WITHDRAWN
WRIGHT STATE UNIVERSITY LIBRARIES

PATTY'S INDUSTRIAL HYGIENE

Fifth Edition

Volume 3
VI LAW, REGULATION, AND MANAGEMENT

PATTY'S INDUSTRIAL HYGIENE

Fifth Edition
Volume 3
VI LAW, REGULATION, AND MANAGEMENT

ROBERT L. HARRIS
Editor

CONTRIBUTORS

R. E. Allan
L. R. Birkner
M. T. Brandt
V. Fiserova-Bergerova
M. R. Flynn
D. K. George
R. J. Harris

A. Hart
T. C. Johnson
S. P. Levine
J. Mraz
R. A. Patnoe
D. J. Paustenbach
C. F. Phillips
C. F. Redinger

C. A. Rice
K. Schmidt
L. K. Simkins
J. F. Stockman
R. G. Tardiff
B. Toeppen-Sprigg
M. R. Zavon

A Wiley-Interscience Publication
JOHN WILEY & SONS, INC.
New York / Chichester / Weinheim / Brisbane / Singapore / Toronto

This book is printed on acid-free paper. ∞

Copyright © 2000 by John Wiley & Sons, Inc. All rights reserved.

Published simultaneously in Canada.

No part of this publication may be reproduced, stored in a retrieval system or transmitted in any form or by any means, electronic, mechanical, photocopying, recording, scanning or otherwise, except as permitted under Sections 107 or 108 of the 1976 United States Copyright Act, without either the prior written permission of the Publisher, or authorization through payment of the appropriate per-copy fee to the Copyright Clearance Center, 222 Rosewood Drive, Danvers, MA 01923, (978) 750-8400, fax (978) 750-4744. Requests to the Publisher for permission should be addressed to the Permissions Department, John Wiley & Sons, Inc., 605 Third Avenue, New York, NY 10158-0012, (212) 850-6011, fax (212) 850-6008, E-Mail: PERMREQ @ WILEY.COM.

For ordering and customer service, call 1-800-CALL-WILEY.

Library of Congress Cataloging in Publication Data:

Patty's industrial hygiene / [edited by] Robert L. Harris. — 5th ed.
   [rev.]
      v. 〈 〉 cm.
     Fourth ed. published as: Patty's industrial hygiene and toxicology.
     Includes bibliographical references and index.
     ISBN 0-471-29753-4 (Vol. 3) (cloth : alk. paper); 0-471-29784-4 (set)
      1. Industrial hygiene.   I. Harris, Robert L., 1924– .
  II. Patty, F. A. (Frank Arthur), 1897–   Industrial hygiene and toxicology.
RC967.P37 2000
613.6′2—dc21   99-32462

Printed in the United States of America.

10 9 8 7 6 5 4 3 2 1

Contributors

Ralph E. Allan, JD, CIH, University of California at Irvine, Irvine, California

Lawrence R. Birkner, MBA, CIH
McIntyre, Birkner and Associates,
Thousand Oaks, California

Michael T. Brandt, Ph.D, Los Alamos National Laboratory
Los Alamos, New Mexico

Vera Fiserova-Bergerova, Ph.D
Department of Anesthesiology,
University of Miami, School of Medicine
Miami, Florida

Michael R. Flynn, Sc.D, CIH
Department of Environmental Sciences and Engineering, University of North Carolina, Chapel Hill, North Carolina

Dennis K. George, Ph.D, CIH
Department of Engineering Technology
Western Kentucky University
Bowling Green, Kentucky

Robert J. Harris, JD, Capital District Law Offices, Raleigh, North Carolina

Alan Hart, Ph.D, CIH, CSP
Hudson, Ohio

Ted C. Johnson, MSPH, CIH, CSP, The Stockman Group, Whittier, California

Steven P. Levine, Ph.D, CIH, University of Michigan, Erb Environmental Management Institute
Ann Arbor, Michigan

Jaroslav Mraz, Ph.D, Division of Occupational Health
Prague, Czech Republic

Richard A. Patnoe, Ph.D
Boulder, Colorado

Dennis A. Paustenbach, Ph.D, CIH, DAPT, Exponent
Menlo Park, California

Carolyn F. Phillips, CIH
Houston, Texas

Charles F. Redinger, MPA, CIH, Ph.D
Redinger and Associates
San Francisco, California

Charlotte A. Rice, CIH
Wheat Ridge, Colorado

Kathleen Schmidt, RN, COHN-S
Akron, Ohio

Lisa K. Simkins, PE, CIH
Alamo, California

Judith F. Stockman, RN, CANP, COHN
The Stockman Group
Whittier, California

Robert G. Tardiff, Ph.D, ATS
Vienna, Virginia

Barbara Toeppen-Sprigg, MD, MPH
North Canton, Ohio

Mitchell R. Zavon, MD, Lewiston, New York

Preface

Industrial hygiene is an applied science and a profession. Like other applied sciences such as medicine and engineering, it is founded on basic sciences such as biology, chemistry, mathematics, and physics. In a sense it is a hybrid profession because within its ranks are members of other professions—chemists, engineers, biologists, physicists, physicians, nurses, and lawyers. In their professional practice all are dedicated in one way or another to the purposes of industrial hygiene, to the anticipation, recognition, evaluation, and control of work-related health hazards. All are represented among the authors of chapters in these volumes.

Although the term "industrial hygiene" used to describe our profession is probably of twentieth century origin, we must go further back in history for the origin of its words. The word "industry," which has a dictionary meaning, "systematic labor for some useful purpose or the creation of something of value," has its English origin in the fifteenth century. For "hygiene" we must look even earlier. Hygieia, a daughter of Aesklepios who is god of medicine in Greek mythology, was responsible for the preservation of health and prevention of disease. Thus, Hygieia, when she was dealing with people who were engaged in systematic labor for some useful purpose, was practicing our profession, industrial hygiene.

Industrial Hygiene and Toxicology was originated by Frank A. Patty with publication of the first single volume in 1948. In 1958 an updated and expanded Second Edition was published with his guidance. A second volume, Toxicology, was published in 1963. Frank Patty was a pioneer in industrial hygiene; he was a teacher, practitioner, and manager. He served in 1946 as eighth President of the American Industrial Hygiene Association. To cap his professional career he served as Director of the Division of Industrial Hygiene for the General Motors Corporation.

At the request of Frank Patty, George and Florence Clayton took over editorship of the ever-expanding *Industrial Hygiene and Toxicology* series for the Third Edition of Volume I, *General Principles*; published in 1978, and Volume II, *Toxicology*, published in

1981–1982. The First Edition of Volume III, *Theory and Rationale of Industrial Hygiene Practice*, edited by Lewis and Lester Cralley, was published in 1979 with its Second Edition published in 1984. The ten-book, two-volume Fourth Edition of *Patty's Industrial Hygiene and Toxicology*, edited by George and Florence Clayton, was published in 1991–1994, and the Third Edition of Volume III, *Theory and Rationale of Industrial Hygiene Practice*, edited by Robert Harris, Lewis Cralley, and Lester Cralley, was published in 1994. With the agreement and support of George and Florence Clayton, and Lewis and Lester Cralley, it is a signal honor for me to follow them and Frank A. Patty as editor of the Industrial Hygiene volumes of *Patty's Industrial Hygiene and Toxicology*.

Industrial hygiene has been dealt with very broadly in past editions of *Patty's Industrial Hygiene and Toxicology*. Chapters have been offered on sampling and analysis, exposure measurement and interpretation, absorption and elimination of toxic materials, occupational dermatoses, instrument calibration, odors, industrial noise, ionizing and nonionizing radiation, heat stress, pressure, lighting, control of exposures, safety and health law, health surveillance, occupational health nursing, ergonomics and safety, agricultural hygiene, hazardous wastes, occupational epidemiology, and other vital areas of practice. These traditional areas continue to be covered in this new edition. Consistent with the past history of *Patty's*, new areas of industrial hygiene concerns and practices have been addressed as well: aerosol science, computed tomography, multiple chemical sensitivity, potential endocrine disruptors, biological monitoring of exposures, health and safety management systems, industrial hygiene education, and other areas not covered in earlier editions.

Although industrial hygiene has been practiced in one guise or another for centuries, the most systematic approaches and the most esoteric accomplishments have been made in the past fifty or sixty years—generally in the years since Frank Patty published his first book. This accelerated progress is due primarily to increased public awareness of occupational health and safety issues and need for environmental control as is evidenced by Occupational Safety and Health, Clean Air, and Clean Water legislation at both federal and state levels.

Industrial hygienists know that variability is the key to measurement and interpretation of workers' exposures. If exposures did not vary, exposure assessment could be limited to a single measurement, the results of which could be acted upon, then the matter filed away as something of no further concern. We know, however, that exposures change. But not only do exposures change—change is characteristic of the science and practice of our profession as well. We must be alert to recognize new hazards, we must continue to evaluate new and changing stresses, we must evaluate performance of exposure controls and from time to time upgrade them. These volumes represent the theory and practice of industrial hygiene as they are understood by their chapter authors at the time of writing. But, as observed by the Greek philosopher Heracleitus about 2500 years ago, "There is nothing permanent except change." Improvements and changes in theory and practice of industrial hygiene take place continuously and are generally reported in the professional literature. Industrial hygienists, the practitioners, the teachers, and the managers, must stay abreast of the professional literature. Furthermore, when an industrial hygienist develops new knowledge, he/she has what almost amounts to an ethical obligation to share it in our journals.

PREFACE

One cannot ponder the rapid changes and advancements made in recent decades in science and technology, and in our own profession as well, without wondering at what the next two or three decades will bring. Developments in computer technology and information processing and exchange have greatly influenced manufacturing (robotics, computer controlled machining) and the general conduct of commerce and business in the past one or two decades. This change will only accelerate with computer speeds and capacities doubling every 18 months or so, and processing units approaching microsize. The possibility for continuously monitoring and computer storage of exposures of individual workers may become reality within the next decade. The human genome project holds promise for prevention and cure of many diseases, including some associated with conditions of work. World population continues to increase geometrically and is expected to be about eight billion in the year 2020; with improvements in preventive health care this will be an increasingly older population. Genetic engineering and highly effective pesticides are already improving yields of agricultural commodities; if all goes well in this area, and if we can avoid set-backs as might be associated with potential endocrine disruptors, feeding the expanding human population may not be a limiting factor. Globalization of manufacturing and commerce has already begun to reduce manufacturing employment in the United States and in Europe, and to expand opportunities for expanding populations in some developing nations. The United States and other developed nations are on their way to becoming world centers of information and innovation.

How will all of this affect the future practice of industrial hygiene? In the Preface to the Fourth Edition of *Patty's,* George and Florence Clayton suggested that the future of industrial hygiene is limited only by the narrowness of vision of its practitioners. More recently, Lawrence Birkner, past president of the American Academy of Industrial Hygiene, and his co-worker and spouse, Ruth McIntyre Birkner, in writing about the future of the occupational and environmental hygiene profession, say much the same thing. (See "The Future of the Occupational and Environmental Hygiene Profession" in *A.I.H.A.* Journal, pp. 370–374, 1997) Larry and Ruth report that we must be aware of the changes likely to take place in the next couple of decades, and must develop strategies now to assure the profession's full participation in protecting the health and safety of workers, and the environment, of tomorrow.

ROBERT L. HARRIS

Raleigh, North Carolina

Contents

VI. Law, Regulation, and Management

34. Job Safety and Health Law — 1595
Robert J. Harris, JD

35. Compliance and Projection — 1651
Robert J. Harris, JD

36. Industrial Hygienist's Liability Under Law — 1685
Ralph E. Allan, JD, CIH

37. Litigation in Industrial Hygiene Practice — 1697
Ralph E. Allan, JD, CIH

38. Odor: A Legal Overview — 1725
Ralph E. Allan, JD, CIH

39. Hazard Communication and Worker Right to Know Programs — 1735
Lisa K. Simkins, CIH, PE and Charlotte A. Rice, CIH

40. Pharmacokinetics and Unusual Work Schedules — 1787
Dennis J. Paustenbach, Ph.D., CIH, DABT

41.	**The History and Biological Basis of Occupational Exposure Limits for Chemical Agents**	1903
	Dennis J. Paustenbach, Ph.D., CIH, DABT	
42.	**Biological Monitoring of Exposure to Industrial Chemicals**	2001
	Vera Fiserova-Bergerova, Ph.D. and Jaroslav Mraz, Ph.D.	
43.	**Cost-Effectiveness of a Multidisciplinary Approach to Loss Control and Prevention Programs**	2061
	Judith F. Stockman, RN, CANP, COHN and Ted C. Johnson, MSPH, CIH, CSP	
44.	**Industrial Hygiene Surveys, Records and Reports**	2085
	Carolyn F. Phillips, CIH	
45.	**Data Automation**	2113
	Richard A. Patnoe, Ph.D.	
46.	**Risk Analysis for the Workplace**	2151
	Robert G. Tardiff, Ph.D., ATS	
47.	**Health Surveillance Programs In Industry**	2199
	Mitchell R. Zavon, MD	
48.	**Health Promotion in the Workplace**	2221
	B. Toeppen-Sprigg, MD, MPH, K. Schmidt, RN, COHN-S, and A. Hart, Ph.D., CIH, CSP	
49.	**Occupational Health and Safety Management Systems**	2247
	Charles F. Redinger, Ph.D., CIH and Steven P. Levine, Ph.D., CIH	
50.	**Business Analysis for Health and Safety Professionals**	2303
	Lawrence R. Birkner, MBA, CIH and Michael T. Brandt, Ph.D.	
51.	**Industrial Hygiene Education, Training, and Information Exchange**	2343
	Dennis K. George, Ph.D., CIH and Michael R. Flynn, Sc.D, CIH	

Index   2375

USEFUL EQUIVALENTS AND CONVERSION FACTORS

1 kilometer = 0.6214 mile
1 meter = 3.281 feet
1 centimeter = 0.3937 inch
1 micrometer = 1/25,4000 inch = 40 microinches = 10,000 Angstrom units
1 foot = 30.48 centimeters
1 inch = 25.40 millimeters
1 square kilometer = 0.3861 square mile (U.S.)
1 square foot = 0.0929 square meter
1 square inch = 6.452 square centimeters
1 square mile (U.S.) = 2,589,998 square meters = 640 acres
1 acre = 43,560 square feet = 4047 square meters
1 cubic meter = 35.315 cubic feet
1 cubic centimeter = 0.0610 cubic inch
1 cubic foot = 28.32 liters = 0.0283 cubic meter = 7.481 gallons (U.S.)
1 cubic inch = 16.39 cubic centimeters
1 U.S. gallon = 3,7853 liters = 231 cubic inches = 0.13368 cubic foot
1 liter = 0.9081 quart (dry), 1.057 quarts (U.S., liquid)
1 cubic foot of water = 62.43 pounds (4°C)
1 U.S. gallon of water = 8.345 pounds (4°C)
1 kilogram = 2.205 pounds

1 gram = 15.43 grains
1 pound = 453.59 grams
1 ounce (avoir.) = 28.35 grams
1 gram mole of a perfect gas ≎ 24.45 liters (at 25°C and 760 mm Hg barometric pressure)
1 atmosphere = 14.7 pounds per square inch
1 foot of water pressure = 0.4335 pound per square inch
1 inch of mercury pressure = 0.4912 pound per square inch
1 dyne per square centimeter = 0.0021 pound per square foot
1 gram-calorie = 0.00397 Btu
1 Btu = 778 foot-pounds
1 Btu per minute = 12.96 foot-pounds per second
1 hp = 0.707 Btu per second = 550 foot-pounds per second
1 centimeter per second = 1.97 feet per minute = 0.0224 mile per hour
1 footcandle = 1 lumen incident per square foot = 10.764 lumens incident per square meter
1 grain per cubic foot = 2.29 grams per cubic meter
1 milligram per cubic meter = 0.000437 grain per cubic foot

To convert degrees Celsius to degrees Fahrenheit: °C (9/5) + 32 = °F
To convert degrees Fahrenheit to degrees Celsius: (5/9) (°F − 32) = °C
For solutes in water: 1 mg/liter ≎ 1 ppm (by weight)
Atmospheric contamination: 1 mg/liter ≎ 1 oz/1000 cu ft (approx)
For gases or vapors in air at 25°C and 760 mm Hg pressure:
    To convert mg/liter to ppm (by volume): mg/liter (24,450/mol. wt.) = ppm
    To convert ppm to mg/liter: ppm (mol. wt./24,450) = mg/liter

CONVERSION TABLE FOR GASES AND VAPORS[a]
(Milligrams per liter to parts per million, and vice versa; 25°C and 760 mm Hg barometric pressure)

Molecular Weight	1 mg/liter ppm	1 ppm mg/liter	Molecular Weight	1 mg/liter ppm	1 ppm mg/liter	Molecular Weight	1 mg/liter ppm	1 ppm mg/liter
1	24,450	0.0000409	39	627	0.001595	77	318	0.00315
2	12,230	0.0000818	40	611	0.001636	78	313	0.00319
3	8,150	0.0001227	41	596	0.001677	79	309	0.00323
4	6,113	0.0001636	42	582	0.001718	80	306	0.00327
5	4,890	0.0002045	43	569	0.001759	81	302	0.00331
6	4,075	0.0002454	44	556	0.001800	82	298	0.00335
7	3,493	0.0002863	45	543	0.001840	83	295	0.00339
8	3,056	0.000327	46	532	0.001881	84	291	0.00344
9	2,717	0.000368	47	520	0.001922	85	288	0.00348
10	2,445	0.000409	48	509	0.001963	86	284	0.00352
11	2,223	0.000450	49	499	0.002004	87	281	0.00356
12	2,038	0.000491	50	489	0.002045	88	278	0.00360
13	1,881	0.000532	51	479	0.002086	89	275	0.00364
14	1,746	0.000573	52	470	0.002127	90	272	0.00368
15	1,630	0.000614	53	461	0.002168	91	269	0.00372
16	1,528	0.000654	54	453	0.002209	92	266	0.00376
17	1,438	0.000695	55	445	0.002250	93	263	0.00380
18	1,358	0.000736	56	437	0.002290	94	260	0.00384
19	1,287	0.000777	57	429	0.002331	95	257	0.00389
20	1,223	0.000818	58	422	0.002372	96	255	0.00393
21	1,164	0.000859	59	414	0.002413	97	252	0.00397
22	1,111	0.000900	60	408	0.002554	98	249.5	0.00401
23	1,063	0.000941	61	401	0.002495	99	247.0	0.00405
24	1,019	0.000982	62	394	0.00254	100	244.5	0.00409
25	978	0.001022	63	388	0.00258	101	242.1	0.00413
26	940	0.001063	64	382	0.00262	102	239.7	0.00417
27	906	0.001104	65	376	0.00266	103	237.4	0.00421
28	873	0.001145	66	370	0.00270	104	235.1	0.00425
29	843	0.001186	67	365	0.00274	105	232.9	0.00429
30	815	0.001227	68	360	0.00278	106	230.7	0.00434
31	789	0.001268	69	354	0.00282	107	228.5	0.00438
32	764	0.001309	70	349	0.00286	108	226.4	0.00442
33	741	0.001350	71	344	0.00290	109	224.3	0.00446
34	719	0.001391	72	340	0.00294	110	222.3	0.00450
35	699	0.001432	73	335	0.00299	111	220.3	0.00454
36	679	0.001472	74	330	0.00303	112	218.3	0.00458
37	661	0.001513	75	326	0.00307	113	216.4	0.00462
38	643	0.001554	76	322	0.00311	114	214.5	0.00466

CONVERSION TABLE FOR GASES AND VAPORS (*Continued*)
(Milligrams per liter to parts per million, and vice versa; 25°C and 760 mm Hg barometric pressure)

Molecular Weight	1 mg/liter ppm	1 ppm mg/liter	Molecular Weight	1 mg/liter ppm	1 ppm mg/liter	Molecular Weight	1 mg/liter ppm	1 ppm mg/liter
115	212.6	0.00470	153	159.8	0.00626	191	128.0	0.00781
116	210.8	0.00474	154	158.8	0.00630	192	127.3	0.00785
117	209.0	0.00479	155	157.7	0.00634	193	126.7	0.00789
118	207.2	0.00483	156	156.7	0.00638	194	126.0	0.00793
119	205.5	0.00487	157	155.7	0.00642	195	125.4	0.00798
120	203.8	0.00491	158	154.7	0.00646	196	124.7	0.00802
121	202.1	0.00495	159	153.7	0.00650	197	124.1	0.00806
122	200.4	0.00499	160	152.8	0.00654	198	123.5	0.00810
123	198.8	0.00503	161	151.9	0.00658	199	122.9	0.00814
124	197.2	0.00507	162	150.9	0.00663	200	122.3	0.00818
125	195.6	0.00511	163	150.0	0.00667	201	121.6	0.00822
126	194.0	0.00515	164	149.1	0.00671	202	121.0	0.00826
127	192.5	0.00519	165	148.2	0.00675	203	120.4	0.00830
128	191.0	0.00524	166	147.3	0.00679	204	119.9	0.00834
129	189.5	0.00528	167	146.4	0.00683	205	119.3	0.00838
130	188.1	0.00532	168	145.5	0.00687	206	118.7	0.00843
131	186.6	0.00536	169	144.7	0.00691	207	118.1	0.00847
132	185.2	0.00540	170	143.8	0.00695	208	117.5	0.00851
133	183.8	0.00544	171	143.0	0.00699	209	117.0	0.00855
134	182.5	0.00548	172	142.2	0.00703	210	116.4	0.00859
135	181.1	0.00552	173	141.3	0.00708	211	115.9	0.00863
136	179.8	0.00556	174	140.5	0.00712	212	115.3	0.00867
137	178.5	0.00560	175	139.7	0.00716	213	114.8	0.00871
138	177.2	0.00564	176	138.9	0.00720	214	114.3	0.00875
139	175.9	0.00569	177	138.1	0.00724	215	113.7	0.00879
140	174.6	0.00573	178	137.4	0.00728	216	113.2	0.00883
141	173.4	0.00577	179	136.6	0.00732	217	112.7	0.00888
142	172.2	0.00581	180	135.8	0.00736	218	112.2	0.00892
143	171.0	0.00585	181	135.1	0.00740	219	111.6	0.00896
144	169.8	0.00589	182	134.3	0.00744	220	111.1	0.00900
145	168.6	0.00593	183	133.6	0.00748	221	110.6	0.00904
146	167.5	0.00597	184	132.9	0.00753	222	110.1	0.00908
147	166.3	0.00601	185	132.2	0.00757	223	109.6	0.00912
148	165.2	0.00605	186	131.5	0.00761	224	109.2	0.00916
149	164.1	0.00609	187	130.7	0.00765	225	108.7	0.00920
150	163.0	0.00613	188	130.1	0.00769	226	108.2	0.00924
151	161.9	0.00618	189	129.4	0.00773	227	107.7	0.00928
152	160.9	0.00622	190	128.7	0.00777	228	107.2	0.00933

CONVERSION TABLE FOR GASES AND VAPORS (Continued)
(Milligrams per liter to parts per million, and vice versa; 25°C and 760 mm Hg barometric pressure)

Molecular Weight	1 mg/liter ppm	1 ppm mg/liter	Molecular Weight	1 mg/liter ppm	1 ppm mg/liter	Molecular Weight	1 mg/liter ppm	1 ppm mg/liter
229	106.8	0.00937	253	96.6	0.01035	277	88.3	0.01133
230	106.3	0.00941	254	96.3	0.01039	278	87.9	0.01137
231	105.8	0.00945	255	95.9	0.01043	279	87.6	0.01141
232	105.4	0.00949	256	95.5	0.01047	280	87.3	0.01145
233	104.9	0.00953	257	95.1	0.01051	281	87.0	0.01149
234	104.5	0.00957	258	94.8	0.01055	282	86.7	0.01153
235	104.0	0.00961	259	94.4	0.01059	283	86.4	0.01157
236	103.6	0.00965	260	94.0	0.01063	284	86.1	0.01162
237	103.2	0.00969	261	93.7	0.01067	285	85.8	0.01166
238	102.7	0.00973	262	93.3	0.01072	286	85.5	0.01170
239	102.3	0.00978	263	93.0	0.01076	287	85.2	0.01174
240	101.9	0.00982	264	92.6	0.01080	288	84.9	0.01178
241	101.5	0.00986	265	92.3	0.01084	289	84.6	0.01182
242	101.0	0.00990	266	91.9	0.01088	290	84.3	0.01186
243	100.6	0.00994	267	91.6	0.01092	291	84.0	0.01190
244	100.2	0.00998	268	91.2	0.01096	292	83.7	0.01194
245	99.8	0.01002	269	90.9	0.01100	293	83.4	0.01198
246	99.4	0.01006	270	90.6	0.01104	294	83.2	0.01202
247	99.0	0.01010	271	90.2	0.01108	295	82.9	0.01207
248	98.6	0.01014	272	89.9	0.01112	296	82.6	0.01211
249	98.2	0.01018	273	89.6	0.01117	297	82.3	0.01215
250	97.8	0.01022	274	89.2	0.01121	298	82.0	0.01219
251	97.4	0.01027	275	88.9	0.01125	299	81.8	0.01223
252	97.0	0.01031	276	88.6	0.01129	300	81.5	0.01227

[a] A. C. Fieldner, S. H. Katz, and S. P. Kinney, "Gas Masks for Gases Met in Fighting Fires," *U.S. Bureau of Mines, Technical Paper No. 248,* 1921.

PATTY'S INDUSTRIAL HYGIENE

Fifth Edition

Volume 3
VI LAW, REGULATION, AND MANAGEMENT

CHAPTER THIRTY-FOUR

Job Safety and Health Law

Robert J. Harris, JD

The principal legislation relating to job safety and health is the federal Occupational Safety and Health Act of 1970 (OSHA). However, certain industries and/or portions of industries are subject to regulation by federal statutes other than OSHA. In addition, OSHA does not entirely preclude regulation of job safety and health by the states or their political subdivisions. Nevertheless, OSHA is clearly the most comprehensive legislative directive relating to workplace safety and health; accordingly, this chapter and the next focus primarily on developments and requirements pursuant to this act.

1 OVERVIEW OF SOURCES OF LAW

The actions of Congress and the executive-branch agencies that carry out its legislative directives are subject to judicial review in the federal courts of the United States. Other than the actions of Congress and the agencies themselves, the decisions of the federal courts and other quasijudicial bodies that rule on challenges to these actions are the primary source of developments in the field of job safety and health law.

Because judicial review of legislative and agency actions lies directly with the federal courts of appeal, these courts, for practical purposes, are the most important judicial source of developments in job safety and health law. There are twelve regional United States Courts of Appeal. These twelve federal appellate courts and their respective regions of jurisdiction are as follows:

District of Columbia Circuit: District of Columbia

First Circuit: Maine, Massachusetts, New Hampshire, Rhode Island, and Puerto Rico

---

Patty's Industrial Hygiene, Fifth Edition, Volume 3. Edited by Robert L. Harris.
ISBN 0-471-29753-4 © 2000 John Wiley & Sons, Inc.

Second Circuit: Connecticut, New York, and Vermont

Third Circuit: Delaware, New Jersey, Pennsylvania, and Virgin Islands

Fourth Circuit: Maryland, Virginia, West Virginia, North Carolina, and South Carolina

Fifth Circuit: Louisiana, Mississippi, and Texas

Sixth Circuit: Tennessee, Kentucky, Ohio, and Michigan

Seventh Circuit: Indiana, Illinois, and Wisconsin

Eighth Circuit: Arkansas, Missouri, Iowa, Minnesota, Nebraska, South Dakota, and North Dakota

Ninth Circuit: Arizona, California, Nevada, Oregon, Washington, Idaho, Montana, Alaska, Hawaii, Guam, and Northern Mariana Islands

Tenth Circuit: New Mexico, Utah, Colorado, Kansas, Oklahoma, and Wyoming

Eleventh Circuit: Florida, Georgia, and Alabama

Oftentimes the rulings of the various federal appellate courts are in conflict with one another, so that the law on a particular issue may be different in one region from another. Sometimes the United States Supreme Court takes for review issues that have been ruled upon by one or more federal appellate courts, called "granting certiorari". When the Supreme Court rules on such issues, it generally settles any conflicts that have developed among the circuits; however, unless and until the Supreme Court rules, the circuits can and often do remain in conflict on certain issues.

When a federal appellate court has ruled on a particular issue, its opinion is "binding" authority only in the region that the court covers. If a circuit has not ruled on a particular issue, then the decisions of other circuits can be "persuasive" authority on that particular issue in the circuit that has not ruled, but the decisions of the other circuits are not "binding" in that circuit.

1.1 The Administrative Procedure Act

The Administrative Procedure Act was enacted by Congress in 1946 to impose some coherent system of procedural regularity on the growing regulatory bureaucracy of the federal government. The act applies generally to all federal agencies. It provides for administrative "rule making" and administrative "adjudication," among other things, and it prescribes the methods by which agencies are to go about "rule making" and "adjudication." With regard to agency rule making, the act prescribes that the agency is to provide notice of its intent to make a particular rule and is to receive public comment and conduct hearings on the matter before promulgating a final rule. The act also sets up the process by which an aggrieved party can appeal any final agency action, including a final rule making, which generally involves going through an administrative appeal process with further appeal rights to the federal courts of appeal (1).

2 LEGISLATIVE HISTORY AND BACKGROUND OF OSHA

The Occupational Safety and Health Act of 1970 was enacted by Congress on December 17, 1970, and became effective on April 28, 1971. It represents the first job safety and health law of nationwide scope (2). Passage of the act was preceded by a dramatic and bitter labor-management political fight. The legislative history of OSHA is summarized in The Job Safety and Health Act of 1970 (2).

Congress enacted OSHA for the declared purpose of assuring "so far as possible every working man and woman in the Nation safe and healthful working conditions" [§ 2(b)]. The act is intended to prevent work-related injury, illness, and death.

Since its enactment, OSHA has received very little amendment. In 1998, however, President Clinton signed two Republican bills that amended OSHA for the first time in more than two decades (3). The enactment of this legislation was the first break in a stalemate on OSHA reform between Congressional Republicans and Democrats that had existed since President Clinton's election in 1992.

2.1 Agencies Responsible for Implementing and Enforcing OSHA

The Department of Labor is responsible for implementing OSHA. On the date the act became effective, the Department of Labor created the Occupational Safety and Health Administration (OSH Administration or OSHA) to carry out such responsibilities. The OSH Administration is headed by an assistant secretary of labor for occupational safety and health and is responsible, among other things, for promulgating rules and regulations, setting health and safety standards, evaluating and approving state plans, and overseeing enforcement of the act (4).

Section 12(a) of the act establishes the Occupational Safety and Health Review Commission (OSAHRC) as an independent agency to adjudicate enforcement actions brought by the secretary of labor. The commission is composed of three members appointed by the president for six-year terms. The chairperson of the commission is authorized to appoint such administrative law judges as he or she deems necessary to assist in the work of the commission [§ 12(e)].

Sections 20 and 21 of the act give the secretary of health, education and welfare (HEW) [now Health and Human Services (HHS)] broad authority to conduct experimental research relating to occupational safety and health, to develop criteria for and recommend safety and health standards, and to conduct educational and training programs (5). Section 22 establishes the National Institute for Occupational Safety and Health (NIOSH) to perform the functions of the secretary of health and human services under Sections 20 and 21.

The act specifically directs NIOSH to develop criteria documents that describe safe levels of exposure to toxic materials and harmful physical agents and to forward recommended standards for such substances to the secretary of labor (6). The act also directs NIOSH to publish at least annually a list of all known toxic substances and the concentrations at which such toxicity is known to occur [§ 20(a)(6)].

Section 7(a) establishes a National Advisory Committee on Occupational Safety and Health (NACOSH), whose basic functions are to advise, consult with, and make recommendations to the secretary of labor and the secretary of health and human services on matters relating to the administration of the act. NACOSH consists of 12 members who represent management, labor, occupational safety and occupational health professions, and the public. The members are appointed by the secretary of labor, although four members are to be designated by the secretary of health and human services.

The secretary of labor is authorized by Section 7(b) to appoint other advisory committees to assist him or her in the formulation of standards under Section 6. For example, *ad hoc* advisory committees have been used to assist in developing standards for exposure to asbestos and coke oven emissions. The secretary of labor has also appointed various standing advisory committees (7).

2.2 Scope of OSHA's Coverage

The Occupational Safety and Health Act of 1970 applies to every private employer engaged in a business affecting commerce, regardless of the number of employees (8). It applies with respect to employment performed in a workplace in any of the 50 states, the District of Columbia, Puerto Rico, the Virgin Islands, American Samoa, Guam, the Trust Territory of the Pacific Islands, Wake Island, the Outer Continental Shelf Lands, Johnston Island, and the Canal Zone [§§ 3(5) and 4(a)].

The act's definition of "employer" does not include the states, political subdivisions of the states, or the United States (9). However, Section 19 directs the head of each federal agency to establish and maintain an effective and comprehensive occupational safety and health program that is consistent with the standards required of private employers.

3 REGULATION OF JOB SAFETY AND HEALTH BY FEDERAL STATUTES OTHER THAN OSHA

Section 4(b)(1) of OSHA states that nothing in the act shall apply to working conditions of employees with respect to which other federal agencies exercise statutory authority to prescribe or enforce standards or regulations affecting occupational safety or health. Thus federal agencies other than OSHA that are authorized by statute to regulate employee safety and health can continue to do so after the effective date of OSHA; in fact, the exercise of such authority preempts OSHA from regulating with respect to such working conditions.

Although Section 4(b)(1) seems to be self-defining, it has generated a tremendous volume of litigation. Three major interpretive questions have been raised:

1. What constitutes a sufficient exercise of regulatory authority to preempt OSHA regulation?
2. Does the exercise of authority by another federal agency in substantial areas of employee safety exempt the entire industry from OSHA standards?
3. Must the other federal agency's motivation in acting have been to protect workers?

It appears to be well settled that the mere existence of statutory authority to regulate safety or health is not sufficient to oust OSHA's regulatory scheme; some exercise of that authority is necessary. Furthermore, at least three federal courts of appeals have taken the position that speculative pronouncements of proposed regulations by a federal agency are not sufficient to warrant preemption of OSHA standards. Rather, it has been held that Section 4(b)(1) requires a concrete exercise of statutory authority (10).

The same courts of appeals have also rejected the notion that the exercise of statutory authority by another federal agency creates an industrywide exemption from OSHA regulations. Rather, the courts have agreed that the term "working conditions" in Section 4(b)(1) refers to something more limited than every aspect of an entire industry. Ambiguity remains, however, with respect to the scope of the displacing effect of another agency's regulation of a working condition.

For example, in *Southern Pacific Transportation Company* (10), the Fifth Circuit explained that the term "working conditions" has a technical meaning in the language of industrial relations; it encompasses both a worker's surroundings and the hazards incident to the work. The court stated that the displacing effect of Section 4(b)(1) would depend primarily on the agency's articulation of its regulations (11):

> Section 4(b)(1) means that any FRA [Federal Railroad Administration] exercise directed at a working condition—defined either in terms of a "surrounding" or a "hazard"—displaces OSHA coverage of that working condition. Thus comprehensive FRA treatment of the general problem of railroad fire protection will displace all OSHA regulations on fire protection, even if the FRA activity does not encompass every detail of the OSHA fire protection standards, but FRA regulation of portable fire extinguishers will not displace OSHA standards on fire alarm signaling systems.

The Fourth Circuit defined "working conditions" as "the environmental area in which an employee customarily goes about his daily tasks." The court in *Southern Railway Company* (10) explained that OSHA would be displaced when another federal agency had exercised its statutory authority to prescribe standards affecting occupational safety or health for such an area (12).

The courts of appeals seem to indicate, at least implicitly, that regulation of a working condition by another federal agency need not be as effective or as stringent as an OSHA standard to preempt the OSHA standard (10). But it remains unclear whether a decision by another federal agency that a particular aspect of an industry should not be regulated at all would preempt or preclude OSHA regulation of that same aspect. Resolution of these and other issues involving the scope of the displacement effect under Section 4(b)(1) awaits future litigation or legislation.

Finally, the commission has held that to be cognizable under Section 4(b)(1), "a different statutory scheme and rules thereunder must have a policy or purpose that is consonant with that of the Occupational Safety and Health Act. That is, there must be a policy or purpose to include employees in the class of persons to be protected thereunder" (13). In *Organized Migrants in Community Action, Inc. v. Brennan* (14), the U.S. Court of Appeals for the District of Columbia, although not deciding the issue, implicitly rejected the argument that preemption under Section 4(b)(1) exists only where the allegedly preempting statute was passed primarily for the protection of the employees (15).

4 OVERVIEW OF FEDERAL REGULATORY SCHEMES OTHER THAN OSHA

The following material represents an overview of the major federal regulatory schemes other than OSHA that deal with or relate to job safety and health. The listing is by no means exhaustive, and employers are urged to consult specific statutory schemes in substantive areas relating to their respective industries.

4.1 Mine Safety and Health Legislation

Occupational safety and health matters with respect to the nation's mining industry are regulated pursuant to the Federal Mine Safety and Health Act of 1977, which became effective in March 1978 (16). The Department of Labor is responsible for enforcing that legislation. The secretary of labor has delegated its enforcement authority to the Mine Safety and Health Administration (MSHA).

Prior to the effective date of the 1977 act, safety and health matters with respect to the mining industry were covered by two separate statutes. The Metal and Non-Metallic Mine Safety Act of 1966 covered mines of all types other than coal mines. Safety and health matters concerning coal mines were regulated by the Coal Mine Health and Safety Act of 1969. The 1977 Mine Safety and Health Act is now the single mine safety and health law for all mining operations. The 1977 act directs the secretary of labor to, inter alia, "develop, promulgate, and revise as may be appropriate improved mandatory health or safety standards for the protection of life and prevention of injuries in coal or other mines" (17). The secretary is also empowered to enforce those standards.

4.2 Environmental Pesticide Control Act of 1972

The Federal Environmental Pesticide Control Act of 1972 (FEPCA) regulates the use of pesticides and makes misuse civilly and criminally punishable. The Court of Appeals for the District of Columbia has held that FEPCA authorizes the Environmental Protection Agency (EPA) to promulgate and enforce occupational health and safety standards with respect to farm workers' exposure to pesticides. The EPA has exercised that authority (18). OSHA is thus preempted from regulating in that area [*Organized Migrants in Community Action* (14), (19)].

4.3 Federal Railroad Safety Act of 1970

The Federal Railroad Safety Act of 1970 authorizes the Federal Railroad Administration (FRA) within the Department of Transportation (DOT) to promulgate regulations for all areas of railroad safety, including employee safety. To date, however, DOT has not adopted railroad occupational safety standards for all railroad working conditions or workplaces (20). The Department of Labor (OSHA) retains jurisdiction over safety and health of railroad employees with respect to those "working conditions" for which DOT has not adopted standards (21).

4.4 Federal Aviation Act of 1958

The Federal Aviation Act of 1958, as amended, empowers the Federal Aviation Administration (FAA) within the Department of Transportation to promote safety of flight of civil aircraft in air commerce (22), as well as to establish minimum safety standards for the operation of airports that serve any scheduled or unscheduled passenger operation of air carrier aircraft designed for more than 30 passenger seats (23). If the congressional mandate in that statute is deemed to include the safe working conditions of airline and/or airport employees, OSHA would be precluded from exercising its jurisdiction with respect to the working conditions regulated by the FAA. At least one administrative law judge has determined that the FAA's mandate encompasses the safe working conditions of airline ground crews when performing aircraft maintenance (24).

4.5 Hazardous Materials Transportation Act

The Hazardous Materials Transportation Act (HMTA) authorizes the secretary of transportation to issue regulations governing any safety aspect of the transportation of materials designated as hazardous by the secretary (25). The act encompasses shipments by rail, air, water, and highway. If worker safety is deemed to be a purpose of the HMTA, safety standards promulgated by DOT under the statute would trigger a preemption of OSHA jurisdiction with respect to the working conditions covered by such standards. The Occupational Safety and Health Review Commission, however, has held that DOT regulations promulgated under the act do not pre-empt OSHA standards (26).

4.6 Natural Gas Pipeline Safety Act of 1968

The Natural Gas Pipeline Safety Act of 1968 (NGPSA) authorizes the secretary of transportation to establish minimum federal safety standards for pipeline facilities and the transportation of gas in commerce. In *Texas Eastern Transmission Corp.* (27), the Occupational Safety and Health Review Commission determined that the NGPSA was intended to affect occupational safety and health. Thus employers engaged in the transmission, sale, and storage of natural gas would be exempt from OSHA with respect to working conditions covered by DOT standards promulgated under NGPSA (28).

4.7 Federal Noise Control Act of 1972

Although a health and safety standard adopted pursuant to OSHA governs the level of noise to which a worker covered by the act may be exposed in the workplace (see Section 6.3.15), other federal statutes deal with noise abatement and control as well (29).

The first such enactment, a 1968 amendment to the Federal Aviation Act, required the administrator of the FAA to include aircraft noise control as a factor in granting type certificates to aircraft under the act (30). To the extent that the administrator of the FAA denies certification to an aircraft that produces noise in excess of the standards or prohibits the operation of a certified aircraft in a manner that violates regulations, the general environmental noise level in workplaces covered by OSHA and located adjacent to airports and landing patterns is correspondingly reduced.

The first attempt to deal with noise on a nationwide basis, however, was the federal Noise Control Act of 1972. The control strategy of that act is generally as follows: The administrator of the EPA is required to develop and publish criteria with regard to noise, reflecting present scientific knowledge of the effects on public health and welfare that are to be expected from different quantities or qualities of noise. The administrator then must identify products or kinds of products that in his or her opinion are major sources of environmental noise, and publish noise emission regulations where it is feasible to limit the amount of noise produced by such products (31).

Under the act, the administrator is further charged with publishing regulations identifying "low noise emission products." Such products, once so designated, must thereafter be purchased by federal agencies in preference to substitute products, provided the "low noise emission product" costs no more than 125% of the price of the substitute.

As the administrator of the EPA identifies more and more products as major sources of noise and subjects those products to noise emission standards adopted under the Noise Control Act of 1972, the noise levels found in workplaces covered by OSHA and in which such products are used should decrease.

4.8 Federal Toxic Substances Control Act of 1976

The Toxic Substances Control Act of 1976 establishes a broad, nationwide program for the federal regulation of the manufacture and distribution of toxic substances (32). The act divides all "chemical substances" and "mixtures" into two categories: the old and the new. The administrator of the EPA is charged under Section 8(b) with the gargantuan task of compiling and publishing in the Federal Register an inventory or list of all chemical substances manufactured or processed in the United States.

Manufacturers of substances that appear on that inventory are at liberty to continue to manufacture and distribute such substances unless the administrator by rule promulgated under Section 4 of the act first requires that a designated substance be tested and data from the tests be submitted to the EPA (33). If EPA thereafter makes a determination under Section 6 that the continued manufacture, processing, or distribution in commerce of the substance presents an unreasonable risk of injury to health or the environment, the administrator may either prohibit altogether the manufacture of the chemical substance or may impose restrictions (limitations on the quantity manufactured, the use to which the chemical may be put, the concentrations in which it may be used, the labels and warnings that must accompany its sale, etc.).

Section 5 of the act provides a different treatment with respect to a "new chemical substance" or a "significant new use" of a substance that appears on the Section 8(b) inventory. The manufacturer is not at liberty to commence manufacture or distribution of such a "new" substance but must first submit a notice to the administrator of intention to manufacture such a new substance or to engage in a significant new use. Next, testing data relating to the substance's toxicity and the effect on health and on the environment must be submitted. If the administrator does not act within 90 days, the manufacturer is at liberty to proceed with manufacture or distribution. During the initial period, however, the administrator may extend his or her time for action an additional 90 days. If during the original period (or its extension) the administrator believes that the information available

JOB SAFETY AND HEALTH LAW

is inadequate to make a reasoned finding that the proposed new substance or use does not present an unreasonable risk of injury to health or the environment, he or she may prohibit or limit the manufacture of the substance and obtain an injunction in court for that purpose. It would appear that in the absence of testing data submitted in compliance with Section 4, this injunction against manufacture or distribution of the new substance or use would continue indefinitely. If, however, the administrator finds, based on information provided, that the proposed new substance or use does present an unreasonable risk to health and safety, the administrator must proceed by means of the provisions of Section 6 to prohibit manufacture or to impose restrictive conditions (34).

The administrator of the EPA has promulgated myriad regulations implementing the Toxic Substances Control Act. These regulations can be found at 40 *CFR* Parts 702–766 and 790–799. Besides detailing the procedural and chemical inventory requirements of the act, the regulations set forth guidelines for the manufacture, processing and distribution of polychlorinated biphenyls and fully halogenated chlorofluoroalkanes (35). Moreover, the regulations require local education agencies to identify friable asbestos-containing material in public and private school buildings (Part 763). It is likely that the Toxic Substances Control Act will continue to be a major weapon in the federal health, safety, and environmental arsenal (36).

4.9 Clean Air Act

Like the Toxic Substances Control Act of 1976, the Clean Air Act was not enacted for the primary purpose of protecting workers (37). However, the EPA, the agency charged with administering the mandates of the Clean Air Act, has promulgated regulations designed to protect workers, among others, from accidental chemical releases. Under a final rule published by EPA in 1996, nearly 70,000 facilities that manufacture or handle certain hazardous chemicals were to develop risk management plans by June of 1999 to reduce the likelihood and severity of accidental chemical releases (38).

4.10 Federal Consumer Product Safety Act

The Consumer Product Safety Act of 1972 was drafted to apply only to "consumer products." That term is defined in the act in a manner that serves to exclude most products destined principally for use in workplaces covered by OSHA. Nevertheless, the act promises to provide increased protection of the American worker from hazardous products that by their nature are "consumer products" within the meaning of the act, yet are frequently found in the workplace (39).

The act created a Consumer Product Safety Commission and empowered that agency to promulgate "consumer product safety standards" applicable to consumer products found by the commission to present an unreasonable risk of injury. Such standards may be performance standards or they may require that products not be sold without adequate warnings or instruction. Where no feasible safety standard that could be promulgated would eliminate an unreasonable risk of injury, the commission is empowered to ban the consumer product altogether from interstate sale or distribution.

In addition to publishing safety standards, the commission is empowered to file suit and seek the seizure of a consumer product believed to be "imminently hazardous," regardless of whether the product in question is covered by already promulgated consumer product safety standards. The commission is also authorized to find, after hearing, that a consumer product presents a "substantial product hazard." In the event of such a finding, the commission may order the manufacturer, distributor, or retailer of the product to give public notice of that finding, and to repair, replace, or refund the purchase price of the product affected (40).

4.11 Hazardous Substances Act

The federal Hazardous Substances Act provides a mechanism by means of which the Consumer Product Safety Commission may find that a substance distributed in interstate commerce is "hazardous." After such a finding the commission may either impose packaging and labeling requirements to protect public health and safety or, in the cases of hazardous substances intended for the use of children or likely to be subject to access by children, or substances intended for household use, prohibit distribution altogether ("banned hazardous substance") (41). Insofar as safety in the U.S. workplace is concerned, the effect of the act is that hazardous substances distributed in interstate commerce and utilized by the U.S. worker will arrive safely packaged and accompanied by appropriate warnings.

4.12 The Atomic Energy Act of 1954 and Other Statutory Sources of Radiation Control

The Department of Labor has published occupational health and safety standards regulating exposure to ionizing (i.e., alpha, beta, gamma, X-ray, neutron, etc.) radiation and nonionizing (i.e., radiofrequency, electromagnetic) radiation. However, the primary federal law regulating human exposure to radiation is not OSHA but rather the Atomic Energy Act of 1954, as amended. Exercising power under that statute, the Nuclear Regulatory Commission (NRC) has published "Standards for Protection Against Radiation" (42).

A detailed discussion of those regulations is beyond the scope of this chapter. The operation of the standards can be summarized briefly as follows, however. Any person holding a license issued under the Atomic Energy Act of 1954 and using "licensed material" (i.e., radioactive or radiation-emitting material) may not permit the exposure of individuals within a "restricted area" (i.e., an area in which radioactive materials are being used) to greater doses of radiation than are set forth in the regulations (43).

Although the primary thrust of the NRC regulations is to control ionizing radiation within the "restricted area," the NRC has also published regulations on permissible levels of radiation in unrestricted areas, in effluents discharged into unrestricted areas, and for the disposal of radioactive materials by release into sanitary sewerage systems (44).

The administrator of the EPA, exercising authority under the Atomic Energy Act of 1954 (authority acquired by means of the Reorganization Plan No. 3 of 1970), has also promulgated regulations limiting exposure of the general population to ionizing radiation produced during the operation of nuclear power plants licensed by the NRC (45).

JOB SAFETY AND HEALTH LAW

There are additional federal agencies empowered to set standards for ionizing radiation control within areas under their jurisdiction. The Department of Labor, for example, has promulgated regulations regarding radiation exposure in underground mines (46). The Department of Labor has also issued radiation standards for uranium mining conducted under the Walsh-Healey Public Contracts Act (47).

Radiation generated by devices and products that are not governed by the Atomic Energy Act and licensed by the NRC is regulated by the federal Radiation Control for Health and Safety Act of 1968 (48).

4.13 Outer Continental Shelf Lands Act

The Outer Continental Shelf Lands Act authorizes the head of the department in which the Coast Guard is operating to promulgate and enforce "such reasonable regulations with respect to lights and other warning devices, safety equipment, and other matters relating to the promotion of safety of life and property" on the lands and structures referred to in the act or on the adjacent waters (49). If worker safety is deemed to be within the mandate of this statute, OSHA jurisdiction may be pre-empted with respect to working conditions that are governed by standards issued by the Coast Guard pursuant to this act.

4.14 North American Free Trade Agreement

The treaty, enacted in late 1993 and effective January 1, 1994, includes side pacts on the environment and labor issues. Among the provisions of these side pacts, any party, including government agencies and labor unions, may bring a complaint relating to job health and safety. Trinational arbitration panels made up of representatives from the United States, Mexico and Canada hear complaints and can assess monetary fines and institute trade sanctions against the offending nation if it is found not to have enforced its own laws.

5 REGULATION OF JOB SAFETY AND HEALTH BY THE STATES

One of the primary factors that induced Congress to enact the Occupational Safety and Health Act was the failure of many of the states to regulate workplace safety and health adequately (50). In passing OSHA, Congress hoped to ensure at least a minimum level of regulation of the conditions experienced by workers throughout the country.

The Occupational Safety and Health Act of 1970 preempts state regulation of job safety and health with respect to matters that OSHA regulates, even when a state has a more stringent regulation with respect to a particular hazard (51). However, OSHA does not totally ban the states from developing and enforcing occupational safety and health standards. Pursuant to Section 18(b) of the act, a state may regain jurisdiction over development and enforcement of occupational safety and health standards by submitting to the federal government an effective state occupational safety and health plan. Final approval of a state plan can lead ultimately to exclusive authority by a state over the matters included in its plan.

The process of regaining jurisdiction over the regulation of occupational safety and health begins with the submission of a plan that sets forth specific procedures for ensuring workers' safety and health. According to the regulations of the secretary of labor, the states can submit either of two types of plan: a complete plan or a developmental plan.

A "complete" plan (52) is a plan that, upon submission, satisfies the criteria for plan approval set forth in Section 18(c) of the act, as well as certain additional criteria outlined by the secretary of labor in administrative regulations (53). Complete plans are given "initial" approval by the secretary of labor upon submission. For at least 3 years following the "initial" approval, the secretary of labor will monitor the state plan to determine whether on the basis of the actual operations of the plan, the criteria set forth in Section 18(c) are being applied. If this determination [the "Section 18(e) determination"] is favorable, the state plan will be granted "final approval" and the state will regain exclusive jurisdiction with respect to any occupational safety or health issue covered by the state plan. Federal (i.e., OSHA) standards continue to apply to hazards not covered by the state program; thus, state plans need not address all hazards and yet gaps in protection are avoided.

A "developmental" plan (54) is a plan that, upon submission, does not fully meet the criteria set forth in the statute or in the regulations. A developmental plan may receive initial approval upon submission, however, if the plan contains "satisfactory assurances" that the state will take the necessary steps to bring its program into conformity within three years following commencement of the plan's operation.

If the developmental plan satisfies all the statutory and administrative criteria within the 3-year "developmental period," the secretary of labor will so certify and will initiate an evaluation of the actual operations of the state plan for purposes of making a Section 18(e) determination. The evaluation must proceed for at least one year before such a determination can be made.

Plans that have received final approval will continue to be monitored and evaluated by the secretary of labor pursuant to Section 18(f) of the act, which authorizes the secretary to withdraw approval if a state fails to comply substantially with any provision of the state's plan.

Although a state does not regain exclusive jurisdiction over matters contained in its plan until the plan receives final approval, a state with initial plan approval may participate in the administration and enforcement of the act prior to final approval by satisfying the following four criteria (55):

1. The state must have enacted enabling legislation conforming to that specified in OSHA and the regulations.
2. The state plan must contain standards that are found to be at least as effective as the comparable federal standards.
3. The state plan must provide for a sufficient number of qualified personnel who will enforce the standards in accordance with the state's enabling legislation.
4. The plan's provisions for review of state citations and penalties (including the appointment of the reviewing authority and the promulgation of implementing regulations) must be in effect.

JOB SAFETY AND HEALTH LAW

If the criteria above are met, the state plan is deemed to be "operational." Thereupon the federal government enters into an operational agreement with the state whereby the state is authorized to enforce safety and health standards under the state plan (55). During this period the act permits, but does not require, the federal government to retain enforcement activity in the state (56). Thus during this period an employer could be subject to enforcement activities by both the state and federal authorities. However, the secretary's regulations provide that once a plan (either complete or developmental) becomes "operational," the state will conduct all enforcement activity, including inspections in response to employee complaints, and accordingly, the federal enforcement activity will be reduced and the emphasis will be placed on monitoring state activity (55).

Plans have been approved for the Virgin Islands and 15 states: Alaska, Arizona, Connecticut, Hawaii, Indiana, Iowa, Kentucky, Maryland, Minnesota, New York, South Carolina, Tennessee, Utah, Virginia, and Wyoming. In addition, the secretary of labor has certified the (developmental) plans of Puerto Rico and 8 states: California, Michigan, Nevada, New Mexico, North Carolina, Oregon, Vermont, and Washington (57).

Thirteen states or territories plus the District of Columbia have submitted plans and are awaiting initial approval by the secretary of labor: Alabama, American Samoa, Arkansas, Delaware, Florida, Guam, Idaho, Massachusetts, Missouri, Oklahoma, Rhode Island, Texas, and West Virginia (57).

The following 11 states submitted plans at one time but have withdrawn them: Colorado, Georgia, Illinois, Maine, Mississippi, Montana, New Hampshire, New Jersey, North Dakota, Pennsylvania, and Wisconsin. Five states (Kansas, Louisiana, Nebraska, Ohio, and South Dakota) have never submitted plans to the Department of Labor (57).

Most states that are presently operating approved and/or certified plans have adopted standards that are substantially similar, if not identical, to the federal standards. At least five state plans, however, contain certain provisions that vary from the federal standards and have been approved as being "at least as effective" as the federal standards. The states are California, Hawaii, Michigan, Oregon, and Washington (57). In some instances these states have adopted standards that are more stringent than the analogous federal standards. Employers who are operating in more than one state should be aware that they may have to deal with different regulations, different enforcement procedures, and perhaps different interpretations of similar standards for purposes of complying with the applicable occupational safety and health laws (58).

5.1 State Jurisdiction in Areas Regulated by Federal Legislation Other than OSHA

If state legislation regulates a job safety or health issue that OSHA does not cover, the state may continue to enforce its relevant standards unless other applicable federal law has pre-empted state enforcement.

In some areas a federal regulatory scheme permits concurrent federal and state regulation of job safety and health. For example, the Federal Mine Safety and Health Act of 1977 states that no state law that was in effect on December 30, 1969, or that may become effective thereafter, shall be superseded by any provisions of the federal mine act unless the state law is in conflict with the mine act. State laws and rules that provide for standards more stringent than those of the mine act are deemed not to be in conflict (59).

On the other hand, the Federal Railroad Safety Act of 1970 has essentially pre-empted the states' regulation of railroad safety and health except in cases of a state having a more stringent law because of the need to eliminate or reduce an essentially local safety hazard (60).

6 EMPLOYERS' DUTIES UNDER THE OCCUPATIONAL SAFETY AND HEALTH ACT OF 1970

A private employer's primary duties under the Occupational Safety and Health Act of 1970 are found in Section 5(a), which provides that each employer:

1. Shall furnish to each employee employment and a place of employment that are free from recognized hazards that are causing or are likely to cause death or serious physical harm to employees.
2. Shall comply with occupational safety and health standards promulgated under the act (61).

6.1 The General Duty Clause [§ 5(a)(1)]

The essential elements of the so-called general duty clause of OSHA are the following (62):

1. The employer must render the workplace "free" of hazards that arise out of conditions of the employment.
2. The hazards must be "recognized."
3. The hazards must be causing or likely to cause death or serious physical harm (63).

6.1.1 Failure to Render Workplace "Free" of Hazard

It is fairly well settled that Congress did not intend to make employers strictly liable for the presence of unsafe or unhealthful conditions on the job. The employer's general duty must be an achievable one. Thus the term "free" has been interpreted by the courts and the commission to mean something less than absolutely free of hazards. Instead, the courts and the commission have held that the employer has a duty to render the workplace free only of hazards that are preventable (64).

The determination of whether a hazard is preventable generally is made in the context of an enforcement proceeding under the act when an employer asserts the inability of preventing the hazard as an affirmative defense to a proposed citation. The employer often contends that: (*1*) compliance is impossible or infeasible (65) because either the technology does not exist to prevent the hazard or the cost of the technology is prohibitive; (*2*) compliance with a standard will result in a greater hazard to employees than would noncompliance; or (*3*) the hazard was created by an employee's misconduct that was so unusual that the employer could not reasonably prevent the existence of the hazard.

When technology does not exist to prevent a hazard, OSHA does not require prevention by shutting down the employer's operation. Rather, the Secretary of Labor must be able to show that "demonstrably feasible" measures would have materially reduced the hazard [*National Realty and Construction Company* (64)]. Similarly, it seems that measures that, even though technologically feasible, would have been so expensive as to bankrupt the employer are not "demonstrably feasible." (For further discussion relating to feasibility, see Section 6.2.3 of this chapter.)

To establish the "greater hazard" defense, the employer must show that: (*1*) the hazards of compliance are greater than the hazards of noncompliance; (*2*) alternative means of protecting employees are unavailable; and (*3*) a variance either cannot be obtained or is inappropriate (66).

In addition, the courts and the commission have recognized that certain isolated or idiosyncratic acts by an employee that were not foreseeable by the employer could result in unpreventable hazards for which the employer should not be held liable. For example, in *National Realty and Construction Company*, the court of appeals stated:

> Hazardous conduct is not preventable if it is so idiosyncratic and implausible in motive or means that conscientious experts, familiar with the industry, would not take it into account in prescribing a safety program. Nor is misconduct preventable if its elimination would require methods of hiring, training, monitoring, or sanctioning workers which are either so untested or so expensive that safety experts would substantially concur in thinking the methods infeasible (67).

The court in *National Realty* emphasized, however, that an employer does have a duty to attempt to prevent hazardous conduct by employees. Thus the employer must adopt demonstrably feasible measures concerning the hiring, training, supervising, and sanctioning of employees to reduce materially the likelihood of employee misconduct (68). The Occupational Safety and Health Review Commission has ruled that an employer who has issued only oral reprimands to its employees for their failure to follow the employer's work rules fails to establish the unpreventable employee misconduct defense (69). Likewise, the Commission has ruled that an employer's failure to communicate its work rules by properly training its supervisors voids the defense (70), as does an employer's failure to take steps to discover respirator violations (71).

There is disagreement among the courts of appeals as to who bears the burden of proving employee misconduct. Most courts have held that the burden is appropriately placed on the employer. Others place the burden on the secretary to disprove unforeseeable employee misconduct (72). In 1998, the Secretary of Labor sought to end this split when he asked the United States Supreme Court to review the Fourth Circuit's decision in *Secretary of Labor v. L.R. Willson and Sons Inc.* (73), in which the Fourth Circuit had held that the Secretary must prove that employees' failure to comply with work rules was foreseeable in order to overcome the employee misconduct defense. In asking the Supreme Court for review of this decision, the Secretary asserted that the employee misconduct defense is an "affirmative defense," which must be proven by the employer. However, the Supreme

Court declined to review the decision, thus leaving it intact as the law of the Fourth Circuit and preserving the split among the circuits on this issue.

6.1.2 The Hazard Must Be a "Recognized" Hazard

The general duty clause does not apply to all hazards but only to the hazards that are "recognized" as arising out of the employment. The test for determining a "recognized" hazard is whether the hazard is known by the employer or generally by the industry of which the employer is a part (74). This test involves an objective determination and does not depend on whether the employer is aware in fact of the hazard (75).

With regard to a particular enforcement proceeding, the Second Circuit has held that the Secretary, not the employer, has the burden of proof to show that the employer knew about the safety violations for which it was cited (76).

In *American Smelting and Refining Company v. OSAHRC* (77), the Eighth Circuit held that the general duty clause is not limited to recognized hazards of types detectable only by the human senses but also encompasses hazards that can be detected only by instrumentation.

6.1.3 The Hazard Must Be Causing or Likely to Cause Death or Serious Physical Harm

It is not necessary that there be actual injury or death to trigger a violation of the general duty clause. The purpose of the act is to prevent accidents and injuries. Thus, violation of the general duty clause arises from the existence of a statutory hazard, not from injury in fact (78).

Proof that a hazard is "causing or likely to cause death or serious physical harm" does not require a mathematical showing of probability. Rather, if evidence is presented that a practice could eventuate in serious physical harm upon other than a freakish or utterly implausible concurrence of circumstances, the commission's determination of likelihood will probably be accorded considerable deference by the courts (79).

The term "serious physical harm" is defined neither in the act nor in the secretary's regulations, but OSHA's *Field Operations Manual* defines it to mean:

(i) Permanent, prolonged, or temporary impairment of the body in which part of the body is made functionally useless or is substantially reduced in efficiency on or off the job. Injuries involving such impairment would require treatment by a medical doctor, although not all injuries which receive treatment by a medical doctor would necessarily involve such impairment. Examples of such injuries are amputations, fractures (both simple and compound); deep cuts involving significant bleeding and which require extensive suturing; disabling burns and concussions.

(ii) Illnesses that could shorten life or significantly reduce physical or mental efficiency by inhibiting the normal function of a part of the body, even though the effects may be cured by halting exposure to the cause or by medical treatment. Examples of such illnesses are cancer, silicosis, asbestosis, poisoning, hearing impairment and visual impairment.

6.2 The Specific Duty Clause

Section 5(a)(2) of OSHA imposes on employers a duty to comply with the occupational safety and health standards promulgated by the secretary of labor. These standards constitute the employers' so-called specific duties under the act. Specific promulgated standards preempt the general duty clause, but only with respect to hazards expressly covered by the specific standards (80).

6.2.1 Processes for Promulgating Standards

The act established processes for promulgating three types of occupational safety and health standards: interim, permanent, and emergency.

Interim standards consist of standards derived from (*1*) established federal standards or (*2*) national consensus standards that were in existence on the effective date of OSHA (81). Section 6(a) of the act directed the secretary of labor to publish such standards in the *Federal Register* immediately after the act became effective (i.e., April 28, 1971) or for a period of up to two years thereafter. These standards became effective as OSHA standards upon publication without regard to the notice, public comment, and hearing requirements of the Administrative Procedure Act (82).

The intent of the interim standards provisions was to give the secretary a mechanism by which to promulgate speedily standards with which industry was already familiar and to provide a nationwide floor of minimum health and safety standards (83). The secretary's 2-year authority to promulgate interim standards expired on April 29, 1973.

Pursuant to Section 6(b) of the act, the secretary of labor is authorized to adopt "permanent" occupational safety and health standards to serve the objectives of OSHA. With respect to standards relating to toxic materials or harmful physical agents, Section 6(b)(5) specifically directs the Secretary to set the standard "which most adequately assures, to the extent feasible, on the basis of the best available evidence, that no employee will suffer material impairment of health or functional capacity even if such employee has regular exposure to the hazard dealt with by such standard for the period of his working life."

The promulgation of these "permanent" occupational safety and health standards requires procedures similar to informal rule making under Section 4 of the Administrative Procedure Act. Upon determination that a rule should be issued promulgating such a standard, the secretary must first publish the proposed standard in the Federal Register. Publication is followed by a 30-day period during which interested persons may submit written data or comments or file written objections and requests for a public hearing on the proposed standard. If a hearing is requested, the secretary must publish in the *Federal Register* a notice specifying the standard objected to and setting a time and place for the hearing. Within 60 days after the period for filing comments, or, if a hearing has been timely requested, within 60 days of the hearing, the secretary must either issue a rule promulgating a standard or determine that no such rule should be issued (84). Once a rule is issued, the secretary may delay the effective date of the rule for a period not in excess of 90 days to enable an affected employer to learn of the rule and to familiarize itself with its requirements.

Section 6(c) (1) of OSHA authorizes the secretary to issue emergency temporary standards if he or she determines: (*1*) that employees are exposed to grave danger from exposure to substances or agents determined to be toxic or physically harmful or from new hazards and (2) that such emergency standard is necessary to protect employees from such danger.

An emergency temporary standard may be issued without regard to the notice, public comment, and hearing provisions of the Administrative Procedure Act. It takes effect immediately upon publication in the *Federal Register*.

The key to the issuance of an emergency temporary standard is the necessity to protect employees from a grave danger, as defined in *Florida Peach Growers* (85). After issuing an emergency temporary standard, the secretary must commence the procedures for promulgation of a permanent standard, which must issue within 6 months of the emergency standard's publication in accordance with Section 6(c)(3).

6.2.2 Challenging the Validity of Standards

Any person who may be adversely affected by an OSHA standard may file a petition under Section 6(f) challenging its validity in the United States Court of Appeals in the circuit wherein such person resides or has the principal place of business. The petition may be filed at any time prior to the 60th day after the issuance of the standard. Unless otherwise ordered, the filing of a petition does not operate as a stay of the standard.

Section 6(f) of the act directs the courts to uphold the Secretary of Labor's determinations in promulgating standards if those determinations are "supported by substantial evidence in the record considered as a whole" (86). In practice, the courts have generally declined to apply a strict "substantial evidence" standard of review. Instead, the courts have chosen to apply two different standards depending on whether the agency determination to be reviewed is one of fact or policy. They have essentially taken the position that only the Secretary's findings of fact should be reviewed pursuant to a substantial evidence standard, while the Secretary's policy determinations should be substantiated by a detailed statement of reasons, which are subject to a test of reasonableness (87). The courts have adopted this approach with respect to emergency temporary standards as well as permanent standards [e.g., in *Florida Peach Growers* (84)].

In addition to a direct petition for review under Section 6(f), a majority of the courts of appeals have held that both procedural and substantive challenges to the validity of OSHA standards may be raised in enforcement proceedings under Section 11 (88). In fact, the Third Circuit in *Atlantic & Gulf Stevedores* (88) stated that the validity of a standard may be challenged not only in a federal court of appeals as part of an appeal from an order of the commission but in the commission proceedings themselves. The court explained, however, that in an enforcement proceeding invalidity is an affirmative defense to a citation and the employer bears the burden of proof on the issue of the reasonableness of the adopted standard. To carry its burden the employer must produce evidence showing why the standard under review, as applied to it, is arbitrary, capricious, unreasonable, or contrary to law (89).

6.2.3 Economic and Technological Feasibility of Standards

In enacting OSHA, Congress did not intend to make employers strictly liable for unavoidable occupational hazards. Accordingly, feasibility of compliance is a factor the secretary of labor must consider in developing occupational safety and health standards. As the U.S. Supreme Court explained in *American Textile Mfrs. Inst. Inc. v. Donovan* (90), OSHA's legislative history makes clear that any standard that is not economically or technologically feasible would *a fortiori* not be "reasonably necessary or appropriate" as directed by Section 3(8) of the act (91). Thus, in enacting OSHA, "Congress does not appear to have intended to protect employees by putting their employers out of business" (92).

In analyzing economic feasibility, the Secretary has tried to determine whether proposed standards threaten the competitive stability of an affected industry (93). The Supreme Court has not yet decided whether a standard that actually does threaten the long-term profitability and competitiveness of an industry would be "feasible" (94).

The Supreme Court has expressly decided, however, that in promulgating a toxic material and harmful physical agent standard under Section 6(b)(5), the secretary is not required to determine that the costs of the standard bear a reasonable relationship to its benefits [*American Textile Mfrs. Inst. Inc. v. Donovan* (90)]. Rather, Section 6(b)(5) directs the Secretary to issue the standard that "most adequately assures that no employee will suffer material impairment of health," limited only by the extent to which this is economically and technologically feasible, or, in other words, capable of being done (95). The Supreme Court left open the possibility, however, that cost-benefit analysis might be required with respect to standards promulgated under provisions other than Section 6(b)(5) of the act (96). The Court also left open the question of whether cost-benefit balancing by the secretary might be appropriate for deciding between issuance of several standards regulating different varieties of health and safety hazards (97).

In *Auto Workers v. OSHA* (98), the Court of Appeals for the District of Columbia affirmed somewhat reluctantly OSHA's authority to engage in rule making without using cost-benefit analysis. The case involved a lockout/tagout standard that OSHA initially promulgated in 1989. Upon the standard's initially being challenged, the appeals court remanded it to the agency in 1991 so that OSHA could address whether its promulgation of the standard had violated the doctrine of "nondelegation." The nondelegation doctrine is a bedrock concept in administrative law that, in seeking to preserve the constitutionally required separation of powers between branches of the federal government, requires Congress to provide reasonably specific guidelines to the agencies to which it delegates its authority to implement legislation and requires those agencies to act within these specific guidelines (99). In remanding the standard, the Court expressed concern that, in addressing a perceived job safety hazard, OSHA could take any position in the spectrum from doing nothing to promulgating a standard so burdensome as to bankrupt an industry. OSHA responded by listing six criteria that it viewed as limiting its rulemaking authority: (*1*) finding that the standard will substantially reduce a significant risk of material harm; (*2*) developing a standard that is economically feasible; (*3*) developing a standard that is technologically feasible; (*4*) finding a standard that employs the most cost-effective protective measures; (*5*) listing reasons why a promulgated standard would better effectuate the purposes of the act when the standard differs from an existing national consensus standard; and (*6*) supporting the agency's choice of standard with evidence in the rulemaking record and explaining any inconsistency with prior agency practice. In the appeals court's final decision upholding the standard, it expressed concern that OSHA's six listed criteria, standing alone, still might not pass constitutional muster because they may not ensure that the cost of safety standards are reasonably related to their benefits. However, on a briefing technicality, the court declined to decide whether the act actually requires a reasonable relationship between a standard's costs and its benefits.

In September of 1993, President Clinton issued Executive Order 12866, which stated that federal agencies should assess the costs and benefits of various regulatory approaches and should choose the approach that maximizes the net benefits to society. In January of 1996, the Office of Management and Budget issued guidelines expanding on the executive order, instructing agencies to consider alternative strategies for regulation and advising them to determine whether regulations should require different results for different segments of the regulated population. In March of 1999, bipartisan legislation was introduced in the Senate that would force federal agencies to conduct cost-benefits analyses in "major" rule makings, defined as those having an economic impact of $100 million or more. This legislation would not require an agency to pursue the less costly rulemaking option, but the cost-benefit analyses conducted during the rule making could be used to make an "arbitrary and capricious" challenge to a final rule upon its issuance (100).

In cases of violations of standards caused by employee disobedience or idiosyncratic behavior, the decisions of the courts and the commission have been similar to those rendered under the general duty clause (101) (see also discussion in Section 6.1.1 of this chapter).

For example, in *Brennan v. OSAHRC & Hendrix (d/b/a Alsea Lumber Co.)* (102), an employer was cited for violation of OSHA standards requiring workers to wear certain personal protective equipment. The record established that the violations resulted from individual employee choices, which were contrary to the employer's instructions. The Ninth Circuit affirmed the commission's decision vacating the citations, explaining as follows (103):

> The legislative history of the Act indicates an intent not to relieve the employer of the general responsibility of assuring compliance by his employees. Nothing in the Act, however, makes an employer an insurer or guarantor of employee compliance therewith at all times. The employer's duty, even that under the general duty clause, must be one which is achievable. See *National Realty*, supra. We fail to see wherein charging an employer with a nonserious violation because of an individual, single act of an employee, of which the employer had no knowledge and which was contrary to the employer's instructions, contributes to achievement of the cooperation [between employer and employee] sought by the Congress. Fundamental fairness would require that one charged with and penalized for violation be shown to have caused, or at least to have knowingly acquiesced in, that violation [emphasis added, footnote omitted].

Nevertheless, even though Congress did not intend the employer to be held strictly liable for violations of OSHA standards, an employer is responsible if it knew or, with the exercise of reasonable diligence, should have known of the existence of a violation. Thus, in *Brennan v. Butler Lime & Cement Co. and OSAHRC* (104), the Seventh Circuit drew on general duty clause concepts from the *National Realty* case (64) and explained that a particular instance of hazardous employee conduct may be considered preventable even if no employer could have detected the conduct or its hazardous nature at the moment of its occurrence, where such conduct might have been precluded through feasible precautions concerning the hiring, training, or sanctioning of employees (105).

In *Atlantic & Gulf Stevedores, Inc.* the Third Circuit held that such feasible precautions include disciplining or dismissing workers who refuse to wear protective headgear, even where such employer action could subject the company to wildcat strikes by employees adamantly opposed to the regulation (106).

There are some limits to the employer's obligations, however. In *Horne Plumbing and Heating Company v. OSAHRC and Dunlop*, the Fifth Circuit found that an employer had taken virtually every conceivable precaution to ensure compliance with the law, short of remaining at the job site and directing the employees' operations himself. The court held that the final effort of personally directing the employees was not required by the act, and that such an effort would be a "wholly unnecessary, unreasonable and infeasible requirement" (107).

6.2.4 Environmental Impact of Standards

Section 102(2)(C) of the National Environmental Policy Act of 1969 (NEPA) (108) requires all federal agencies, including OSHA, to prepare a detailed environmental impact statement in connection with major federal actions significantly affecting the quality of the human environment. The Secretary of Labor has identified the promulgation, modification, or revocation of standards that will significantly affect air, water or soil quality, plant or animal life, the use of land or other aspects of the human environment as always constituting such major action requiring the preparation of an environmental impact statement (109). On the other hand, promulgation, modification, or revocation of any safety standard, such as machine guarding requirements, safety lines, or warning signals, would normally qualify for categorical exclusion from NEPA requirements because "[s]afety standards promote injury avoidance by means of mechanical applications of work practices, the effects of which do not impact on air, water or soil quality, plant or animal life, the use of land or other aspects of the human environment" (110). The secretary's regulations regarding the procedure for preparation and circulation of environmental impact statements can be found at 29 CFR § 11.1 et seq.

6.2.5 Variances from Standards

Section 6(d) of OSHA provides that any affected employer may apply to the Secretary of Labor for a variance from an OSHA standard. To obtain a variance, the employer must show by a preponderance of the evidence submitted at a hearing that the conditions, practices, means, methods, operations, or processes used or proposed to be used will provide employees with employment and places of employment that are as safe and healthful as those that would prevail if the employer complied with the standard.

If granted, the Section 6(d) variance may nevertheless be modified or revoked on application by an employer, employees, or by the secretary of labor on his or her own motion at any time after six months from its issuance. Affected employees are to be given notice of each application for a variance and an opportunity to participate in a hearing.

The act also provides mechanisms to enable employers to obtain variances of a more temporary nature than those sought under Section 6(d). Section 6(b)(6)(A) provides for "temporary" variances upon application when an employer establishes, after notice to employees and a hearing, that (*1*) he or she is unable to comply with a standard by its effective date because of unavailability of professional or technical personnel or of materials and equipment needed to come into compliance with the standard or because necessary construction or alteration of facilities cannot be completed by the effective date, or (*2*) he or she is taking all available steps to safeguard employees against the hazards covered by the standard and has an effective program for coming into compliance with it.

Section 6(b)(6)(C) authorizes the secretary to grant a variance from any standard or portion thereof whenever he or she determines, or the secretary of health and human services certifies, that the variance is necessary to permit the employer to participate in an experiment approved by the secretary of labor or the secretary of health and human services designed to demonstrate or validate new and improved techniques to safeguard the health or safety of workers.

Finally, Section 16 permits the secretary of labor, after notice and an opportunity for hearing, to provide such reasonable limitations and rules and regulations allowing "reasonable variations, tolerances, and exemptions to and from any or all provisions of" the act as he or she may find necessary and proper to avoid serious impairment of the national defense.

6.3 Overview of Occupational Safety and Health Standards (111)

The bulk of federal job safety and health standards deal with occupational safety rather than with occupational health (112).

The occupational safety standards promulgated by the secretary of labor pursuant to OSHA are voluminous, encompassing hundreds of pages in the *Code of Federal Regulations*. A comprehensive analysis of these safety standards, many of which were adopted in 1971 as interim standards (113), is accordingly beyond the scope of this chapter (114).

For purposes of simplification, however, OSHA's safety standards can be broken down into the following general categories: (*1*) requirements relating to hazardous materials (e.g., compressed gas, acetylene, hydrogen, oxygen) and related equipment; (*2*) requirements for personal protective equipment and first aid; (*3*) requirements for means of egress, fire protection standards and the national electrical code; (*4*) general environmental controls (e.g., control of hazardous energy or "lockout/tagout"); (*5*) design and maintenance requirements for industrial equipment and walking-working surfaces; and (*6*) operational procedures and equipment utilization requirements for certain hazardous industrial operations such as welding, cutting and brazing, and materials handling. Certain industries are also subject to specialized safety standards (115).

Like the safety standards, the bulk of OSHA's health standards were promulgated in 1971 as interim standards under Section 6(a). At that time, previously established federal standards were used to set workplace exposure limits for approximately 400 chemical and hazardous substances. These were referred to as threshold limit values (TLVs) and were expressed in terms of milligrams of substance per cubic meter of air and/or parts of vapor or gas per million parts of air. TLVs were defined as representing conditions under which it was believed nearly all workers may be repeatedly exposed day after day without adverse effects (116).

Amidst ongoing criticism that the limits set in the original start-up standard were too lenient, in January of 1989, pursuant to Section 6(b), OSHA issued a massive rule revising its air contaminants standard (117). The rule revised permissible exposure limits (PELs) for many of the approximately 400 contaminants that were the subject of the original standard. In addition, PELs were established for substances not previously regulated (118).

Immediately following promulgation, the revised standard on air contaminants was challenged in court by various labor and industry groups. Eleven such cases were consolidated, and in 1992, in a surprise move, the Eleventh Circuit vacated the standard (*AFL-CIO v. OSHA*, 15 OSHC 1729, 965 F.2d 962). The Eleventh Circuit ruled that OSHA had failed to establish that each exposure limit reduced a significant risk to worker health or that each exposure limit was technologically and economically feasible for the affected industries. The Clinton administration decided not to appeal the decision to the United States Supreme Court, and OSHA thus returned to its original 1971 limits for the contaminants at issue (119). The Eleventh Circuit's decision was a major blow to the multiple-substance rulemaking approach, the proponents of which contend that OSHA cannot feasibly undertake rulemaking for each and every regulated contaminant in the depth that the court suggested.

Since the *AFL-CIO v. OSHA* decision, OSHA has moved deliberately to re-establish new PELs for the substances covered in its air contaminants standard. In January of 1996, OSHA announced that it was beginning the rule making process for 20 toxic substances, all of which had their revised PELs struck down in the *AFL-CIO v. OSHA* case. The substances are carbon disulfide, carbon monoxide, choroform, dimethyl sulfate, epichlorohydrin, ethylene dichloride, glutaraldehyde, *n*-hexane, 2-hexanone, hydrazine, hydrogen sulfide, manganese and its compounds, mercury and its compounds, nitrogen dioxide, perchloroethylene, sulfur dioxide, toluene, toluene diisocyanate, trimellitic anhydride, and vinyl bromide (120).

OSHA has promulgated other permanent health standards pursuant to Section 6(b) (e.g., regarding occupational exposure to bloodborne pathogens, discussed below at Section 6.3.13). Each of the permanent standards adopted under Section 6(b) can be viewed as a "complete" standard since each provides not only a specific value for the level of exposure to the toxic substance but also specifications as to monitoring, engineering controls, personal protective equipment, record keeping, medical surveillance, and other matters.

Compliance with promulgated standards is mandatory. In February of 1984, OSHA formally removed 153 provisions that included the advisory word "should" instead of the mandatory "shall" (121). The agency did not replace these provisions with rules promulgated under Section 6(b). Instead, the agency has relied on the general duty clause or mandatory language in general industry standards to enforce the deleted standards (122).

A summary of several of the major health standards under OSHA follows.

6.3.1 Asbestos

OSHA issued a revised asbestos rule in 1986 (123). The revised rule consisted of two simultaneously issued asbestos standards, one for general industry and one for construction. It covered asbestos, tremolite, actinolite and anthophyllite, reducing the 8-hour time-weighted average (TWA) from 2 fibers per cubic centimeter of air to 0.2 fibers per cubic centimeter of air.

The revised standards were challenged by the Asbestos Information Association, North America (AIA), and by the Building and Construction Trades Department (BCTD) of the AFL–CIO. The unions challenged OSHA's refusal to set a lower 8-hour TWA of 0.1 fiber per cubic centimeter of air. OSHA claimed it promulgated the higher limit because the lower limit was not feasible in the entire industry, and OSHA had discretion to decide what industries should be grouped together for regulatory purposes. The Court of Appeals for the District of Columbia disagreed and held that if OSHA was concerned with the administrative problems involved in desegregating industrial sectors for purposes of the TWA, it must make specific findings on the issue. Therefore, the D.C. Circuit remanded to OSHA on this issue, as well as on whether a short-term exposure limit (STEL) should be established (124).

In 1988, OSHA established an asbestos STEL of 1 fiber per cubic centimeter of air over a 30-min sampling period (125). OSHA decided that the 8-hr TWA issue would require additional rule making, and the D.C. Circuit Court approved the OSHA decision.

In August of 1994, OSHA published a final rule lowering the PEL for all forms of asbestos, tremolite, anthophyllite and actinolite dust to 0.1 fibers per cubic centimeter of air (126). The standard also required building owners to disclose the presence of asbestos in their buildings and enacted a new protective scheme requiring stricter controls for operations deemed to be more hazardous. Specifically, negative-pressure enclosures were required for workers removing "high-risk" asbestos-containing materials. The new rule affected general industry, the construction industry and the shipyard industry.

Eighteen different legal challenges to the new standard and to revisions to it that were published in 1995 and 1996 were filed by various industry and labor groups, and the cases were consolidated in the Fifth Circuit Court of Appeals. OSHA then set about settling these challenges with the various parties. In January of 1995, OSHA settled with the National Brake Care Coalition, allowing the use of chemical solvent sprays to clean asbestos-lined brake and clutch assemblies. In March of 1995, several roofing contractors and the Safe Buildings Alliance settled their challenges, with OSHA agreeing to amend the rule to allow employers to rely on information submitted by the roofers' association concerning negative exposure assessment and to allow alternative methods of compliance for installation, removal, repair and maintenance of certain roofing materials. The BCTD of the AFL–CIO settled its dispute with OSHA in exchange for OSHA agreeing to, among other things: (*1*) make certain modifications to the training programs prescribed for construction and shipyard workers; (*2*) require construction and shipyard employers to notify employees who are required to wear respirators under the rule that they may require their employer to provide a powered, air-purifying respirator instead of a negative-pressure one; and (*3*) require that warnings about exposure be communicated to workers in an effective manner (127). OSHA settled with the American Iron and Steel Institute in exchange for OSHA amending the standard to allow, among other things, that: (*1*) a "competent person" simply may be "available", not necessarily at the work site itself, for consultation while Class II, III or IV work is being performed under 29 *CFR* 1926.1101; and (*2*) negative exposure assessments may be made by a competent person on the basis of "professional judgment" (128). All the other challenges to the final asbestos rule were withdrawn, except for one: in July of 1997, in response to a challenge by the AIA, the Fifth Circuit vacated the construction and shipyard standards on asbestos-containing asphalt roof coatings and sealants, stating that the standards were not supported by substantial evidence (129). Specifically, the court held that OSHA had not made a critical distinction between built up roofing, which the court said "poses real risks of exposure", and roofing sealants, which the secretary had not seriously argued posed a significant risk of asbestos exposure. OSHA published revisions to the rule to conform with the Fifth Circuit decision (130), and the litigation over the asbestos standard was over (131).

6.3.2 Vinyl Chloride

On April 5, 1974, the Secretary promulgated an emergency standard for vinyl chloride. One of its requirements was that no worker be exposed to concentrations of vinyl chloride in excess of 50 parts per million parts of air over any 8-hour period. The prior TLV had set an exposure limit of 500 ppm (132).

A permanent standard was promulgated on October 1, 1974. It set a permissible exposure limit for vinyl chloride at no greater than 1 ppm averaged over 8 hours (133). The Court of Appeals for the Second Circuit upheld the standard in *Society of the Plastics Industry*, and it became effective in April 1975 (134).

Generally, the standard requires feasible engineering and work practice controls to reduce exposure below the permissible level wherever possible. Specifically, the employer is required to provide respirator protection and protective garments for employees (135). The employer must provide, for each employee engaged in vinyl chloride operation, a training program "relating to the hazards of vinyl chloride and precautions for its safe use" (136). Entrances to regulated areas must be posted with signs warning of the cancer-suspect nature of vinyl chloride (137). Moreover, the employer must institute a program of medical surveillance for each employee exposed to vinyl chloride in excess of the action level without regard for use of respirators (138). Finally, the employer must maintain accurate medical records to measure employee exposure to vinyl chloride (139).

6.3.3 Carcinogens

The Department of Labor excluded the American Industrial Hygiene Association's (ACGIH's) list of carcinogenic chemicals from its interim standards package, which it promulgated in 1971 (116). After approximately one year of consulting with NIOSH and receiving data and commentary from interested groups, the secretary promulgated emergency temporary standards on May 3, 1973, for a list of 14 chemicals found to be carcinogenic. Permanent standards for the 14 carcinogens were issued on January 29, 1974 (140).

In *Synthetic Organic Chemical Manufacturers Ass'n v. Brennan*, the Third Circuit upheld all the carcinogens standards except one and except for the provisions pertaining to medical examinations and to laboratory usage of said chemicals (141). The standard for 4,4′-methylene bis(2-chloraniline) (MOCA) was subsequently revoked by OSHA (142).

In January of 1980, OSHA promulgated a generic carcinogen policy for identifying, classifying, and regulating workplace carcinogens (143). The policy includes a process for screening chemicals and establishing priorities for rule making. It places substances in two categories: potential carcinogenicity and suggestive potential. Model permanent and emergency temporary standards are included (144). Lists of carcinogen candidates were to be published by OSHA periodically pursuant to the policy, but the listing requirements have been administratively stayed since 1983 (145). Since its issuance and amendment in 1981, little has been done to implement and utilize the policy due to legal challenges, pressure from outside interests to amend the policy, and inactivity at OSHA (146).

6.3.4 Coke Oven Emissions

In October of 1976, the secretary of labor issued a permanent standard regulating workers' exposure to coke oven emissions. The standard defines coke oven emissions as the benzene-soluble fraction of total particulate matter present during the destructive distillation of coal for the production of coke. It limits exposure to 150 micrograms of benzene-soluble fraction of total particulate matter per cubic meter of air averaged over an 8-hour period (147).

The standard mandates specific engineering controls and work practices that were to be in use by January 20, 1980. For example, the employer is required to provide protective clothing and equipment (148), hygienic changing rooms (149), and lunchrooms with a filtered air supply (150). The employer must ensure that in the regulated area food and beverages are not consumed, smoking is prohibited, and cosmetics are not applied (151). The employer must provide training to employees regarding the dangers of the regulated area (152). Additionally, the employer must post precautionary signs and labels (153). The employer must institute a medical surveillance program for all those employed in the regulated area at least 30 days per year (154). The employer also must maintain accurate medical records to measure employee exposure to coke oven emissions (155).

The steel industry challenged the validity of the coke oven standard, but it was upheld by the Third Circuit in *American Iron and Steel Institute v. OSHA* (156).

6.3.5 Lead

The OSH Administration issued a permanent standard regulating occupational exposure to lead in November 1978 (157). The standard set a permissible exposure limit of 50 micrograms per cubic meter of air over an 8-hour period (158). In addition, the standard required, among other things, the use of respirators and protective work clothing and equipment whenever lead exposure exceeds the PEL, compliance with vigorous rules on housekeeping and hygiene, biological monitoring, and medical surveillance. The standard also required a controversial medical removal protection (MRP) provision pursuant to which certain workers must be removed from the exposed workplace without loss of earnings, benefits, or seniority for at least 18 months. The standard further required employers to create safety and health training programs for their workers exposed to lead; to keep detailed records on environmental (workplace) monitoring, biological monitoring, and medical surveillance; and to make those records available to workers, certain of their representatives, and the government.

Virtually every aspect of the lead standard was challenged by the industry and by organized labor. The U.S. Court of Appeals for the District of Columbia rejected those challenges and upheld the standard as to certain industries (159). However, with respect to 38 other industries, the court held that OSHA failed to present substantial evidence or adequate reasons to support the feasibility of the standard for those industries and thus

JOB SAFETY AND HEALTH LAW

remanded to the secretary of labor for reconsideration of the technological and economic feasibility of the standard as to those industries (160).

OSHA subsequently amended the standard to require employers that cannot reach the PEL to reduce exposure only to the lowest feasible level (161), and it also found that the standard of 50 micrograms per cubic meter of air was technologically and economically feasible for all but nine of the remand industries (162). OSHA continued to gather information as to these nine industry sectors, and on July 11, 1989, OSHA issued a statement of reasons as to why the 50-microgram standard was feasible through engineering controls for eight of the remaining nine sectors (163). For the ninth industry sector, nonferrous foundries, OSHA found the 50-microgram standard was technologically but not economically feasible for small foundries and therefore promulgated a bifurcated standard for large and small nonferrous foundries (164).

In March of 1990, the D.C. Circuit upheld the standard with respect to all but six industries (165), and finally, on July 19, 1991, the court lifted its existing stay of the standard's engineering and work practice control requirements with respect to all of these six industries except the brass and bronze ingot manufacturing industry (166). In a notice published in October of 1995, OSHA set the final lead standard for brass and bronze ingot manufacturers at an 8-hour time weighted average of 75 micrograms per cubic meter of air, reflecting an agreement reached between the agency and two trade associations that finally settled all litigation arising out of the 1978 standard. These employers have six years to meet the 75-microgram limit (167).

In May of 1993, in response to legislation signed by President Bush in 1992, OSHA published an interim rule intended to protect construction workers who are exposed to lead on the job. The interim rule reduced the PEL for lead in construction from a TWA of 200 micrograms per cubic meter of air to 50 micrograms, and it established a 30-microgram threshold beyond which exposed workers are to be provided additional protections, including medical surveillance. The interim rule is to remain in effect until OSHA promulgates a final rule on the subject (168).

The EPA has published a final rule setting national training, certification and accreditation standards for people conducting lead-based paint inspections, risk assessments and abatement in certain houses and child-occupied facilities. The rule, which EPA was required to develop under Sections 402 and 404 of the Toxic Substances Control Act, includes a model state program by which states can seek permission from EPA to administer their own training, certification and abatement programs. The agency omitted, however, controversial provisions that had targeted lead-based paint activities in commercial and public buildings (169).

6.3.6 Cotton Dust

In June of 1978, OSHA promulgated its original cotton dust standard (170). Extensively revised in 1985, the standard sets different PELs for different industries: 200 micrograms per cubic meter of air averaged over an 8-hour period for exposures in yarn manufacturing and cotton washing operations; 500 micrograms in textile mill waste house operations and yarn manufacturing from lower-grade washed cotton; and 750 micrograms in slashing and weaving operations (171). The action levels are set at one-half the PELs (172).

The standard commands compliance with the PELs through a mix of engineering and work practice controls, except to the extent that employers can establish that such controls are infeasible (173). Specifically, the standard requires, among other things the provision of respirators, the monitoring of cotton dust exposure, medical surveillance, medical examinations, employee education and training, and the posting of warning signs (174).

Upon revision of the standard in 1985, OSHA exempted the cottonseed processing and cotton waste processing industries from all but the medical surveillance provisions; the cotton warehousing industry was totally exempted. OSHA also retained, in limited form, a provision that guaranteed transfer to another available position having dust exposure at or below the PEL, without loss of wage rate or benefits, for employees unable to wear respirators (175).

6.3.7 DBCP

In March of 1978, OSHA promulgated a permanent standard regarding 1,2-dibromo-3-chloropropane (DBCP) (176). The standard, which became effective April 17, 1978, sets a PEL of one part per billion parts of air over an 8-hour period. Where engineering controls and work practices are not sufficient to reduce exposure to permissible limits, respirators may be used as a supplement in order to achieve the required protection. Protective clothing is required where eye or skin contact may occur. The standard does not apply to the use of DBCP as a pesticide or when it is stored, transported, or distributed in sealed containers (177).

6.3.8 Acrylonitrile

In September of 1978, OSHA issued a permanent standard governing workplace exposure to acrylonitrile (178). The standard, which became effective November 2, 1978 (except as to training programs, which were to be set up by January 2, 1979, and engineering controls, which were to be installed by November 2, 1980), sets a permissible exposure limit of 2 parts per million parts of air over an 8-hour period. The standard also sets a ceiling limit of 10 ppm for any 15-minute period and an action level of 1 ppm. Exposure above the action level triggers periodic monitoring requirements, medical surveillance, protective clothing and equipment requirements, employee information and training, and housekeeping. Skin or eye contact with the substance is prohibited (179).

6.3.9 Benzene

OSHA's first permanent standard for benzene, which limited occupational exposure to 1 part benzene per million parts of air, was invalidated by the Fifth Circuit in 1978. On appeal, the U.S. Supreme Court affirmed, finding that OSHA had failed to show that exposure above the 1 ppm limit presented a "significant risk of material health impairment" to workers (180).

In 1987, OSHA issued a revised benzene standard (181). The revised standard sets the 8-hr TWA for benzene at 1 ppm, and the STEL at 5 ppm for a 15-min sampling period (182). Under the revised standards, the employer must provide a medical surveillance program to monitor the health of employees exposed to benzene in amounts over the action level of 0.5 ppm for more than 30 days per year. Additionally, OSHA requires a medical removal plan for temporary and permanent removal of employees showing adverse health effects from benzene exposure. The employer is required to provide six months of medical removal protection benefits to a removed employee, unless the employee has been transferred to a comparable job with benzene exposure below the action level (183).

Union and industry petitions for review of the new standard were withdrawn from the Third and D.C. Circuits in November 1987, shortly before the effective date of the new standard (184).

6.3.10 Inorganic Arsenic

The OSH Administration issued a permanent standard for inorganic arsenic in May 1978 (185). The standard established a PEL of 10 micrograms per cubic meter of air over an 8-hour period and also specified various other requirements such as engineering and work practice controls, respiratory protection, and employee monitoring and training (186).

The standard was challenged by industry in *ASARCO, Inc. v. OSHA* (187) and was remanded to OSHA by the Ninth Circuit for reconsideration in light of the Supreme Court's decision invalidating the benzene standard in *Industrial Union Dept. v. American Petroleum Institute* (180). In 1983, OSHA published a final risk assessment indicating that the 10-microgram limit reduced the risk of lung cancer by about 98 percent from the previous TLV of 500 micrograms and that such a reduction satisfies the Supreme Court's "significant risk" test (188). The Ninth Circuit later upheld the standard as supported by substantial evidence of a significant health risk (189).

The standard applies to all occupational exposures to inorganic arsenic except employee exposures in agriculture or industries involving pesticide application, the treatment of wood with preservatives, or the use of wood preserved with arsenic (190).

6.3.11 Ethylene Oxide

OSHA issued a final rule for ethylene oxide in 1984. Pursuant to the rule, OSHA established an 8-hour TWA of 1 ppm but deferred setting a STEL (191). Several months later, OSHA concluded that a STEL was not warranted by the available health evidence. This decision was challenged in court, and the D.C. Circuit, while upholding the validity of the standard, ruled that OSHA's decision regarding the STEL should be remanded for further consideration (192). Finally, in 1988, OSHA set an ethylene oxide STEL of 5 ppm for a 15-minute period (193).

The ethylene oxide standard requires the familiar use of engineering and work practice controls, when feasible, as well as medical surveillance, protective clothing and equipment, and employee training and warnings (194). A research study on the effectiveness of the ethylene oxide standard found that the standard was instrumental in significantly reducing worker exposure at a cost much less than what OSHA itself had projected. The study, which was released in 1992, credited the standard with increasing public awareness of the substance's toxicity and in spurring the development of new and inexpensive control technology (195).

6.3.12 Formaldehyde

OSHA's formaldehyde standard, issued in 1987 (196), was significantly revised in 1992 to, among other things, lower the existing PEL from 1 part formaldehyde per million parts air to 0.75 ppm for the 8-hour TWA (197). The revisions were prompted by a decision of the D.C. Circuit to remand the 1987 standard for OSHA's reconsideration of several issues (198).

The standard as amended has medical removal protection (MRP) provisions that supplement its medical surveillance requirements. It also requires specific hazard labeling and annual employee training. The amendments became effective at various dates in 1992 (199).

OSHA has announced that the standard's amendments will provide additional protection primarily to workers in the apparel, furniture, and foundry industries (200).

6.3.13 Bloodborne Pathogens

In one of its most controversial rulemakings, OSHA has promulgated a standard aimed at controlling occupational exposure to bloodborne pathogens such as the human immunodeficiency virus (HIV) and the hepatitis B virus (HBV). The final rule was issued in late 1991 following a comment period of several years (201).

Application of the standard is triggered by an employee's reasonably anticipated contact with blood or other potentially infectious materials (defined by a list of body tissues and fluids). Affected employers must develop written exposure control plans that identify exposed employees and tailor the standard's work practices and engineering controls to the particular work environment. Central to the standard is the mandatory use of "universal precautions," a practice by which all blood and other potentially infectious materials are treated as if known to be infected with bloodborne pathogens. The standard has stringent requirements regarding the disposal of wastes and used needles, the use of personal protective equipment, employee training, and labels and warnings. Employers must offer exposed employees the hepatitis B vaccine free of charge and must also provide free postexposure medical evaluation and testing. Additionally, extensive record keeping, including documented follow-up of exposure incidents, is mandated by the standard. The standard is aimed at protecting over five million workers, many of whom are outside the health care industry (202).

The standard was challenged by two health care industry groups, but the Seventh Circuit upheld the standard in January of 1993. However, the appeals court did agree with Home Health Services and Staffing Association Inc. that OSHA needed to explain better the compliance obligations of home health care employers who do not control their work sites, and the court remanded the rule to the agency for clarification of its application to those employers (203). The United States Supreme Court declined to review the Seventh Circuit's ruling, thus killing the challenge to the rule made by the other challenging group, the American Dental Association, which had argued that dentists and their employees generally were exposed less to blood than other health care providers.

6.3.14 Hazard Communication (204)

OSHA issued a hazard communication standard in 1983 (205). The original standard required manufacturers to establish a method to communicate the hazards of chemicals to those who worked with them, mainly through the use of labels on containers, materials safety data sheets (MSDS), and training programs. The Third Circuit upheld the standard in most respects (206).

On August 24, 1987, OSHA expanded the scope of the standard to encompass non-manufacturing employment (207). The Third Circuit denied subsequent petitions for review of the expanded standard (208). The standard now applies to all industries and is designed to protect 32 million workers from exposure to an estimated 650,000 chemicals (209).

In September of 1993, the Sixth Circuit Court of Appeals, in the context of an appeal from an enforcement proceeding, upheld the secretary's interpretation that the hazard communication standard requires target organ warnings on container labels. The court upheld a violation of 29 *CFR* 1910.1200(f)(1)(ii) against a manufacturer for failure to have "appropriate hazard warnings" on labels of hazardous chemicals. The manufacturer, which produced resins and molding compounds, had placed labels on its products that said, "WARNING! HARMFUL IF INHALED" (210).

OSHA published a final rule clarifying the hazard communication standard in February of 1994. The clarification, which OSHA made under pressure from the Office of Management and Budget, applied to: (*1*) certain exemptions from labeling requirements and other provisions in the standard; (*2*) parts of the provision for a written hazard communication program; (*3*) the duties of distributors, manufacturers and importers of hazardous chemicals to provide MSDSs; and (*4*) provisions regarding the content of safety data sheets (211).

Additionally, OSHA promulgated another final rule in July of 1994 requiring employers who receive shipments of hazardous materials to leave any existing placards or warning labels on the shipments until the materials are removed from their containers (212).

6.3.15 Noise

OSHA's standard for occupational exposure to noise specifies a maximum permissible noise exposure level of 90 decibels (dB) for a duration of 8 hours (213). Employers are required to use feasible engineering or administrative controls, or a combination of both, whenever employee exposure to noise in the workplace exceeds the permissible exposure level. Personal protective equipment may be used to supplement the engineering and administrative controls where such controls are not able to reduce the employee exposure to within the permissible limit.

In 1983, OSHA issued a final hearing conservation amendment to the noise standard (214). Under the amendment, the employer must establish a baseline audiometric measurement from which it must compare subsequent annual measurements to determine if a shift in hearing capability has occurred for any employees whose exposures equal or exceed an 8-hour TWA of 85 dB (215). To determine if the exposure equals or exceeds this action level, employers are required to use personal sampling of noise where factors, such as high worker mobility or significant variation in sound levels, make area monitoring generally inappropriate (216). The employer must make hearing protectors available to all employees whose exposure equals or exceeds an 8-hour TWA of 85 dB and must ensure that hearing protectors are worn by any employee who has experienced a shift in hearing capability (217). The Fourth Circuit has upheld the amendment (218).

In 1987, OSHA sought public comments as to whether and to what extent the information collection requirements of the noise standard could be reduced without lessening the standard's effectiveness in preventing hearing loss. OSHA later announced, however, that no such changes would be made (219).

6.3.16 Cadmium

OSHA issued its final rule for cadmium exposure in September of 1992 (220). Cadmium is a heavy metal that has been linked to lung cancer and kidney disease. The rule set a PEL of 5 micrograms of cadmium per cubic meter of air, to be achieved primarily by engineering controls. Immediately, eight different lawsuits challenging the standard were filed by various manufacturers and industry groups. By June of 1993, OSHA had settled all but one of these challenges by addressing what specific measures the individual manufacturers needed to take to comply with the standard. In the remaining challenge, the D.C. Circuit Court of Appeals in December of 1993 stayed the operation of the rule as applied to the cadmium pigments industry (221). The trade association had argued that cadmium pigments do not have the carcinogenic effects found in other forms of cadmium because the pigments are less soluble than other forms. With the stay, cadmium pigments manufacturers became bound only by the earlier cadmium standard, which dictated a PEL of 50 micrograms per cubic meter with an action level of 25 micrograms. Later, in the same case, the Eleventh Circuit ordered OSHA to restudy the technological and economic feasibility of the final cadmium standard for the cadmium pigments industry (222).

6.3.17 Methylene Chloride

The final OSHA standard for exposure to methylene chloride was initially published in January of 1997 (223). Methylene chloride is used as a solvent in the furniture stripping industry and as an adhesive in the manufacture of flexible foam; it has been linked to cancer, increased risk of heart attack and nerve damage. The new final standard, which was to be phased in over the following three years, reduced the PEL from 500 parts per million to 25 ppm over an eight-hour, time-weighted period. Other provisions required employers to provide training, medical surveillance and exposure monitoring. The United Auto Workers and the Halogenated Solvents Industry Alliance challenged the rule (224). The trade association's suit sought review of the standard under the Small Business Regulatory Enforcement Fairness Act of 1996 (225), GOP-sponsored legislation that sought to give small businesses "more influence over the development of regulations; additional compliance assistance for federal rules; and new mechanisms for addressing enforcement actions by agencies." The trade association argued for an alternative limit of 50 ppm; the UAW sought inclusion of medical removal protection in the rule. In response to these challenges, OSHA revised its final rule to extend the dates for certain employers to be in compliance with engineering controls and respiratory protection and to provide medical removal protection (226).

OSHA also faced opposition over the standard from Congress and from the American College of Occupational and Environmental Medicine. Congress directed OSHA to give its highest priority to consulting with smaller employers with regard to the standard and barred OSHA from enforcing the standard against employers who were found to have violated it because of their own technical or economic infeasibility of complying (227). OSHA issued a new compliance instruction in response to this Congressional directive (228). The physicians' group sought reopening of the methylene chloride rulemaking, arguing that the standard's allowing required medical surveillance to be conducted by health care professionals other than licensed doctors was a "radical departure" from other OSHA standards and put employees at risk of incorrect diagnosis and treatment (229). However, OSHA denied ACOEM's petition.

6.3.18 Confined Spaces

In January of 1993, OSHA published its final rule on confined spaces, requiring employers to establish permit systems and training programs for workers (230). The standard was designed to protect workers, who enter storage tanks and other confined spaces nearly five million times each year, from toxic atmospheres, lack of oxygen and other hazards. The rule requires employers to identify all confined spaces at their work sites that could pose a hazard and to establish policies preventing unauthorized entry into them; it also requires employers to provide trained rescue teams, although off-site emergency response personnel can be used to satisfy this requirement. OSHA issued a compliance directive stating that violations of the confined space rule generally will be classified as serious because of the danger that any violation can lead directly to death or serious injury (231). Settling several legal challenges, OSHA issued an amendment to the final confined space rule in December of 1998 that provides for greater participation by employees in the permitting system, allows employees to observe testing or monitoring of confined spaces and clarifies employers' duties in providing the rescue program (232).

OSHA has also promulgated a separate final confined space standard for the shipyard industry (233), which requires shipyard employers: (*1*) to designate "competent" individuals to make initial evaluations of confined spaces; (*2*) to ensure that detailed records of atmospheric testing of confined spaces are kept; and (*3*) to label unsafe confined spaces.

6.3.19 Butadiene

In November of 1996, after a negotiated rulemaking process involving OSHA, labor and industry groups, a final rule for exposure to 1,3-butadiene was published (234). Butadiene is a flammable and carcinogenic chemical used in the production of synthetic rubber. The standard calls for reduction in the PEL from 1,000 parts per million to 1 ppm over an eight-hour period and for a STEL of 5 ppm. If exposures reach 0.5 ppm, additional engineering controls are to be implemented, and the rule also covers medical surveillance, exposure monitoring and other issues.

6.3.20 Respirators

OSHA published its final respiratory protection rule in January of 1998 (235). The standard requires employers to develop and implement written respiratory protection programs. Such programs are to list all circumstances in the employees' working environment for which respirators are necessary to protect their health and are to contain workplace-specific procedures for respiratory protection. OSHA estimated that five million workers at over a million workplaces would be protected by the final standard, which is considered to be a "building block standard" because future standards that call for respirator use in working with specific toxic substances will refer back to the final respiratory protection rule (236). The respiratory protection standard is currently being challenged by several industry groups (237).

6.3.21 Personal Protective Equipment

OSHA published a final rule setting out employers' requirements to provide their workers with protective equipment, including gloves, goggles, helmets and safety shoes, in April of 1994 (238). The rule was intended to provide improved eye, face, hand, head and foot protection for nearly 11.7 million employees (239).

In October of 1997, the Review Commission ruled that employers need not pay for the equipment required under the personal protective equipment standard (240). Rather than appealing this decision, OSHA decided to revise its final rule to make clear that it intended for employers to pay for the required equipment. In March of 1999, the agency published a proposed rule that would require employers to pay for most required personal protective equipment (241).

6.3.22 Proposed Standards

Perhaps the most controversial rulemaking OSHA has ever engaged in is the proposed ergonomics standard. As of early 1999, the standard was still in the proposal stage, with a draft proposal having been released on February 19, 1999. The draft rule would require employers to evaluate potential workplace hazards for employees engaged in manufacturing or manual handling operations and, if a hazard were discovered, to implement training and medical management (242). OSHA released the draft proposal after getting the go-ahead from Congress, which had prohibited any action on an ergonomics rulemaking since the GOP takeover in 1994. In allowing OSHA to move forward, however, Congress attached the condition that the link between musculoskeletal disorders and workplace exposure to repetitive stress would have to undergo further study by the National Academy of Sciences (243). In March of 1999, bipartisan legislation was introduced in the House to block OSHA from issuing an ergonomics standard before completion of the NAS study; according to the chief sponsor, the bill does not prevent the agency from issuing a proposed rule, though (244). Without an ergonomics standard, OSHA has been using the general duty clause of the act to force abatement of alleged repetitive stress hazards, with mixed results (245). The 1990s have seen considerable developments on the ergonomics front in the courts (246) and in state agencies (247). These developments are beyond the scope of this chapter.

In March of 1994, as part of a larger rulemaking on indoor air quality in the workplace, OSHA proposed a rule that would require virtually all employers to ban smoking or to provide separately ventilated areas for smokers (248). As of early 1999, OSHA had not taken final action on the indoor air proposal, and an outright smoking ban would probably join the proposed ergonomics standard among OSHA's most controversial actions; however, several states and agencies have enacted workplace smoking bans (249), and, of course, litigation of smoking-related claims has flourished (250).

OSHA proposed a tuberculosis standard in October of 1997 that would cover an estimated 5.3 million health care workers (251). The rule, as proposed, would require covered employers to develop written exposure control and response plans, install engineering controls, offer employees free TB skin tests and provide respiratory protection for workers exposed to certain conditions.

7 EMPLOYEE RIGHTS AND DUTIES UNDER JOB SAFETY AND HEALTH LAWS

7.1 Employee Duties

Section 5(b) of OSHA requires employees to "comply with occupational safety and health standards and all rules, regulations, and orders issued pursuant" to the act that are applicable to their own actions and conduct.

The act does not, however, expressly authorize the secretary of labor to sanction employees who disregard safety standards and other applicable orders. The U.S. Court of Appeals for the Third Circuit has held that although Section 5(b) would be devoid of content if not enforceable, Congress did not intend to confer on the secretary or the commission the power to sanction employees (252).

Section 110(g) of the Federal Mine Safety and Health Act of 1977 not only requires mine employees to comply with health and safety standards promulgated under that statute, it also authorizes the imposition of civil penalties on miners who "willfully violate the mandatory safety standards relating to smoking or the carrying of smoking materials, matches, or lighters."

7.2 Employee Rights

The Occupational Safety and Health Act grants numerous rights to employees and/or their authorized representatives. The most fundamental employee right is the right set forth in Section 5(a) to a safe and healthful employment and place of employment. Other significant rights of employees and/or their authorized representatives include the following:

1. The right to request a physical inspection of a workplace and to notify the Secretary of Labor of any violations that employees have reason to believe exist in the workplace [§ 8(f)].
2. The right to accompany the Secretary during the physical inspection of a workplace [§ 8(e)] (253).
3. The right to challenge the period of time fixed in a citation for abatement of a violation of the act and an opportunity to participate as a party in hearings relating to citations [§ 10(c)].
4. The right to be notified of possible imminent danger situations and the right to file an action to compel the secretary to seek relief in such situations if he or she has "arbitrarily and capriciously" failed to do so [§§ 13(c) and 13(d)].

5. Various rights, including the right to notice, regarding an employer's application for either a temporary or permanent variance from an OSHA standard [§ 6(b)(6) and § 6(d)].
6. The right to observe monitoring of employee exposures to potentially toxic substances or harmful physical agents, the right to records thereof, and the right to be notified promptly of exposures to such substances in concentrations which exceed those prescribed in a standard [§ 8(c)].
7. The right to petition a court of appeals to review an OSHA standard within 60 days after its issuance [§ 6(f)].

Section 11(c) of OSHA makes it unlawful for any person to discharge or in any manner discriminate against an employee because the employee has exercised his or her rights under the act. This provision is designed to encourage employee participation in the enforcement of OSHA standards (254).

Employees who believe they have been discriminated against in violation of Section 11(c) must file a complaint with the secretary of labor within 30 days after such violation has occurred. The secretary is to investigate the complaint, and if he or she determines that Section 11(c) has been violated, he or she is authorized to bring an action in federal district court for an order restraining the violation and for recovery of all appropriate relief, including rehiring or reinstatement of the employee to his or her former position with back pay. The statute authorizes only the secretary of labor to bring an action for violation of Section 11(c) (255).

The Federal Mine Safety and Health Act of 1977 also contains a broad antiretaliation provision and grants other rights to mine employees as well (256).

An employee has no explicit right, under OSHA, to refuse a work assignment because of what he or she feels is a dangerous working condition. The Secretary of Labor, however, has issued an administrative regulation that interprets the act as implying such a right under certain limited circumstances (257). The U.S. Supreme Court, in *Whirlpool Corp. v. Marshall* (258), has held the promulgation of that regulation to be a valid exercise of the Secretary's authority under the act. The Court observed that despite the detailed statutory scheme for speedily remedying dangerous working conditions, circumstances may arise when an employee justifiably believes that the statutory scheme will not sufficiently protect him:

> [S]uch a situation may arise when (*1*) the employee is ordered by his employer to work under conditions that the employee reasonably believes pose an imminent risk of death or serious bodily injury, and (*2*) the employee has reason to believe that there is not sufficient time or opportunity either to seek effective redress from his employer or to apprise OSHA of the danger (259).

In holding that the regulation conformed to the fundamental objective of the act to prevent occupational deaths and injuries, the Court observed that the regulation also served to effectuate the general duty clause of Section 5(a)(1)(260).

The First Circuit has ruled that OSHA is immune from liability in claims that employees were injured as a result of negligent inspections by the agency's compliance officers (261).

JOB SAFETY AND HEALTH LAW

It is fairly well settled that OSHA does not create a private right of action for damages suffered by an employee as the result of an employer's violation of the act (262). Finally, Section 4(b)(4) states that the act does not supersede or in any manner affect any worker's compensation law (263).

ACKNOWLEDGMENTS

Authors of this Chapter in the previous Edition of *Patty's Industrial Hygiene and Toxicology* are Martha Hartle Munsch, JD and Jacqueline A. Koscelnik, JD. It is respectfully acknowledged that much of this Chapter remains their work from the previous Edition. Some general orientation text and the updating with listings and discussion of post 1992 materials and events are by the current author.

BIBLIOGRAPHY

1. See generally H. Linde and G. Bunn, *Legislative and Administrative Processes*, Foundation Press, Mineola, NY, 1976, p. 814. The Administrative Procedure Act is codified at Title 5 of the United States Code (USC) §§ 551 et seq.
2. See Bureau of National Affairs (BNA), *The Job Safety and Health Act of 1970*, Washington, D.C., 1971, pp. 13–21. The Occupational Safety and Health Act is codified at 29 U.S.C. §§ 651–678. References to the act in this chapter and the next are to the appropriate section of the statute itself and do not include a corresponding citation to the United States Code.
3. P.L. 105-297, cited as "The Occupational Safety and Health Administration Compliance Assistance Authorization Act of 1998," amended Section 21 of OSHA to add, as subsection (d), language specifically authorizing OSHA's voluntary employer consultation programs, which were created under the Reagan administration. P.L. 105-298 amended Section 8 of OSHA to add, as subsection (h), a one-sentence directive prohibiting citation "quotas" from being used to evaluate the performance of OSHA inspectors.
4. N. Ashford, *Crisis in the Workplace*, M.I.T. Press, Cambridge, MA, 1976, pp. 141, 236–237.
5. See American Bar Association, *Report of the Committee on Occupational Safety and Health Law*, ABA Press, Chicago, 1975, p. 107.
6. For a discussion of the weight to be accorded by the secretary of labor to the NIOSH recommendations, see *Industrial Union Department, AFL-CIO v. Hodgson*, 499 F.2d 467, 476–77; 1 OSHC 1631 (D.C. Cir. 1974). [References to "F.2d" designate the volume (e.g., 499) and page (e.g., 467) of the *Federal Reporter*, Second Series, which contains the official reported decisions of the U.S. Courts of Appeals as published by the West Publishing Company. References to "OSHC" designate the same decision as reported in the "Occupational Safety and Health Cases" published by the Bureau of National Affairs (BNA).
7. See Ashford, *Crisis in the Workplace* (Ref. 4), pp. 249–251. Section 27(b) of the act established a National Commission on State Workmen's Compensation Laws, which was directed to study and evaluate such laws to determine whether they provide an adequate, prompt, and equitable system of compensation for injury or death arising out of, or in the course of, employment. The commission's tasks were completed in July 1972, and the commission was disbanded. For

a description of the activities of the commission and an evaluation of its work, see Ashford, *Crisis in the Workplace*, pp. 246, 289–292.
8. The Review Commission has held that the Secretary may even cite individuals or companies who have gone out of business and fine them for violating OSHA regulations; such authority is intended to encourage future compliance with job safety rules. *Secretary of Labor v. Yandell*, OSHRC, No. 94-3080, March 12, 1999; *Secretary of Labor v. Kenny Niles Construction Co.*, OSHRC, No. 95-1539, March 12, 1999.
9. In 1998, President Clinton signed legislation extending OSHA coverage to U.S. Postal Service facilities. BNA, OSHR, *Current Report* for September 30, 1998. As of early 1999, legislation had been introduced on the House floor to extend OSHA coverage to state, municipal and county employers. BNA, OSHR, Current Report for March 3, 1999.
10. *Southern Pacific Transportation Co. v. Usery and OSAHRC*, 539 F.2d 386; 4 OSHC 1693 (5th Cir. 1976), certiorari denied, 434 U.S. 874; 5 OSHC 1888 (1977); *Southern Railway Company v. OSAHRC and Brennan*, 539 F.2d 335; 3 OSHC 1940 (4th Cir.), *certiorari denied*, 429 U.S. 999; 4 OSHC 1936 (1976); *Baltimore & Ohio Railroad Co. v. OSAHRC*, 548 F.2d 1052; 4 OSHC 1917 (D.C. Cir. 1976).
11. 539 F.2d at 391; 4 OSHC at 1696.
12. The Third Circuit has also adopted this definition of "working conditions." See *Columbia Gas of Pennsylvania, Inc. v. Marshall*, 636 F.2d 913; 9 OSHC 1135 (3rd Cir. 1980).
13. *Fineberg Packing Co.*, 1 OSHC 1598, 1599 (Rev. Comm. 1974).
14. 520 F.2d 1161; 3 OSHC 1566, 1572 (D.C. Cir. 1975).
15. American Bar Association, *Report of the Committee on Occupational Safety and Health Law*, ABA Press, Chicago, 1976, pp. 247–248.
16. The Federal Mine Safety and Health Act of 1977 is codified at 30 USC §§ 801 et seq.
17. 30 USC § 811 (a). Comprehensive regulations have been promulgated by the Mine Safety and Health Administration. Those regulations appear in title 30 of the *Code of Federal Regulations* at Parts 1 to 100. The *Code of Federal Regulations* is hereinafter cited as "*CFR*." The mandatory health standards can be found at 30 *CFR* Parts 70 et seq.
18. For example, the EPA announced in May of 1996 that farm workers working with molinate would be required to wear protective clothing and engage in other risk-reducing practices in order to minimize adverse male reproductive effects from exposure to the herbicide. BNA, OSHR, *Current Report* for May 15, 1996.
19. The FEPCA is codified at 7 USC §§ 136 et seq. It is a comprehensive revision of the Federal Insecticide, Fungicide and Rodenticide Act of 1970, 7 USC §§ 135 et seq. (1970). The regulations promulgated by EPA to protect farm workers from toxic exposure to pesticides are found in 40 *CFR* §§ 170.1 et seq.
20. See, for example, *PBR, Inc. v. Secretary of Labor*, 643 F.2d 890, 896; 9 OSHC 1357, 1361 (1st Cir. 1981). The Secretary of Labor has acknowledged that under the Federal Railroad Safety Act the Department of Transportation (DOT) has authority to regulate all areas of employee safety for the railway industry. *Southern Railway Company v. OSAHRC* (10), 539 F.2d at 333; 3 OSHC at 1941. However, the scope of DOT's statutory authority to regulate matters relating to worker health is still unsettled. *Southern Pacific Transportation Company v. Usery* (10), 539 F.2d at 389; 4 OSHC at 1694, n. 3. The Federal Railroad Safety Act is codified at 45 USC §§ 421 et seq. Other federal statutes dealing with railway safety include the Safety Appliance Acts, 45 USC §§ 1–16; the Train Brakes Safety Appliance Act, 45 USC § 9; the Hours of Service Act, 45 USC §§ 61 et seq.; and the Rail Passenger Service Act, 45 USC §§ 501 et seq. Department of Transportation regulations relating to railway safety can be

JOB SAFETY AND HEALTH LAW 1633

found in 49 *CFR*, Chapter II. The Department of Transportation also has statutory authority to regulate safety in modes of transportation other than rail. For example, the secretary of transportation may prescribe requirements for "qualifications and maximum hours of service of employees of, and safety of operation and equipment of, a motor carrier." 49 USC § 3102 (Revised Special Pamphlet 1992). For additional areas of DOT jurisdiction, see Sections 4.4–4.6 of this chapter.

21. See Section 3 of this chapter.
22. The provisions of the statute regarding safety regulation of civil aeronautics are codified at 49 USC §§ 1421 et seq.
23. See 49 USC § 1432(a).
24. See decision in *Usery v. Northwest Orient Airlines, Inc.*, 5 OSHC 1617 (E.D.N.Y. 1977). ["E.D.N.Y." refers to the federal district court in the Eastern District of New York. Reported decisions of federal district courts dealing with job safety and health matters are reported in BNA's Occupational Safety and Health Cases (OSHC), and many are also reported in West Publishing Company's *Federal Supplement* (F. Supp.)]. The administrative law judge determined that the FAA had exercised its authority by requiring each air carrier to maintain a maintenance manual that must include all instructions and information necessary for its ground maintenance crews to perform their duties and responsibilities with a high degree of safety.
25. The HMTA is codified at 49 USC App. §§ 1801 et seq. A table of materials that have been designated as hazardous by the Secretary can be found at 49 CFR Part 172. The Secretary's regulations prescribe the requirements for shipping papers, package marking, labeling, and transport vehicle placarding applicable to the shipment and transportation of those hazardous materials.
26. *Secretary of Labor v. Yellow Freight Systems, Inc.*, 17 OSHC 1699 (Rev. Comm. 1996).
27. 3 OSHC 1601 (Rev. Comm. 1975).
28. The Secretary of Transportation has exercised statutory authority under NGPSA to promulgate safety standards for employees at natural gas facilities. These regulations can be found at 49 *CFR* Part 192. The NGPSA is codified at 49 USC App. §§ 1671 et seq. See also *Columbia Gas of Pennsylvania, Inc. v. Marshall* (12) (regulation by DOT requiring operators to take steps to minimize danger of accidental ignition of gas while employees were performing a "hot tap" on existing gas main while installing auxiliary natural gas pipeline preempted authority of OSHA over the matter).
29. There also exist thousands of state and local laws regulating noise, discussion of which is beyond the scope of this work. See Compilation of State and Local Ordinances on Noise Control, 115 Cong. Rec. 32178 (1969).
30. See 49 USC App. § 1431. In carrying out this task the administrator of the FAA has adopted and published aircraft noise standards and regulations. See 14 *CFR* Part 36. Section 7 of the Noise Control Act of 1972, discussed below, amended the noise abatement and control provision of the Federal Aviation Act to provide generally that standards adopted with respect to aircraft noise must have the prior approval of the administrator of the EPA. In addition, the Noise Control Act was amended by the Quiet Communities Act of 1978, P.L. 95-609, 92 Stat. 3079, to provide for, among other things, a unified effort among state, local, and federal authorities to develop an effective noise abatement control program with respect to aircraft noise associated with airports.
31. The Noise Control Act is codified at 42 USC §§ 4901 et seq. The products with which the administrator is statutorily authorized to deal must come from among four categories: construction equipment, transportation equipment (including recreational vehicles and related equip-

ment), any motor or engine (including any equipment of which an engine or motor is an integral part), and electrical or electronic equipment. The administrator has promulgated comprehensive regulations under the Noise Control Act, which appear at 40 *CFR* Subchapter G, Parts 201–211. To date these regulations cover interstate rail carriers (Part 201), motor carriers engaged in interstate commerce (Part 202), procedure and criteria for determining low-noise-emission products (Part 203), construction equipment including portable air compressors (Part 204), and transportation equipment, including medium and heavy trucks (Part 205). Pursuant to Section 8 of the Noise Control Act, the administrator has also published product noise labeling requirements, which can be found at 40 CFR Part 211.

32. The Toxic Substances Control Act of 1976 is codified at 15 USC §§ 2601 et seq. Prior to 1976 there existed no enactment that authorized the federal government to regulate toxic substances generally. Although regulations published under OSHA did govern worker exposure to toxic substances in the workplace (see Section 6.3, this chapter), federal law lacked a general authority to prohibit or restrict the manufacture of such substances.

33. For example, the EPA published a final rule in 1996 (61 *Federal Register* 7421, February 28, 1996) amending 40 *CFR* 712 to require manufacturers of 28 designated alkylphenols and alkylphenol ethoxylates to submit to EPA reports on volume, end use and exposure as well as past, current and prospective unpublished health and safety studies, so that such information could be used to test the substances further. (The *Federal Register* is published by the federal government every business day and is hereinafter cited as "*Fed. Reg.*").

34. Section 7 of the act empowers the administrator of EPA to commence a civil action in federal court for the purpose of seizing an "imminently hazardous chemical substance," defined in the act as one that "presents an imminent and unreasonable risk of serious or widespread injury to health or the environment."

35. Polychlorinated biphenyls (PCBs) are the only group of chemicals with which the administrator is statutorily obligated to deal. See Section 6(e) of the act.

36. The regulations issued by the EPA in 1979 dealing with PCBs were judicially reviewed by the U.S. Court of Appeals for the District of Columbia in *Environmental Defense Fund v. EPA*, 636 F.2d 1267 (D.C. Cir. 1980). In that proceeding the Environmental Defense Fund (EDF) challenged: (1) the determination by the EPA that certain uses of PCBs were "totally enclosed" and hence exempt from regulation under the Toxic Substances Control Act; (2) the applicability of the regulations to materials containing concentrations of PCBs greater than 50 parts per million (50 ppm); and (3) the decision of the EPA to authorize the continued availability of 11 nontotally enclosed uses of PCBs. The court upheld the regulations regarding the continued availability of the 11 nontotally enclosed PCB uses, but it set aside the regulations classifying certain PCB uses as "totally enclosed" and the 50-ppm cutoff figure for materials containing PCBs. In response to this decision the EPA revised and amended the Part 761 regulations on PCBs and published new regulations at 47 Fed. Reg. 37342-60 (August 25, 1982).

37. The Clean Air Act is codified at 42 U.S.C. §§ 7401 et seq. National primary and secondary ambient air quality standards are published at 40 CFR Part 50.

38. 61 *Fed. Reg.* 31668 (June 20, 1996). The rule is published at 40 CFR Part 68.

39. The Consumer Product Safety Act, as amended, is codified at 15 USC §§ 2051 et seq.

40. Provision is made in the act for suit by "any person who shall sustain injury by reason of any knowing (including willful) violation of a consumer product safety rule or any other rule or order issued by the Commission" against the responsible party in a federal district court. This right to sue is in addition to existing common law, federal, and state remedies. Comprehensive

regulations promulgated by the Consumer Product Safety Commission can be found at 16 CFR Parts 1000 et seq.

41. The Hazardous Substances Act is codified at 15 USC §§ 1261–1276. Regulations published under this statute can be found at 16 CFR Subchapter C, Parts 1500 et seq.
42. The Atomic Energy Act is codified at 42 USC §§ 2011 et seq. The Nuclear Regulatory Commission, an independent executive commission, was created by the Energy Reorganization Act of 1974, 42 USC § 5841(a), and all licensing and related regulatory functions of the Atomic Energy Commission were then transferred to the NRC. See 42 USC § 5841(f) and (g). The NRC's "Standards for Protection Against Radiation" can be found at 10 *CFR* Part 20.
43. The permissible dosage per calendar quarter within such a "restricted area" is 1.25 rems to the whole body, head, and trunk, active bloodforming organs, lens of the eyes, or gonads; 18.75 rems to hands and forearms, feet and ankles; 7.5 rems to the skin of the whole body. Dosage standards are also set forth for the inhalation of radioactive substances. See 10 *CFR* §§ 20.101–103. Detailed personnel monitoring and reporting requirements are also included in the regulations.
44. See 10 *CFR* §§ 20.105, 20.106, and 20.303.
45. See 40 *CFR* Part 190 et seq.
46. See 30 *CFR* Part 57 et seq. These regulations are revisions of regulations previously promulgated by the Secretary of Interior under the Metal and Non-Metallic Mine Safety Act, which was repealed by the Federal Mine Safety and Health Amendments Act of 1977 (see Section 4.1 of this chapter).
47. The Walsh-Healey Act is codified at 41 USC §§ 35 et seq. The relevant regulations can be found in 41 CFR § 50-204. These standards were later promulgated by the secretary of labor as established federal standards under Section 6(a) of OSHA. See Section 6.2.1 of this chapter.
48. That statute amended the Public Health Service Act and is codified at (21 USC §§ 360gg-ss). The regulations promulgated by the Food and Drug Administration under this statute can be found at 21 CFR Parts 1000–1050 (Radiological Health). See also 21 CFR § 1020.30, the standard applicable to diagnostic X-ray systems.
49. The Outer Continental Shelf Lands Act is codified at 43 USC §§ 1331 et seq.
50. See Ashford, *Crisis in the Workplace* (Ref. 4), pp. 47–51.
51. See D. Currie, "OSHA," *Am. Bar Found. Res. J.*, pp. 1107, 1111 (1976). Section 18(a) of OSHA explicitly directs, however, that the states may assert jurisdiction under state law with respect to occupational safety or health issues for which no federal standard is in effect. See Section 5.1 of this chapter.
52. A so-called complete plan is described in an administrative regulation issued by the secretary of labor and codified at 29 *CFR* § 1902.3.
53. Pursuant to Section 18(c), the state plan must: a. Designate a state agency or agencies as the agency or agencies responsible for administering the plan throughout the state. b. Provide for the development and enforcement of safety and health standards relating to one or more safety or health issues, which standards (and the enforcement of which standards) are or will be at least as effective in providing safe and healthful employment and places of employment as the standards promulgated under Section 6 of OSHA, which relate to the same issues. c. Provide for a right of entry and inspection of all workplaces subject to this chapter, which is at least as effective as that provided in Section 8 of OSHA and include a prohibition on advance notice of inspections. d. Contain satisfactory assurances that such agency or agencies have or will have the legal authority and qualified personnel necessary for the enforcement of such standards. e. Give satisfactory assurances that such state will devote adequate funds to the admin-

istration and enforcement of such standards. f. Contain satisfactory assurances that such state will, to the extent permitted by its law, establish and maintain an effective and comprehensive occupational safety and health program applicable to all employees of public agencies of the state and its political subdivisions, which program is as effective as the standards contained in an approved plan. g. Require employers in the state to make reports to the secretary in the same manner and to the same extent as if the plan were not in effect. h. Provide that the state agency will make such reports to the secretary in such form and containing such information as the secretary shall from time to time require. The additional criteria outlined in the secretary's regulations can be found at 29 *CFR* §§ 1902.3 and 1902.4.

54. 29 *CFR* § 1902.2(b).
55. 29 *CFR* § 1954.3, 1954.10.
56. The commission has held that OSHA is not precluded from exercising its own enforcement authority during this period. *Par Construction Co., Inc.*, 4 OSHC 1779 (Rev. Comm. 1976); *Seaboard Coast Line Railroad Co. and Winston-Salem Southbound Railway Co.*, 3 OSHC 1767 (Rev. Comm. 1975).
57. A chart on the status of state plans can be found in BNA, OSHR, Reference File at 81:1003.
58. See P. Hamlar, "Operation and Effect of State Plans," *Proceedings of the American Bar Institute on Occupational Safety and Health Law*, ABA Press, Chicago, 1976, pp. 42–45.
59. See Section 303(e) of the Federal Mine Safety and Health Act of 1977.
60. See Section 205 of the Federal Railroad Safety Act.
61. An employer's general duty to provide a safe workplace may not necessarily be discharged simply by the employer's compliance with specific OSHA standards. In *UAW v. General Dynamics*, 815 F.2d 1570, 1577; 13 OSHC 1201, 1206-07 (D.C. Cir.), *certiorari denied*, 484 U.S. 976 (1987), the court stated: [I]f . . . an employer knows a particular safety standard is inadequate to protect his workers against the specific hazard it is intended to address, or that the conditions in his place of employment are such that the safety standard will not adequately deal with the hazards to which his employees are exposed, he has a duty under section 5(a)(1) to take whatever measures may be required by the Act, over and above those mandated by the safety standard, to safeguard his workers. In sum, if an employer knows that a specific standard will not protect his workers against a particular hazard, his duty under section 5(a)(1) will not be discharged no matter how faithfully he observes that standard.
62. There is language in the legislative history of OSHA to indicate that the general duty clause merely restates the employer's common law duty to exercise reasonable care in providing a safe place for his employees to work. However, the courts have generally characterized such statements as "misleading." For example, in *REA Express v. Brennan*, 495 F.2d 822, 825; 1 OSHC 1651, 1653 (2d Cir. 1974), the Second Circuit could not "accept the proposition that common law defenses such as assumption of the risk or contributory negligence will exculpate the employer who is charged with violating the Act."
63. At a multiemployer construction worksite, this duty may extend to hazardous conditions that the employer neither creates nor fully controls, under what have become known as the Anning-Johnson/Grossman rules, unless the employer can show that it took realistic or reasonable measures to protect its employees or that it neither knew or reasonably could have known of the violation. *D. Harris Masonry Contracting, Inc. v. Dole*, 876 F.2d 343; 14 OSHC 1034 (3d Cir. 1989); *Dun-Par Engineered Form Co. v. Marshall*, 676 F.2d 1333, 1335–1336; 10 OSHC 1561, 1562 (10th Cir. 1982); *Electric Smith, Inc. v. Secretary of Labor*, 666 F.2d 1267, 1268–1270; 10 OSHC 1329, 1330–1332 (9th Cir. 1982); *DeTrae Enterprises, Inc. v. Secretary of Labor*, 645 F.2d 103, 104; 9 OSHC 1425, 1426 (2d Cir. 1981); *Bratton Corp. v. OSAHRC*,

590 F.2d 273, 275; 7 OSHC 1004, 1005 (8th Cir. 1979). Conversely, the Review Commission has held that an employer who creates a hazardous condition in violation of federal job safety standards may be held liable even if the only workers exposed to the hazard belong to another employer. *Secretary of Labor v. Flint Engineering and Construction Co.* 15 OSHC 2052 (Rev. Comm. 1992). In *IBP v. Secretary of Labor*, 18 OSHC 1353 (1998), the D.C. Circuit vacated an OSHA citation against a meatpacking company for a subcontractor's lockout/tagout violation; the court did not strike down the multi-employer theory of liability, but it did question its validity, finding that the Review Commission's definition of "control" as the ability to fire a subcontractor was "irrational." OSHA has drafted a revised citation policy to address multi-employer work sites that would require compliance officers to determine whether general contractors have overall responsibility for enforcing safety and health requirements before they are fined for violations. As of early 1999, this draft policy was being reviewed by OSHA's Advisory Committee on Construction Safety and Health. BNA, OSHR *Current Report* for November 11, 1998.

64. *National Realty and Construction Company, Inc. v. OSAHRC*, 489 F.2d 1257; 1 OSHC 1422 (D.C. Cir. 1973); *Brennan v. OSAHRC and Canrad Precision Industries*, 502 F.2d 946; 2 OSHC 1137 (3d Cir. 1974).

65. In *Dun-Par Engineered Form Co.*, 12 OSHC 1949, 1956 (Rev. Comm. 1986), the review commission modified the defense of impossibility to one of infeasibility ("We overrule Commission precedent that requires employers to prove that compliance with a standard is 'impossible' rather than 'infeasible.'"). In deciding the subsequent appeal, the Eighth Circuit upheld the commission on this point, although the court did reverse the commission's analysis with respect to allocation of the burden of proof following a showing of infeasibility. *Brock v. Dun-Par Engineered Form Co.*, 843 F.2d 1135; 13 OSHC 1652 (8th Cir. 1988).

66. See *Dole v. Williams Enterprises, Inc.*, 876 F.2d 186, 188; 14 OSHC 1001, 1003 (D.C. Cir. 1989); *Donovan v. Williams Enterprises, Inc.*, 744 F.2d 170, 178, n. 12; 11 OSHC 2241 (D.C. Cir. 1984); *True Drilling Co. v. Donovan*, 703 F.2d 1087, 1090; 11 OSHC 1310, 1311 (9th Cir. 1983); *Carlyle Compressor Co. v. OSAHRC*, 683 F.2d 673, 677; 10 OSHC 1700 (2d Cir. 1982); *PBR, Inc. v. Secretary of Labor*, 643 F.2d 890, 895; 9 OSHC 1357, 1360 (1st Cir. 1981).

67. 489 F.2d at 1266; 1 OSHC at 1427.

68. 489 F.2d at 1266–1267; 1 OSHC at 1427; see also *Capital Electric Line Builders of Kansas, Inc. v. Marshall*, 678 F.2d 128, 130; 10 OSHC 1593, 1594 (10th Cir. 1982); *H. B. Zachry Co. v. OSAHRC*, 638 F.2d 812, 818; 9 OSHC 1417, 1421–1422 (5th Cir. 1981); *General Dynamics Corp. v. OSAHRC*, 599 F.2d 453, 458; 7 OSHC 1373, 1375 (1st Cir. 1979). See also the discussion in Section 6.2.3 of this chapter.

69. *Secretary of Labor v. GEM Industrial Inc.*, 17 OSHC 1861 (Rev. Comm. 1996).

70. *Secretary of Labor v. Halmar Corp. and DeFoe Corp.*, 18 OSHC 1014 (Rev. Comm. 1997).

71. *Secretary of Labor v. American Sterilizer Co.*, 18 OSHC 1082 (Rev. Comm. 1997).

72. See, for example, *Brock v. L.E. Myers Co.*, 818 F.2d 1270; 13 OSHC 1289 (6th Cir.) (burden on employer to prove employee misconduct), certiorari denied, 484 U.S. 989 (1987); *D.A. Collins Construction Co. v. Secretary of Labor*, 17 OSHC 2099, 117 F.3d 691 (2nd Cir. 1997) (burden on employer) and *Pennsylvania Power & Light Co. v. OSHRC*, 737 F.2d 350, 11 OSHC 1985 (3d Cir. 1984) (burden on secretary). As of early 1999, six federal courts of appeals, the First, Second, Fifth, Sixth, Eighth and Eleventh, had held that the burden of proof in the employee misconduct defense was on the employer; the Third, Fourth and Tenth Circuits had held that the burden of proof was on the Secretary.

73. 4th Cir., No. 97-1492, January 28, 1998.

74. *McLaughlin v. Union Oil of California*, 864 F.2d 1039, 1044; 13 OSHC 2033, 2036 (7th Cir. 1989); *Kelly Springfield Tire Co. v. Donovan*, 729 F.2d 317, 321; 11 OSHC 1889, 1891 (5th Cir. 1984); *Pratt & Whitney Aircraft v. Secretary of Labor*, 649 F.2d 96, 100; 9 OSHC 1554, 1557 (2d Cir. 1981); *Continental Oil Co. v. OSHRC*, 630 F.2d 446, 448; 8 OSHC 1980, 1981 (6th Cir. 1980), *certiorari denied*, 450 U.S. 965 (1981); *Brennan v. OSAHRC and Vy Lactos Laboratories*, 494 F.2d 460, 464; 1 OSHC 1623, 1625 (8th Cir. 1974); *National Realty* (Ref. 55), 489 F.2d at 1265 n. 32; 1 OSHC at 1426.
75. *National Realty* (64); *Pratt & Whitney Aircraft* (74); and *Kelly Springfield Tire* (74).
76. *New York State Electric & Gas Corp. v. Secretary of Labor*, 17 OSHC 1650, 88 F.3d 98 (2nd Cir. 1996). In June of 1994, the United States Supreme Court ruled that the Department of Labor's "true doubt" rule, which dictated that when the evidence was "evenly balanced" the petitioner won on its claim, violated the provisions of the Administrative Procedure Act by improperly shifting the burden of proof to the responding party. *Office of Workers' Compensation Programs v. Greenwich Collieries*, 16 OSHC 1825, 114 S.Ct. 2251 (U.S. Sup. Ct. 1994). The "true doubt" rule had been relied upon for decades in cases involving black lung disease, for example.
77. 501 F.2d 504; 2 OSHC 1041 (8th Cir. 1974).
78. *Brennan v. OSAHRC and Vy Lactos Laboratories* (74), 1 OSHC at 1624; R. Morey, "The General Duty Clause of the Occupational Safety and Health Act of 1970," *Harv. Law Rev.* 86, 988, 991 (1973). The same is true for violations of the specific duty clause. The fact that a violation is based on the existence of a hazard rather than an injury precludes an employer from arguing lack of "proximate cause." In *Dye Construction Co. v. OSAHRC*, 698 F.2d 423; 11 OSHC 1104 (10th Cir. 1983), the court rejected a defense by an employer that the alleged hazard was not the proximate cause of an employee's injury. The court held that "[t]he relevant inquiry is not the proximate cause of this particular accident but the risk of accident or injury as a result of the alleged violations and the seriousness of the potential injuries." 698 F.2d at 426; 11 OSHC at 1106.
79. *National Realty* (64), 489 F.2d at 1265; 1 OSHC at 1426, n. 33; *Babcock & Wilcox Co. v. OSAHRC*, 622 F.2d 1160, 1165; 8 OSHC 1317, 1319 (3d Cir. 1980); *Illinois Power Co. v. OSAHRC*, 632 F.2d 25, 28; 8 OSHC 1512, 1514–1515 (7th Cir. 1980); *Titanium Metals Corp. of America v. Usery*, 579 F.2d 536, 541; 6 OSHC 1873, 1876–1878 (9th Cir. 1978). "[T]he 'likely to cause' test should be whether reasonably foreseeable circumstances could lead to the perceived hazard's resulting in serious physical harm or death—or more simply, the proper test is plausibility, not probability." Morey, "The General Duty Clause of the Occupational Safety and Health Act of 1970" (78), pp. 997–998. In *Kelly Springfield Tire Co. v. Donovan* (74), 729 F.2d at 324; 11 OSHC at 1894, the Fifth Circuit rejected an argument to change the "likely to cause" requirement in general duty cases to the more lenient "significant risk of harm" test in specific duty cases.
80. 29 *CFR* § 1910.5(f).
81. An "established Federal standard" is defined in Section 3(10) of the act as "any operative occupational safety and health standard established by any agency of the United States and presently in effect, or contained in any Act of Congress in force on the date of enactment of this Act." Section 4(b)(2) of the act listed several federal statutes from which established federal standards were to be derived, including the *Walsh-Healey Act*, 41 USC §§ 35–45, the Service Contract Act of 1965, 41 USC §§351–357, and the National Foundation on Arts and Humanities Act, 20 USC §§ 951–960. A "national consensus" standard is defined in Section 3(9) of the act as any occupational safety and health standard, which "(1) has been adopted and promulgated by a nationally recognized standards producing organization under procedures,

whereby it can be determined by the Secretary that persons interested and affected by the scope or provisions of the standard have reached substantial agreement on its adoption, (2) was formulated in a manner which afforded an opportunity for diverse views to be considered, and (3) has been designated as such a standard by the Secretary, after consultation with other appropriate Federal agencies." The principal sources for national consensus standards were the American National Standards Institute (ANSI) and the National Fire Protection Association. *American Federation of Labor v. Brennan*, 530 F.2d 109, 111 at n. 2; 3 OSHC 1820, 1821 at n. 2 (3d Cir. 1975).

82. The courts have held that OSHA does not have the right to change advisory national consensus standards ("should") to mandatory standards ("shall") upon adoption as OSHA standards without following formal rule-making procedures. Absent such rule making, citations issued to employers pursuant to these standards have been vacated. *Usery v. Kennecott Copper* Corp., 577 F.2d 1113, 1117–1118; 6 OSHC 1197, 1199 (10th Cir. 1977); *Marshall v. Pittsburgh-Des Moines Steel Co.*, 584 F.2d 638, 644; 6 OSHC 1929, 1933 (3d Circ. 1978). See also *Marshall v. Anaconda Co.*, 596 F.2d 370, 376–377; 7 OSHC 1382, 1385–1386 (9th Cir. 1979).

83. The Job Safety and Health Act of 1970, BNA, Washington, D.C., 1971, p. 23.

84. See *Florida Peach Growers Association v. U.S. Department of Labor*, 489 F.2d 120, 124; 1 OSHC 1472 (5th Cir. 1974) and Sections 6(b) (1) to 6(b) (4) of OSHA. In *National Congress of Hispanic American Citizens v. Usery*, 554 F.2d 1196; 5 OSHC 1255 (D.C. Cir. 1977), the Court of Appeals for the District of Columbia held that the statutory deadlines in Sections 6(b) (1) to 6(b) (4) for the promulgation of permanent standards were discretionary rather than mandatory as long as the secretary's exercise of discretion was honest and fair.

85. *Florida Peach Growers Association* (84), 489 F.2d at 124; 1 OSHC at 1475; see also *Industrial Union Dept. AFL-CIO v. American Petroleum Institute*, 448 U.S. 607, 651 n. 59; 8 OSHC 1586, 1602 (1980).

86. The "substantial evidence" standard of judicial review is traditionally conceived of as suited to adjudication or formal rule making. OSHA, however, calls for informal rule making, which under the Administrative Procedure Act generally entails judicial review pursuant to the less stringent "arbitrary and capricious" test. This apparent anomaly can be explained historically as a legislative compromise. The Senate OSHA bill called for informal rule making, but the House version specified formal rule making and substantial evidence review. The House receded on the procedure for promulgating standards, but the substantial evidence standard of review was adopted. *Industrial Union Department, AFL-CIO v. Hodgson* (6), 499 F.2d at 473; 1 OSHC at 1635. For a more detailed discussion of these legislative events, see *Associated Industries of New York State, Inc. v. U.S. Department of Labor*, 487 F.2d 342, 1 OSHC 1340 (2d Cir. 1973).

87. See B. Fellner and D. Savelson, "Review by the Commission and the Courts," *Proceedings of the American Bar Association Institute on Occupational Safety and Health Law*, ABA Press, Chicago, 1976, pp. 113–114. This approach has been summarized as one requiring the reviewing court to determine whether the agency (1) acted within the scope of its authority; (2) followed the procedures required by statute and by its own regulations; (3) explicated the bases for its decision; and (4) adduced substantial evidence in the record to support its determination. *United Steelworkers of America, AFL-CIO, v. Marshall and Bingham*, 647 F.2d 1189, 1206; 8 OSHC 1810, 1816 (D.C. Cir. 1980), certiorari denied, 453 U.S. 913 (1981). See also *Texas Independent Ginners Assoc. v. Marshall*, 630 F.2d 398, 404–405; 8 OSHC 2205, 2209–2210 (5th Cir. 1980); *American Iron and Steel Institute v. OSHA*, 577 F.2d 825, 830–31; 6 OSHC 1451, 1455 (3d Cir. 1978); *Society of the Plastics Industry, Inc. v. OSHA*, 509 F.2d 1301, 1304; 2 OSHC 1496, 1498 (2d Cir. 1975).

88. *Atlantic & Gulf Stevedores, Inc. v. OSAHRC*, 534 F.2d 541; 4 OSHC 1061 (3d Cir. 1976); *Arkansas-Best Freight Systems, Inc. v. OSAHRC* and Secretary of Labor, 529 F.2d 649; 3 OSHC 1910 (8th Cir. 1976); *Deering Milliken, Inc. v. OSAHRC*, 630 F.2d 1094, 1099; 9 OSHC 1001, 1004 (5th Cir. 1980); *Marshall v. Union Oil Co. and OSAHRC*, 616 F.2d 1113, 1117–1118; 8 OSHC 1169, 1173 (9th Cir. 1980); *Daniel International Corp. v. OSAHRC and Secretary of Labor*, 656 F.2d 925, 928–930; 9 OSHC 2102, 2104–2106 (4th Cir. 1981). But see *National Industrial Contractors v. OSAHRC*, 583 F.2d 1048, 1052–1053; 6 OSHC 1914, 1916–1917 (8th Cir. 1978), in which the Eighth Circuit held that procedural challenges to an OSHA standard must be brought in a pre-enforcement proceeding pursuant to Section 6(f) within 60 days from the challenged standard's effective date.
89. *Atlantic & Gulf Stevedores* (88), 534 F.2d at 550–552; 4 OSHC at 1067–1068.
90. 452 U.S. 490; 9 OSHC 1913 (1981).
91. 452 U.S. at 513, n. 31; 9 OSHC at 1922, n. 31. Section 3(8) of OSHA contains the general definition of an occupational safety and health standard. It provides as follows:

The term "occupational safety and health standard" means a standard which requires conditions, or the adoption or use of one or more practices, means, methods, operations, or processes, *reasonably necessary or appropriate* to provide safe or healthful employment and places of employment. (Emphasis added.)

For standards dealing with toxic materials or harmful physical agents, Section 6(b)(5) imposes the following additional requirements:

The Secretary, in promulgating standards dealing with toxic materials or harmful physical agents under this subsection, shall set the standard which most adequately assures, *to the extent feasible*, on the basis of the best available evidence, that no employee will suffer material impairment of health or functional capacity, even if such employee has regular exposure to the hazard dealt with by such standard for the period of his working life. (Emphasis added.)
92. *Industrial Union Department, AFL-CIO v. Hodgson* (6), 499 F.2d at 478; 1 OSHC at 1639, cited approvingly by the Supreme Court in *American Textile Mfrs. v. Donovan* (90), 452 U.S. at 513, n. 31; 9 OSHC at 1922, n. 31. In *Industrial Union Department*, the Court of Appeals for the District of Columbia applied Section 6(b)(5) in a case challenging OSHA's standard for exposure to asbestos dust and held that the secretary, in promulgating the standard, could properly consider problems of both economic and technological feasibility.
93. *American Textile Mfrs. v. Donovan* (90), 452 U.S. at 530, n. 55; 9 OSHC at 1928–1929, n. 55; *United Steelworkers of America, AFL-CIO v. Marshall and Bingham* (87), 647 F.2d at 1265; 8 OSHC at 1864.
94. *American Textile Mfrs. v. Donovan* (90), 452 U.S. at 530, n. 55; 9 OSHC at 1928–1929, n. 55.
95. *American Textile Mfrs. v. Donovan* (90), 452 U.S. at 509. *American Textile Mfrs.* involved a challenge by the textile industry to OSHA's standard governing occupational exposure to cotton dust. The industry contended, among other things, that the act required OSHA to demonstrate that its standard reflected a reasonable relationship between the costs and benefits associated with the standard. The Supreme Court rejected that argument and upheld the validity of the entire cotton dust standard except for a wage guarantee requirement for employees who are transferred to another position when they are unable to wear a respirator.
96. *American Textile Mfrs. v. Donovan* (90), 452 U.S. at 513, n. 32. In *Donovan v. Castle & Cooke Foods and OSAHRC*, 692 F.2d 641; 10 OSHC 2169 (9th Cir. 1982), the Ninth Circuit held that the Supreme Court's holding concerning cost-benefit analysis in the *American Textile Manufacturers* case applied only to standards promulgated under Section 6(b) of OSHA and did not apply to standards promulgated under Section 6(a), such as the noise standard at issue in *Castle*

& *Cooke*, which had been originally promulgated as an established federal standard under the *Walsh-Healey Act*. But compare *Sun Ship, Inc.*, 11 OSHC 1028 (Rev. Comm. 1982), where the Review Commission applied the reasoning of *American Textile Manufacturers* to the noise standard and rejected the application of cost-benefit analysis in enforcing that standard. See BNA, OSHR, Current Report for February 3, 1983, pp. 735–736.
97. *American Textile Mfrs. v. Donovan* (90) 452 U.S. at 509, n. 29.
98. 16 OSHC 2065, 37 F.3d 665 (D.C. Cir. 1994).
99. See, e.g., *Panama Refining Co. v. Ryan*, 293 U.S. 388 (1935).
100. The Regulatory Improvement Act of 1999 (S.746), BNA, OSHR, Current Report for March 31, 1999.
101. Ashford, *Crisis in the Workplace* (4), p. 169.
102. 511 F.2d 1139; 2 OSHC 1646 (9th Cir. 1975).
103. 511 F.2d at 1144–1145; 2 OSHC at 1650–1651. See also *Daniel International Corp. v. OSAHRC and Secretary of Labor*, 683 F.2d 361; 10 OSHC 1890 (11th Cir. 1982).
104. 520 F.2d 1011; 3 OSHC 1461 (7th Cir. 1975).
105. Accord, *H. B. Zachry Company v. OSAHRC*, 638 F.2d 812; 9 OSHC 1417 (5th Cir. 1981) (defense of employee negligent misconduct fails because of the employer's inability to establish to the satisfaction of the fact-finder that it effectively communicated and enforced work rules, which were necessary to ensure compliance with OSHA standards).
106. *Atlantic & Gulf Stevedores* (88), 534 F.2d at 555; 4 OSHC at 1068–1069. See Note, "Employee Noncompliance with OSHA Safety Standards," 90 Harv. Law Rev. 1041 (1977).
107. 528 F.2d 564, 570; 3 OSHC 2060, 2064 (5th Cir. 1976).
108. 42 USC § 4332.
109. 29 *CFR* § 11.10(a)(3).
110. 29 *CFR* § 11.10(a)(1).
111. Health and safety standards promulgated by the Secretary of Labor pursuant to OSHA can be found at 29 *CFR* Part 1910. A standards digest (*OSHA Publication 2201*) outlining the basic applicable standards is published in BNA, OSHR, Reference File at 31:4001. Federal health and safety standards for the construction industry were initially promulgated under the Contract Work Hours and Safety Standards Act, 40 USC §§ 327 et seq. These standards were incorporated by reference under OSHA, are enforceable under both laws, and can be found at 29 *CFR* Part 1926.

A standards digest (*OSHA Publication 2202*) outlining the basic applicable construction standards is published in BNA, OSHR, Reference File at 31:3001. Health and safety standards for ship repairing, shipbuilding, shipbreaking, and longshoring were initially promulgated pursuant to the Longshoremen's and Harbor Worker's Compensation Act, 33 USC §§ 901 et seq. These standards were incorporated by reference by OSHA, are enforceable under both laws, and can be found in 29 *CFR* Parts 1915 and 1918. Health and safety standards originally promulgated under the Walsh-Healey Public Contracts Act, the McNamara-O'Hara Service Contract Act of 1965, and the National Foundation on the Arts and Humanities Act of 1965 can be found in 41 *CFR* Part 50-204, 29 *CFR* Part 1516, and 20 *CFR* Part 505. These were adopted and are enforceable by OSHA. Standards promulgated under the aforementioned statutes will be superseded if corresponding standards that are promulgated under OSHA are determined by the secretary of labor to be more effective. See Section 4(b)(2) of OSHA. Federal health and safety standards for coal mines were promulgated by the Department of Interior pursuant to the federal Coal Mine Health and Safety Act of 1969. Standards under the 1969 act were adopted without change by the Federal Mine Safety and Health Act of 1977. CCH, Empl. Safety and Health Guide ¶¶ 5924, 5931. Health standards for underground coal mines can be found in 30 *CFR* Part 70; health standards for surface work areas of underground coal mines and surface coal mines are codified in 30 *CFR* Part 71. Requirements for approval of coal mine dust personal sampler units designed to determine the concentrations of respirable dust in coal mine atmospheres can be found in 30 *CFR* Part 74; health standards for coal miners with evidence of pneumoconiosis are codified in 30 *CFR* Part 90. The safety standards for underground coal mines can be found in 30 *CFR* Part 75, and the safety standards for surface coal mines and surface work areas of underground coal mines are codified in 30 *CFR* Part 77. The secretary of interior promulgated health and safety standards for metal and nonmetallic mines pursuant to the federal Metal and Non-Metallic Mine Safety Act of 1966. Mandatory standards adopted under the 1966 act were adopted without change by the Federal Mine Safety and Health Act of 1977. However, advisory metal and nonmetallic standards did not become mandatory standards under the 1977 act. A committee later reviewed these advisory standards and made recommendations for conversion to mandatory standards, which MSHA accepted in August 1979 by converting scores of noncoal advisory standards to mandatory standards. CCH, Empl. Safety and Health Guide §§ 5924, 5931. The standards for surface metal and nonmetal mines and for underground metal and nonmetal mines are found at 30 CFR Parts 56 and 57, respectively.

112. Safety standards generally focus on the time that an employee is actually working. The harm created by a safety hazard is generally immediate and violent. An occupational health hazard, on the other hand, is slow acting, cumulative, irreversible, and complicated by nonoccupational factors. Ashford, *Crisis in the Workplace* (Ref. 4), pp. 68–83.
113. For a listing of the initial package of national consensus and established federal standards, see 36 *Fed. Reg.* 10466–10714 (1971).
114. In 1995, pursuant to a presidential order, OSHA pledged to eliminate over 1,000 of the over 3,000 pages dedicated to the agency in the Code of Federal Regulations. Most of the standards to be done away with were included in 29 *CFR* 1910, which sets job safety and health provisions for general industry, and OSHA pledged also to revise and simplify the 1971 consensus standards. On March 7, 1996, OSHA published a final rule (61 *Fed. Reg.* 9228), characterized as a "down payment" on OSHA's pledge, that eliminated 275 of OSHA's CFR pages.
115. Industries covered by specific OSHA regulations include pulp, paper and paperboard mills, textiles, bakery equipment, laundry machinery and operations, sawmills, pulpwood logging, telecommunications, agriculture, grain handling facilities, commercial diving, construction, ship repairing, shipbuilding, shipbreaking, marine terminal employment, and longshoring. The mining industry is subject to comprehensive safety regulations issued pursuant to the Federal Mine Safety and Health Act of 1977 (formerly the Federal Coal Mine Health and Safety Act of 1969 and the Federal Metal and Non-Metallic Mine Safety Act of 1966). See also Section 3 of this chapter.
116. The TLVs had been developed principally in 1968 by the American Conference of Governmental Industrial Hygienists (ACGIH) and were subsequently incorporated into the Walsh-Healey Act. See Ashford, *Crisis in the Workplace* (4), p. 154. The Secretary of Labor did not include the ACGIH's carcinogen standards in his Section 6(a) package, but instead preferred to develop his own standards regarding carcinogens. Ashford, pp. 154, 247–248.
117. 54 *Fed. Reg.* 2332 (January 19, 1989).
118. The PELs established by OSHA can be found at 29 *CFR* § 1910.1000, Tables Z-1 Z-2, and Z-3. Pursuant to the January 1989 rule making, OSHA made 212 PELs more protective, established 162 new PELs for previously unregulated substances, and left other PELs unchanged.

In so doing, OSHA stated that it relied heavily on the widely accepted 1987–1988 TLVs published by ACGIH and the Recommended Exposure Limits (RELs) developed by the National Institute for Occupational Safety and Health (NIOSH). 54 *Fed. Reg.* 2332, 2333–35 (January 19, 1989).

119. OSHA formally revoked the 1989 PELs in June of 1993 with publication of a notice at 58 *Fed. Reg.* 35338 (June 30, 1993). In so doing, OSHA republished the less stringent limits that were in effect prior to 1989.
120. 61 *Fed. Reg.* 1947 (January 24, 1996). In March of 1996, OSHA announced that the styrene industry had agreed to a voluntary reduction in the PEL for styrene—considered to be a nervous system toxin with narcotic effects—to the 1989 levels of 50 parts per million as an eight-hour time-weighted average (TWA) and a short term exposure limit (STEL) of 100 ppm. Under the voluntary program, employers were to assess compliance with the standard and provide data to OSHA, although individual employers would not be identified in the data.
121. 49 *Fed. Reg.* 5318 (1984).
122. BNA, OSHR, Current Report for March 17, 1983, p. 894 (stating that OSHA intended to issue general duty clause citations for serious hazards to replace the revoked standards).
123. 51 *Fed. Reg.* 22612 (June 20, 1986). The original asbestos standard was issued on June 7, 1972, in a Section 6(b) rule making that retained the emergency temporary standard and limited the 8-hour TWA of airborne concentration of asbestos to 5 fibers greater than 5 micrometers in length per milliliter of air (the "five-fiber standard") for 4 years, then required 2 fibers. 37 *Fed. Reg.* 11322 (1971). In *Industrial Union Department, AFL-CIO v. Hodgson*, 449 F.2d 467; 1 OSHC 1631 (D.C. Cir. 1974), the Court of Appeals for the District of Columbia upheld the standard with two exceptions: OSHA had to reconsider the effective date of the two-fiber standard and the standard's record-keeping provision. In response, OSHA initiated a new rule-making proceeding and issued a proposed revised asbestos standard in 1975, but this proposal was never promulgated. OSHA issued an emergency temporary standard for asbestos in 1983 [48 *Fed. Reg.* 51086 (1983)], and this was vacated by the Fifth Circuit in 1984. *Asbestos Information Ass'n v. OSHA*, 727 F.2d 415; 11 OSHC 1817 (5th Cir. 1984).
124. *Building & Construction Trades Dep't, AFL-CIO v. Secretary of Labor*, 838 F.2d 1258 (D.C. Cir. 1989).
125. The STEL is now codified at 29 *CFR* § 1910.1001(c)(2).
126. 59 *Fed. Reg.* 40964, August 10, 1994. The rule was codified at 29 *CFR* 1910.1001, 1915.1001 and 1926.1101 (formerly 1926.58), and the several revisions that followed were written into those sections.
127. BNA, OSHR, *Current Report* for May 1, 1996. These changes were published at 61 *Fed. Reg.* 43454, August 23, 1996.
128. BNA, OSHR, *Current Report* for June 4, 1997.
129. *Asbestos Information Association/North America v. Secretary of Labor*, 17 OSHC 2089, 117 F.3d 891 (5th Cir. 1997).
130. 63 *Fed. Reg.* 35137, June 29, 1998. The changes were made to 29 *CFR* Parts 1915 and 1926. The asbestos provisions in 29 *CFR* Part 1910, the general industry asbestos standard, remained unchanged.
131. Of course, there remains plenty of litigation over asbestos exposure in general. The United States Supreme Court, in a case involving pipe fitters in Grand Central Terminal who were covered in asbestos-containing material so thick they were dubbed "the snowmen of Grand Central," ruled that workers may not recover damages under the Federal Employers' Liability Act for fear of cancer alone when the workers are asymptomatic. *Metro-North Commuter*

Railroad Co. v. Buckley, 17 OSHC 2153 (U.S. Sup. Ct. 1997). The Supreme Court has also struck down a $1.3 billion class-action settlement of future asbestos claims against some 20 companies because the settlement failed to satisfy the criteria for class certification under the Federal Rules of Civil Procedure. *Amchem Products Inc. v. Windsor*, U.S. Sup. Ct., No. 96-270, June 25, 1997. The Fifth Circuit held the *Amchem* precedent to be inapplicable to another mass tort asbestos exposure settlement, however, when it upheld a $1.5 billion class-action settlement in *In re Asbestos Litigation: Flanagan v. Ahearn*, 5th Cir., No. 95-40635, January 27, 1998.

132. See *Society of the Plastics Industry v. OSHA* (87), 509 F.2d at 1306; 2 OSHC at p. 1500.
133. The standard can be found at 29 *CFR* § 1910.1017.
134. *Society of the Plastics Industry v. OSHA* (87), 509 F.2d at 1311; 2 OSHC at 1504.
135. 29 *CFR* § 1910.1017(h)(1) (ii).
136. 29 *CFR* § 1910.1017(j).
137. 29 *CFR* § 1910.1017(l).
138. 29 *CFR* § 1910.1017(k).
139. 29 *CFR* § 1910.1017(m).
140. See *Dry Color Manufacturers' Association, Inc. v. U.S. Department of Labor*, 486 F.2d 98; 1 OSHC 1331, 1332 (3d Cir. 1973). The 14 chemicals included in the carcinogen standards were 4-nitrobiphenyl, alpha-naphthylamine, 4,4'-methylene bis (2-chloroaniline), methyl chloromethyl ether, 3,3'-dichlorobenzidine (and its salts), bis-chloromethyl ether, beta-naphthylamine, benzidine, 4-aminodiphenyl, ethyleneimine, beta-propiolactone, 2-acetylaminofluorene, 4-dimethylaminoazobenzene, and *N*-nitrosodimethylamine. The permanent standards were codified at 29 *CFR* §§ 1910.1003–1910.1016.
141. 500 F.2d 385; 2 OSHC 1402 (3d Cir. 1974), *certiorari denied*, 423 U.S. 830 (1975).
142. 29 *CFR* § 1910.1005 was deleted in 41 *Fed. Reg.* 35184 (August 20, 1976).
143. OSHA's Cancer Policy is found at 29 *CFR* Part 1990.
144. CCH, *Empl. Safety and Health Guide* ¶ 1031.
145. 48 *Fed. Reg.* 242 (January 4, 1983); see also CCH, Empl. Safety and Health Guide ¶ 1031.
146. BNA, OSHR, *Current Report* for December 23, 1987.
147. The text of the coke ovens standard can be found at 29 *CFR* § 1910.1029.
148. 29 *CFR* § 1910.1029(h). This includes, but is not limited to, flame-resistant pants, jackets, gloves, footwear, and face shields.
149. 29 *CFR* § 1910.1029(i)(1).
150. 29 *CFR* § 1910.1029(i)(3).
151. 29 *CFR* § 1910.1029(i)(5).
152. 29 *CFR* § 1910.1029(k).
153. 29 *CFR* § 029(l).
154. 29 *CFR* § 1910.1029(j).
155. 29 *CFR* § 1910.1029(m).
156. 577 F.2d 825; 6 OSHA 1451 (3d Cir. 1978), *certiorari dismissed*, 448 U.S. 917 (1980).
157. CCH, *Empl. Safety and Health Guide* ¶ 1198. Before issuance of the separate standard, occupational exposure to lead was regulated pursuant to the standard on air contaminants.
158. The lead standard can be found at 29 *CFR* § 1910.1025. For a detailed early history of the standard, see *United Steelworkers of America, AFL-CIO v. Marshall*, 647 F.2d 1189; 8 OSHC 1810 (D.C. Cir. 1980), *certiorari denied*, 453 U.S. 913 (1981).

… JOB SAFETY AND HEALTH LAW   1645

159. *United Steelworkers of America v. Marshall* (158).
160. For a listing of the industries as to which the standard was remanded, as well as for a summary of the court's order, see *United Steelworkers of America v. Marshall* (158), 647 F.2d at 1311; 8 OSHC at 1901–1902.
161. See BNA, OSHR, *Current Report* for December 17, 1981, pp. 539–540.
162. 46 *Fed. Reg.* 60758 (December 11, 1981).
163. 54 *Fed. Reg.* 29142 (July 11, 1989).
164. 55 *Fed. Reg.* 3146 (January 30, 1990). OSHA set the standard for small foundries at 75 micrograms and at 50 micrograms for large foundries.
165. CCH, *Empl. Safety and Health Guide* ¶ 1198. Those six industries were: nonferrous foundries, secondary copper smelters, brass and bronze ingot manufacturers, independent collectors and processors of scrap lead (including independent battery breakers), leaded steelmaking operations, and lead chemical manufacturers.
166. BNA, OSHR, *Current Report* for July 24, 1991, pp. 228–229. The court remanded the standard as to the brass and bronze industry for an economic feasibility determination of the 50-microgram standard.
167. 60 *Fed. Reg.* 52856, October 11, 1995.
168. 58 *Fed. Reg.* 26590, May 4, 1993.
169. 61 *Fed. Reg.* 45778, August 29, 1996. The rule is codified at 49 *CFR* Part 745 Subparts L and Q.
170. The cotton dust standard is found at 29 *CFR* § 1910.1043. The original standard was generally upheld by the Supreme Court in *American Textile Mfrs. Inst. v. Donovan*, 452 U.S. 490; 9 OSHC 1913 (1981). Cotton that has been washed in accordance with outlined processes is exempt from the standard. 29 *CFR* § 1910.1043(n). Some forms of cotton dust exposure not covered by the separate cotton dust standard are regulated pursuant to the air contaminants standard, 29 *CFR* § 1910.1000.
171. 29 *CFR* § 1910.1043(c) (1).
172. 29 *CFR* § 1910.1043(c) (2).
173. 29 *CFR* § 1910.1043(e).
174. 29 *CFR* § 1910.1043(f), (h), (i) and (j).
175. CCH, *Empl. Safety and Health Guide* ¶ 1202.
176. The DBCP standard can be found at 29 *CFR* § 1910.1044.
177. CCH, *Empl. Safety and Health Guide* § 1203.
178. The acrylonitrile standard can be found at 29 *CFR* § 1910.1045.
179. See CCH, *Empl. Safety and Health Guide* § 1203.
180. *Industrial Union Dept. v. American Petroleum Institute*, 448 U.S. 607; 8 OSHC 1586 (1980). OSHA was required to review other standards (e.g., the arsenic standard) in light of this Supreme Court decision.
181. 52 *Fed. Reg.* 34460 (September 11, 1987). The benzene standard is codified at 29 *CFR* § 1910.1028.
182. The 1-ppm limit is identical to that set in the 1978 standard. The 1987 standard, however, is followed by a risk assessment analysis from which OSHA determined the benzene level that posed a significant risk. 52 *Fed. Reg.* 34460 (September 11, 1987). The 1978 standard was based on a feasibility determination after OSHA simply assumed that no level of benzene exposure would be safe for humans. 43 *Fed. Reg.* 5918, 5946–5947 (1978).

183. 29 *CFR* § 1910.1028(i).
184. See CCH, *Empl. Safety and Health Guide* ¶ 1199.
185. The inorganic arsenic standard can be found at 29 *CFR* § 1910.1018.
186. There is no ceiling limit because of technical inability to control occasional exceedances. See CCH, *Empl. Safety and Health Guide* § 1197.
187. 647 F.2d 1; 9 OSHC 1508 (9th Cir. 1981).
188. 48 *Fed. Reg.* 1864 (January 14, 1983).
189. The arsenic standard was validated in *Asarco, Inc. et al. v. OSHA*, 746 F.2d 483 (9th Cir. 1984).
190. CCH, *Empl. Safety and Health Guide* ¶ 1197.
191. 49 *Fed. Reg.* 25734 (June 22, 1984). The ethylene oxide standard is codified at 29 *CFR* § 1910.1047.
192. *Public Citizen Health Research Group v. Tyson*, 796 F.2d 1479; 12 OSHC 1905 (D.C. Cir. 1986).
193. 53 *Fed. Reg.* 1724 (January 8, 1988). The STEL appears at 29 *CFR* § 1910.1047(c) (2).
194. CCH, *Empl. Safety and Health Guide* ¶ 1204. The standard prohibits rotating employees in and out of exposure areas as a means of exposure control, because of the risk to other employees.
195. BNA, OSHR, *Current Report* for July 22, 1992, p. 245. The study analyzed the use of ethylene oxide in hospitals, which is the largest industry sector affected because ethylene oxide is used as a sterilant.
196. 52 *Fed. Reg.* 46168 (December 4, 1987). The standard appears at 29 *CFR* § 1910.1048.
197. 57 *Fed. Reg.* 22290 (May 27, 1992).
198. *United Auto Workers v. Pendergrass*, 878 F.2d 389; 14 OSHC 1025 (D.C. Cir. 1989).
199. CCH, *Empl. Safety and Health Guide* ¶ 1205.
200. BNA, OSHR, *Current Report* for June 3, 1992, p. 7.
201. 56 *Fed. Reg.* 64004 (December 6, 1991). The standard is codified at 29 *CFR* § 1910.1030.
202. BNA, OSHR, *Current Report* for December 11, 1991, p. 875.
203. *American Dental Association v. Secretary of Labor*, 15 OSHC 2097, 984 F.2d 823 (5th Cir. 1993).
204. A federal district court has ruled, in the context of a product liability suit brought by a worker against a chemical maker and distributor, that OSHA's hazard communication standard does not pre-empt a failure-to-warn claim. *Wickham v. American Tokyo Kasei Inc.*, D.C. Ill., No. 95 C 50014, May 17, 1996.
205. The hazard communication standard for general industry is codified at 29 *CFR* § 1910.1200.
206. *United Steel Workers of America v. Auchter*, 763 F.2d 728; 12 OSHC 1337 (3d Cir. 1985).
207. 52 *Fed. Reg.* 31852 (August 24, 1987).
208. *Associated Builders and Contractors, Inc. v. OSHA*, 862 F.2d 63; 13 OSHC 1945 (3d Cir. 1988).
209. BNA, OSHR, *Current Report* for February 9, 1994, p. 1188.
210. *Secretary of Labor v. American Cyanomid Co.*, 16 OSHC 1369, 8 F.3d 980 (6th Cir. 1993).
211. 59 *Fed. Reg.* 6126, February 9, 1994.
212. 59 *Fed. Reg.* 36695, July 19, 1994.
213. The noise standard can be found at 29 *CFR* § 1910.95.

JOB SAFETY AND HEALTH LAW 1647

214. 29 *CFR* § 1910.95(c)-(p). The amendment exempts employers engaged in oil and gas well drilling and service operations. 29 *CFR* § 1910.95(o).
215. 29 *CFR* § 1910.95(g). The agency review commission has ruled that the action level of 85 decibels excludes "impulse noise," defined as sharp noise peaks lasting less than one second and spaced more than one second apart. *Collier-Keyworth Co.*, 13 OSHC 1208, remanded 13 OSHC 1269 (Rev. Comm. 1987), on remand 13 OSHC 1940 (Rev. Comm. 1988). An employer's failure to make audiometric tests available to its employees is properly classified as a "serious" OSHA violation. *Secretary of Labor v. Miniature Nut and Screw Corp.*, 17 OSHC 1557 (Rev. Comm. 1996).
216. 29 *CFR* § 1910.95(d) (1) (ii).
217. 29 *CFR* § 1910.95(i) (1), (2) (ii) (B). Notwithstanding the fact that an employer provides hearing protectors, employers may receive citations for failing to ensure their use. See, for example, *United States Container Corp.*, 13 OSHC 1415 (Rev. Comm. 1987).
218. *Forging Indus. Ass'n v. Donovan*, 773 F.2d 1436 (4th Cir. 1985).
219. 53 *Fed. Reg.* 26437 (July 13, 1988).
220. 57 *Fed. Reg.* 42101. The standard for general industry is codified at 29 CFR 1910.1027, and the standard for construction is codified at 29 CFR 1926.63.
221. *Color Pigments Manufacturers Association v. OSHA*, D.C. Cir., No. 92-3057, December 3, 1993.
222. *Color Pigments Manufacturers Association v. OSHA*, 16 OSHC 1665, 16 F.3d 1157 (11th Cir. 1994).
223. 62 *Fed. Reg.* 1494. The standard is codified at 29 CFR 1910.1052.
224. *UAW v. OSHA*, D.C. Cir., No. 97-1040, January 17, 1997; *Benco Sales Inc. v. OSHA*, 5th Cir., No. 97-60037, January 17, 1997.
225. P.L. 104-121, 110 Stat. 857–874.
226. 63 *Fed. Reg.* 50712.
227. BNA, OSHR, *Current Report* for April 29, 1998.
228. CPL 2-2.68.
229. BNA, OSHR, *Current Report* for June 18, 1997.
230. The standard is codified at 29-*CFR* 1910.146. The Review Commission has held that OSHA's definition of a "confined space" is not unenforceably vague. *Secretary of Labor v. CBI Services, Inc.*, 15 OSHC 2046 (Rev. Comm. 1992).
231. CPL 2.100.
232. 63 *Fed. Reg.* 66018, December 1, 1998.
233. 58 *Fed. Reg.* 37816, July 25, 1994. The standard is codified at 29 *CFR* Part 1915, Subpart B.
234. 61 *Fed. Reg.* 56746, November 4, 1996. The standard is codified at 29 *CFR* § 1910.1051.
235. 63 *Fed. Reg.* 1152, January 8, 1998. The standard is codified at 29 *CFR* 1910.134.
236. BNA, OSHR, *Current Report* for January 7, 1998.
237. *AISI v. OSHA*, 11th Cir., No. 98-6146.
238. 59 *Fed. Reg.* 16334 (April 6, 1994). The standard is codified at 29 CFR 1910.132.
239. BNA, OSHR, *Current Report* for April 13, 1994.
240. *Secretary of Labor v. Union Tank Car Co.*, 18 OSHC 1067 (Rev. Comm. 1997).
241. BNA, OSHR, *Current Report* for March 31, 1999.
242. BNA, OSHR, *Current Report* for February 24, 1999.

243. BNA, OSHR, *Current Report* for October 21, 1998.
244. BNA, OSHR, *Current Report* for March 10, 1999.
245. The Review Commission finally handed OSHA a victory in its efforts to litigate the question of the use of the general duty clause to cite ergonomics violations when, in April of 1997, it ruled that lifting and repetitive motion hazards can be cited under the general duty clause. *Secretary of Labor v. Pepperidge Farm Inc.*, 17 OSHC 1993 (Rev. Comm. 1997). In the *Pepperidge Farm* case, a Commission administrative law judge had earlier ruled that OSHA could not force employers to "experiment" to lessen ergonomic hazards without promulgating specific standards. In another closely-watched ergonomics case, a Commission ALJ dismissed a citation for an ergonomics violation, holding that the secretary had failed even to prove the first element of a general duty clause violation: that the employees suffered harm. *Secretary of Labor v. Dayton Tire, Bridgestone/Firestone*, 18 OSHC 1225 (Rev. Comm. ALJ 1998). Applying the *Pepperidge Farm* precedent, the ALJ reasoned that the existence of a "hazard"—the first element to be proven for a general duty clause violation—requires "actual or potential physical harm, and a sufficient causal connection between that harm and the workplace." The ALJ then held that OSHA had failed to prove actual or potential harm. OSHA did not appeal the result.
246. For example, a federal court jury awarded $5.3 million in compensatory damages to a former secretary in her failure-to-warn product liability suit against a keyboard manufacturer. The plaintiff suffered extensive carpal tunnel damage to both her wrists, neck injuries and severe wasting of her upper arm muscles. *Geressy v. Digital Equipment Corp.*, E.D.N.Y., No. 94-1427, December 4, 1996. However, the award was struck down in April 1997 because of newly discovered medical evidence, and another jury in the same court later ruled in favor of the keyboard maker. *Gonzalez v. Digital Equipment Corp.*, E.D.N.Y. 92-CV-5230, June 16, 1998. The Americans with Disabilities Act (42 U.S.C. § 12101 et seq.) has been used to attempt to force employers to accommodate their employees who have repetitive stress injuries and/or to recover damages from those employers who allegedly refuse to make such accommodations. See, e.g., *Feliberty v. Kemper Corp.*, 7th Cir., No. 95-1724, October 16, 1996. Although the scope and application of ADA law to provide redress for aggrieved employees is beyond the scope of this chapter, it has been a rapidly evolving area of the law during the 1990s; generally, if an employee can show that he is "substantially limited in a major life activity," then his employer must provide him with "reasonable accommodation" if doing so will enable the disabled employee to perform the essential functions of his or her job.
247. For example, in November of 1996, after a lengthy campaign by the state's organized labor interests, California became the first state to adopt a statewide ergonomics standard. BNA, OSHR, *Current Report* for November 20, 1996. The measure, which exempted employers with fewer than nine employees, required covered employers to conduct work site evaluations, control exposures to repetitive motion hazards and provide training whenever two workers performing identical tasks at the same work site were diagnosed with repetitive motion injuries within a 12-month period. The standard continues to be litigated, and a state trial court has struck down the small employer exemption. *Pulaski v. California Occupational Safety and Health Standards Board*, Calif. Sup. Ct., No. 95-CS-00362, September 5, 1997.
248. BNA, OSHR, March 30, 1994. While the proposed smoking ban would apply to nearly all of the six million employers under OSHA's jurisdiction, the broader indoor air quality provisions would apply to only non-industrial employers and would require those employers to ensure that their heating, air-conditioning and ventilation systems are operated and maintained to ensure healthful indoor air.
249. For example, after the defeat of Proposition 188 in the 1994 election, California in 1995 implemented legislation that generally prohibits smoking in any workplace statewide. BNA,

OSHR, *Current Report* for November 23, 1994. Maryland, a state that administers its own occupational safety and health enforcement program, promulgated a workplace smoking ban in 1994, thus becoming the first state to make such a rule under its own job safety authority; the state's highest court ruled in favor of the ban in February of 1995. *Fogle v. H&G Restaurant*, 17 OSHC 1099, 654 A.2d 449 (Md. Ct.App. 1995). The Defense Department banned smoking in its civilian and military installations worldwide in 1994. BNA, OSHR, *Current Report* for March 9, 1994. Later, in August of 1997, President Clinton signed an executive order banning smoking in all federal buildings operated by the executive branch.

Some question about the efficacy of workplace smoking bans was raised, however, by the decision of a North Carolina federal district court judge in July of 1998, who rejected the EPA's landmark 1993 report on secondhand smoke as failing to establish a statistically significant association between secondhand smoke and lung cancer. *Flue-Cured Tobacco Cooperative Stabilization Corp. v. EPA*, M.D.N.C., No. 6:93-DV-370, July 17, 1998. The EPA's 1993 report concluded that secondhand tobacco smoke is a Class A carcinogen. In September of 1998, the EPA appealed the district court's decision to the 4th Circuit Court of Appeals. BNA, OSHR, *Current Report* for September 23, 1998.

250. For example, asthmatic workers have succeeded in making claims against their employers under the Americans with Disabilities Act for failing to accommodate their asthma by providing smoke-free environments. See *Bell v. Elmhurst Chicago Stone Co.*, N.D. Ill., No. 95-CV-5686, March 12, 1996; *Muller v. Costello*, N.D.N.Y., No. 94-CV-842, April 16, 1996.

251. 62 *Fed. Reg.* 54160, October 17, 1997.

252. *Atlantic & Gulf Stevedores* (88), 534 F.2d at 553; 4 OSHC at 1069–1070.

253. The Seventh Circuit has held that an employee on strike has the same right as a nonstriking employee to accompany an OSHA compliance officer during an inspection. *In re Establishment Inspection of Caterpillar, Inc.*, 17 OSHC 1243, 55 F.3d 334 (7th Cir. 1995).

254. *Dunlop v. Trumbull Asphalt Company, Inc.*, 4 OSHC 1847 (E.D. Mo. 1976).

255. See *Powell v. Globe Industries, Inc.*, 431 F. Supp. 1096; 5 OSHC 1250 (N.D. Ohio 1977). The National Labor Relations Board (NLRB) has concurrent jurisdiction over Section 11(c) cases. In 1975 the general counsel of the NLRB and the Secretary of Labor entered into an understanding for the procedural coordination of litigation arising under Section 11(c) of OSHA and Section 8 of the National Labor Relations Act, to avoid duplicate litigation. See J. Irving, "Effect of OSHA on Industrial Relations and Collective Bargaining," in *Proceedings of the ABA National Institute on Occupational Safety and Health Law*, ABA Press, Chicago, 1976, pp. 125–127. See also, 40 *Fed. Reg.* 26083 (June 20, 1976). In February of 1999, a former employee who alleged she was fired for refusing to violate OSHA rules asked the United States Supreme Court to review the Fifth Circuit's dismissal of her retaliatory discharge claim. *Dixon v. Boise Cascade Corp.* U.S., No. 98-1349, February 10, 1999. The Fifth Circuit had upheld the dismissal of her claim on the basis that the OSH Act itself does not create a private right of action for retaliatory discharge.

256. The general anti-retaliation provision in the 1977 Mine Safety and Health Act is set forth in Section 105(c) of the statute. The 1977 mine act also provides for immediate inspection of a coal mine at the request of a miner [§ 103(g)], the right of employees to accompany the inspector on his walk-around inspection of the coal mine [§ 103(f)], and limited payments to miners when a safety violation closes the mine [(§ 111)]. Black lung (coal worker's pneumoconiosis) benefits are provided to totally disabled coal miners and surviving dependents of coal miners whose deaths were due to black lung disease [§§ 401 et seq.].

257. 29 *CFR* § 1977.12(b) (2).

258. 445 U.S. 1; 8 OSHC 1001 (1980).
259. 445 U.S. at 10–11; 8 OSHC at 1004. The Supreme Court has also declined to review a ruling by the D.C. Circuit that effectively overturned a test developed by the National Labor Relations Board that allowed workers to strike over unsafe working conditions only when they reasonably believed that an inherently dangerous workplace had changed "significantly for the worse." *TNS, Inc. v. Oil, Chemical & Atomic Workers International Union*, U.S. Sup. Ct., No. 94-2067, certiorari denied October 2, 1995.
260. With respect to work stoppages over safety disputes in the context of collective bargaining agreements and the National Labor Relations Act, see *Gateway Coal Company v. United Mine Workers*, 414 U.S. 368; 1 OSHC 1461 (1974) and *National Labor Relations Board v. Tamara Foods, Inc.*, 692 F.2d 1171 (8th Cir. 1982).
261. *Irving v. U.S.*, 1st Cir., No. 96-2368, December 13, 1998.
262. See, for example, *Skidmore v. Travelers Ins. Co.*, 483 F.2d 67; 1 OSHC 1294 (5th Cir. 1973). See also, *Taylor v. Brighton Corp.*, 616 F.2d 256; 8 OSHC 1010 (6th Cir. 1980). Also, the great majority of federal appeals courts that have considered the issue have held that a company's violation of an OSHA standard does not, absent independent proof, conclusively establish that the company was negligent. See, e.g., *Elliott v. S. D. Warren Co.*, 1st Cir., No. 97-1848, January 13, 1998; *Jones v. Spentonbush-Red Star Co.*, 2nd Cir., No. 97-9586, September 14, 1998.
263. Generally, state and federal workers' compensation statutes, in a trade-off for the relative certainty of workers' compensation benefits, bar negligence claims by employees against their employers for injuries employees sustain in the course and scope of their employment. See, e.g., *Building and Construction Trades Department; AFL-CIO v. Rockwell International Corp.*, 10th Cir., No. 91-1163, October 26, 1993 (current and former Rocky Flats Nuclear Weapons Plant employees who sought medical monitoring for cancer were barred from doing so by Colorado's workers' compensation law). In order to overcome these "exclusive remedy" provisions, an injured employee generally has to make a high-threshold showing of conduct by the employer that approaches or amounts to conduct by the employer that is *intended* to injure the employee.

CHAPTER THIRTY-FIVE

Compliance and Projection

Robert J. Harris, JD

1 INVESTIGATIONS AND INSPECTIONS

With the enactment of the Occupational Safety and Health Act of 1970 (OSHA), Congress authorized the secretary of labor to enter, inspect, and investigate places of employment to discover possible violations of the employer's general and specific duties under the act.

Section 8(a) authorizes the Secretary, upon presenting appropriate credentials to the owner, operator, or agent in charge:

1. To enter without delay and at reasonable times any factory, plant, establishment, construction site, or other area, workplace or environment where work is performed by an employee or employer.
2. To inspect and investigate during regular working hours and at other reasonable times, and within reasonable limits and in a reasonable manner, any such place of employment and all pertinent conditions, structures, machines, apparatus, devices, equipment, and materials therein, and to question privately any such employer, owner, operator, agent, or employee.

The Federal Mine Safety and Health Act of 1977 directs authorized representatives of the Secretary to make frequent, unannounced inspections and investigations in coal or other mines each year. The purposes of these visits include determining whether an imminent danger exists and whether there is compliance with the mandatory health and safety standards issued under that statute (1).

Patty's Industrial Hygiene, Fifth Edition, Volume 3. Edited by Robert L. Harris.
ISBN 0-471-29753-4   © 2000 John Wiley & Sons, Inc.

1.1 Inspection Procedures—Warrants

Section 8(a) of OSHA on its face gives the secretary of labor the unqualified right to enter and inspect, in a "reasonable manner" any place of employment upon the presentation of credentials. The United States Supreme Court in 1978 held Section 8(a) unconstitutional insofar as it authorized nonconsensual warrantless inspections at an employer's establishment. In *Marshall v. Barlow's, Inc.* (2), the Court held that an employer may refuse entry to an OSHA compliance officer unless a warrant is obtained, reasoning that employers have a reasonable expectation of privacy in their commercial property and are therefore guaranteed the right, under the Fourth Amendment, to be free from unreasonable official intrusions (3).

The Supreme Court explained, however, that the secretary need not make a showing of probable cause in the criminal sense to obtain a warrant. Instead, probable cause authorizing an administrative inspection may be based either on specific evidence of a violation or on a showing that reasonable legislative or administrative standards for conducting an inspection are satisfied with respect to a particular establishment (4). By way of illustration of the latter basis, the Court stated that a warrant could properly be issued upon a showing that a particular business was chosen for inspection pursuant to a general enforcement plan derived from "neutral sources" (5). An inspection may proceed without a warrant, of course, if the employer or an authorized third party consents to the inspection (6). It is OSHA's policy to seek preinspection warrants only when it is likely that employers will not allow OSHA inspectors access to areas or records (7).

1.1.1 Probable Cause Necessary for Issuance of Warrants

Following the *Barlow's* decision, federal courts have addressed the question of what constitutes a showing of probable cause necessary to obtain an OSHA inspection warrant. The Supreme Court stated in *Barlow's* that probable cause authorizing an OSHA inspection may be based on evidence that a specific violation exists at the establishment to be inspected. Thus, probable cause for warrant issuance may be established by OSHA's receipt of an employee complaint (8), although courts of appeals have held that the nature of the violation complained of must be described in the warrant application so the magistrate issuing the warrant may make an independent determination that probable cause exists (9). In addition, the majority of federal courts addressing the issue have concluded that a warrant based on specific employee complaints is overly broad if it purports to authorize a "wall-to-wall" inspection of the entire plant. Instead, the scope of the warrant and resulting inspection must bear a reasonable relationship to the specific violations complained of (10).

There appears to be some question whether a past history of OSHA violations satisfies the requirement of probable cause for issuance of a warrant. The Second Circuit Court of Appeals has upheld the issuance of a warrant based on an employer's record of past violations (11). The Seventh Circuit has upheld a warrant based on a showing that the employer had been cited for a violation at an old plant and that the employer had tried to abate the violation by moving the unsafe operations to a new plant (12). The same court, however, has stated that a history of past violations, standing alone, is insufficient to establish probable cause (13).

Finally, the Supreme Court in *Barlow's* stated that an inspection warrant could be issued pursuant to an inspection plan based on "neutral criteria." Under this standard, the warrant application must describe the administrative plan being used, so the magistrate can make an independent determination that it is based on "neutral criteria" (14). Thus, warrant applications describing OSHA's general inspection program, in which employers are randomly chosen for inspection from a list of firms in industries with above-average lost workday rates, have been held to satisfy probable cause (15).

In addition, the Seventh Circuit has held that an inspection warrant was properly issued on the basis of OSHA's National Emphasis Program targeting high-hazard industries, even without evidence in the warrant application showing why a particular firm within the industry was selected (16). On the other hand, district courts in Pennsylvania and New Jersey have ruled that an employer's involvement in a high-hazard industry is insufficient, by itself, to establish probable cause for issuance of a warrant. Instead, these courts have held that the warrant application must contain information from which the magistrate can conclude that the particular firm within the high-hazard industry was chosen for inspection on the basis of "neutral criteria" (17).

1.1.2 Ex Parte Warrants

Additional questions raised by the Supreme Court's decision in *Barlow's* involve the proper procedure for obtaining, enforcing, and contesting OSHA inspection warrants. Although the Fifth Circuit Court of Appeals has held that the U.S. district courts do not have jurisdiction to issue injunctions compelling employers to submit to OSHA inspections (18), it is well settled that both the district courts and the U.S. magistrates have the authority to issue administrative inspection warrants (19). A more difficult question has been presented regarding OSHA's right to obtain an *ex parte* inspection warrant: that is, a warrant obtained without prior notice to the employer. In *Barlow's*, the Supreme Court suggested that the secretary could, by appropriate regulation, provide for *ex parte* warrants (although the regulation then in force called instead for "compulsory process") (20). When the original regulation (21) was held not to include *ex parte* warrants (22), an "interpretive rule" was issued stating that the term "compulsory process" was intended to include *ex parte* warrants (23). In *Cerro Metal Products, Division of Marmon Group, Inc. v. Marshall* (24), the Third Circuit held the "interpretive rule" invalid because it was inconsistent both with OSHA's prior interpretations of the "compulsory process" regulation and with the Supreme Court's dictum in *Barlow's*. Thus the court held that *ex parte* warrants were unavailable to the secretary. The Seventh, Ninth, and Tenth Circuits disagreed, holding that the secretary's "interpretive rule" properly authorized *ex parte* warrants (25). The controversy became moot when OSHA subsequently promulgated a new regulation (26) that clearly states that OSHA may obtain an inspection warrant without prior notice to the employer (27).

1.1.3 Challenging Validity of Warrant

A final question that faced the federal courts as a result of the *Barlow's* decision was the proper procedure for an employer to use in challenging the validity of an OSHA inspection warrant. In general, it appears that an employer who wishes to contest a warrant has two choices: (*1*) He or she may refuse to obey the warrant and may move to quash it in district court or (2) he or she may allow the inspection to proceed and challenge the warrant's validity before the Review Commission if the employer contests any citations issued pursuant to the inspection.

In *Babcock & Wilcox Company v. Marshall,* the Third Circuit indicated that an employer may obtain a district court hearing on a warrant's validity prior to inspection by refusing to obey the warrant (thereby risking contempt), moving to quash the warrant, and promptly appealing if the motion is denied (28). However, the Third Circuit held that if the employer obeys the warrant, he must "exhaust his administrative remedies" before the Review Commission prior to obtaining judicial review of his objections to the warrant (29).

The overwhelming majority of federal appellate courts have agreed with the Third Circuit that the doctrine of exhaustion of remedies precludes an employer from challenging an executed warrant in federal court (30). In contrast, the Seventh Circuit has held that in certain circumstances an employer may obtain district court review of a warrant even after an inspection has been completed (31). The same court, however, modified its stance on postexecution challenges when it joined the other circuits in requiring employers who challenge completed OSHA inspections on Fourth Amendment grounds to go to the Review Commission first before turning to the federal courts (32).

1.2 Other Inspection Matters

It is well settled that OSHA inspections must be made at reasonable times, in a reasonable manner, and within reasonable limits pursuant to the act (33). The act also requires the inspector to present his or her credentials to the employer before beginning the inspection (34). The Fifth Circuit Court of Appeals has held, however, that even if the inspector fails to present credentials, such failure cannot operate to exclude evidence obtained in the inspection when there is no showing that the employer was prejudiced thereby in any way (35). At least one federal district court has enjoined OSHA to provide employers within the court's jurisdiction with copies of employee complaints at the time of any employee complaint-based inspections (36).

Section 8(e) of OSHA requires that a representative of the employer and an authorized employee representative be allowed to accompany the OSHA inspector during the physical inspection of any workplace. In *Chicago Bridge & Iron Company v. OSAHRC and Dunlop* (37), the Seventh Circuit held the directives of Section 8(e) to be mandatory rather than merely directory. The court refused, however, to hold that the absence of a formalized offer of an opportunity to accompany the compliance officer on his inspection rendered the citations for violations observed during that inspection void *ab initio.* Rather, the court explained that when there has been substantial compliance with the mandate of the act regarding walk-around rights and the employer is unable to demonstrate that prejudice resulted from his nonparticipation in the inspection, citations issued as a result of the inspection are valid (38).

In *Accu-Namics* (35), the Fifth Circuit did not reach the question of whether the language of Section 8(e) was mandatory or directory, but it did refuse to adopt a rule that would exclude all evidence obtained illegally, no matter how minor or technical the government's violation and no matter how egregious or harmful the employer's safety violation.

The Ninth and Tenth Circuits have also concluded that minor violations of Section 8(e)'s inspection procedures do not justify dismissing a citation (39) or suppressing evidence gained from an inspection (40), at least as long as there has been substantial compliance with the act and the employer's defense on the merits has not been prejudiced.

The Fourth and Eighth Circuits have gone even further, holding as a matter of law that regardless of whether there has been substantial compliance by the inspector, Section 8(e) violations do not affect the validity of citations issued or evidence obtained unless the employer can show that he or she has been prejudiced in preparing or presenting a defense on the merits. In *Pullman Power Products, Inc. v. Marshall and OSAHRC* (41), the Fourth Circuit held that, in the absence of such prejudice, the validity of citations was not affected by an inspector's alleged failure properly to present his credentials, conduct opening and closing conferences, and provide walk-around rights. In so holding, the court declined to reach the issue of substantial compliance, stating broadly that the employer's inability to show prejudice bars its attack on the validity of the citations. The Eighth Circuit reached a similar conclusion in *Marshall v. Western Waterproofing Co., Inc.* (42). Although stating that the requirements of Section 8(e) are not merely directory, the court nevertheless held that, in the absence of prejudice to the employer, an inspector's failure to comply with these requirements will not justify suppression of evidence, regardless of whether there has been substantial compliance (43).

Applicable regulations authorize OSHA consultants to make inspections at work sites in order to help employers comply with OSHA standards (44). The consultants do not report violations to the compliance officers. There is also a voluntary protection program (VPP), under which OSHA recognizes exemplary workplaces by removing participating employers from OSHA's general inspection lists (45). The VPP was introduced in 1982 (46). Legislation aimed at codifying the VPP as part of the OSH Act was introduced in the House in March of 1999 (47).

2 RECORD KEEPING AND REPORTING

Section 8(c)(1) of OSHA requires employers to make, keep, and preserve such records regarding their OSHA-related activities as the secretary of labor, in cooperation with the secretary of health and human services (HHS), may prescribe as necessary or appropriate for the enforcement of the act or for developing information on the causes and prevention of occupational accidents and illnesses. These records must also be made available to the secretaries of labor and/or HHS (48).

Section 8(c)(2) more specifically directs the secretary of labor, in cooperation with the secretary of HHS, to issue regulations requiring employers to maintain accurate records of, and to make periodic reports on, work-related deaths, injuries, and illnesses (other than minor injuries requiring only first aid treatment and not involving medical treatment, loss of consciousness, restriction of work or motion, or transfer to another job).

Regulations promulgated by the secretary of labor to implement Sections 8(c)(1) and (2) require employers to keep the following records (or the equivalents thereof):

OSHA Form 200   A log and annual summary of all recordable occupational injuries and illnesses.
OSHA Form 101   A supplementary record for each occupational injury or illness (49).

The regulations further require that a copy of Form 200, summarizing the year's occupational illnesses and injuries, be posted in each establishment in a conspicuous place or places where notices to employees are customarily posted. The required records must be retained in each establishment for 5 years following the end of the year to which they relate (50).

Employers must report incidents involving employee fatalities or the hospitalizations of three or more workers to OSHA within eight hours. The rule requires the employer's representative to make such a report verbally by telephone or in person to the nearest OSHA office, or by using OSHA's toll-free telephone number (51). This rule has been extended to federal agencies whose employees are covered by OSHA (52).

In addition, Section 24 of the act directs the secretary of labor, in consultation with the secretary of HHS, to develop and maintain a program of collection, compilation, and analysis of occupational safety and health statistics. The secretary of labor has given the commissioner of the Bureau of Labor Statistics (BLS) the authority to develop and maintain such a program. This program requires employers to participate in periodic surveys of occupational injuries and illnesses. The survey form is OSHA Form 200S, and an employer who receives such a form has a duty to complete and return it promptly (53).

Small employers are exempt from many, but not all, of OSHA's record-keeping requirements. For example, employers who had no more than 10 employees at any time during the calendar year preceding the current one are exempt from the requirements of keeping OSHA Forms 200 and 101. Such employers, however, are not exempt from the requirement of reporting accidents resulting in fatalities or multiple hospitalizations. In addition, small employers may be selected to participate in the BLS periodic surveys and, if selected, must maintain a log and summary on Form 200 for the survey year, and must make the required reports on survey Form 200S (54).

Finally, with respect to toxic materials or harmful physical agents, the secretary of labor, in cooperation with the secretary of HHS, is directed by Section 8(c)(3) of the act to issue regulations requiring employers to maintain accurate records of employee exposure to potentially toxic materials or harmful physical agents that are required to be monitored or measured under Section 6. These regulations must guarantee employees or their representatives an opportunity to observe the required monitoring or measuring, and to have access to the records of employee exposure. Further, the regulations must ensure employees and former employees access to such records as will indicate their own exposure to such substances (55).

These statutory directives have been implemented by the secretary of labor in a rule governing employee exposure and medical records (56). This rule requires employers to maintain exposure and medical records pertaining to their employees' exposure to toxic substances and harmful physical agents. The exposure records must be made accessible to exposed and potentially exposed employees, as well as to designated employee representatives and to OSHA. Access to medical records must also be ensured for the employee and for OSHA; because of the privacy interests involved, however, an employee's medical records are open to a collective bargaining representative only with the employee's consent (57).

3 SANCTIONS FOR VIOLATING SAFETY AND HEALTH LAWS

3.1 Citations

The Occupational Safety and Health Act authorizes the secretary of labor to issue citations and proposed penalties to employers who are believed to have violated the act or its implementing regulations. Section 9(a) directs the secretary to issue citations "with reasonable promptness" following an inspection or investigation. According to Section 9(c), no citation may be issued after the expiration of 6 months from the occurrence of any violation (58).

Each citation is to be in writing and must "describe with particularity the nature of the violation," including a reference to the provision of the statute, standard, rule, regulation, or order alleged to have been violated (59). Section 9(a) states that each citation must also fix a reasonable time for the abatement of the violation (60).

Section 9(b) requires employers to post each citation prominently at or near each place where a violation referred to in the citation occurred. The mechanics of how, when, where, and how long to post the citations are set forth in regulations issued by the secretary (61).

3.2 Penalties

Within a reasonable time after a citation has been issued, the secretary is directed by Section 10(a) to notify the employer by certified mail of the penalty, if any, that will be assessed for the violation. Violations fall into the following general categories: serious, nonserious, de minimis, willful, repeated, and criminal, and the penalty will be based at least in part on the nature of the violation (62).

With the Omnibus Budget Reconciliation Act of 1990, Congress amended OSHA to provide for dramatically increased penalties for occupational safety and health violations, primarily to raise revenue. In general, maximum penalty levels were increased sevenfold, and a $5000 statutory minimum penalty for willful violations was established (63).

3.2.1 *Serious Violations*

Section 17(k) of OSHA defines a serious violation as follows:

> [A] serious violation shall be deemed to exist in a place of employment if there is a substantial probability that death or serious physical harm could result from a condition which exists, or from one or more practices, means, methods, operations, or processes which have been adopted or are in use, in such place of employment unless the employer did not, and could not with the exercise of reasonable diligence, know of the presence of the violation.

The *probability* of an *accident* occurring need not be shown to establish that a violation is serious (64). Rather, as the Ninth Circuit court ruled in *California Stevedore and Ballast* (64), a serious violation exists if any accident that could result from the violation would have a substantial probability of resulting in death or serious physical harm (65). No actual death or physical injury is required to establish a serious violation (66).

Employer knowledge is clearly an element of a serious violation. The knowledge requirement in Section 17(k) deals with actual or constructive knowledge (67) of practices or conditions that constitute violations of the act; it is not directed to knowledge of the law (68). The burden of proof is on the secretary to prove knowledge (69) as well as the other elements of a serious violation of the act.

Since 1990, Section 17(b) provides that an employer who has received a citation for a serious violation of the act must be assessed a civil penalty of up to $7000.

3.2.2 Nonserious Violations

The original Senate version of the occupational safety and health bill treated all violations as "serious." As finally enacted, however, OSHA incorporated a House proposal for violations "determined not to be of a serious nature" (70).

The statute does not describe the elements of a nonserious violation and provides no guidelines for determining when a violation is not serious. The Fifth Circuit, however, has described nonserious violations as violations that do not create a substantial probability of serious physical harm (71). The commission has explained that serious and nonserious violations are distinguished on the basis of the seriousness of injuries that experience has shown are reasonably likely to result when an accident does arise from a particular set of circumstances (72). At least one federal court of appeals has held that employer knowledge is an element of a nonserious violation (73).

When a violation is determined not to be serious, the assessment of a penalty is discretionary rather than mandatory (74). Section 17(c) states that the employer may be assessed a civil penalty of up to $7000 for each nonserious violation.

Section 110(a) of the Federal Mine Safety and Health Act of 1977 does not allow discretionary penalties for so-called nonserious violations, but instead requires the secretary of labor to assess a civil penalty of up to $50,000 for each violation of a mandatory health or safety standard under the act.

3.2.3 De Minimis Violations

If noncompliance with an OSHA provision or standard presents no direct or immediate threat to the safety or health of employees, the violation is *de minimis* and the secretary of labor may issue only a notice—not a citation—to the employer (75). The notice contains no proposed penalty (76). The secretary has explained that a violation should be considered *de minimis* if: (a) an employer complies with the intent of a standard but deviates from its particular requirements in a way that has no direct or immediate relationship to safety and health; or (b) an employer complies with a proposed amendment to a standard and the amendment provides equal or greater safety and health protection than the standard itself; or (c) an employer's workplace is "state of the art"—that is, it is technically advanced beyond the requirements of a standard and provides equal or greater safety and health protection (77).

3.2.4 Willful or Repeated Violations

The Occupational Safety and Health Act provides more stringent civil penalties for employers who "willfully or repeatedly" violate the act or any regulations promulgated pursuant thereto. Under Section 17(a) of the act, willful or repeated violations are subject to penalties of up to $70,000 for each violation. Under an OSHA policy announced in June of 1994, the minimum penalty for most willful violations became $25,000. Under the policy, willful violations that are deemed to be serious are divided into categories of high, medium and low gravity, with maximum penalties of $70,000, $55,000 and $40,000, respectively. Up to 10% may be excused based on the employer's history of OSHA violations, although no penalty may be reduced below the $25,000 minimum. Under the policy, non-serious willful violations are penalized at a minimum of $5,000. State OSHA programs were to raise their minimum penalties as well under the policy (78). Based largely on concerns raised by state OSHA programs, OSHA later scaled back its minimum penalty for willful violations for employers with 50 or fewer employees from $25,000 to $5,000 (79).

The act contains no definition of either "willful" or "repeated" as applied to violations. Thus it is not surprising that there has been some disagreement on the elements of these types of violations.

In *Frank Irey Jr., Inc. v. OSAHRC* (80), the Third Circuit initially held that "[w]illfulness connotes defiance or such reckless disregard of consequences as to be equivalent to a knowing, conscious, and deliberate flaunting of the Act. Willful means more than merely voluntary action or omission—it involves an element of obstinate refusal to comply." The majority of the circuits, as well as the Review Commission, declined to follow the *Frank Irey* definition of "willfulness." The First and Fourth Circuits interpreted a willful action as a "conscious, intentional, deliberate voluntary decision," regardless of venial motive (81). The Review Commission agreed that no showing of malicious intent is necessary to establish "willfulness," (82) defining a willful violation as one "committed with either an intentional disregard of, or plain indifference to, the Act's requirements" (83). Most courts of appeals have either adopted the Review Commission's standard or have embraced similar definitions that do not require a showing of a bad motive (84).

Thus, a conflict developed between the Third Circuit's view, as expressed in *Frank Irey*, and the majority approach. The Court of Appeals for the District of Columbia Circuit characterized this conflict as more apparent than real. In *Cedar Construction Co. v. OSAHRC and Marshall* (85) that court indicated that the two approaches were likely to yield the same results in particular cases, since there is little practical difference between "obstinate refusal to comply" and "intentional disregard" of the act. The Third Circuit later agreed when it again addressed the "willfulness" question. In *Babcock & Wilcox Co. v. OSAHRC* (86), the court explained its holding in *Frank Irey* as follows (87):

> [T]he supposed conflict among the circuits on this point has been generated by several courts of appeals reading into our *Irey* definition a requirement that the employer act with "bad

purpose." Read in this fashion, *Irey* has not been followed by some circuits. To our way of thinking, an "intentional disregard of OSHA requirements" differs little from an "obstinate refusal to comply;" nor is there in context much to distinguish "defiance" from "intentional disregard." We also believe, as does the District of Columbia Circuit, that the same results would likely be reached in various cases, including the one here, regardless of the verbiage utilized. It is not unusual that different words are used to describe the same basic concept.

This clarification by the Third Circuit thus resulted in general agreement among the circuits that the Review Commission's "intentional disregard" standard is the correct one, and that no malicious intent need be shown to establish a willful violation.

The interpretation of "repeated" violation has also generated disagreement among employers, the courts, and the commission. Once again, the controversy has centered on a Third Circuit ruling. In *Bethlehem Steel Corp. v. OSAHRC and Brennan* (88), the Third Circuit held that the commission can find a repeated violation only when the evidence shows the employer consciously ignored or "flaunted" the requirements of the act and was cited for a similar violation on at least two prior occasions. The court further explained that in applying the "repeatedly" portion of the act, the commission must determine that the acts themselves "flaunt" the requirements of the statute, but need not determine whether the acts were performed with an intent to "flaunt" the requirements of the statute.

Most other federal courts refused to adopt the Third Circuit's interpretation of a "repeated" violation. The Fourth Circuit, in *George Hyman Construction Company v. OSAHRC* (89), reasoned that the requirement of flagrant misconduct to establish a repeated violation fails to "recognize a meaningful distinction between willful and repeated violations." Other circuits have agreed that an employer need not have a particular state of mind or motive for "flaunting" the act, nor otherwise exhibit an aggravated form of misconduct to be guilty of a repeated violation (90). Most of these courts also rejected the Third Circuit's conclusion that a repeated violation must be based on at least two prior violations, requiring instead only one prior and substantially similar infraction.

The Review Commission also agreed that no "flaunting" of the act need be shown to establish a repeated violation. In *Potlach Corporation* (91), the commission defined a "repeated" violation as follows (92):

A violation is repeated under section 17(a) of the Act if, at the time of the alleged repeated violation, there was a Commission final order against the same employer for a substantially similar violation (93).

With respect to the length of time for which a previous violation may serve as the basis for a subsequent, "repeated" violation, it is OSHA policy to issue a citation as a repeated violation if:

a. The citation is issued within 3 years of the final order of the previous citation, or
b. The citation is issued within 3 years of the final abatement date of that citation, whichever is later (94).

The Commission has also ruled that an employer's substantial history of OSHA violations can outweigh the gravity of a particular cited violation and thus skew the penalty for that violation upward (95).

The Federal Mine Safety and Health Act of 1977 [Section 110(d)] does not provide more stringent civil penalties for willful violations of that act's safety and health standards, but does impose criminal liability on mine operators who are found guilty of such willful conduct. The act does impose [Section 110(g)] a civil penalty on miners who willfully violate the safety standards relating to smoking or the carrying of smoking materials.

3.2.5 Criminal Sanctions

Job safety and health laws also provide criminal sanctions for certain specified conduct (96). The most stringent criminal sanctions are those set forth in the Federal Mine Safety and Health Act of 1977. Section 110(d) states that a mine operator shall be subjected to a fine of up to $25,000 or a prison term of up to one year (or both) for willfully violating mandatory mine health or safety standards or for knowingly failing to comply with certain orders issued under that statute. That section further provides for increased maximum penalties of $50,000 and 5 years in prison for a second conviction (97).

Under OSHA, willful violations (98) of the act (or of relevant standards, rules, regulations or orders) that cause death to an employee shall be punishable upon conviction by a fine of up to $10,000 or imprisonment for up to 6 months or both. Second convictions carry maximum penalties of $20,000 and one year in jail (or both) (99).

Criminal penalties may also be imposed for: (a) knowingly making any false statement, representation, and so forth, in any document filed pursuant to or required to be maintained by OSHA or by the Mine Safety and Health Act of 1977 (100); (b) giving advance notice of an OSHA or Mine Safety and Health inspection without the authority of the secretary of labor (101); (c) knowingly distributing, selling, and so on, in commerce any equipment for use in coal or other mines that is represented as complying with the Mine Safety and Health Act and does not do so (102); (d) killing an OSHA inspector or investigator on account of the performance of his or her duties (103).

3.2.6 Failure to Abate a Violation

The Occupational Safety and Health Act does not specify fixed periods within which violations must be remedied, but Section 9(a) does require that each citation "fix a reasonable time for the abatement of the violation" (104). The abatement period does not begin to run until the date of the final order of the commission affirming the citation, as long as the review proceeding, if any, initiated by the employer was in good faith and not solely for delay or avoidance of penalties (105). An employer who fails to correct a violation within the period specified in the citation may receive an additional citation pursuant to Section 10(b) for failure to abate. Failure to abate may result in the assessment of civil penalties of not more than $7000 per day for each day the violation continues (106).

OSHA published a final rule in March of 1997 requiring employers to certify to OSHA that they have corrected cited violations and to inform their employees of the abatement actions taken (107). Prior to the final rule, employer compliance with requests to provide showings of abatement was voluntary, and follow-up inspections often were necessary in order to determine whether abatement had occurred (108).

OSHA instituted a nationwide "Quick Fix" abatement incentive program in August of 1996. Under the program, employers may be given a 15% penalty reduction for correcting a cited workplace hazard within 24 hours of the inspection. The penalty reduction does not apply to violations linked to fatal injury or illness, to serious incidents resulting in serious injuries to employees, nor to high- or medium-gravity serious, willful, repeat or failure-to-abate violations (109).

Notices of violations under Sections 104(a) and 104(b) of the Federal Mine Safety and Health Act of 1977 must similarly specify time periods for abatement of violations. Failure to abate under that statute can result in an order directing all persons to be withdrawn from the affected area of the mine until a representative of the secretary determines that the violation has been abated.

3.3 Contesting Citations and Penalties

Section 10(a) of OSHA and the regulations promulgated thereunder provide a means for contesting citations and proposed penalties. After the employer has been notified of the penalty proposed by the secretary, the employer has 15 working days to notify the secretary that he or she wishes to contest the citation or the proposed assessment of penalty. A failure to notify the secretary within 15 days of intent to contest the citation or proposed penalty will render the citation or penalty "a final order of the Commission and not subject to review by any court or agency" (110).

The secretary's regulations (111) instruct the employer that "[e]very notice of intention to contest shall specify whether it is directed to the citation or to the proposed penalty, or both." Similarly, the courts have construed the OSHA enforcement scheme as mandating a distinction between contesting a citation and contesting a proposed penalty (112). Thus, the Fifth Circuit held (113) that an employer's letter that contested the proposed penalty but failed to contest the citation (in fact, the letter affirmatively admitted the violation) constituted waiver of the employer's right to challenge the citation on appeal. However, the commission has stated that it will construe notices of contest that are limited to the penalty to include a contest of the citation as well if the cited employer indicates later that it was his or her intent to contest the citation (114).

If an employer files a timely notice of contest (or if within 15 working days of the issuance of a citation, a representative of the employees files a notice challenging the period of abatement specified in the citation), the secretary must immediately advise the Occupational Safety and Health Review Commission of the intent to contest. The commission then must afford an opportunity for an administrative hearing (115).

A commission hearing is conducted pursuant to the Administrative Procedure Act and is presided over by a single administrative law judge employed by the commission. After taking testimony, the judge writes an opinion, which is subject to review by the full three-member commission at its discretion. An aggrieved party may petition for discretionary review before the full commission (116), and any commission member may direct review of a case on his or her own motion (117). If no commissioner directs review, or if a timely petition for review is not filed, the administrative law judge's decision becomes a final order of the commission (118).

Section 10(c) authorizes the commission to review either the citation or the proposed penalty or both:

> The Commission shall thereafter [i.e., after hearing] issue an order, based on findings of fact, affirming, modifying or vacating the Secretary's citation or proposed penalty, or directing other appropriate relief, and such order shall become final thirty days after its issuance.

Section 17(j) further empowers the commission to assess appropriate civil penalties, giving due consideration to the size of the business of the employer being charged, the gravity of the violation, the good faith of the employer, and the history of previous violations (119). The commission has taken the position that it may exercise its power to increase the secretary's proposed penalty after considering the factors outlined above, and courts of appeals have expressed the view that the commission may act in this manner (120). The Commission has also held that it does not have to give "substantial weight" to the Secretary's proposed penalties (121).

Courts have also sanctioned the commission's right to increase the degree of a violation from nonserious to serious (122). The commission has taken the view that it can reduce the degree of a violation as well (123).

In November of 1994, the United States Supreme Court denied an employer's request to review the question of whether the Commission had lost its neutrality in adjudicating disputes between the Secretary and employers (124). In the petition for certiorari, the employer had maintained that the Commission had devolved into no more than a ratifier of OSHA's enforcement policies and practices.

The final stage of an OSHA enforcement proceeding is review in a court of appeals (and thereafter discretionary review by the Supreme Court). Any person adversely affected or aggrieved by the commission's disposition (125) may obtain review in an appropriate court of appeals pursuant to Section 11(a) of the act (126). Section 11(b) provides that the secretary of labor may also obtain review or enforcement of any final order of the commission by filing a petition for such relief in the appropriate court of appeals. The reviewing court is bound by Section 11(a) to apply the "substantial evidence test" to the commission's findings of fact (127). The same section empowers the court to direct the commission to consider additional evidence if the evidence is material and reasonable grounds existed for a party's failure to admit it in the hearing before the commission. Regarding the penalty imposed by the commission, the reviewing court may inquire only whether the commission abused its discretion because the assessment of a penalty is not a finding of fact but rather the exercise of a discretionary grant of power (128). The Seventh Circuit has held that interest on penalties assessed against an employer by the Commission accrues from the date the Commission's order is entered, even when an employer appeals to the federal courts (129).

Following a general trend in the law during the 1990s, the Review Commission in February of 1999 published an interim final rule establishing a one-year pilot alternative dispute resolution program (130). Under the program, parties in cases appealed to the Commission with proposed penalties exceeding $200,000 must participate in formal settlement talks, known as the "Settlement Part," before a trial is scheduled. The program is designed to stem the tide of complex cases from going to trial and thus to help clear the clogged Review Commission docket. It provides for an administrative law judge other than the one assigned to hear the case to act as the "settlement part judge," requires the attendance at the settlement conference of officials from both sides with authority to settle and provides for the inadmissibility in subsequent proceedings of any statements made during the negotiations.

The Mine Safety and Health Act of 1977 has administrative review procedures that are similar to the OSH Act's for contesting citations and penalties (131). The United States Supreme Court has held that an employer cannot circumvent the MSHA's administrative law procedures but rather must exhaust the administrative steps before seeking redress in federal court (132). Presumably, the same analysis would hold were an employer to attempt to sidestep OSHA's administrative adjudication process.

3.4 Imminent Danger Situations

Section 13(a) of OSHA confers jurisdiction on the U.S. district courts, upon petition of the secretary of labor, to restrain hazardous employment conditions or practices if they create an imminent danger of death or serious physical harm that cannot be eliminated through the act's other enforcement procedures.

As originally reported out of the House committee, the act contained a provision that would have permitted an OSHA inspector to close down an operation for up to 72 hours without a court order if he or she found that an imminent danger existed. The original Senate version of the bill also contained a provision allowing an inspector to close down an operation for 72 hours, but this provision was revised so that no shutdown may occur unless a federal district judge grants an application for a temporary restraining order (133).

The Federal Mine Safety and Health Act of 1977, however, permits coal or other mine operations to be shut down without a restraining order from a court. Section 107(a) of the act provides that when a federal inspector finds that an imminent danger is present in a mine, he or she shall order the withdrawal of all other persons from a part or all of that mine until the imminent danger no longer exists (134).

3.5 Challenges to the OSHA Enforcement Scheme

Not surprisingly, the citation and penalty scheme of OSHA has been subject to challenges on several fronts. For example, in *Atlas Roofing Company, Inc. v. OSAHRC* (135), a cited employer contended that the act was constitutionally defective because: (a) civil penalties under OSHA are really penal and call for the constitutional protections of the Sixth Amendment and Article III; (b) even if the penalties are civil, OSHA violates the Seventh Amendment because of the absence of a jury trial for fact finding; (c) the act denies the employer his or her right to a Fifth Amendment "prejudgment" due process hearing since commission orders are self-executing unless the employer affirmatively seeks review; and (d) the overall penalty structure of OSHA violates due process because it "chills" the employer's right to seek review of the citation and penalty. The Fifth Circuit rejected all the employer's constitutional contentions. The Supreme Court granted a petition for certiorari in that case, limited to the Seventh Amendment issue (136), and subsequently upheld the act's provision for imposition of civil penalties without fact finding by a jury (137).

There have been other constitutional challenges to various OSHA enforcement policies. In March of 1998, the Seventh Circuit rejected an employer's contention that its being fined administrative penalties even after it was criminally prosecuted for the same offenses under the Act violated the double jeopardy clause of the Fifth Amendment (138). The double jeopardy ban prohibits the government from prosecuting a person twice for the same offense.

In a due process challenge, the Review Commission has held that OSHA did not violate an employer's due process rights when it issued separate sets of citations, instead of only one citation, for alleged violations at nine different trenching sites on the same construction project (139).

3.5.1 Egregious Case Policy

In what some described as a significant loss for OSHA's enforcement policy, the Fifth Circuit ruled in April of 1997 that OSHA may not cite employers for per-employee violations of the general duty clause under the agency's "egregious case" policy, which OSHA had used since 1986 to cite alleged violations on an instance-by-instance basis in particularly bad cases (140). The case grew out of a 1992 explosion at a urea reactor. OSHA charged that during the explosion and the 24-hour period leading up to it, a total of 87 employees had been willfully exposed to a recognized hazard (the catastrophic failure of a pressure vessel), and the agency used its egregious case policy to cite the employer 87 times with a proposed penalty of $50,000 per instance, for a total proposed penalty of $4.3 million. The Review Commission's administrative law judge, characterizing the use of the egregious case policy in a general duty case as "inappropriate," vacated 86 of the 87 citation items and all but $50,000 of the proposed penalty. On review, the Commission affirmed the ALJ, holding that the Secretary is not authorized to use the egregious case policy in general duty clause cases (141). The Secretary appealed the Commission's decision to the Fifth Circuit, and, for the first time, a federal appeals court was asked to rule on the legality of the egregious case policy. The Fifth Circuit ruled that the plain meaning of the general duty clause allows OSHA to cite only for each violative condition in a workplace, not for each employee exposed to the violative condition. Viewing the Fifth Circuit's holding as a rather narrow one, because it affects only general duty clause cases, OSHA decided not to appeal the ruling to the Supreme Court (142).

Arcadian related only to OSHA's authority to use the egregious case policy under the general duty clause, which OSHA uses when there is no specific standard under which to cite an employer. The egregious case policy remains viable with regard to specific standards, so long as the standard itself allows for stacking of penalties (143).

3.5.2 Cooperative Compliance Program

In November of 1997, OSHA announced that it would expand its Cooperative Compliance Program (CCP)—designed to combine aspects of enforcement and partnerships between OSHA and employers—to all 29 states under federal OSHA jurisdiction. The program was to use data collected in nationwide illness and injury surveys to target specific employers that showed higher-than-average rates of illness or injury, then give those employers the choice of whether to cooperate with OSHA to attempt to improve safety at their workplaces (in exchange for a lower probability that such workplaces would be inspected) or be placed on OSHA's primary inspection list (144).

Several industry groups, including the United States Chamber of Commerce, filed a petition for review challenging the CCP with the District of Columbia Court of Appeals in January of 1998 (145). The challenging groups complained that the CCP—billed as the "coercive compliance program" by some—constitutes a standard that was not properly promulgated through the notice-and-comment process and that the program's method of targeting specific employers violates the Fourth Amendment prohibition of unreasonable search and seizure. The D.C. Circuit stayed the operation of the CCP in February of 1998 pending a decision on the merits, and, in response, OSHA instituted an interim inspection plan in which whole industries with higher-than-average lost workday rates are targeted, not specific employers (146). The court approved the interim plan's methodology over the plaintiffs' objections, and OSHA subsequently announced that, during the pendency of the litigation, the interim enforcement plan would take priority over the agency's other specific enforcement programs (147).

The D.C. Circuit has expressed concern that the CCP may not be a "standard" at all, and the court thus may not have jurisdiction to rule on the merits of the case (148). If the court of appeals refuses to consider the merits and dismisses the challenge on jurisdictional grounds, the plaintiffs would then have to renew their challenge in federal district court. As of early 1999, a ruling was pending.

4 THE FUTURE OF JOB SAFETY AND HEALTH LAW

Overall, employers can expect OSHA's complementary strategies of cooperation with willing employers and aggressive enforcement with others to continue. While "egregious case" proceedings will probably remain uncommon, the number of "significant" cases—defined by OSHA as cases involving proposed penalties of more than $100,000—will probably increase owing to OSHA's strategy of targeting employers (or industries) with higher-than-average injury and illness rates (149). Of course, much of OSHA's overall strategy could be impacted by rulings on the challenges to the agency's Cooperative Compliance Program.

Observers can expect a long and bitter struggle over OSHA's proposed ergonomics rulemaking. Other rulemakings on tap include those for indoor air quality, tuberculosis exposure and an update of the PEL for silica dust, as well as a final recordkeeping regulation. In addition, OSHA has described as the "centerpiece" of its latest regulatory agenda a proposed rule that would require employers to develop comprehensive safety and health programs for their work sites (150).

Legislatively, OSHA plans to lobby vigorously for improved protections for job safety whistleblowers. Provisions that have been discussed include extending the filing deadline for complaining workers who are fired or otherwise discriminated against from 30 days to six months and providing "make-whole" relief to whistleblowers who prevail in their cases (151). Legislation to this effect was introduced in the Senate in March of 1999 (152).

Other OSHA-reform legislation that had been introduced or was expected to be introduced as of early 1999 included bills to require cost-benefit analyses on regulations that are projected to have an annual impact on the economy of $100 million or more (153), to give employers relief from penalties if they voluntarily hire consultants certified by OSHA to audit their workplaces (154), to shield employers' safety self-audits from OSHA (155), and even to repeal most of the OSHA Act of 1970 in favor of an agency that would simply assist employers to improve worker safety and health (156). However, in the wake of the 1998 elections, the strengthened Democratic majority in Congress is expected to make it tough going for OSHA reform legislation, which has largely been GOP-driven (157).

ACKNOWLEDGMENTS

Authors of this Chapter in the previous Edition of *Patty's Industrial Hygiene and Toxicology* are Martha Hartle Munsch, JD, and Jacqueline A. Koscelnik, JD It is respectfully acknowledged that much of this Chapter remains their work from the previous Edition. Some general orientation text and the updating with listings and discussion of post 1992 materials and events are by the current author.

BIBLIOGRAPHY

1. See Section 103(a) of the Federal Mine Safety and Health Act of 1977. For a discussion of "imminent danger," see Section 3.4 of this chapter. Underground coal or other mines must be inspected at least four times each year. Each surface coal or other mine must be inspected at least twice a year.
2. 436 U.S. 307; 6 OSHC 1571 (1978). Several state courts have likewise held the inspection provisions of their respective state occupational safety and health acts unconstitutional or have held that inspections are permissible under those provisions only pursuant to a warrant. For example: *Woods & Rohde, Inc., d/b/a Alaska Truss & Millwork v. State*, 565 P.2d 138; 5 OSHC 1530 (Alaska Sup. Ct. 1977) (provision authorizing warrantless inspections violates state constitution); *State v. Albuquerque Publishing Co.*, 571 P.2d 117; 5 OSHC 2034 (New Mexico Sup. Ct. 1977), *certiorari denied*, 435 U.S. 956; 6 OSHC 1570 (1978) (nonconsensual inspection pursuant to state occupational safety and health act requires warrant satisfying administrative standards of probable cause); *Yocom v. Burnette Tractor Co., Inc.*, 566 S.E.2d 755; 6 OSHC 1638 (Kentucky Sup. Ct. 1978).
3. By contrast, the Supreme Court has upheld the constitutionality of warrantless inspections under the Federal Mine Safety and Health Act of 1977 on the grounds that mining has long been a pervasively regulated industry, in which a businessman can have no reasonable expectation of privacy against official inspections. In *Donovan v. Dewey*, 452 U.S. 594 (1981), the Court reasoned further that the Federal Mine Safety and Health Act inspection provisions are more narrowly drawn and provide for less administrative discretion concerning inspections than does OSHA, thus making such inspections reasonable under the Fourth Amendment.
4. 436 U.S. at 320–21; 6 OSHC at 1575–76. Several state courts have held that this relaxed administrative standard of probable cause is sufficient to support an inspection warrant pursuant to their respective occupational safety and health acts. See, for example, *Yocom v. Burnette*

Tractor Co., Inc., (2); *State v. Keith Manufacturing Co.*, 6 OSHC 1043 (Oregon Ct. of Appeals 1977) (provision of Oregon Safe Employment Act authorizing inspection warrant based on administrative standard of probable cause is constitutional); *State v. Kokomo Tube Co.*, 426 N.E.2d 1338; 10 OSHC 1158 (Indiana Ct: of Appeals 1981) (administrative probable cause standard articulated in *Barlow's* is applicable to inspection warrants under Indiana OSH Act); *Salwasser Mfg. Co. v. Occupational Safety and Health Appeals Bd.*, 14 OSHC 1278 (California Ct. of Appeal 1989).

5. 436 U.S. at 321; 6 OSHC at 1575–76. Such "neutral sources," according to the Court, could include statistics indicating "dispersion of employees in various types of industries across a given area, and the desired frequency of searches in any of the lesser divisions of the area . . ." 6 OSHC at 1576.

6. See, for example, *Donovan v. A. A. Beiro Constr. Co.*, 746 F.2d 894; 12 OSHC 1017 (D.C. Cir. 1984) (all contractors at the site consented to the inspection of a site owned by the District of Columbia); *J. L. Foti Constr. Co. v. Donovan*, 786 F.2d 714; 12 OSHC 1737 (6th Cir. 1986) (general contractor at work site consented); *National Eng'g & Contracting Co. v. Department of Labor*, 687 F. Supp. 1219; 13 OSHC 1793 (S.D. Ohio 1988) (U.S. Army Corps of Engineers consented to inspection), *affirmed*, 902 F.2d 34; 14 OSHC 1621 (6th Cir.), *certiorari denied*, 111 S. Ct. 344; 14 OSHC 1920 (1990). See also *Secretary of Labor v. LaForge & Budd Construction Co.*, 16 OSHC 2002 (Rev. Comm. 1994), in which the Review Commission held that a warrantless inspection of a construction site, during which police were called after workers and management blocked a federal inspector, was constitutional because the employer had agreed in its contract with the municipal authority to grant common authority over the work site to the municipal authority, and the municipal authority's consent to the search alone was therefore sufficient.

7. OSHA Instruction CPL 2.45B.III.B.3.a (June 15, 1980), amended by CPL 2.45 *OSHR*, Reference File at 77:2301.

8. See *Marshall v. Chromalloy American Corp. (Gilbert & Bennett Manufacturing Co.)*, 589 F.2d 1335; 6 OSHC 2151 (7th Cir.), *certiorari denied*, 444 U.S. 884; 7 OSHC 2238 (1979); *Marshall v. W and W Steel Co., Inc.*, 604 F.2d 1322; 7 OSHC 1670 (10th Cir. 1979); *Burkart Randall Div. of Textron, Inc. v. Marshall*, 625 F.2d 1313; 8 OSHC 1467 (7th Cir. 1980); *Martin v. Intern. Matex Tank Terminals—Bayonne*, 928 F.2d 614, 14 OSHC 2153 (3d Cir. 1991). On the other hand, it has been held that the occurrence of an accident on the employer's premises is insufficient evidence of a specific violation to establish probable cause for an inspection. See, for example, *Donovan v. Federal Clearing Die Casting Co.*, 655 F.2d 793; 9 OSHC 2072 (7th Cir. 1981) (newspaper reports of accident do not satisfy "specific evidence of existing violation" standard); *Marshall v. Pool Offshore Co.*, 467 F. Supp. 978; 7 OSHC 1179 (W.D. La. 1979) (while actual fatalities do not constitute evidence of specific violation, an OSHA policy of investigating all fatalities may satisfy administrative "neutral criteria" standard).

9. *Weyerhaeuser Co. v. Marshall*, 592 F.2d 373; 7 OSHC 1090 (7th Cir. 1979); *Marshall v. Horn Seed Co., Inc.*, 647 F.2d 96; 9 OSHC 1510 (10th Cir. 1981). The Seventh Circuit has concluded that, since probable cause in the strict criminal sense is not required in OSHA inspection cases, the warrant application need not establish the reliability of the complainant or the basis of his complaint. *Marshall v. Chromalloy American Corp. (Gilbert & Bennett Manufacturing Co.)* (Ref. 8); *Burkart Randall Div. of Textron, Inc., v. Marshall* (Ref. 8). The Tenth Circuit, on the other hand, held in *Horn* that a warrant application based on an employee complaint must contain evidence showing the complaint was actually made by a complainant who was sincere in asserting a violation existed and who had some plausible basis for his belief. *Accord, Martin v. Inter. Matex Tank Terminals—Bayonne* (Ref. 8) (citing *Horn Seed* test).

10. *Marshall v. North American Car Co.*, 626 F.2d 320; 8 OSHC 1722 (3d Cir. 1980) (rejecting the secretary's argument that 29 *CFR* § 1, which provides that an inspection based on an employee complaint is not limited to matters complained of, permits a wall-to-wall inspection); *Donovan v. Sarasota Concrete Co.*, 693 F.2d 1061; 11 OSHC 1001 (11th Cir. 1982); *Establishment Inspection of Asarco, Inc.*, 508 F. Supp. 350; 9 OSHC 1317 (N.D. Tex. 1981); *Trinity Industries, Inc. v. OSHRC*, 16 OSHC 1609, 16 F.3d 1455 (6th Cir. 1994) (OSHA Instruction CPL 2.45A, which widens an employee-complaint based inspection to a full-scope inspection in certain circumstances, was invalid; the court stated that a limited warrant should issue based on the employee complaint, and, if the limited inspection leads OSHA to suspect that further inspections are necessary, then it should apply for a second warrant). However, in *In re Establishment Inspection of Cerro Prods. Co.*, 752 F.2d 280, 283; 12 OSHC 1153, 1155 (7th Cir. 1985), a full-scope warrant based on an employee complaint was upheld where: (1) there was no evidence of harassment in the filing of the complaint; (2) there was evidence that the employer had a high-hazard workplace in a high-hazard industry (based on a high industry lost workday incident or LWDI rate); (3) there had been no inspections in the previous fiscal year; and (4) OSHA's limited resources would be conserved by allowing a full-scope inspection. The Seventh Circuit thus followed the approach of the Eighth Circuit, which, in *Appeal of Carondelet Coke Corp.*, 741 F.2d 172; 11 OSHC 2153 (8th Cir. 1984), held that whether a general warrant should issue must be determined on a case-by-case basis. A wall-to-wall inspection may also be permissible if the complained-of violation affects the entire plant. For example, see *In re Establishment Inspection of Seaward International, Inc.*, 510 F. Supp. 314 (W.D. Va. 1980), *affirmed without opinion*, 644 F.2d 880 (4th Cir. 1981) (complaint involving exposure to carcinogens); *Hern Iron Works, Inc. v. Donovan*, 670 F.2d 838; 10 OSHC 1433 (9th Cir.), *certiorari denied*, 103 S. Ct. 69 (1982) (complaint pertaining to ventilation system).

11. *Marshall v. Northwest Orient Airlines, Inc.*, 574 F.2d 119; 6 OSHC 1481 (2d Cir. 1978).

12. *Pelton Casteel, Inc. v. Marshall*, 588 F.2d 1182; 6 OSHC 2137 (7th Cir. 1978).

13. *Marshall v. Chromalloy American Corp. (Gilbert & Bennett Manufacturing Co.)* (Ref. 8). *Accord, Marshall v. Weyerhaeuser Co.*, 456 F. Supp. 474; 6 OSHC 1920 (D.N.J. 1978) (past violations do not establish probable cause when a follow-up inspection of those violations revealed they had been corrected).

14. *In re Establishment Inspection of Northwest Airlines, Inc.*, 587 F.2d 12; 6 OSHC 2070 (7th Cir. 1978); *Marshall v. Weyerhaeuser Co.* (13).

15. *Peterson Builders, Inc.*, 525 F. Supp. 642; 10 OSHC 1169 (E.D. Wis. 1981); *Donovan V. Athenian Marble Corp.*, 10 OSHC 1450 (W.D. Okla. 1982); *Erie Bottling Corp.*, 539 F. Supp. 600; 10 OSHC 1632 (W.D. Pa. 1982); *Urick Foundry Co. v. Donovan*, 542 F. Supp. 82; 10 OSHC 1765 (W.D. Pa. 1982); *Brock v. Gretna Mach. & Ironworks*, 769 F.2d 1110; 12 OSHC 1457 (5th Cir. 1985); *Donovan v. Trinity Indus.*, 824 F. 2d 634, 636; 13 OSHC 1369, 1371 (8th Cir. 1987); *Pennsylvania Steel Foundry & Mach. Co. v. Secretary of Labor*, 831 F.2d 1211; 13 OSHC 1417 (3d Cir. 1987); *Industrial Steel Prods. Co. v. OSHA*, 845 F.2d 1330; 13 OSHC 1713 (5th Cir.), *certiorari denied*, 488 U.S. 993 (1988).

16. *Marshall v. Chromalloy American Corp. (Gilbert & Bennett Manufacturing Co.)* (8). *Accord, Marshall v. Multi-Cast Corp.*, 6 OSHC 1486 (N.D. Ohio 1978); *The Fountain Foundry Corp. v. Marshall*, 6 OSHC 1885 (S.D. Ind. 1978).

17. *Urick Foundry*, 472 F. Supp. 1193; 7 OSHC 1497 (W.D. Pa. 1979) (warrant would be sufficient if it showed, e.g., that the particular firm chosen for inspection had a history of prior OSHA violations, or that it was selected at random from the list of firms in the high-hazard industry); *Marshall v. Weyerhaeuser Co.* (13) (fact that employer was member of high-hazard industry

does not establish probable cause in the absence of additional facts explaining why this particular employer was selected).
18. *Marshall v. Gibson's Products, Inc. of Plano*, 584 F.2d 668; 6 OSHC 2092 (5th Cir. 1978). Based on its decision in *Gibson's Products*, the Fifth Circuit has further held that the district court has no jurisdiction to issue an injunction enforcing a warrant that has already been issued. *Marshall v. Shellcast Corp.*, 592 F.2d 1369; 7 OSHC 1239 (5th Cir. 1979). In *Marshall v. Pool Offshore Co.* (8), however, the district court held that, despite *Gibson's Products*, it had jurisdiction to impose sanctions on an employer for its refusal to permit an inspection because the proceeding was for civil contempt for failure to obey a warrant rather than for an injunction to compel an inspection.
19. There are statutory bases, for example, 28 U.S.C. §§ 1337 and 1345, that generally give the district courts jurisdiction. Some federal courts have held that the district courts (and the U.S. magistrates) have the authority to issue warrants simply because it would be inconsistent with the rationale of *Barlow's* to hold otherwise. For example, *The Fountain Foundry Corp. v. Marshall* (16); *Marshall v. Weyerhaeuser Co.* (13); *Marshall v. Huffhines Steel Co.*, 478 F. Supp. 986; 7 OSHC 1850 (N.D. Tex. 1979). The Seventh Circuit has held that federal magistrates have the authority to issue OSHA inspection warrants pursuant to 28 U.S.C. § 636, which permits district judges to assign to magistrates any duties not inconsistent with the constitution or federal law. See *Marshall v. Chromalloy American Corp. (Gilbert & Bennett Manufacturing Co.)* (8); *Pelton Casteel, Inc. v. Marshall* (Ref. 12). Accord, *Marshall v. Multi-Cast Corp.* (16).
20. 436 U.S. at 316–317, 6 OSHC at 1575.
21. The regulation permitting the secretary to obtain "compulsory process" to compel an inspection was published at 29 *CFR* § 1903.4.
22. See *Cerro Metal Products v. Marshall*, 467 F. Supp. 869, 872; 7 OSHC 1125, 1126–1127 (E.D. Pa. 1979), affirmed, 620 F.2d 964; 8 OSHC 1196 (3d Cir. 1980).
23. 43 *Fed. Reg.* 59838 (1978); 29 *CFR* § 1903.4(d). See discussion in *Cerro Metal Products* (Ref. 22), 7 OSHC at 1127.
24. 620 F.2d 964; 8 OSHC 1196 (3d Cir. 1980). Accord, *Marshall v. Huffhines Steel Co.*, 7 OSHC 1910 (N.D. Tex. 1979), *affirmed without opinion sub nom. Donovan v. Huffhines Steel* Co., 9 OSHC 1762 (5th Cir. 1981).
25. *Rockford Drop Forge Co. v. Donovan*, 672 F.2d 626; 10 OSHC 1410 (7th Cir. 1982); *Stoddard Lumber Co. v. Marshall*, 627 F.2d 984; 8 OSHC 2055 (9th Cir. 1980); *Marshall v. W & W Steel Co., Inc.* (8).
26. 45 *Fed. Reg.* 65916–65924 (1980).
27. The new regulation, specifically held valid by the court in *Donovan v. Blue Ridge Pressure Castings, Inc.*, 543 F. Supp. 53; 10 OSHC 1217 (M.D. Pa. 1981), reads as follows [29 *CFR* § 1903.4(d)]:

    (d) For purposes of this section, the term compulsory process shall mean the institution of any appropriate action, including ex parte application for an inspection warrant or its equivalent. Ex parte inspection warrants shall be the preferred form of compulsory process in all circumstances where compulsory process is relied upon to seek entry to a workplace under this section.
28. 610 F.2d 1128, 1135–1136; 7 OSHC 1880, 1884 (3d Cir. 1979). An employer's right to a pre-inspection hearing in federal court on a warrant's validity was reaffirmed by the Third Circuit in *Cerro Metal Products v. Marshall* (24), holding an employer could bring an action in district court, prior to inspection, seeking to enjoin OSHA from obtaining an *ex parte* warrant. The

COMPLIANCE AND PROJECTION 1671

Seventh Circuit has also held an employer may, in limited circumstances, raise a pre-inspection warrant challenge in district court. See *Blocksom & Co. v. Marshall*, 582 F.2d 1122; 6 OSHC 1865 (7th Cir. 1978) (employer may raise invalidity of warrant as a defense in action seeking to hold it in contempt for failure to obey warrant).

29. 610 F.2d at 1135–1137; 7 OSHC at 1883–1885. Accord, *Marshall v. Whittaker Corp., Berwick Forge & Fabricating Co.*, 610 F.2d 1141; 6 OSHC 1888 (3d Cir. 1979); *Establishment Inspection of the Metal Bank of America, Inc.*, 700 F.2d 910; 11 OSHC 1193 (3d Cir. 1983).

30. In re *Worksite Inspection of Quality Products, Inc.*, 592 F.2d 611; 7 OSHC 1093 (1st Cir. 1979); *Marshall v. Central Mine Equipment Co.*, 608 F.2d 719; 7 OSHC 1907 (8th Cir. 1979); *Baldwin Metals Co., Inc. v. Donovan*, 642 F.2d 768; 9 OSHC 1568 (5th Cir.), *certiorari denied sub nom. Mosher Steel Co. v. Donovan*, 454 U.S. 893 (1981); In re *J. R. Simplot Co.*, 640 F.2d 1134 (9th Cir. 1981), *certiorari denied*, 455 U.S. 939 (1982); *Robert K. Bell Enters. v. Donovan*, 710 F.2d 673 (10th Cir. 1983), *certiorari denied*, 464 U.S. 1041 (1984); *Establishment Inspection of Gould Publishing Co.*, 934 F.2d 457; 15 OSHC 1073 (2d Cir. 1991).

31. *Weyerhaeuser Co. v. Marshall*, 592 F.2d 373; 7 OSHC 1090 (7th Cir. 1979); *Federal Casting Div., Chromalloy American Corp. v. Donovan*, 684 F.2d 504; 10 OSHC 1801 (7th Cir. 1982) (following Weyerhaeuser).

32. In re *Establishment Inspection of Kohler Co.*, 935 F.2d 810 (7th Cir. 1991).

33. *Dunlop v. Able Contractors, Inc.*, 4 OSHC 1110 (D. Mont. 1975), *affirmed sub nom. Marshall v. Able Contractors, Inc.*, 573 F.2d 1055; 6 OSHC 1317 (9th Cir.), *certiorari denied*, 439 U.S. 826 (1978). In *Hamilton Fixture v. Secretary of Labor*, 16 OSHC 1889 (6th Cir. 1994), the employer challenged the reasonableness of OSHA's conducting a wall-to-wall inspection of a manufacturing facility during a work slowdown prompted by labor unrest. The employer claimed that a work slowdown was not a reasonable time to do the inspection, because it was unable to follow its normal safety procedures under those circumstances. The appeals court, however, rejected this argument and held the inspection to be reasonable.

The Seventh Circuit has held that it is not unreasonable for an inspection warrant to allow videotaping of the workplace. *In re Establishment Inspection of Kelly-Springfield Tire Co.*, 7th Cir., No. 93-1082, January 18, 1994. Going further, the Fourth Circuit has held that worksite videotaping not only does not violate the act, but it also does not violate the Fourth Amendment's right to privacy because the employer "left the construction site open to observation from vantages outside its control." *L. R. Willson and Sons, Inc. v. OSHRC*, 4th Cir., No. 97-1492, January 28, 1998. See also *Tri-State Steel Construction, Inc. v. Secretary of Labor*, D.C. Cir., No. 92-1614, June 17, 1994 (applying the "open fields doctrine," employer "lacked a reasonable expectation of privacy in the open areas of [its] construction site.")

34. See Section 8(a). The secretary's regulations provide in relevant part [29 *CFR* § 1903.7(a)]:

At the beginning of an inspection, Compliance Safety and Health Officers shall present their credentials to the owner, operator, or agent in charge at the establishment; explain the nature and purpose of the inspection; and indicate generally the scope of the inspection and the records . . . which they wish to review.

By regulation, employers are entitled to a copy of the requests for inspection if the inspection follows an employee's complaint [29 *CFR* § 1903.11]. Employers also have the right not to have trade secrets disclosed as a result of the inspection [29 *CFR* § 1903.9]. The secretary's regulations further provide for a conference at the conclusion of the inspection. During this conference the compliance safety and health officer advises the employer of any apparent safety or health violations disclosed by the inspection, and the employer is afforded an opportunity to bring to the attention of the officer any pertinent information regarding conditions in the workplace [29 *CFR* § 1903.7(e)].

35. *Accu-Namics, Inc. v. OSAHRC and Dunlop*, 515 F.2d 828; 3 OSHC 1299 (5th Cir.), rehearing denied, 521 F.2d 814 (5th Cir. 1975), *certiorari denied*, 425 U.S. 903 (1976). See also *Secretary of Labor v. GEM Industrial Inc.*, 17 OSHC 1184 (Rev. Comm. 1995), in which the Review Commission held that, although the compliance officer observed alleged violations prior to presenting his credentials, the employer's Section 8(a) rights do not apply unless a violation of the Fourth Amendment right against unreasonable searches and seizures is shown; the Commission also held that, in order to show a violation of Section 8(a) or 8(e), the employer must show that it was "actually prejudiced in the preparation or presentation of its defense of the merits.".

    Another controversial question involving the exclusion of evidence has arisen as a result of the Supreme Court's decision in the *Barlow's* case: whether the exclusionary rule, borrowed from the field of criminal law, applies in OSHA hearings to prohibit the use of evidence obtained without a valid search warrant. The Ninth Circuit, in *Todd Shipyards Corp. v. Secretary of Labor*, 586 F.2d 683; 6 OSHC 2122 (9th Cir. 1978), suggested that the rule should not apply because OSHA hearings are civil rather than criminal in nature. The Tenth Circuit indicated the contrary in *Savina Home Industries, Inc. v. Secretary of Labor* and OSAHRC, 594 F.2d 1358; 7 OSHC 1154 (10th Cir. 1979), holding that the rule should apply in order to deter improper OSHA inspections. The Eleventh Circuit, while not deciding whether the exclusionary rule must be applied in OSHA enforcement proceedings, has stated that the OSHA Review Commission is free to apply the rule if it sees fit, and need not allow an exception permitting the use of evidence obtained by OSHA "in good faith." *Donovan v. Sarasota Concrete Co.* (10). On the other hand, the Seventh Circuit held in *Donovan v. Federal Clearing Die Casting Co.* and OSAHRC, 695 F.2d 1020; 11 OSHC 1014 (7th Cir. 1982), that the commission must apply a "good faith" exception to the exclusionary rule where an OSHA search was made pursuant to a warrant upheld by the district court, even though the warrant was later invalidated on appeal. In 1984, in *INS v. Lopez-Mendoza*, 468 U.S. 1032, the Supreme Court held that the exclusionary rule does not apply in one civil context, deportation proceedings. Subsequently, in *Pennsylvania Steel Foundry and Machine Co. v. Sec'y of Labor*, 831 F.2d 1211; 13 OSHC 1417 (3d Cir. 1987), the Third Circuit cited *Lopez-Mendoza* in finding the exclusionary rule did not apply to an OSHA proceeding when the invalidity of the warrant was not based on constitutional violations but merely on a violation of OSHA regulations. Similarly, in *Smith Steel Casting Co. v. Brock*, 800 F.2d 1329; 12 OSHC 2121 (5th Cir. 1986), the Fifth Circuit stated that the exclusionary rule should not be used to exclude evidence obtained through an invalid warrant for purposes of preventing the secretary from ordering correction of OSHA violations involving unsafe working conditions; however, the illegally obtained evidence could be excluded for purposes of assessing penalties to punish past violations (subject to the good faith exception). The Sixth Circuit has followed the Fifth Circuit's decision in *Smith Steel Casting. Trinity Industries, Inc. v. OSHRC*, (10). As these conflicting decisions indicate, the applicability of the exclusionary rule in OSHA enforcement proceedings remains an unresolved question.

36. *Jones v. OSHA*, 17 OSHC 1276 (W.D.Mo. 1995). In *Jones*, the federal district court for the Western District of Missouri, in response to a Freedom of Information Act challenge, ordered OSHA thenceforth to comply with the requirements of 29 *CFR* 1903. 11(a), which states that copies of employee complaints must be provided to the employer "no later than at the time of inspection." The court agreed that employees' confidentiality must be maintained in order to encourage employees to report suspected violations, but it ruled that OSHA, in failing to comply with its own regulation, had "overstepped its boundaries." The court noted that, in

order to attempt to preserve confidentiality, OSHA could retype the complaints for production to employers.

37. 535 F.2d 371; 4 OSHC 1181 (7th Cir. 1976).
38. 535 F.2d at 377; 4 OSHC at 1185. The Seventh Circuit found substantial compliance in the *Chicago Bridge & Iron Co.* case because of the on-site employer representative was informed of the pending inspection and was given literature setting forth the directives of the act.
39. See *Marshall v. C.F. & I. Steel Corp.* and OSAHRC, 576 F.2d 809; 6 OSHC 1543 (10th Cir. 1978), in which the inspector failed to give the employer formal notice that its facilities—in addition to those of its subcontractor—were the target of the inspection. Because the employer's personnel manager did in fact accompany the inspector on his rounds, the court held that dismissal of the citation would be a grossly excessive sanction in relation to the inspector's minor violations of Section 8(e).
40. See *Hartwell Excavating Co. v. Dunlop*, 537 F.2d 1071; 4 OSHC 1331 (9th Cir. 1976), in which the court found substantial compliance even though the employer's superintendent was not notified of the inspection until it had been partially completed, because the inspector had made an unsuccessful attempt to locate the supervisor earlier.
41. 655 F.2d 41; 9 OSHC 2075 (4th Cir. 1981).
42. 560 F.2d 947; 5 OSHC 1732 (8th Cir. 1977).
43. In *Leone v. Mobil Oil Corp.*, 523 F.2d 1153 (D.C. Cir. 1975), the District of Columbia Circuit held that neither OSHA nor the Fair Labor Standards Act of 1938, 29 U.S.C. § 203(o), requires an employer to pay wages for time spent by employees in accompanying OSHA inspectors on walk-around inspections of the employer's plant. However, in 1977, OSHA announced an amendment of its administrative regulations (29 *CFR* § 1977.21) to reflect a new policy that employees should be paid by their employers for time spent on walk-arounds. The Chamber of Commerce of the United States filed suit challenging the policy, and the Court of Appeals for the District of Columbia Circuit ultimately held the regulation invalid because OSHA had failed to promulgate it properly under the notice and comment requirements of the Administrative Procedure Act. *Chamber of Commerce of United States v. OSHA*, 636 F.2d 464; 8 OSHC 1648 (D.C. Cir. 1980). The regulation was subsequently revoked [45 Fed. Reg. 72118 (1980); BNA, OSHR, Current Report for October 30, 1980, pp. 593–594], and a new one was proposed and issued [45 *Fed. Reg.* 75232 (1980); 46 *Fed. Reg.* 3852 (1981); BNA, OSHR, Current Reports for: November 20, 1980, pp. 653–658; January 22, 1981, pp. 845–853]. After the Reagan administration took office, however, implementation of the new walk-around pay rule was delayed in response to a memorandum from the president [BNA, OSHR, Current Report for February 5, 1981, p. 1225], and was ultimately withdrawn [46 *Fed. Reg.* 28842 (1981); BNA, OSHR, Current Report for June 4, 1981, pp. 21–24].
44. 29 *CFR* § 1908.6.
45. See OSHA Instruction TED 8.1 (Nov. 10, 1986), in BNA, OSHR, Reference File at 77:4001.
46. See 47 *Fed. Reg.* 29025 (July 2, 1982); OSHA Instruction CPL 5.1.II.
47. BNA, OSHR, Current Report for March 17, 1999.
48. Record-keeping and reporting requirements under the Federal Mine Safety and Health Act of 1977 are set forth in subsections 103(c)–103(e) and 103(h) of that statute.
49. These regulations are found in 29 *CFR* Part 1904. The secretary has defined "recordable occupational injuries or illnesses" as those that result in [29 *CFR* § 1904.12(c)]:

1. Fatalities, regardless of the time between the injury and death, or the length of the illness; or
2. Lost workday cases, other than fatalities, that result in lost workdays; or
3. Nonfatal cases without lost workdays, which result in transfer to another job or termination of employment, or require medical treatment (other than first aid) or involve: loss of consciousness or restriction of work or motion. This category also includes any diagnosed occupational illnesses that are reported to the employer but are not classified as fatalities or lost workday cases.

   Employers should resolve doubts about whether an injury is recordable—for example, because it is unclear whether an injury required first aid or medical treatment or whether the injury is occupationally related—in favor of recording. See, for example, *General Motors Corp., Inland* Div., 8 OSHC 2036 (Rev. Comm. 1980); *General Motors Corp. Warehousing and Distribution Division*, 10 OSHC 1844 (Rev. Comm. 1982).
50. 29 *CFR* § 1904.6. Effective January 1, 1991, the responsibility for establishing record-keeping requirements for occupational illnesses and injuries was transferred from the Bureau of Labor Statistics (BLS) to OSHA and its newly created Office of Recordkeeping and Data Analysis [BNA, OSHR, Current Report for January 23, 1991, pp. 1269–1270]. Throughout the 1990s, OSHA has been in the process of revising its record-keeping regulations. As of early 1999, a final rule making substantial revisions to OSHA's record-keeping requirements was to be published in June of 1999 (BNA, OSHR, Current Report for January 20, 1999). The rule was to address the controversial issue of how OSHA defines whether an injury is work-related and thus must be recorded, among other issues.
51. 59 *Fed. Reg.* 15594 (April 1, 1994). The rule is codified at 29 CFR 1904.8.
52. 60 *Fed. Reg.* 18994 (April 14, 1995).
53. 29 *CFR* §§ 1904.20–1904.21. In addition, since 1996, OSHA has conducted an annual survey that requires about 80,000 employers to submit injury and illness records directly to OSHA. The regulation creating this annual survey (29 *CFR* 1904.17) was invalidated by a federal district court in *American Trucking Associations v. Secretary of Labor*, 17 OSHC 1881, 955 F.Supp. 4 (D.C.D.C. 1997), and OSHA responded by promulgating a final rule to preserve its authority to collect the survey (62 *Fed. Reg.* 6434, February 11, 1997).
54. 29 *CFR* § 1904.15.
55. Subsection 103(c) of the Federal Mine Safety and Health Act of 1977 contains a similar requirement guaranteeing employees access to toxic materials exposure records.
56. 29 *CFR* § 1910.1020. The medical records rule is augmented by the specific permanent standards issued by the secretary of labor for certain toxic materials and harmful physical agents. See, for example, the OSHA standard regarding employee exposure to coke oven emissions, 29 *CFR* § 1910.1029(m).
57. Despite these privacy interests, OSHA is specifically granted access to employee medical records. However, additional regulations require OSHA to observe administrative procedures designed to protect the confidentiality of this information. 29 *CFR* Part 1913. In a case involving similar privacy concerns, the Supreme Court refused to review the Sixth Circuit's ruling that NIOSH may subpoena employee medical records in the course of a health hazard inquiry. In *General Motors Corp. v. NIOSH*, 636 F.2d 163; 9 OSHC 1139 (6th Cir. 1980), *certiorari denied*, 454 U.S. 877; 10 OSHC 1032 (1981), the court held that, as long as there is no public disclosure of this medical information, NIOSH may obtain access to it without violating the employees' privacy rights. At least one federal district court has reached a similar conclusion regarding OSHA's medical records access rule. *Louisiana Chemical Assoc. v. Bingham*, 550 F. Supp. 1136; 10 OSHC 2113 (W.D. La. 1982), affirmed, 731 F.2d 280, 11 OSHC 1992 (5th Cir. 1984).

58. The Review Commission has held that this 6-month limitation period does not apply if the secretary's inability to discover the violation was caused by the employer's failure to report a fatal accident as the secretary's regulations require [29 *CFR* §1904.8]. The commission reasoned that allowing an employer to escape a citation because of its own failure to comply with OSHA's reporting requirements would reward it for its own wrongdoing. *Yelvington Welding Service*, 6 OSHC 2013 (Rev. Comm. 1978). Also, the commission has held that OSHA may issue a citation for an alleged violation even though it may have had the opportunity during an earlier inspection to discover the same alleged violation and cite it—the holding was based on the idea that the purpose of Section 9(c) is simply to ensure that violations are prosecuted while events are still relatively fresh and that this purpose is not compromised so long as the subsequent violative conduct is prosecuted within six months. *Secretary of Labor v. Safeway Store No. 914*, 16 OSHC 1504 (Rev. Comm. 1993).

59. See Section 9(a). Interpretations of the "particularity" requirements have generally dealt with: (a) the precision of the reference to the standard allegedly violated, and (b) the adequacy of the description of the alleged violation. Several federal courts have held that, to meet the particularity requirement, a citation must provide the employer with "fair notice" of the violation sufficient to enable it both to prepare its defense and to correct the cited hazard. *Marshall v. B.W. Harrison Lumber Co.* and OSAHRC, 569 F.2d 1303; 6 OSHC 1446 (5th Cir. 1978); *Whirlpool Corp. v. OSAHRC and Marshall*, 645 F.2d 199; 7 OSHC 2059 (9th Cir. 1980), *certiorari denied*, 457 U.S. 1132 (1982); *Noblecraft Industries, Inc. v. Secretary of Labor* and OSAHRC, 614 F.2d 199; 7 OSHC 2059 (9th Cir. 1980); *Whirlpool Corp. v. OSAHRC and Marshall*, 645 F.2d 1096; 9 OSHC 1362 (D.C. Cir. 1981). See also *Otis Elevator Co.*, 13 OSHC 1791 (Rev. Comm. 1988) [holding that the citation was not too vague because (1) the basic elements of the charge describing the hazard were present, and (2) there was extensive pretrial discovery], affirmed, 871 F.2d 155; 13 OSHC 2085 (D.C. Cir. 1989).

60. The Fifth Circuit has held that an employer may raise a citation's lack of particularity in a later failure-to-abate action by the Secretary. In so holding, the court reasoned that a citation that is too vague to give notice of the action necessary to correct the cited hazard also makes it impossible for the Review Commission to determine whether the hazard has been abated. See *Marshall v. B.W. Harrison Lumber Co. and OSAHRC* (59).

61. The secretary's regulations can be found at 29 *CFR* § 1903.16. The employer must post notice of OSHA citations at or near the place of the violation immediately upon receipt of the notice [29 *CFR* §§ 1903.14, 1903.16]. If it is not possible to post the notice of citation at the violation site, employers must post it in a "prominent place where it will be readily observable by all affected employees" [29 *CFR* § 1903.16(a)]. If the employer settles with OSHA (to reduce the fines), a notice of settlement must also be posted [29 *CFR* § 2200.100(c)]. Section 17(i) mandates penalties of up to $7000 for each violation of the posting requirements.

62. Penalties are prescribed by Section 17 of OSHA. Applicable regulations on proposed penalties are found at 29 *CFR* § 1903.15. In addition to the gravity of the violation, the size of the business, the employer's good faith, and the employer's history of previous violations are also taken into account in determining the amount of the proposed penalty [29 *CFR* § 1903.15(b)].

63. Following passage of the new law, OSHA amended its Field Operations Manual with respect to its instructions on penalties. See BNA, OSHR, Current Report for January 30, 1991, p. 1289. The penalties section of the *Field Operations Manual* [OSHA Instruction CPL 2.46B, as amended] is found in BNA, OSHR, Reference File at 77:2701.

64. *Dorey Electric Co. v. OSAHRC*, 553 F.2d 357; 5 OSHC 1285 (4th Cir. 1977); *California Stevedore and Ballast Co. v. OSAHRC*, 517 F.2d 986; 3 OSHC 1174 (9th Cir. 1975); *Shaw Construction, Inc. v. OSAHRC*, 534 F.2d 1183, 1185 & n. 4; 4 OSHC 1427, 1428 & n. 4 (5th

Cir. 1976). The Review Commission has held that the probability of an accident, while not necessary to establish a serious violation, is a factor to be considered in determining the gravity of the violation for penalty assessment purposes. Thus, an employer's accident-free history, while not a defense to a citation for a serious violation, may justify reducing the proposed penalty. *George C. Christopher & Sons, Inc.*, 10 OSHC 1436, 1446 (Rev. Comm. 1982).

65. *Accord, Usery v. Hermitage Concrete Pipe Co. and OSAHRC*, 584 F.2d 127; 6 OSHC 1886 (6th Cir. 1978); *Kent Nowlin Construction Co., Inc. v. OSAHRC*, 648 F.2d 1278; 9 OSHC 1709 (10th Cir. 1981); *Central Brass Mfg. Co.*, 13 OSHC 1609, 1611 (Rev. Comm. 1987).
66. *Brennan v. OSAHRC and Vy Lactos Laboratories Inc.*, 494 F.2d 460; 1 OSHC 1623 (8th Cir. 1974).
67. An employer has an obligation to inspect the workplace to discover and prevent possible hazards, and its failure to do so can support a finding that it should have known of the violations it failed to discover. *Joseph J. Stolar Construction Co., Inc.*, 9 OSHC 2020 (Rev. Comm.), appeal denied, 681 F.2d 801; 10 OSHC 1936 (2d Cir. 1981). However, the existence of an effective employer inspection and maintenance program, designed to discover and remedy hazardous conditions, may preclude a finding that the employer should reasonably have known of violations it actually failed to discover. *Cullen Industries, Inc.*, 6 OSHC 2177 (Rev. Comm. 1978). See, however, *East Texas Motor Freight, Inc. v. OSAHRC and Donovan*, 671 F.2d 845; 10 OSHC 1457 (5th Cir. 1982) (because employer's inspection program was not effective in discovering and repairing defective machinery, employer was charged with knowledge of undiscovered defects).
68. *Mid-Plains Construction Co.*, 3 OSHC 1484 (Rev. Comm. 1975); *Southwestern Acoustics & Specialty, Inc.*, 5 OSHC 1091 (Rev. Comm. 1977); *Cleveland Consol.*, 13 OSHC 1114 (Rev. Comm. 1987); *B. B. Riverboats, Inc.*, 13 OSHC 1350 (Rev. Comm. 1987).
69. *Brennan v. OSAHRC and Hendrix (d/b/a Alsea Lumber Co.)*, 511 F.2d 1139, 1144; 2 OSHC 1646, 1648–1649 (9th Cir. 1975). *Accord, Ocean Electric Corp. v. Secretary of Labor and OSAHRC*, 594 F.2d 396; 7 OSHC 1149 (4th Cir. 1979); *Diversified Industries Div., Independent Stave Co. v. OSAHRC and Marshall*, 618 F.2d 30, 31 n. 8; 8 OSHC 1107, 1108 n. 8 (8th Cir. 1980).
70. *Conference Report 91-1765, 1970, U.S. Code Congressional and Administrative News*, p. 5237; OSHA § 17(c).
71. *Ryder Truck Lines, Inc. v. Brennan*, 497 F.2d 230, 233; 2 OSHC 1075, 1077 (5th Cir. 1974).
72. *Standard Glass and Supply Co.*, 1 OSHC 1223–1224 (Rev. Comm. 1973). See also *Consolidated Rail Corp.*, 10 OSHC 1564, 1568 (Rev. Comm. 1982) (where the evidence did not show the violation created a substantial probability of serious injury in case of an accident, the citation was affirmed as nonserious rather than serious); *Pace Constr. Co.*, 12 OSHC 1830, 1831 (Rev. Comm. 1986) (because the operators near an unguarded hand grinder machine wore goggles and stood at a distance from the machine and there had been no prior accidents, the violation was only nonserious); *Cuthers Corp. d/b/a Woodland Constr.*, 13 OSHC 1906 (Rev. Comm. 1988) (there was no probability of death because the machine moved very slowly).

In determining whether a violation is serious or nonserious, the commission has held that a number of nonserious violations may be grouped together to form a serious violation if the cumulative effect of the violations could result in an accident causing death or serious injury. *H. A. S. & Associates*, 4 OSHC 1894, 1897–1898 (Rev. Comm. 1976). When the violation poses a risk of illness through cumulative exposure to a toxic substance, the relevant question is whether the secretary can show a substantial probability of serious harm resulting from the degree and length of actual employee exposure. *Bethlehem Steel Corp.*, 11 OSHC 1247, 1252 (Rev. Comm. 1983); *Texaco, Inc.*, 8 OSHC 1758, 1761 (Rev. Comm. 1980).

73. *Brennan v. OSAHRC and Hendrix (d/b/a Alsea Lumber Co.)* (69); *National Steel and Shipbuilding Co. v. OSAHRC*, 607 F.2d 311, 315–316 n. 6; 7 OSHC 1837, 1840 n. 6 (9th Cir. 1979). See also *Dunlop v. Rockwell International*, 540 F.2d 1283, 1291; 4 OSHC 1606, 1611–1612 (6th Cir. 1976), in which the court upheld a divided Review Commission's affirmance of a nonserious, rather than a serious, violation where employer knowledge was not established. The court, in dictum, approved of the reasoning of the Ninth Circuit in the *Alsea Lumber* case. Contra, *Arkansas-Best Freight Systems, Inc. v. OSAHRC*, 529 F.2d 649, 655 n. 11; 3 OSHC 1910, 1913 n. 11 (8th Cir. 1976); *Brennan v. OSAHRC and Interstate Glass Co.*, 487 F.2d 438, 442 n. 19; 1 OSHC 1372, 1375 n. 19 (8th Cir. 1973).

74. Section 17(i) of the act requires that a penalty be assessed for violation of OSHA's posting requirements (61). However, the commission has stated that such a violation is nevertheless considered "nonserious." *Thunderbolt Drilling, Inc.*, 10 OSHC 1981 (Rev. Comm. 1982).

75. *Lee Way Motor Freight, Inc. v. Secretary of Labor*, 511 F.2d 864, 869; 2 OSHC 1609, 1612 (10th Cir. 1975). See § 9(a) of the act and the regulations promulgated thereunder at 29 *CFR* § 1903.14.

76. It is OSHA policy to document *de minimis* violations in the same way as other violations except that *de minimis* violations are not included on the citation. OSHA Instruction 2.45 B, as amended, is found at BNA, OSHR, Reference File at 77:2501. The passage pertaining to *de minimis* violations is found at 77:2512–13.

77. OSHA Instruction 2.45 B, as amended, BNA, OSHR, Reference File at 77:2512–13. For some examples of *de minimis* violations: *Cleveland Consol. Inc.*, 13 OSHC 1114, 1118 (Rev. Comm. 1987) (failure to post hospital telephone numbers was a *de minimis* violation where hospital was located within 300 yards of the work site); *Daniel Constr. Corp.*, 13 OSHC 1128, 1131 (Rev. Comm. 1986) (failure to properly label an air receiver in compliance with the Boiler and Pressure Vehicle Code was a *de minimis* violation because it had no bearing on the health and safety of employees).

78. BNA, OSHR, *Current Reports* for June 15, 1994 and June 22, 1994.

79. BNA, OSHR, *Current Report* for March 8, 1995.

80. 519 F.2d 1200, 1207; 2 OSHC 1283, 1289 (3d Cir. 1974), *affirmed on other points sub non. Atlas Roofing Co., Inc. v. OSAHRC*, 430 U.S. 442 (1977).

81. *Messina Construction Corp. v. OSAHRC*, 505 F.2d 701; 2 OSHC 1325 (1st Cir. 1974); *Intercounty Construction Co. v. OSAHRC*, 522 F.2d 777, 780; 3 OSHC 1337, 1339 (4th Cir. 1975), *certiorari denied* 423 U.S. 1072 (1976).

82. *Kent Nowlin Construction, Inc.*, 5 OSHC 1051, 1055 (Rev. Comm. 1977), *affirmed in relevant part* 593 F.2d 368, 369; 7 OSHC 1105, 1108 (10th Cir. 1979).

83. *Kus-Tum Builders, Inc.*, 10 OSHC 1128, 1131 (Rev. Comm. 1981); *A. Schonbek & Co.*, Inc., 9 OSHC 1189, 1191 (Rev. Comm. 1980), *affirmed* 646 F.2d 799; 9 OSHC 1562 (2d Cir. 1981). Several state tribunals have also adopted the "intentional disregard and indifference" test for "willfulness." See *Monadnock Fabricators, Inc.* (Docket No. RB491), a ruling of the Vermont Occupational Safety and Health Review Board (summarized in BNA, OSHR, *Current Report* for June 24, 1982, p. 96). The New Mexico Health and Safety Review Commission applied the same standard in *Environmental Improvement Division v. Stearns-Roger, Inc.* (Docket No. 82-3) (summarized in BNA, OSHR, *Current Report* for November 18, 1982, pp. 491–492).

84. *A. Schonbek & Co., Inc. v. Donovan and OSAHRC*, 646 F.2d 799, 9 OSHC 1562 (2d Cir. 1981), affirming, 9 OSHC 1189 (Rev. Comm. 1980); *Mineral Industries & Heavy Construction Group (Brown & Root, Inc.) v. OSAHRC and Marshall*, 639 F.2d 1289, 9 OSHC 1387 (5th

Cir. 1981); *Georgia Electric Co. v. Marshall and OSAHRC*, 595 F.2d 309; 7 OSHC 1343 (5th Cir. 1979); *Empire-Detroit Steel Div., Detroit Steel Corp. v. OSAHRC and Marshall*, 579 F.2d 378; 6 OSHC 1693 (6th Cir. 1978); *Western Waterproofing Co., Inc. v. Marshall and OSAHRC*, 576 F.2d 139; 6 OSHC 1550 (8th Cir.), certiorari denied, 439 U.S. 965 (1978); *National Steel & Shipbuilding Co. v. OSAHRC* (63); *Kent Nowlin Construction Co. v. OSAHRC and Marshall*, 593 F.2d 368; 7 OSHC 1105 (10th Cir. 1979), affirming in relevant part 5 OSHC 1051 (Rev. Comm. 1977).

85. 587 F.2d 1303; 6 OSHC 2010, 2012 (D.C. Cir. 1978).
86. 622 F.2d 1160; 8 OSHC 1317 (3d Cir. 1980).
87. 622 F.2d at 1167–1168; 8 OSHC at 1322 (footnotes omitted).
88. 540 F.2d 157; 4 OSHC 1451 (3d Cir. 1976).
89. 582 F.2d 834, 840–41; 6 OSHC 1855, 1859 (4th Cir. 1978).
90. See *Todd Shipyards Corp. v. Secretary of Labor*, 586 F.2d 683; 6 OSHC 2122 (9th Cir. 1978); *Bunge Corp. v. Secretary of Labor and OSAHRC*, 638 F.2d 831; 9 OSHC 1312 (5th Cir. 1981); *J.L. Foti Construction Co. v. OSAHRC and Donovan*, 687 F.2d 853; 10 OSHC 1937 (6th Cir. 1982); *Dun-Par Engineered Form Co. v. Marshall and OSAHRC*, 676 F.2d 1333; 10 OSHC 1561 (10th Cir. 1982).
91. 7 OSHC 1061 (Rev. Comm. 1979).
92. 7 OSHC at 1063. Except for those within the Third Circuit, Review Commission judges continue to find repeated violations after only one prior citation. For example: *Carl Thomas Constr. Corp.*, 13 OSHC 1671 (Rev. Comm. 1988); *Bat Masonry Co.*, 13 OSHC 1876, 1877 (Rev. Comm. 1988). Compare *Mellon Stuart Co.*, 13 OSHC 1025, 1026 (Rev. Comm. 1987).
93. The definition of "substantially similar" violations and the proper allocation of the burden of proving such similarity is not settled. For a discussion of these problems, see *Bunge Corp. v. Secretary of Labor and OSAHRC* (90). In Chapter IV of its Field Operations Manual (pertaining to citing violations), OSHA sets forth its position on various issues as to what constitutes a repeated violation (e.g., whether the violations must be of the identical standard; whether the violations must have occurred at the same site for a single employer with multiple workplace locations; and whether the time between violations affects a "repeated" finding). See BNA, OSHR, Reference File at 77:2501, 2511–12.
94. Chapter IV of OSHA's *Field Operations Manual*, printed in BNA, OSHR, Reference File at 77:2501, 2512.
95. *Secretary of Labor v. Quality Stamping Products Co.*, 16 OSHC 1927 (Rev. Comm. 1994). In this case, the employer had a history of violation throughout the 1970s and 1980s, and the Commission found that the administrative law judge was thus justified in fining the employer $25,750 for more than 30 cited violations in which the gravity ranged from "quite low" to high.
96. In what is believed to have been the first application of a state's criminal homicide laws to an occupational death, three former officials of a film recycling operation pleaded guilty to involuntary manslaughter in September of 1993 in the 1983 cyanide poisoning death of an employee. *Illinois v. Kirschbaum*, Ill. Cir. Ct., No. 84-5064, September 9, 1993. The three had originally been convicted of murder in 1985, but an Illinois appellate court reversed the convictions on a narrow technical ground and remanded the case for retrial.

The U.S. Supreme Court in 1991 decided not to review the issue of whether OSHA preempts state criminal proceedings for health and safety violations. In *Pymm v. New York*, the state convicted two corporate officers on charges of conspiracy, falsifying business records, assault, and reckless endangerment. 151 A.D.2d 133, 546 N.Y.S.2d 871; 14 OSHC 1297 (1989), affirmed, 76 N.Y.2d, 561 N.Y.S.2d 687, 563 N.E.2d 1; 14 OSHC 1833 (1990), *certiorari denied*, 111 S. Ct. 958 (1991). The verdict was set aside by the trial court on the grounds that federal OSHA preempted state prosecution. The appellate court and the state high court reinstated the verdict, concluding that the OSH act does not preempt state criminal prosecutions. New York, Illinois, Texas, Maine, Wisconsin and Michigan courts all have ruled that the OSH Act does not pre-empt the use of state criminal laws to prosecute employers for endangering their employees. BNA, OSHR, *Current Report* for September 15, 1993.

97. 30 U.S.C. § 820(d).
98. For an interpretation of "willful" in the context of Section 17(e) of OSHA, 29 U.S.C. § 666(e), see *United States v. Dye Construction Co.*, 510 F.2d 78 (10th Cir. 1975); *United States v. Dye Constr. Co.*, 510 F.2d 78; 2 OSHC 1510 (10th Cir. 1975).
99. 29 U.S.C. § 666(e).
100. See Section 17(g) of OSHA, 29 U.S.C. § 666(g) and Section 110(f) of the Federal Mine Safety and Health Act of 1977, 29 U.S.C. § 820(f).
101. See Section 17(f) of OSHA, 29 U.S.C. § 666(f) and Section 110(e) of the Federal Mine Safety and Health Act of 1977, 29 U.S.C. § 820(e).
102. See Section 110(h) of the Federal Mine Safety and Health Act of 1977.
103. See the United States Criminal Code, 18 U.S.C. § 1114.
104. 29 U.S.C. § 658(a). In determining a reasonable abatement period, OSHA's *Field Operations Manual* advises consideration of the following factors: (1) the gravity of the alleged violation; (2) the availability of needed equipment, material and/or personnel; (3) the time required for delivery, installation, modification, or construction; and (4) training of personnel. BNA, OSHR, Reference File at 77:2325.
105. 29 U.S.C. § 659(b).
106. 29 U.S.C. § 666(d).
107. 62 *Fed. Reg.* 15324, March 31, 1997.
108. BNA, OSHR, *Current Report* for June 18, 1997.
109. OSHA Instruction CPL 2.112.
110. Despite the requirements of § 10(a), the Review Commission may in limited circumstances entertain a late notice of contest pursuant to Federal Rule of Civil Procedure 60(b). This rule permits a federal court to grant relief from a final order for a number of reasons, including a party's mistake, surprise or excusable neglect, the presence of newly discovered evidence, the fraud or misconduct of an adverse party, and so on. Rule 60(b), which was held applicable to the Review Commission by the Third Circuit in *J. I. Hass Co. v. OSAHRC*, 648 F.2d 190; 9 OSHC 1712 (3d Cir. 1981), has been viewed by the commission as affording a possible basis for considering an untimely notice of contest. *Special Coating Systems of New Mexico, Inc.*, 10 *OSHC* 1671 (Rev. Comm. 1982). But see *J. F. Shea Co.*, 15 OSHC 1092 (Rev. Comm. 1991) (refusing to allow a notice of contest that had been filed eight days late due to an administrative mishap in the company's mailroom); *Medco Plumbing, Inc.*, 15 OSHC 1325 (OSHRC J. 1991) [dismissing notice of contest that was one day late because illness of secretary and being busy with other matters were not considered excusable neglect under Rule 60(b)].
111. 29 *CFR* § 1903.17.
112. *Brennan v. OSAHRC and Bill Echols Trucking Co.*, 487 F.2d 230; 1 OSHC 1398 (5th Cir. 1973); *Dan J. Sheehan Co. v. OSAHRC and Dunlop*, 520 F.2d 1036; 3 OSHC 1573 (5th Cir. 1975), *certiorari denied*, 424 U.S. 956 (1976).

113. *Dan J. Sheehan* (112), 520 F.2d at 1038–1039; 3 OSHC at 1575.
114. *Turnbull Millwork Co.*, 3 *OSHC* 1781 (Rev. Comm. 1975); *State Home Improvement Co.*, 6 OSHC 1249 (Rev. Comm. 1977). The Seventh and Eighth Circuits have expressed their approval of the commission's *Turnbull* rule, at least in situations in which the employer is a layman proceeding *pro se* (without an attorney). *Penn-Dixie Steel Corp. v. OSAHRC and Dunlop*, 553 F.2d 1078; 5 OSHC 1315 (7th Cir. 1977); *Marshall v. Gil Haughan Construction Co. and OSAHRC*, 586 F.2d 1263; 6 OSHC 2067 (8th Cir. 1978). The commission has gone further and has applied the rule even where the employer is represented by counsel. *Nilsen Smith Roofing & Sheet Metal Co.*, 4 OSHC 1765 (Rev. Comm. 1976). However, the commission has made it clear that the same lenient policy does not apply to an employer who understands the difference between contesting a penalty and a citation, and who nevertheless limits his notice of contest to the penalty. *F. H. Sparks of Maryland, Inc.*, 6 OSHC 1356 (Rev. Comm. 1978). The Commission has applied its Rule 41(b), which states that sanctions may be set aside "for reasons deemed sufficient by the Commission or Judge," to vacate an ALJ's dismissal of a *pro se* employer's notice of contest because there was no intentional disregard of the Commission's rules by the employer. *Secretary of Labor v. Arkansas Abatement Services Corp.*, 17 OSHC 1163 (Rev. Comm. 1995).
115. See Section 10(c) of OSHA. The rules governing practice before the commission can be found at 29 *CFR* §§ 2200.1–2200.212. In addition to the extensive, formal rules of practice, these rules include a subpart (at §§ 2200.200–2200.212) to permit any party to request simplified proceedings before an ALJ in cases that do not involve alleged general duty violations or alleged violations of certain enumerated standards. Procedures are simplified in several ways: the pleadings are limited, discovery is generally not permitted, the Federal Rules of Evidence do not apply, and interlocutory appeals are not permitted. In May of 1995, the Commission released a publication designed for non-lawyers called the *Guide to Commission Procedures*, which was to be sent to each employer, employee or union that contests an OSHA citation (BNA, OSHR, *Current Report* for May 17, 1995). In August of 1995, the Review Commission issued a final rule (60 *Fed. Reg.* 41805, August 14, 1995) creating an "E–Z" trial system to speed the adjudication of simple disputes. Cases designated as "E–Z" have relatively few alleged violations involved, generally have a total proposed penalty of not more than $10,000, have no allegation of willful conduct, involve a hearing expected to last less than two days, and/or involve small employers. Under the rule, which was originally to expire on September 30, 1996, a Commission administrative law judge can unilaterally assign cases to the "E–Z" track without the consent of either party. After the average time necessary to reach a decision dropped from 423 days to 141 days, the Commission later removed the sunset provision of the rule, thus leaving it in place, with the added provision that cases with aggregate proposed penalties of up to $20,000 are now eligible for "E–Z" designation (62 *Fed. Reg.* 40933, July 31, 1997).
116. The commission's rules of procedure provide that a petition for discretionary review must state specific grounds for relief. The commission normally limits review to cases in which a party asserts that: (a) a finding of material fact is not supported by a preponderance of the evidence; (b) the ALJ's decision is contrary to law or to the rules or decisions of the commission; (c) an important question of law, policy, or discretion is involved; (d) the administrative law judge committed a prejudicial procedural error or an abuse of discretion; or (e) review by the full commission will resolve a question about which individual ALJs have rendered differing opinions. 29 CFR § 2200.91(d).
117. 29 CFR §§ 2200.91 and 2200.92. Any member of the commission clearly has the authority to direct review of a case on his own motion, even if no party has requested review. *GAF Corp.*,

8 OSHC 2006 (Rev. Comm. 1980). However, the commission's rules limit the grounds on which a member may independently order review. Except in extraordinary circumstances, review is limited to the issues raised by the parties before the administrative law judge, 29 CFR § 2200.92(c). Thus, it is normally preferable for a party to obtain review by filing a petition. Indeed, a party's failure to file a petition may foreclose later judicial review of any objections to the administrative law judge's decision. 29 CFR § 2200.91(f). See also *Keystone Roofing Co., Inc. v. OSAHRC* and *Dunlop*, 539 F.2d 960; 4 OSHC 1481 (3d Cir. 1976); *McGowan v. Marshall*, 604 F.2d 885; 7 OSHC 1842 (5th Cir. 1979).

118. 29 CFR § 2200.90(d).

119. The Review Commission's failure to adequately consider these statutory penalty assessment factors can result in the Court of Appeals remanding the case to the commission for more particularized findings regarding the penalty. *Astra Pharmaceutical Products, Inc. v. OSAHRC and Donovan*, 681 F.2d 69; 10 OSHC 1697 (1st Cir. 1982).

120. *REA Express v. Brennan*, 495 F.2d 822; 1 OSHC 1651 (2d Cir. 1974); *Brennan v. OSAHRC and Interstate Glass Co.*, 487 F.2d 438; 1 OSHC 1372 (8th Cir. 1973); *California Stevedore & Ballast Co. v. OSAHRC* (64). Reduction of the secretary's proposed penalty is also a matter within the commission's discretion. *Western Waterproofing Co., Inc. v. Marshall and OSAHRC* (84). Compare *Dale M. Madden Construction, Inc. v. Hodgson*, 502 F.2d 278; 2 OSHC 1236 (9th Cir. 1974), where the court held that the commission has no authority to modify settlements made by the secretary and the cited employer. See also *Marshall v. Sun Petroleum Products Co. and OSAHRC*, 622 F.2d 1176; 8 OSHC 1422 (3d Cir. 1980), *certiorari denied*, 449 U.S. 1061, holding that the commission may review and modify settlements, but only for the limited purpose of determining the reasonableness of the abatement period when it has been challenged by employees.

121. *Secretary of Labor v. Hern Iron Works, Inc.*, 16 OSHC 1619 (Rev. Comm. 1994).

122. See *California Stevedore* (64).

123. See, for example, *Dixie Roofing and Metal Co.*, 2 OSHC 1566 (Rev. Comm. 1975). See also *Consolidated Rail Corp.*, 10 OSHC 1564, 1568 (Rev. Comm. 1982), in which the commission reduced the degree of a violation from serious to nonserious without specifically discussing its authority to do so. Federal appeals courts have upheld the Commission's authority to downgrade non-serious violations to de minimis status. See, e.g., *Secretary of Labor v. OSHRC and Erie Coke Corp.*, 16 OSHC 1241, 998 F.2d 134 (3rd Cir. 1993) (in which the court noted that the Commission had exercised this authority for over 20 years and that ruling in favor of the Secretary would create an "undesirable inter-circuit conflict"). The distinction between a "non-serious" violation and a de minimis one is important because a "non-serious" designation requires the employer to cease the violative practice, while a de minimis designation does not.

124. *Piping of Ohio, Inc. v. Secretary of Labor*, U.S. Sup. Ct., No. 94-411, cert. denied November 28, 1994.

125. Although § 11(a) permits "any person" aggrieved by a commission order to obtain judicial review, the Third Circuit has held that the role of employees or their representatives in a proceeding initiated by the employer is strictly limited to challenging the reasonableness of the abatement period pursuant to § 10(c) of the act. In *Marshall v. Sun Petroleum Products* (120), the court indicated that this limitation should also apply to the right of employees to initiate judicial review. On the other hand, the District of Columbia Circuit has concluded that employees, although prohibited from instituting a commission action on matters other than the reasonableness of the abatement period, may nevertheless participate fully as parties in employer-initiated proceedings. The right of employees to participate in enforcement proceed-

ings, the court reasoned, must include the right to appeal from an unfavorable commission decision. *Oil, Chemical, and Atomic Workers v. OSAHRC*, 671 F.2d 643; 10 OSHC 1345 (D.C. Cir.), *certiorari denied sub nom. American Cyanamid and Atomic Workers*, 459 U.S. 905 (1982).

126. Petitions for review of a commission order must be filed within 60 days of issuance of the order. Review is obtained in either the court of appeals where the violation is alleged to have occurred, where the employer has its principal office, or in the Court of Appeals for the District of Columbia Circuit. 29 U.S.C. § 660(a).

127. The Sixth Circuit has defined "substantial evidence" in the context of OSHA proceedings as "such relevant evidence as a reasonable mind might accept as adequate to support a conclusion." *Martin Painting & Coating Co. v. Marshall*, 629 F.2d 437; 8 OSHC 2173, 2174 (6th Cir. 1980), *certiorari denied*, 449 U.S. 1062, quoting *Dunlop v. Rockwell International*, 540 F.2d 1283, 1287; 4 OSHC 1606, 1608 (6th Cir. 1976). While the commission's factual findings are conclusive if supported by substantial evidence, its interpretations of statutory language are legal conclusions, which are not accorded the same deference on review. *Usery v. Hermitage Concrete Pipe Co. and OSAHRC* (65). The commission's resolution of questions of credibility, on the other hand, are insulated from reversal on appeal unless they are contradicted by "uncontrovertible documentary evidence or physical facts." *Super Excavators, Inc. v. OSAHRC and Secretary of Labor*, 674 F.2d 592; 10 OSHC 1369, 1370 (7th Cir. 1981), *certiorari denied*, 457 U.S. 1133; 11 OSHC 1304 (1982), quoting *International Harvester Co. v. OSAHRC*, 628 F.2d 982, 986; 8 OSHC 1780, 1783 (7th Cir. 1980).

128. *Secretary v. OSAHRC and Interstate Glass* (73), 487 F.2d at 442; 1 OSHC at 1375.

129. *Secretary of Labor v. Sea Sprite Boat Co.*, 17 OSHC 1331, 64 F.3d 332 (7th Cir. 1995).

130. 64 *Fed.* Reg. 8243, February 19, 1999. The pilot program is currently scheduled to run through February 22, 2000.

131. 30 *CFR* Part 100. Section 105 of the Act created the Mine Safety and Health Review Commission to hear citation contests.

132. *Thunder Basin Coal Co. v. Secretary of Labor*, 16 OSHC 1553, 114 S.Ct. 711 (U.S. Sup.Ct. 1994). In this case, Thunder Basin attempted to obtain injunctive relief in federal district court against MSHA enforcement action, but the Supreme Court disallowed this approach.

133. See *Usery v. Whirlpool Corp.*, 416 F. Supp. 30, 34; 4 OSHC 1391, 1392–1393 (N.D. Ohio 1976), *reversed on other points sub nom. Marshall v. Whirlpool Corp.*, 593 F.2d 715; 7 OSHC 1075 (6th Cir. 1979), affirmed, 445 U.S. 1 (1980). The U.S. Supreme Court in the *Whirlpool* case was faced with the question whether an employee is protected from retaliation by his employer if he walks off the job in an imminent danger situation. The secretary of labor, in 1973, promulgated a regulation that protected employees from discrimination if they refused in good faith to work under life-threatening conditions. The rule applied only if the employee reasonably believed that a real danger of death or serious injury existed, and that the danger was too immediate to be eliminated through the act's normal enforcement channels. 29 *CFR* § 1977.12(b) (2). In *Whirlpool Corp.*, the Supreme Court upheld § 1977.12(b) (2) as a valid exercise of the secretary's authority under the act. Thus, employees are afforded a limited right to refuse to work in imminent danger situations. See also *G. Scarzafava and F. Herrera, Jr., Workplace Safety–The Prophylactic and Compensatory Rights of the Employee*, 13 St. Mary's L.J. 911, 931–933 (1982).

134. Imminent danger is defined by Section 3(j) of the Federal Mine Safety and Health Act of 1977 as "the existence of any condition or practice in a coal or other mine that could reasonably be expected to cause death or serious physical harm before such condition or practice can be

abated." The definition of "imminent danger" in the Coal Mine Health and Safety Act of 1969 was almost identical to that which now appears in the 1977 act. In a case arising under the 1969 provision, the Seventh Circuit affirmed the Board of Mine Operations Appeals' holding that an "imminent danger" situation exists if, in a reasonable man's estimation, it is at least as probable as not that continuation of normal coal extraction operations in the disputed area will result in the occurrence of the feared accident or disaster before the danger is eliminated. *Freeman Coal Mining Co. v. Interior Board of Mine Operations Appeals*, 504 F.2d 741; 2 OSHC 1308 (7th Cir. 1974). *Accord, Old Ben Coal Corp. v. Interior Board of Mine Operations Appeals*, 523 F.2d 25; 3 OSHC 1270 (7th Cir. 1975); *Eastern Assoc. Coal Corp. v. Interior Board of Mine Operations Appeals*, 491 F.2d 277 (4th Cir. 1974).

135. 518 F.2d 990; 3 OSHC 1490 (5th Cir. 1975); *affirmed*, 430 U.S. 442; 5 OSHC 1105 (1977).

136. The Supreme Court also granted certiorari in *Frank Irey, Jr., Inc. v. OSAHRC and Brennan*, 519 F.2d 1215; 3 OSHC 1329 (3d Cir. en banc 1975) to review the same issue. The *Atlas Roofing and Frank Irey* cases were decided by the Court in consolidated proceedings.

137. 430 U.S. 442; 5 OSHC 1105 (1977).

138. *S. A. Healy v. Occupational Safety and Health Review Commission*, 18 OSHC 1193 (7th Cir. 1998). The Seventh Circuit's decision was a reversal of its previous holding on the matter, which was vacated by the Supreme Court. On remand from the Supreme Court, the Seventh Circuit held that the administrative penalty was "civil" rather than "criminal" and thus did not constitute a separate criminal proceeding for purposes of double jeopardy.

139. *Secretary of Labor v. Andrew Catapano Enterprises, Inc.*, 17 OSHC 1776 (Rev. Comm. 1996). In this case, the employer argued that all nine trench sites should have been treated as a single workplace and the violations thus treated as a single case, but the Commission ruled that it was within OSHA's discretion to prosecute the violations at each trench site as a separate case.

140. *Secretary of Labor v. Arcadian Corp.*, 17 OSHC 1929, 110 F.3d 1192 (5th Cir. 1997). The "egregious case" policy is referenced in the *Field Operations Manual*'s Chapter VI, relating to penalties, at section B.9.d. (BNA, OSHR, Reference File at 77:2704). According to that section, OSHA considers as egregious cases those "willful, repeated and high gravity serious citations and failures to abate." The policy and its implementation are explained in detail in OSHA Instruction CPL 2.80 (October 1, 1990), reprinted in BNA, OSHR, Reference File at 21:9649.

141. *Secretary of Labor v. Arcadian Corp.*, 17 OSHC 1345 (Rev. Comm. 1995).

142. BNA, OSHR, *Current Report* for July 30, 1997.

143. In a case decided by the Commission the same day as *Arcadian*, the Commission disallowed OSHA's use of the egregious case approach with regard to a specific fall-protection standard, ruling that the application of the policy to this particular standard was improper:

> . . . where the Act and standard clearly allow the Secretary to consider each failure to comply with a standard as a discrete violation, he also has the discretion to group them for penalty purposes as if they were one violation . . . Here, however, the Secretary seeks the discretion to expand what is clearly a single violation into multiple violations based on the number of employees exposed, even where the standard calls for abatement by the performance of a single discrete action.

Secretary of Labor v. Hartford Roofing Co., 17 OSHC 1361 (Rev. Comm. 1995). Thus it appears that, while the egregious case policy remains in OSHA's enforcement repertoire, its use with regard even to specific standards will be scrutinized on review. For example, in March of 1997, the Commission directed for review an ALJ's decision to reduce the total penalties in an "egregious case" involving 100 violations of the lockout/tagout standard from $7.5

million to $518,000. *Secretary of Labor v. Dayton Tire*, OSHRC, No. 94-1374, direction for review March 28, 1997. The Commission decision in this case is pending.
144. BNA, OSHR, *Current Reports* for October 29, 1997, December 3, 1997 and December 9, 1998. The CCP was outlined in OSHA Instruction CPL 2-0.119.
145. *Chamber of Commerce of the United States v. U.S. Department of Labor*, D.C. Cir., No. 98-1036, January 22, 1998.
146. BNA, OSHR, *Current Report* for March 4, 1998.
147. OSHA Instruction CPL 2.98-1.
148. BNA, OSHR, *Current Report* for December 9, 1998.
149. BNA, OSHR, *Current Report* for January 13, 1999.
150. BNA, OSHR, *Current Report* for November 11, 1998.
151. BNA, OSHR, *Current Report* for January 13, 1999.
152. BNA, OSHR, *Current Report* for March 24, 1999.
153. BNA, OSHR, *Current Reports* for July 29, 1998 and January 27, 1999.
154. BNA, OSHR, *Current Report* for February 10, 1999.
155. BNA, OSHR, *Current Report* for March 31, 1999.
156. BNA, OSHR, *Current Report* for March 24, 1999.
157. BNA, OSHR, *Current Report* for January 13, 1999.

CHAPTER THIRTY-SIX

Industrial Hygienist's Liability Under Law

Ralph E. Allan, MS, JD, CIH

1 INTRODUCTION

The complex area of toxic-substance law and regulation is one of the fastest growing areas of the law, and industrial hygiene activity is an inherent component of this exploding legal area. Society is mandating that irresponsibility in reference to toxic substance management be dealt with immediately in order to protect the current and future health of the earth's population. The occupational and environmental health issues are complex and in many instances rest largely upon inferences derived at the very frontiers of science and technology. Lawyers and courts assume their roles in adjudicating complex social issues not easily resolvable whether dealing in the areas of administrative, civil, or criminal law. It is therefore important that the industrial hygienist in the process of anticipation, recognition, evaluation, and control of health hazards in the workplace and community understand their responsibility in the legal system in order to assist more favorably in unraveling some of the most complicated legal issues that have ever needed resolution in our society.

No two words in the English language strike more fear into the hearts of any professional than "professional liability". In this day and age of many liability suits being filed, professionals may well be reminded of the old saying, "An ounce of prevention is worth a pound of cure", that is, giving no cause for a liability suit is better than defending oneself in a lawsuit. In past years, industrial hygienists may have thought that the only professionals who had to worry about being sued for liability were the physicians and nurses. However, "no man is so high that he is above the law" (1).

Patty's Industrial Hygiene, Fifth Edition, Volume 3. Edited by Robert L. Harris.
ISBN 0-471-29753-4 © 2000 John Wiley & Sons, Inc.

The practice of industrial hygiene has dramatically changed since the early days of the profession. At one time, the industrial hygienist had to be concerned only about being able to recognize the factors and stresses associated with work and work operations and to understand their effects on humans and their well-being; to evaluate, on the basis of experience and with the aid of quantitative measurement techniques, the magnitude of these stresses in terms of the ability to impair health and well-being; and to prescribe methods to eliminate, control, or reduce such stresses when necessary to alleviate their adverse effects. Presently, environmental and safety responsibilities have added to traditional industrial hygiene practice, creating some uncertainty in the basic definition of industrial hygiene practice. In order to assist in clarifying the definition of industrial hygiene practice, the American Industrial Hygiene Association is launching a broad project to document guidelines and standards of good practice aimed at helping industrial hygiene professionals in their everyday practice (2).

Presently, not only must an industrial hygienist perform his or her duties with the ever-present thought that whatever is done may become the subject of litigation, but he or she also must deal with an ever-expanding definition of industrial hygiene functions and responsibilities. Needless to say, it is not sufficient to do just an adequate job. Industrial hygienists are in a position to practice one of the most interesting areas of professional interaction in an ever-changing professional environment. In addition to the actual performance of the work and the understanding of the legal parameters related to the work, elements of importance include the definition, standards of practice, and the Code of Ethics for industrial hygiene.

2 ETHICAL RESPONSIBILITIES

The question of professional liability cannot be dealt with effectively without a consideration of ethics. In the practice of industrial hygiene, the Code of Ethics is a consideration of industrial hygiene responsibility that, in turn, can reflect on industrial hygiene liability. The American Academy of Industrial Hygiene, a nonincorporated, voluntary, nonprofit, professional society activated by the American Board of Industrial Hygiene, initiated the development of standards of ethical conduct (3) to be followed by industrial hygienists. The American Industrial Hygiene Association (4) and the American Conference of Industrial Hygienists (5) have subscribed to its canons. These canons provide standards of ethical conduct for industrial hygienists as they practice their profession and exercise their primary mission, to protect the health and well-being of working people and the public from chemical, microbiological, and physical health hazards present at, or emanating from, the workplace. Under the canons of ethical conduct and interpretive guidelines, industrial hygienists shall:

1. Practice their profession following recognized scientific principles with the realization that the lives, health and well-being of people may depend upon their professional judgment and that they are obligated to protect the health and well-being of people.

- Industrial Hygienists should base their professional opinions, judgments, interpretations of findings and recommendations upon recognized scientific principles and practices which preserve and protect the health and well being of people.
- Industrial Hygienists shall not distort, alter or hide facts in rendering professional opinions or recommendations.
- Industrial Hygienists shall not knowingly make statements that misrepresent or omit facts.

2. Counsel-affected parties factually regarding potential health risks and precautions necessary to avoid adverse health effects.
   - Industrial Hygienists should obtain information regarding potential health risks from reliable sources.
   - Industrial Hygienists should initiate appropriate measures to see that the health risks are effectively communicated to the affected parties.
   - Parties may include management, clients, employees, contractor employees, or others dependent on circumstances at the time.

3. Keep confidential personal and business information obtained during the exercise of industrial hygiene activities, except when required by law or overriding health and safety considerations.
   - Industrial Hygienists should report and communicate information which is necessary to protect the health and safety of workers in the community.
   - If their professional judgment is overruled under circumstances where the health and lives of people are endangered, industrial hygienists shall notify their employer or client or other such authority, as may be appropriate.
   - Industrial Hygienists should release confidential personal or business information only with the information owners' express authorization, except when there is a duty to disclose information as required by law or regulation.

4. Avoid circumstances where a compromise of professional judgement or conflict of interest may arise.
   - Industrial Hygienists should promptly disclose known or potential conflicts of interest to parties that may be affected.
   - Industrial Hygienists should not solicit or accept financial or other valuable consideration from any party, directly or indirectly, which is intended to influence professional judgement.
   - Industrial Hygienists shall not offer any substantial gift, or other valuable consideration, in order to secure work.
   - Industrial Hygienists should not accept work that negatively impacts the ability to fulfill existing commitments.
   - In the event that this Code of Ethics appears to conflict with another professional code to which industrial hygienists are bound, they will resolve the conflict in the manner that protects the health of affected parties.

5. Perform services only in the area of competence.
   - Industrial Hygienists should undertake to perform services only when qualified by education, training or experience in the specific technical fields involved, unless sufficient assistance is provided by qualified associates, consultants or employees.
   - Industrial Hygienists shall obtain appropriate certifications, registrations and/or licenses as required by federal, state and/or local regulatory agencies prior to providing industrial hygiene services, where such credentials are required.
   - Industrial Hygienists shall affix or authorize the use of their seal, stamp or signature only when the document is prepared by the Industrial Hygienist or someone under their direction and control.
6. Act responsibly to uphold the integrity of the profession.
   - Industrial Hygienists shall avoid conduct or practice which is likely to discredit the profession or deceive the public.
   - Industrial Hygienists shall not permit the use of their name or firm name by any person or firm which they have reason to believe is engaging in fraudulent or dishonest industrial hygiene practices.
   - Industrial Hygienists shall not knowingly permit their employees, their employers or others to misrepresent the individuals' professional background, expertise or services which are misrepresentations of fact.
   - Industrial Hygienists shall not misrepresent their professional education, experience or credentials.

5 THEORIES OF POTENTIAL LIABILITY

5.1 Torts

The 1980s could well be classified as the decade of torts. Simply speaking, a tort is a civil wrong for which a remedy may be obtained, usually in the form of damages (6). Because of the increase in occupational health tort litigation, more and more lawyers are becoming educated in environmental and occupational health issues. The complexity of the issues and subsequent effects of litigation regarding environmental and occupational health matters are nearly overwhelming.

Tort litigation concerning an environmental or occupational health matter can involve the issue of malpractice. Although the issue of malpractice has been used in litigation, its exact definition is somewhat unclear. The dictionary defines malpractice as professional misconduct or unreasonable lack of skill. The term has been employed in a broad sense; an Ohio court has defined malpractice as any professional misconduct, unreasonable lack of skill or fidelity in professional or fiduciary duties, evil practice, or illegal or immoral conduct (7). In professional malpractice, the injured party generally files suit under the theory of negligence.

5.1.1 Negligence

The Restatement (second) of Torts (8) defines negligence as "conduct which falls below the standard established by law for the protection of others against unreasonable risk of harm." The elements of a cause of action in negligence are:

- A duty or obligation recognized by the law requiring the actor to conform to a certain standard of conduct for the protection of others against unreasonable risks or harm
- A failure to conform to the standard required
- A reasonably close causal connection between the conduct and the resulting injury, commonly known as "legal cause" or "proximate cause"
- Actual loss or damage to the injured party

Before a plaintiff can get to a jury on the issue of negligence, he or she must establish all elements of negligence. As a matter of law, the first element to be proven relates to the establishment of a duty or standard of conduct. Some negligent conduct involves unintentional harm resulting from lack of care. The duty or conduct is judged on average carefulness, what the ordinary reasonable prudent person (industrial hygienist) would have done under the same or similar circumstances. This duty or standard of care is then compared with what was actually done by the individual (industrial hygienist).

For professionals, the standard of conduct that needs to be followed can only be clearly defined by other professionals in the field. Therefore, associate professionals become the resource for determining the standard of care. The standard of care for all professions would seem to be basically the same, that is, the general average of professionally acceptable conduct or the learning and skill ordinarily possessed and experienced by the profession.

For example, an industrial hygienist has contracted to perform an industrial hygiene review of a degreasing operation including identification of any toxic substances, evaluation of exposures and control of the exposures, if necessary. The contract or agreement establishes the duty of the industrial hygienist to perform the indicated services. This represents the first element involved in the cause of action for negligence-duty. If the industrial hygienist did not perform the duty indicated with a standard of care expected as a reasonable and prudent industrial hygienist would, under the same or similar circumstances, there would be a breach of the duty; therefore the second element of the cause of action for negligence would be fulfilled.

Another element related to the negligence cause of action is sufficient causal connection between the defendant's conduct and the injured party's damages. Establishing the causal relationship between the defendant's conduct and the plaintiff's damages can be difficult in toxic tort litigation. For instance, the long latent periods, the unforeseen effect of conduct undertaken years ago by persons unknown, the exposure of the plaintiff to many harmful elements over the years, and loss or spoliation of evidence tend to obscure causation.

In the previously presented example, above, if the industrial hygienist failed to identify excessive exposure to the toxic substance present at the degreasing operation, and thereby breaches the recognized duty, there may be no clear causation element because the adverse health effects of the excessive exposure may occur due to other causes, that is, other environmental exposure not related to the degreasing operation or other disease consequences. Particularly in toxic tort litigation, there is often little proof of cause and effect, and courts tend to be skeptical of statistical correlation. In time, these problems of proof may be solved by science. The test of causation does not require absolute medical certainty.

Thus the newly developing field of toxic torts may leave the legal system no option but to be unfair on the causation issue. For example, cancers caused by a toxic chemical or agent (like ionizing radiation) may not be scientifically distinguished from the many background cases of naturally occurring cancers.

Sometimes the deleterious exposure is easy to prove; the real challenge comes in connecting the exposure to the illness. Where the disease itself is poorly understood or is of unknown etiology, the likelihood of establishing causation is slim, even with strong evidence of chronic chemical exposure. For example, what caused the plaintiff's shortness of breath? Asbestos from long-term exposure to asbestos fibers, or chronic obstructive pulmonary disease, from a history of smoking two packs a day? Or what about exposure to formaldehyde? It may have caused the plaintiff's running nose and tearing eyes, but did it cause gastritis or asthma?

Many treating doctors are not trained in occupational medicine and epidemiology; therefore, a treating doctor may not be able to identify toxic substances as factors in a disease process with sufficient particularity to substantiate the injured party's complaint. In the absence of many similar cases exhibiting the same symptoms, establishing causation can be difficult. If the evidence is compelling and the disease is serious, however, decisions are easier resolved.

The absence of readily identifiable acute symptoms in a person with a chronic illness after long-term exposure may not rule out a link between the original exposure and the resulting illness. If the chemical is a carcinogen to which there is no known safe level of exposure, for example, the worker may have been exposed at a level long enough to produce noticeable acute warning signs of exposure, like a rash or blurred vision. Yet that same low-level exposure may ultimately contribute to illness years later. Even where chronic disease has not appeared, a number of jurisdictions now recognize the damage element of the cause of action based upon fear of future illnesses. If such a claim is to be pursued, the lawyer must establish that there was a deleterious exposure to a substance capable of inducing chronic illness.

The following scenario is typical in trying to prove negligence in a toxic tort setting. First, the defendant had a duty and usually did something or failed to do something that has caused harm to the plaintiff. Second, there has been at least some recognition by medical authorities that exposure to the alleged substance causes some physical harm. For example, chest doctors generally agree that exposure to asbestos can cause asbestosis, mesothelioma, and lung cancer. Third, the defendant usually has reason to know of the contaminant's harmful effects. The plaintiff proves this through discovery based on the complaint or through articles in the medical literature or in trade journals. Fourth, the plaintiff has been exposed to the contaminant, and the plaintiff claims that the exposure caused some injury or disease.

INDUSTRIAL HYGIENIST'S LIABILITY UNDER LAW

The etiology of the plaintiff's physical harm becomes a major battleground. Etiology can become the plaintiff's nemesis, however, because the plaintiff has the burden of proof on this issue, he or she must also fight the strategy of a resourceful defense tactician.

When dealing with causation, it is important to review the medical history of the plaintiff, the family background, and when the plaintiff realized that health problems might be related to the chemical exposure. It is important to amass a complete work history, ascertain whether the plaintiff is involved with certain hobbies in which chemicals are used, and find out if there was more than one job involved.

The fourth and last element of a negligence cause of action is that the plaintiff must suffer actual loss or damage. Damages can include medical expenses, economic losses such as loss of income and pain and suffering.

With regard to the negligence cause of action, the duty of an industrial hygienist is to recognize, evaluate, and control environmental and occupational stresses by virtue of special studies and training in a competent manner. The duties include keeping abreast of current statutes and regulations, using state-of-the-art equipment when measuring environmental and workplace operations, and evaluating data in a competent manner. After evaluating the situation with or without data, the industrial hygienist would have to act appropriately to identify potential hazards, including the use of proper personal protection, if applicable. The duty of the industrial hygienist could extend toward anyone whose interests may be harmed by the failure of the industrial hygienist to exercise the standard of care in a competent manner.

5.1.2 Intentional and Other Tort Actions

Intentional or fraudulent misrepresentation has been identified with the common law action of deceit. In the typical case, one will find that the plaintiff has parted with money or property of value in reliance upon the defendant's representations. The court relies upon the following elements when dealing with intentional misrepresentation (9):

- A false representation ordinarily of fact made by the defendant
- Knowledge or belief on the part of the defendant that the representation is false
- An intention to induce the plaintiff to act or refrain from acting in reliance upon the misrepresentation
- Justifiable reliance upon the representation on the part of the plaintiff in taking action or refraining from it
- Damage to the plaintiff resulting from such reliance

If a professional is liable for negligent performances of his or her duties, liability may also be found for fraudulent misrepresentation. The issue of fraudulent misrepresentation concerning information exchange goes to the very essence of ethical and moral responsibilities of industrial hygienists. Failure to inform adequately on issues concerning industrial hygiene can result in allegation of fraudulent misrepresentation.

An example of a situation in which fraudulent misrepresentation may apply is as follows: One of the problems of many industrial hygiene projects is that multiple parties are involved with the project. The fact that an industrial hygienist is in a position to supervise the project creates a possibility for a conflict of interest with the client, for example, a building owner. A client building owner may employ an industrial hygiene firm without knowing that the firm was employed by another contractor on the project; the court may find liability if the contractor did a poor job under the industrial hygienist's supervision, causing the building owner to pay additional money to remedy the situation.

At times, industrial hygienists will be asked to recommend another industrial hygienist or contractor for a particular project. The industrial hygienist should be careful not to make defamatory statements to a third person either by libel (written words) or by slander (usually oral).

Generally, the theory of strict liability, or liability without fault, has been limited to litigation involving products and the abnormally dangerous commercial activities of industry. A court will probably be reluctant to apply strict liability in a toxic-tort case involving professionals. Strict liability will generally be found in cases involving product liability. In LaCossa v. Scientific Design (10), the court stated that "those who hire (experts) are not justified in expecting infallibility, but can expect only reasonable care and competence. They purchase service, not insurance."

5.1.3 Contracts

A contract is defined as a promissory agreement between two or more parties creating obligations that are enforceable or otherwise recognizable under law (11). The same act by an individual can be, and very often is, both negligence and a breach of contract.

It is important to know under what theory of law the underlying suit is being filed. For instance, the statute of limitations can be different for contract versus tort law; the defenses the defendant may raise are generally altogether different in contract than in tort law, and the damages are different as well.

An industrial hygienist who occupies a specific position by contract could be held liable to anyone injured as a result of his or her failure to discharge that duty with due care. In cases in which an individual is injured, the courts may look to the contract to find the existence of a duty on the part of the industrial hygienist to protect the plaintiff from injury. Thus, industrial hygienists who voluntarily, and for a fee, make it part of their business to supervise work and enforce safety regulations may be responsible for failure to discharge those duties carefully.

5.2 Workers' Compensation and Individual Liability

Historically, workers' compensation cases have been filed against an employer when an employee sustained an injury. Regardless of how the injury was sustained, whether the fault of the employer or employee, workers' compensation was the exclusive remedy. Because many states provide inadequate benefits, plaintiffs in an occupational disease case may try to find a theory that will circumvent the exclusivity provisions of the Workers' Compensation Act.

The doctrine of exclusivity limits the common law remedy that the injured employee would have had if the Workers Compensation Act did not apply. In return for agreeing to forego other legal remedies, workers are guaranteed a swift and sure payment, although circumscribed, which covers lost wages and medical payments. To avoid the limitation of remedies, workers may seek to avoid the exclusivity provisions.

Most large companies employ full-time industrial hygienists. The industrial hygienists may be employed by the employer, the employer's parent companies, or an industrial hygiene consulting firm. Plaintiffs can circumvent the workers' compensation ban on suing employers of coemployees for negligence by asserting that the person causing the injury was not a coemployee but an independent contractor.

The injured worker may attempt to sue a coemployee or employer by reclassifying the employer as another entity or "dual capacity." The claim presents the theory that the employer is functioning toward the worker in a manner no different from the way the employer acts toward the general public. It is important that a company employing industrial hygienists have a carefully worded job description establishing employment status; an established industrial hygiene protocol and a limitation on performance of services to others outside the company; or a carefully worded description of policy where such actions are performed so as to limit dual-capacity exceptions.

A contract-based negligence action against an industrial hygienist also might be raised in situations in which workers compensation insurance companies provide safety inspections to their clients. The action might occur where the insurance company renders loss-control prevention as a service to its client and not for the protection of the insurance company.

6 CRIMINAL SANCTIONS

Senior level officials at the U.S. Environmental Protection Agency (EPA) and the U.S. Department of Justice have assigned high priority to the criminal enforcement of environmental laws. The number of cases referred to the Department of Justice by the EPA has increased dramatically. Many of the indictments have included management-level employees, directors, presidents, vice presidents, or owner-operators.

Environmental protection statutes have been created to address widespread public recognition of the need to control pollution. The environmental statutes seek to establish a balance among business activity, protection of the public health and welfare, and preserving natural resources. In the United States, Congress recognized the public's concern for violations of environmental laws by providing criminal sanctions and incarceration to ensure that the goals of the legislature are achieved (12).

6.1 History

Federal programs to prosecute environmental crimes first began in 1982. Although the programs have been in existence for a relatively short period of time, an enforcement presence has been established in nearly every geographic region of the country. Investigators and prosecutors hired to enforce environmental laws have been trained in both the complex technical and scientific areas of law.

Before initiation of the criminal program, most of the cases referred to the Department of Justice for prosecution were declined usually because they lacked merit, were insufficiently investigated, or did not receive the staff support necessary to bring the violators to trial. The initiation of the criminal program has changed the situation. The two organizations largely responsible for the success of the program are the Land and Natural Resources Division of the U.S. Department of Justice and the EPA.

The EPA began to hire its first criminal investigators in 1982. Many of the investigators had little environmental experience but were experienced criminal investigators. The Land and Natural Resources Division of the Department of Justice ultimately organized an Environmental Crimes Unit staffed by attorneys with both criminal and environmental law experience. One of the unit's main purposes has been to prosecute cases and set substantial penalties.

The majority of convictions have been against managerial level officials acting illegally in their capacity. The conduct of the typical defendant in the cases under prosecution is no different and no less serious than the conduct of one who has been convicted of traditional "white-collar" or "street-crime" felonies. The acts of the defendant have shown some degree of intent whether it be willful, deliberate, or premeditated over a period of time.

The prosecutor's objective is clearly to deter illegal conduct by implementing strong penalties including incarceration. Initially, financial profit may motivate an illegal act; thus fines must exceed by a substantial amount the illegal gain of the violator. If not, noncompliance will be viewed as simply a cost of doing business (13).

6.2 The Statutes

A number of federal statutes create specific sanctions for violators of their provisions and the regulations promulgated pursuant to them. Prosecutions may be brought under statutes such as: The Resource Conservation and Recovery Act (RCRA) (14); The Comprehensive Environmental Response, Compensation, and Liability Act (CERCLA) (15); The Hazardous Substance Act (16); Occupational Safety and Health Act (OSHA) (17); The Clean Air Act (18); and the Water Pollution Control Act (Clean Water Act) (19).

The criminal enforcement provisions of the aforementioned statutes define a liable person to include an individual and generally require that the liable person have an awareness of wrongful conduct.

Local district attorney's offices are also filing criminal charges against individuals violating health and safety statutes. The County of Los Angeles, Office of the District Attorney, Occupational Safety and Health Section was the first of its kind in the country created specifically to protect workers' rights to a safe and healthy workplace. This objective is pursued in a number of ways: aggressive prosecution of cases referred by the California Division of Occupational Safety and Health; identification and investigation of other possible cases involving safety and health violations through the assistance of employees, organized labor, and health professionals; and strengthening regulations, legislation, and the practices and policies of the relevant regulatory agencies relating to worker health and safety issues.

Violations of laws relating to occupational safety and health are prosecuted either criminally or civilly. Most of the cases are multiple defendants, generally with the corporation and corporate officers and managers charged. Complex factual issues and a substantial amount of scientific or medical testimony are involved (20).

6.3 The U.S. v. Park

The general rule concerning officers, directors, or agents of a corporation is that they may be criminally liable individually for their acts on behalf of the corporation. They cannot, in the absence of a statute, be held liable for acts either in which they have not actively participated or which they have not directed or permitted. But the Supreme Court has broadened the traditional theory of individual criminal liability under welfare statutes by holding corporate officers personally culpable even when they do not have knowledge of the unlawful acts.

In U.S. v. Park (21), the Supreme Court upheld the chief executive officer's conviction for violation of the Food, Drug and Cosmetics Act (FDCA) for allowing unsanitary conditions to exist in a company warehouse. The court stated that the "government establishes a prima facie case when it introduces evidence sufficient to prove that the defendant had, by reason of his position in the corporation, responsibility and authority either to prevent the unsanitary conditions in the first instance, or promptly to correct the violation complained of, and that he failed to do so." The failure to fulfill the duty imposed by the interaction of the corporate agent's authority, and the statute, furnished a sufficient causal link.

Although the application of the Park holding is only beginning to see its effect in environmental statutes, the potential liability for officers and employees of firms involved in the handling of toxic materials is clear: the Public Welfare statutes may be the vehicle for convicting employers using the responsible corporate officer principle.

8 CONCLUSION

It is clear that professional industrial hygienists are faced with more complex legal issues today than in past years, and the future holds more of the same. The steady growth of applicable statutes plus the expansion in the area of tort law requires industrial hygienists to conduct their activities completely and competently. Industrial hygienists must perform their jobs with the utmost of care, be ethical, keep abreast of the current statutes and regulations, and learn to work with the legal system as an ally rather than as an enemy. It is important to remember that an attorney does not testify in a court proceeding. All information for deliberation in a court proceeding is presented through witnesses. It may be up to you to provide the facts that will determine the outcome of the court proceedings.

BIBLIOGRAPHY

1. United States v. Lee, 106 U.S. 196, 220 (1882). See also Carey v. Piphus, 435 U.S. 247 (1978) ("[O]ver the centuries the common law of torts has developed a set of rules to implement the

principle that a person should be compensated fairly for injuries caused by the violation of his legal rights.").
2. One Small Step for Industrial Hygiene: Developing Practice Standards and Guidelines. *The Synergist* December 15, 25, (1998).
3. American Board of Industrial Hygiene (Membership Roster), Code of Ethics for the Practice of Industrial Hygiene, pp 2–3, 1999.
4. *http://www.aiha.org/ethics.html.*
5. American Conference of Governmental Industrial Hygienists (Worldwide), Membership Directory and Information Guide, pp. 229–230, 1999.
6. B. A. Gardner, Ed., *Black's Law Dictionary*, 7th ed., West Group, St. Paul, Minn., 1999, p. 1498.
7. Mathews v. Walker, 34 Ohio App ed. 128. 1996 N.E. ed. 569, 630, and 208.
8. Restatement (2nd) of Torts.
9. Prosser, Law of Torts, p. 31, 1982.
10. LaCossa v. Scientific Design, 402 F 2nd 937 (3d Cir 1968).
11. B. A. Gardner, Ed., *Black's Law Dictionary*, 7th ed., West Group, St. Paul, Minn., 1999, p. 318.
12. American Bar Association, National Institute on Environmental Compliance, Los Angeles, May 7–8, 1987.
13. J. W. Starr, "Countering Environmental Crimes," *Boston College Environ. Affairs Law Rev.* **13**(3), 103 (1986).
14. 42 U.S.C. S 6901 et seq.
15. 42 U.S.C. S 9601 et seq.
16. 15 U.S.C. S 1261 et seq.
17. 29 U.S.C. S 651 et seq.
18. 42 U.S.C. S 7401 et seq.
19. 33 U.S.C. S 1251 et seq.
20. From Department Mission Statement, Office of the District Attorney, County of Los Angeles, Los Angeles, California, 1985.
21. 421 U.S. 658, 95 S.Ct, 9903, 1975.

CHAPTER THIRTY-SEVEN

Litigation in Industrial Hygiene Practice

Ralph E. Allan, JD, CIH

1 INTRODUCTION

Lawyers and industrial hygienists do share a common goal of problem solving. The initial aspect of problem solving is understanding the problem in order to provide the optimum in decision making. Therefore, understanding the legal system, the vehicle for legal problem solving, becomes especially important for the industrial hygienist as a precursor in order to clear the way for efficient attorney–industrial hygiene team effort for toxic substance issue resolution.

It is the purpose of the following material to provide more awareness of the law and its interaction with industrial hygiene activity.

2 COMMUNICATIONS

2.1 Potential Problems

One area of concern is the communication problems that can exist between the legal community and industrial hygienists. A good industrial hygienist can do wonders to aid an attorney in synthesizing the knowledge of specialists into a usable and understandable body of general knowledge to solve problems. The industrial hygienists who wish their work to have an impact must be capable of communicating their knowledge in an understandable and usable form. This becomes complicated because certain dissimilarities between the disciplines of law and science do exist. Many writers have emphasized the contrast between the inductive process utilized by science and the deductive process utilized by law. But this is an oversimplification; by no means does it explain the conflicts one hears about. In any event, both the industrial hygienist and the attorney should recognize and respect the differences in the way each approaches and thinks about a problem. Each needs to learn more of, and to understand better, the objectives and techniques of the other.

2.2 Role of the Industrial Hygienist in Litigation

The dissatisfaction that industrial hygienists feel when they are called upon to participate as witnesses in legal proceedings also stems from another contrast—the contrast between the role that they visualize for themselves and the role they often actually play. Too often industrial hygienists consider themselves as giving highly specific answers to highly specific questions. As they visualize the scene, they make their appearances in the courtroom surrounded by an aura of science; they state their pieces with great authority; and then they depart. The effect is dramatic and impressive because difficult controversies are resolved by their testimony.

But, things being the way they are, the drama often plays out altogether differently. Industrial hygienists may blame the attorneys for any mishaps and the attorneys may blame the industrial hygienists. A major part of the problem is a communications gap, if not a communications barrier. This is partly due to increasing specialization within the sciences. Even the field of industrial hygiene has become so compartmentalized that its practitioners may suffer from a sort of professional myopia.

Sometimes an industrial hygienist chafes under the usual requirement that evidence must be given by question and answer. The result thus may depend more on the skill of the questioner than that of the witness. The questioner is dominant. The form of the questions conditions the form and often the content of the answers. Unless the questioner really understands the subject matter, he or she may phrase questions in ways that restrict the scientific witness or misplace the emphasis. At worst, he or she may ask questions that are unanswerable because, to the witness, they have no rational content.

To get an intelligent answer, of course, one has to ask an intelligent question. If the answer is to be highly technical, the question itself must be technical in content and must be clearly phrased to bring out the pertinent answer. With all due respect to the legal skill of the attorney, he or she may find it difficult to feel at home with the highly specialized language of another field. It is a most distressing experience for an industrial hygienist to be subjected, from a supposedly friendly attorney, to a barrage of questions that really do not mean what the attorney thinks they mean. In fact, they may be technologically unintelligible, while at the same time sounding like the best English. The obvious solution is good communication.

In any communication between an attorney and an industrial hygienist, in or out of court, precise language must be employed, or at the very least there must be enough of exchange of views so that the language has an agreed meaning. Since the attorney and the industrial hygienist view problems from different vantage points, it is quite possible that they will see different objectives. These can be reconciled if recognized in time, but the danger is that they may go their separate ways, leaving both unaware that they are working at cross purposes. To be effective, the precision of the industrial hygienist and the legal skill of the attorney should focus on the same target.

It should go without saying that the attorney should seek the advice of the industrial hygienist on the wording of questions before they are asked in court, and the attorney should thoroughly understand and heed this advice. Then both will be on target. Too often, however, time limitations on the part of the attorney or the industrial hygienist, or both, prevent such careful collaboration. But both should insist on it as far as they can.

3 TYPES OF LAWSUITS

There are two general types of lawsuits: civil and criminal. A civil action is a lawsuit that is brought by a person or group of persons against another person or group of persons for losses, injuries, or damages they have suffered as the result of another's acts. The causes of these losses can run the gamut from negligence, intentional tort, breach of contract, libel, slander, or invasion of one's right of privacy to the violation of local, state, or federal law. In a civil case the suit is brought in the name of the person (John Smith) who alleges that he has suffered a loss at the hands of another (James Jones). Hence, the case is identified as *Smith v. Jones*.

A criminal case is one that involves the breach of some local, state, or federal law. In a criminal case the suit is filed in the name of the "people," and not by the individual against whom the crime was committed. This is because our legal system is based on the principle that all crimes are said to be against the public interest. The local, state, or federal legislative bodies that represent the "people" enact laws to protect these "people." Therefore, any breach of such a law is against the "people" and not the individual. If a local law has been violated (a crime has been committed), the lawsuit is brought in the name of the "People of the City" by the local officer in charge of this type of case. Usually this is the city attorney or county prosecutor from the jurisdiction where the crime allegedly occurred. If the offense upon which the case is founded is a violation of state law, the lawsuit is brought in the name of the "People of the State" by the state's attorney of the county in which the crime is alleged to have been committed. If violation of a federal law has been charged, the case is brought in the name of the "People of the United States" by the U.S. attorney for the Federal District Court in which the crime is alleged to have taken place. The defendant in either a civil or criminal case is the person, persons, or company charged with the infraction of violating a particular law.

3.1 Civil Cases

When an attorney determines that an individual has been harmed, for whatever reason, the first step is to draw up certain papers required by law. The first and most important paper is called a complaint in most states. It informs the opposing party (defendant) of the facts involved in the particular case, tells why the plaintiff is suing, and lists the amount of damages (i.e., money sought, etc.). After the complaint is filed in court, a summons is issued by the sheriff, which is then served on the defendant. The defendant then has a time period in which to respond. This response is called an answer. In his or her answer the defendant either admits or denies each paragraph of the complaint. Depending on how the defendant answers the complaint determines what becomes of the issues, which are the facts to be decided by the court and jury. If the defendant admits any charge in the complaint, this does not come before the court or jury since it is no issue on that particular point and does not need to be proved by the plaintiff.

Once the issues have been established, the next step is the trial itself. The time it takes for a case to come to trial depends on the individual case and the jurisdiction where the case has been filed. Each of the parties has a right to request a jury trial. Usually it is the plaintiff who will make this request. If neither party requests a jury, then the case will be heard by the judge alone, who is then the sole judge of the facts and the law and makes the decisions concerning all the issues of trial. In a jury trial, the jury determines the facts and the judge has the responsibility of interpreting the law. The witnesses present the facts to the jury whereas the judge gives them the law of the jurisdiction where the case is being heard. It is the responsibility of the jury to decide the reliability of the witnesses and weigh all the evidence presented during the trial. They must then return only one of two verdicts: in favor of the plaintiff or in favor of the defendant. If the verdict is for the plaintiff, the jury must assess the amount of the damages, which is included on the verdict sheet.

In addition to instructing the jury what law should be applied by them to the case, the judge has several other duties. The judge must overturn the jury's verdict if one of the following reasons is present:

1. If the verdict resulted from passion and prejudice (which is seldom found).
2. If there were insufficient facts presented to justify a verdict under the law.
3. Because of legal error during the trial.

The granting of a new trial is the exception rather than the rule. Another of the judge's responsibilities is to see that the trial is conducted by accepted rules and regulations and that the attorneys and witnesses strictly adhere to the rules of the court. In addition, the judge is the gatekeeper in allowing the admissibility of expert testimony. In this day and age, an industrial hygienist could be called to testify in a wide variety of types of cases ranging from criminal or civil cases before a judge to an OSHA hearing before an administrative law judge. For those who have never been to a trial and whose only courtroom experience is exposure to scenes from television's "Peoples' Court" or "Judge Judy" it might be a good idea to outline the proceeding in the conduct of a trial.

After the issues have been formed in a case, the first step of a trial is the selection of a jury if one has been requested. In most jurisdictions 12 persons form a jury; in some, 6-person juries are used. The jury is picked in most instances by the attorneys for the respective parties and is done after questioning the jurors to determine fairness and absence of prejudice. Actually, each attorney tries to pick jurors who will be sympathetic to his or her side of the case, but it cannot be predetermined that a favorable jury has been chosen.

Once the jury has been selected, the trial actually starts. The attorney who represents the plaintiff/prosecution (the side that has the burden of proof) will make his or her opening statement first. After the opening statement has been made by the plaintiff, or prosecution, then the attorney for the defendant will make his or her opening statement. The purpose of the opening statement is to have the attorneys clarify the issues in the particular case by telling the jury what they believe their respective evidence will show. This enables the jury to hear something about the case before the testimony begins. The judge will instruct the jury that the attorney's opening remarks should not be considered as evidence by the jury. The only evidence that is official is that testimony that comes to the jury from witnesses who have been sworn to tell the truth and testify from the witness stand. Documentary evidence that has been properly submitted to the court may be used as additional evidence in conjunction with the oral testimony.

After the opening statement, there is generally a motion by one of the attorneys to exclude witnesses from the courtroom. This means that all witnesses other than the plaintiff or defendant must leave and cannot hear testimony until each has personally testified. Thus, an expert seldom hears the testimony in any case in which he or she is a witness. Most attorneys prefer that witnesses not listen either before or after testifying because, if they do, the jury may receive the impression that the witnesses are interested in the outcome of the case and may be prejudiced in their testimony.

The next step in a trial is for the side that has the burden of proof, namely the plaintiff in a civil case or the prosecution in a criminal case, to begin presentation of its case through its witnesses. The attorney for the side that has the burden of proof must keep in mind what the particular issues are in the case so that the necessary proof that is required under the law is presented. If he or she fails to prove the elements of his or her case, it is within the prerogative of the judge to enter a directed verdict or a finding in favor of the defendant. A good attorney will always attempt to present witnesses in a good logical and chronological sequence to make it easy for the jury to understand the evidence and to follow the case.

After the evidence has been presented for the side that has the burden of proof, the attorney for the defendant may present a written motion for a verdict directed in his or her client's favor. This is done when the attorney feels that the plaintiff/prosecutor has not met the burden of proof necessary for a favorable verdict. If an attorney does this, it is done outside the presence of the jury. If the judge sustains this motion, the trial is over. In a civil case the term directing a verdict is the one most commonly used. In a criminal case the term finding for the defendant is the term that is generally used. In essence they both mean the same thing and have the same effect. If the judge overrules the motion for a directed verdict, it means that the attorney for the defendant must present his or her side of the case.

After the plaintiff/prosecutor has presented all of his evidence, and either no motions were presented or if presented they were denied by the judge, the attorney for the defendant presents his or her evidence. It should be pointed out that the same rules of evidence apply equally to both plaintiff/prosecutor and defendant witnesses. If a case involves more than one defendant, the defendant whose name appears first on the complaint will present witnesses, who will be followed by the witnesses for the later named defendant(s).

After all evidence has been presented, the plaintiff/prosecutor and/or defendant have an opportunity to present a written motion for a verdict directed in his or her favor. This is another time that the phrase "directing a verdict," comes into being. This term can be confusing for most laypeople. In directing a verdict, the judge takes the case away from the jury's consideration and directs the jury to find the decision that he or she tells them. As mentioned earlier, each attorney is given the opportunity to present a written motion requesting that the judge direct the verdict in favor of his or her client. To have the judge take the case away from the jury when it is the duty of the jury to decide issues of the case does not make a whole lot of sense to the average layperson. The judge does not do this very often, but it does occur often enough so that one should be aware of it.

Why would a judge direct a verdict? It occurs if the plaintiff/prosecutor has not proved his or her case under the law or the defendant has not made a satisfactory defense under the law. Remember, each side must prove its case by proving certain elements through the evidence presented through its witnesses. As an example, in a civil personal injury case the plaintiff must prove:

1. Negligence on the part of the defendant; a duty must be owed and a breach must be shown.
2. Freedom from negligence on the plaintiff's part.
3. That the injuries involved had a direct causal connection with the defendant's negligence.

If any of these essential elements is not proven through the evidence presented by witnesses, the plaintiff is not entitled to recover, and the judge will direct a verdict for the defendant and order the jury to sign a verdict finding for the defendant. Conversely, if the defendant fails to provide a defense to the plaintiff's action in which the essential elements were proven, the judge will direct a verdict for the plaintiff. In the latter case, the jury will not decide whether the defendant was liable or not. Their duty as a jury will be to arrive at a decision involving only the issue of damages, which in most cases deals with the amount of money the plaintiff is entitled to receive.

Before a judge directs a verdict in favor of either of the parties, he or she is required, under the law, to construe the evidence most strongly in favor of the party against whom judgment is sought. Generally speaking, the evidence must be overwhelming before the judge will enter a directed verdict. The law is very specific about this and, where possible, wants to give each citizen the benefit of the doubt before the judge directs the verdict. This action is taken only when the case is unfounded under the law.

Considering the number of cases that are heard by judges and juries, there are not very many directed verdicts. If a judge enters a directed verdict, the party against whom the verdict has been directed has the right to appeal to a higher court.

Once all the evidence has been presented, and if there is no directed verdict, the next step in the trial is the final arguments. Attorneys for both sides are allotted the same amount of time, but the plaintiff's/prosecutor's argument usually is in two parts, as he or she has the burden of proof. The plaintiff's/prosecutor's attorney goes first. He or she decides how much time to use initially in the final argument; what is not used can be reserved for use as part of the rebuttal after the defendant's attorney presents his or her final argument. This rebuttal argument of the plaintiff's/prosecutor's attorney is an answer to the final argument of the defendant's attorney.

When the final arguments for both sides have been given, the court will instruct the jury as to the law in that particular case. Sometimes the jury is given written instructions, which they take to the jury room as an aid in their deliberations. After the judge gives the jury his or her instructions, the jury will retire to the jury room where they will choose a foreperson and then deliberate the verdict. After they have reached their decision, they will complete a form that lists all the issues that had to be decided in the case. This is given to the clerk of the court who gives it to the judge to review. When this review is finished, the judge will instruct the clerk to read the verdict to everyone present in the courtroom.

The procedure applies to a case tried before a jury. Some cases are tried before only a judge. The procedures are the same for both with the only difference being that at the end of the trial the judge gives his or her decision without input from anyone else. The evidence is the same in both cases; the main difference is how the attorneys present the evidence since, in the absence of a jury, they do not have to worry about how the jury is going to perceive the witnesses and the arguments.

3.2 Criminal Cases

The trial procedures in a criminal case are a little different from those in a civil case. The action is brought by the "People of the State" in which the action is originated or in the name of the federal government if it is filed in a federal court. One difference is in the document that originates the case. In a criminal case this document is called an indictment, or the information, either of which is comparable to a complaint in a civil action. This document gives the court the authority to have the defendant apprehended. Once the defendant is taken into custody, he or she is arraigned. At the time of the arraignment the defendant will either plead guilty to the charge, plead not guilty, or may remain mute. If the defendant remains mute, the court will enter a plea of not guilty. When a plea of not guilty is entered, it creates the issue of denial of the charges in the indictment, or the information. No paper or written pleading is filed denying the charges as in a civil case.

Criminal cases may be tried by a court or a court and a jury. Only the defendant is allowed to ask for a jury in a criminal case. The big difference between a criminal case and a civil case is the level of the burden of proof. In a criminal case the burden of proof is greater than that in a civil case because the prosecution must prove beyond a reasonable doubt in order to sustain its verdict. The remainder of the trial is basically conducted in the same way as a civil trial as discussed above.

4 EXPERT WITNESS

An expert witness is one who possesses special knowledge and experience on matters in issue in a lawsuit. More and more industrial hygienists are being asked to perform this role. For those who have never experienced this, it is not surprising that they are hesitant to do so. One basic reason for hesitation is the fear of not knowing what might confront them when they come to court.

An expert witness is in a special class. The court has placed the expert witness in a little different position than the average witness in an effort to be more scientific.

The selection of an expert witness involves trying to match the education and experience of an individual with the specific facts associated with the case in question. The function of the expert witness is to assist the court and the jurors in arriving at a correct conclusion upon matters that are not familiar to their everyday experiences so that as part of their deliberations they can arrive at an intelligent understanding of the issues that are before them.

An expert may testify only about a matter that is a proper subject for expert testimony. The trial court is entrusted with the responsibility of determining whether the proffered testimony is a proper subject for an expert. As the Supreme Court has noted, the trial court has "broad discretion in the matter of the admission or exclusion of expert evidence, and his action is to be sustained unless manifestly erroneous" (1).

Federal Rule of Evidence 702 provides that expert testimony is admissible if "scientific, technical, or other specialized knowledge will assist the trier of fact to understand the evidence or to determine a fact in issue" (2). According to the federal drafters:

> The rule is broadly phrased. The fields of knowledge which may be drawn upon are not limited merely to the "scientific" and "technical" but extend to all specialized knowledge. Thus within the scope of the rule are not only experts in the strictest sense of the word, e.g., physicians, physicists, and architects, but also the large group sometimes called "skilled" witnesses, such as bankers or landowners testifying to land values (3).

The standard adopted by Federal Rule of Evidence 702—whether expert testimony will "assist the trier of fact"—is a more liberal formulation of the subject matter requirement than that found in many common law opinions, which often phrased the requirement as whether the subject was beyond the comprehension of a layperson (4). Under Rule 702, "the test is not whether the jury could reach some conclusion in the absence of the expert evidence, but whether the jury is qualified without such testimony to determine intelligently and to the best possible degree the particular issue without enlightenment from those having a specialized understanding of the subject" (5). This test is consistent with Wigmore's formulation of the test for expert testimony: "On this subject can a jury receive from this person appreciable help?" (6).

With respect to expert witnesses the trial judge has the burden of determining whether or not expert testimony should be admitted to prove an issue as well as determining whether or not the expert witness possesses the knowledge and experience to permit him or her to testify on the particular issues involved in that trial. The judge must also rule on whether or not an expert witness can give an opinion upon a given subject matter and whether or not the opinion is relevant and material to the issues in the case.

4.1 Types of Testimony

An expert witness can provide testimony that can be divided into two classifications: (*1*) where he or she has personal knowledge of the facts and (*2*) where all or part of his or her knowledge of the case comes from other sources. In the latter situation the evidence will have to be given through the use of hypothetical questions. Courts have relied principally on two different tests to determine the admissibility of innovative scientific evidence. One approach treats the validity of the underlying principle and the validity of the technique as aspects of relevancy. The relevancy approach is to treat scientific evidence in the same way as other evidence, weighing its probative value against countervailing dangers and considerations. In a 1954 text, Professor C. McCormick wrote:

> "General scientific acceptance" is a proper condition upon the court's taking judicial notice of scientific facts, but not a criterion for the admissibility of scientific evidence. Any relevant conclusions which are supported by a qualified expert witness should be received unless there are other reasons for exclusion. Particularly, its probative value may be overborne by the familiar dangers of prejudicing or misleading the jury, unfair surprise and undue consumption of time (7).

This approach, which dovetails with the Federal Rules of Evidence, requires a three-step analysis: (*1*) ascertaining the probative value of the evidence, (*2*) identifying any countervailing dangers or considerations, and (*3*) balancing the probative value against the identified dangers.

The second approach, which required the proponent of a novel technique to establish its general acceptance in the scientific community, was based on *Frye v. United States* (8), decided in 1923. In Frye the D.C. Circuit Court considered the admissibility of polygraph evidence as a case of first impression. The court wrote:

> Just when a scientific principle or discovery crosses the line between the experimental and demonstrable stages is difficult to define. Somewhere in this twilight zone the evidential force of the principle must be recognized, and while the courts will go a long way in admitting expert testimony deduced from a well recognized scientific principle or discovery, the thing from which the deduction is made must be sufficiently established to have gained general acceptance in the particular field in which it belongs (9).

In a unanimous decision on June 28, 1993 (*Daubert v. Merrel Dow Pharmaceuticals, Inc.*, US SupCt, No. 92-102, Blackmun, J.) (10), the U.S. Supreme Court held that the "general acceptance" standard for the admission of scientific evidence adopted in *Frye v. U.S.* no longer governs trials conducted under the Federal Rules of Evidence. The Supreme Court held that Fed. R. Ev. 702, which provides a more liberal rule for admitting relevant scientific evidence, supersedes the common law rule announced in Frye.

In the Daubert case, two children alleged that they suffered from birth defects caused by their *in utero* exposure to the drug Bendectin. They presented a number of well-qualified experts who were prepared to testify based on animal studies and reanalysis of epidemiological studies that Bendectin causes the type of birth defects suffered by the plaintiffs. The defendant, Merrell Dow, had equally well-credentialed experts who would have testified, based solely on epidemiological studies, that Bendectin is not a teratogen capable of causing malformations in fetuses.

The U.S. District Court for the Southern District of California granted summary judgment for the defendant, saying that scientific evidence is admissible only if the principle on which it is based is generally accepted in the field to which it belongs. The court found that the plaintiffs' evidence did not meet this standard. Given the vast body of epidemiological data concerning Bendectin, the court held, expert opinion that is not based on epidemiological evidence is not admissible to establish causation.

The U.S. Court of Appeals for the Ninth Circuit affirmed. Citing Frye, the court stated that expert opinion based on a scientific technique is inadmissible unless the technique is "generally accepted" as reliable in the relevant scientific community. Reanalysis, the court contended, is generally accepted by the scientific community only when it is subjected to verification and scrutiny by others in the field. Since the reanalysis on which the plaintiffs' experts based their opinions had not been subjected to such peer review, the court concluded that this evidence was not admissible and, therefore, that the plaintiffs could not satisfy their burden of proving causation at trial.

Although the Frye standard has dominated the admissibility of scientific evidence for 70 years, its merits have been the subject of much debate. However, the plaintiffs in this case did not attack the standard's scope or application; rather, they argued that the Frye test was superseded by the adoption of the Federal Rules of Evidence in 1975. The U.S. Supreme Court agreed. Fed.R.Ev. 402 provides the "baseline" for admissibility. "All relevant evidence is admissible ... Evidence which is not relevant is not admissible." Evidence is relevant if it has "any tendency to make the existence of any fact that is of consequence to the determination of the action more probable than it would be without the evidence." Rule 401.

In *U.S. v. Abel* (11), the court previously considered the pertinence of background common law in interpreting the Federal Rules of Evidence. The court noted that the rules occupy the field, but that the common law could still serve as an aid to their application. The common law precept at the issue in that case was found to be consistent with Rule 402's general requirement of admissibility. In *Bourjaily v. U.S.* (12), on the other hand, the court was unable to find a particular common law doctrine in the rules, and so held it superseded.

In Daubert, the court found, there is a specific rule governing the contested issue. Rule 702 provides:

> If scientific, technical, or other specialized knowledge will assist the trier of fact to understand the evidence or to determine a fact in issue, a witness qualified as an expert by knowledge, skill, experience, training, or education, may testify thereto in the form of an opinion or otherwise.

The court found that there was nothing in the text of this rule that establishes "general acceptance" as an absolute prerequisite to admissibility. Nor was there any evidence presented that the rule was intended to incorporate a general acceptance standard. In fact, the rigid general acceptance test is at odds with the "liberal thrust" of the rules and their "general approach of relaxing the traditional barriers to 'opinion' testimony," the court ruled, quoting *Beech Aircraft Corp. v. Rainey* (13).

The court held that "Given the Rules" permissive backdrop and their inclusion of a specific rule on expert testimony that does not mention "general acceptance," the assertion that the Rules somehow assimilated Frye is unconvincing. Frye made 'general acceptance' the exclusive test for admitting expert scientific testimony. That austere standard, absent from and incompatible with the Federal Rules of Evidence, should not be applied in federal trials," the court concluded (14).

Where this is all headed is open to conjecture. Everyone generally agrees that Daubert changed the rules for introducing expert scientific evidence. But agreement pretty much stops there. Justice Harry A. Blackmun, who wrote the opinion for the unanimous court, also wrote in a part joined by only six other justices that the federal rules do place some restrictions on expert testimony. The rules especially Rule 702, assign to trial judges the task of ensuring that an expert's testimony rests on a reliable foundation and also is relevant. According to this majority of the court, a trial judge, when faced with expert proof, must determine whether the expert's proffered testimony is based on scientific knowledge that will assist the trier of fact to understand or determine a fact in issue. Rule 702, "which clearly contemplates some degree of regulation of the subjects and theories about which an expert may testify (15), is the primary locus of this obligation, Blackmun wrote. Parsing the rule, he said the subject of an expert's testimony must be "scientific . . . knowledge." The adjective "scientific" implies a grounding in the methods and procedures of science, the court found, while the word "knowledge" connotes more than subjective belief or unsupported speculation. "Proposed testimony must be supported by appropriate validation—i.e., 'good grounds,' based on what is known. In short, the requirement that an expert's testimony pertain to 'scientific knowledge' establishes a standard of evidentiary reliability," Blackmun wrote. Additionally, "Rule 702's 'helpfulness' standard requires a valid scientific connection to the pertinent inquiry as a precondition to admissibility" (16).

It is the position of the majority that the trial judge, when faced with a proffer of expert scientific testimony, must determine at the outset whether the expert is proposing to testify in accordance with these principles. While not "presuming to set out a definitive checklist or test," the majority made some "general observations" about how a court should go about this task.

The majority said that the lower courts should look at whether the scientific knowledge being presented has been tested, whether it has been subjected to peer review and publication, what the evidence's known rate of error is, and whether the evidence has a particular degree of acceptance in the relevant community. The majority further cautioned that the inquiry must remain a flexible one. The traditional measures of judicial control—such as vigorous cross-examination, screening unqualified experts or irrelevant evidence, and granting summary judgment and directed verdicts where warranted—are adequate to contain unreliable testimony.

Chief Justice William H. Rehnquist and Justice John Paul Stevens concurred with the majority's holding that the Frye rule is no longer good law. In their dissent from the remainder of the court's opinion, they said that the majority should not have gone beyond its determination that the Federal Rules of Evidence has supplanted Frye. They recognized that "general observations by this Court customarily carry great weight with lower federal courts," and pointed out that "the ones offered here suffer from the fatal flaw common to most such observations—they are not applied to deciding whether or not particular testimony was or was not admissible, and therefore they tend to be not only general, but vague and abstract" (17).

Chief Justice Rehnquist went even further as he wrote, "I do not doubt that Rule 702 confides to the judges some gatekeeping responsibility in deciding questions of the admissibility of proffered expert testimony. But I do not think it imposes on them either the obligation or the authority to become amateur scientists in order to perform that role. I think the Court would be far better advised in this case to decide only the questions presented, and to leave the further development of this important area of the law to future cases" (18).

According to the seven-member majority, the rules contain no general acceptance test. The "austere standard" of Frye "absent from and incompatible with the Federal Rules of Evidence, should not be applied in federal trials" (19). The court also said that there is a reliability component in the scientific knowledge component. "Under the rules the trial judge must ensure that any and all scientific testimony or evidence admitted is not only relevant, but also reliable. The requirement that an expert's testimony pertain to 'scientific knowledge' establishes a standard of evidentiary reliability," the opinion said. The judge then must determine whether the reasoning or methodology underlying the testimony is scientifically valid. The court cautioned that the judge's examination must be on the expert's methods, not conclusions.

The following are the key points of the Daubert decision:

- The Frye Rule or "general acceptance standard," has been supplanted by the Federal Rules of Evidence, which were enacted in 1975.
- The Federal Rules of Evidence do not contain a requirement that scientific evidence be generally accepted in the field in order to be admissible at trial.
- The inquiry envisioned by Fed. R. Evid. 702 is a flexible one.
- Fed. R. Evid. 702 requires that a trial judge ensure that an expert's testimony is based on scientific knowledge and will assist the trier of fact.
- A standard of evidentiary reliability is established by the requirement that an expert's testimony relate to "scientific knowledge."
- A trial judge, under Fed. R. Evid. 104(a), must determine whether the expert is proposing to testify to scientific knowledge that will assist the trier of fact.
- Factors to be applied in arriving at the above include:
  whether a theory or technique can be and has been tested.
  whether it has been subject to peer review or publication.
  the known or potential error rate.
  whether the theory of technique has gained general acceptance in the field.

The "general acceptance" standard is gone. The question based upon this decision, for those involved in the selection of expert witnesses, is "what impact does this ruling have with regard to witness preparation and jury persuasion?" In the past, attorneys chose experts based on their credentials and conclusions. Daubert shifts the emphasis to the reliability of the expert's data and conclusions.

This ruling promotes the novel and innovative. It results in expert witnesses needing to be prepared to clearly explain how their testimony relates to the dispute. Will Daubert have an effect on the way attorneys prepare experts to testify? Currently, experts undergo preparation for two distinct stages of the trial: the deposition and courtroom testimony. There is a significant difference.

There are four major differences: (*1*) the purpose for the testimony, (*2*) the audience for the testimony; (*3*) the rules governing the testimony considering the purpose and audience; and (*4*) the tools employed to achieve the purpose. In deposition, the expert sits on a pile of information, which the one taking the deposition wants. The purpose for the testimony is to give out only the information directly requested. The audience is the opposition who wants to secure information that can be used to either disqualify the expert or to challenge the credibility of the conclusions or the person. The rules are specific that the deponent only has to answer the questions asked. When a question is asked, however, the answer given is to be done cooperatively, fully, and honestly but that does not mean that the deponent has to volunteer any answers over and above the questions asked. The tools are supplied by the one taking the deposition and usually are documents.

At trial, the purpose of the testimony is to aid the fact-finder (the audience) to understand certain key elements. The rules are relatively simple—the witness testifies solely for the benefit of the fact-finder (judge or jury) whether he or she is questioned on direct or cross-examination. The expert's sole purpose in testifying is to be the objective supplier of the truth and not an advocate, which means that the expert talks to the judge/jury freely, openly, and honestly. Preparation of the experts should emphasize the need to understand the judge/jurors, their backgrounds, their basic level of understanding, and the best way to present information to help them do their job. The tools in this arena can include demonstrative exhibits, documents, drawings, photographs, tables of data, graphs, and so on—whatever it takes to aid their understanding. Clearly, the expert has both a right and an obligation to convince the judge that reliable principles, derived from a reliable, valid, and trustworthy scientific method, form the basis for testimony.

The effect Daubert has on litigation continues to develop. Meanwhile another recent Supreme Court case, *Kumho Tire Co. v. Carmichael* (20), provides an insight into the application of Daubert to the testimony of engineers and other experts who are not scientists. The Court in Kumho Tire finds that the trial judge's general "gatekeeping" responsibility applies not only to testimony based upon "scientific" knowledge but also to testimony based on "technical" and "other specialized" knowledge.

The implications of Daubert and Kumho Tire will continue to develop and influence the presentation of scientific expert testimony in the federal courts, but only in the federal courts does it apply. Technically, Daubert and Kumho Tire are binding only in the federal court system.

State courts, however are initiating their own methods to control bogus theories presented by science experts. Georgia, for example, once an expert has met the requirements to qualify as an expert witness, allows the expert to freely present their opinions before the jury and specifically rejects the Dalbert test (21). A Georgia statute provides that the opinions of experts on any question of science . . . shall always be admissible . . . (22). However, the Georgia jurisdiction does not allow a qualified expert free license before a jury. The Georgia Court of Appeals in the Motorola case, in order to protect against unreliable expert opinions, held that an expert must do more than just establish his credentials and give his opinion. He must also present how he related the facts of the case and applied his expertise to them to reach his opinion.

California has also rejected the Daubert decision in favor of the Kelly/Frye rule (23) which is an exclusionary rule of evidence that allows courts to exclude expert testimony that is based on controversial scientific techniques. State courts, therefore, have different bases for scientific expert witness testimony to be presented before a jury.

4.2 Evidence

In general, a witness is permitted to testify to anything that he or she has actually seen or experienced, which is material, that is, pertinent, to the issues involved in the case. The witness is permitted to testify to anything he or she has heard in the presence of the opposing party. He or she can testify to the statements made by one of the parties to the lawsuit that may be against the interest of the party making the statement. The witness is allowed to testify to anything within his or her knowledge that concerns the issues. The witness is not permitted to testify to anything he or she heard from a third party or to facts learned from someone other than the parties to the case. This is what is known as hearsay evidence, which is inadmissible.

There are many exceptions to the general rules of evidence, but the focus in this discussion will be on those that pertain to the expert witness presenting testimony in the average case. Opinion evidence and hypothetical questions are both exceptions to the hearsay rule. If a witness says, "It is my opinion that . . .," or "I think . . .," there will be an immediate objection by the opposing attorney. This objection will be sustained by the court. The reason is that when a witness tries to give an opinion or tell what he or she thinks, he or she is trying to decide one of the issues of the case. This is the role of the jury, and thus such evidence invades the province of the jury, which the court will not permit. The ordinary witness is entitled to give the facts and let the jury decide the case from the facts admitted into evidence. The expert witness is given some latitude as the law permits the testimony of the witness to include opinions upon a given set of facts in evidence, and then permits the jury to decide whether or not it wishes to adopt the opinions given by the expert witness. Expert testimony is not utilized to substitute the opinion of the expert for that of the jury. It is just an aid, but in some cases it can be the deciding factor in the outcome of the particular lawsuit.

4.3 Opinion Questions

As was mentioned, the expert witness is allowed to render opinions in two instances: (*1*) where he or she has personal knowledge of the facts and (*2*) where he or she knows none of the facts. An expert witness can render an opinion on facts about which he or she has personal knowledge and on which there is sufficient data on which to base the opinion. This is always done in response to questions presented by the attorney. This could include tests or experiments performed by the witness. It should be pointed out that facts must be within the witness' personal knowledge and not on what someone else did or told him. If he uses other information his opinion will be stricken from the consideration of the jury as being invalid.

Typically, the expert witness will testify to all the facts that he or she has personal knowledge of and upon which he or she can form an opinion. The witness will then be asked an opinion question that must be in his or her specialty or it is not admissible in evidence. The opinion question must be based on a reasonable degree of certainty to be admissible in law. It cannot be speculative. If the witness is an industrial hygienist, the question must be based on a reasonable degree of scientific certainty. It is not necessary that it be a certainty. The word reasonable is all that is required upon which to formulate an opinion. This is far different than the requirements needed to form an opinion in other fields. It is much less stringent and less exacting. It would be unjust for an expert witness to apply a more strict interpretation to the particular problem than is required under law. What is this thing called reasonable degree of certainty? Plain and simple, it is a legal fiction. It has been interpreted differently on various occasions. The key word is reasonable. A typical definition would be that which would induce a person of ordinary prudence to believe it under the circumstances.

For some to use this standard might prove uncomfortable. However, it might help to view the alternative. Typically the people who make up the jury will not have any technical or scientific knowledge on which to base an opinion, and they are looking to the expert witness to provide them with some assistance. The role of the expert witness is to give his or her opinion based on the facts as presented, using the training and experience that he or she has to arrive at conclusions.

4.4 Hypothetical Questions

The hypothetical question is actually an opinion question. It was stated above that opinion questions were asked of the expert witness where he or she has personal knowledge of the facts. In a hypothetical question situation the expert witness is giving an opinion based on facts from other sources. This is due to the fact that the expert witness was not present when the situation occurred so he or she could not have all the facts needed to express an opinion. Since the expert witness cannot render an opinion unless he or she knows certain facts, the law permits the attorney to ask a hypothetical question in which all the necessary facts that have been proven by witnesses during the trial will be given. This will be done in a hypothetical form so that the expert witness can give his or her opinion on these facts as they are applicable to his or her special knowledge. From this hypothetical question the expert witness may give as many opinions as are applicable to the issues of the particular case. It is up to the jury to use this information to come to its decision.

The hypothetical question has two principal advantages. First, it informs the jury of the facts upon which the expert's opinion is based. Second, if the question contains assumed facts that are not supported by evidence admitted at trial, it provides the opposing party with an opportunity to object before an opinion is expressed.

The witness must be made to appreciate the underlying premise of a hypothetical question. He or she must understand that the evidence postulated in the question is assumed to be true.

Therefore, a hypothetical question is one in which an attorney gives the expert witness an assumed set of facts that have been given by other witnesses in the case so that the expert witness can give an opinion on that particular set of facts. Generally, most hypothetical questions will begin with "assume," "suppose," or "consider." More often than not, the attorney will generally start the question with "Assume that a hypothetical person

. . ." This should alert the expert witness that the question being asked falls into this special category.

How the hypothetical question will be put together must begin long before it is put to the expert in the courtroom. The length of the question is governed by the atmosphere of the trial and the attorney's own sense of effective timing. It would be poor practice to propound a question so long that it needlessly confuses either the expert or the jury. Hopefully, the question will be framed so that it will not become clouded in the jury's minds by frequent objections from opposing counsel.

The expert witness must be able to determine what are the salient facts from among the many that will be presented in the hypothetical question. Typically, it will contain facts that are not needed to render an opinion, but have been included so that the attorney can impress the jury with certain facts that have been brought out in the testimony. Many times the attorney will take this opportunity to present the facts logically and lucidly in a narrative form. The expert witness must keep in mind that regardless of what is being presented, his or her answer must be limited to the facts given in the hypothetical question. It is the expert witness's responsibility to assimilate all the data in the hypothetical question, apply his or her special learning and experience, that is, expertise, to the facts, and then render an opinion based on reasonable certainty.

After an expert witness has given his or her opinion, the attorney should ask the reasons why he or she arrived at this opinion. This gives the expert witness the opportunity to discuss all the facts used to arrive at his or her decision and is one of the most important parts of the testimony. Since the purpose of his or her testimony is to instruct the jury about a technical matter that they do not fully understand, the expert witness should clearly itemize and specify the basis for the opinion so that the jury can follow the logic of the analysis. The importance of this part of the testimony cannot be overemphasized. The expert witness is given all the time needed to thoroughly explain his or her position. If the jury is to responsibly perform its duties, then it is up to the expert witness to educate them so that they can understand the technical problem that is before them.

5 WORK PRODUCT

Federal Rule 26(b) (3) sets out the general principles governing the discoverability of materials (i.e., materials prepared in anticipation of litigation or for trial). It states that relevant "documents and tangible things" that were prepared by or for a party or his or her "representative," "in anticipation of litigation or for trial," are not discoverable, except on a showing of "undue hardship" and "substantial need" (24).

As the Sixth Circuit Court of Appeals stated in In re Grand Jury Subpoena dated November 8, 1979:

> Work product consists of tangible and intangible material which reflects an attorney's efforts at investigating and preparing a case, including one's pattern of investigation, assembling of information, determination of the relevant facts, preparation of legal theories, planning of strategy, and recording of mental impressions (25).

A wide variety of persons have been held to fall within the work product immunity. Materials prepared for an attorney are, of course, covered (26). When a person, such as an industrial hygienist, does work specifically at the direction of an attorney, that person should assume that what he or she is doing will be protected under the work product privilege and should conduct him or herself accordingly. If there are any changes to be made in this relationship, let them be left to the attorney.

The Supreme Court in *United States v. Nobles* (27) held the work product privilege applicable to both the pretrial and trial stages. In cases involving scientific evidence, the work product rule may preclude discovery of reports prepared by expert witnesses. In Nobles the Court wrote:

> At its core, the work product doctrine shelters the mental processes of the attorney, providing a privileged area within which he can analyze and prepare his client's case. But the doctrine is an intensely practical one, grounded in the realities of litigation in our adversary system. One of those realities is that attorneys often must rely on the assistance of investigators and other agents in the compilation of materials in preparation for trial. It is therefore necessary that the doctrine protect material prepared by agents for the attorney as well as those prepared by the attorney himself (28).

The Court, however, also held that the calling of a witness at trial is a waiver of the privilege (29).

6 ATTORNEY–CLIENT PRIVILEGE

The attorney–client privilege is designed to protect confidential communications between a client and his or her attorney. The Supreme Court has noted: "its purpose is to encourage full and frank communication between attorneys and their clients and thereby promote broader public interests in the observance of law and administration of justice. The privilege recognizes that sound legal advice or advocacy depends upon the attorney being fully informed by the client" (30).

Application of the attorney–client privilege to expert witnesses sometimes arises in cases involving industrial hygienists. It is important to distinguish between the two ways in which an industrial hygienist might be used as an expert witness. First, he or she may be retained for the purpose of testifying at trial. When this happens, the privilege is waived (31). A "party ought not to be permitted to thwart effective cross-examination of a material witness whom he will call at trial merely by invoking the attorney–client privilege to prohibit pretrial discovery" (32).

Second, an industrial hygienist might be retained as a consultant to provide the attorney with information concerning a particular case. This could be for the plaintiff or the defendant. Many times an attorney will rely on this information to determine whether or not it is feasible to continue with a particular case. A number of courts have held that the attorney–client privilege covers communications made to an attorney by an expert retained for the purpose of providing information necessary for proper representation (33). There are times when the attorney needs to obtain expert advice because of the technical issues of a particular case.

If an industrial hygienist is retained by an attorney, whether his or her input will be covered under the attorney–client privilege is not that industrial hygienist's concern. However, the industrial hygienist should conduct him or herself as though the attorney–client privilege does apply and not discuss the case outside the attorney's office. The confidentiality of the industrial hygienist's role must be honored. That way the industrial hygienist will not inadvertently waive the privilege if, in fact, one does exist.

7 DISCOVERY

Discovery is a pretrial process whereby an attorney attempts to gather information pertaining to the trial for the following purposes:

1. One of the most basic purposes is simply the obtaining of information. By exchanging facts through the discovery process, each side is provided with a great deal more information to help prepare their cases than they would have if required to obtain all information on their own.
2. Another purpose is to create a record of testimony prior to the trial by using depositions and document production. Sometimes attorneys use interrogatories and requests for admissions to aid them in trying to determine what their opponent does or does not know or to locate documents or people who have relevant knowledge.
3. Generally the requests for admissions helps to define and limit the issues at trial. Discovery, more than likely, will eliminate certain issues and focus the parties attention on others.
4. As a result of the exchange of information, all parties are able to evaluate the strengths and weaknesses of their cases for potential settlement discussions. It gives the attorneys an opportunity to evaluate the kind of witness an individual will make at trial in terms of demeanor, credibility, sympathy, and the other intangible factors, which, absent the discovery process, would leave the parties to their peril in the courtroom.
5. Sometimes the pretrial discovery gives the parties additional information to allow them to supplement their pleadings.
6. Sometimes the information gathered during the pretrial discovery process will form the basis of motions for summary judgment and other fact-based motions that may be dispositive of the case in whole or in part.

What can be discovered? Any matter relevant to the subject matter involved in the pending litigation so long as it is not privileged and not trial preparation material. As a matter of right, the discovery pertaining to expert witnesses that the attorney is entitled to is through the use of interrogatories [Federal Rule of Civil Procedure 26(b) (4)]. The information available under the rules pertaining to interrogatories includes the following:

1. The identity of the expert witness who is expected to be used at trial.
2. The subject matter of that expert's testimony.

LITIGATION IN INDUSTRIAL HYGIENE PRACTICE

3. The substance of the facts and the opinions to which the expert will testify.
4. A summary of the grounds for each opinion that the expert will give.

8 DEPOSITIONS

A deposition is a record of testimony, and the one giving a deposition is a witness. Testimony is a statement made or evidence given by a witness under oath or affirmation in a legal proceeding and more properly refers to oral evidence than to documentary evidence. The reasons for taking a deposition include:

1. To discover the theory of the case.
2. To "freeze" the testimony of the witnesses on some crucial points.
3. To discover information that could be used at trial or may lead to information that can be used at trial.
4. To secure admissions from adverse witnesses.
5. To perpetuate the testimony of a friendly witness or an eyewitness who may not be available for trial.

Generally, a motion must be filed to take the deposition of an expert although in some cases it may be done by agreement. The need for a deposition is especially manifest if the answers to interrogatories posed to an expert regarding his or her opinion are not complete or are evasive.

If you, as an industrial hygienist, are ever called to give a deposition, you should make sure that your attorney adequately prepares you. The following points may be used as your guide:

1. Your attorney should discuss with you the manner in which you answer the questions. You must remember that what you have to contribute is your credibility.
2. You should not admit that any material with which you are not totally familiar is authoritative.
3. There may be occasions when your attorney will need to object to a question or may instruct you to not answer a question. In order to give him or her time to do so, you need to learn to pause before you answer a question.
4. Be sure to take with you to the deposition only those documents or materials that you have seen before and with which you are totally familiar. If you use a document at a deposition, assume that it will be turned over to the other side.
5. Know that the other attorney will question your credentials.
6. If interrogatories were propounded to you, be familiar with your responses to them because the other attorney will question you about them. He or she will look for what might have been left out of the answers or to supplement them.
7. The opposing attorney will probably ask you who you recognize as authoritative in your field.

8. Your attorney should conduct a full cross-examination of you prior to the deposition. This means that he or she simulates the type of questioning you can anticipate from the opposing attorney.
9. Tell the truth at all times.
10. Only answer the questions that are asked. There is a natural tendency to think out loud and run on with explanations. Do not volunteer any information beyond the scope of inquiry.

9 WITNESS PREPARATION

The attorney should tell the expert witness what his or her expected role will be in the case in which he or she will be involved. The expert witness should be told who he or she is representing and what evidence he or she will be expected to give in court. If you, as an industrial hygienist, are called to be an expert witness in a trial, most attorneys will give you this information, and they will meet with you and discuss the following:

1. Why your testimony is necessary.
2. When you will be expected to testify.
3. How long the attorney thinks that your testimony will take.
4. Where your testimony fits into the overall picture.
5. What procedure will be followed in court.
6. What testimony you will be expected to give during the trial.
7. What you might expect from the opposing attorney during cross-examination.

If the attorney who engages you does not give you this information, be sure to question him or her about it. If you are not satisfied with what you think is expected of you and what you should expect while you are testifying, do not hesitate to ask questions of the attorney. On direct examination, most trial attorneys will attempt to anticipate matters that are likely to be raised on cross-examination. The attorney can avoid many stressful moments for the industrial hygienist by placing an abundance of information in the record concerning the sufficiency of the industrial hygienist's qualifications and the adequacy of the examination upon which the opinion, if there is one, will be based. Matters concerning the compensation of the expert, or any other possible motives he or she may have in testifying, should also be brought out in direct examination in order to avoid the impact such facts may have on the trier of fact if they are forced out of the witness by opposing counsel.

9.1 Clarity of Testimony

One of the problems of communication in expert testimony occurs because experts are not used to talking to laypersons about their knowledge and judgments. An expert witness must remember that he or she is more familiar with the terminology than the attorney, and

LITIGATION IN INDUSTRIAL HYGIENE PRACTICE 1717

especially the jurors. Even if the terms are clear, an expert witness must be careful that the explanation is not too confusing, which it can be if the witness makes it too complex and it involves situations unfamiliar to the jury. A good rule of thumb to remember is that an expert witness is, in essence, teaching the jury, often from scratch.

9.2 Qualifications

The attorney must qualify an expert witness carefully in order to demonstrate the credibility of his or her testimony (opinion) to the judge and jury. The witness's background and standing in his or her profession should be set out in detail, including university degrees, affiliation with professional societies, writings in the field, tests and experiments conducted, and relevant experience generally.

The following is a sample qualification of an industrial hygienist in an intentional tort case involving a former employee who has charged his employer with intentionally causing his injury:

Q. Mr. _____ what is your profession?

A. I am an industrial hygienist.

Q. Where did you receive your formal education?

A. I took my degree, Bachelor of Science in Chemical Engineering, at the University of _____ in 1975, and I received a Master of Science in Public Health degree from the University of _____ in 1977.

Q. Do you hold a license to practice engineering in this state?

A. Yes, I am licensed as a professional engineer in this and four neighboring states.

Q. Are you certified in industrial hygiene?

A. Yes, I am a Certified Industrial Hygienist (CIH) by the American Board of Industrial Hygiene. I received my certification in 1981.

Q. Upon completion of your formal education, did you become associated with any firm in the practice of your profession?

A. Yes. The _____ Consulting Company in New York City.

Q. What was your position with that firm and how long were you associated with it?

A. I have been with that firm since 1977. I started as Industrial Hygienist I and have been promoted several times. I presently am Manager of the Industrial Hygiene Department.

Q. Of what professional societies are you a member?

A. I am a member of the American Industrial Hygiene Association, the American Academy of Industrial Hygiene, the American Public Health Association, and the American Institute of Chemical Engineers.

Q. Have you written any papers, given any speeches, or taught any classes in your area of practice?

A. Yes I have.
Q. Would you please identify them?
A. In September 1977, I . . .

10 RECORDKEEPING

10.1 General

Many practicing industrial hygienists collect samples and prepare reports on a regular basis. In determining what and how many reports and records should be kept, it is prudent to consider legal requirements and liability matters. Some safety and health regulations require recordkeeping as a form of accountability. These recordkeeping requirements include such documents as fire-extinguishing equipment inspection records, local exhaust ventilation system testing records, water-quality testing records, and hazardous waste removal records. Some recordkeeping is not obligatory but is considered to be a state-of-the-art requirement in a particular industry. For example, deluge shower and eye wash equipment testing records are considered necessary in the chemical industry.

One of the most important responsibilities of an industrial hygienist is to keep good records. Generally, industrial hygienists probably maintain records on assessment of hazards, exposure measurements, development and maintenance of exposure controls, audit procedures, training and education, to name a few. It is now becoming increasingly important to be able to track where an employee worked so that his or her exposures can be calculated based on places of employment for various time periods. Maintaining a file of job descriptions that can be linked to job title records of individual employees can be of immeasurable value.

An industrial hygienist's records not only serve the primary function of helping document efforts and activities in the recognition, evaluation, and control of employee exposures, but they also could serve as evidence in a workers' compensation claim, health-related grievance, or arbitration case or might be used by the plant physician to evaluate an environment that might be suspect as related to an employee's exposure and illness. In the long term these records are needed in epidemiological studies. Finally, they may be of great importance in defense of a professional liability lawsuit.

A logical question that needs to be addressed is: "What constitutes adequate records?" Suffice it to say there is not one "adequate" record-keeping system that will work for every situation. Typically, there should be sufficient detail that if an industrial hygienist has to go back to them in the future he or she can generate the information that the records represent. For example, if the records are of a sampling activity, it is important to record not only the sampling data completely and accurately but also to document the methods used to obtain the results. An industrial hygienist should assume that his or her records will be reviewed as historical data by someone else and that they should be self-explanatory.

In this day and age it makes good sense to develop a system of recordkeeping that can be loaded into a computer database. Regardless of how the data are maintained, the system should allow for a systematic storage and retrieval of the data. Data collected in the field should be kept as a permanent record. It is extremely important to include all notes that are pertinent. It may be of some value to develop a standardized form for collecting the data. When conducting the field work, be sure to write down every detail that could have any value so that when the data are being used some time later all the information needed to do a proper evaluation, make reasonable recommendations, and prepare a comprehensive report will be available. It is just as important to record all data pertinent to a sample as it is to collect the sample. A good rule of thumb is to provide enough information so another industrial hygienist could duplicate the exercise without further assistance from the industrial hygienist making the record.

Be sure to include any calibration information that is pertinent. This should be part of the permanent record. Specific forms can be developed or adopted for the various types of data collection. Forms with labeled spaces for the basic information help to ensure that essential information is documented at the time the work is done. The use of standard forms will promote consistent data collection and documentation.

10.2 Reports

As a general rule an industrial hygiene report should present the facts, show analysis of the data, interpret the findings, and contain conclusions and recommendations. All of the ways in which a report may be used cannot be anticipated at the time it is prepared; an industrial hygienist should make sure that each report includes sufficient information that it can be understood by others who read it later.

The makeup of a report can take various forms, but a common makeup for reports that will be useful in future testimony or litigation would include the following sections:

1. A summary at the beginning that concisely describes the work performed, why it was done, sufficient background information so that a reader can understand what was done, the conclusion(s) drawn, and the recommendations made.
2. A description of the reported exercise or project, which could include illustrations, tables, graphs, diagrams, and photographs. It should tell what was done, why it was done, what was observed at the workplace or otherwise that would be pertinent to the reported exercise, what data were collected, and how they were collected.
3. There should be complete documentation of the sampling and analytical procedures and equipment calibration.
4. There should be discussion of the results and an interpretation of them. Reference should be made to any other pertinent studies.
5. Conclusions should be drawn based on the findings.
6. Recommendations should be identified. All proposed changes or actions should be listed and supported by the conclusions drawn.
7. Detailed data should be included as an addendum or appendix.
8. The report author(s) and principal participants in the project should be identified.

Do not rely on memory. Reports written from memory tend to be incomplete, and there may be a question as to their admissibility should they be needed as evidence in a court proceeding. Generally, unless the original notes from a survey or other project can be produced, the testimony may not be allowed.

Every inspection program must define how inspection results will be handled, filed, classified, and followed up. A plant or area inspection generally results in a determination of what should be done and is very seldom a complete result in and of itself. Usually further action, assistance, reinspection, or correction of some sort is called for. The results of the inspection must be forwarded to the people who have an interest—the supervisor of the area, especially if something should be done, the person in charge of the industrial hygiene program, the medical department, and so forth.

Because supervisors are key persons in safety and health programs, it is important to communicate to them the results, and, if possible, recommendations arising from any industrial hygiene project. For example, in an industrial hygiene survey it is determined that the ventilation hood at a grinding operation has a capture velocity that is less than the minimum recommended for good practice. The mere recording of the potential hazard represents communication. Chances are that the supervisor does not know that this is a problem and may need assistance in process change and design as well as budget support. Whoever inspects the operation should make every effort to communicate at the supervisory level as well as at the management level. Some mechanism should be set up so that problems can be addressed as soon as they are recognized.

There should be a system established for setting priorities for safety and health needs and corrections, and for ways to communicate these needs to management. Sometimes corrections require considerable expenditures of funds that are not immediately available. This situation can sometimes be dealt with by implementing interim programs such as the use of effective personal protective equipment.

One question an industrial hygienist may ask is: "How will the information I have gathered and reported stand up to scrutiny in a courtroom?" Whenever an industrial hygienist is involved in any project that involves sampling, interpreting, and report writing, he or she should proceed on the basis that whatever is being done is going to end up in trial. That way he or she can be prepared for most contingencies.

There are numerous rules that control the admissibility of records during a trial. These are generally found in the Federal Rules of Evidence. The details of this are beyond the scope of this discussion and suffice it to say that your attorney should be very familiar with how the records can be used.

11 CHAIN OF CUSTODY

Regardless of the type of case at issue, the term chain of custody basically has the same meaning. When one is dealing with scientific evidence, it is often necessary to show that the sample in question is the sample that was collected at a particular location at a particular time, was brought to the laboratory for analysis, and is the same sample that produced the result that is being introduced at trial. When a chain of custody is required, it is necessary to establish where the chain begins and ends. Between these points are the "links" in the chain of custody.

11.1 Links in the Chain

The links in the chain of custody are those persons who have had physical custody of the object. Failure to account for the sample(s) during possession by a custodian may constitute a break in the chain of custody. The critical point is that while a custodian in the chain of possession need not testify under all circumstances, the sample(s) should be accounted for while in that custodian's control. The most important part of this function is to set up standard operating procedures with sufficient documentation to be able to reproduce the steps involved from the beginning to the end.

11.2 Burden and Standard of Proof

The party offering the evidence has the burden of proving the chain of custody (34). Prior to having the Federal Rules of Evidence adopted, the courts described the standard of proof in a variety of ways. There were several phrases used to describe the standard: "reasonable certainty" (35), "reasonable assurance" (36), and the most common expression in which the offering party had to establish the identity and condition of the exhibit by a "reasonable probability" (37). The reasonable probability standard appears to require no more than the "preponderance of evidence" or "more probable than not" standard (38), and some courts have explicitly expressed the standard in those terms (39).

Contrast this with Federal Rule 901(a), which requires only that the offering party introduce "evidence sufficient to support a finding that the matter in question is what its proponent claims." Thus, the trial court does not decide finally or exclusively whether the item has been identified; rather, the court decides only whether sufficient evidence has been introduced from which a reasonable jury could find the evidence identified (40).

12 STANDARDIZED PROCEDURES

As documentation for possible future use in litigation, and for other purposes as well, every routine function or operation that an industrial hygienist performs should be described in detail in writing as standard operating procedures (SOPs). These procedures should describe the minimum acceptable requirements for the function or operation. For example, if a routine function is to evaluate performance of ventilation systems, SOPs are needed that describe what equipment is to be used, describe how it is to be used (including calibration), describe what information is needed (including forms for standardizing the information), describe in detail the entire procedure to be conducted, and define what reports are required.

Standard operating procedures must be in writing and show date or dates of approval and be signed by someone who has authority and responsibility for industrial hygiene activities. Periodic review is required to determine that the procedures are still adequate and that they reflect the way operations are performed.

13 SUMMARY

Even though the words "professional liability" may cause an industrial hygienist to feel uncomfortable, there is little need to be concerned if he or she conducts his or her affairs as a true professional. Industrial hygienists should use procedures and test methods generally recognized in the profession as reliable and accurate. As a general rule, if industrial hygienists live by the Code of Ethics for the Practice of Industrial Hygiene (41), everything will take care of itself with regard to professional liability.

ACKNOWLEDGMENTS

This Chapter relies upon Chapter 19, "Professional Liability and Litigation," by Robert B. Weidner, J. D., which appears in *Patty's Industrial Hygiene and Toxicology*, Vol. IIIA, 3rd ed., 1994.

BIBLIOGRAPHY

1. *Salem v. United States Lines Co.*, 370 U.S. 31, 35 (1962).
2. See, generally, 3 D Louisell & Co. Mueller, *Federal Evidence* § 382 (1979); 3 J. Weinstrin & Berger, *Weinsteins' Evidence*, 702[01] (1982).
3. Advisory Committee's Note, Fed. R. Evid. 702.
4. For example, *Fineberg v. United States*, 393 F.2d 417, 421 (9th Cir. 1968) ("beyond the knowledge of the average layman"); *Jenkins v. United States*, 307 F.2d 637, 643 (D.C. Cir. 1962) ("beyond the ken of the average layman").
5. *State v. Chapple*, 135 Ariz. 281, 660 P.2d 1208, 1219–20 (1963) [quoting Ladd, "Expert Testimony," 5 Vand. L. Rev. 414, 418 (1952)].
6. J. Wigmore, *Evidence* § 1923. at 29 (Chadboum rev. 1978). See also Ladd, supra note 9, at 418: "There is no more certain test for determining when experts may be used than the common sense inquiry whether the untrained laymen would be qualified to determine intelligently and to the best possible degree the particular issue without enlightenment from those having a specialized understanding of the subject involved in the dispute."
7. C. McCormick, (1954). *Evidence*, West Publishing Co., St. Paul, MN, 1954, pp. 363–364.
8. 293 F. 1013 (D.C. Cir. 1923).
9. 293 F. at 1014.
10. 61 *Law Week* 4805.
11. 460 US 45 (1984).
12. 483 US 171 (1987).
13. 488 US 169 (1988).
14. 61 *Law Week* 4808.
15. 61 *Law Week* 4808.
16. 61 *Law Week* 4808.
17. 61 *Law Week* 4810.

18. 61 *Law Week* 4810.
19. 61 *Law Week* 4808.
20. *Kumho Tire Company, LTD., et al, Petitioners v. Patrick Carmichael, etc., et al.* 131 F.3 1433, March 23, 1999.
21. *Motorola, Inc. v. Ward*, 223 Ga. App. 678, 478 S.E. 2d 465 (1996).
22. *Orkin Exterminating Co. v. McIntosh*, 215 Ga. App. 587, 592 S.E.2d 159, 165 (1994).
23. *People v. Kelly*, 17 Cal. 3d 24, 130 Cal Rptr. 144, 549 P.2d 1240; *Frye v. United States*, 293 F. 1013 (D.C. Cir. 1923)
24. Rule 25(b)(3) of the *Federal Rules of Civil Procedure* reads as follows:
    Trial Preparation: Materials. Subject to the provisions of subdivision (b)(4) of this rule, a party may obtain discovery of documents and tangible things otherwise discoverable under subdivision (b)(1) of this rule and prepared in anticipation of litigation or for trial by or for another party or by or for that other party's representative (including his attorney, consultant, surety, indemnitor, insurer, or agent) only upon a showing that the party seeking discovery has substantial need of the materials in the preparation of his case and that he is unable without undue hardship to obtain the substantial equivalent of the materials by other means. In ordering the discovery of such materials when the required showing has been made, the court shall protect against disclosure of the mental impressions, conclusions, opinions, or legal theories of an attorney or other representative of a party concerning the litigation.
25. 622 F.2d at 935.
26. *United States v. Nobles*, 422 U.S. 225, 95 S. Ct. 2160; 45 L. Ed.2d 141 (1975) (work product doctrine "necessarily" applies to materials prepared for an attorney).
27. 422 U.S. 225 (1975). See generally Feldman, "Work Product in Criminal Practice and Procedure," 50 U. Cin. L. Rev. 495 (1981).
28. Id. See also *People v. Collie*, 30 Cal.3d 43, 59, 634 P.2d 534, 543, 177 Cal. Rptr. 458, 467 (1981) (work product doctrine applies to criminal cases and protects the work product of defense investigators).
29. 422 U.S. at 239.
30. *Upjohn Co. v. United States*, 449 U.S. 383, 389 (1981), See also *Fisher v. United States*, 425 U.S. 391, 403 (1976) ("The purpose of the privilege is to encourage clients to make full disclosure to their attorneys").
31. See *United States v. Alvarez*, 519 F.2d 1036, 1046–47 (3d Cir. 1975); *Pouncy v. State*, 353 So.2d 640, 642 (Fla. App. 1977); *State v. Mingo*, 77 N.J. 576, 584, 392 A.2d 590, 595 (1975); see also *United States v. Nobles*, 422 U.S. 225, 239 (1975) ("Respondent, by electing to present the investigator as a witness, waived the [work product] privilege with respect to matters covered in his testimony"); *Miller v. District Court*, 737 P.2d 834 (Colo. 1987) (psychiatrist).
32. Friedenthal, "Discovery and Use of an Adverse Party's Expert Information," 14 Stan. L. Rev. 455, 464–65 (1962).
33. See *United States v. Alvarez*, 519 F.2d 1036, 1046–47 (3d Cir. 1975) (psychiatrist); *United States v. Kovel*, 296 F.2d 918, 921–22 (2d Cir. 1961) (accountant); *United States v. Layaton*, 90 F.R.D. 520, 525 (N.D. Cal. 1981); *Baily v. Meister Brau, Inc.*, 57 F.R.D. 11, 13 N.D. (Ill. 1972) (financial expert); *Houston v. State*, 602 P.2d 784, 791 (Alaska 1979) (psychiatrist); *People v. Lines*, 13 CAl. 3d 500, 614–15, 531 P.2d 793, 802–03, 119 Cal. Rptr. 225, 234–235 (1975); *Pouncy v. State*, 353 So.2d 640, 642 (Fla. App. 1977); *People v. Knippenberg*, 66 Il.2d 276, 283–84, 362 N.E.2d 681, 684 (1977) (investigator); *State v. Pratt*, 284 Md. 516, 520–22, 396 A.2d 421, 423–24 (1979) (psychiatrist); *People v. Hilliker*, 29 Mich. App. 543, 546–47, 185 N.W.2d 831, 833–

34 (1971); *State v. Kociolek*, 23 N.J. 400, 129 A.2d 417 (1957); *State v. Hitopoulus*, 297 S.C. 549, 309 S.E.2d 747 (1983).
34. See *United States v. Santiago*, 534 F.2d 768. 770 (7th Cir. 1976); I. Wigmore, *Evidence* § 18, at 841 (Tillers rev. 1983).
35. See *United States v. Jones*, 404 F. Supp. 529, 543 (E.D. Pa. 1975); *Sorce v. State*, 88 Nev. 350, 352–53, 497 P.2d 902, 903 (1972); *State v. Tillman*, 208 Kan. 954, 958–59, 494 P.2d 1178, 1182 (1972).
36. See *State v. Cress*, 344 A.2d 57, 61 (Me. 1975); *State v. Baines*, 394 S.W.2d 312, 316 (Mo. 1965), *cert. denied*, 384 U.S. 992 (1966); *People v. Julian*, 41 N.Y.2d 340, 344, 360 N.E.2d 1310, 1313, 392 N.Y.S.2d 610, 613 (1977).
37. For example, *United States v. Brown*, 482 F.2d 1226, 1228 (8th Cir. 1973) ("reasonable probability the article has not been changed in any important respect"); *United States v. Robinson*, 447 F.2d 1215, 1220 (D.C. Cir. 1971), rev'd on other grounds, 414 U.S. 218 (1973); *United States v. Capocci*, 433 F.2d 155, 157 (1st Cir. 1970); *Gass v. United States*, 416 F.2d 767, 770 (D.C. Cir. 1969); *West v. United States*, 359 F.2d 50, 55 (8th Cir.), *cert. denied*, 385 U.S. 867 (1966); *Gallego v. United States*, 276 F.2d 914, 917 (9th Cir. 1960); *United States v. S.B. Penick & Co.*, 136 F.2d 413, 415 (2d Cir. 1943); *State v. Johnson*, 162 Conn. 215, 232, 292 A.2d 903, 911–912 (1972); *Doye v. State*, 16 Md. App. 511, 519, 299 A.2d 117, 121 (1973).
38. See *People v. Riser*, 47 Cal.2d 566, 580–81, 305 P.2d 1, 10 ("The requirement of reasonable certainty is not met when some vital link in the chain of possession is not accounted for, because then it is as likely as not that the evidence analyzed was not the evidence originally received"), *appeal dismissed*, 358 U.S. 646 (1959); *State v. Serl*, 269 N.W.2d 785, 788–89 (S.D. 1978).
39. See *State v. Henderson*, 337 So.2d 204, 206 (La. 1976); *State v. Sears*, 298 So.2d 814, 821 (La. 1974); *State v. Williams*, 273 So.2d 280, 281 (La. 1973) ("clear preponderance").
40. See *Zenith Radio Corp. v. Matsusshita Elect. Indus. Co.*, 505 F. Supp. 1190, 1219 (E.D. Pa. 1980) ["The Advisory Committee Note to Rule 104(b) makes plain that preliminary questions of conditional relevancy are not determined solely by the judge, for to do so would greatly restrict the function of the jury."], rev'd on other grounds, 723 F.2d 238 (3d Cir. 1983).
41. The *Code of Ethics for the Practice of Industrial Hygiene* (appears in annual membership rosters of the Academy of Industrial Hygiene, American Conference of Governmental Industrial Hygienists and the American Industrial Hygiene Association).

CHAPTER THIRTY-EIGHT

Odor: A Legal Overview

Ralph E. Allan, CIH, JD

1 INTRODUCTION

The practice of industrial hygiene has changed tremendously since the passage of the Occupational Safety and Health Act of 1970. With the issues of indoor air pollution generally and tight building syndrome specifically becoming a part of an industrial hygienist's daily involvement, odors are and will continue to be more and more an issue of contention. Is odor an index of hazard potential? Can it be such an index? How should odors be handled in reference to industrial hygiene activity? What are the legal issues considering an odor present without a quantitative determination of hazard potential? Currently we do not have all the answers available to answer fully all questions posed. However, history sometimes reveals indications of the future.

2 HISTORICAL REVIEW

Since ancient times, it has been indicated that pleasing odors preserve health and unpleasant odors are injurious. In the early days this thinking provided the basis for the application of aromatic eau de cologne and of pomanders stuffed with balsam for health preservation. Alternatively, diseases have been attributed to atmospheric "miasmas." In fact, the word malaria is related to the Italian translation of "bad air," that is, malaria. Because of this thinking, prior to the nineteenth century people avoided crowded places because it was generally believed that crowded locations, with the accompanying odors, were responsible for the breeding of disease, and the odors specifically were responsible for the spread of infection. Even as new information developed into the twentieth century, it was difficult for people to accept the premise that contagion is primarily a fingerborne phenomenon and not airborne. In fact, it was a convenient notion to associate foul smelling air with unhealthy situations. What better way to provide a basis for freshening the air! Church activities that included the burning of incense supported the idea that a pleasant odor represents healthy air. This concept was also supported by pre-nineteenth century medical practice. As the twentieth century approached, scientists' experiments tended to conclude that there was not consistent hazard associated with occupancy odor (1).

In 1923, a New York State Commission focused on ventilation considerations, including such functions and indexes as comfort, body temperature, intellectual performance, motivation, respiration, metabolism, condition of the nasal mucosa, frequency of colds, blood pressure, hematocrit, appetite, and rate of physical work. There was no cause for medical concern found under normal conditions of occupancy. The study determined that control of occupancy odor had to stand on the basis of comfort rather than ill health (2).

As understanding of odors progressed, and more and more substances became part of everyday living through the industrial revolution and into the twentieth century, it became important further to organize scientific thinking concerning odors and categorize them according to known properties.

One such classification follows (3):

- Odorous substances that have been well established as toxic to humans.
- Odorous substances that have produced well-defined pathological changes in animals that have not been identified in humans.
- Odorous substances that have not been identified as toxic to humans, but that evoke violent and alarming physical symptoms in a substantial fraction of an exposed population whenever odor intensity is high and exposure more than fleeting.
- Odorous substances that have not been identified as toxic to humans, but that are capable of evoking violent and alarming physical symptoms in a small number of people even when exposure is moderate and fleeting.
- Odorous substances that have not been identified as toxic to humans (or are present at concentrations substantially below a well-established toxic threshold), but that produce more than passing vexation by the continuing or frequent presence of their unpleasant odor.
- Odorous substances that have no known toxic properties and are universally recognized as pleasant or neutral, but that produce vexation in a substantial fraction of an exposed population because of unusual intensity or persistence.
- Odorous substances of no known toxicity that are sensed to the point of conscious recognition, but that evoke only pleasant or indifferent sensations.

The above classification is indeed important to nurture and develop especially considering the need for unambiguous classification, identification and definition of responses to odors, and their effect on human populations. A clear and unequivocal as possible assessment is important in order to minimize litigation and provide guidance for the development of laws, regulations, and standards. It therefore becomes desirable to attempt to deal with the effect of odorous substances on humans by means of a series of precisely defined categories, rather than with a continuum of effects that stretches from highly pleasing (however, what may be highly pleasing to one may be antagonistic to another!) to violently repugnant (perhaps to all?) and then ultimately merges into the clearly toxic category. The sole unifying factor through the continuum remains the ability of each substance to stimulate or act on the olfactory system in some fashion.

3 HUMAN RESPONSE

Odors affect the well-being of an individual by precipitating unpleasant feelings, by initiating harmful responses and other physiological interactions, and by adjusting olfactory activity. Some objective responses to unpleasant odors include vomiting, nausea, headache, shallow breathing and cough, sleeplessness, stomach problems, interference with appetite, and eye-, nose-, and throat irritation; emotional upset resulting in lack of sense of well-being and interference with the enjoyment of home, food, and general environment; and irritability and depression. Physiological effects of unpleasant odor include decreased heart rate, constriction of the blood vessels of the skin and muscles, release of epinephrine, and changes in the size and condition of the cells in the olfactory bulbs of the brain. Science, however, has not provided us an understanding of the relationships between the symptoms and the intensity or duration of the exposure to the odor. Indication, therefore, of changes of olfactory function or sensitivity in populations exposed under controlled conditions are not available to provide a clear insight into the phenomenon of attributable incidence. There is no doubt that certain physiological phenomena are related; that is, respiratory and cardiovascular responses are elicited by stimulation of receptors in the nasal mucosa. These effects have been documented in various animal species and include sneezing, bronchodilation, decrease in breathing rate, decrease in heart rate, increase in arterial blood pressure, decrease in cardiac output, and vasoconstriction in various parts of the body. All information, regarding adverse reactions occurring in humans as a result of environmental odors, however, has come from complaints and surveys, which are difficult to verify and evaluate scientifically (3).

4 LEGAL ASPECTS

The law, meanwhile, considering this continuum, does indeed have difficulty in dealing with odor problems, somewhat as it does with the definition of noise as unwanted sound: what is noise to some is music to others' ears; similarly what is odiferous to one (or even toxic to one) is not to another. The cattleman's "pleasant" animal odor is foul smelling to the uninterested, or perhaps to those who are unacclimatized to the economic aspects of the cattle industry. Differently from noise, however, where there is a definite measurement of hearing loss accruing from high-intensity sound (whether or not it is classified as noise), there are no easy identifying biological markers for odor intensity unless it tends to trip into a frankly toxic effect. Odors therefore do indeed present a difficult issue to handle from the legal viewpoint because of differences of perspective, insufficiency of data, or ambiguities in data. Despite the difficulty, there is no doubt that malodor should be regulated by one means or another, as a matter of public policy. There are many cases in the legal archives relating to odor. In California, the South Coast Air Quality Management District (SCAQMD) has indicated that objectionable odors are the No. 1 source of complaints made to the regulatory agency. In 1997, there were 5,983 complaints or 63% of all air quality complaints received by the SCAQMD were due to odors (4). Whether it be a case concerning the odors emanating from a rendering plant (5) or a chicken processing plant (6), the issue of public or private nuisance becomes involved. Cases that arise under private or public nuisance have developed through a long line of judicial precedents and are codified in some situations by local ordinances. Although more scientific approaches to odor control are becoming more involved, nuisance law is the oldest and strongest source of law for controlling odors in our society.

4.1 Nuisance

Nuisance law is divided into public nuisance and private nuisance. A public nuisance is created when an act invades a right common to all members of the public. *Black's Law Dictionary* (7) defines nuisance as follows:

> That which annoys and disturbs one in possession of his property, rendering its ordinary use or occupation physically uncomfortable to him. *Yaffe v. City of Ft. Smith*, 178 Ark, 406, 10 S.W.2d 886, 890, 61 A.L.R. 1138. Everything that endangers life or health, gives offense to senses, violates the laws of decency, or obstructs reasonable and comfortable use of property. *Hall v. Putney*, 291 Ill. App. 508, 10 N.E. 2d 204, 207. Annoyance; anything which essentially interferes with enjoyment of life or property. *Holton v. Northwestern Oil Co.*, 201 N.C. 744, 161 S.E. 391, 393. That class of wrongs that arise from the unreasonable, unwarrantable, or unlawful use by a person of his own property, either real or personal, or from his own improper, indecent, or unlawful personal conduct, working an obstruction of or injury to the right of another or of the public and producing such material annoyance, inconvenience, discomfort, or hurt, that the law will presume resulting damage. *City of Phoenix v. Johnson*, 51 Ariz. 115, 75 P.2d 30; Wood, Nuis t 1; *District of Columbia v. Totten*, 55 App. D.C. 312, 5F.2d 374, 380, 40 A.L.R. 1461. Anything that unlawfully worketh hurt, inconvenience, or damage. 3 Bl. Comm. 216; *City of Birmingham v. Hood-McPherson Realty Co.*, 233 Ala. 352, 172 So. 114, 120, 108 A.L.R. 1140. Anything which is injurious to health, or is indecent or offensive to the senses, or an obstruction to the free use of property, so as to interfere with the conferrable enjoyment of life or property, or which unlawfully obstructs the free passage or use, in the customary manner, of any navigable lake or river, bay, stream, canal, or basin, or any public park, square, street, or highway, is a nuisance. Civ. Code Cal. t 3479; *Veazie v. Dwinel*, 50 Me. 479; *Bohan v. Port Jervis Gaslight Co.*, 122 N.Y. 18, 25 N.E. 246, 9 L.R.A. 711; *Baltimore & P.R. Co. v. Fifth Baptist Church*, 137 U.S. 568 11 S.Ct. 185, 34 L.Ed. 784; Ex parte Foote, 70 Ark. 12, 65 S.W. 706, 91 Am. St. Rep. 63.
>
> In determining what constitutes a "nuisance," the question is whether the nuisance will or does produce such a condition of things as in the judgment of reasonable men is naturally productive of actual physical discomfort to persons or ordinary sensibility and ordinary tastes and habits. *Meeks v. Wood* 66 Ind. App. 594, 118 N.E. 591, 592.
>
> Nuisances are commonly classed as public and private, and mixed. A public nuisance is one which affects an indefinite number of persons, or all the residents of a particular locality, or

all people coming within the extent of its range or operation, although the extent of the annoyance or damage inflicted upon individuals may be unequal. *Burnham v. Hotchkiss*, 14 Conn. 317; *Chesbrough v. Com'rs*, 37 Ohio St. 508; *Lansing v. Smith*, 4 Wend., N.Y., 30, 21 Am. Dec. 89. A private nuisance was originally defined as anything done to the hurt or annoyance of the lands, tenements, or hereditaments of another. 3 Bl. Comm. 216; *Whittenmore v. Baxter Laundry Co.*, 181 Mich. 564, 148 N.W. 437, 52 L.R.A., N.S., 930, Ann. Cas. 1916C, 818. As distinguished from public nuisance, it includes any wrongful act which destroys or deteriorates the property of an individual or of a few persons or interferes with their lawful use or enjoyment thereof, or any act which unlawfully hinders them in the enjoyment of a common or public right and causes them a special injury different from that sustained by the general public. Therefore, although the ground of distinction between public and private nuisances is still the injury to the community at large or, on the other hand to a single individual, is evident that the same thing or act may constitute a public nuisance and at the same time a private nuisance. *Heeg v. Licht*, 80 N.Y. 582, 36 Am. Rep. 654; *Baltzeger v. Carolina Midland R. Co.*, 54 S.C. 242, 32 S.E. 358, 71 Am. St. Rep. 789; *Wilcox v. Hines*, 100 Tenn. 538, 46 S.W. 297, 41 L.R.A. 278; *Harris v. Poulton*, 99 W. Va. 20, 127 S.E. 647, 650, 651, 40 A.L.R. 334. A mixed nuisance is of the kind last described; that is, it is one which is both public and private in its effects-public because it injures many persons or all the community, and private in that it also produces special injuries to private rights. *Kelley v. New York*, 27 N.Y.S. 164, 6 Misc. 516.

A private nuisance, therefore, involves an invasion of a private party's interest in the use and enjoyment of his property.

To result in fulfilling the elements of a private nuisance cause of action, odors complained of must be judged a substantial annoyance based upon standards established by the ordinary reasonable person living in that specific locality. If the odor were located in an industrial community, a highly sensitive person may find it impossible to establish a cause of action on the basis of odor pollution because (*1*) the odor is a common characteristic of the industrial community and (*2*) the odor is considered harmless by most residents of the area. Contrarily, if a foundry were located within a residential community where the average homeowner is not used to the smell of amines and other decomposition products from the pouring process, it would be extremely difficult to defend against a complaint based upon private nuisance. It must, however, be shown by the plaintiff that the odors produced were unreasonable—the liability is not automatic. The plaintiff will have to show that the harm to him is greater than he should be required to bear without compensation. Elements of consideration by the trier of fact can include the economic viability consideration of the community. If the suit is successful at all, the usual remedy is an award of damages, rather than an injunction that would force the defendant to abate the odor.

Other legal aspects in addition to the difficulty of obtaining injunctions also limit the role of private litigation as a technique for regulating odorants. If one literally moves into the vicinity of an odor-emitting source, a defense for the defendant can arise. Called appropriately "coming to the nuisance," this defense can prevent recovery by a plaintiff. In *McLung v. Louisville* (8) plaintiffs purchased land near an old railroad and were denied injunction after the railroad recommenced active operation. Sometimes delay in presenting legal rights can cause a detrimental change in the plaintiff's position and ultimately bar the nuisance action altogether. Further, from the legal standpoint, the private nuisance is tied to interests in land; therefore an action in nuisance cannot be maintained by an employee or by any person who has no property right in the affected land. Finally, because private litigation is a costly and uncertain route, plaintiffs only rarely have the resources available to pursue remedies available to them under the law of nuisance.

Despite any regulation or standard implementation, whether based upon federal legislation or regulation and subsequent compliance, these basic nuisance actions would not be explicitly preempted and would remain valid avenues for seeking control of undesirable odors (1).

4.2 Air Pollution Regulations

As presented earlier, whether or not odor presents a public health hazard is a difficult question to answer in many situations. Thus federal government intervention concerning odor control occurs only indirectly, as air pollution regulation of basic air pollutants provides control requirements for gas, vapor, and particulate emissions. There is, therefore, no specific nationwide effort directed at controlling odors per se.

Many states and local authorities, however, have implemented odor control regulations as part of a total program for air quality maintenance. In attempts to answer the question what constitutes an "acceptable" level of odor, local jurisdictions have used the basic public nuisance criteria that include community consensus as an important factor in defining acceptable limits for odor perception. As indicated earlier, pleasantness or unpleasantness of odor can be determined by one's interest in the source of emission; that is, a farmer finds odors of the farm at least unobjectionable (if not pleasant) because the odors reflect and relate to a vital personal economic interest. On the other hand, a nonfarmer, commuter resident living near the farm may indeed associate the odors with unpleasantness. An example of a local air pollution regulation relating to odor control is as follows:

> No person shall discharge ... one or more air contaminants (including odors) or combinations thereof in such concentrations and of such duration as are ... injurious to ... human health or welfare, animal life, vegetation or property ... (1).

The violation of the regulation is proved primarily on the basis of testimony from affected residents of the community. This type of regulation essentially codifies the traditional public nuisance cause of action. More specific regulatory direction has been developed in some jurisdictions for designated industries. The regulations are directed toward reducing highly odorous emissions from sources such as rendering plants by requiring incineration control methods. In addition, regulations, such as Rule 1179 (9) of the California South Coast Air Quality Management District relating to Publicly Owned Treatment Works Operations (POTWs), require the operator of the POTW to submit and implement an Emissions Inventory Plan which quantifies odor related emissions, sources of odorous emissions, odor-related citizen complaints and requires recommendations for abatement or elimination of the odorous emissions.

Another approach at establishing regulatory control is based on sensory evaluation of odors in the general environment. Using a dilution system by mixing portions of the subject-contaminated air with portions of clean air, a violation is determined by assessing whether or not an odor exists after continued dilution of the odorous air with the clean air. There are, however, obvious drawbacks to this direction of regulatory control including the following possible criticisms: a questionable link between dilution factors and community annoyance, the problem associated with obtaining ambient air samples, and questions concerning the reliability of the odor panel(s) involved in the assessment.

Effort by First (10) and by Copley International Corporation (11) resulted in presentation of model ordinances with accentuation on a recognition that a number of variables relate to the acceptability of an odor condition.

First's proposal considers factors such as odor intensity and quality, duration, and frequency of the odor and includes the time and the day of the week as well as the wind direction. By appropriately considering the weighting of these factors, an "odor perception index" is used to assess the magnitude of the annoyance condition and thereby determine whether or not a nuisance requiring control is present.

Problems are associated with this approach; however, in that the numerical values assigned can be open to challenge based on arbitrary and capricious establishment of the index system. In addition, given the subjective nature of the establishment of the standard, a procedure is necessary in order to maintain the rights of offenders to due process of law, yet at the same time provide the moving party (agency) the opportunity to present a violation even though the specific index may not have been exceeded.

The Copley International Corporation system is based on a public attitude survey that includes the elements of establishment of the presence of a community odor problem and prescribes odor control requirements as appropriate. A subsequent public attitude survey assesses the sufficiency of abatement. The legal system could have difficulty dealing with the Copley approach because the surveys may be considered hearsay unless additional information supports the public attitude survey as evidence is presented to the court.

4.3 Audit

Schroeder, in "Industrial Odor Technology Assessment" (12), has developed a checklist for evaluation of an air pollution odor situation that can provide an insight into the status of legal significance for interested parties. The checklist follows:

1. What is the nature of the problem?
2. Who in the company has authority to control it?
3. Is the situation covered by any law or ordinance covering odors or air pollution, such as the Occupational Safety and Health Act or general nuisance law? Exactly what do pertinent provisions require?
4. Does the problem involve an accident, periodic occurrence, or constant emission? If a violation has been charged, is it substantiated or capable of being substantiated? Could there have been mistakes in measurement, procedures, interpretation, or conclusions?

5. Background history:
   a. Have there been previous complaints, by private persons or a public agency, by workers, by the community? (Has the scope of the community changed since then?)
   b. Company responses.
   c. Resolution of problem.
6. Present company measures or equipment to curb odors, if any. Can this be documented?
7. What new measures or equipment would be needed to prevent such an episode or abate the odor?
   a. Cost.
   b. Lead time to get equipment.
8. If a complaint has been filed, how bad is the effect on complainant or complainants?
   a. Mere annoyance.
   b. Health effects.
   c. So severe as to support a preliminary injunction to cease operations pending trial, or compliance with statute or ordinance.
9. Consequences of shutdown cost in
   a. Jobs.
   b. Production.
   c. Start-up costs.
   d. Inventory spoilage.
   e. Financial stability of company.
10. What court or government agency has enforcement jurisdiction?
11. What is the purpose of the action?
    a. Close down the plant.
    b. Have company control the odor.
    c. Money damages for actual injuries, punitive damages, and exemplary damages?
12. What are the steps of the legal procedure?
    a. Who has the burden of proof?
    b. Avenues of appeal.
    c. Possibilities of delaying or expediting procedure.
13. Estimate of costs in alternative courses of action: paying damages, settlement, legal fees, control equipment costs, production interruption.
14. Is there any relationship between this action and other court cases or enforcement actions in this industry or political subdivision that would affect the outcome?
    a. Court decisions on what constitutes a nuisance.
    b. Decisions on availability of abatement technology.
    c. Admissions on economically feasible control measures contained in content decrees.

15. Is legislation pending that will require compliance anyway? Will defiance only assure passage of the legislation? Or will good faith efforts to control the odor give legislators an example of what is economically and technically feasible and thereby possibly help achieve reasonable legislation?

5 SUMMARY

Based upon the historical and scientific basis for odor evaluation and control, the legal system continues to struggle with appropriate legal basis for objective assessment of alleged odor problems. It is the nature of the issue, as with most toxic substance issues, that an objective, convenient judicial basis for noncontroversial problem resolution is difficult to achieve. Resolution of odor problems through the science-law mechanism will therefore be slow to develop into efficient, rapid processes; however, continued progress from the scientific standpoint will lead the way to a more objective legal basis of resolution in this most difficult area of community interest.

BIBLIOGRAPHY

1. National Research Council-Committee on Odors from Stationary and Mobile Sources, National Academy of Sciences, Washington, DC, 1979.
2. Committee on Indoor Pollutants, National Research Council, National Academy Press, Washington, DC, 1981.
3. Committee on Odors from Stationary and Mobile Sources, Odors from Stationary and Mobile Sources, National Academy of Sciences, Washington, DC, 1979.
4. (South Coast) AQMD News. (California) March 31, 1998. http://www.aqmd.gov/news1/Archives/odors.html
5. *Cox v. Schlachter*, 147 Ind. App. 530, 262 N.E. 2d 550, 1 ERC 1681 (1970).
6. *Ozark Poultry v. Garman*, 251 Ark. 389, 472 S.W.2d 714, 3 ERC 1545 (1971).
7. H. C. Black, *Black's Law Dictionary*, West Publishing Company, St. Paul, MN, 1968.
8. *McLung v. Louisville and N.R. Co.*, 255 Ala. 302, 51 Sr.2d 371 (1951).
9. South Coast Air Quality Management District (California), Regulation 11 (Source Specific Standards), Rule 1179, *Publicly Owned Treatment Works Operations*, Amended March 6, 1992. (http://www.aqmd.gov/rules/html/r1179.html)
10. M. W. First, "A Model Odor Control Ordinance" in H. M. Englund and W. T. Perry, eds. *2d International Clean Air Congress Proceedings*, Academic Press, New York, 1971, pp. 1255–1259.
11. Copley International Corporation, "A Study of the Social and Economic Impact of Odors. Phase III. Development and Evaluation of a Model Odor Control Ordinance," A report to U.S.E.P.A. EPA Publication #650/5-73-001. Contract #68-02-0095. Washington, DC, Feb. 1973.
12. P. N. Cheremisinoff, P. E. and R. A. Young, *Industrial Odor Technology Assessment*, Ann Arbor Science, Ann Arbor, MI, 1975.

CHAPTER THIRTY-NINE

Hazard Communication and Worker Right-To-Know Programs

Lisa K. Simkins, CIH, PE and Charlotte A. Rice, CIH

1 INTRODUCTION

More than half a million chemical products are in use today with hundreds of new chemicals introduced every year. Many of these chemicals find significant commercial application, exposing an estimated 32 million U.S. workers to one or more potentially hazardous chemicals. Although not all of these chemicals pose a serious threat, some can cause harm to human health and the environment, if handled improperly.

Much of the original concern about chemical safety was with the flammable and explosive nature of chemicals and not with their toxicity and carcinogenicity. Since the early 1970s, there has been increased emphasis on health and environmental issues and support for the right of employees to know the nature of the hazards associated with chemicals they encounter on the job.

The Occupational Safety and Health Administration (OSHA) believed that providing information to workers on chemical hazards was a critical factor in protecting workers. This was reflected in the following quotation from the Occupational Safety and Health Act of 1970: (1)

> Any standard promulgated under this subsection shall prescribe the use of labels or other appropriate forms of warning as are necessary to insure that employees are apprised of all hazards to which they are exposed, relevant symptoms and appropriate emergency treatment, and proper conditions and precautions of safe use or exposure (2).

Patty's Industrial Hygiene, Fifth Edition, Volume 3. Edited by Robert L. Harris.
ISBN 0-471-29753-4   © 2000 John Wiley & Sons, Inc.

OSHA met these requirements in the standards promulgated for specific chemicals. However, since only a handful of chemicals was regulated by specific standards, OSHA identified the need for a "generic" chemical information standard, and proposed the Hazard Communication Standard (HCS) in March of 1982 (3).

Chemical manufacturers, labor unions, health and safety professionals, and various governmental agencies supported the idea of a national, generic chemical information and labeling standard. This support arose in part because of the proliferation of state and community right-to-know laws that were creating a confusing patchwork of overlapping and, in some cases, contradictory regulations. However, despite this broad support, the rule underwent numerous court battles and revisions after its introduction. With approximately 25 changes, modifications, amendments and clarifications to the HCS, it is no wonder that there is confusion regarding the requirements and application of the standard. This confusion may, in part, explain why the HCS is the most frequently cited OSHA standard in recent years.

The HCS, as originally proposed, applied only to manufacturing business sectors. In 1987 it was extended to virtually all employers. However, OSHA was prevented from enforcing the rule in construction, and prevented from enforcing requirements dealing with MSDSs on multi-employer worksites, coverage of consumer products, and drugs in the nonmanufacturing sectors. These issues were subsequently resolved. A hazard communication standard is now part of the OSHA Longshoring, Shipyard Employment, Construction, and Agriculture standards (4). Clarifications and modifications to the language of the HCS were made in the final rule that was published in the *Federal Register* in 1994 (5). These clarifications and modifications have been summarized in Section 8.1 of this chapter. The HCS now covers approximately 32 million workers, in approximately 3.5 million locations (6).

The HCS preempts state right-to-know regulations, except in states with approved OSHA plans. This ensures similar minimum requirements in all states. Some local regulations that exceed, but do not conflict with the HCS, are still applicable. Employers in a state or municipality with "right-to-know" regulations should contact local authorities for interpretation on how federal, state, and local regulations apply.

Since its introduction, the basic provisions have not changed. A written program document must explain how compliance is achieved. Material Safety Data Sheets (MSDSs) must be developed by chemical manufacturers and maintained by users for all hazardous chemicals. A list of chemicals must be maintained at the worksite. Employees must receive information and training regarding the hazards of the chemicals. These may seem like simple requirements. However, changes in chemicals, materials, processes, hazard determinations, and/or employees can make continuous compliance challenging.

To add to the challenge, the HCS is a performance standard. The requirements are specific, but the implementation methods are at the discretion of the employer. This allows a great deal of flexibility. However, the yardstick typically used to measure the effectiveness of a hazard communication program is whether employees recognize the hazards posed by chemicals and understand how to protect themselves. It is not always easy to determine if this is true.

The relationship between the HCS and other OSHA standards can be complex. The standards covering specific chemicals, for example, have requirements for labeling, train-

HAZARD COMMUNICATION AND WORKER RIGHT-TO-KNOW PROGRAMS

ing, information, and controls. In general, the requirements of these more specific standards take precedence over the generic approach of the HCS. However, it is wise to study the standards carefully to determine which requirements take precedence.

The tragic release of methyl isocyanate in Bhopal, India in 1984 resulted in public outrage and concern that the incident could be repeated with an equally devastating outcome. Citizens throughout the United States began to question whether safety precautions and emergency plans would be effective in preventing a similar incident in their own communities. This concern fueled demands that information about chemicals used in industry be provided to communities.

Three years after OSHA proposed the HCS the U. S. Congress passed the Emergency Planning and Community Right-to-Know Act of 1986 (EPCRA). This law, also known as SARA (Superfund Amendments and Reauthorization Act) Title III attempted to address community concern regarding hazardous materials. It requires the inventory and reporting of hazardous materials for the purpose of providing information to the community and planning for emergencies. Although this law will not be covered in any depth in this chapter, the relationship between it and the HCS will be discussed briefly. A brief discussion, however, does not imply that it is less important than HCS. A thorough understanding of EPRCA is critical.

In 1995, the Hazard Communication Workgroup to the National Advisory Committee on Occupational Safety and Health (NACOSH) convened to identify ways to improve chemical hazard communication and the "right to know" in the workplace. Its report, transmitted to OSHA and NIOSH in 1996, contained recommendations related to MSDSs, electronic management of MSDSs, labeling, training, enforcement of the HCS, harmonization of hazard communication requirements and misinformation about the HCS (7). Some of these recommendations are presented within relevant sections of this chapter.

The purpose of this chapter is to present the requirements of the HCS and to provide some tools and strategies for the implementation of a program. In Section 2 the essential elements of a hazard communication program are reviewed and explained. Training and labeling are discussed in Sections 3 and 4. The informational systems that may be used are described in Section 5. Program management strategies are discussed in Section 6. Related regulations, interpretation, and enforcement are covered briefly in Sections 7 and 8. Section 9 contains recommendations for program auditing.

2 HAZARD COMMUNICATION REQUIREMENTS

This section is designed primarily for users, rather than manufacturers of chemicals. While it contains a discussion of the requirements for conducting hazard determinations and distributing MSDSs to down stream users, it does not present detailed discussion of MSDS development.

2.1 Application of Hazard Communication

The HCS applies to chemical manufacturers, importers, and distributors of hazardous chemicals, as well as all employers who use hazardous chemicals. A "hazardous chemical" is broadly defined in the HCS as any chemical that is a physical hazard or a health hazard.

There are some substances to which the requirements of the standard do not apply, including:

- Any hazardous waste subject to the regulations issued under the Resource Conservation and Recovery Act (RCRA).
- Any hazardous substance subject to regulations issued under the Comprehensive Environmental Response, Compensation, and Liability Act (CERCLA).
- Tobacco or tobacco products.
- Wood or wood products, including lumber which will not be processed, where the manufacturer or importer can establish that the only hazard they pose to employees is the potential for flammability or combustibility.
- Articles (defined as "a manufactured item which (a) is formed to a specific shape or design during manufacture, (b) has an end-use function dependent in whole or in part upon its shape or design during end use, and (c) does not release more than very small quantities of a hazardous chemical and does not pose a physical hazard or health risk to employees").
- Food or alcoholic beverages which are sold, used or prepared in a retail establishment, and foods intended for personal consumption by employees while in the workplace.
- Any drug, as that term is defined by the Federal Food, Drug, and Cosmetic Act, when it is in solid, final form for direct administration to the patient; drugs which are packaged by the chemical manufacturer for sale to consumers in a retail establishment; and drugs intended for personal consumption by employees while in the workplace.
- Cosmetics which are packaged for sale to consumers in a retail establishment, and cosmetics intended for personal consumption by employees while in the workplace.
- Any consumer product or hazardous substance, as those terms are defined in the Consumer Product Safety Act and Federal Hazardous Substance Act respectively, where the employer can show that it is used in the workplace for the purpose intended by the chemical manufacturer or importer of the product, and the use results in a duration and frequency of exposure which is not greater than the range of exposures that could reasonably be experienced by consumers when used for the purpose intended.
- Nuisance particulates where the manufacturer or importer can establish that they do not pose any physical or health hazard.
- Ionizing and nonionizing radiation.
- Biological hazards.

Some chemicals are subject to labeling requirements of other laws or regulations and are, therefore, exempt from hazard communication labeling requirements. Chemicals exempt from labeling requirements of the HCS include:

- Pesticides subject to the labeling requirements of the Federal Insecticide, Fungicide, and Rodenticide Act (FIFRA) and pursuant regulations.

HAZARD COMMUNICATION AND WORKER RIGHT-TO-KNOW PROGRAMS 1739

- Chemical substances or mixtures subject to the labeling requirements of the Toxic Substances Control Act (TSCA) and pursuant regulations.
- Foods, food additives, color additives, drugs, cosmetics, or medical or veterinary devices or products (including materials intended for use as ingredients in such products) subject to the labeling requirements of the Federal Food, Drug and Cosmetic Act or Virus-Serum-Toxin Act and pursuant regulations.
- Distilled spirits (beverage alcohols), wine, or malt beverages intended for nonindustrial use that are subject to the labeling requirements of the Federal Alcohol Administration Act and pursuant regulations.
- Consumer products or hazardous substances as defined in the Consumer Product Safety Act and Federal Hazardous Substances Act, respectively, when subject to a consumer product safety standard or labeling requirement of these Acts and pursuant regulations.
- Agricultural or vegetable seed treated with pesticides and labeled in accordance with the Federal Seed Act and pursuant regulations.

There is a provision in the HCS addressing work situations in which employees handle only sealed containers which are not opened under normal conditions of use, such as retail sales, warehousing, and cargo handling. For these types of operations employers are required to:

- Ensure that labels on incoming containers are not removed or defaced.
- Maintain copies of MSDSs received with incoming shipments of sealed containers.
- Obtain the MSDS for a chemical in sealed containers as soon as possible if an employee asks for one.
- Ensure that MSDSs are readily accessible during each work shift.
- Provide training and information to employees as required by the HCS (except for the location and availability of the written hazard communication program) to the extent necessary to protect them in the event of a leak or spill.

Laboratories, as defined within the HCS, are exempt from some of the provisions. Those that do apply to laboratories include the following:

- Labels on incoming containers of hazardous chemicals must not be removed or defaced.
- Employers must maintain MSDSs that are received with incoming shipments of hazardous chemicals and ensure that they are readily accessible during each workshift to laboratory employees.
- Employers must ensure that laboratory employees are provided information and training as required by the HCS (except for the location and availability of the written hazard communication program).
- Laboratories that ship hazardous chemicals are considered to be either a chemical manufacturer or distributor and thus must ensure that containers leaving the laboratory are labeled and that a MSDS is provided to distributors and other employers.

2.2 Program Elements—Chemical Manufacturer, Supplier, and Importer Responsibilities

The basic responsibilities of chemical manufacturers, suppliers, or importers of hazardous substances in a hazard communication program are described below.

2.2.1 Hazard Determination

Hazard determination is the responsibility of chemical manufacturers and importers. When evaluating chemicals, the chemical manufacturer or importer must identify and consider the available scientific evidence concerning such hazards. For health hazards, hazard determination criteria are specified in Appendix B of the HCS. This appendix specifies criteria for (a) carcinogenicity, (b) human data, (c) animal data, and (d) adequacy and reporting of data.

Chemicals listed in the following sources must be treated as hazardous:

- **29** *CFR* part 1910, subpart Z, Toxic and Hazardous Substances, Occupational Safety and Health Administration (OSHA), or
- *Threshold Limit Values for Chemical Substances and Physical Agents in the Work Environment*, American Conference of Governmental Industrial Hygienists (ACGIH) (latest edition). The chemical manufacturer or importer is still responsible for evaluating the hazard associated with the chemicals in these source lists.

Determination of whether a chemical is a carcinogen or a potential carcinogen is based on findings by the National Toxicology Program (NTP), International Agency for Research on Cancer (IARC), and/or OSHA. A positive determination on carcinogenicity by any of these groups is considered conclusive. The *Registry of Toxic Effects of Chemical Substances* published by the National Institute for Occupational Safety and Health (NIOSH) indicates whether a chemical has been found by NTP or IARC to be a potential carcinogen.

When human data such as epidemiological studies and case reports of adverse health effects are available, they must be considered in a hazard evaluation. Because human evidence of health effects is not generally available for many chemicals, animal data must often be relied upon. The results of toxicology testing in animals must often be used to predict the health effects that may occur in exposed workers. This reliance is particularly evident in the definition of some acute hazards that refer to specific animal data, such as median lethal dose (LD_{50}) or median lethal concentration (LC_{50}).

The results of any positive studies that are designed and conducted according to established scientific principles, and have statistically significant conclusions regarding health effects of a chemical, are considered sufficient basis for hazard determination and reporting on a MSDS.

The HCS specifies the following categories for hazard determination in its mandatory Appendix A.

- Carcinogen—a cancer causing substance based on determination and/or listing by IARC, NTP or OSHA.

- Corrosive—a chemical that causes visible destruction of, or irreversible alterations in, living tissues by chemical action at the site of contact.
- Highly toxic—a chemical with:
    a. LD_{50} of less than or equal to 50 milligrams per kilogram of body weight (mg/kg) by oral ingestion.
    b. LD_{50} of less than or equal to 200 mg/kg by dermal contact, or
    c. LC_{50} of less than or equal to 200 parts per million (ppm) or 2 milligrams per liter (mg/L) by inhalation.
- Irritant—a chemical which causes a reversible inflammatory effect on living tissue by chemical action at the site of contact.
- Sensitizer—a chemical that causes a substantial proportion of exposed people or animals to develop an allergic reaction in normal tissue after repeated exposure.
- Toxic—a chemical with:
    a. LD_{50} of more than 50 mg/kg but not more than 500 mg/kg by oral ingestion,
    b. LD_{50} of more than 200 mg/kg but not more than 1000 mg/kg by dermal contact, or
    c. LC_{50} of more than 200 ppm but not more than 2000 ppm or more than 2 mg/L but not more than 20 mg/L by inhalation;
- Target organ effects—chemicals that affect specific organs. Examples of target organ effect categories include: (a) hepatotoxins (liver), (b) nephrotoxins (kidney), (c) neurotoxins (nervous system), (d) agents that act on the blood or hematopoietic (blood forming) system, (e) agents that damage the lung, (f) reproductive toxins, (g) cutaneous hazards (skin), and (h) eye hazards.

2.2.2 *Material Safety Data Sheets*

Chemical manufacturers and importers are required to obtain or develop an MSDS for each hazardous chemical they produce or import. Chemical manufacturers and importers must then ensure that distributors and purchasers of hazardous chemicals are provided with an MSDS with the initial shipment of each hazardous chemical and with the first shipment following an MSDS update. Distributors must ensure that MSDSs and updated information are provided to distributors and purchasers of hazardous chemicals.

The preparer of the MSDS must ensure that the information reported is accurate and reflects scientific evidence used in making the hazard determination. If the MSDS preparer becomes newly aware of significant information regarding the hazards of a chemical or ways to protect against the hazards, the new information must be added to the MSDS within three months.

The MSDS may be in any format, but must provide specific information regarding the hazardous chemical. Required information on an MSDS includes:

1. *Identity of the material.* The identity on the MSDS must match the identity on the label. The MSDS must also provide the chemical and common name(s) of the hazardous chemical. In the case of a mixture, the chemical and common name(s) of the ingredients must be listed. If the mixture has been tested as a whole to

determine its hazards, the chemical and common name(s) of any ingredients contributing to the known hazards must be listed. If the hazardous chemical has not been tested as a whole, the chemical and common name(s) of all hazardous ingredients that comprise 1% or greater of the composition must be listed, except for carcinogens, which must be listed if they comprise 0.1% or greater of the composition. If there is evidence that a hazardous ingredient could be released from the mixture in concentrations which would exceed an established OSHA permissible exposure limit (PEL) or ACGIH Threshold Limit Value (TLV) or could present a health risk to employees, then the ingredient must be listed even if it comprises less than 1% of the mixture. In all cases, the chemical and common name(s) of ingredients that present a physical hazard when present in the mixture must be listed. In some states, the Chemical Abstract Services (CAS) number is also required for each component. The standard does include provisions for protecting trade secrets of chemical manufacturers. The specific identity of hazardous chemicals can be withheld from the MSDS if this information is a trade secret; however, all other information regarding the properties and effects of the chemical must be provided.

2. *Physical and chemical characteristics.* This information includes characteristics such as vapor pressure, flash point, etc.
3. *Physical hazards.* This includes information regarding potential for fire, explosion, and reactivity.
4. *Health hazards.* Health hazard information must include signs and symptoms of exposure and any medical conditions that are generally aggravated by exposure to the chemical.
5. *Primary route(s) of entry into the body.* Typical routes of entry are inhalation, ingestion, skin contact, and skin absorption.
6. *Exposure limits.* These limits include the OSHA PEL, the ACGIH TLV, and any other exposure limit used or recommended.
7. *Carcinogenicity.* The MSDS must state whether the chemical has been determined to be a carcinogen or potential carcinogen by the NTP Annual Report on Carcinogens, the IARC Monographs, or OSHA.
8. *Precautions for safe handling and use.* This includes precautions that are generally applicable including hygienic practices, protective measures during repair and maintenance of contaminated equipment, and procedures for cleanup of spills and leaks.
9. *Control measures.* These include engineering controls, work practices, or personal protective equipment.
10. *Emergency and first aid procedures.*
11. *Date of preparation or latest change.*
12. *Name, address, and telephone number.* This is required for the chemical manufacturer, importer, employer, or other responsible party preparing or distributing the MSDS. This is provided in case the user needs additional information on the hazardous chemical and emergency procedures.

HAZARD COMMUNICATION AND WORKER RIGHT-TO-KNOW PROGRAMS

If the MSDS preparer cannot find any information for one of these categories to include on the MSDS, then that section should be marked as not applicable or a notation that no information was found should be made. There must be no blanks on an MSDS.

OSHA has an optional form that may be used to prepare an MSDS. This OSHA form is available from OSHA.

The American National Standards Institute (ANSI) has developed a detailed standard, ANSI Z400.1-1993, which provides further guidance on the preparation of MSDSs (8).

2.2.3 Labels

The chemical manufacturers, importers, or distributors must ensure that each container of hazardous chemicals leaving their workplaces is labeled. The label on a "shipped" container must include:

- Identity of the hazardous chemical.
- Appropriate hazard warning.
- Name and address of the chemical manufacturer, importer, or other responsible party.

The identity on the label must match the MSDS chemical identity. The label is meant to provide an immediate warning. The appropriate hazard warning should cover the major effects of exposure; however, it will not include all of the detailed information provided by the MSDS. The label must warn of specific acute and chronic health hazards as well as physical hazards. A precautionary statement such as "caution," "harmful," or "harmful if inhaled" is not adequate since it does not provide information on the actual hazard. The hazard warning should include target organ effects information, such as "causes lung damage when inhaled," when a specific target organ effect is known. Some manufacturers include additional information, such as emergency first aid procedures, that can prove useful, but is not required for compliance.

2.3 Program Elements—Employer Responsibilities

An employer that uses hazardous substances must have a hazard communication program that includes the components described below.

2.3.1 Written Hazard Communication Program

The employer must develop, implement, and maintain at each workplace, a written hazard communication program that explains how compliance with the standard is achieved. The written program must include a description of how the employer complies with labeling, MSDS, and employee information and training requirements. The written program must also include a list of hazardous chemicals known to be present at the workplace. When OSHA inspects a workplace, the OSHA compliance officer will ask to see the written program and will look for specific items to ensure that each of the elements are properly addressed.

For labeling, the following items should be described in the written program:

- The person(s) responsible for ensuring labeling of in-plant containers.
- The person(s) responsible for ensuring labeling of any shipped containers.
- Description of the labeling system used.
- Description of written alternatives to labeling of in-plant containers if used.
- Procedures to review and update label information when necessary.

For MSDSs, the following items should be described in the written program:

- The person(s) responsible for obtaining and maintaining MSDSs.
- How MSDSs are maintained in the workplace and how employees can obtain access to them in their work areas during their work shift.
- Procedures to follow when an MSDS is not received at the time of the first shipment of a hazardous chemical.
- For producers, procedures to update the MSDS when new and significant health information is found.
- Description of alternatives to actual MSDSs in the workplace, if used.

For employee information and training, the following elements should be described in the written program:

- The person(s) responsible for conducting training.
- Format of the training program to be used.
- Elements of the training program.
- Procedures to train new employees at the time of their initial assignment to work with a hazardous chemical and to train employees when a new hazard is introduced.

A written program must also include the following:

- A list of hazardous chemicals known to be present at the workplace. The identity on the list must match the identity used on the MSDS and label. The list can be for the entire facility or for individual work areas.
- The methods an employer will use to inform employees of the hazards of nonroutine tasks, such as cleaning reactor vessels, and the hazards associated with chemicals contained in unlabeled pipes in their work areas.

For multiemployer workplaces where employers produce, use, or store hazardous chemicals at a workplace in such a way that the employees or other employers may be exposed (e.g., contractor working on-site), the written program must also include the following:

- Methods to be used to provide the other employers on-site access to MSDSs for each hazardous chemical the other employers' employees may be exposed to while working.

HAZARD COMMUNICATION AND WORKER RIGHT-TO-KNOW PROGRAMS

- Methods to be used to inform the other employers of any precautionary measures that need to be taken to protect employees during the workplace's normal operating conditions and foreseeable emergencies.
- Methods to be used to inform the other employers of the labeling system used in the workplace.

Employers must make the written program available, upon request, to employees and their representatives. If employees travel between workplaces during a workshift, the written program can be kept at the primary workplace.

2.3.2 Material Safety Data Sheets

The employer may rely on the chemical manufacturer, importer, or distributor for the hazard determination and MSDS preparation. The employer is responsible for maintaining copies of the MSDSs in the workplace and ensuring that MSDSs are readily accessible during each work shift to employees when they are in their work areas. MSDSs can be made available through electronic access, microfiche, and other alternatives to maintaining paper copies provided that no barriers to employee access in each workplace are created by the alternative system. Where employees travel between workplaces during a workshift, the MSDSs may be kept at the primary workplace facility; however, employees need to be able to immediately obtain required information in an emergency.

Material Safety Data Sheets may be kept in any form, including operating procedures, and may be designed to cover groups of hazardous chemicals in a work area where it may be more appropriate to address the hazards of a process rather than individual hazardous chemicals. However, the employer shall ensure that in all cases the required information is provided for each hazardous chemical, and is readily accessible during each work shift to employees when they are in their work area(s).

2.3.3 Labels

The employer must ensure that each container of hazardous chemicals in the workplace is labeled, tagged, or marked with (a) the identity of the hazardous chemical and (b) appropriate hazard warning.

The labels provided by suppliers on original containers are typically adequate for those containers. The employer must label in-plant containers into which hazardous chemicals are transferred from original containers. Labels similar to those used on original containers can be used, or an alternate method can be employed. Signs, placards, process sheets, batch tickets, operating procedures, or other such written materials may be used in lieu of actual labels on individual stationary process containers provided the method conveys the required label information.

Employers are not required to label portable containers into which hazardous chemicals are transferred from labeled containers, and which are intended for immediate use by the employee who performs the transfer. Immediate use is typically interpreted to mean within the same work shift that the transfer was done.

Existing labels on original containers of hazardous chemicals must not be defaced or removed, unless the required information is immediately marked on the container again.

Labels must always be present or available in English. It is acceptable and often advantageous to supplement the English labels with hazard information in other languages, where employees speak other languages.

2.3.4 Employee Information and Training

Employers must provide specific information and training to employees to comply with the HCS. Employees must be trained on hazardous chemicals in their work area at the time of their initial assignment and whenever a new hazard is introduced into the work area. Information and training may be designed to cover categories of hazards (e.g., flammability, carcinogenicity) or specific chemicals. Labels and MSDSs must always be available to provide chemical-specific information.

Information that must be provided to employees includes:

- Requirements of the HCS.
- Operations in the work area where hazardous chemicals are present.
- Location and availability of the written hazard communication program, including the list(s) of hazardous chemicals and MSDSs.

Employee training must include, at a minimum, the following:

- Methods and observations that may be used to detect the presence or release of a hazardous chemical in the work area.
- Physical and health hazards of chemicals in the work area.
- Measures employees can take to protect themselves from hazards.
- Procedures the employer has implemented to protect employees from exposure to hazardous chemicals, such as work practices, emergency procedures, and personal protective equipment to be used.
- Details of the hazard communication program developed for the facility, including an explanation of the labeling system, the MSDS, and how employees can obtain and use appropriate hazard information.

2.4 Trade Secrets

A trade secret is defined in the HCS as follows:

> "Trade secret means any confidential formula, pattern, process, device, information or compilation of information that is used in an employer's business, and that gives the employer an opportunity to obtain an advantage over competitors who do not know or use it" (9).

Chemical manufacturers, importers, or employers may withhold the specific chemical identity from a material safety data sheet if the information is a trade secret provided that:

HAZARD COMMUNICATION AND WORKER RIGHT-TO-KNOW PROGRAMS

- The trade secret claim can be supported.
- The MSDS contains information regarding the properties and effects of the hazardous chemical.
- The MSDS indicates that the specific chemical identity is being withheld as a trade secret.
- The specific chemical identity is made available to health professionals, employees, and designated representatives as required by the HCS.

The specific chemical identity must be immediately disclosed to a treating physician or nurse during a medical emergency if the information is necessary for emergency or first aid treatment. This disclosure must be made regardless of the absence of a written statement of need from emergency personnel or a confidentiality agreement, which may be required from the physician or nurse later as soon as circumstances permit.

Trade secret chemical identity information must also be disclosed in nonemergency situations provided it is requested in writing and meets a specific occupational health need as detailed in the standard. A written confidentiality agreement is typically required by the manufacturer, importer, or employer.

3 TRAINING

The HCS specifies who must be trained, when and what topics must be covered. As explained above, employees must be trained on hazardous chemicals in their work area at the time of their initial assignment and whenever a new hazard is introduced into the work area. Training must include general information about the HCS and the employers' program, as well as information specific to the work area, as explained in Section 2.3.4.

The performance-oriented approach of this standard allows the employer to determine how training will be conducted. This provides employers with a great deal of flexibility to implement a training program that is both effective and practical for their organizations. Different approaches may be applied depending on the number of employees, number of worksites, number of chemicals, projected turnover of personnel, and the need for refresher or update training. Companies with large numbers of chemicals and those in which chemicals change frequently may choose to train on the categories or types of physical and health hazards. Those having fewer chemicals may find it more efficient to train on specific materials.

As noted above, training must be provided initially and when a new hazard (not a new chemical) is introduced into the work area. If a new chemical is brought in whose hazards are essentially the same as an existing chemical and training has already been done, no additional training would be required.

Routine retraining is not required by the federal standard. However, it may be required by state regulation. In any case, an employer may find it advisable to refresh the basic information on chemical hazards on a routine basis.

3.1 Training Techniques

There are a variety of training techniques and tools available. Typically, a combination of techniques will enhance the learning process and produce more effective training. For example, the general information on the standard and the program may be presented in a large group lecture. The area-specific information may then be presented in smaller groups using more interactive techniques. The following are some of the most common training techniques:

- *Group Lectures.* This method can be used when all members of a group need the same information and the information is conducive to mass display. With very large groups there is little opportunity for interaction or participation, unless the participants can be organized into several smaller groups.
- *Group Problem Solving or Role-Play.* Group members work together to solve problems or play roles in hypothetical situations. Using real-life situations can help students relate principles they have learned to actual practice. This method may be used to simulate emergency response or procedures that must be performed within a critical time frame. It can be an effective learning method if coupled with a group lecture and hands-on equipment training. However, it is time consuming, requires more training staff than a group lecture and involves significant preparation time.
- *Equipment Demonstrations and Hands-On Practice.* Students learn by watching demonstrations and practicing behavior with fellow class participants. Group demonstrations can help to familiarize students with equipment. However, individual hands-on practice is generally considered to be essential for students to be able to use equipment properly.
- *Self-Paced Instruction.* The student reads information and responds to questions or situations independently. Immediate feedback and explanation of the correct answers should be provided.
- *Interactive Computer Programs.* The information is presented to the student individually in the form of an interactive program at a computer terminal. The student may be quizzed or tested to help assure comprehension of the material before going on. Documentation of successful completion may be electronically recorded. This type of training may lack the benefit of interaction with other students and the trainer; however, it assures consistent presentation of information, can be an efficient way to train large numbers of employees, and may facilitate record keeping.

3.2 Training Tools

For any of the above techniques a variety of training tools may be used such as those described below:

- *Videotape.* A number of excellent instructional videos are available. Well-produced videos will help ensure consistency in the information presented and can hold the interest of students. Unless videos are produced on-site, they will be very general. Relying exclusively on purchased videotapes for instruction may produce training that is ineffective and may not be sufficiently work-area specific. Videotapes of facility operations, labeling, on-site scenarios, and real-life simulations can be an excellent addition to a training program, allowing students to relate the information that has been presented with their work areas and jobs. Poorly produced videotapes can be noisy, difficult to watch because of inadequate lighting, an unsteady camera, or badly selected vantage point. Videotapes can be incorporated into customized computer-based training programs.

- *Slides and Overheads.* Text, charts and diagrams may be used during a lecture to summarize information and to emphasize important points. Overheads may be edited or changed more easily than slides, however flipping them by hand is distracting. This problem can be alleviated by the use of projectors that display computer presentation programs directly on the screen.

  Inserting slides of facility operations, labels, familiar people and locations can help to keep the students' interest and stimulate questions and interaction. Slides can also be used to engage students in critique of actual labels, storage, personal protective equipment, and chemical handling. Quality site-specific photographs can often be more easily produced that videos. Photographs can be incorporated into customized computer-based training programs.

- *Flip Charts.* Developing information on flip charts during the lecture lends spontaneity to the presentation and helps to retain student interest. This same technique can be used on overheads for larger audiences.

3.3 Training Records

The purpose of training documentation is to demonstrate that employees have been trained prior to initial assignment and whenever a new hazard is introduced into the work area. As a minimum, records should be kept for each hazard communication training session that describe the content of the training, who was in attendance, and when the training occurred. The OSHA HCS does not specify the content of the records or the length of time they must be retained.

3.4 Training Effectiveness

The typical measure of Hazard Communication training effectiveness is that employees understand the hazards of their work area, know how to obtain information about the hazards, and are able to protect themselves. A number of different evaluation methods may be used including those summarized in the following paragraphs:

- *Comprehension Testing.* Testing for comprehension is not required by the HCS. However, it can provide valuable feedback on how much of the information was understood. Testing will not provide information on long-term retention.

- *Random Comprehension Sampling.* An auditor, health and safety professional or supervisor may verbally quiz employees during field visits to test for comprehension of basic information. OSHA compliance officers often use this method.

- *Work Practice Observation.* The employee's work practices and utilization of personal protective equipment may be observed to determine whether principles taught during training are being implemented in the workplace.

When hands-on practice and interaction is included in a training program, it is more likely the information will be retained. Reinforcing the concepts following initial training can promote long-term retention; for example, periodic discussions of chemical hazards, labeling and personal protection at the job site. Integrating the principles learned in hazard communication training with other health and safety training demonstrates the importance of the concepts and shows how they relate to all other aspects of health and safety.

In addition, actively involving employees in the development and presentation of the training program can enhance their comprehension of the information and increase the acceptance of the training by all employees.

4 LABELING

The HCS allows flexibility in the format and content of labels used by both manufacturers and employers, provided they include the required information.

4.1 Manufacturer's Labels

As discussed in Section 2.2.3, manufacturers must ensure that each container of hazardous chemicals leaving the workplace is labeled with (*1*) the identity of the hazardous chemical, (*2*) an appropriate hazard warning, and (*3*) the name and address of the chemical manufacturer, importer, or other responsible party. For substances shipped in a tank truck or rail car, the appropriate label or label information may either be posted on the vehicle or attached to the shipping papers.

There are numerous labeling systems used in industry. Labeling systems that rely on numerical or alphabetic codes to define hazards typically do not provide target organ effect information and are not appropriate for shipped containers unless additional narrative information is included on the container label. Some chemicals regulated by OSHA in a substance-specific health standard require a specific warning label. Examples of these are asbestos, benzene, and ethylene oxide.

4.2 Labels in the Workplace

As discussed in Section 2.3.3, employers are responsible for ensuring that all containers in their facility are properly labeled with the identity of the hazardous chemical, and an appropriate hazard warning.

The employer has more flexibility than the manufacturer regarding labeling containers since they can integrate their labeling, training, and written programs. In the case of individual stationary process containers, employers have the option of using signs, placards, process sheets, batch tickets, operating procedures, or other written materials in lieu of affixing labels directly to the container. The alternative method used must identify the container to which it applies and convey the required information. The written materials must be accessible to employees throughout their work shift.

Labeling systems that include numerical or alphabetic codes to convey hazards may be permissible for in-plant labeling provided employees are adequately trained on their use and they are described in the written program. The target organ effects must be communicated to employees in some manner. Proper training on a specific labeling system is essential to help ensure its effectiveness in promoting safe handling and use of chemicals.

5 MATERIALS/SYSTEMS/EQUIPMENT

5.1 Hazardous Material Information Systems

Hazard communication programs for medium to large manufacturers and employers require gathering and communication of large volumes of data. Because the HCS is performance oriented, the system used is left to the discretion of the individual company and a multitude of options is available.

5.1.1 MSDS Preparation

The Hazard Communication Standard provides manufacturers with flexibility in MSDS format and presentation. The only strict requirement for MSDS preparation is that all of the required elements as discussed in Section 2.2.2 be included on the form. This flexibility has prompted most manufacturers of chemicals and many suppliers and distributors to develop computer-generated MSDSs for their products. Computer generation allows for easier updating of MSDSs as new technical information becomes available and provides a convenient method for storing and retrieving MSDSs for a wide variety of chemicals.

Some individual publishers have developed databases of thousands of MSDSs for sale to manufacturers and users of chemicals. These databases typically include MSDSs for pure chemicals only. Therefore, they are not a complete answer to MSDS preparation, but they can provide a starting point. Because the manufacturers, suppliers, distributors, or importers will have the name of their companies on the MSDS and are ultimately responsible for hazard determination and other contents, they should verify all information prior to distribution of MSDSs.

5.1.2 MSDS Management

Managing MSDSs can present a challenge to employers who use many different chemicals in their workplaces. As discussed in Section 2.3.2, employers are required to maintain MSDSs such that they are readily accessible to employees during each work shift. There are several options for maintaining MSDSs. The simplest method is to maintain hard-copy MSDSs in an accessible file location at the worksite. This type of system works well for a workplace where all the employees are located in one building or within reasonable proximity to one another and MSDSs can, therefore, be kept in just one or two locations. A manual system is usually only practical when the types and numbers of hazardous materials are somewhat limited.

Another option used by some larger employers is to enter information manually from MSDSs into a database, which can then provide MSDS information to the user. This allows MSDS information to be available online at computer terminals located throughout a large facility. Employees must, of course, be trained to use the computer system or have access to an individual with appropriate training on all shifts. Manual entry of MSDS information can be time consuming and presents the risk of incorrect entry.

If a uniform MSDS format is desired, MSDSs received from manufacturers can be reformatted for computer input. A major advantage of a system with a uniform format is the ease in use. The MSDS user need only understand one MSDS format to access the information easily through a computer terminal. A disadvantage to this type of system is the increased potential liability the end user incurs by changing the format of, and interpreting information from, the original MSDS.

Another option used by some employers is to scan the original MSDSs into a computer system to allow for online access. This option provides the advantage of online access with much less time required to input data. Typically a scanned image is treated as a graphics file and is therefore identical to the original MSDS. This eliminates the problem of incorrect data entry, which can occur with systems where MSDS data is entered into the system. As graphics files, MSDSs will be in a variety of formats that cannot be manipulated, making their online use somewhat more difficult for the user than with the uniform format option.

The capabilities of commercially available MSDS software products have expanded over the past several years. As of 1997, there were 64 MSDS software products available commercially. Over half of the MSDS database products allow for using scanners to enter MSDSs. About one-third of the software packages allow users to add MSDSs to the database via electronic transfer (e.g., e-mail or on-line). Many of the packages run under Microsoft Windows, client/server applications, which is often useful to companies with multiple locations. Most of the packages will allow entry of multiple MSDS formats. Some of the packages also offer conversion from one language to another (10).

A major advantage of the electronic storage of MSDSs is the ability to search the database and easily retrieve information. The majority of commercially available systems allow the user to search the database for chemicals by chemical name, product name, CAS number, date, date range, words, phrases, manufacturer or vendor. This can be useful to health and safety professionals as well as other users.

5.1.3 Hazardous Material Tracking

Employers who use hazardous chemicals are faced with the task of managing very large volumes of paper, particularly MSDSs. Once MSDSs are received and stored for existing hazardous chemicals, the employer is still confronted with the task of receiving, tracking, and storing MSDSs for new chemicals and updating MSDSs for existing chemicals. The employer will also often typically track "approved" hazardous materials that may be purchased for use in the facility.

Many approaches are available for tracking MSDSs, ranging from a completely manual system to a completely automated system. Except in cases where very few chemicals are used, some form of automated tracking is typically needed. In addition to the advantages discussed in the previous section, electronically entering and tracking MSDSs allows for (a) cross-referencing and updating of hazardous chemical list, (b) tracking the date of the latest MSDS revision, (c) checking chemicals and suppliers for pre-approval, and (d) tracking departments using each chemical for training.

A database system facilitates communication between users and purchasing and health/safety professionals regarding chemicals used and approved. Materials approved by either a health/safety professional or review committee can be entered into the database as "approved" so that purchasers and users of the chemical know that the material is acceptable for use. This can help avoid duplication of effort in reviewing substances and speed the purchasing process for commonly used materials. The hazardous materials tracking system will also allow tracking of departments that use each chemical. This information can be used to determine training requirements within the department. An enhancement to the system may include tracking the training (initial and update) of employees within each department. When updated MSDSs are received, the departments using the chemical can be alerted to provide updated information to their employees on that chemical.

5.1.4 Related Programs

An information system set up for hazard communication compliance should be flexible enough to use in complying with other related regulatory requirements, such as those discussed in Section 7. Additionally, other state or local regulations may require information tracking that could be easily incorporated into a hazardous material tracking system set up for hazard communication.

An example of another use of the hazard communication tracking information is the requirement of Section 311 of SARA Title III to produce copies of MSDS chemicals. In addition, Section 312 of SARA Title III requires the submission of an inventory of hazardous chemicals with additional information regarding quantities present at a facility, daily use, locations, and storage information. This information can be submitted either as an aggregate by OSHA categories of health and physical hazards (Tier I) or by individual chemicals (Tier II). Including the information required for producing SARA Title III reports with the hazard communication database helps avoid duplication of efforts. The hazard communication database may also be useful as a starting point for preparing a Toxic Chemical Release Form as required by Section 313 of SARA Title III.

If the information tracking system includes a provision for tracking training, other training programs may be added to the system for tracking purposes. This may be particularly useful when training programs are combined. For example, a hazard communication training program may be expanded to include training of employees who handle hazardous waste, as required by RCRA regulations. The training may be combined, but records must be kept to show compliance with both sets of regulations. An automated system for tracking this is helpful, particularly when many employees require training.

6 PROGRAM MANAGEMENT

The objective of hazard communication is to provide information to employees on the chemical hazards in the workplace. If employees understand this information, they will be better able to use chemicals safely and reduce the likelihood of injury or illness. Keeping this in mind when designing a hazard communication program can help to maintain program focus. Assembling the essential elements, such as MSDSs, labels, a written program, and training, although vital to the program, will not, in itself, accomplish this objective.

There is a danger is that even a compliant program can quickly become stagnant with ever-growing collections of MSDSs clogging file cabinets, and "canned" training programs shown over and over again. A hazard communication program must be a dynamic process that includes continuous updating of MSDSs, chemical lists, and training. This will not happen without adequate resources. However, immature programs that require too much professional staff support can exhaust resources and cause weakening of management commitment. The challenge in program management is to continually explore ways to improve efficiency as well as quality. This will help to ensure that resources and funding do not wither over time, or with changes in philosophy and leadership. This section will provide some suggestions for designing and managing an effective hazard communication program, including planning, program responsibilities, policies and procedures, program costs, communication, quality control, program improvement and recordkeeping.

In this section the term program "maturity" is used. For the purposes of this section, a mature program is one that has become part of doing business. Characteristics of a mature program are that costs tend to be stable or decline over time. Professional staff time is focused and efficiently used. Maintenance is minimal and updating occurs routinely. Stable systems support essential functions. Continuous improvement is an integral part of the program. Mature programs tend to be less vulnerable to loss of management commitment because the use of resources becomes more efficient over time, and because program support and ownership reside in many parts of the company, not just one department.

6.1 Planning

Early in the program development, the primary concern will be to achieve compliance. At this stage, considerable professional time will be needed to organize and launch the program. The responsibility and support for the program may be within one group, possibly one or two individuals.

When compliance is attained, the focus should switch to activities that will encourage program maturation. Such activities may include: reviewing existing processes for opportunities to capture and use existing systems; establishing mechanisms to identify and track program costs and support staff time; and identifying all of the departments or individuals within the company that have a stake in or a contribution to hazard communication.

Whether designing an initial program or upgrading an existing program, an activity-based or task-based approach will provide a clear path from task responsibilities back to program requirements. This approach to program development and redesign provides structured information for cost analysis and resource allocation decisions. As the program matures cost savings, for example, can be tracked and directly related to program elements and regulatory requirements. Although other methods are available, this one will be used as an organizational tool throughout this section.

HAZARD COMMUNICATION AND WORKER RIGHT-TO-KNOW PROGRAMS

A program plan is an excellent tool for program development or upgrading. It will describe how each HCS element will be implemented. The first step is to develop a task list. Costs, responsibilities and systems will then be directly related to the implementation of each program element.

The plan should address items such as:

- Tasks associated with each element.
- Responsibility (current/potential) for each task.
- Systems that will be used to collect and maintain data and records.
- Information and training delivery systems.
- Policies, procedures and/or work instructions that will support HCS program elements.
- Mechanisms to evaluate, upgrade, and measure the effectiveness of the program.
- Cost analysis and budgeting.
- Time lines for implementation.

This planning document will evolve as decisions are made regarding key items such as program responsibilities and delivery systems (MSDS and training). Portions can then be extracted to develop or revise the HCS written program.

Table 39.1 provides examples of some basic tasks that may be needed to implement each element of the program. The exact tasks and the level of detail may differ from those listed. This same list will be used for discussion of program costs in Section 6.4.

6.2 Program Responsibilities

As mentioned earlier in this section, it can be beneficial to distribute the responsibilities for some program elements across organizational boundaries. Only in smaller companies can one or two people effectively drive or manage the entire program. In many cases, programs become more efficient if tasks are integrated into existing jobs. Table 39.2 contains some examples of typical departments and the HCS functions that they might perform.

The list in Table 39.2 is not comprehensive, and is not meant to imply that certain department or functional groups within an organization should perform the duties indicated. It is intended as a guide to help identify groups that may participate in some aspect of the program. The task assignments will depend on organization structure, size, labor agreements and other factors.

A word of caution: Although ownership of the program should be shared, a program that is too diffused within an organization may disappear altogether. Ensure that the responsibility for the overall program is clear, and that ownership, accountability, and support accompanies each task assignment.

6.3 Policies and Procedures

Company policies, procedures and work instructions reflect the support structure for programs. If a company is large, written procedures may address most key functions. Smaller

Table 39.1. HCS Program Elements and Example Tasks

Key Elements	Task
Written program	Meet with key personnel to plan strategy
	Develop HCS policies/procedures
	Develop written program/update program
	Review/revise existing HCS-related policies
Chemical list	Conduct inspection to identify chemicals used
	Develop data management system for chemical lists and/or MSDS[a]
	Compile and update list(s) of chemicals
	Input list(s) into data management system[a]
	Organize and distribute chemical list(s)
	Check lists against chemical purchasing records
MSDS	Develop and update MSDSs for products (chemical manufacturer)
	Acquire, organize and distribute new and revised MSDSs
	Review MSDSs for completeness and readability
	Match MSDSs with chemical list
	Contact manufacturer for missing MSDSs
	Input MSDSs into management system[a]
Labeling	Develop and update labels for products (chemical manufacturer)
	Check labels of incoming containers, isolate unlabeled containers at receiving stations, and obtain replacement labels from supplier, if appropriate
	Develop/implement in-house labeling system for transfer and secondary containers
	Apply labels to secondary containers, transfer containers, vessels, etc.
	Review piping systems, vessels and information that may be used in lieu of labels
Training	Assemble or develop training materials
	Train the trainers
	Train current employees
	Train new hires/transfer employees
	Train upon introduction of new hazard/changes in processes
Contractors	Provide information to contractors on hazardous materials at worksite
	Request information on materials contractors will bring onsite
	Obtain documentation that contractors comply with HCS
Recordkeeping	Develop/revise records retention policy and document retrieval systems for HCS
	Maintain training records
	Maintain chemical usage information (plant and contractor)
	Maintain MSDS documentation
	Maintain audit/inspection records
	Maintain trade secret information files
Program auditing[b]	Develop schedule and description of type and frequency of program review and audit
	Conduct reviews/inspections/audits

[a]These tasks would represent entire projects if an electronic data management and delivery system for MSDSs were chosen. Projects would include field-testing of systems for accessibility and accuracy.
[b]Not specifically required by HCS, but part of effective program management.

HAZARD COMMUNICATION AND WORKER RIGHT-TO-KNOW PROGRAMS

Table 39.2. Possible HCS Functions for Typical Departments

Department	Possible HCS Functions
Purchasing	Request MSDSs from chemical manufacturers
	Develop purchasing procedures that incorporate HCS requirements
	Collect documentation of contractors' compliance with the HCS
Receiving	Check packaging integrity of incoming chemicals
	Verify that MSDS accompanies incoming chemicals, or MSDS is on file
	Verify same name on label, chemical list, and MSDS
	Check containers for complete, readable labels
	Hold materials if labels are defective and make necessary contacts to correct
	Apply internal labels, if needed
Operations, production, maintenance, construction groups	Check labels of containers as received in area and periodically as used
	Check chemicals against chemical list and MSDS file when received
	Update list of chemicals, if appropriate, or notify responsible party
	Apply additional labeling for transfer and secondary containers
	Ensure new employees and transfer employees are trained
	Ensure training is received by all employees when a new hazard is introduced
	Ensure that employee questions and concerns about chemical usage or safety receive prompt response
	Maintain departmental HCS records (training, chemical usage, PPE, etc.)
	Ensure that employees have and use proper personal protective equipment
	Conduct periodic inspections of work area (labeling, MSDSs availability, personal protective equipment availability and usage, and training records)
	Conduct periodic review of employees' knowledge of chemical safety
	Participate in review of new and reformulated chemicals
Health and safety professionals	Develop and manage the Hazard Communication Program
	Participate in review of new and reformulated chemicals
	Participate in the review of chemical usage; provide information on safer alternatives where possible
	Conduct and/or coordinate HCS compliance reviews and audits
	Review incoming MSDSs for completeness and accuracy
	Recommend labeling practices and participate in the selection of secondary container labels or in-house labeling system, if used

Table 39.2. (continued)

Department	Possible HCS Functions
Laboratories	Develop or review content and effectiveness of training
	Conduct assessments, including air sampling to determine potential employee exposure and assess effectiveness of controls
	Assist in the development of procedures for safe handling of chemicals in routine and nonroutine tasks
	Provide guidance in the selection of personal protective equipment
	Develop, review and update existing product MSDSs (chemical manufacturers)
	Review accident and injury records related to chemical usage; identify improvements needed in personal protective equipment usage and training
	Facilitate physical, chemical, and toxicological testing to provide data for MSDSs (chemical manufacturers)
	Review test data from outside laboratories (chemical manufacturers)
	Research potential substitute chemicals
Engineering	Participate in development or review of product MSDSs
	Ensure that HCS requirements are included in project development and review
	Ensure that introduction of new or reformulated chemicals, including test batches and samples, follows HCS requirements
	Ensure that on-site contractors have fulfilled safety requirements, including HCS, before beginning work
	Implement regular project safety reviews with contractors
Legal department	Participate in initial and periodic review of HCS written program
	Provide legal interpretation of local and federal regulations related to HCS
	Participate in review of product MSDSs (chemical manufacturers)
	Assist in the negotiation of confidentiality agreements
Medical department	Participate in review of product MSDSs (chemical manufacturers)
	Review MSDSs of products in use in the plant
	Ensure ready access of MSDS for emergency care providers and hospitals
	Contact manufacturers for additional information when required in cases of trade secret formulations
	Alert appropriate personnel of exposure and injury incidents, and possible failures in PPE or exposure controls
Training department	Assist in the selection of and development of training delivery programs
	Support training efforts with equipment and technical advice
	Conduct training or train the trainers

companies often rely to some extent on verbal procedures. The philosophical approach to health and safety is usually stated in one or more company policies. Either a general or specific statement of top management support for Hazard Communication is very desirable. The tactical program support is generally in procedures and/or work instructions.

Effective procedures clearly state:

- Responsibility for each key task and the overall program.
- Interaction of responsible parties.
- Methods that will be used to maintain information and records.
- Associated job tasks needed to support those functions.
- Mechanisms to evaluate task completion.
- Methods to measure the effectiveness of the program.
- Process to be used for continuous improvement.

When key individuals or groups have been identified, negotiations should take place to define and assign program responsibilities. Issues will include funding, impact on staffing, level of responsibility, integration with other departments, reporting, updating, quality control, and so forth. Some examples of organizational responsibilities are provided in the previous section. Procedures and work instructions will be written or revised to reflect agreements.

Some system for the development, revision, review and acceptance of policies and procedures is generally part of the organizational process. Responsibility for specific program tasks may be covered in a variety of procedures, work instructions, and identified within the written HCS program itself. However, in many organizations, management and employees are held accountable only for responsibilities as described in the organization's published and approved procedures. Employees might not be held accountable for those tasks described only in program documents. Human Resources professionals or legal counsel can provide clarification and guidance.

6.4 Program Costs

The costs associated with a hazard communication program may be reflected in the budgets of many departments within a company. Realistic planning for and tracking of these costs will help to ensure the continued effectiveness of the program. Actual costs will, of course, depend on the size of the company and means chosen to implement the program.

An activity-based cost accounting system that includes elements and tasks such as those described in Table 39.3 will relate actual costs to program elements and regulatory requirements. This information can be used to (*1*) support budget and resource requests, (*2*) correlate to individual and group performance measures, (*3*) reveal opportunities for cost savings, and (*4*) demonstrate the effectiveness of cost saving and process improvement initiatives. It provides the basis for clear and understandable presentations to upper management by relating program requirements with cost and staffing implications, or demonstrating the effect of alternative implementation strategies. For example, the actual costs

Table 39.3. HCS Program Elements, Example Tasks and Associated Cost Factors

		Start-Up Costs			Program Maintenance Costs		
Key Elements	Task	Hours (Direct/ Indirect)	Equipment (Purchase/ installation)	Outside Services	Hours (Direct/ Indirect)	Equipment (Maintenance)	Outside Services
Written program	Meet with key personnel to plan strategy						
	Develop HCS policies/procedures						
	Develop written program/update program						
	Review/revise existing HCS-related policies						
Chemical list	Conduct inspection to identify chemicals used						
	Develop data management system for chemical lists and/or MSDS[a]						
	Compile and update list(s) of chemicals						
	Input list(s) into data management system[a]						
	Organize and distribute chemical list(s)						
MSDS	Check lists against chemical purchasing records						
	Develop and update MSDSs for products (chemical manufacturer)						
	Acquire, organize and distribute new and revised MSDSs						
	Review MSDSs for completeness and readability						
	Match MSDSs with chemical list						
	Contact manufacturer for missing MSDSs						
	Input MSDSs into management system[a]						
Labeling	Develop and update labels for products (chemical manufacturer)						
	Check labels of incoming containers, isolate unlabeled containers at receiving stations, and obtain replacement labels from supplier, if appropriate						

Training	Develop/implement in-house labeling system for transfer and secondary containers
	Apply labels to secondary containers, transfer containers, vessels, etc.
	Review piping systems, vessels and information that may be used in lieu of labels
	Assemble or develop training materials
	Train the trainers
	Train current employees
	Train new hires/transfer employees
Contractors	Train upon introduction of new hazard/changes in processes
	Provide information to contractors on hazardous materials at worksite
	Request information on materials contractors will bring onsite
	Obtain documentation that contractors comply with HCS
Recordkeeping	Develop/revise records retention policy and document retrieval systems for HCS
	Maintain training records
	Maintain chemical usage information (plant and contractor)
	Maintain MSDS documentation
	Maintain audit/inspection records
	Maintain trade secret information files
Program auditing	Develop schedule and description of type and frequency of program review and audit
	Conduct reviews/inspections/audits

[a] Associated with electronic MSDS and chemical management systems.

of using "stand-up" training may be contrasted with the projected costs of a proposed computer-based training delivery system.

For each element and task, start-up and program maintenance costs should be projected for new programs and budget cycles, then actual costs tracked. Cost breakdown should include, at a minimum: direct and indirect hours; purchase, installation and maintenance of equipment and systems (e.g., computer-based training, information tracking and reporting, etc.); and outside services. Many organizations have financial systems in place to support activity-based program cost accounting.

6.5 Flow of Communication

The smooth flow of information from the source (the chemical manufacturer, importer, or distributor) to the employee is critical to the program. This transfer of information utilizes MSDSs and labels as the source of information and training as the means of communication. Training, although it is the key step in this transfer of information, is not the final step. Feedback from employees, key individuals and departments can provide information to identify problems and streamline the program. Consider the following points to help avoid communication "sinks" and "traps."

6.5.1 Employee Concerns

Response to employee concerns should be quick, accurate and consistent, but need not be given on the spot. It is often better to admit that the answer is not immediately available, than to "shoot from the hip."

Questions may be addressed in a number of ways. They may be referred to in-house staff. Chemical manufacturers or distributors can sometimes be used as a resource. Additional information sources may be provided to the employee. Failure to address employees' questions will result in loss of confidence in the program and possibly more serious employee relations problems. Inconsistency or inaccuracy in responses will also result in loss of confidence. Investigating these questions can often provide fresh insight into hazards or controls that can then be incorporated into training.

Concerns regarding the level of exposure to chemicals, the efficiency of personal protective equipment, or the effectiveness of engineering controls may arise in training classes, safety committees, employee participation groups, directly to area supervisors, or to safety and health professionals. Regardless of where the concern is voiced, key people should be trained to respond and follow-up. Employee concerns should be taken seriously. Employees should be informed what actions will be taken to address their concerns. An approach that involves the participation of the employees, their representatives, and management is often effective in obtaining resolution.

6.5.2 Communication Between Key Personnel

The efficiently functioning program thrives on feedback from all active participants. Collecting and managing this information helps to maintain compliance. More importantly, however, it can improve employee safety and health, and may prevent costly mistakes. Critical information needed to make sound decisions may reside with a number or indi-

HAZARD COMMUNICATION AND WORKER RIGHT-TO-KNOW PROGRAMS

viduals or departments. The following example illustrates a situation in which input from key personnel is essential to arrive at a truly safe and economical decision.

As a cost savings initiative, a purchasing department proposed that a chemical be purchased in a more concentrated form, and in a larger unit volume. The cost savings appeared to be substantial. Considering only direct savings may lead to erroneous conclusions. In this example, significant hidden costs could be associated with any one of the following factors:

- The increased health and safety risk associated with the distribution and handling of more a concentrated chemical may required additional controls. Additional personal protective equipment and spill response equipment may be needed. Additional health, safety and environmental training may be required. A larger number of employees may potentially be exposed.
- The increased health risk to the surrounding community, air, soil, and water in the event of a release or spill may necessitate additional expenditure to manage risk. There may be additional environmental reporting responsibilities and increased waste disposal costs.
- Existing receiving and distribution systems may need upgrading or redesign.
- Existing storage facilities and leak containment may be inadequate for larger volumes or more concentrated chemicals.
- Leak or spill response capabilities may be inadequate for the increased volume, or the more concentrated chemical.

Anticipating the costs associated with this change can help to present a more accurate risk assessment and financial picture. A review process, such as a chemical control committee, involving key multidisciplinary personnel will help ensure that all significant factors are evaluated before changes to chemicals and processes are made. If the review process is not balanced, it may be perceived as a strategy to resist change. This perception may detrimental to the program. Changes such as those in the example above, can and should be made when the risks are identified and found to be acceptable and manageable.

The example provided above shows one way in which communication between key personnel could improve safety and possibly save money. The paragraphs below provide more examples of the contributions that can be made by key personnel within an organization, and ways in which the hazard communication program can benefit.

- Health and safety professionals evaluate employee exposure, the effectiveness of personal protective equipment, and engineering controls. They are typically knowledgeable of illness and injury experience. This information can be used to adjust the emphasis of the program, and can be used to evaluate and improve training.
- Supervisors, lead persons and employees will have input on how personal protective equipment or control devices are used (or misused) in the workplace. They will also have insight into accidents or injuries involving chemicals. Near misses or misuse of controls may indicate a lack of hazard awareness. This information may influence decisions regarding changes in the purchase and use of chemicals and it may be used

to refocus training. Real life examples derived from these experiences can make training more relevant and effective.
- Trainers will hear questions and concerns voiced by employees during training sessions. All such questions should receive quick and rapid response, as indicated above. However, the nature of the questions may suggest a need for alternative training methods or other program changes. They may also provide vital insight to the application of chemical handling principles in the field.
- Legal staff may have input on the hazard communication program in light of court decisions or regulatory trends.
- Physicians' and nurses' feedback on medical cases involving chemicals can provide additional insight regarding the effectiveness of personal protective equipment, level of employee knowledge, and handling procedures.
- Engineering personnel will be involved in changes to processes and materials. This information is key in ensuring that information and training are current and accurate. Engineering personnel may also be involved in the installation, testing, and functional evaluation of controls, such as ventilation.

Valuable information from a variety of sources is needed to ensure continuing compliance with the HCS and to fine-tune and redirect the program, as necessary.

6.6 Quality Control and Program Improvement

Part of the program planning should be devoted to quality control and continuous improvement. Both should be designed into the program. Although auditing is important, it should not be exclusively relied on for quality control. Program appraisals, record reviews, and performance assessments are essential parts of quality control. They are covered in Section 9. A formal, routine process for developing, evaluating and implementing program improvements is essential. Many organizations have process improvement or quality control programs that can serve as a model.

Introducing periodic improvements and program upgrades helps to stimulate interest and contributes to overall effectiveness. The following paragraphs provide some useful ideas for improvements. Hazard communication program elements are listed followed by program upgrade suggestions. Some of these suggestions go beyond the minimal requirements of the HCS. Therefore, evaluate the potential benefit to the program within the organizational environment, then choose and implement upgrades as resources permit.

Written Program

- Review and analyze applicability of HCS to the workplace; for example, laboratories and areas in which employees handle only sealed containers. Labs may qualify for some exceptions to the HCS or may be covered under the OSHA Lab Standard. A correct determination of which requirements apply may help to streamline the HCS program.
- Review written program annually. Audits of compliance, management systems and performance by key personnel may also be useful.

Chemical List

- Implement periodic, documented inspections of chemical usage areas.
- Review processes routinely to verify that the hazardous materials currently used and generated are in the program and covered in training. These reviews may already be part of an existing program, such as a configuration or change control program.
- Check chemical list(s) periodically against chemical purchasing records, MSDS file, and materials physically located in work areas.
- Refine procedures for continuous updating of the chemical list.
- Check the chemical list against lists kept for EPA SARA reporting.
- Evaluate the potential benefits of using a single chemical tracking system.
- Establish multidisciplinary team(s) to review chemical usage, chemical purchasing, the feasibility of substituting less hazardous process materials, and the benefits of reducing the number or volume of chemicals purchased.
- Identify ways chemicals may enter workplace outside normal purchasing channels (e.g., blanket purchase orders, off-the-shelf purchases, and product samples). Develop procedures to improve control of the entry of chemicals into the workplace.

MSDS

- Purge active MSDS files of duplicates and materials no longer onsite. Archive MSDSs according to record retention procedures.
- Compare MSDSs from manufacturers supplying similar products. Identify and research major differences. Address discrepancies between MSDSs in training.
- Test accessibility to MSDSs, chemical list, and written program in the workplace.
- Evaluate the potential benefit of electronic MSDS systems.
- Conduct a spot check of MSDSs for completeness of information.

Labeling

- Conduct routine spot checks of containers for legibility and completeness of labels. Report problems to supplier and purchasing agent.
- Conduct routine spot checks of secondary containers for correct labeling.
- Evaluate the potential benefits of using a computer program to generate secondary labels.
- Consider prelabeling frequently used transfer containers.
- If batch tickets or process flow diagrams are used in lieu of labels, test the accessibility and accuracy of this information.

Training

- Review training materials periodically for technical content.
- Compare training content with current hazards.

- Evaluate language skill level and accessibility of training to sight and hearing impaired employees.
- Review and upgrade audio-visual materials, consider inserting "real life" examples into training.
- Review accident/injury experience; utilize this information to refocus training, if appropriate.
- Request employee evaluations.
- Review test questions and scores, if comprehension tests are used.
- Provide skills development for trainers.
- Consider including area employees in training presentation and review.
- Evaluate alternative training methods, such as interactive computer-based training.
- Consider varying the training methods to fit diverse groups (e.g., operations groups using limited numbers of chemicals vs. maintenance groups using a large variety of chemicals).
- Interview employees to assess knowledge of chemical hazards and controls.
- Routinely spot check use of PPE. Lack of use or misuse may indicate the need to refocus training.
- Consider integrating portions of HCS training with other required safety and environmental training, such as RCRA or hazardous waste, or on-the-job training.
- Provide periodic refresher training for all employees.

Contractors
- Obtain acknowledgment that contractors receive the information on worksite hazardous materials.
- Review contractor training to ensure that on-site hazard training is adequate.
- Review service contract language with legal counsel and contract negotiators to determine if compliance with Hazard Communication (and other safety, health and environmental regulations) is adequately covered.

6.7 Record Keeping

Record keeping can be a challenging aspect of hazard communication program management. The mechanisms and systems used (paper, electronic, or a combination) depends on the size of the company, the location of the worksites, and the volume of records. Decisions regarding how and where the records will be kept, and how files will be updated and purged should be made during planning. Both security and accessibility must be addressed. The major record keeping needs are outlined in this section. Document retention policy and practice should be reviewed with legal counsel.

6.7.1 MSDSs and Chemical Lists

Current MSDSs, the written hazard communication program, and the chemical list must be immediately accessible to employees on all shifts. However, additional record keeping is required. As explained in Section 7.1.1, *29 CFR* 1910.20 (Access to Employee Exposure and Medical Records), requires that exposure and medical records be maintained for 30 years. Chemical lists are considered exposure documentation. All of this information must meet the requirements for employee access. Please note that some other OSHA standards have record retention requirements longer than 30 years. Develop a record retention schedule to manage these documents and ensure that records are kept for the correct period of time and allow the required accessibility.

When MSDSs are developed, the source of all information provided on the MSDS should be documented and retained. This holds true for chemical manufacturers as well as organizations that choose to prepare in-house MSDSs.

Written and verbal attempts to obtain MSDSs, requests for new and updated MSDSs or for replacements of MSDSs that are incomplete or unreadable should also be documented and retained. The results of the contact should be included.

6.7.2 Trade Secret Disclosures

If trade secret information is requested from a chemical supplier, files should be kept that include the agreement under which the information is provided, as well as the specific information. Other information that should be in the files includes the requests for information, the responses, and the outcomes. Legal counsel may be needed to provide guidance on obtaining trade secret information, negotiating agreements, securing access, and maintaining the files.

Once the supplier has provided the information, it is prudent to keep a copy of the agreement with the trade secret information. The agreement(s) will specify how information is to be safeguarded and who within the organization may have access. Implement controls on access to the filing system that are appropriate for the level of confidentiality required by the agreements. Controls are needed which are sufficient to prevent not only inappropriate but also *inadvertent* duplication or distribution of confidential information. Breach of these contractual agreements could result in legal action.

It is generally advisable to store trade secret information in a separate filing system, not in the same system as the chemical lists or the MSDSs. However, there should be some way to identify materials for which there is trade secret information. In this way authorized persons will know that the information is available and will be able to access the information, if the need arises.

6.7.3 Training Records

Although not required under the HCS, it is prudent to document training. These records should include:

- Content of training (course curriculum), date, person conducting the training, materials used, sample of test administered, training evaluations.
- Names of employees trained, department or work area, job classification or position.
- Follow-up training: chemicals or hazards covered in training, persons conducting training, date, materials used, persons trained, specific department or work area.
- Qualification of persons conducting the training and train-the-trainer courses.

6.7.4 Contractors

There are no requirements within the HCS for the retention of contractor records. However, information on chemicals used by another employer at a worksite may qualify as exposure records for the hiring company employees under **29** *CFR* 1910.20. It is advisable to document the information provided by the contractor, as well as that provided to the contractor by the hiring organization. This documentation should include, at a minimum:

- A list of the chemical materials used and copies of the MSDSs.
- Documentation that onsite hazardous materials information was provided to the contractor.
- Safety and health concerns raised during the work, including how those concerns were resolved.
- Exposure records such as air sampling data, if available.

Most companies keep records of contract work. The above information may be easier to organize as part of a single set of files on the contract. Organizing the information in this way, however, will make it difficult to retrieve information in response to a specific employee request.

7 RELATED REGULATIONS

A variety of federal, state and local regulations have been promulgated which cover the safety, health and environmental issues involved in managing hazardous chemicals from manufacture to disposal. Many of these regulations have requirements for training, hazard information, and labeling. Although each regulation deals with a slightly different aspect of this problem, they are related and, in some cases, overlap. This section provides a summary of some of these regulations and briefly discusses how they relate to the HCS.

7.1 OSHA Regulations

The relationship between the HCS and other OSHA standards can be complex. The standards covering specific chemicals, for example, may have requirements for labeling, training, information, and controls. In general, the requirements of these more specific standards take precedence over the generic approach of the HCS. However, it is wise to study the standards carefully to determine which requirements apply under specific conditions. This section will briefly present some key OSHA regulations that may impact elements of hazard communication.

7.1.1 Access To Employee Exposure and Medical Records (29 CFR 1910.20)

The purpose of this regulation is to provide employees and their designated representatives access to relevant exposure and medical records. MSDSs are identified in the regulation as one of the types of information that constitutes an employee exposure record. As such, they must be available to employees, their designated representatives, and OSHA.

Exposure records must be maintained for 30 years; however, it is not necessary to keep MSDSs for materials which are no longer used, provided records are kept which identify chemicals, provide the original formulation, describe where is was used, and when it was used. Some organizations choose to archive all old MSDSs and chemical lists to help assure compliance with this regulation.

It is not necessary to keep two lists. The requirements for maintaining a list of chemicals in the HCS can also meet the requirements for this regulation, if the list includes information on where and when (time period, not each application) each chemical is used. Purchasing records and annual chemicals inventories can be the source of this information.

OSHA has noted that some employers have confused the accessibility requirements of the 1910.20 with those of the HCS (11). For example, some employers have mistakenly believed that they have 15 days to produce a copy of the written hazard communication program at the worksite. This is not the case. A copy of the program must be at the worksite and accessible to employees on all work shifts.

7.1.2 Air Contaminants (29 CFR 1910.1000)

When the permissible exposure limits (PELs) were revised in 1989, limits for 376 chemicals were added or changed. Currently, these changes have been stayed. However, when PELs are changed, or new specific standards are promulgated, exposure limits and/or hazard information must be amended on MSDSs within three months of the change.

7.1.3 Ethylene Oxide (29 CFR 1910.1047)

Requirements for labeling are found within this standard. When there is a more specific regulation covering a chemical, such as for ethylene oxide, employers must comply with the specific regulation. In this case, even though the HCS requires the labeling of all hazardous chemical containers, ethylene oxide containers do *not* have to be labeled unless they meet the requirements specified in the ethylene oxide regulation (12). When labeling *is* required under this regulation they supersede those of the HCS.

7.1.4 Occupational Exposure To Hazardous Chemicals in Laboratories (29 CFR 1910.1450)

This standard takes precedence for laboratories that fit the laboratory use and scale definition in **29** *CFR* 1910.1450. However, there are laboratories, such as some quality control laboratories, which do not fit the definition in the Laboratory Standard. For these worksites, the HCS applies. Determine first, which definition fits your laboratories. If a laboratory is *not* a "1910.1450 lab", it may fit the definition within the HCS. Laboratories, as described within the HCS, are exempted from certain provisions of the HCS; however, facilities that may be referred to as "labs" may not fit this definition. Pilot plants, dental, photo-finishing, and optical labs, for example, are considered to be manufacturing operations, not laboratories. Careful assessment of how a facility fits these definitions is critical because it determines exactly which requirements will apply: the Laboratory Standard, all aspects of HCS, or the laboratory exemptions within the HCS.

7.1.5 Occupational Exposure to Benzene (29 CFR 1910.1028)

There are references to the HCS within the benzene standard, under sections dealing with the communication of benzene hazards to employees. Training and information requirements must meet those of the HCS when benzene is present. However, if exposures exceed the action level, training and information must be provided at least annually. The labeling requirements for benzene also supersede those of the HCS. Consider the HCS as a minimum requirement. If benzene exposure fits those specified within the benzene standard, the more specific standard prevails.

7.1.6 Process Safety Management (29 CFR 1910.119)

Specific highly hazardous toxic, reactive, flammable, or explosive chemicals are listed in the Process Safety Management (PSM) standard. If these materials are used at threshold quantities, the provisions of the PSM standard apply. The requirements for providing employees with information on these materials go well beyond those of the HCS. MSDSs may be used to provide a portion of the information, but will not provide all the information required.

7.1.7 Hazardous Waste Operations and Emergency Response (29 CFR 1910.120)

This standard covers both hazardous waste handling and emergency response. Most of the confusion with HCS occurs in the interpretation of the training requirements for emergency procedures.

Where the emergency procedures involve evacuation of the work areas and notification to emergency responders, the training required includes emergency procedures, emergency notification (alarms, etc.), and evacuation routes. Where employees are required to take additional actions, the level of training required would be dictated by **29** *CFR* 1910.120. At a minimum, it would include: emergency procedures, spill and leak cleanup procedures, personal protective equipment, decontamination procedures, shut-down procedures, recognizing and reporting incidents and evacuation. Careful reading of both standards and possibly consultation with legal counsel or regulators may be needed to understand how they apply to a given situation.

7.1.8 Specific Industry Standards

The General Industry HCS was adopted for Longshoring (**29** *CFR* 1918.90), Shipyard Employment (**29** *CFR* 1915.99), Construction (**29** *CFR* 1926.59), and Agriculture (**29** *CFR* 1928.21) under the specified Code of Federal Regulations citations.

7.2 EPA Regulations

7.2.1 Superfund Amendments and Reauthorization Act

On October 17, 1986, the Superfund Amendments and Reauthorization Act of 1986 (SARA) was enacted into law. Title III of SARA: the Emergency Planning and Community Right-to-Know Act of 1986 (13) establishes requirements for federal, state, and local governments and industry regarding emergency planning and community right-to-know reporting on hazardous and toxic chemicals. When SARA Title III was enacted, the HCS applied only to the manufacturing sector. However, in 1987 the HCS was expanded to cover all employers. As a result, SARA Title III requirements were also extended to all employers. Community Right-to-Know reporting requirements as detailed in **40** *CFR* 370 require facilities that must prepare or have MSDSs available under the OSHA HCS to submit copies of MSDSs or a list of the MSDS chemicals above minimum threshold levels to:

- The Local Emergency Planning Committee.
- The State Emergency Response Commission.
- The local fire department.

If a list of chemicals is submitted, it must include the chemical name or common name of each substance and any hazardous components as provided on the MSDS. The list must be organized in categories of health and physical hazards as set forth in OSHA regulations unless modified by EPA.

Minimum threshold levels for submitting MSDSs are greater than or equal to 500 pounds of extremely hazardous chemicals or equal to or greater than 10,000 pounds of hazardous substances or the threshold planning quantity (TPQ). Facilities required to have MSDSs must also prepare a hazardous chemical inventory (Tier I or Tier II forms) for the emergency response groups listed above. These requirements are also subject to minimum threshold levels.

7.2.2 Resource Conservation and Recovery Act

The Resource Conservation and Recovery Act (RCRA) enacted in 1976 (14) regulates the management of hazardous wastes. One of the requirements of this regulation is that employees involved in the handling of hazardous wastes receive training. Although materials covered under this regulation are specifically exempt from the provisions of the HCS, the requirements for training are similar. Because employees handling hazardous materials must receive training under the HCS, combining this training in some cases can save time.

7.2.3 Worker Protection Standard for Agricultural Pesticides

This standard issued on August 21, 1992 (15) regulates the exposure of farm workers to pesticides. In an effort to coordinate regulatory jurisdiction, OSHA has agreed not to cite employers who are covered under this rule with respect to the HCS. Agricultural operations otherwise covered by OSHA regulations will still be required to have a Hazard Communication program for chemicals other than pesticides. An appropriations rider prevents OSHA from enforcing standards on farms with ten or fewer employees.

7.2.3 Clean Air Act Amendments

On November 15, 1990, the United States Congress enacted amendments to the United States Clean Air Act (16). One of the amendments authorized the creation of an independent federal agency, the Chemical Safety and Hazard Investigation Board.

As defined by Congress, the Board's mission is to provide industries that manufacture, use or otherwise handle chemicals with information to enable identification and mitigation of operational conditions that compromise safety. Congress directed the Board to accomplish its mission by:

- Conducting investigations of chemical accidents both at fixed facilities and "on the road."
- Evaluating and advising Congress on duplication of effort in the prevention of industrial chemical accidents among 14 other federal agencies, including the EPA and OSHA.
- Conducting special studies.
- Developing and communicating recommendations to improve the safety of operations involved in the production, transportation, and industrial handling, use and disposal of chemicals.

Congress also encouraged the Board to engage in international outreach efforts by offering investigative assistance to other countries.

8 INTERPRETATION

8.1 HCS Final Rule

In 1994 OSHA reopened the record and requested comments in several subject areas. This was in part to clarify sections of the standard about which there had been persistent questions or misinterpretations and in part to address criticism. The revised rule was published in February of 1994 (17). The following paragraphs highlight some of the issues clarified or modified.

Employee Information and Training. Requirements have not changed since the original promulgation of the HCS in 1983. The final rule clarified the language of the standard to emphasize the fact that re-training is not required when new chemicals are introduced. It is required when a new *hazard* is introduced.

MSDSs. OSHA has revised the rule on the "automatic distribution" of MSDS with respect to retail distributors in paragraph (g) (7).

- Retail distributors selling hazardous chemicals to employers having a commercial account: provide MSDS on request and post a sign or inform them that MSDS is available. A commercial account is defined within the HCS final rule. Other aspects of MSDS requirements remain unchanged.
- Wholesale distributors selling hazardous chemicals to employers over-the-counter: provide MSDSs upon request at the time of purchase and post a sign or inform them that an MSDS is available.
- Employers without a commercial account purchase hazardous chemical from a retail distributor not required to have MSDS: provide the employer, upon request, with the

HAZARD COMMUNICATION AND WORKER RIGHT-TO-KNOW PROGRAMS

name, address, and phone number of the chemical manufacturer, importer, or distributor from which MSDS can be obtained.

Electronic Access to MSDS. OSHA has acknowledged that electronic access, microfiche and other alternatives to paper MSDS copies are acceptable as long as no barriers to immediate employee access are created by such options.

Labels. When employers choose to use "in-house" labeling systems, employee training may require augmentation to ensure that employees understand and can readily discern the hazards. For example, color-coding or number systems must be explained and target organ information may need to be added, if it is not included on the labels.

In the preamble to the final standard, OSHA commented on the use of the ANSI labeling system. OSHA states that compliant labels can be generated using the system, and that widespread use of consistent labels may generally improve quality; however, they did express concern. HCS requires that the inherent hazard of the material be addressed on the label and the MSDS. The ANSI standard states that the hazard reflected on the label should be based on hazardous exposure resulting from customary and reasonable foreseeable occupational use, misuse, handling and storage. OSHA's conclusion was that it would be possible to generate labels that were not compliant with the HCS, if ANSI labeling guidelines were used exclusively.

Labeling Limitation for Certain Shipments. Solid metal may be considered an "article" and therefore exempt from the labeling requirements. When metal is not considered an article, a change has been made to the standard allowing shippers to send the label once to the customer, as long as the material does not change. OSHA modified the final rule to include similar provisions for wood, plastic, and whole grain.

Multi-employer Worksite Provision. Additional requirements have been included in the standard for employers on a multi-employer worksite producing, using or storing chemicals. Where the employees of other employers at the work site may be exposed, the employer must include in their Hazard Communication program:

- The methods used to provide or make MSDSs available to other employers.
- The methods used to inform other employers of any necessary precautionary measures to protect employees under normal operating conditions and in foreseeable emergencies.
- The methods used to inform other employers of the labeling system used in the workplace.

Other requirements for on-site availability of MSDSs remain unchanged.

Written Hazard Communication Program. OSHA revised the final rule to permit employers of employees who travel between worksites during a shift to maintain MSDSs at the primary work place. This is permissible as long as the MSDSs are accessible in the event of an emergency and accessible to employees at the primary work site. In addition, OSHA revised paragraph (e) (5) of the HCS to allow employers to keep the written Hazard Communication Program at the primary worksite as well.

Articles. "Article" has been redefined and expanded slightly in the final rule. The new definition includes items that release minute or trace amounts of a hazardous chemical, provided that the items do not pose a health risk to employees.

Food, Drugs, Cosmetics, and Alcoholic Beverages. OSHA clarified the exemption on food and alcoholic beverages in retail establishments to include those used, sold or prepared. HCS labeling requirements do not apply to containers of drugs dispensed by a pharmacy to health care providers for direct administration to a patient. OSHA has clarified this issue in (f) (7) of the HCS.

Consumer Products. OSHA clarified the language of the exemption of "consumer products" as defined by the Consumer Product Safety Commission.

Laboratory Coverage. The section regarding laboratory employee training has been revised to emphasize the fact that employees must receive information and training. OSHA has adopted an amendment to specify a laboratory's duties when shipping or transferring a chemical out of the laboratory. In summary, the laboratory must comply with hazard determination, labeling and MSDS requirements as would a chemical manufacturer or distributor.

Labeling Exemptions. Veterinary biological products have been specifically exempted from labeling requirements in the revised final standard. OSHA has likewise exempted seeds labeled in accordance with the Federal Seed Act and substances labeled in accordance with the Toxic Substances Control Act.

Hazardous Waste. The HCS totally exempts hazardous waste regulated under the Resource Conservation and Recovery Act. The exemption has been modified to include hazardous substances regulated by EPA under the Comprehensive Environmental Response, Compensation, and Liability Act.

Wood Dust. Wood and wood products are currently exempt under the HCS. This exemption has been modified to indicate that wood *dust* is not exempt. OSHA recognizes that labeling is impractical; however, MSDS and training requirements apply. In the preamble to the revised standard, OSHA states that the responsibility for generating the MSDS resides with the first employer who handles or processes the raw material such that the hazardous chemical (wood dust) is released into the environment. That would generally be the sawmill. This same logic is applied to grain dust. MSDS generation in that case would probably be the responsibility of the grain elevator.

8.2 Recommendations of the Hazard Communication Workgroup to the National Advisory Committee on Occupational Safety and Health (NACOSH)

"The New OSHA: Reinventing Worker Safety and Health" was issued as part of the Clinton Administration National Performance Review in 1995. One of the recommendations was the formation of a National Advisory Committee on Occupational Safety and Health (NACOSH) workgroup on hazard communication. The Workgroup issued its report to OSHA and NIOSH on September 12, 1996 (18). This section summarizes the twenty-three recommendations in that report.

HAZARD COMMUNICATION AND WORKER RIGHT-TO-KNOW PROGRAMS

MSDSs

OSHA should:

- Endorse the order and section titles described in the ANSI Z400.1-1993.
- Endorse the addition of a statement by the MSDS preparer which indicates whether the product is regulated under the HCS and its hazard type.
- Actively participate in future ANSI Z400.1-1993 revisions.
- Partner with industry, labor and professional association to develop a guidance document for conducting a hazard determination.

Labeling

- OSHA should not mandate the use of symbols for MSDSs and labels without validation studies; OSHA should insist on validation studies prior to adopting symbols in the U.S. as part of an internationally harmonized system; and, if the U.S. adopts symbols, OSHA should require training on their meaning.
- OSHA should not mandate the use of color-coding as the sole method of communicating hazards on labels. As with symbols, the U.S. should insist on validation studies prior to adopting such a system as part of an internationally harmonized system and should require training.
- NACOSH supported the HCS labeling requirement for containers and its requirement of MSDS accessibility for employees.
- NACOSH recommended that OSHA endorse ANSI Z129.1-1994 for precautionary labeling. (See Section 8.1 for OSHA's concerns regarding the application of this ANSI standard.)

Electronic Management of MSDSs

OSHA should require that:

- Electronic devices be readily accessible in the workplace at all times;
- Employees be trained in the use of these devices; and
- Employers must have an adequate backup plan in the event of service disruption.
- Use of an off-site MSDS management service does not relieve the employer of MSDS accessibility requirements or other program requirements.

Employee Training

OSHA should:

- Develop "model" employee training programs and evaluation criteria to assess the overall effectiveness of the training.
- Clarify what portion of the training should stay with the employee who changes employers.

Enforcement of the HCS

OSHA should:
- Adopt an enforcement policy which provides an alternative to citation where minor deficiencies that do not affect the safety and health of employees and are abated within 24 hours.
- Develop ways to address enforcement related to the accuracy of MSDSs.
- Develop an outreach system to address implementation inconsistencies.
- Reemphasize the role of Regional Hazard Communication Coordinators to improve consistency in enforcement and interpretation.
- Increase internal communication on the HCS implementation and to improve consistency in enforcement and interpretation.
- Review compliance memoranda and correct confusion and inconsistencies to ensure that the standard does not change through interpretation.
- Review guidance provided to compliance officers on the application of HCS to consumer products and distribute an information bulletin.
- Use HCS to develop partnerships with industry, labor and professional associations to provide better communication.

Harmonization of Hazard Communication Requirements

OSHA should:
- Continue active pursuit of international harmonization, North American harmonization and harmonization of federal agency requirements for hazard classification, labeling and MSDSs.
- Review substance-specific standards and take action to ensure that future standards are consistent with the HCS.

Misinformation about the HCS

OSHA should implement an outreach effort to systematically address misinformation and misconceptions about the HCS.

9 PROGRAM AUDIT

9.1 Program Appraisal

Every hazard communication program should be periodically reviewed to determine its effectiveness and degree of compliance with the OSHA standard. The best form of review is a formal audit, performed either by an independent outside party, such as a consultant, or an internal knowledgeable individual, such as an industrial hygienist or safety professional.

The audit should include a review of records to determine the adequacy of the documentation of the program, as well as a determination of how well the program is being performed.

9.2 Record Review

A key part of every OSHA compliance officer's visit is a review of an employer's written hazard communication program. As discussed earlier, the written hazard communication program provides an explanation of all the elements of an employer's hazard communication program and how they are implemented. Therefore, a logical first step of any audit is a review of the written hazard communication program for the required elements, as discussed in Section 2.3.1.

The MSDS is an essential ingredient of every hazard communication program. For manufacturers, backup information should be available for MSDS preparation. The source of hazard determination and protective measure information should be available. In some cases each individual MSDS may warrant auditing.

For an employer using chemicals, the availability of MSDSs to employees should be audited. Documentation to be reviewed includes (*1*) MSDS copies in the workplace, (*2*) written requests to suppliers for missing or inadequate MSDSs, and (*3*) tracking method for new and updated MSDSs.

Hazard communication training should be well documented and therefore a review of training records is an important part of the record review. The records should include a description of (*1*) what was covered at each training session, (*2*) who was in attendance and (*3*) when the training occurred. Training of new employees and employees transferred to a department with different chemicals should also be audited. Documentation of follow-up training for new hazards in the workplace should also be reviewed. Refresher training is required at specific time intervals in some states. The documentation should be reviewed for compliance with state as well as federal OSHA requirements.

Specific labeling systems should be audited for their compliance with OSHA's requirements. Manufacturers must include (*1*) identity of the material, (*2*) hazard warnings, and (*3*) the name and address of the manufacturer, supplier, importer, or other responsible party. Containers used at a specific facility, such as bulk tanks or small containers used at workstations, must include the first two items. In both cases, the labels, particularly the hazard warning, should contain specific information regarding target organ effects, when applicable.

9.3 Performance

The HCS is a "performance standard." Thus employers have much latitude in determining how to comply with specific provisions of the standard.

The most important test of its effectiveness is whether a program actually results in a more aware worker. The OSHA compliance officer tests this by actually interviewing employees at random and asking them about the chemicals in their workplace (19). A good audit will also include this spot check. It is not feasible to interview every worker, but a few employees in several different work settings should be interviewed. For example, maintenance, process control, in-house construction workers, and outside contractors should be included in the interview (when possible), in addition to production employees with routine exposure. Employees should be asked key questions about hazard communication such as:

- What are the potential hazards of that chemical?
- What protective equipment do you use when working with that chemical?
- Where would you find the MSDS?
- What would you do if some of that chemical spilled?
- Did you receive any training regarding chemicals?

Other elements of the program, such as MSDS and labeling, should be checked in the workplace for (*1*) compliance with OSHA standards, (*2*) implementation in the workplace, and (*3*) actual performance. This part of the audit should be through a random check, as with training. A number of chemicals should be selected in the workplace. The adequacy of the labels on the containers should be checked at the workstation. The MSDS should then be located and checked for completeness and accessibility. The protective measures and air monitoring data should also be reviewed as appropriate.

9.4 Checklist

The checklist shown in Figure 39.1 may be used a guideline for a hazard communication audit. This checklist may be augmented to include state and local regulations. It also can be supplemented to include related regulations such as SARA Title III. The checklist is prepared in two parts, one for employers and one for suppliers. The appropriate portions may be used for specific audits.

10 CONCLUSION

The HCS has been called a performance-oriented standard, in that the specific means of accomplishing the requirements of the standard are left to the employer. The measure of an effective hazard communication program, therefore, is whether employees recognize the nature and extent of the hazards posed by chemical substances and understand what steps must be taken to protect themselves.

All components of an effective safety and health program should be integrated into the overall management structure. The level of commitment, the implementation strategies, the measurement of effectiveness, funding, and responsibility for program implementation should be considered when developing or redesigning a hazard communication program.

Employers with an effective hazard communication program will reap the benefits of enhanced employee and community relations. Overall, OSHA expects this standard to benefit the workforce in the ways indicated below:

- Reduction of accidents, injuries, and illness related to exposure to hazardous materials.
- Reduction of workers' compensation cost.
- Reduction of lost time accidents.
- Improved productivity.
- Improved employee morale.

HAZARD COMMUNICATION AND WORKER RIGHT-TO-KNOW PROGRAMS

Part 1

Hazard Communication Audit - Employers

WRITTEN PROGRAM

A. Statement of Purpose (*Optional*)

  [ ] Purpose of Hazard Communication Program

  Comments: _____

B. Location:

  [ ] Available for review upon request to employees and OSHA

  Comments: _____

C. Labeling:

  [ ] Who is responsible for checking and maintaining original labels?

  [ ] Who is responsible for making and maintaining in-plant labels?

  [ ] Description of in plant labeling system for secondary containers.

  [ ] Description of labeling system for bulk storage tanks (*if applicable*)

  [ ] Description of alternative systems used for batch processes (*if applicable*)

  [ ] Description of fixed labels (*if applicable*)

  Comments: _____

D. MSDS

  [ ] Who is responsible for requesting MSDS and keeping files up to date?

  [ ] How are MSDS obtained for every Hazardous Substance?

  [ ] Where are MSDS kept?

  [ ] Accessibility to employees

  [ ] Review of MSDS for omissions (checklist) (optional).

  [ ] How are missing MSDSD requested?

  [ ] MSDS alternatives (if used).

  Comments: _____

Figure 39.1. Checklist as a guideline for a hazard communication audit.

E. **Training**

   [ ] Who is responsible?

   [ ] Initial training plan.

   [ ] New employee training.

   [ ] Refresher training plan. (optional)

   [ ] Update of training for new hazards.

   [ ] Documentation of training.

   Comments: _____

F. **List of Hazardous Substances**

   [ ] Included in written program.

   [ ] Identities match MSDS and labels.

   [ ] Procedure to update list.

   [ ] Locations listed by work area.

   Comments: _____

G. **Non-Routine Task and Unlabeled Pipes**

   [ ] How employees will be informed.

   [ ] Who is responsible for informing employees?

   [ ] Incorporation of existing procedures.

   Comments: _____

H. **Contractors**

   [ ] How they will be informed of hazards and protective measures.

   [ ] How they will provide MSDSs for hazardous substances they bring onsite.

   [ ] Who is responsible for informing contractors and obtaining MSDs?

   Comments: _____

Figure 39.1. Continued.

I. **Multi-Employer Workplaces**

    [ ] Methods to provide onsite access to MSDSs to other employers.

    [ ] Methods to inform other employers of precautionary measures.

    [ ] Methods to inform other employers of labeling system.

    Comments: _____

J. **Hazard Determination (*optional*)**

    [ ] How chemicals are evaluated (e.g., rely on supplier or do own evaluation).

    [ ] Who is responsible for evaluating chemicals?

    Comments: _____

LABELING

A. **Labeling System**

    [ ] Incude product identity and hazard warning (health and physical hazards).

    [ ] Labels displayed and legible.

    [ ] Method for replacing damaged labels.

    [ ] Checking labels on incoming containers.

    Comments: _____

Figure 39.1. Continued.

MATERIAL SAFETY DATA SHEETS

[ ] MSDS available for each hazardous substance.

[ ] Accessible location to employees on all shifts.

[ ] Requests for missing MSDS documented.

[ ] Requests for update of incomplete MSDS documented.

[ ] Review of MSDS for completeness. (optional)

[ ] MSDS in all files up-to-date.

[ ] Method for updating MSDS files.

[ ] Provision for providing MSDS to employees, physicians, and/or representatives upon request.

Comments: _____

INFORMATION AND TRAINING

A.   Training Curriculum Include:

[ ] Explanation of MSDS and information it contains.

[ ] MSDS contents for each substance or class of substances used.

[ ] Explanation of requirements of Hazard Communication Standard.

[ ] Location and availability of written program.

[ ] Location of operations where hazardous substances are used.

[ ] Observation and detection methods for hazardous substance presence or release.

[ ] Physical and health hazards of substances used in work area.

[ ] Protective measures for work area.

[ ] Details of employers hazard communication program, including labeling and MSDS program.

Comments: _____

Figure 39.1.   Continued.

B. Training Update

[ ] Periodic refresher training. (optional)

[ ] Provision for informing employees within 30 days of receiving a new or revised MSDS which indicates significantly increased risks or protective measures.

[ ] Provisions for training employees when a new hazard is introduced into the workplace.

Comments: _____

C. Training Documentation

[ ] Attendance lists.

[ ] Tests of comprehension (optional).

Comments: _____

Part II

Hazard Communication Audit - Suppliers

[ ] Who is responsible for hazard determination?

[ ] List of information sources used.

[ ] Written procedure for evaluating all products.

[ ] Plan for getting updated information.

Comments: _____

MATERIAL SAFETY DATA SHEETS

[ ] Who is responsible for:

    [ ] Preparing and updating MSDS.

    [ ] Distributing MSDS.

    [ ] Replying to MSDS requests from customers.

Figure 39.1. Continued.

[ ] Complete and accurate MSDS written for each product.

[ ] Copies of MSDS provided to each product.

[ ] Method of providing MSDS: (e.g. with bill of lading, mailed to purchaser).

[ ] Procedure for updating MSDS when new information becomes available.

Comments: _____

LABELS

[ ] Who is responsible for:

    [ ] Determining label wording.

    [ ] Updating label wording.

    [ ] Making and maintaining labels.

[ ] Product containers labeled with:

    [ ] Name and address of manufactuer or supplier.

    [ ] Product identity (same as on MSDS).

    [ ] Hazard warning (specific physical and health hazards).

TRADE SECRETS (if applicable)

[ ] Who is responsible for making trade secret determination?

[ ] Justification documentation.

[ ] Procedure for applying to OSHA for trade secret status.

[ ] Procedure for providing trade secret chemical identities in emergencies.

[ ] Procedure for providing trade secret chemical identities in non-emergencies under specified conditions.

[ ] Standard Confidentiality Agreement? (optional)

Figure 39.1. Continued.

These benefits notwithstanding, communicating technical information is not without risk. Communication on any subject can be plagued with misinterpretation. It is doubly so for issues that have become emotionally charged. However, the consequences of non-compliance with the HCS include OSHA citations, and penalties, adverse publicity, and increased liability for failure to inform employees of hazards that, in extreme cases, can involve criminal prosecution. Failure to warn employees of a potential health hazard may be interpreted as a negligent violation of the workers' rights. This may permit civil actions against the employer outside the protection of the workers' compensation system.

BIBLIOGRAPHY

1. The Occupational Safety and Health Act of 1970 (P.L. 91-596, 91st Cong., S.2193, Dec. 29, 1970) is codified at **29** *USC* 651 et seq.
2. *Fed. Reg.* **47** FR 12092.
3. *Code of Federal Regulations* **29** *CFR* 1910.1200.
4. Longshoring (**29** *CFR* 1918.90), Shipyard Employment (**29** *CFR* 1915.99), Construction (**29** *CFR* 1926.59), and Agriculture (**29** *CFR* 1928.21).
5. *Fed. Reg.* **59** FR 6169-84.
6. *Fed. Reg.* **59** FR 6126.
7. National Advisory committee on Occupational Safety and Health (NACOSH), *Report of the Hazard Communication Workgroup to the National Advisory Committee on Occupational Safety and Health*, September 12, 1996, transmitted to the Occupational Safety and Health Administration and the National Institute for Occupational Safety and Health.
8. American National Standard for Hazardous Industrial Chemicals, Material Safety Data Sheet Preparation, ANSI Z400.1-1993, ANSI, New York, 1993.
9. **29** *CFR* 1910.1200 (c).
10. "Whole Lotta Shakin' Goin' On in MSDS Software." *Environmental Software Report* **VII** (8). (Summer 1997).
11. U.S. Department of Labor, Occupational Safety and Health Administration, *OSHA Instruction CPL 2-2.38D*, March 20, 1998, pages 19–20.
12. *Ibid.*, p. 19.
13. Enacted by Public Law 99-499, Oct. 17, 1986, **42** *USC* δδ 11001–11050.
14. **42** *USC* δδ 6901-6992k, EPA regulations in **40** *CFR* Parts 124, 260–280.
15. **57** FR 38102 (Aug. 21, 1992) *Worker Protection Standard for Agricultural Pesticides*.
16. Public Law 101-549.
17. **59** FR 6126-6169, **29** *CFR* Part 1910, et al., Hazard Communication: Final Rule.
18. National Advisory committee on Occupational Safety and Health (NACOSH), *Report of the Hazard Communication Workgroup to the National Advisory Committee on Occupational Safety and Health*, September 12, 1996, transmitted to the Occupational Safety and Health Administration and the National Institute for Occupational Safety and Health.
19. *OSHA Instruction CPL 2-2.38D*, March 20, 1998, pages 14–15.

GENERAL REFERENCES

Several of the books or journals listed below are also cited in the Bibliography; other books, journals, and articles are listed because they provide excellent background information on the topic.

U.S. Department of Labor, Occupational Safety and Health Administration, Directive Number CPL 2-2.38D, *Inspection procedures for the Hazard Communication Standard*, **29** *CFR 1910.1200, 1915.99, 1917.28, 1918.90, 1926.59 and 1928.21*, March 20, 1998, Washington, DC, 1998.

U.S. Department of Labor, Occupational Safety and Health Administration, Office of Training and Education, *Hazard Communication: A Key to Compliance*, Washington, DC, 1994.

U.S. Department of Labor, Occupational Safety and Health Administration, OSHA 3084, *Chemical Hazard Communication*, Washington, DC, 1994.

U.S. Department of Labor, U.S. Department of Labor Program Highlights, Fact Sheet No. OSHA 93-26, *Hazard Communication Standard, U.S. GPO: 1993-342-523/72520*, Washington, DC, 1993.

AIHA Publications, *Hazard Communication: An AIHA Protocol Guide*, Fairfax, VA, 1995.

United States General Accounting Office Report to Congressional Requesters, *Occupational Safety & Health: OSHA Action Needed to Improve Compliance with Hazard Communication Standard*, GAO/HRD-92-8, Washington, DC, 1991.

United States General Accounting Office Report to Congressional Requesters, *Occupational Safety & Health: Employers' Experience in Complying With the Hazard Communication Standard*, GAO/HRD-92-63BR, Washington, DC, 1992.

M. L. Goldstein, "Hazard Communication in the Workplace," *Hofstra Labor Law Journal*, **7.2**, 303–340 (1990).

R. Piccioni, "Industry's Response to OSHA's Hazard Communication Standard: Is the Law Working as Intended?" *Chemical Times & Trends* 31–38, (April, 1990).

T. G. Robins, M. K. Hugentobler, M. Kaminski, and S. Klitzman, "Implementation of the Federal Hazard Communication Standard: Does Training Work?" *J. Occupational Med.* **32**, 1133–1140 (Nov. 1990).

Occupational Safety and Health Reporter, The Bureau of National Affairs, Inc., 1992–1996.

Hazard Communication Compliance Manual for the Society for Chemical Hazard Communication, The Bureau of National Affairs, Washington, D.C., 1995.

CHAPTER FORTY

Pharmacokinetics and Unusual Work Schedules

Dennis J. Paustenbach, Ph.D., CIH, DABT

1 INTRODUCTION

The concern about the adverse health effects of night shift work has existed for over 100 years. In 1860, for example, some scientists were worried that bakers might be at increased risk of physical and emotional illness because they always worked at night and, subsequently, some effort was made to try to regulate their work hours and their working conditions (1). Since then, it has been shown that some persons who have been very productive while working standard 8-hr/day, 40-hr/week daytime schedules can become especially fatigued, unhappy, less productive, and perhaps even more susceptible to the effects of chemical agents and physical agents after they are placed on shift work. Even though many studies have been conducted, the degree to which shift work effects a worker's capabilities, longevity, mortality, and overall wellbeing is still not fully understood.

So-called unusual work shifts and work schedules have been implemented in a number of industries in an attempt to eliminate or at least reduce, some of the problems caused by normal shift work that require three work shifts per day. These unusual or "other-than-normal" shifts have been termed odd, novel, extended, extraordinary, compressed, non-normal, nonroutine, prolonged, exceptional, nonstandard, unusual, peculiar, weird, and nontraditional (2). In 1981 the AIHA established a committee to address the potential occupational health aspects of shift work and the need to adjust exposure limits for persons who work schedules that were markedly different than the "normal" workweek, which consists of five consecutive 8-hr daylight workdays followed by two days off. At the committee's first meeting, it was agreed that the term *unusual work shift* should be used,

---

Patty's Industrial Hygiene, Fifth Edition, Volume 3. Edited by Robert L. Harris.
ISBN 0-471-29753-4   © 2000 John Wiley & Sons, Inc.

for sake of consistency, to describe these other-than-normal shifts. The committee completed its work by 1990 and it was retired.

In general, most unusual work schedules will involve workdays markedly longer than 8 hr in duration, however, because many persons are regularly exposed to high concentrations of xenobiotics for very short periods during shifts and, because TWA occupational exposure limits were not necessarily intended for use during short exposures, these exposure periods were also classified as "unusual" by the committee. The assessment of the health aspects of unusual shifts is complex since some of these schedules require the worker to alternate between night and day work every few days (rapid rotation). These schedules have been called rapidly rotating, fast, or simply rapid-roto shifts (3–6).

In order for the health professional to ensure protection of workers who are exposed to airborne chemicals during unusually long or unusually short work periods, he or she should be familiar with the toxicology and pharmacokinetics of the chemical of interest as well as understand the rationale for its occupational exposure limit, since most limits were established to prevent injury following exposure during a normal 8-hr/day, 5-day/workweek schedule.

1.1 Background on Shift Work

Traditionally, a fixed work schedule is established by the employer and consists of five 8-hr days each week starting at 8 A.M. and ending at 5 P.M. Initially, the need to operate certain manufacturing processes 24 hr/day served as the impetus for using 3 shifts of workers to cover the 24-hr workday. Later, shift work was implemented because the economics of having idle equipment 70% of the week was prohibitive. Of course, the use of three shifts per day allows uninterrupted production throughout the year so that a process never has to shut down. Many continuous process operations, such as those found in oil refineries, chemical plants, steel and aluminum mills, pharmaceutical manufacturing, glass plants, and paper mills, cannot be shut down without causing serious production and financial losses so they require round-the-clock manning (2, 4).

In many professions adherence to the 40-hr/week schedule is unusual. For example, craftspeople and equipment repair persons frequently work beyond 8 hr/day (overtime) to accommodate routine equipment repair, fill in for absent workers, or deal with seasonal fluctuations in demand (4). For some persons, such as those in the railroad industry and the military, a workday of 16–24 hr is not uncommon. In certain industries, very complicated work schedules that have involved both short and long periods of work have been implemented for a wide number of reasons. The advantages and disadvantages of alternative (unusual) work schedules, as well as the number of persons involved in various types of work schedules have been discussed (7–10).

Dozens of alternatives to traditional work schedules have been developed and are now used in a number of industries. As observed during the recession of the 1980s and early 1990s, the length of time worked by a person was often altered by the use of part-time employees and many workweeks were shortened or lengthened due to wide fluctuations in the demand for a given product. Overall, the high cost of manufacturing equipment, increased foreign competition, and the inability to halt certain chemical and physical processes after they have begun has forced industry to make shift work and modified work schedules a permanent part of manufacturing (4).

The shift worker employed in a 24-hr/day operation, usually on a rotating, 8-hr schedule, faces many disruptive events that arise due to his or her work schedule. Overall, the primary complaint of these workers surrounds the unsatisfactory impact shift work has on their social lives (3, 4, 10, 11). When they are home, most other persons are asleep, at work, or at school. In general, these shift workers have only one full weekend per month during which they are not at work. As a result, shift workers and members of their family are frequently disappointed with the quantity and quality of time that they spend together (4, 13–16).

Researchers who have studied the effects of shift work on human health have noted that shift workers may have trouble sleeping, often feel fatigued, often feel overly tired during days off work, and are often chronically irritable (13, 16, 17). Other undesirable effects of shift work that are more easily measured have included constipation, gastritis, gastroduodenal ulcers, peptic ulcers, high absenteeism, and lessened productivity. As a consequence, numerous kinds of unusual work schedules have been developed and implemented by companies who are seeking alternatives to standard shift work. Throughout this chapter, standard shift work is always defined as a workweek of 8 hr/day and 5 days that occurs during daylight hours, usually between 8 A.M. to 5 P.M., and is followed by 2 days off work.

2 UNUSUAL WORK SCHEDULES

One kind of work schedule classified as unusual is the type involving work periods longer than 8 hr and varying (compressing) the number of days worked per week (e.g., a 12-hr/day, 3-day workweek). Another type of unusual work schedule that involves a series of brief exposures to a chemical or physical agent during a given work schedule (e.g., a schedule where a person is exposed to a chemical for 30 min, 5 times per day with 1 hr between exposures). Another type of unusual schedule is that involving the "critical case" wherein persons are continuously exposed to an air contaminant (e.g., spacecraft, submarine).

Compressed workweeks are a type of unusual work schedule that has been used primarily in nonmanufacturing settings. It refers to full-time employment (virtually 40 hr/week) that is accomplished in less than 5 days/week. Many compressed schedules are currently in use, but the most common are (a) 4-day workweeks with 10-hr days; (b) 3-day workweeks with 12-hr days; (c) 4.5-day workweeks with four 9-hr days and one 4-hr day (usually Friday); and (d) the 5/4, 9 plan of alternating 5-day and 4-day workweeks of 9-hr days (8–10).

Over the past two or three decades, unusual work shifts and schedules including some form of the compressed workweek have been implemented in many manufacturing facilities (2, 18). Examples of the types of schedules and one type of industry that has used them include four 10-hr workdays per week (chemical); a 6-week cycle of three 12-hr workdays for 3 weeks followed by four 12-hr workdays for 3 weeks (pharmaceutical); a 6-hr per day, 6-day workweek (rubber); a 56/21 schedule involving 56 continuous days of work of 8 hr per day followed by 21 days off (petroleum); a 14/7 schedule involving 14 continuous days of work of 8–12 hr per day followed by 7 days off (petroleum); a 3/4 schedule involving only three 12-hr workdays in one week followed by a week of four 12-hr workdays (pharmaceutical); four 12-hr workdays followed by three 10-hr workdays, followed by five 8-hr workdays then 4 days off; five 8-hr workdays, followed by two 12-hr days, then 5 more 8-hr workdays followed by 5 days off (petrochemical); a 2/3 schedule involving 18 hr of work for 2 days then 3 days off (military); and numerous other variations of these (2, 4). Of all workers, those on unusual schedules represent only about 5 percent of the working population (9). Of this number, only about 50,000–200,000 Americans who work unusual schedules are employed in industries where there is routine exposure to significant levels of airborne chemicals. In Canada, the percentage of chemical workers on unusual schedules is thought to be greater than in most other countries due to the number of persons employed by the petroleum, paper, and chemical industries.

3 CHEMICAL PHARMACOKINETICS

Pharmacokinetics is defined as the study of the rate processes of absorption, distribution, metabolism, and excretion of drugs and toxicants in intact animals (19). Many of the approaches or models that will be described in the following sections for adjusting exposure limits for both a series of short and very long or continuous exposure periods are based on the pharmacokinetics of the toxicant.

Pharmacokinetic modeling is the science of describing, in mathematical terms, the time course of drug and metabolite concentrations in body fluids and tissues (20, 21). Most often, the modeling of a chemical's behavior is based on the blood concentration of a contaminant versus the time course of its concentration in the blood following exposure. It is generally assumed that blood is in dynamic equilibrium with all tissues and fluids of the body; consequently, it is the biologic medium of choice for monitoring. In some cases it may be necessary or useful to measure the output of its parent chemical or drug and the metabolites in the urine, breath, and feces and then model this profile mathematically.

A pharmacokinetic model is regarded as being accurate if it correctly predicts the concentration of a substance or its metabolite in the blood, urine, breath, or feces for as long as these substances can be measured analytically. The model may be tested by comparing its predictions with experimental data after exposure to the drug by intravenous, oral, or inhalation administration. The effects of various dosing regimens can also be modeled. Usually, recovery of $+95\%$ of the administered dose(s) as unchanged drug and metabolites in urine, breath, and feces, and agreement between the recovered amounts and the pharmacokinetic model is considered confirming evidence that the model accounts for the fate of the absorbed chemical or drug (22).

The principles of industrial pharmacokinetics (i.e., absorption, distribution, metabolism, and excretion) are discussed here with emphasis on those principles necessary to understand the various models for adjusting occupational exposure limits for unusual periods of exposure.

3.1 Absorption of Chemicals

For workers, the two predominant routes of entry for industrial chemicals are absorption through the respiratory tract and the skin. Absorption that takes place through the lungs can be roughly assessed through analysis of the air that is breathed. The results can be interpreted by comparison with an exposure limit. Absorption via the skin, on the other hand, evades this method of measuring exposure. Uptake by the skin cannot be directly accounted for except by biological monitoring. It is acknowledged that methods for analyzing most chemicals in various excretory pathways or fluids have not been developed for most chemicals; therefore, when it is anticipated that a significant portion of the daily dose may be due to skin absorption, it can be assumed that a proportionally greater amount will be absorbed by persons who work on shifts longer than 8 hr/day. Indirect methods such as those used in health risk assessment are another approach to estimating dermal uptake (23).

Absorption of organic vapors may take place both in the upper and in the lower parts of the respiratory tract. Two basic processes may be observed here depending on the physical properties of the toxic substance. If the toxic substance is present in air in the form of an aerosol, absorption will often be preceded by deposition of the substance in the upper respiratory tract. When very small particles are present (less than 10 μm diameter), the deposition will be in the alveolar region.

The basic absorption mechanism for most industrial chemicals is gas diffusion, and the factor that predicts the efficiency of this process is the partition coefficient between air and blood. Values for the coefficient between air and blood have been estimated for perhaps fifty or more volatile chemicals (24). The coefficients for partition between air and water are available for a far greater number of compounds. Usually, these have been obtained while working out sampling procedures involving absorption of a gaseous substance in water. By knowing the oil/water partition coefficient, one can often predict the relative efficiency of absorption through the skin and the lung. In addition, the partition coefficient and degree of solubility in water of a substance can give some insight as to the degree of distribution among the various tissues and the likely rate of elimination. For chemicals that have not been studied pharmacokinetically, these physical properties (especially the partition coefficient) can be very useful for predicting the chemical's behavior during unusually long exposures (25).

The respective values of the air/water partition coefficients for organic compounds may vary by several orders of magnitude: from about 10 for carbon disulfide, through 10^{-3} for acetone, acrylonitrile, and nitrobenzene, to 10^{-5} for aniline and toluidine. The degree of pulmonary absorption (i.e., retention of vapors in the lungs) increases with a decreasing partition coefficient for air/blood (water); however, variation of the retention is much less than the variation of the respective partition coefficients. For example, lung retention of aniline vapor (partition coefficient 10^{-5}) is about 90 percent; nitrobenzene (partition coefficient 10^{-3}) up to 80%; benzene (partition coefficient 10^{-1}) up to about 50–75%; and finally, carbon disulfide (partition coefficient 10^0) up to about 40% (25).

3.2 Concept of Half-Life

Most biologic processes follow first-order kinetics. In other words, the rate of elimination or metabolism is known to be a function of the concentration of all reactive species and, where there is only one of these, the reaction is termed first-order and the rate is directly proportional to the concentration:

$$\text{Rate} = dC/dt = kC \tag{1}$$

This is the differential form of a first-order reaction and is of little practical use since dC/dt, although it can be found from the tangent of the time-concentration curve, is very difficult to measure precisely.

The integrated form of Eq. 1 is more useful:

$$C_t = C_0 e^{-kt} \tag{2}$$

C_0 is the initial concentration and is sometimes designated a; it is not unusual to find C_t, the concentration at time t, designated $a - x$, where x is the amount that has disappeared at the time t. Thus Eq. 2 sometimes appears as

$$\ln \frac{a}{a - x} = kt \tag{3}$$

By definition, the half-life ($t_{1/2}$) for any first-order kinetic process is the time taken for the original amount or concentration to be reduced by one-half. So from Eq. 3, if $C_t = \tfrac{1}{2}C_0$ at time $t_{1/2}$, then

$$\begin{aligned}
\ln \frac{C_0}{C_0/2} &= kt_{1/2} \\
\ln 2 &= kt_{1/2} \\
t_{1/2} &= \frac{\ln 2}{k} = \frac{2.303 \times 0.3010}{k} \quad \text{(e.g., min, hr)}
\end{aligned} \tag{4}$$

Note that $t_{1/2}$ is independent of concentration for a first-order process and that the larger the half-life, the slower the elimination or reaction rate, that is, the smaller the rate coefficient k. The fundamental concepts are illustrated in Figure 40.1.

When describing the elimination of a chemical from humans or experimental animals, it is often convenient to use the term *biologic half-life*. The biologic half-life of a chemical is the time needed to eliminate 50% of the absorbed material, either as the parent compound or one of its metabolites. As will be shown, the biologic half-life is the most important criterion for assessing whether a TLV should be adjusted for either very short or very long durations of exposure. The second key factor is the type of adverse effect or hazard posed by the chemical. The biologic half-life and the exact exposure regimen (time exposed and recovery time) are the two factors that determine the degree of adjustment needed to provide equal protection.

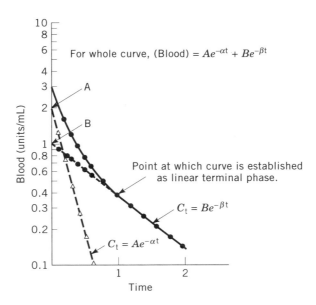

Figure 40.1. Semilogarithmic plot of blood concentration versus time after intravenous administration illustrating biphasic elimination (two compartments) and the use of "curve stripping" to resolve the A intercept.

3.3 Concept of Steady State

The concept of steady state can be a difficult one for persons to comprehend. It is easy to visualize how persons who are exposed to a chemical with a short half-life will rapidly eliminate the chemical so that the blood and tissue levels of the chemical return to zero before returning to work the next day (e.g., carbon disulfide). The other extreme is represented by chemicals such as the 2,3,7,8-tetrachlorodibenzo-p-dioxin (TCDD), which have extremely long half-lives in humans (whole body biologic half-life = 7 to 9 years). Here, the body burden never returns to zero before reexposure occurs so each successive dose adds to the existing burden until the steady state concentration is reached in the deep tissue compartment. These phenomena are shown in Figures 40.2 and 40.3.

The goal of the approach to adjusting occupational limits, however, is to identify a dose that ensures that the daily peak body burden or weekly peak body burden does not exceed during unusual work schedules that which occurs during a normal 8-hr/day, 5-day/week shift. The potential for the body burden to exceed normal levels during unusually long work periods exists whenever the biologic half-life for the chemical in man is in the range of 3–200 hr (see Fig. 40.3). The key point to understand here is that for any chemical, a steady-state plasma or tissue level will eventually be achieved following regular exposure to any work schedule and even during continuous exposure. At first glance, most persons might think that with continuous exposure, the body burdens will continuously increase as long as exposure is maintained. However, this is not true. For any chemical, a steady-

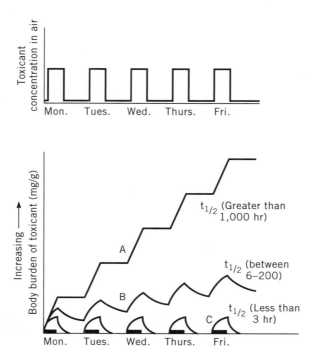

Figure 40.2. Body or tissue burden as a function of a chemical's uptake and elimination rate (i.e., biologic half life).

state blood or tissue level of the contaminant will be achieved when the rate of elimination of the drug equals the rate of absorption (19, 27). (See Figure 40.4.) A rule-of-thumb is that steady-state body burdens occur when exposure occurs after a period greater than five biologic half-lives. For example, for TCDD, the steady-state concentrations in adipose tissue occurs after about 35 years (5×7 yr).

During continuous inhalation exposure to volatile workplace gases and vapors, the concentration of the chemical in the blood increases toward an equilibrium between absorption, on the one hand, and metabolism and elimination, on the other. This is accompanied by a decreasing retention of the absorbed gases and vapors during each breath. This decrease of retention during the early periods of continuous inhalation exposure may be observed in practice for compounds whose air/blood (water) partition coefficient is of the order of 10^{-3} or greater. Decreasing retention is characteristic for carbon disulfide, tri- and tetrachloroethylene, benzene, toluene, nitrobenzene, and other chemicals (carbon disulfide is illustrated in Fig. 40.5). If metabolism and distribution are rapid, this phenomenon is less pronounced and, for example, has not been observed for aniline (low partition coefficient) or for styrene (25). In the case of styrene, analogous to acrylonitrile, the explanation for lack of time-dependent decrease of retention seems to lie not so much in the magnitude of the physical partition coefficient as in the metabolism of the chemical.

PHARMACOKINETICS AND UNUSUAL WORK SCHEDULES

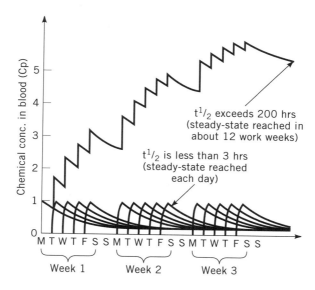

Figure 40.3. The principle of graphic summation assuming regular weekly periods free of exposure (From Ref. 26). Illustrates that biologic half-life determines the degree of day-to-day accumulation and the time to steady-state.

The overall or average retention (R) of an organic vapor has been studied on human volunteers in chamber-type experiments, where it can be determined for a particular exposure period directly from the ratio of concentrations of the chemical in the inhaled and expired air:

$$R = \frac{C_i - C_e}{C_i} \tag{5}$$

C_i and C_e denote the concentrations in inhaled and exhaled air, respectively.

It is often useful for the industrial hygienist to know how to calculate an individual's anticipated daily uptake (absorbed dose) of a toxicant. One of the key determinants or elements in these calculations is the ventilation rate. In chamber-type experiments, in which volunteers are usually exposed in a sitting position and not subject to additional physical effort, the ventilation rate is usually of the order of 0.3–0.4 and 0.4–0.5 m^3/hr, in females and males, respectively (25). Differences are due to the differences in body weights and heart rates. It can be assumed that for people engaged in light work, corresponding to a slow walk, the rate is at least doubled. Specific ventilation rates have been determined for hundreds of tasks and should be used when appropriate (29).

In calculating the amount of a substance absorbed through the pulmonary tract during industrial work activities, various authors have used different average ventilation rates as typical for workers performing light work: from about 0.8 m^3/hr to about 1.25 m^3/hr. The

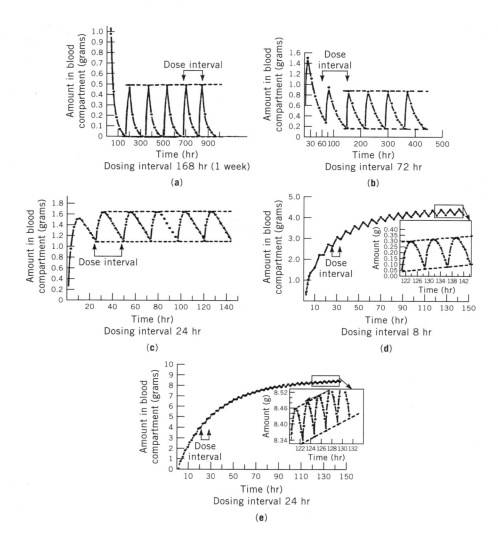

Figure 40.4. Multiple-dosing curves arising from one-compartment pharmacokinetic model with first-order absorption and elimination. These curves illustrate that the steady-state blood level and the time to reach steady state are dependent on the period of time between exposures and the biologic half-life of the substance. Reprinted with permission from Ref. 20.

latter figure (10 m³ per 8-hr working shift) is most often used by industrial health professionals in the United States since it is almost always higher than actual levels and, therefore, health protective.

Taking the above factors into account, the total amount of a substance absorbed through the respiratory tract over a period of exposure would be the product of air concentration

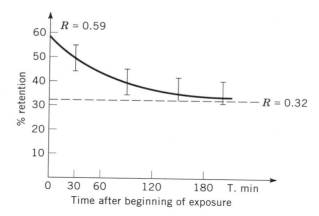

Figure 40.5. Retention of carbon disulfide vapors by the respiratory tract with continuous exposure. (Based on work of Jakubowski, ref. 28).

(C), duration of exposure (T), ventilation rate (V), and average fractional retention rate (R) for the time of exposure:

$$C(\text{mg/m}^3) \times T(\text{hr}) \times V(\text{m}^3/\text{hr}) \times R = \text{absorbed dose} \qquad (6)$$

This simple formula is useful for estimating the acceptability of an airborne concentration of a toxicant in the workplace if the person is to be exposed for periods markedly longer than 8 hr/day or 40 hr/week. Even if the industrial hygienist knows little else about the chemical's biologic or physical properties, he or she can limit the absorbed dose during the longer workday to that expected for a normal workday by calculating the air concentration for the nonnormal condition, which would yield the same daily dose allowed for the normal schedule.

4 PHARMACOKINETIC MODELS

The dynamic behavior of toxic substances can be described in terms of mathematical compartments in which a compartment represents all of the organs, tissues, and cells for which the rates of uptake and subsequent clearance of a toxicant are sufficiently similar to preclude kinetic resolution. This simplification of the body introduces few errors in the final analysis and is much more realistic than attempting to describe the fate of a chemical in each major tissue (e.g., liver, heart, lung, etc.). The mathematical description of how a chemical behaves in a living organism requires a model.

In general, most models divide the body into from one to five compartments. About twenty-five years ago, Fiserova-Bergerova et al. (30) and others suggested an approach to compartment categorization based on perfusion, ability to metabolize the inhaled substance, and the solubility of the substance in the tissue. Lung tissue, functional residual air, and arterial blood form the central compartment LG (lung group) in which pulmonary uptake and clearance take place. The partial pressure of inhaled vapor equilibrates with four peripheral compartments. Vessel-rich tissues form two peripheral compartments: BR (blood rich) compartment includes brain, which lacks capability to metabolize most xenobiotics, and is treated as a separate compartment because of its biological importance and the toxic effect of many vapors and gases on the central nervous system. VRG (vessel-rich group) compartment includes tissues with sites of vapor metabolism such as liver, kidney, glands, heart, and tissues of the gastrointestinal tract. Muscles and skin form compartment MG (muscle group), and adipose tissue and white marrow form compartment FG (fat group). The FG compartment is treated separately, since the dumping of lipid-soluble vapors in this compartment has a smoothing effect on concentration variation in other tissues, and these variations caused by changes in exposure concentrations, minute ventilation, and exposure duration would be more dramatic if not for the buffering effect of adipose tissue. Although the organs placed in various compartments have changed over the years when specific chemicals were evaluated, the conceptual approach remains the same (31).

An example of a pharmacokinetic model that describes the possible fate of an inhaled substance is depicted in Figure 40.6. An illustration of the possible fate of a common solvent like tetrachloroethylene in the various imaginary compartments is shown in Figures 40.7 and 40.8. The key point illustrated here is that fat- or lipid-soluble chemicals will quickly reach equilibrium in the highly perfused tissues, and they will also be removed

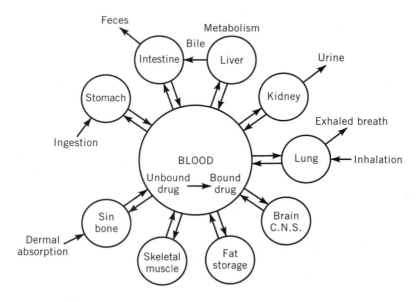

Figure 40.6. A diagram illustrating the potential distribution of a chemical following exposure. The differences in the rates of absorption or elimination among these tissues is the reason for using multicompartment pharmacokinetic models.

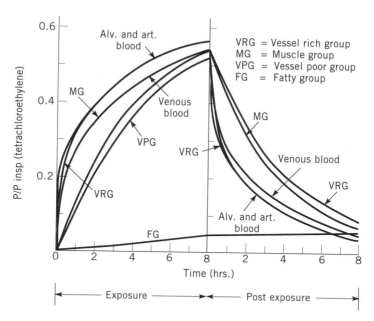

Figure 40.7. Predicted partial pressure of tetrachloroethylene in alveolar air, mixed venous blood, and tissue groups during and after 8 hr of exposure to a constant air concentration of tetrachloroethylene. Partial pressures in alveoli, blood, and tissues (P) are expressed as a fraction of the constant partial pressure in ambient air during exposure (P_{insp}). From Ref. 34.

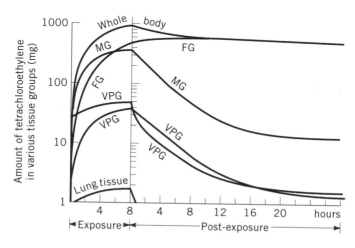

Figure 40.8. Predicted distribution of tetrachloroethylene to the tissue groups during and after 8 hr exposure to 100 ppm. In contrast with many other volatile chemicals, due to its relatively high lipid solubility, tetrachloroethylene demonstrates some persistence in fatty tissues. From Ref. 34.

quickly following cessation of exposure. However, the less perfused tissues, which are usually high in lipid, do not reach saturation quickly and, more importantly, they take a great deal of time to reach background levels (32, 33).

4.1 One-Compartment Model

The simplest pharmacokinetic model is the one-compartment open model illustrated in Figure 40.9. Here, the body is viewed as a single homogeneous box with a fixed volume. A drug or chemical entering the systemic circulation by any route is instantaneously distributed throughout the body. Although the absolute concentrations in all body tissues and fluids are not identical, it is assumed that they all rise and fall in parallel as drug is added and eliminated.

Overall elimination of drug from the body by urinary excretion and/or metabolism usually obeys first-order kinetics; that is, the rate of elimination is proportional to drug concentration in blood or plasma, Cp, as shown by the following equation:

$$\frac{dCp}{dt} = -k_{el}Cp \qquad (7)$$

Rearrangement and integration of Eq. (7) gives

$$\log Cp = -\frac{k_{el}(t)}{2.3} + \log Cp^0$$

which says that following an intravenous dose, a semilog plot of plasma concentrations versus time will be a straight line with a slope of $-k_{el/2.3}$ and an initial plasma concentration of Cp^0.

One simple way to determine whether a one-compartment model is appropriate for a given drug is to plot the intravenous (IV) plasma concentrations against time on semilog paper and see if the plot is a straight line (see Fig. 40.10). This approach permits determination of the two key parameters of the model: k_{el} the overall elimination rate constant, and V_D, the apparent volume of distribution of the drug in the body. Computer programs are available for quickly conducting such analyses.

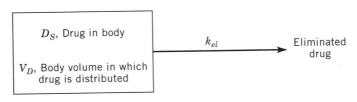

Figure 40.9. One-compartment open pharmacokinetic model.

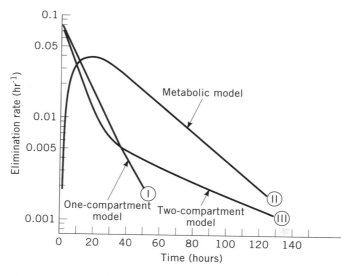

Figure 40.10. Three basic types of elimination curves following a single instantaneous exposure (IV) at time equals zero. From Ref. 25.

4.2 Two-Compartment Model

Often, it is clear from the blood plasma curves that there is a slower distribution to some tissues than others. In this case the kinetic model that will allow the interpretation of the observed experimental data is a two-compartment model, illustrated in Figure 40.11. In the two-compartment model the body is essentially reduced to an accessible compartment, the blood, and a second less accessible and diffuse compartment, the tissues. In the physiological sense, as soon as the molecules in the administrated dose mix with the blood, they will be rapidly carried to all parts of the body and brought into intimate contact (by perfusion) with organs, tissues, fat depots, and even bone. Some tissues, like the liver and the kidney, are very well perfused; others, such as muscle and fat, may be poorly perfused. Molecules that are carried to the perfused organs and tissues will then diffuse from the

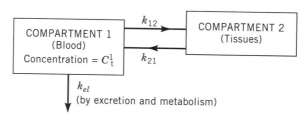

Figure 40.11. Schematic of the two-compartment pharmacokinetic model.

blood across cellular membranes and a dynamic equilibrium will usually be established (Fig. 40.12) (20, 35).

The distribution of a chemical as described by a two-compartment model is readily apparent in a plot of blood versus time following exposure. A blood concentration-time curve following an intravenous bolus administration, which exhibits the characteristics of distribution, is shown in Figure 40.13. An inspection of this kind of plot reveals an initial rapid (steep slope) nonlinear portion followed after a time (which depends on the nature and characteristics of the administered substance) by a slower linear portion. This linear portion, sometimes referred to as the terminal phase, has a definite slope from which a rate coefficient can be evaluated. If the rate coefficient for this linear terminal phase is defined as β, and the intercept of the line extrapolated back to the ordinate axis is B, then this part of the relationship can be described by

$$C_t^1 = Be^{-\beta t} \tag{8}$$

From the initial curved portion of the elimination curve, the rate at which the central compartment releases the toxicant is the slope of the feathered line. This can be determined easily from the method of residuals (19). Having determined A, the intercept of the feathered line, and α, the slope, the whole experimentally observed curve can be described by the biexponential equation

$$C_t^1 = Ae^{-\alpha t} + Be^{-\beta t} \tag{9}$$

where C_t^1 is the concentration of toxicant in blood at any time following exposure.

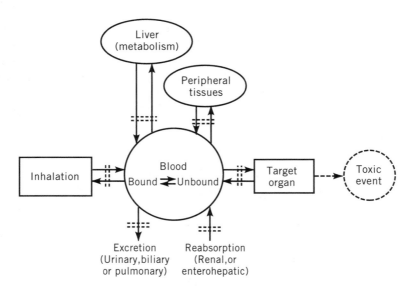

Figure 40.12. The physiologic model for toxic action proposed by Withey (20) modified to emphasize the fate of inhaled substances. The dashed lines represent diffusion barriers.

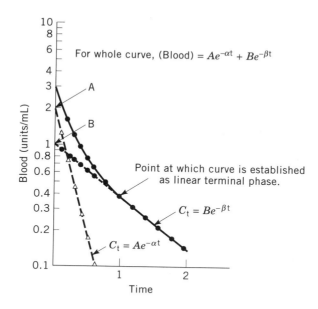

Figure 40.13. Semilogarithmic plot of blood concentration against time after intravenous administration illustrating biphasic elimination (two compartments) and the use of "curve stripping" to resolve the A intercept.

It is essential to note that in this case $Ae^{-\alpha t}$; that is, the initial phase of the curve, dominated by the rate coefficient α, is essentially over and completed when $e^{-\alpha t}$ approaches 0 and thereafter the curve is described by $C_t = Be^{-\beta t}$.

The individual rate coefficients α and β, obtained empirically from the plotted data, and the intercepts A and B can then be used to determine the individual rate coefficients of the model (20):

$$k_{21} = \frac{A\alpha - B\beta}{\alpha - \beta}$$

$$k_e = \frac{\alpha\beta}{k_{21}}$$

$$k_{12} = \alpha + \beta - k_{21} - k_e \qquad (10)$$

where k_{21} = rate transfer coefficient for the movement of molecules from compartment 2 to compartment 1
k_{12} = rate transfer coefficient for the movement of two molecules from compartment 1 to compartment 2
k_e = rate transfer coefficient for the elimination of the chemical from the system

Curve-fitting computer programs are available which require only rough estimates of the four parameters, that is, the slopes α and β and the intercepts A and B, to allow more precise calculation of the individual rate coefficients from the animal or human data.

To consider the body as one or two compartments involves considerable simplification of reality. To account for each of the 10 compartments shown in Figure 40.6 would involve considerable experimental and mathematical difficulty and could conceivably require 9 different equations to resolve the chemical's behavior. A 3-compartment model, as shown in Figure 40.14, is often adequate for most data analysis. Due to the minimal practical benefits of performing more rigorous mathematical analysis, combined with the comparatively less quantitative basis by which most occupational exposure limits are established, toxicologists should use Occam's razor in that the simplest model describing the available data is always the best one!

4.3 Accumulation of Chemicals in the Body

Cumulation of a substance in the body is a process wherein the concentration of a particular substance increases following repeated or continuous exposure. When using biologic monitoring to evaluate occupational exposure, cumulation is likely when the concentration of a substance in the analyzed media (urine, feces, blood, or expired air) increases with each day or week of exposure. This phenomenon has been observed even during normal work schedules when the airborne concentration of the toxicant was below the TLV (37) and the biologic half-life of the chemical was longer than 8 hrs. Therefore, for these types of chemicals exposure during unusually long work periods might be expected to increase the cumulation. Steady increases in the concentration of the inhaled contaminant in adipose tissue is indicative of cumulation.

Cumulation, if it occurs, results from a slow turnover of the substance. Thus it may take place under conditions of every kinetic model, provided the elimination rate constant is low (long biologic half-time). From theoretical considerations, it can be expected that the highest value of the rate constant (single compartment model) at which cumulation might occur would not exceed 0.1 hr^{-1} (38). The tendency of a chemical to cumulate in the body with repeated exposure can be due to several factors. For substances that are eliminated mainly unaltered in the breath, the cause of slow turnover may be the low air/

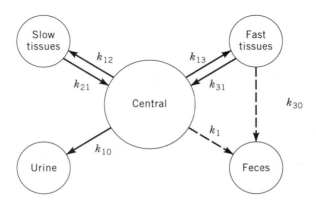

Figure 40.14. Schematic illustrating a three-compartment open model.

PHARMACOKINETICS AND UNUSUAL WORK SCHEDULES

blood partition coefficient or deposition in poorly perfused compartments such as fat; for substances excreted by the kidney, a low clearance may result from poor glomerular filtration, or intensive tubular reabsorption, or both. In practical situations, two other events may be of significance, namely slow biotransformation (when excretion takes place in metabolized form) and deposition in adipose tissue. These factors can act in combination, as has been shown for nitrobenzene and DDT (25).

The likelihood that a chemical will cumulate in the body or a key tissue is dependent on the overall rate at which a chemical is absorbed, metabolized, and excreted. The time required for the concentration of the parent chemical or its metabolite to be reduced by one-half in a particular medium (e.g., blood) or a tissue (e.g., fat), is called the biologic half-life. It must be emphasized that the biologic half-life varies for each substance and is dependent on the species studied and maybe influenced by the route of exposure. As a result, identification of the appropriate biologic half-life of a substance in humans based on animal data usually requires considerable effort. Estimates of the human biologic half-life can be made using physiologically-based pharmacokinetic (PB-PK) model.

As shown in Figure 40.15, chemicals with moderate half-lives (greater than 3 hr and less than 200 hr) are likely to show some degree of day-to-day cumulation during the workweek even at exposures at or near the TLV (19). Cumulation of a chemical during an exposure regimen (workweek or year) is not necessarily detrimental as long as the peak or steady-state tissue levels do not reach levels that are above the threshold concentration for various adverse effects as cytotoxicity, behavioral toxicity, and so forth. This weekly increase in body burden in workers periodically exposed has been demonstrated for some chemicals (37).

One of the primary concerns with unusual work schedules is the possibility that day-to-day increases in the toxicant concentration at the site of action might occur during

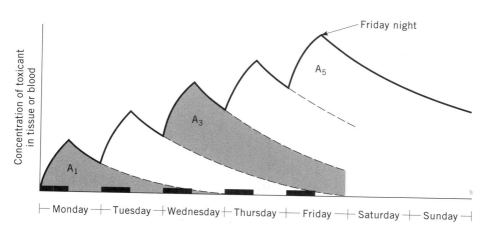

Figure 40.15. Day-to-day increase of toxicant concentration in tissue or blood following exposures of 8 hr/day, 5 day/week at TLV levels. For this type of cumulation to occur, the half-life of the air contaminant or its metabolite would be about 6 → 200 hrs., as shown in Figures 40.6 and 40.7.

unusually long work shifts to a much greater degree than that which occurs during standard 8 hr/day schedules. This potential problem is illustrated in Figure 40.16 where it is shown that a chemical with a 24-hr biologic half-life would clearly provide a greater peak body burden following 4 days of exposure on a 10-hr/day schedule than that observed in the workers exposed 8 hr/day for 5 days unless the chemical concentration in air was not lowered. The point here is that even when the number of hours of exposure per week are the same, for some chemicals, the peak body burden may be different when the work schedules vary. In this figure, C_T represents the concentration in the blood following a certain period of exposure and C_s is the concentration in the blood at the saturation level attained following continuous exposure to that same air concentration. In this figure t represents a schedule involving five 8-hr workdays per week and t' represents a workweek consisting of four workdays: a 10 hr/day, 8 hr/day, 12 hr/day, and another 10 hr/day followed by 3 days off work. Even though this unusual work schedule involves 40 hr of

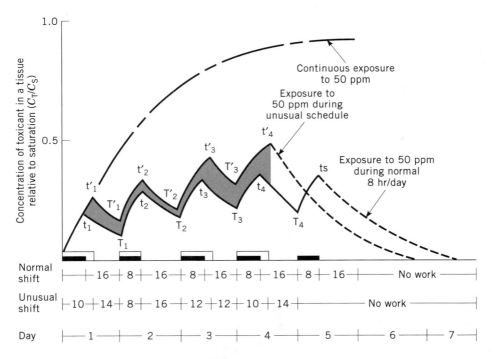

Figure 40.16. The accumulation of a substance during irregular periods of intermittent exposure that could occur during unusual work shifts and overtime. This plot illustrates that even when the total weekly dose (40 hr × 50 ppm) is unchanged, the peak body burden for two different shift schedules may not be equivalent. To help ensure that the peak tissue concentration for the unusual exposure schedule does not exceed the presumably "safe" level of the normal schedule, the air concentration of toxicant in the workplace may have to be reduced. K_{elim} has a value of 0.03 hr^{-1} in this illustration. (Courtesy of Dr. J. Walter Mason.)

exposure to 50 ppm of a chemical vapor, exposed workers are likely to have greater peak body burdens than persons who are also exposed for 40 hr per week but only 8 hr per day.

4.4 Nonlinear Pharmacokinetic Models

All the models discussed so far have been based on the assumption that all pharmacokinetic pathways can be described by linear differential equations with constant coefficients and first-order rate constants. This may not always be true and it is important to recognize that in some cases nonlinear behavior may occur. This can have a great deal of effect on the estimation of risks associated with exposure (20, 39–41). So, in general, the concentration of an air contaminant at or near the occupational exposure limit will not be sufficient to saturate the blood or bring about nonlinear behavior (42). However, this could occur if exposures were continuous or at high concentrations for short periods, even though the acceptable 8-hr TWA concentration was not exceeded. This phenomenon must be understood by the hygienist and toxicologist when interpreting some oncogenicity studies wherein a maximum tolerated dose (MTD) must be administered to at least one group of animals. In certain cases it is only when saturation occurs that an carcinogenic response is plausible and, in these cases, the likelihood (although remotely small) that Michaelis-Menten kinetics (see Fig. 40.17) will exist in exposed workers should be considered (32, 43).

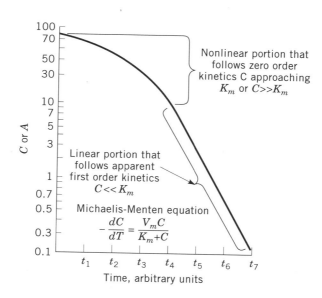

Figure 40.17. Simulated plasma concentration of a chemical (C) or the amount in the body (A) as a function of time for a chemical that displays dose-dependent or nonlinear pharmacokinetics described by the Michaelis-Menten equation: dc/dT is the change in concentration with time, V_m is the maximum rate of the process, and K_m is the Michaelis constant.

A dose-dependent change in pharmacokinetic behavior may arise from several causes. At saturation, the capacity of some biological system in the overall process can become overloaded. Some examples of chemicals that exhibit dose-dependent pharmacokinetic at concentrations well in excess of the TLV include vinyl chloride, vinylidene chloride, methylene chloride, methanol, benzene, and ethylene dichloride. Dose-dependent effects are of special interest to the toxicologist since some toxic effects are associated only with large or excessive doses of drugs. Alcohol pharmacokinetics is one example of dose-dependent kinetics. As is frequently the case, the dose dependence is due to metabolic saturation of the particular enzyme; in this case alcohol dehydrogenase (25). When this enzyme system is saturated, the person will be exposed to higher levels of unmetabolized alcohol via the blood than would normally occur at lower levels of alcohol ingestion.

In many of the cases cited in the literature, the observed dose dependence has been attributed to changes in the metabolic reaction involved. The metabolism of the drug may be due to an enzymatic reaction and thus would behave according to the classical Michaelis-Menten model of enzyme kinetics (Fig. 40.17). Such a system involves reaction of drug as substrate with enzyme to produce a metabolite via an intermediate enzyme-drug complex. At low drug concentrations, the reaction is overall first-order with a rate constant, $k_m = V_m/K_m$. At high drug concentrations, the capacity of the limited amount of enzyme present for reaction is exceeded and the reaction rate becomes constant (V_m). Thus, the production of metabolite becomes a constant rate or zero-order process.

The transformation from first-order to zero-order reaction rate is not an abrupt jump but a graded transition. During this transition the kinetics may remain apparently first-order but the elimination constant changes. Thus, both types of dose-dependent effects mentioned earlier may be explicable by a model of the Michaelis-Menten type. For those chemicals for which the rationale for the TLV or PEL is based on effects that occur only when the metabolic process is overloaded or saturated, such as vinyl chloride's carcinogenicity, it is suggested that no adjustment to the TLV is probably needed to protect workers on long shifts as long as the average weekly exposure throughout the month is about 40 hr. This suggestion is based on the observation that the occupational exposure limits are set at air concentrations far below those at which saturation and dose-dependent kinetics are likely to occur (43). Some common causes of dose-dependent pharmacokinetics are presented in Table 40.1 (20).

5 WHY ADJUST EXPOSURE LIMITS FOR UNUSUAL WORK SCHEDULES

It has been speculated by a number of researchers that some workers may be at increased risk of injury during unusual work schedules. In particular, industrial hygienists and occupational physicians have been concerned that unusual work schedules and the potential effects on the circadian rhythm might eventually compromise a worker's capacity to cope with exposure to airborne toxicants. In some cases, it is clear that adjustments to the TLV or any other occupational exposure limit (OEL) are necessary to provide workers on long workdays the same degree of protection given those persons on normal 8-hr workdays (2). Establishing acceptable limits of exposure for unusual work schedules is a difficult task

Table 40.1 Possible Causes of Dose-Dependent Pharmacokinetics[a]

Metabolic saturation	a. Self-inhibition by excess drugs
	b. Insufficient metabolic enzyme
	c. Competition for coenzyme or cosubstrate
Excretory saturation	a. Competition for tubular secretion or resorption mechanism
	b. Changed drug distribution
Changed drug distribution	a. Protein and tissue binding effect
	b. Overflow into new volumes of distribution

[a]From Ref. 2.

since, at best, each person's susceptibility to stressors is dependent on many factors that are unique to that individual (44).

Because tens of thousands of persons are exposed to airborne substances for unusually short or long periods of time, several organizations have set limits for these situations. Pennsylvania promulgated short-term exposure limits soon after the OSHA regulations took effect in 1971. Brief and Scala (2) and OSHA (46) have developed simple, easily applied models aimed at adapting exposure limits to unusual situations. The National Institute for Occupational Safety and Health has applied its recommended standard for chloroform (47) and benzene to both 8 and 10-hr work shifts. Dozens of states and communities have established limits of short term exposure to airborne toxicants.

The National Research Council has described the need to adjust exposure limits for long-term continuous exposure, such as might be encountered in space exploration (45). ACGIH has incorporated short-term exposure limits into its TLV list (50). Numerous pharmacokineticists and other researchers (18, 38, 51–56) have proposed complex models for adjusting limits and a few toxicologists have begun to investigate the differences in toxicologic response between administration of the same daily dose over particularly short and long periods of exposure. These will be discussed later in the chapter.

Two questions regarding the possible occupational health hazards of the 12-hr schedules surfaced shortly after its commencement. One of the initial concerns of occupational physicians and industrial hygienists was whether the recommended limit for exposure to noise in the workplace, 85 dBA for 8 hr, was sufficiently low to protect workers on the 12-hr shift. There was speculation that regulatory agencies such as OSHA might arbitrarily lower the standard by 3–5 dBA for workers on these shifts. Second, since most federal standards for occupational exposure to airborne toxicants are based a work schedule of 8 hr per day, 5 days per week, there was concern that a marked decrease in the concentration of air contaminants in the workplace might be necessary to protect workers. The concern that the existing exposure limits might not protect those on unusual work schedules was understandable since both the noise and air contaminant limits were based on the results of either animal testing, which was usually conducted for 6.0 hr or less per day, or the epidemiological experience of workers who were exposed during normal, 8-hr/day work shifts.

5.1 Limits for Exposure to Noise During Unusual Shifts

Currently, ACGIH recommends that exposures to sound be limited to 85 dBA for periods of 8 hr per day and to no more than 82 dBA if exposed 16 hr per day (50). The setting of a special guideline for those persons working 12-hr shifts has not been specifically addressed by the ACGIH, but one can infer that some noise level between 82 and 85 dBA would be appropriate. The regulatory agency responsible for setting standards that protect worker health, OSHA, currently enforces a legal limit for occupational exposure to noise of 85 dBA time-weighted average. This standard applies to all persons who work 8-hr/day or longer, but the total exposure should not exceed 50 hr per week.

Although ACGIH has embraced this approach for more than a decade, there remains some level of disagreement among experts in the field of noise-induced hearing loss that routine exposure to a 12-hr/day, 3- or 4-day/week work schedule requires a special noise limit. Many believe that as long as the workers are exposed throughout the year an average of no more than 40 hr per week then the increased health risk is insignificant. This seems reasonable since there is evidence that injury due to exposure to certain physical agents such as noise or radiation is solely a function of the intensity and duration of exposure. In short, for most physical agents, the risk is dictated by the total dose over a month, year, or lifetime rather than each day or week, as long as daily threshold for acute effects are excepted.

5.2 Exposure Limits for Air Contaminants During Unusual Schedules

In the preface of the ACGIH publication *Threshold Limit Values For Chemical Substances and Physical Agents and Biological Exposure Indices in the Work Environment with Intended Changes for 1999*, it states that the TLVs refer to airborne concentrations of substances and represent conditions under which it is believed that nearly all workers may be repeatedly exposed day after day without adverse effect (50). The values refer to a TWA concentration for a normal 8-hr workday and a 40-hr workweek (50). Implicit in this evaluation is an assumption that a balance exists between the accumulation of a contaminant in the body while exposed at work, and the elimination of the contaminant while away from work (during which time there is presumably no exposure).

It is readily apparent to a toxicologist or pharmacokineticist that some modification of the TLVs may be necessary if these limits are to provide an equivalent amount of protection to persons working unusual shifts. For example, a 12-hr work shift involves a period of daily exposure that is 50% greater than that of the standard 8-hr workday and the period of recovery before reexposure is shortened from 16 to 12 hr (2). For certain systemic toxins having half-lives between 5 and 500 hr, it can be envisioned that shifts longer than 8 hr/day would present a correspondingly greater hazard than that incurred during normal workweeks if the exposure (i.e., dose) were not reduced by some albeit small amount.

The very basis of occupational exposure limits for inhaled toxicants is that there is a maximum concentration of the air contaminant in the workplace to which persons can safely be exposed and suffer no adverse effect. Consequently, it can be envisioned that there is a corresponding maximum body burden at which an adverse effect is not likely. TLVs represent conditions under which it is believed that nearly all workers may be exposed day after day without adverse effect, so exposure to a substance at its TLV should result in some maximum burden of the substance in a particular target organ of the body (28). Therefore, it seems logical that TLVs for certain chemicals should be modified if persons are exposed more than 8 hr/day so as to prevent this maximum body burden from being exceeded (see Fig. 40.18).

Since the rationales for the more than 700 ACGIH TLVs vary, identifying adjusted TLVs for some, if not most, of the listed chemical is not a trivial task. Several factors should be considered. For example, the need to lower the average exposure limit for certain industrial chemicals would be especially important in those situations where the safety factor in the TLV is small, toxicity data are limited, the toxic effect is serious, accumulation of the chemical is possible following several days of repeated exposure, or where there is the possibility for extreme variability in worker response to a toxicant.

6 MODELS FOR ADJUSTING OCCUPATIONAL EXPOSURE LIMITS

Several researchers have proposed mathematical formulas or models for adjusting occupational exposure limits (PELs, TLVs, etc.) for use during unusual work schedules, and these have received a good deal of interest in the industrial and regulatory arenas (2, 18, 38, 51–56). Although OSHA has not officially promulgated specific exposure limits applicable to unusual work shifts, it has published guidelines for use by OSHA compliance officers for adjusting exposure limits (46, 59). These generally apply to work shifts longer than 8 hr/day.

Under the General Duty Clause of OSHA, Section 5A1, employers that have employees who, during long shifts, are exposed to workplace air concentrations in excess of adjusted exposure limits are citeable for noncompliance. All of the models that an OSHA compli-

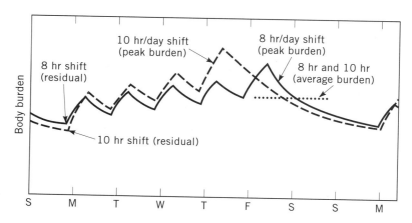

Figure 40.18. Comparison of the peak, average, and residual body burdens of an air contaminant following exposure during a standard (8-hr/day) and unusual (10-hr/day) workweek. In this case the weekly average body burdens are the same for both schedules since each involved 40 hr/week. The residual (Monday morning) body burden of the 8-hr shift worker, however, is greater than the 10-hr shift worker and the peak body burden of the person who worked the 10-hr shift is higher than the 8-hr worker. Based on Hickey (51).

ance officer might use in determining the acceptability of an air contaminant level are presented and discussed in this chapter. Each of the models has advantages and disadvantages. The shortcomings in determining modified exposure limits for unusual schedules have been discussed (57). It should be noted that OSHA has, over the past 20 years, cited few firms that have unusually long shifts and are not in compliance with the limits suggested by these models.

6.1 Brief and Scala Model

In the early 1970s, due to the increasingly large number of workers who had begun working unusual schedules, the Exxon Corporation began investigating approaches to modifying the occupational exposure limits for their employees on 12-hr shifts. In 1975, the first recommendations for modifying TLVs and PELs were published by Brief and Scala (2), wherein they suggested that TLVs and PELs should be modified for individuals exposed to chemicals during novel or unusual work schedules.

They called attention to the fact that, for example, in a 12-hr workday the period of exposure to toxicants was 50% greater than in the 8-hr workday, and that the period of recovery between exposures was shortened by 25 percent, or from 16 to 12 hr. Brief and Scala noted that repeated exposure during longer workdays might, in some cases, stress the detoxication mechanisms to a point that accumulation of a toxicant might occur in target tissues, and that alternate pathways of metabolism might be initiated. Although theoretically possible, it has generally been held that due to the margin of safety in the TLV (albeit small), there was little potential for frank toxicity to occur due to unusually long work schedules if one is in compliance with the TLV.

Brief and Scala's approach was simple but important since it emphasized that unless worker exposure to some systemic toxicants was lowered, the daily dose would be greater, and due to the lesser time for recovery between exposures, peak tissue levels might be higher during unusual shifts than during normal shifts. This concept is illustrated in Figure 40.18. Their formulas (Eqs. 11 and 12) for adjusting limits are intended to ensure that this will not occur during novel work shifts:

$$\text{TLV reduction factor (RF)} = \frac{8}{h} \times \frac{24h}{16} \tag{11}$$

where h is hours worked per day.

For a 7-day workweek, they suggested that the formula be driven by the 40-hr exposure period; consequently, they developed Eq. 12, which accounts for both the period of exposure and period of recovery:

$$\text{TLV reduction factor (RF)} = \frac{40}{h} \times \frac{168h}{128} \tag{12}$$

where h = hours exposed per week.

One advantage of this formula is that the biologic half-life of the chemical and the mechanism of action do not need to be understood in order to calculate a modified TLV. Such a simplification has shortcomings since the reduction factor for a given work schedule is the same for all chemicals even though the biologic half-lives of different chemicals vary widely. Consequently, this formula overestimates the degree to which the limit should be lowered.

Brief and Scala (2) were cautious in describing the strength of their proposal and offered the following guidelines for its use. These should be considered when applying this model and also the others:

1. Where the TLV is based on systemic effect (acute or chronic), the TLV reduction factor will be applied and the reduced TLV will be considered as a time-weighted average (TWA). Acute responses are viewed as falling into two categories: (a) rapid with immediate onset and (b) manifest with time during a single exposure. The former are guarded by the C notation and the latter are presumed time and concentration dependent, and hence, are amenable to the modifications proposed. Number of days worked per week is not considered, except for a 7-day workweek discussed later.

2. Excursion factors for TWA limits (Appendix D of the 1974 TLV publication) will be reduced according to the following equation:

$$EF = (EF_8 - 1)RF + 1 \qquad (13)$$

where EF = desired excursion factor
EF_8 = value in Appendix D for 8-hr TWA.
RF = TLV reduction factor

3. Special case of 7-day workweek. Determine the TLV Reduction Factor based on exposure hours per week and exposure-free hours per week.

4. When the novel work schedule involves 24-hour continuous exposure, such as in a submarine, spacecraft or other totally enclosed environment designed for living and working, the TLV reduction technique cannot be used. In such cases, the 90-day continuous exposure limits of the National Academy of Science should be considered, where applicable limits apply.

5. The techniques are not applicable to work schedules less than seven to eight hours per day or $\leq$ 40 hours per week.

Brief and Scala (2) also noted that:

The RF value should be applied (a) to TLV's expressed as time-weighted average with respect to the mean and permissible excursion and (b) to TLV's which have a C (ceiling) notation except where the C notation is based solely on sensory irritation. In this case the irritation response threshold is not likely to be altered downward by an increase in number of hours worked and modification of the TLV is not needed.

In short, the Brief and Scala formula is dependent solely on the number of hours worked per day and the period of time between exposures. For example, for any systematic toxicant, this approach recommends that persons who are employed on a 12-hr/day, 3- or 4-day workweek should not be exposed to air concentrations of a toxicant greater than one-half that of workers who work on an 8-hr/day, 5-day schedule.

In their publication, Brief and Scala acknowledged the importance of a chemical's biologic half-life when adjusting exposure limits, but because they believed this information was rarely available, they were comfortable with their proposal. They noted that a reduction in an occupational exposure limit is probably not necessary for chemicals whose primary untoward effect is irritation since the threshold for irritation response is not likely to be altered downward by an increase in the number of hours worked each day (2), that is, irritation is concentration rather than time dependent. Although this appears to be a reasonable assumption, some researchers believe that it may not be the case for all chemical irritants since duration of exposure could possibly be a factor in causing irritation in susceptible individuals who are not otherwise irritated during normal 8-hr/day exposure periods (60).

6.1.1   Illustrative Example 1 (Brief and Scala Model)

Refinery operators often work a 6-week schedule of three 12-hr workdays for 3 weeks, followed by four 12-hr workdays for three weeks. What is the adjusted TLV for methanol (assume TLV = 200 ppm) for these workers? Note that the weekly average exposure is only slightly greater than that of a normal work schedule.

Solution

$$RF = \frac{8}{12} \times \frac{24 - 12}{16} = 0.5$$

$$\text{Adjusted } TLV = RF \times TLV$$
$$= 0.5 \times 200 \text{ ppm}$$
$$= 100 \text{ ppm}$$

Note: The TLV reduction factor of 0.5 applies to the 12-hr workday, whether exposure is for 3, 4, or 5 days per week.

6.1.2   Illustrative Example 2 (Brief and Scala Model)

What is the modified TLV for tetrachloroethylene (assume TLV = 25 ppm) for a 10-hr/day, 4-day/week work schedule if the biologic half-life in humans is 144 hr?

Solution

$$RF = \frac{8}{10} \times \frac{24 - 10}{16} = 0.7$$

$$\text{New } TLV = 0.7 \times 25 = 17.5 \text{ ppm}$$

Note: This model and the one used by OSHA *do not consider* the pharmacokinetics (biologic half-life) of the chemical when deriving a modified TLV. Other models to be discussed later *do* take this information into account.

6.1.3 Illustrative Example 3 (Brief and Scala Model)

In an 8-hr day, 7-day workweek situation, such as the 56/21 schedule, persons work 56 continuous days followed by 21 days off. What is the recommended TLV for H_2S (assume TLV = 10 ppm) for this special case of a 7-day workweek? Assume that the biologic half-life in humans for H_2S is about 2 hr and the rationale for the limit is the prevention of irritation and systemic effects.

Solution. Exposure hr per week = $8 \times 7 = 56$ hr
Exposure-free hr per week = $(24 \times 7) - 56 = 112$ hr

$$RF = \frac{40}{56} \times \frac{112}{128} = 0.625$$

Adjusted $TLV = RF \times 10$ ppm = 6 ppm.

6.1.4 Illustrative Example 4 (Brief and Scala Model)

Ammonia has a TLV of 25 ppm and is an upper respiratory tract irritant. What is the modified TLV for this chemical for a work schedule of 14 hr/day for 3 day/week?

Solution. Since the rationale for the limit for ammonia is the prevention of irritation, *no* adjustment (lowering) of the limit is needed.

6.2 OSHA Model

Most toxicologists believe that, in general, the intensity of a toxic response is a function of the concentration that reaches the site of action (30, 63). This principle is simplistic and, while it may not apply to irritants, sensitizers, or carcinogens, it is clearly true for the systemic toxics. This assumption is the basis for the OSHA model for modifying PELs for unusual shifts (46). The originators of the model assumed that for chemicals that cause an acute response, if the daily uptake (concentration × time) during a long workday was limited to the amount that would be absorbed during a standard workday, then the same degree of protection would be given to workers on the longer shifts. For chemicals with cumulative effects (i.e., those with a long half-life), the adjustment model was based on the dose imparted through exposure during the normal workweek (40 hr) rather than the normal workday (8 hr).

OSHA recognized that the rationale for the occupational exposure limits for the various chemicals was based on different types of toxic effects. After OSHA adopted the same 500 TLVs of 1968 as PELs (**29** *CFR* 1910.1000), and attempted to do it again in 1989, they placed each of the chemicals into one of six "Work Schedule" categories (Table 40.2) to assure that an appropriate adjustment model would be used by their hygienists. As can be seen in Examples 5, 6, and 7, the degree to which an exposure limit is to be adjusted, if at all, is based to a large degree on the work schedule category (and type of adverse effect) in which a chemical is placed (Table 40.3).

The rationale that OSHA and its expert consultants used when categorizing the various chemicals was based on the primary type of health effect to be prevented, biologic half-

Table 40.2 Work Schedule Categories Listed by OSHA[a]

Category 1A. Ceiling Limit Standard

Substances in this category (e.g., butylamine) have ceiling limit standards that were intended never to be exceeded at any time, and so, are independent of the length of frequency of work shifts. The ceiling PELs for substances in this category should not be adjusted.

Category 1B. Standards Preventing Mild Irritation

Substances in this category have a PEL designed primarily to prevent acute irritation or discomfort (e.g., cyclopentadiene). There are essentially no known cumulative effects resulting from exposures for extended periods of time at concentration levels near the PEL. The PELs for substances in this category should not be adjusted.

Category 1C. Standards Limited by Technologic Feasibility

The PELs of substances assigned to this category have been set either by technologic feasibility (e.g., vinyl chloride) or good hygiene practices (e.g., methyl acetylene). These factors are independent of the length or frequency of work shifts. The PELs for substances in this category should not be adjusted.

Category 2. Acute Toxicity Standards

a. The substances in this category have PELs that prevent excessive accumulation of the substnace in the body during 8 hr of exposure in any given day (e.g., carbon monoxide).
b. The following equation determines a level that ensures that employees exposed more than 8 hr/day will not receive a dosage (i.e., length of exposure × concentration) in excess of that intended by the standard.

$$\text{Equivalent PEL} = \text{8-hr PEL} \times \frac{\text{8 hr}}{\text{hours of exposure in 1 day}} \text{ (daily adjustment)}$$

c. The industrial hygienist should normally conduct sampling for the entire shift minus no more than 1 hr for equipment setup and retrieval (e.g., at least 9 hr of a 10-hr shift). In situations where an employee works multiple shifts in a day (e.g., two 7-hr shifts), and the industrial hygienist can document sufficient cause to expect exposure concentrations to be similar during the other shifts, the sampling should be done during only one shift.

Category 3. Cumulative Toxicity Standards

a. Substances assigned to this category present cumulative hazards (e.g., lead, mercury, etc.). The PELs for these substances are designed to prevent excessive accumulation in the body resulting from many days or even years of exposure.
b. The following equation ensures that workers exposed more than 40 hr/week will not receive a dosage in excess of that intended by the standard.

$$\text{Equivalent PEL} = \text{8-hr PEL} \times \frac{\text{40 hr}}{\text{hours of exposure in 1 week}} \text{ (weekly adjustment)}$$

Table 40.2 (continued)

c. It is the responsibility of the industrial hygienist to conduct sufficient sampling to document exposure levels for the entire week when evaluating conditions on the basis of this equivalent PEL. For most operations, the industrial hygienist will be able to sample during one shift only and then document sufficient cause to predict exposure concentrations during the other shifts.

Category 4. Acute and Cumulative Toxicity Standards

Substances in this category may present both an acute and a cumulative hazard. For this reason, the PELs of these substances should be adjusted by either equation; i.e., whichever provides the greatest protection.

Refined Adjustment Equations for Specific Standards

The adjustment equation presented for Categories 2 and 3 reflect an oversimplification of the actual accumulation and removal of a toxic agent from the body. Additional research, however, is needed in order to apply more complex equations to estimate resulting body burden and health risk due to prolonged exposure periods.

[a]From Ref. 46.

Table 40.3 Prolonged Work Schedule Categories[a]

Category[b]	Classification	Adjustment Criteria
1A	Ceiling standard	None
1B	Irritants	None
1C	Technologic limitations	None
2	Acute toxicants	Exposed—8 hr/day
3	Cumulative toxicants	Exposed—40 hr/week
4	Both acute and cumulative	Exposed—8 hr/day and/or Exposed—40 hr/week

[a]From Ref. 46.
Note: The health effects and classification of violation sections of this chapter have been reviewed by a panel of toxicologists and industrial hygienists from NIOSH using policy guidelines established by OSHA.
[b]This column indicates the code designation for prolonged work schedules that may require an adjustment to the PEL.

life (if known), and the rationale for the limit. The categories include (a) ceiling limit, (b) prevention of irritation, (c) technological feasibility limitations, (d) acute toxicity, (e) cumulative toxicity, as well as (f) acute and cumulative toxicity. A review of the degree of adjustment required for each type of chemical can be found in Table 40.3. Table 40.4 contains names of the categories of the different health effects into which the 700 chemicals with PELs were placed. Table 40.5 presents some examples of how chemicals are classified according to primary adverse health effects. The complete list can be found in the *Federal Register* (62). Beginning in 1997, the ACGIH TLV committee also began to note the primary target organ for each chemical in their annual booklet (50).

Table 40.4 OSHA Substance Toxicity Table (Rationale for Placing a Chemical into One of the Categories Listed in the OSHA *Officers Field Manual*)[a]

Health Code Number	Health Effect
1	Cancer—Currently regulated by OSHA as carcinogens; chiefly work practice standards
2	Chronic (cumulative) toxicity—Suspect carcinogen or mutagen
3	Chronic (cumulative) toxicity—Long-term organ toxicity other than nervous, respiratory, hematologic or reproductive
4	Acute toxicity—Short-term high hazards effects
5	Reproductive hazards—Fertility impairment or teratogenesis
6	Nervous system disturbances—Cholinesterase inhibition
7	Nervous system disturbances—Nervous system effects other than narcosis
8	Nervous system disturbances—Narcosis
9	Respiratory effects other than irritation—Respiratory sensitization (asthma)
10	Respiratory effects other than irritation—Cumulative lung damage
11	Respiratory effects—Acute lung damage/edema
12	Hematologic (blood) disturbances—Anemias
13	Hematologic (blood) disturbances—Methemoglobinemia
14	Irritation—eye, nose, throat, skin—Marked
15	Irritation—eye, nose, throat, skin—Moderate
16	Irritation—eye, nose, throat, skin—Mild
17	Asphyxiants, anoxiants
18	Explosive, flammable, safety (no adverse effects encountered when good housekeeping practices are followed)
19	Generally low-risk health effects—Nuisance particulates, vapors or gases
20	Generally low-risk health effects—Odor

[a]From Refs. 46, 59.

Irrespective of the model that will be used to make the adjustments, including the pharmacokinetic models to be discussed, the table in the *Federal Register* (62) or the ACGIH TLV booklet should be consulted before the hygienist begins the task of modifying an exposure limit. The use of this table, combined with some professional judgment, will prevent hygienists from requiring control measures when they are unnecessarily restrictive as well as minimize the risk of injury or discomfort from overexposure during an unusual exposure schedule. For example, as noted by OSHA, substances in Category 1A, 1B, and 1C do not require adjustment during long shifts due to the rationale for those limits (Table 40.3).

As discussed briefly, OSHA has proposed two simple equations for adjusting occupational health limits. These equations are offered to their compliance officers as an alternative to the more complex models of Brief and Scala (2) or Hickey and Reist (18). The first equation, which appears in Chapter 13 of the *OSHA Field Manual* (46), is to be used with chemicals whose primary hazard is acute injury (Category 2, Table 40.3). It should be noted that not all later editions of the OSHA field manual contained this information.

Table 40.5 Example of OSHA Classification of Chemicals by Primary Adverse Health Effect[a]

Substance	Health Code No.	Health Effects	Work Category
Abate	6	Cholinesterase inhibition	3
Acetaldehyde	14	Marked irritation—eye, nose, throat, skin	1B
Acetic acid	14	Marked irritation—eye, nose, throat, skin	1B
Acetic anhydride	14	Marked irritation—eye, nose, throat, skin	1B
Acetone	16, 8	Mild irritation—eye, nose, throat/narcosis	1B
Acetonitrile	16, 4	Mild irritation—eye, nose, throat/acute toxicity (cyanosis)	4
2-Acetylaminofluorene	1	Cancer	1C
Acetylene	18, 17	Explosive/simple asphyxiation	1C
Acetylene tetrabromide	3, 10	Cumulative liver and lung damage	4
Acrolein	14	Marked irritation—eye, nose, throat, lungs, skin	1B
Acrylamide—skin	7, 3	Polyneuropathy, dermatitis/skin, eye irritation	4
Acrylonitrile—skin	2, 5	Suspect carcinogen	4
Aldrin—skin	2, 3	Suspect carcinogen/cumulative liver damage	4
Allyl alcohol—skin	4, 14	Eye damage/marked irritation—eye, nose, throat, bronchi, skin	1B
Allyl chloride	3, 14	Liver damage/Marked irritation—eye, nose, throat	4
Allyl glycidyl ether (AGE)—skin	14	Contact skin allergy/Marked irritation—eye, nose, throat, bronchi, skin	1B
Allyl propyl disulfide	14	Marked irritation—eye, nose, throat	1B
Aluminum oxide	18, 19	Nuisance particulate	1C
4-Aminodiphenyl—skin	1	Cancer	1C
2-Aminopyridine	4, 7	CNS stimulation/headache/increased blood pressure	4
Ammonia	11, 14	Marked irritation—eye, nose, throat, bronchi, lungs	1B
Ammonium chloride (fume)	16	Mild irritation—eye, nose, throat	1B
Catechol pyrocatechol	14, 3	Eye and skin irritation/kidney damage	4
Cellulose (paper fiber)	19	Nuisance particulate	1C
Cesium hydroxide	15	Moderate irritation—eye, nose, throat, skin	1B
Chlordane—skin	3, 2	Cumulative liver damage/suspect carcinogen	4

Table 40.5 (continued)

Substance	Health Code No.	Health Effects	Work Category
Chlorinated camphene (toxaphene)—skin	3	Cumulative liver damage	3
Chlorinated diphenyl oxide	3	Cumulative liver damage/dermatitis	3
Chlorine	11, 14	Lung injury/marked irritation—eye, nose, throat, bronchi	1B
Chlorine dioxide	11, 14	Lung injury/marked irritation—eye, nose, throat, bronchi	1B
Chlorine trifluoride	11, 14	Marked irritation—eye, nose, throat, bronchi, lungs	1A
Chloroacetaldehyde	14	Marked irritation—eye, nose, throat, lungs, skin	1A
alpha-Chloroacetophenone (phenacyl chloride)	14	Marked irritation—eye, nose, throat, bronchi, lungs, skin	1B
Chlorobenzene (monochlorobenze)	3, 8	Cumulative systemic-toxicity/narcosis	4
o-Chlorobenzylidene malonitrile—skin	14	Marked irritation—eye, nose, throat, skin	1B
2-Chloro-1,3-butadiene	(See Chloroprene)		
Chlorobromomethane	3, 8	Cumulative liver damage/narcosis	3
Chlorodifluoromethane (F-22)	18	Good housekeeping practice	1C
Chlorodiphenyl (42% Cl)—skin	2, 3	Suspect carcinogen/Chloracne/cumulative liver damage	4
Chlorodiphenyl (54% Cl)—skin	2, 3	Suspect carcinogen/chloracne/cumulative liver damage	4
Chloroform trichloromethane	2, 3, 8	Suspect carcinogen/cumulative liver and kidney damage/narcosis	4
bis-Chloromethyl ether	1	Cancer (lung)	1C
1-Chloro-1-nitropropane	15	Moderate irritation—eye, nose, throat, skin	1B
Chloropicrin	14, 11	Marked irritation—eye, nose, throat, bronchi, lungs, skin	1B
Chloroprene—(2-chloro-1,3 butadiene)—skin	5, 3, 2	Reproductive hazard/systemic toxicity/suspect mutagen	4
Chloropyrifos (Dursban[R])—skin	6	Cholinesterase inhibition	4
o-Chlorostyrene	3	Cumulative liver, kidney damage	3
o-Chlorotoluene—skin	2, 15	Mild irritant—eye, skin	1B
2-Chloro-6-trichloromethyl pyridine (N-serve[R])	18	Good housekeeping practice	1C
Chromates, certain insoluble forms (as Cr)	2, 10, 3	Suspect carcinogen/cumulative lung damage/dermatitis	4
Chromic acid and chromates	2, 10, 3	Suspect carcinogen/cumulative lung damage/nasal preforation, ulceration	4
Chromium, soluble chromic, chromous salts (as Cr)	10, 3	Cumulative lung damage/dermatitis	3

Table 40.5 (continued)

Substance	Health Code No.	Health Effects	Work Category
Clopidol (Coyden^R)	18	Good housekeeping practice	1C
Coal dust	10	Pneumoconiosis	3
Coal tar pitch volatiles	2, 10	Suspect carinogen/cumulative lung changes	4
Cobalt, metal, fume and dust (as Co)	9, 10, 3	Asthma/cumulative lung changes/dermatitis	3
Coke oven emissions	1, 3	Cancer—lungs, bladder, kidney/skin sensitization	1C
Copper dusts and mists (as Cu)	16	Mild irritation—eye, nose, throat, skin	4
Copper fume (as Cu)	15, 11	Moderate irritation—eye, nose, throat, lung	4
Corundum (Al_2O_3)	19	Nuisance particulate	1C
Cotton dust (raw)	9, 10	Asthma/cumulative lung damage (bysinosis)	4
Crag^R herbicide	3	Cumulative liver damage	3
Cresol (all isomers)—skin	14, 4, 3	Marked irritation—eye, skin/acute toxicity (CNS), liver and kidney damage	1B
Cristobalite	10	Pneumoconiosis	3
Crotonaldehyde	14	Marked irritation—eye, nose, throat, lungs	1B
Crufomate^R	6	Cholinesterase inhibition	3
Lindane—skin	7, 3, 2	Cumulative CNS and liver damage/suspect carcinogen	4
Lithium hydride	14, 11, 7	Marked irritation—eye, nose, throat, skin/lung damage/CNS effects	1B
LPG (liquefied petroleum gas)	18, 17, 8	Explosive/asphyxiant/narcosis	2
Magnesite	19	Nuisance particulate/accumulation in lungs	1C
Magnesium oxide fume	11	Lung effects (fume fever)	2
Malathion—skin	6	Cholinesterase inhibition	3
Maleic anhydride	14, 9, 2	Marked irritation—eye, nose, throat, lungs (edema), skin/asthma	2
Manganese and compounds (as Mn)	7, 10	Cumulative CNS damage/lung damage	1A
Manganese cyclopentadienyl tricarbonyl (as Mn)—skin	4, 7, 3	Acute CNS and blood effects/cumulative kidney damage	4
Marble	19	Nuisance particulate/accumulation in lungs	1C
Mercury, (organo) alkyl compounds, (as Hg)—skin	7, 3, 14	Acute and cumulative CNS damage/marked skin irritation	4

Table 40.5 (continued)

Substance	Health Code No.	Health Effects	Work Category
Mercury, inorganic (as Hg)—skin	7, 3, 2	Acute and cumulative CNS damage/gastrointestinal effects/gingivitis/suspect carcinogen	4
Mesityl oxide	16	Mild irritation—eye, nose, throat	1B
Methane	18, 17	Explosive/simple	1C
Methanethiol	(See Methyl mercaptan)		
Methomyl (Lannate^R)—skin	6	Cholinesterase inhibition	3
Methoxychlor	3	Cumulative kidney damage	3
Methyl acetate	16, 8, 7	Mild irritation–nose, throat, lungs/narcosis/CNS effects	4
Methyl acetylene (propyne)	18, 8	Explosive/narcosis	1C
Methyl acetylene–propadiene mix (MAPP)	18	Flammable	1C
Methyl acrylate—skin	14, 4, 3	Marked irritation—eye, nose, throat, skin/acute lung damage	1B
Tetrahydrofuran	15, 8	Moderate irritation—eye, nose, throat/narcosis	2
Tetramethyl lead (as Pb)—skin	3, 7, 4	Cumulative liver, CNS and kidney damage/acute CNS effects	3
Tetramethyl succinonitrile—skin	4	Acute systemic toxicity (CNS)—headache, nausea, convulsions	4
Tetranitromethane	14, 4, 3	Marked irritation—eye, nose, throat/acute CNS and lung effects (edema)/cumulative systemic damage	4
Tetryl (2,4,6-Trinitrophenyl Methylnitramine)—skin	3	Contact dermatitis, skin sensitization/cumulative systemic toxicity	3
Thallium (soluble compounds)—skin (as Tl)	3	Cumulative systemic toxicity	3
4,4'-Thiobis (6-tert-butyl-*m*-cresol)	19	Apparent low toxicity	1C
Thiram^R	4, 5	Acute systemic toxicity (antabuselike effects)/suspect teratogen	4
Tin (Inorganic compounds, except oxide) (as Sn)	4, 3	Acute and chronic systemic toxicity	4
Tin (Organic compounds) (as Sn)	14, 3	Marked irritation—skin/cumulative systemic toxicity	3
Tin oxide	10	Pneumoconiosis (stannosis)	3
Titanium dioxide	19	Nuisance particulate (accumulation in lungs)	1C
Toluene—skin	15, 8	Moderate irritation—eye, nose, throat/narcosis	2

Table 40.5 (continued)

Substance	Health Code No.	Health Effects	Work Category
Toluene-2,4-diisocyanate (TDI)	9, 14, 3	Asthma/marked irritation—eye, nose, throat, bronchi, lungs/dermatitis	4
o-Toluidine—skin	13, 4, 2	Methemoglobinemia/acute systemic effects/suspect carcinogen	4
Toxphene	(See Chlorinated camphene)		
Tributyl phosphate	15, 7	Moderate irritation—nose, throat, lungs/headache	4
1,1,2-Trichloroethane—skin	3, 8	Cumulative liver damage/narcosis	4
Trichloroethylene	8, 3, 2	Narcosis/cumulative systemic toxic effects/suspect carcinogen	4

a From Refs. 46, 63.

In these cases, the objective is to modify the limit for the unusually long shift to a level that would produce a dose (mg) that would be *no greater* than that obtained during 8 hr of exposure at the PEL. Examples of chemicals with exclusively acute effects include carbon monoxide or phosphine. The following equation is recommended by OSHA for calculating an adjustment limit (equivalent PEL) for these types of chemicals:

$$\text{Equivalent PEL} = \text{8-hr PEL} \times \frac{8 \text{ hr}}{\text{hours of exposure per day}} \qquad (14)$$

The other formula recommended by OSHA applies to chemicals for which the PEL is intended to prevent the cumulative effects of repeated exposure. For example, PCBs, PBBs, mercury, lead, and DDT are considered cumulative toxins because repeated exposure is usually required to cause an adverse effect, and the overall biologic half-life is clearly in excess of 10 hr. The goal of PELs in this category is to prevent excessive cumulation in the body following many days or even years of exposure. Chemicals whose rationale is based on cumulative toxic effects are placed in Category 3 in Table 40.2. Accordingly, Eq. 15 is offered to OSHA compliance officers as a viable approach for calculating a modified limit for chemicals whose half-life would suggest that not all of the chemical will be eliminated before returning to work the following day. Its intent is to ensure that workers exposed more than 40 hr/week will not eventually develop a body burden of that substance in excess of persons who work on normal 8-hr/day, 40-hr/week schedules.

$$\text{Equivalent PEL} = \text{8-hr PEL} \times \frac{40 \text{ hr}}{\text{hours of exposure in one week}} \qquad (15)$$

The specific approaches that an OSHA compliance officer can use to evaluate a workplace using unusual work shifts is described in detail in the *OSHA Field Operations Manual* (46, 59). The OSHA models, although less rigorous than the pharmacokinetic models that will be discussed, have certain advantages due to their simplicity, e.g., they do account for the kind of toxic effect to be avoided, require no pharmacokinetic data, and tend to be more conservative than the pharmacokinetic models.

It is interesting that in the first OSHA occupational health regulation that discusses the long workday (unusual shifts), they prohibited the use of its own adjustment scheme to establish acceptable levels of exposure (55). Because regulatory agencies must make decisions based on political, social, and scientific information, and, especially because they must survive legal scrutiny, it is important that the industrial hygienist, toxicologist, or physician understand that regulatory guidelines have to consider each and every exposure condition. Consequently, professionals should take the time to become familiar with the rationales for the TLV, PEL, AIHA WEEL, European MAC, and any other occupational exposure limit before adjusting it for unusually short or long periods of exposure. In this respect, Tables 40.2, 40.3, 40.4, and 40.5 as well as the various books that document the rationales for the various limits, are important aids. Beginning in 1997, the ACGIH began to place the various chemicals in the TLV booklet into categories and these are useful for deciding whether adjustments are needed.

6.2.1 Illustrative Example 5 (OSHA Model)

An occupational exposure limit of 1 µg/m³ has been suggested by NIOSH for polychlorinated biphenyls (PCBs). Studies of exposed workers indicate that the biologic half-life of PCBs is as long as several years. What adjustment to the occupational exposure limit might be suggested by NIOSH for workers on the standard 12-hr work shift involving 4 days of work per week if they adopted the simple OSHA formulas?

Solution

$$\text{Recommended limit} = 8 \text{ hr PEL} \times \frac{8 \text{ hr}}{12 \text{ hr}}$$

$$\text{Recommended limit} = 1 \text{ µg/m}^3 \times 0.667 = 0.667 \text{ µg/m}^3$$

Note: Since PCBs (chlorodiphenyl) are listed as both cumulative and acute toxicants (Category 4) in Table 40.5, Eq. 14 rather than Eq. 15 should be used since it yields the more conservative results.

6.2.2 Illustrative Example 6 (OSHA Model)

Many industries such as boat manufacturing are seasonal in their workload. During the months of January, February, March, and April, the builders of boats work 5 days per week, 14 hr per day and could be exposed to concentrations of toluene diisocyanate (TDI) at the TLV of 0.005 ppm. What occupational exposure limit is recommended for TDI for a person who works 14 hr per day for 5 days per week but only works 8 weeks per year?

Solution. No adjustment is made.

Note: TDI is categorized in Table 40.5 as a Category 1A chemical (i.e., one that has a ceiling limit). Substances in this category have limits that should never be exceeded and consequently the limits are independent of the length or frequency of exposure. Exposure limits for chemical irritants such as these are currently thought not to require adjustment. Until more is known about human response to irritants during unusually long durations of exposure, the physician, nurse, and hygienist should make note of the employee tolerance to the presence of irritants at levels at or near the TLV. Eventually, human experience will tell us whether irritation is a time-dependent phenomenon and whether the response varies with the different chemicals.

6.2.3 Illustrative Example 7 (OSHA Model)

Assume the permissible exposure limit for elemental mercury is 50 µg/m³. It has a half-life in humans in excess of several days. What adjustment to the limit would be recommended by OSHA for workers on a shift involving 4 days at 12 hr/day followed by 3 days of vacation, then 3 days of 12 hr/day followed by 4 days off?

Solution

$$\text{Equivalent PEL} = 8 \text{ hr PEL} \times \frac{40 \text{ hr}}{48 \text{ hr}}$$

$$= 50 \text{ µg/m}^3 \times 0.833 = 40 \text{ µg/m}^3$$

Note: Since elemental mercury is classified as a cumulative toxin (Category 3) according to Table 40.5, Eq. 15 should be used. The 48-hr workweek was used in this example since it yields a more conservative adjustment than the 36-hr work-week. It could be argued that a smaller adjustment factor based on the average number of hours worked every 2 weeks more accurately reflects the exposure [i.e., (48 + 36)/2 = 42] since the chronic effects of mercury are due to many weeks or years of excess exposure. A more detailed discussion of how to deal with toxins that tend to accumulate can be found in Hickey (55).

One major drawback to the simple formulas suggested by Brief and Scala (2) as well as OSHA is that they are conservative (i.e., they suggest a modified TLV or PEL that is lower than that predicted by more accurate pharmacokinetic models). This occurs because they *do not* take into account (quantitatively) the toxicant's overall biologic half-life (i.e., metabolism and elimination). As will be shown later in this chapter, pharmacokinetic models usually recommend a smaller degree of reduction in the air contaminant limit and, therefore, the achievement of the adjusted limit should be less costly yet still provide adequate protection.

6.3 Iuliucci Model

Robert Iuliucci of Sun Chemical proposed a formula (64) for adjusting limits for long workdays that is similar to Brief and Scala's except that it accounts for the number of days worked each week as well as the number of hours worked each day. It is mentioned only for the sake of completeness since it poses no particular advantages over Brief and Scala's approach and because it is limited only to the schedule Iuliucci described (12-hr/day,

4-day/week schedule). For the work schedule, Iuliucci recommended the following equation for modifying the TLV:

$$TLV_x = TLV_s \times \frac{8 \text{ hr worked}}{12 \text{ hr worked}} \times \frac{12 \text{ hr recovery}}{16 \text{ hr recovery}} \times \frac{4 \text{ day workweek}}{5 \text{ day workweek}} \quad (16)$$

In addition to this formula, he recommended that exposure to carcinogens during 12-hr shifts always be reduced by 50%. Although this may be prudent, it would seem to be unnecessarily strict based on pharmacokinetic considerations.

6.4 Pharmacokinetic Models

Pharmacokinetic models for adjusting occupational limits have been proposed by several researchers (18, 38, 51–56, 65–67). These models acknowledge that the maximum body burden arising from a particular work schedule is a function of the biological half-life of the substance. Pharmacokinetic models, like the other models, generate a correction factor based on the pharmacokinetic behavior of the substance as well as the number of hours worked each day and week, and this is applied to the standard limit in order to determine a modified limit. Unlike the OSHA, Brief and Scala, and Iuliucci approaches, by accounting for a chemical's pharmacokinetics, these models can also identify those exposure schedules where a reduction in the limit is not necessary.

The rationale for a pharmacokinetic approach to modifying limits is that during exposure to the TLV for a normal workweek, the body burden rises and falls by amounts governed by the biological half-time of the substance (Figs. 40.2 and 40.3). A general formula provides a modified limit for exposure during unusual work shifts so that the peak body burden accumulated during the unusual schedule is no greater than the body burden accumulated during the normal schedule. This is the goal of all of the pharmacokinetic models that have, thus far, been developed (18).

It is worthwhile to note that the maximum body burden arising from continuous uniform exposure under the standard 8-hr/day work schedule always occurs at the end of the last work shift before the 2-day weekend. On the other hand, the maximum body burden under an extraordinary work schedule may not occur at the end of the last shift of that schedule (see Example 15). This is especially true when the duration and spacing of work shifts that precede the last shift differ markedly from the standard week. Because unusual work schedules can be based on a 2-week, 3-week, 4-week, or even 11-week cycle and the work shift may be 10, 12, 16, or even 24 hr in duration, no generalization regarding the time of peak body burden can be offered. The time of peak tissue burden for unusual schedules must therefore be calculated for each specific schedule and chemical.

6.4.1 Mason and Dershin Model

Mason and Dershin (38) were the first to propose a pharmacokinetic model for adjusting exposure limits for unusual exposure schedules. Apparently, due to the manner in which the model was presented, their publication did not receive the attention and use that later researchers enjoyed. In spite of this, their approach is entirely accurate. Like the other pharmacokinetic models to follow, it accounted for the biologic half-life of the chemical and the number of hours of exposure per day and per week.

The pharmacokinetic model they developed accounted for a number of factors known to influence the rate of accumulation of a chemical. These factors included the toxicant concentration to which the individual is exposed, the physicochemical form of the material, the rate of metabolism and excretion, as well as the distribution of the material in the body following absorption. The major drawback of the model is that it assumes the body acts as a single compartment. The one-compartment approach (discussed previously) assumes the chemical to be uniformly distributed throughout blood and aqueous body fluids without significant storage in specific tissues except where such tissues may constitute the rate limiting step. In Mason and Dershin's model, the overall respiratory exchange, metabolism, and renal excretion were accounted for by using a single effective clearance constant, k.

In their study they noted that for a simple single compartment model, Ruzic (68) had shown that the overall rate of change in accumulation can be expressed as

$$\frac{d[A]}{dt} = k_i^*[M] - k_c[A] \tag{17}$$

where [M] = concentration of the contaminant in the environment (alveolar spaces) mg/L
[A] = concentration in the compartment, mg/L
k_i^* = effective rate constant for uptake, hr^{-1}
$k_c[A]$ = overall clearance constant, hr^{-1}

Note: The effective rate constants may also include other factors: for example, changes in vital capacity, minute volume, membrane permeability, or absorption from other sources as the cutaneous absorption and loss of carbon disulfide.

In a cyclic pattern of exposure and recovery, Eq. 17 has a general solution following the final period of recovery of

$$[A](t_m) = \frac{k_i^*}{k_c}[M][e^{-k_c(t_m - t_{m-1})} - e^{-k_c(t_m - t_{m-2})} + e^{-k_c(t_m - t_{m-3})} \cdots - e^{-k_c(t_m - t_{m-i})} \quad \text{for } t_{(m-1)} \geq 0 \tag{18}$$

where k^*, k, and M are held constant and
$t_{(m)}$ = time elapsed from the onset of the initial exposure, hours, including recovery following the last exposure
$t_{(m-1)}$ = time elapsed from the onset of exposure through the completion of the ith phase of uptake or recovery (loss).

and the initial body burden is assumed to be negligible.

Similarly, Eq. 18 may be solved for the body burden obtained at the close of the last exposure by substituting $t_{(m-1)}$ for $t_{(m)}$ so

$$[A]_{(t_{m-1})} = \frac{k_i^*}{k_c} [M][1 - e^{-k_c(t_{m-1}-t_{m-2})} + e^{-k_c(t_{m-1}-t_{m-3})} \cdots - e^{-k_c(t_{m-1}-t_{m-i})}] \quad (19)$$

As noted by the authors, in intermittent exposure and recovery, an upper limit to accumulation should be achieved for many of the typical, lipid-soluble solvents, within six days, provided that the periods of recovery and rate of excretion are sufficiently large. The fraction of the saturation value (from continuous exposure) attained at the close of a series of cycles of exposure and recovery varies with the clearance coefficient k_c and the pattern and duration of exposure according to the function

$$[e^{-k_c(t_m - t_{m-1})} - e^{-k_c(t_m - t_{m-2})} + e^{-k_c(t_m - t_{m-3})} \cdots - e^{-k_c(t_m - t_{m-i})}] \quad (20)$$

which is dependent on both the concentration of exposure and final equilibrium position. Using Henderson and Haggard's (69) definition of the equilibrium distribution coefficient,

$$D = \frac{C}{C_1} \quad (21)$$

where D = distribution coefficient, dimensionless
 C = concentration of contaminant in the fluid phase (mg/L)
 C_1 = concentration of contaminant in the vapor phase of the alveolar air (mg/L)

and solving Eq. 21 at equilibrium,

$$\frac{[A]}{[M]} \approx \frac{k_i^*}{k_c} \approx D \quad (22)$$

This completes the data requirements for calculation of an expected body burden due to a series of intermittent exposures.

The authors acknowledged the limitations inherent in the assumption regarding the use of a one-compartment model but noted that in most cases this limitation had little practical significance with respect to adjusting exposure limits. Other kineticists, although not all of them, would probably agree with this observation (22). In addition, Mason and Dershin supported the prior recommendations of Brief and Scala wherein a modeling approach should not be used to adjust limits whose goal is to minimize the likelihood of irritation, sensitization, or a carcinogenic response. Illustrative Examples 8, 9, and 10 demonstrate the use and general applicability of the model.

6.4.1.1 Illustrative Example 8 (Mason and Dershin Model).

Several workers exposed to methanol in a printing operation complained that their exposure left them dizzy and with optic neuritis at the end of work on Wednesday, 56 hr after reporting to work Monday morning. If the distribution coefficient for methanol at saturation is 1700 to 1 for body water over the concentration in alveolar air; and the concentration in workroom air is 350 ppm (TLV = 200 ppm), what concentration would a worker have obtained in blood at (*1*) the end of the last exposure and (*2*) the time at which he or she would report to work on Thursday morning?

A schematic diagram illustrating the behavior of methanol of this situation is shown in Figure 40.19.

Part I. What is the saturation fraction at the close of the last shift, 56 hr after initial exposure? Shifts are 8 hr in length and are separated by 16 hr without exposure (recovery).

A.

$$C_T/C_S = [1 - e^{-k(T)} + e^{-k(T-t_1)} - e^{-k(T-T_1)} + e^{-k(T-t_n)} - e^{-k(T-\tau_n)}]$$

where $t_1 = 8$ hr, $\tau_1 = 24$ hr, $t_2 = 32$ hr, $\tau_2 = 48$ hr, $t_3 = 56$ hr $= T$ (which is quitting time on Wednesday), k for methanol (MeOH) $= 0.03$ hr^{-1},

$$C_T/C_S = [1 - e^{-k(56)} + e^{-k(56-8)} - e^{-k(56-24)} + e^{-k(56-32)} - e^{-k(56-48)}]$$

$$C_{56}/C_S = [1 - e^{-0.03(56)} + e^{-0.03(48)} - e^{-0.03(8)}]$$

$$C_{56}/C_S = 0.3678$$

B. To calculate the tissue concentration (whole body) at saturation:

$$C_S = DC_{\text{MeOH}} = 1700 \times 0.44 \text{ mg/L} = 747 \text{ mg/L (body water) @ } 37°C$$

where D is the distribution coefficient for methanol (body fluids/alveolar air) and C_{MeOH} the ambient methanol concentration

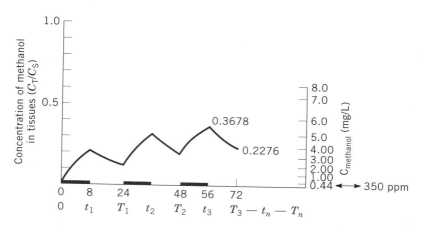

Figure 40.19. Graphical illustration of the behavior of methanol following the exposure schedule described in Example 8. Day-to-day increase in body burden were predicted by pharmacokinetic models. Exposure begins at $t = 0$, the 1st period ends at T_1, followed by 1st period of recovery (beginning at t_1), which ends at T_1; and repeats through t_n if the last point (T) is during exposure, and t_n if in recovery. (Courtesy of Dr. J. Walter Mason.)

$$C_{56} = 747 \text{ mg/L} \times 0.3678 = 275 \text{ mg/L}$$

Note: If this was the only exposure to MeOH, exposure to $C_{8.5}/C_{\text{Uns}} \times \text{TLV}_{8.5}$ would not allow an increase in the tissue concentration over that experienced under the 8-hr day, 5-day week. This provides the general solution for adjusting standards to unusual shifts as

$$\text{TLV}_{\text{unusual}} = C_{5.8}/C_{\text{Uns}} \times \text{TLV}_{5.8}$$

where

$$C_{5.8}/C_{\text{Uns}} = F$$

Then,

$$\text{TLV}_{\text{unusual}} = F \times \text{TLV}_{5.8}$$

Part II. What is the saturation fraction at the end of the final period of recovery 72 hr after the onset of the first exposure? Shifts are 8 hr as before.

A.

$$C_T/C_S = [-e^{-k(T)} + e^{-k(T-t_1)} - e^{-k(T-\tau_1)} + e^{-k(T-t_2)} - e^{-k(T-\tau_2)} + e^{-k(T-t_n)}]$$

where $T = 72$, $t_1 = 8$, $\tau_1 = 24$, $t_2 = 32$, $\tau_2 = 48$, $t_3 = 56$, $\tau_3 = 72 = T$ (which is the starting time for work on Thursday), k for methanol = 0.03 hr^{-1},

$$C_T/C_S = [-e^{-k(72)} + e^{-k(72-8)} - e^{-k(72-24)} + e^{-k(72-32)} - e^{-k(72-48)} + e^{-k(72-56)}]$$

$$C_{72}/C_S = [0 - e^{-0.03(72)} + e^{-0.03(64)} \cdots + e^{-0.03(16)}]$$

$$C_{72}/C_S = 0.2276$$

B. The tissue concentration is then found by

$$C_S = DC_{\text{MeOH}} = 747 \text{ mg/l (body water) @ 37°C},$$

$$C_{72} = 747 \text{ mg/l} \times 0.2276 = 170 \text{ mg/l}$$

6.4.1.2 Illustrative Example 9 (Mason and Dershin Model). **Part I.** Workers are exposed to methanol for 8 hr/day for 5 days/week. What is the body burden of those workers at the end of the fifth day knowing that the distribution coefficient (blood/air) is 1700, the effective clearance constant, k_c, is $0.03 h^{-1}$ ($t_{1/2} = 24$ hr) and the concentration to which they are exposed (1993 TLV) is 0.26 mg/1 (200 ppm)?

Solution. Setting $1700 = D$, $k_c = 0.03 \text{ hr}^{-1}$, and the periods of exposure and recovery at 8 and 16 hr, respectively; for which $t_m = 5(24) = 120$, $t_{(m-1)} = 120 - 16 = 104$,

PHARMACOKINETICS AND UNUSUAL WORK SCHEDULES

$t_{(m-2)} = 104 - 8 = 96$ and so forth; the body burden remaining at the close of recovery on the fifth day may be calculated as a function of the concentration of exposure.

$$[A]_{(t_m)} = 1700[M][e^{-0.03(16)} - e^{-0.03(24)} + e^{-0.03(40)} - e^{-0.03(48)} + e^{-0.03(64)} - e^{-0.03(72)} \cdots]$$

$$[A]_{(t_m)} = (1700M)(0.249) = 420M$$

Similarly, if the concentration at the end of the last exposure is of interest t_m is set equal to $t_{(m-1)}$ and

$$[A]_{(t_{m-1})} = 1700[M][1 - e^{-0.03(104-96)} + e^{-0.03(104-80)} \cdots - e^{-0.03(104-72)}]$$

$$A_{(t_{m-1})} = 1700M(0.401) = 680M \text{ (peak burden during 8 hr/day schedule)}$$

Part II. Having calculated the body burden (peak) at the end of 5 days of exposure during a normal 8-hr workday, what modified TLV would be recommended for a 14-hr workday and 4-day workweek?

Solution. If the modified exposure limit is chosen so that the tissue concentration attained at the close of the last work phase under standard conditions is equal to the accumulation allowed under the novel shift arrangement:

$$[A]_{(t_{m-1})_m} = [A]_{(104)_5}$$

where $[A]_{(t_{m-1})_m}$ = concentration that would be obtained at the close of the final exposure period in a novel shift arrangement
$[A]_{(104)_5}$ = concentration obtained after a 5-day workweek under standard conditions

The accumulation at the close of the last exposure may be obtained directly with Eq. 22 or from tables constructed with the exponential term of the same equation. The final form is then reduced to

$$[M]_n = \frac{[A]_{(104)_5}}{D} \times \frac{1}{\text{saturation fraction for } (t_{m-1})_n}$$

where $[M]_n$ = alveolar concentration of the contaminant resulting in a body burden equal to that attained by exposure under standard conditions
$TLV_n = [M]_n$ adjusted to ambient conditions

Therefore

$[A]_{95-56}$ = body burden of toxicant following four days of exposure during 14 hr/day schedule (i.e., 10hr/day or recovery)

By substitution

$$[A]_{96} = 1700M[1 - e^{-0.03(96-82)} + e^{-0.03(96-72)} - e^{-0.03(96-58)} + e^{-0.03(96-48)} \cdots]$$
$$[A]_{96} = (1700M)[1 - 0.657 + 0.487 - 0.320 + 0.237 - 0.156 + 0.115 - 0.75]$$
$$[A]_{96} = (1700M)0.630 = 1070M \text{(peak burden during 14-hr/day schedule)}$$

To calculate the modified TLV for the 4-day, 14 hr/day schedule:

$$F[A]_{(t_{m-1})_m} = [A]_{(104)_5}$$
$$F = \frac{A_{(120)}}{A_{(96)}} = \frac{680M}{1070M} = 0.636$$

Adjusted TLV = 0.636 (8 hr TLV) = 0.636 (0.26 mg/L) = 0.165 mg/L

Conclusion. The concentration of airborne methanol for the 14-hr/day schedule should be reduced from 0.260 to 0.165 mg/L (37% lower) in order to have the same peak body burden as that noted during the standard workweek.

Figure 40.20 shows how a series of curves can be generated for a particular chemical which would permit the rapid identification of an adjustment factor for a number of sched-

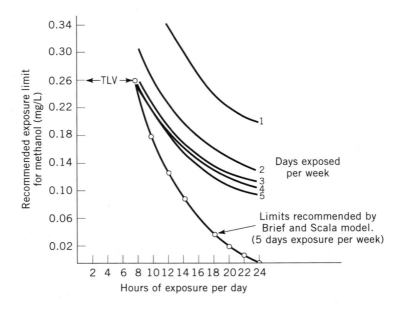

Figure 40.20. Modified exposure limits for methanol for various work schedules as determined by Mason and Dershin's formula. The dotted line illustrates the limits recommended by the Brief and Scala models. (From Ref. 38)

ules. The factor suggested by Mason and Dershin's formula is compared to that recommended by Brief and Scala (dotted line) in this figure.

6.4.1.3 Illustrative Example 10 (Mason and Dershin).
In the oil producing regions of Canada and the North Sea, work schedules can become very complex. Calculate a modified exposure limit for a 2-week work schedule where persons will be exposed to cyclohexane (assume biologic half-life in humans of 23 hr) for four 10-hr workdays, followed by 4 days off, then four 12-hr workdays followed by 2 days off, to complete two calendar weeks, which involved 88 hr of exposure.

Equation A

$$[A]_{(t_m)} = \frac{k_i^*}{k_c} [M][e^{-k_c(t_m - t_{m-1})} - e^{-k_c(t_m - t_{m-2})} + e^{-k_c(t_m - t_{m-3})} \cdots - e^{-k_c(t_m - t_{m-i})}]$$

Equation B

$$[A]_{(t_{m-1})} = \frac{k_i^*}{k_c} [M][1 - e^{-k_c(t_{m-1} - t_{m-2})} + e^{-k_c(t_{m-1} - t_{m-3})} = e^{-k_c(t_{m-1} - t_{m-i})}] \quad \text{for } t_{(m-i) \geq 0}$$

where k_i^*, k_c and $[M]$ are held constant and
t_m = time elapsed since initial onset of exposure ($t = 0$) through the last phase of recovery = 336 hr
t_{m-1} = time elapsed to the point at which work ends = 276 hr

This schedule and the day-to-day increase in blood plasma concentration are shown in Figure 40.21.

The solid line was calculated by iteration [as a check against Eq. A]. But, the point at 252 and 336 hr were obtained via Eqs. B and A, respectively. Data points for the iteration are:

$A(t)/A_{\text{sat}}{}^a$	t, hr	$A(t)/A_{\text{sat}}{}^a$	t, hr	$A(t)/A_{\text{sat}}{}^a$	t, hr
0	0	0.3132	96	0.3178	240
0.2592	10	0.1525	120	0.5240	252
0.1703	24	0.0742	144	0.3656	264
0.3854	34	0.0361	168	0.5574	276
0.2532	48	0.0176	192	0.3889	288
0.4468	58	0.3146	204	0.4893	312
0.2935	72	0.2195	216	0.0921	336
0.4766	82	0.4555	228		

^{a}This is the "saturation fraction."

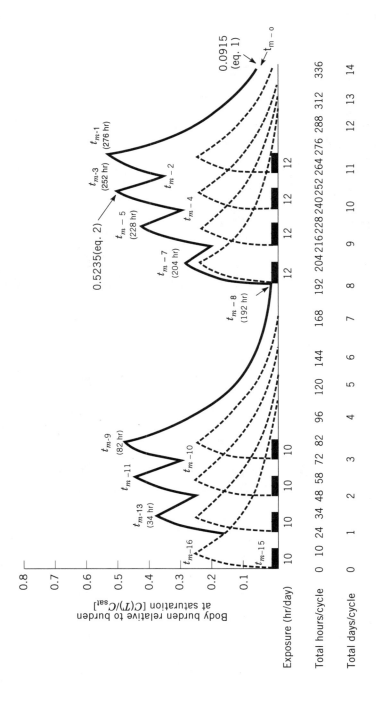

Figure 40.21. The likely body burden of cyclohexane in persons who work an unusual schedule involving four 10-hr days followed by four 12-hr days as described in Example 10. The dashed line represents the behavior of the chemical during each day of exposure and the dark lines represent the overall behavior of the chemical due to repeated exposure.

PHARMACOKINETICS AND UNUSUAL WORK SCHEDULES

The construction of a table is the easiest way to solve Eq. A and B. For $A(t_m)$ use Eq. A $[e^{-k\Delta t} - e^{-k\Delta t} \ldots]$. For $A(t_{m-1})$ use Eq. 6 $[1 - e^{-k\Delta t} \ldots]$.

t_m	t_{m-n}		Δt	$-k$	$-k\Delta t$	$e^{-k\Delta t}$
336 −	276₁	=	60	−0.03	−1.80	+0.1653
336 −	264₂	=	72	−0.03	−2.16	−0.1153
336 −	252₃	=	84	−0.03	−2.52	+0.0805
336 −	240	=	96	−0.03	−2.88	−0.0561
336 −	228	=	108	−0.03	−3.24	+0.039200
336 −	216	=	120	−0.03	−3.60	−0.027300
336 −	204	=	132	−0.03	−3.96	+0.019100
336 −	192	=	144	−0.03	−4.32	−0.013300
336 −	82	=	254	−0.03	−7.62	+0.0005
336 −	72	=	264	−0.03	−7.92	−0.0004
336 −	58	=	278	−0.03	−8.34	+0.0024
336 −	48	=	288	−0.03	−8.64	−0.000177
336 −	34	=	302	−0.03	−9.06	+0.000116
336 −	24	=	312	−0.03	−9.36	−0.000086
336 −	10₁₅	=	326	−0.03	−9.78	+0.000057
336 −	0₁₆	=	336	−0.03	−10.08	−0.000042

$$\Sigma e^{-k\Delta t} = 0.0915$$

A diagram comparing the behavior of the body (blood) levels for the work schedule described in this example with that of a standard schedule is shown in Figure 40.21. Each day's exposure or "daily additions" are represented by the dashed lines at the bottom of the plot. The solid line, which leads up to point t, represents the sum of the dashed line values for the same point. Piotrowski and others have used this approach to illustrate the principle of summation (25).

Examples 8 through 10 illustrate one of the advantages of the Mason and Dershin model in that some persons feel that it is more flexible than the other models for calculating a modified exposure limit for complex work schedules. Specifically, it is useful whenever there is no fixed number of hours worked each day or a fixed number of days worked per week. In short, the best aspect of this approach is that the period in the cycles need not be of equal duration or number.

6.4.2 Hickey and Reist Model

In 1977, Hickey and Reist published a paper describing a general formula approach to modifying exposure limits that was equivalent to that of Mason and Dershin. However, the benefits of their work were manifold. They confirmed the soundness of the previous model but, equally importantly, they also validated it to some extent by comparing the results with published biological data. In addition, they proposed broader uses of the pharmacokinetic approach to modifying limits and presented a number of graphs that could be used to adjust exposure limits for a wide number of exposure schedules. The graphs were based on (a) the biologic half-life of the material, (b) hours worked each day, and (c) hours worked per week.

Over the next 3 years they published studies that illustrated how their model could be used to set limits for persons on overtime (65) and for seasonal workers (54). Hickey's treatment of the topic of adjusting exposure limits is quite thorough, and his publications are primarily responsible for most of the interest and research activity in this area.

As discussed, it is clear that for any schedule, the degree of toxicant accumulation in tissue is a function of the biologic half-life of the substance. Figure 40.22 illustrates how a toxicant might behave in a biologic system or a tissue following repeated exposure to a particular average air concentration during a typical work schedule. Note that the peak body burden, rather than the average (Ba) or residual body burden (Br), is the parameter of interest. The biologic half-life not only dictates the level to which a chemical accumulates with repeated exposure, it dictates the time at which steady state will be reached for any given exposure regimen (normal, unusual, or continuous). For example, for moderately volatile substances (e.g., solvents) that have half-lives in the range of 12–60 hr, and for most work schedules, the steady-state tissue burden will be reached after approximately 2–6 weeks of repeated exposure. For most volatile chemicals (low-molecular-weight solvents) with shorter half-lives, the steady-state blood levels will be reached after about 2–4 workdays. Under conditions of continuous uniform exposure, most chemicals will be within 10 percent of the steady-state levels following about 4 times the biologic half-life of the chemical, and after 7 half-lives it will be within 1 percent of the plateau (steady-state) levels (19).

Several indices of body or tissue burden could have been chosen as the basis for predicting equal protection for any two different exposure regimens. These indices are the peak, residual, and average body or tissue burden of a substance. Figure 40.20 illustrates these three potential criteria from which one must choose in order to build a mathematical model. As in Mason and Dershin's model. Hickey and Reist selected the peak body burden

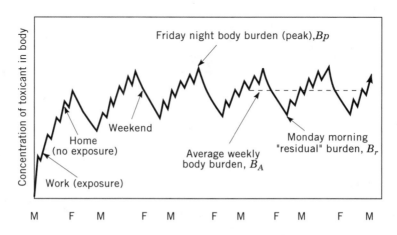

Figure 40.22. Illustration of the weekly fluctuation of body burden resulting from occupational exposure to an inhaled substance. Peak (B_p), residual (B_r), and average body burdens (B_A) are shown. From Ref. 18.

as the criterion since it is a more conservative approach to predicting the occurrence of a toxic effect than either the average or residual tissue concentrations. A thorough discussion of the rationale for selecting the peak burden rather than the residual or average for building the models can be found in Hickey's dissertation (51).

Other choices for predicting safety are problematic. For most chemicals, the residual body burden goes to virtual zero for most chemicals after a weekend away from exposure. Consequently, modeling to control this criterion would not prevent excessive peak burdens. The use of the average burden reduces the model to Haber's law (63). This, of course, would allow high tissue burdens to occur for long periods even though the TWA burden might be acceptable. Peak burden, therefore, is the best criterion, however, it may not be appropriate when the goal of an exposure limit is to avoid a carcinogenic hazard. In these cases, control of the average weekly or daily exposure (at the TLV or PEL) should generally be adequate to prevent any significant risk.

The Hickey and Reist model can, like Mason and Dershin's approach, be used to determine a special exposure limit for workers on extraordinary schedules, which will prevent peak tissue or body burdens from being greater than that observed during standard shifts. This special limit is expressed as a decimal adjustment factor that, when multiplied by the appropriate exposure limit, would yield the "modified" limit. It is worthwhile to note that all of the researchers have been careful to note that they *did not assert that currently prescribed or recommended occupational health limits are safe, but only that the special limit that can be predicted from their models should yield "equal protection" during a special exposure situation*! Examples 11 through 15 illustrate the use of the Hickey and Reist model.

One limitation of the models of Hickey and Reist (18, 51), Mason and Dershin (38), and Roach (67) is that they assume that the body acts as one compartment. Although this simplification may not pose many shortcomings for the task of adjusting occupational exposure limits, it is well known that many, if not most, chemicals do not exhibit one-compartment behavior (36, 71). This is not surprising since after a substance (particulate, gaseous, or vapor) is inhaled in air, it is taken up by the body, distributed, perhaps metabolized, and excreted by complex processes. Even though these processes can now be modeled quite well through the use of complex mathematical models, the one-compartment model has been used by these scientists since, in most cases, even this simple approach will yield results similar to those that incorporate more complex approaches.

As discussed by Hickey, the one-compartment model is the simplest one and it assumes that the body is a homogeneous mass, comparable to a room or compartment containing a clean fluid such as air. More of the air, bearing a contaminant, enters and flows continuously through the compartment, mixing en route with the air therein in a process analogous to inhalation. If contaminated air continues to enter, the contaminant concentration in the compartment increases until it reaches equilibrium with that of the incoming air; that is, as much contaminant is leaving as entering (see Fig. 40.23). When the contaminated air supply is replaced with clean air, the process is reversed, and the contaminant concentration in the compartment decreases exponentially (18). An analogy can be shown using fluid in tanks, as shown in Figure 40.24. The mass transfer phenomena are described by the following equations:

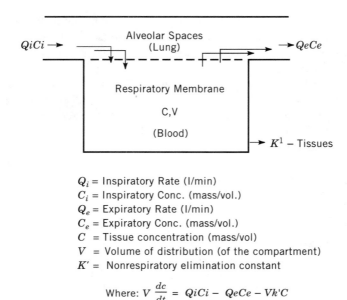

Q_i = Inspiratory Rate (l/min)
C_i = Inspiratory Conc. (mass/vol.)
Q_e = Expiratory Rate (l/min)
C_e = Expiratory Conc. (mass/vol.)
C = Tissue concentration (mass/vol)
V = Volume of distribution (of the compartment)
K' = Nonrespiratory elimination constant

Where: $V \dfrac{dc}{dt} = Q_iC_i - Q_eC_e - Vk'C$

| Change in concentration of toxicant in body | = | Respiratory input | − | Respiratory output | − | Metabolism and nonrespiratory elimination |

Figure 40.23. Diagram illustrating the driving forces for the pharmacokinetic behavior of an inhaled air contaminant. This simple description is the basis of the models developed by Mason and Dershin (38), as well as, Hickey and Reist (18).

A: Inhaled contaminant
B: Exhaled contaminant
D: Relative capacity of body tissues F, S1, S2, and M
R: Alveolar air-pulmonary blood compartment
F: "Fast" compartment
S: "Slow" compartments 1 and 2
M: Compartment with metabolism
D: Relative blood flow to a compartment, or metabolism rate

Figure 40.24. Simulation of body uptake of an air contaminant using fluid in tanks as analogy. Reprinted from Hickey (51) with permission.

For uptake:

$$B_t = CWK(1 - e^{-kt}) + B_0(e^{-kt}) \qquad (23)$$

For excretion:

$$B_r = B_t(e^{-kt_r}) \qquad (24)$$

where B_t = body burden of substance at time t (mass)
B_0 = initial body burden of substance at time zero (mass)
B_r = residual body burden of substance at time t (mass)
C = substance concentration in air (mass/volume)
K = ratio of the substance's equilibrium solubility in the body to that in air, or "partition coefficient" (dimensionless)
W = volume of body
k = uptake and excretion rate of substance in the body, equal to L/WK, in which L is the flow rate of air to the body time (time, hr^{-1})
t = time of exposure to substance in air (hr)
t_r = time since cessation of exposure to substance in air (hr).

It should be noted that k may also be expressed in terms of half-life or half-time of the substance in the body, $t_{1/2}$, where $k = (\ln 2)/t_{1/2}$.

Hickey and Reist have noted that while the predictive capability of the pharmacokinetic models is limited by the shortcomings caused by simplification to a one-compartment system, there are practical circumstances that minimize these draw-backs (51). First, many of the body tissues that are important targets for inhaled substances ("critical tissues") are highly perfused (19, 25, 27, 34, 42, 61, 68) and the concentration of the contaminant in these tissues may follow that of the arterial blood closely, thus in effect, becoming part of the lung-arterial blood compartment. The opposite case occurs when the buildup of contaminant in the critical or target tissue is extremely slow compared to the buildup in the rest of the body. In such a case, the remainder of the body, or more specifically the arterial blood may be assumed to reach saturation relatively quickly and remain at a virtually constant concentration. In effect, the body (except for the critical tissue) becomes part of the ambient environment, and the critical tissue becomes the one-compartment body (18).

Figure 40.21 illustrates how the body takes up and excretes an inhaled air contaminant as described according to Eq. 23 and 24, where exposure to contaminated air occurs during working hours and clean air is inhaled during nonworking hours. The body takes up the contaminant according to rate k during periods of exposure, and during nonworking hours the body excretes the contaminant according to negative rate $-k$ (18). The rate constant k is assumed to remain unchanged for each chemical regardless of the duration of exposure or whether there are repeated exposures. Small changes in k have been reported following repeated exposure to unusual shifts, but these are not usually large enough to justify mathematical correction. It should be remembered that for a given exposure schedule and any k, the body will eventually reach some equilibrium level with the contaminated air after continuous or repeated exposures (see Fig. 40.4).

The variation in body burden upon exposure to an air contaminant for five work-days per week for a period long enough to reach equilibrium (steady state) is also illustrated in Figure 40.21. Equilibrium implied that the "Monday morning" body burden (B_r) remains the same from week to week for a given exposure schedule. Each schedule also has a characteristic Friday afternoon peak body burden (B_p), and average body burden (B_a). This is illustrated in Figure 40.17 where two different exposure schedules and the resulting body burdens are described for a chemical with a moderately long half-life.

In spite of the fact that nearly all volatile chemicals will demonstrate some degree of two-and three-compartment behavior, the one-compartment assumption is generally satisfactory (if the half-life has been properly calculated). Dittert (22) has noted that in many, if not most situations, simplification to one-compartment behavior poses a minimal source for error in most calculations.

In their publications, Hickey and Reist (18) described the derivation and use of the following equation for adjusting limits:

$$(1 - e^{-kT_s})\left[1 - \exp(-kt_n) + \exp\left(-k \sum_{i=n-1}^{n} t_i\right)\right.$$
$$\left. - \cdots + \cdots - \exp\left(-k \sum_{i=1}^{n} t_i\right)\right]n$$
$$(1 - e^{-kl_n})\left[1 - \exp(-kt_s) + \exp\left(-k \sum_{j=s-1}^{s} t_j\right)\right.$$
$$\left. - \cdots + \cdots - \exp\left(-k \sum_{j=1}^{s} t_j\right)\right]s \qquad (25)$$

in which the t values represent duration of sequential work and rest periods in cycle T for normal (n) and special (s) exposure schedules. The authors noted that in their model, the use of ratios causes many of the imponderable and unknown terms to cancel, leaving only the special work schedule, which will be known, and the substance half-life (or uptake/excretion rate), which may or may not be known.

This equation can be used to determine a modified TLV or PEL for any exposure schedule since it accounts for the number of hours worked per day, days worked per week, time between exposures, and biologic half-life of the toxicant.

Where the special or extraordinary work cycle uses normal days and weeks, Eq. 25 can be simplified to the following form:

$$F_p = \frac{(1 - e^{-8k})(1 - e^{-120k})}{(1 - e^{-hk})(1 - e^{-24dhk})} \qquad (26)$$

in which, using hours as the time unit,
  F_p = TLV or PEL reduction factor
  k = uptake and excretion rate of the substance in the body (biologic half-life)
  h = length of special daily work shift
  d = number of workdays per "workweek" in the special schedule

PHARMACOKINETICS AND UNUSUAL WORK SCHEDULES

The general Eq. 25 for regular repetitive schedules simplifies to

$$F_p = \frac{[1 - e^{-kt_{1n}}][1 - e^{-k(t_{1n}+t_{2n})n}][1 - e^{-kT_s}][1 - e^{-k(t_{1s}+t_{2s})}]}{[1 - e^{-kt_{1s}}][1 - e^{-k(t_{1s}+t_{2s})m}][1 - e^{-kT_n}][1 - e^{-k(t_{1n}+t_{2n})}]} \qquad (27)$$

where t_{1n} = length of normal daily work shift (8 hr)
  t_{2n} = length of normal daily nonexposure periods (16 hr)
$t_{1n} + t_{2n}$ = length of normal day (24 hr)
  T_n = length of normal week (168 hr)
  n = number of workdays per normal week (5),
  t_{1s} = length of special "daily" work shift (hr)
  t_{2s} = length of special nonexposure periods between shifts (hr)
$t_{1s} + t_{2s}$ = length of basic work cycle, analogous to the "day" (hr)
  T_s = length of periodic work cycle, analogous to the "day" (hr)
  m = number of work "days" per work "week" in the special schedule

The model may be used to predict the permissible level and duration of exposure necessary to avoid exceeding the normal peak body burden during intrashift, short, high-level exposures. The model does this by establishing excursion limits that will provide equal protection for these situations. Equation 28 is used to do this:

$$F_p = \frac{1 - e^{-kt_n}}{1 - e^{-kt_e}} \qquad (28)$$

where $k = \ln 2 / t_{1/2}$
  t_e = exposure time (hr)
  t_n = normal shift length (8 hr)

Hickey and Reist have noted that for substances with short biologic half-lives, less than 3 hr, no adjustment needs to be applied for workers on most extraordinary work shifts since there is no opportunity for accumulation. In Figure 40.25, the normal workweek is compared to workweeks from one to seven 8-h days. It can be seen that exposure limits may not be increased, even if exposure is for only one day per week, unless the substance half-life is greater than 6 hr. Similarly, limits need not be decreased for 6- or 7-day workweeks involving exposures of 8 hr/day unless the substance half-life is greater than about 16 hr. For substances with very long half-lives, those in excess of 40 hr, F_p is simply proportional to the number of hours worked per week, as compared to 40 hr.

In an effort to simplify the process of adjusting limits for unusual shifts, Hickey and Reist developed a number of graphs, shown later in Figures 40.26, 40.27 and 40.34. Many health professionals have found these to be very useful when estimating safe levels of exposure for chemicals for which they have little or no pharmacokinetic data. In these graphs, the adjustment factor, F_p is usually plotted as a function of substance half-life, $t_{1/2}$ for a particular work schedule(s). For example, Figure 40.26 shows the difference in the occupational exposure limit between a normal workweek and a workweek of four 10-hr days, a workweek of three 12-hr days, and a single 40-hr shift per week.

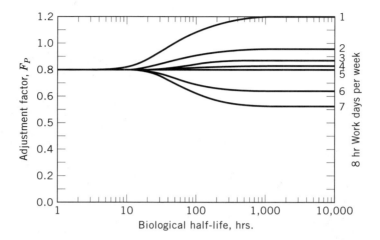

Figure 40.25. Adjustment factor (F_p) as a function of substance half-life ($t_{1/2}$) for various workweeks. From Ref. 18.

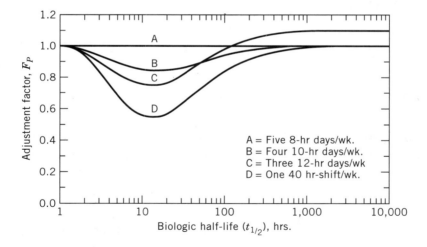

Figure 40.26. Adjustment factor (F_p) as a function of substance half-life ($t_{1/2}$) for various exposure regimens (shift schedules). From Ref. 18.

As is shown in Figure 40.26, for substances with very short half-lives, one hour or less, no correction is needed since the peak body burden is reached very quickly and is the same for a normal workweek as for any longer schedule. Therefore, if B_p is chosen as the predictor of equal protection, no reduction in OSHA limits is necessary for longer-than-normal work shifts as long as the weekly exposure is less than 40 hr.

PHARMACOKINETICS AND UNUSUAL WORK SCHEDULES

For substances with very long half-lives in the body, the adjustment factor is proportional merely to the number of hours exposed, not the daily or weekly exposure schedule. Thus, all 40-hr weeks have a special exposure limit for such substances equal to the normal limit, or an F_p of unity. Since three 12-hr days total only 36 hr per week, F_p for that schedule is 40/36, or 1.1, for a substance with a very long half-life.

Hickey offered sound advice when he noted that one need not resort to the conservative approaches of Brief and Scala, OSHA, or Iuliucci when the biologic half-life of the substance is not known. By assuming that the chemical has a half-life that would cause the greatest degree of day-to-day accumulation for that particular work schedule, the worst-case F_p can be calculated for any exposure schedule. Some of these worst-case values of F_p for selected schedules are shown in Figure 40.26. For example, since F_p varies as a function of the half-life, the worst-case condition is 0.84 for four 10-hr days, 0.75 for three 12-hr days, and 0.54 for the single 40-hr shift. Where the half-life is not known, the worst-case F_p can be used. Consequently, the pharmacokinetic models can be used to accurately protect workers on any schedule even when the pharmacokinetic behavior of the specific chemical is not known.

Other curves may be generated from Eq. 25 or 27 to compare any two schedules. In Figure 40.27, the correction factor recommended for continuous exposures for several periods of time from one to 1024 days is presented. A rest period equal to three times the exposure periods is needed before reexposure should occur. Again, for substances with very short half-lives, no adjustment to exposure limits is necessary when B_p is the criterion.

6.4.2.1 Illustrative Example 11 (Hickey and Reist Model).
In cities where commuting distances are a burden to the worker, one of the more frequent work schedules is the four day, 10-hr/day "compressed workweek." Assuming that this workweek is used in the

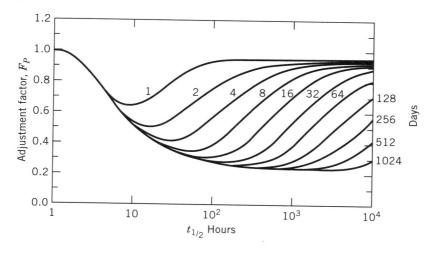

Figure 40.27. Adjustment factor (F_p) as a function of substance half-life ($t_{1/2}$) for continuous exposure schedules (days). From Ref. 18.

textile industry and that persons are routinely exposed to aniline at the PEL of 5 ppm, what adjusted occupational exposure limit would be recommended? Assume that aniline has an overall (beta phase) half-life of about 2 hr in humans.

Solution. Since the workweeks are equal and both schedules have all workdays consecutive, the simplified form of the general Eq. 26 can be used.

$$F = \frac{C_s}{C_n} = \frac{(1 - e^{-8k})(1 - e^{120k})}{(1 - e^{-10k})(1 - e^{-96k})} \quad \text{for } t_{1/2} = 2 \text{ hr}$$

$$k = \ln 2/2 \text{ hr} = 0.347 \text{ hr}^{-1}$$

$$F = 0.9677 \equiv 0.97$$

$$C_s = C_n F = 5 \times 0.97 = 4.85 \approx 5.0 \text{ ppm}$$

Note: No change is needed for chemicals with a half-life this short unless exposure is for 24 hr/day for several days.

6.4.2.2 Illustrative Examples 12 (Hickey and Reist Model).

Some persons in the petrochemical industry will routinely work 12-hr shifts for five consecutive days before having a 4-day period of no work; then they return for three 12-hr days again to be followed by 4 days off. This 5/4, 3/4, 12-hr schedule requires some adjustment of the normal exposure limit for certain systemic toxins if the peak body burden for this unusual shift is not to exceed the normal shift.

Based on Piotrowski's work (25), the overall half-life (β phase) in the human for trichlorethylene and its metabolites is about 9 hr. What modifications of the 1999 TLV of 50 ppm for trichloroethylene would be recommended for the workers on this new shift schedule during their first week of five consecutive 12-hr days of work?

Solution. Begin by determining the body burden for a normal week (B_{p_n}). For

$$t_{1/2} = 9 \text{ hr}, \quad k = 0.077$$

$$B_{p_n} = C_n WK \frac{(1 - e^{-8k})(1 - e^{-120k})}{(1 - e^{-168k})(1 - e^{-24k})}$$

$$B_{p_n} = 0.546 C_n WK$$

The first-week special body burden, B_{p_t}, (after switching schedules) is not to exceed B_{p_n} (normal burden), so

$$B_{p_s}(\text{week 1}) = C_s WK(1 - e^{-12k})$$
$$\cdot (1 + \underset{\underset{\text{day 5}}{\uparrow}}{e^{-124k}} + \underset{\underset{\text{day 4}}{\uparrow}}{e^{-48k}} + \underset{\underset{\text{day 3}}{\uparrow}}{e^{-72k}} + \underset{\underset{\text{day 2}}{\uparrow}}{e^{-96k}}) + \underset{\underset{\substack{\text{residual left} \\ \text{from last week} \\ \text{of old schedule}}}{\uparrow}}{B_{p_n}(e^{-178k})}$$

Set $B_{p_s} = B_{p_n}$ so that the body burden from the "special" schedule (B_{p_s}) is to be equal to the body burden for the normal schedule (B_{p_n}).

$$B_{p_n} - B_{p_n}(e^{-178k}) = B_{p_n}(1 - e^{-178k}) = 0.546 C_n WK(1 - e^{-178k})$$
$$0.546 C_n WK(1 - e^{-178k}) = C_s WK(1 - e^{-12k})(1 + e^{-96k})$$
$$\cdot (1 + e^{-96k} + e^{-72k} + e^{-48k} + e^{-24k})$$

$$\frac{C_s}{C_n} = \frac{\text{TLV special}}{\text{TLV normal}} = \frac{0.546(1 - e^{-178k})}{(1 - e^{-12k})(1 + e^{-96k} + e^{-72k} + e^{-48k} + e^{-24k})}$$
$$= \frac{0.546}{0.716} = 0.76$$

$$C_s = C_n \times 0.76$$
$$C_n = 50 \text{ ppm trichloroethylene}$$

The recommended TLV for trichloroethylene for a 12-hr/day, 5-day schedule = 38 ppm.
Note: Due to the short half-life and 40-hr/week schedule, a short cut approach would yield the same result:

$$F = \frac{1 - e^{-8k}}{1 - e^{-12k}} = 0.76$$

6.4.2.3 Illustrative Example 13 (Hickey and Reist Model).
In an 8-hr day, 7-day workweek situation, such as the 56/21 or 14/7 schedules, what should the TLV for H_2S be (assume TLV = 100 ppm)? This is the special case of a 7-day workweek. Biologic half-life in humans is 2 hr, but the rationale for the standard is based on systemic effects and irritation.

Solution. There is no need to reduce limits to prevent excess irritation, but for system effects it should be. For 14 days on and 7 days off:

$$F_p = \frac{(1 - e^{-8k})(1 - e^{-120k})/(1 - e^{-168k})(1 - e^{-24k})}{(1 - e^{-8k})(1 - e^{-(14 \times 24)k})/(1 - e^{-(3 \times 168)k})(1 - e^{-24k})}$$

where total week = 21 days or 168×3 hr
workweek = 14 days or 14×24 hr

$$F_p = \frac{(1 - e^{1-120k})/(1 - e^{-168k})}{(1 - e^{-336k})/(1 - e^{-504k})} = 1$$

With a $t_{1/2}$ of 2 hr, virtually all of the chemical is lost during the 16 hr of recovery each day, so there is no need to lower the TLV. Also, even though the average hours worked per week is 37.3, the TLV may not be raised by 40/37.3 or by 1.07X. For a 56/21 schedule,

$$F_p = \frac{(1 - e^{1-120k})/(1 - e^{-168k})}{(1 - e^{-(56\times 24)k})/(1 - e^{-(7\times 168k)})} = 1$$

Likewise, with 56/21 there is no need to reduce TLV, even through the average hr/week is 40.7.

6.4.3 Roach Model

Roach (58) also proposed a mathematical model for use during extraordinary work shifts. His model, although developed independently, was virtually identical to that proposed by Mason and Dershin, as well as Hickey and Reist. His general equation is shown below:

$$R = \frac{(1 - e^{-8a})(1 - e^{-120a})(1 - e^{-la})}{(1 - e^{-24a})(1 - e^{-168a})(1 - e^{-ma})\Sigma e^{-na}} \quad (29)$$

In this formula the shifts included are those in one complete work cycle prior to the shift end in question and

- l = total number of hours for a complete work cycle
- m = number of hours duration of the work cycle
- n = number of hours from a prior work shift end to the shift end in question
- e = the exponent of natural logarithms, 2.718
- a = log $2/t_{1/2}$ = $0.693/t_{1/2}$
- $t_{1/2}$ = biologic half-time in hours

The minimum value of this ratio, R_{min}, is the value of R obtained for the particular work shift in the cycle in which the maximum body burden occurs.

Persons who wish to use this model are referred to the original article (53). Since it is functionally the same as the prior models, no examples of its use are provided. Table 40.6, however, was developed by Roach and can serve as a useful guide for quickly approximating the modified exposure limit for a number of types of unusual shifts where the biologic half-life ($t_{1/2}$) is known. Roach has shown that for any given work schedule, no matter how complex, a generalized graphical solution that yields the adjustment factor formula for any chemical can be developed. To illustrate this point, Roach developed Figure 40.28 for the particular complex work schedule shown.

Like previous writers, Roach noted that the limits for substances which (a) have very short biologic half-lives, (b) irritants, or (c) are carcinogens may require no alteration when

PHARMACOKINETICS AND UNUSUAL WORK SCHEDULES

Table 40.6 Examples of Exposure Limit Adjustment Factors for a Variety of Unusual Work Shifts Based on the Chemical's Biologic Half-Life ($t_{1/2}$) in Humans[a]

Work Shifts/Week	Hour/Work Shift	R_{min} when $t_{1/2}$ is			
		1 hr	10 hr	100 hr	1000 hr
4	10	1.00	0.85	0.94	0.99
5	9	1.00	0.92	0.89	0.89
5	10	1.00	0.85	0.81	0.80
5	12	1.00	0.75	0.68	0.67
6	6	1.01	1.26	1.18	1.12
6	8	1.00	1.00	0.89	0.84
6	10	1.00	0.85	0.72	0.67
7	6	1.01	1.26	1.09	0.97
7	8	1.00	1.00	0.82	0.73

[a]Calculated by Roach (53) using a pharmacokinetic model.

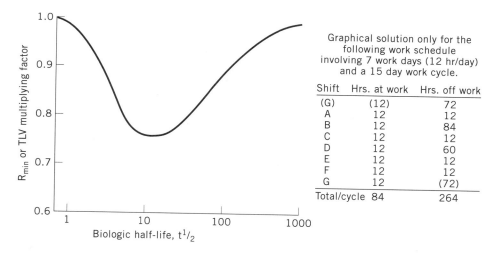

Figure 40.28. Graph showing the adjustment factor for any chemical to which workers are exposed during this specific work schedule. This approach can be quickly used by industrial hygienists who must set limits for dozens of chemicals for a given shift schedule. From Ref. 53.

the standard work schedule is altered. Roach suggested that the TLV for substances that have a very long biological half-time, such as mineral dusts, should be modified in proportion to the average hours worked per week, which has also been the recommendation of OSHA, Mason and Dershin (38) and Hickey and Reist (18). With such substances, the duration of any practical work cycle is short in comparison with their biological half-time, therefore it is appropriate that the limit would be unaltered so long as persons only worked an average of 40 hr per week. Roach suggested that as an additional precaution appropriate medical surveillance to detect any adverse effects would be advisable if the work schedule is such that R_{min} is less than 0.5 or greater than 2.0.

6.5 Determining a Chemical's Biologic Half-life

Effective half-lives of chemicals are difficult to precisely determine because of the complex manner in which many behave in the body. As discussed previously, the body can be envisioned to be made up of many compartments, and these different parts can take up and excrete substances at different rates. Consequently, each compartment has a different half-life for each substance. The difficulties in determining the half-life in a tissue such as the liver is the reason why overall or apparent half-lives that represent clearance of contaminants from the blood are used. Figure 40.29 illustrates how the degree of perfusion of various tissues and the exposure schedule will influence the peak concentration of toxicant that will be achieved, as well as the biologic half-life of the chemical in that tissue.

Effective half-lives for human uptake and excretion have been determined for many substances. Roach (52) and Hickey (51) have listed the half-lives of several industrial chemicals that they compiled from various sources. Table 40.7 presents a slightly more comprehensive list of substances for which half-lives have been determined in humans. Where available, half-lives from human studies should be applied to the models, and where the half-life is not known, the worst-case approach described by Hickey and Reist is convenient. Whenever both the α and β phase half-lives are known, the β phase half-life should be used in the various models since the terminal or β phase half-life conservatively describes the time course of elimination of the chemical. There are numerous differences

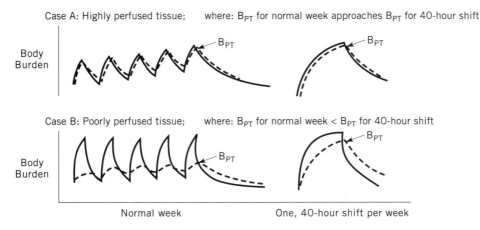

Figure 40.29. Variation in tissue uptake as a function of exposure schedule and blood flow (B_{PT} is peak tissue concentration of toxicant). From Ref. 51.

PHARMACOKINETICS AND UNUSUAL WORK SCHEDULES

Table 40.7 Estimated Half-Lives of Various Chemicals or Their Metabolites in Humans

Substance	Compartment (Media Collected)	Half-Life	Reference
Acetone	Overall (blood)	3 hr	25
Aniline	Overall (urine)	2.9	25
Benzene	Overall (blood)	3.0 hr	25, 74, 75
Benzidine	Overall (urine)	5.3 hr	25
Carbon monoxide	Overall (breath)	1.5 hr	76
Carbon disulfide	Overall (breath)	0.9 hr	25
Carbon tetrachloride	Fast (breath)	20 min	79
	Slow (breath)	3.0 hr	
DDT	Overall (blood)	1.5 yr	77, 78
Dichlorofluoromethane	Overall (blood)	9.4 min	80
Dieldrin	Overall (blood)	1.0 yr	77, 78
Dimethyl formamide	Overall (urine)	3.0 hr	81
2,3,7,8-TCDD	Overall (blood)	7.5 yr	77, 78
Ethyl acetate	Overall (breath)	2.0 hr	74
Ethyl alcohol	Overall (breath)	1.5 hr	74
Ethyl benzene	Overall (urine)	5.0 hr	25
Hexane	Overall (breath)	3.0 hr	74
Methanol	Overall (urine)	7.0	25
Methylene chloride	Overall (blood)	2.4 hr	82
Nitrobenzene	Overall (urine)	86.0 hr	25
PCB	Overall (blood)	1.5 yr	77, 78
Phenol	Overall (urine)	3.4 hr	83
p-Nitrophenol	Overall (urine)	1.0 hr	25
Styrene	Overall (urine)	8.0 hr	84, 85
Tetrachloroethylene	Overall (breath)	70 hr	86, 87
1,1,1,-Trichloroethane	Overall (urine)	8.7 hr	88
Trichloroethylene	Fast (breath)	30 min	89
	Slow (breath)	24 hr	90
Trichlorofluoroethane	Overall (blood)	16 min	80
Toluene	Fast (urine)	4 hr	91
	Slow (urine)	12 hr	92
Xylene	Overall (urine)	3.8 hr	91

and pitfalls in accurately determining biologic half-life, and these have been reviewed by Gibaldi and Weintraub (73).

6.6 Comparing the Various Models

The different models that modify exposure limits without consideration of pharmacokinetic behavior are those of OSHA (46) and Brief and Scala (2). The Brief and Scala model equations are

$$RF = \frac{8}{h} \times \frac{24 - h}{16} \quad \text{on a daily basis} \tag{30}$$

$$RF = \frac{40}{H} \times \frac{168 - H}{128} \quad \text{on a weekly basis} \tag{31}$$

$$EF = [(EF_8 - 1) \times RF + 1] \quad \text{for excursions} \tag{32}$$

where RF = reduction factor to be applied to the TLV or OSHA limit
h = hours worked per day
H = hours worked per week
EF = adjusted excursion factor
EF = normal excursion factor

Note: ACGIH abandoned the use of excursion factors in 1976. OSHA never adopted them.

6.6.1 Illustrative Example 14 (Hickey and Reist Model)

The NIOSH recommended occupational exposure limit for PCB is 1 µg/m³. In humans it has been found that the biologic half-life of PCBs is roughly 2.5 years. What adjustments to the occupational exposure limit would you recommend for workers on the standard 12-hr shift involving four days of work followed by three days of vacation, then three days of work followed by four days off, etc.

Solution

$$F_P = \frac{(1 - e^{-8k})(1 - e^{-120k})/(1 - e^{-168k})(1 - e^{-24k})}{(1 - e^{-12k} + e^{-24k} - e^{-36k} + e^{-48k} \cdots e^{-228k})/(1 - e^{-336k})}$$

$$= \frac{0.2381}{0.2501} = 0.9524$$

As pointed out earlier, as $t_{1/2}$ gets very large,

$$F_p = \frac{40}{\text{hr/wk special schedule}}$$

$$= \frac{40}{84/2} = 0.9525$$

Adjusted limit = TLV_s = 0.95 × 1 µg/m³ (essentially no change is needed).

6.6.2 Illustrative Example 15 (Hickey and Reist Model)

In Canada, unusual work shifts have become more commonplace than in the United States. In one industry, the unions and management decided that a combination of the 8-hr/day and 12-hr/day work schedule best fit their needs. What modified occupational exposure

PHARMACOKINETICS AND UNUSUAL WORK SCHEDULES

limit would be indicated for benzene (proposed TLV of 1.0 ppm) if the limit imposed in the plant were the most rigorous one to which they must adhere during the month (i.e., 12 days of repeated exposure)?

The exact schedule used in this industry involved five days of exposure for 8 hr/day followed by two days of 12 hr/day then five days of exposure for 8 hr/day shift followed by six days off work. The schedule then repeats itself so that workers average only 40 hr per week each 4-week cycle. Figure 40.30 illustrates the qualitative behavior of benzene in the body during this exposure schedule.

Solution. Two cycles must be examined to determine which gives the lower F_p:

a. Cycle of 12 on and 6 off.
b. Cycle of 5 on, 6 off and 7 on, with 6th and 7th days having 12-hr shifts.
c. Benzene $T_{1/2} = 10$ hr, $k = \ln 2/10$

$$F_p = \frac{(1 - e^{-8k})(1 - e^{-120k})/(1 - e^{-168k})(1 - e^{-24k})}{(1 - e^{-8k} + e^{-24k} - e^{-32k} + e^{-48k} \cdots e^{-272k})/(1 - e^{-432k})}$$
$$= \frac{0.52502}{0.52524} = 1.0$$

Note: The 12 on, 6 off cycle, gives an F_p of unity.

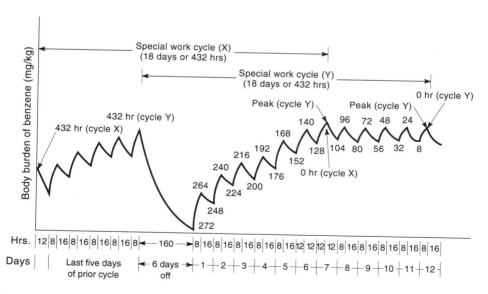

Figure 40.30. Graphical illustration of the likely fluctuations of the body burden of benzene (or its metabolites) following repeated exposure to the complex 8-hr/day and 12-hr/day work schedule described in Example 15.

Why? The residual levels from the two extra exposures from the previous Saturday and Sunday add only 0.0002 *CWK* to the 0.52502 *CWK* from the normal exposure burdens. To confirm, check the burden at the end of the 12-hr Sunday shift.

$$F_p = \frac{0.52502 \leftarrow \text{normal week}}{(1 - e^{-12k} + e^{-24k} - e^{-36k} + e^{-52k} \cdots - e^{-420k})/(1 - e^{-432k})}$$

$$= \frac{0.525}{0.686} = 0.765$$

$$\text{TLV}_s = 0.765 \times 1 \text{ ppm} = 0.765 \text{ ppm}$$

The peak body burden occurs at the end of the second 12-hr shift. The moral is that the time of peak burden must be chosen correctly. Otherwise, as could have occurred here, the incorrect factor would be applied to the exposure limit. If it is not obvious, more than one peak time must be tested. (This example problem was developed by Dr. John Hickey.)

As noted by Hickey (51), when applying the reduction factor to determine allowable exposure limits for nonnormal schedules, Brief and Scala give separate weightings to increased exposure time and decreased recovery time between exposures. Their model incorporated the concept of the "excursion factor" and reduces the TLV by a factor proportional to the decrease in the allowable exposure limit. The Brief and Scala model was not intended to be applied to shorter-than-normal exposures, only to longer-than-normal daily or weekly exposures. It cannot be applied to continuous exposures, as it devolves to zero at this point. Their model does not take into account substances with very short half-lives, but it is evident that rapidity of toxic response was considered in its development.

By comparison, the OSHA model [Eqs. 14 and 15], if applied to other than 8-hr shifts, could be recast in its simplest form using the Brief and Scala symbols:

$$RF = \frac{8}{h} \tag{33}$$

This represents the OSHA model as it would be used to determine limits for a substance (with no peak or ceiling limits) if a single uniform exposure occurred for h hours. For example, if a person were exposed only for 4 hr of an 8-hr shift, the adjustment factor would be 2; that is, a concentration of up to twice the TWA limit would meet OSHA regulations. Assuming application of OSHA regulations to 10 hr shifts, the adjustment factor for a 10-hr exposure would be 0.8; that is, the OSHA 8-hr limit would be reduced to 0.8 of its value to meet regulations.

Neither of the OSHA or Brief and Scala models accounts for the uptake rate; the reduction (or adjustment) factors derived from these models can be compared only to the worst-case adjustment factor for the model under development. It must also be assumed that OSHA regulations would apply to exposures longer than 8-hr. With these qualifications, several comparisons are made in Table 40.8.

Table 40.8. Comparison of Adjustment Factors Derived from Different Models[a]

Condition	Worst Case Pharmacokinetic Models, F_p	Brief and Scala, RF	OSHA Current Practice	OSHA Adjustment Factor for Longer Shifts
Five 8-hr days per week	1	1	1	1
Four 10-hr days per week	0.84	0.7	1	0.8
Three 12-hr days per week	0.75	0.5	1	0.67
Five 16-hr days biweekly	0.56	0.25	1	0.5
Alternating weeks of three and four 12-hr days	0.72	0.5	1	0.67
Four hrs of exposure daily[b]	1	c	2	c
Two hours of exposure daily[b]	1	c	4	c
One-half hour exposure daily[b]	1	c	16	c

[a]This chart was developed by Hickey (51).
[b]For substances with only TWA limits.
[c]Method not applicable.

It is readily apparent that in every case, the pharmacokinetic models call for less reduction in OEL for longer-than-normal exposure periods and for more reduction in short-term exposure limits than those required by the OSHA model. As noted by Hickey (51), this is a direct reflection of the pharmacokinetic model's use of peak body burden rather than average burden as a criterion to predict equal protection and of the fact that it takes into account the uptake and excretion rate of chemicals.

Even considering the limitations of the pharmacokinetic models, the adjustment factors derived by the model reflect more realistically the necessary protection from exposure than does the OSHA model, and to be a further extension of the concept of the Brief and Scala model (2, 51).

The various mathematical models proposed each have advantages and disadvantages. For chemicals with acute or chronic toxicity, the OSHA model restricts the daily dose or the weekly dose, respectively, during an unusual work shift to the same amount as that obtained during a standard 8-hr shift. It does not acknowledge the lesser recovery period or biologic half-life of the compound. The Brief and Scala model does not permit the daily dose of the toxicant under a novel work shift to be greater than that for a standard shift and it also accounts for the lessened time for elimination, however, it does not consider biologic half-life. Consequently, the Brief and Scala model is the most conservative of all of the models. In contrast, the Mason and Dershin, Hickey and Reist, and Roach models account for biologic half-life, increased daily dose, lessened recovery between exposures as well as the weekly dose. As a result these pharmacokinetic models yield less conservative results and they are presumed to be more accurate. Table 40.9 contains some of the general guidelines regarding the adjustment of occupational exposure limits for persons who work unusually long work shifts. Example 16 further illustrates some of the differences in the results of the various models.

Table 40.9. Rules of Thumb for Adjusting Occupational Exposure Limits for Persons Working Unusual Shifts

1. Where the goal of the occupational exposure limit is to minimize the likelihood of a systemic effect, the concentration of toxicant to which persons can be exposed should be less than the TLV if they work more than 8 hr/day or more than 40 hr/week and the chemical has a half-life between 4 and 400 hr.
2. Exposure limits whose goals are to avoid excessive irritation or odor will, in general, not require modification to protect persons working unusual work shifts.
3. Adjustments to TLVs or PELs are not generally necessary for unusual work shifts if the biological half-life of the toxicant is less than 3 hr or greater than 400 hr.
4. The biologic half-life of a chemical in humans can often be estimated by extrapolation from animal data.
5. The four most widely accepted approaches to modifying exposure limits will recommend adjustment factors that will vary. In order of conservatism, the Brief and Scala model will recommend the lowest limit and the kinetic models will recommend the highest.

$$\text{Brief and Scala} > \text{OSHA} > \text{ACGIH} > \text{Pharmacokinetic}$$

6. Whenever the biologic half-life is unknown, a "safe" level can be estimated by assuming that the chemical has a biologic half-life of about 20 hr. (Note: This will generally yield the most conservative adjustment factor for typical 8-, 10-, 12-, and 14-hr workdays.)

6.6.3 Illustrative Example 16 (Comparing the Models)

Assuming that 1,1,2-trichloroethane has a biologic half-life of 16 hr in humans, what modified TLV or PEL would be appropriate for persons who wished to work a 3-day, 12-hr/day workweek? Note that the dose for the unusual workweek (360 ppm-hr) would be less than for the normal 8-hr/day, 5-day workweek (500 ppm). The present PEL and TLV for 1,1,2-trichloroethane is 10 ppm.

$$\text{OSHA model: Modified PEL} = 10.0 \text{ ppm} \times \frac{8 \text{ hr}}{12 \text{ hr worked/day}}$$

$$\text{Modified PEL} = 6.66 \text{ ppm}$$

$$\text{Brief and Scala model: Modified TLV} = \frac{8 \text{ hr}}{12 \text{ hr}} \times \frac{24 - 12}{16} \times 10.0 \text{ ppm}$$

$$\text{Modified TLV} = 5.0 \text{ ppm}$$

$$\text{Hickey and Reist model: Modified TLV} = 10.0 \text{ ppm} \times \frac{(1 - e^{-8k})(1 - e^{-120k})}{(1 - e^{-t_1 k})(1 - e^{-t_2 k})}$$

$$\text{Modified TLV} = 10.0 \text{ ppm} \times \frac{(1 - e^{-8(0.04)})(1 - e^{-120(0.04)})}{(1 - e^{12(0.04)})(1 - e^{72(0.04)})}$$

$$\text{Modified TLV} = 7.5 \text{ ppm}$$

Note:

$$k = \frac{\ln 2}{t_{1/2}} = \frac{0.693}{16} = 0.04$$

t_1 = hr worked per day on unusual schedule
t_2 = 24 × days worked per week on unusual schedule

It is apparent from Examples 11 and 12 that the various models can recommend markedly different limits of exposure. In all cases, the pharmacokinetic approach recommends a less strict TWA limit than that generated by models that do not consider the biologic half-life of the chemical.

6.7 A Generalized Approach to the Use of Pharmacokinetic Models

As discussed, four or five different researchers have proposed pharmacokinetic models for adjusting limits and since they are based on the same assumptions, the results will essentially be the same. Roach (53) and Hickey and Reist (18) have presented charts that can be used to quickly determine the adjustment factor for many common shifts. However, for all other situations, the industrial hygienist must begin with the basic equations in order to calculate a modified limit.

Because most persons have found Hickey and Reist's approach and their publications to be most easily understood, the following generalized scheme for determining the adjustment factor for any schedule will be based on their equations. The limits derived from their model will be virtually identical to those obtained by use of the other models. Where the nonnormal work exposure schedule does not fit a curve derived by Hickey and Reist, one of several equations may be used.

1. For any regular weekly schedule, Eq. 27.
2. For a sporadic schedule, Eq. 25.
3. For an excursion in a normal shift, Eq. 28.
4. Where continuously rising exposure is expected, Eq. 34, Figure 7.31.

$$F_p = \frac{(1 - e^{kt_e})4k}{kt_e - (1 - e^{-kt_e})} \quad (34)$$

5. For discrete variations in exposure levels, Eq. 35.

$$F_p = \frac{1 - e^{-kt}}{f_c(1 - e^{-kt_c}) + f_b(1 - e^{-kt_b})(e^{-kt_c}) + f_a(1 - e^{-kt_a})(e^{-kt_b})(e^{-kt_c})} \quad (35)$$

where t = total exposure time period
f_i = air concentration of substance as fraction of special exposure level (concentrations f_a, f_b, f_c)
t_i = time of exposure at f_i (periods t_a, t_b, t_c)
k = clearance factor, $\ln 2/t_{1/2}$

As noted by Hickey (51), this equation has a drawback in that F_p is calculated on the assumption that peak body burden occurs at the end of a shift. If the actual peak occurs within a shift, F_p must be determined on that basis. Thus, to use Eq. 35 correctly, it must be known or calculated in advance at which exposure level the peak body burden will occur. If exposure levels do not decrease during a shift, the peak burden may be predicted to occur at shift's end. Where levels of contaminant decrease during the shift, F_p may be determined for the end of each discrete period and the lowest one applied.

It should be noted that different exposure regimens can affect how the body accumulates a chemical during a particular work shift. To illustrate how the pharmacokinetic models can predict variations in body burden due to unusual exposure periods, Hickey (51) has offered the following examples (17 and 18).

6.7.1 Illustrative Example 17 (Evaluating Short-Term Exposures)

Assume that three work schedules involve exposure to trichloroethylene (TCE). In situation 1, the worker is exposed to the TWA limit for 8 hr per day (normal). In situation 2, the TCE concentration rises linearly from 0 to 200 ppm during the shift. In situation 3, the workweek is the same, but the worker is exposed to a worst-case situation: a discretely rising TCE concentration with peaks of 300 ppm for 5 min every 2 hr. These situations are depicted in Figure 40.32.

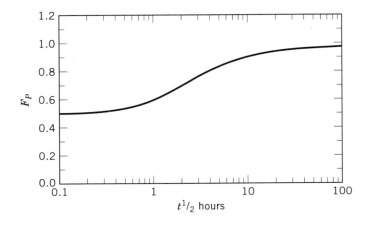

Figure 40.31. Adjustment factor (F_p) for continuously rising contaminant levels during work shift. From Ref. 51.

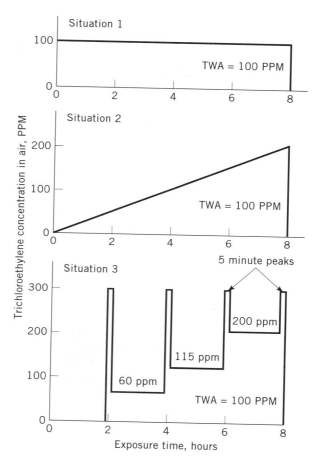

Figure 40.32. Comparison of three different trichloroethylene exposure situations, wherein the 8-hr TWA concentration is always 100 ppm, but the resulting peak body burden could vary between the exposure schedules. Based on Example 17. From Ref. 51.

What can be said about compliance to the various OSHA limits? Would one expect the peak body burdens to vary with the different schedules even though the absorbed dose (ppm-hr) is the same for all three? Is the peak body burden for any of these short-term exposure schedules likely to exceed the peak body burden observed during continuous 8-hr exposure to 100 ppm?

Note: Assume that TCE has an OSHA PEL of 100 ppm (TWA), a ceiling limit of 200 ppm, and an OSHA peak limit of 300 ppm of 5 min every 2 hr. TCE has a fast compartment biologic half-life of 15 min and a slow compartment biologic half-life of 7.6 hr. (This example was developed by Dr. John Hickey.)

Solution

PART I. In each case, exposures are within OSHA limits, since all have TWA averages of 100 ppm and none of the short-term exposures exceed an OSHA limit for short exposure periods.

PART II. Using Eq. 34 for situation 2 and Eq. 35 for situation 3, the F_p values for these two exposure regimens can be calculated. F_p is the adjustment applied to the TWA limit in a special exposure situation that will result in a predicted peak body burden equal to that resulting from a normal exposure at the TWA limit. When the adjustment is not made, and C_s is left equal to C_n, then R_p can be determined. R_p is the predicted ratio of special and normal body burdens (B_{p_s}/B_{p_n}), and is the reciprocal of the predicted F_p.

In the situation at hand, both F_p and R_p values are predicted for situations 2 and 3 relative to normal situation 1. This is shown in Table 40.10 for both the fast compartment of TCE ($t_{1/2}$ = about 15 min) and its slow compartment ($t_{1/2}$ = 7.6 hr). Results are interpreted as follows:

If the slow compartment is the critical tissue, the predicted peak tissue burdens will not exceed the peak burden for normal exposure if the OSHA TWA limit (C_n) is reduced to 0.89 C_n (or 89 ppm) in situation 2, and to 0.86 C_n (or 86 ppm) in situation 3. Stated in terms of R_p, if the concentration is not reduced but is left at C_n (100 ppm), the predicted peak body burden will be 1.12 times greater than desired in situation 2 and 1.16 times greater in situation 3. It is clear that departures from normal exposure have only a small effect on the accumulated burden in a compartment with a low uptake rate (long half-life) for TCE. Also, the additional short peak exposures in situation 3 have little further effect on peak burden over that resulting from situation 2.

PART III. If the fast compartment is the critical tissue group, the OSHA TWA limit must be reduced from 100 to 56 ppm for situation 2 and to 48 ppm for situation 3, if the predicted

Table 40.10. Adjustment Factors (F_p) and the Predicted Ratio of Body Burdens (R_p) for the Three Exposure Schedules Described in Example 17[a]

Predictive Index	Compartment[b]	Exposure Situation[c]		
		Situation 1	Situation 2	Situation 3
F_p	Slow	1	0.89	0.86
	Fast	1	0.56	0.48
R_p	Slow	1	1.12	1.16
	Fast	1	1.8	2.1

[a]This example readily illustrates how the manner in which one is exposed to a particular airborne toxicant can affect the peak body burden even though the total amount of toxicant absorbed each workday remains the same for all three situations.
[b]Fast compartment $t_{1/2}$ = 15 min; slow compartment $t_{1/2}$ = 7.6 hr.
[c]Situations 2 and 3 are compared to situation 1 (normal).

PHARMACOKINETICS AND UNUSUAL WORK SCHEDULES

peak compartment burden is not to exceed that resulting from normal exposure. Failure to reduce the concentration will result in a peak burden of 1.8 times greater than normal in situation 2 and 2.1 times greater than normal in situation 3. Note that in the fast compartment, the brief peaks in situation 3 add considerably (about 16%) to the burden accumulated in situation 2. Part of this increase is due to the longer exposure at the "ceiling" limit of 200 ppm in Situation 3.

This example illustrates how various exposure regimens can result in widely different body accumulations of a substance, and how the model predicts adjustment factors.

6.7.2 Illustrative Example 18 (Predicting Peak Body Burden)

To illustrate how pharmacokinetic models can predict equal peak body burdens for different shift lengths, let us assume two work situations with exposure to TCE. In situation 1, workers are exposed for five 8-hr days per week at the OSHA TWA limit of 100 ppm, and in situation 2, workers are exposed to the same level for four 10-hr days per week (illustrated in Figure 40.18).

What adjustment factor (F_p) must be applied to the normal limit (100 ppm) so that the exposure in situation 2 will result in a body burden no greater than that resulting from exposure in situation 1?

Solution

PART I. Using Eq. 26 or Figure 40.26, it can be determined that for the fast compartment ($t_{1/2}$ = 15 min), F_p = 1, and for the slow compartment ($t_{1/2}$ = 7.6 hr), F_p = 0.87. Thus, if the fast compartment is critical, no reduction in the OSHA limit is required in situation 2 to avoid an increase in fast-compartment accumulation. Since the peak burden is reached in the fast compartment in a few hours, the peak tissue concentration should be no greater with 10 hr exposure than with 8 hr.

PART II. If the slow compartment is critical, the OSHA limit must be reduced to 0.87 of the normal limit, or 87 ppm, to avoid a higher predicted peak in situation 2 than in situation 1. The prudent course would be to make the reduction where there is doubt as to which compartment is critical. If the $t_{1/2}$ values are unknown or in doubt, a worst-case F_p of 0.84, or 84 ppm, can be determined from Figure 40.26.

As shown in the previous examples, the use of the pharmacokinetic approach has a good deal of flexibility in that it can predict modified exposure limits for both short and long periods of exposure.

6.8 Special Application to STELs

Hickey (51) has discussed the use of the pharmacokinetic approach to setting acceptable limits for very short periods of exposure. Figure 40.33 illustrates the model's predicted F_p values for exposure to only a single excursion daily, as a function of substance half-life and the short-term exposure time. Hickey (51) has suggested an approach to determining the effect on F_p of the spacing of multiple excursions during a shift.

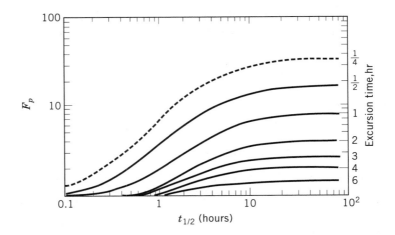

Figure 40.33. Adjustment factor (F_p) as a function of substance half-life ($t_{1/2}$) for three excursion schedules. Based on exposure schedule shown later in Fig. 40.37. From Ref. 51.

Example 17 illustrates how F_p can be calculated for short exposures spaced at intervals other than 60 min. From this example, it is readily seen (Fig. 40.34) that depending on the chemical's half-life, the spacing of the exposures can influence the recommended short term limit.

6.8.1 Illustrative Example 19 (Accounting for Sporadic Peak Exposures)

To illustrate how the time between high exposures (above limit) can influence the degree of adjustment, solve F_p for the following two situations.

CASE A. Assume that a person works in a foundry and is exposed to carbon monoxide for only 15 min, 4 times per day when he opens an oven. There is a 1-hr interval between the times he opens the oven (shown in Figure 40.35, Case A). What adjustment factor (F_p) would be suggested according to the Hickey and Reist model?

CASE B. Assume that the worker who opens the ovens can space the times between exposures at 1.75 hr rather than 1.0 hr (Fig. 40.35 Case B). What F_p is needed to show the level of protection as when the exposures last 1 hr?

Answer. As shown in Figure 40.34 F_p for Case A and Case B will vary with the time between short periods of exposure even though the total dose (ppm-hr) remains the same day. In the case of carbon monoxide, which has a biologic half-life in humans of 3.5 hr, the F_p for the 4 excursions with a 1-hr interval is 5.7. When the rest interval is 1.75 hr, the F_p is 6.7.

Case A of Figure 40.35 illustrates air concentrations of a substance over an 8-hr shift, using a 1-hour respite short-term exposure schedule, as compared to normal exposure.

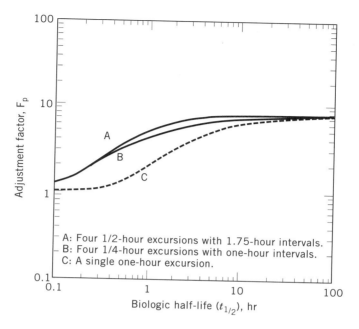

Figure 40.34. Adjustment factor (F_p) as a function of substance half-life ($t_{1/2}$) and excursion time (hours). Dotted line is used to express the high level of uncertainty involved in the prediction. From Ref. 51.

Similarly, Case B of Figure 40.35 depicts four 15-min excursions, but spaced equally over the entire shift, with 1.75-hr rest periods between exposures.

Hickey (51) has noted that generally, as shown in Figure 40.34, it makes little difference in resultant peak body burden whether the excursions are separated by 1-hr or by 1.75-hr intervals of nonexposure. The difference increases markedly, however, as the rest intervals between excursions are diminished, culminating in a single 1-hr exposure (four 15-min excursions with zero rest time between). This situation is shown in the dotted curve in Figure 40.34 for contrast, and is the same as the 1-hr excursion curve in Figure 40.33. Note also that the model's predictions reach a limiting STEL of 8.0, thus satisfying the ACGIH statement that the TLV-TWA may not be exceeded. *The curves in Figure 40.34 indicate once again that for long half-life substances, the exposure schedule is of little consequence when establishing STELs.*

Hickey (51) pointed out that the pharmacokinetic models predict equal protection at somewhat higher STELs than many of those recommended by the ACGIH. This is because many of the ACGIH STEL values continue to be based generally on the excursion limits. By contrast, the model bases STELs solely on first-order uptake rates.

The pharmacokinetic model may be used to predict the permissible level and duration of exposure necessary to avoid exceeding the normal peak burden during short, high-level exposures. The model does this by establishing excursion limits that will predict equal

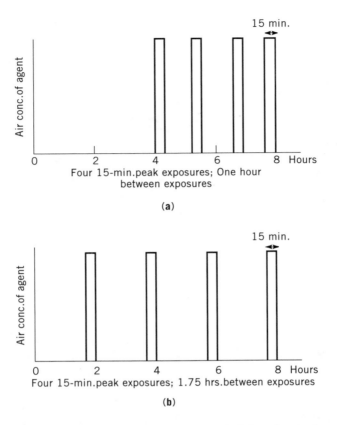

Figure 40.35. Comparison of two short-term exposure schedules wherein the peak and TWA concentration(s) are the same but the peak body burden could in some cases be different. Consequently, a lower short-term exposure limit might be needed. Based on Example 19. From Ref. 51.

protection for these situations. Assuming exposure is limited to a *single daily excursion* of duration t_e, Eqs. 25 and 28 reduce to

$$F_p = \frac{1 - e^{-k(8)}}{1 - e^{-kt_e}} \tag{36}$$

Adjustment factors derived from Eq. 36 have been plotted in Figure 40.34 as a function of substance half-life and excursion time. To illustrate, if a worker is exposed for one 30-min period during an 7-hr shift to a substance with a half-life in the body of 4 hr ($k = 0.17$), the adjustment factor would be determined as follows:

$$F_p = \frac{1 - e^{-(0.17)8}}{1 - e^{-(0.17)1/2}} = 9.0$$

The model would thus predict a limit of 9 times the OSHA TWA limit. This value may be read from Figure 40.34. If the OSHA cumulative exposure formula (46) were used to determine a limit, the limit would be 16 times the OSHA TWA limit, assuming no ceiling or peak limits for the substance. In Figure 40.33, the 15-min curve is dotted because the one-compartment model is not precise for very short exposure times.

6.9 Other Applications of the Pharmacokinetic Modeling Approach

Thus far, the use of modeling has been restricted to only inhaled gases and vapors. Hickey and Reist (18) have suggested that limits for particulates, reactive gases, and vapors can also be adjusted using their approach. For these substances and other situations, peak body burden B_p is still deemed to be the last criterion on which to develop a scheme to provide equal protection for the unusual work shift. The following sections describe these other situations and are based almost exclusively on the work of Hickey (51) and Mason and Hughes (91). The usefulness of kinetic modeling for very short durations, for seasonal workers, for exposures off-the-job, and for carcinogens are also discussed.

For many chemicals, only the results of animal studies will be available. These data can be quite useful for estimating the biologic half-life of specific chemicals in humans. It is, however, always inappropriate to assume that the half-life of a chemical obtained in a mouse, rat, hamster, dog, rabbit, or even monkey will describe its likely fate in humans. Unfortunately, animal data have often been directly used to adjust exposure limits for unusual work schedules. In most cases, the biologic half-life of an industrial chemical in an animal will be much less than in humans.

Quantitative approaches for extrapolating animal metabolism and excretion data to humans has been developed and could be used in the existing approaches. As shown in Table 40.11, it is readily apparent why the biologic half-life of a chemical in the smaller animals can be much less than in humans. Exceptions to this rule-of-thumb occur when humans cannot metabolize the parent compound (but the animal can). Fortuitously, this will rarely

Table 40.11. Interspecies Scaling of Hexane Clearance Based on Alveolar Ventilation Rates[a]

Species	Body Weight (g)	V_{alv}[b] (1/hr/kg)	Expected CL_{hexane}[c,d] (1/hr/kg)	Ratio (Species/human)
Mouse	30	35.9	11.8	6.9
Rat	250	21.1	7.4	4.1
Rabbit	3,000	11.4	3.8	2.2
Human	70,000	5.2	1.7	1.0

[a] Ref. 92
[b] $V_{alv} = [0.084 \text{ l/hr (b.wt.)}^{0.75}]/\text{(body wt.)}$.
[c] For these clearance calculations, it is assumed that S_b for n-hexane is about equal from species to species and that E_i at low inhaled concentrations is 0.25 in all species.
[d] Rate of metabolism is CL times C_{inh}. Relative ratio is given by the ratio of CL in animals/CL in humans.

occur for the majority of industrial chemicals. In general, the rate of metabolism or elimination is a function of the alveolar ventilation rate and the cardiac rate. Table 40.11 illustrates the difference in the ventilation rate and the expected rate of clearance of hexane from four species including humans (94).

Lastly, there are three additional caveats that need to be recognized regarding the determination of a biologic half-life for a substance. First, the half-life in the urine, breath, or feces is not the same as that for blood. Second, unless only one of these routes of elimination is predominant, none may correlate to the blood. Therefore, when available, the blood plasma half-life is the one that should be used. Third, the biologic half life of a particular chemical in animals and humans can vary with repeated exposure. That repeated exposure can vary the biologic half-life was noted in the work of Paustenbach et al. (93) that addressed the effects of repeated exposure of rats for periods of 11.5 hr/day and in the work of O'Flaherty et al. (35) who noted the progressive changes in the half-life of lead in exposed workers with increasing years of exposure. The potential pitfalls in calculating the biologic half-life of a chemical have been discussed elsewhere.

6.9.1 Particulates

6.9.1.1 Chemical. This entire section on particulates and reactive gases was published in Hickey and Reist (18). In that paper, they noted that the modeling approach for particulates is likely to be similar to vapors. Some researchers have suggested that more data needs to be gathered to validate the reasonableness of their model for particulates. Because it seemed appropriate given our state of knowledge, Hickey and Reist asserted that the one-compartment model as applied to particulates is likely to be analogous to inert gases and vapors with one exception; deposition of particulates is presumed to occur in the body at some linear rate proportional to air concentration. In their derivation, Hickey (51) assumed that the clearance of particulates is presumed to conform to a first-order exponential.

For uptake:

$$B_t = (CLf/k)(1 - e^{-kt}) + B_0(e^{-kt}) \tag{37}$$

For clearance:

$$B_r = B_t(e^{-kt_r})$$

where L = flow rate of air to the body (volume/time)
  f = fraction of particulates deposited
  k = clearance rate of deposited particulates ($k = \ln 2/t_{1/2}$)

Note: The other symbols are as defined in Eqs. 23 and 24.

As is the case for inert gases, when F_p is determined, L, f, and the k in the denominator cancel, leaving F_p for particulates identical to that for inert gases. The mechanisms are different, but the model, being a ratio, requires only that $L, f,$ and k be the same for normal as for special schedules.

The retention of inhaled particulates has traditionally been considered to vary directly with their concentration in air. The ICRP Task Group on Lung Dynamics has shown that the processes are independent of the air concentration of particulates, whether deposition is by inertia, gravity, or diffusion (96).

Retention varies significantly with other factors, however. The simple expression Lf in Eq. 37 masks a complex combination of variables including inhalation rate, which in turn affects air velocity in respiratory passages; particulate size, size distribution, density, and shape; respiratory frequency; and breathing habits, such as depth of breathing and mouth or nose breathing. The model does not consider any of these factors in predicting equal peak burdens, as they are canceled out by the assumption that they do not change with exposure time (shift length). In spite of these shortcomings, the assumption of a linear deposition rate appears valid.

As noted by Hickey (51), the assumption of first-order clearance is tenuous. However, the lung clearance rate is generally thought to vary with the magnitude of lung burden, although there is no general agreement on this point. It is thus not definite that clearance follows a single exponential, although half-lives for particulate clearance have been published. Because of the uncertainty of the half-life or half-lives for body clearance of any particulate substance, Hickey and Reist have suggested that the use of the worst case would seem to be prudent in using the model for adjusting particulate limits to predict equal protection.

When the model is applied to short exposures to particulates at high concentrations, there is the implicit assumption that the predicted allowable higher concentration limit is not so high as to overwhelm the deposition or clearance mechanisms of the body. This can occur at very high concentrations, and application of the model to particulates is limited to this extent.

6.9.1.2 Microbial Aerosols. Particulate aerosols may contain viable microorganisms and the use of the models for this hazard have been discussed by Hickey (51). These aerosols behave physically as any other airborne particulates until deposition in the host. There they may exhibit the unique characteristic of being able to multiply, either at the deposition site or some secondary site in the host. The resultant adverse effect may be an infection.

The number of viable organisms that must reach the host in order to initiate an infection depends on many factors, including host susceptibility, deposition site, and organism virulence. However, if one presumes that such a number exists, and that this number is analogous to peak body burden, the model may be applied to viable particulates.

At this time, there are no TLVs or OSHA limits or any other occupational exposure limits for specific microorganisms or viable particulates in air. There are recommended limits applicable to particular locations, such as hospital areas. However, these have not been correlated with, nor are they claimed to be based on, infectious dose. As Hickey (51) noted, in the absence of such limits, the determination of F_p (or F_a or F_r) for microorganisms in air becomes academic, as there is no limit to adjust. However, the potential for application of the model exists and awaits the development of relevant limits.

6.9.2 Reactive Gases and Vapors

6.9.2.1 Systemic Poisons. Two types of reactive substances are discussed in relation to the model: those that are both metabolized and excreted through respiration and those that are only metabolized. The equations of substances that are both metabolized and expired have been derived as follows:

For uptake:

$$B_t = CWK[1 - e^{-(k_1+k_2)t}][k_1/(k_1 + k_2)] + B_0[e^{-(k_1+k_2)t}]$$

For excretion:

$$B_r = B_t[e^{-(k_1+k_2)t_r}] \tag{38}$$

where k_1 = uptake and excretion rate by respiration
k_2 = rate by metabolism

The other symbols are as described in Eqs. 36 and 37.
  Again, when F_p is calculated, the additional factor, $k_1/(k_1 + k_2)$, cancels, and F_p is identical to that for inert gases and vapors, except that the effective half-life is described by $k_1 + k_2 = \ln 2/t_{1/2}$. Use of the prepared curves, such as Figure 40.26, would require knowledge of the combined effective half-life of a substance or use of the worst case, as before. For substances that are only metabolized and not exhaled through the lungs, Eq. 37 applies.

6.9.2.2 Local Irritants and Allergens. As noted by Brief and Scala, Mason and Dershin, Hickey and Reist, and Roach, the models *do not* appear amenable to deriving adjustment factors for exposure to primary irritants or allergens. The actions of these substances appear to be based on such a small local compartment, as contrasted to the entire body, that the predicted equal protection approach may not be applicable. It is also likely that B_p is unsuitable as a criterion for predicting response. By the same token, however, a prolonged exposure, beyond 8 hr, might not require any reduction in exposure limits. Fiserova-Bergerova (97) has suggested that a predictive model could be derived for irritants.

6.9.3 Radioactive Material

The mechanism of uptake and excretion of radioactive substances has been studied thoroughly by many researchers and its discussion is outside the scope of this chapter. However, the fate of inhaled radioactive gases and particulates has been modeled thoroughly and maximum allowable concentrations in air for them have been published, based in a large part on effective half-life. One marked difference here is that for many radioactive substances, the actual dose to which one is exposed is much easier to calculate and measure biologically, thus making the modeling and validation much more straightforward.

6.9.4 Mixtures

Hickey (51) has discussed mixtures and he noted that OSHA regulations state that exposure limits for mixtures of substances in air (except for some dusts) shall be such that

$$C_1/L_1 + C_2/L_2 + \cdots + C_n/L_n \leq 1 \qquad (39)$$

in which C_i is the concentration of a substance in air and L_i is its OSHA limit. The equation presumes strictly an additive effect of inhaled substances.

The model does not take into account potentiation or the possible synergistic effects of mixtures. However, if it is assumed that additive effects exist in proportion to peak body burdens, the model may be modified to accommodate mixtures. Instead of equating peak body burden for a special schedule ($B_{p/s}$) to that for a normal one ($B_{p/a}$), as in Eq. 25, the model would set

$$B_{p/s} = B_{p/n}(C_i/L_i) \qquad (40)$$

and the adjustment factor for any substance in a mixture, $F_{p/m}$, would be its F_p acting alone reduced by whatever factor is needed to meet the limits of Eq. 39 or

$$F_{p/m} = F_p(C_i/L_i) \qquad (41)$$

In practice, F_p would merely be calculated from Eqs. 25 through 27 or read from graphs, but applied as an adjustment to the reduced limit, C_i, as determined for the OSHA formula (Eq. 15), rather than to the normal OSHA limit. Example 20 is provided to illustrate Hickey's conceptual approach to adjusting limits for exposure to mixtures.

6.9.4.1 Illustrative Example 20. (Exposure to Mixtures). The pharmaceutical industry was the first to pioneer the use of the 12-hr work shift. In many firms the typical workweek involves 4 days. In many firms the typical workweek involves 4 days, 12 hr/day, then 3 days off followed by 3 days of 12 hr/day followed by 4 days off. Every 2 weeks, everyone will have worked about 40 hr/week.

Acknowledging that persons are often exposed to more than one chemical at a time, what modification would be recommended for exposure to isoamyl alcohol (TLV = 100 ppm) and carbon tetrachloride (TLV = 5 ppm) for this type of shift?

It is assumed that the biologic half-life in humans for isoamyl alcohol is 12 hr and for carbon tetrachloride it is about 5 hr. Note that each is a systemic toxin and that the intent of the TLV for isoamyl alcohol is to prevent CNS depression and liver toxicity, while the intent of the carbon tetrachloride standard is primarily to prevent liver toxicity.

Solution
Part A. For both isoamyl alcohol and carbon tetrachloride, the following approach could be used:

$$F = \frac{(1 - e^{-8k}(1 - e^{-120k})/(1 - e^{-168k})(1 - e^{-24k})}{[1 - e^{-12k} + e^{-24k} - e^{-36k} + e^{-48k} - e^{-60k} + e^{-72k} - e^{-84k} + e^{-192k} + e^{-204k} + e^{-216k} - e^{-228k} + e^{-240k} - e^{-252k}/(1 - e^{-336k})]}$$

(42)

Part B. For CCl_4, $t_{1/2} = 5$ hr; so $k = 0.139$ hr^{-1}. By substitution,

$$F = \frac{0.695}{0.841} = 0.83$$

Therefore, $TLV_s = TLV_N(0.83) = 5 \times 0.83 = 4.2$ ppm for CCl_4.

Part C. Using the same rationale, the modified TLV for isoamyl acetate is found using $k = \ln 2/12$ hr, or $k = 0.058$ hr^{-1}. By substitution,

$$F = \frac{0.493}{0.664} = 0.74$$

Therefore, special TLV = 74 ppm for isoamyl alcohol.

Part D. This does not consider the potential additive or synergistic effect of simultaneous exposures to both chemicals. If it is desired to apply the ACGIH approach for additive effects of exposure to both chemicals, then one could apply their approach for assessing mixtures. The modified TLVs for this situation would be

$$\frac{C_1}{4.2} + \frac{C_2}{74} \leq 1.0$$

and the concentration of one or both chemicals should be reduced until the equation is less than, or equal to, unity.

6.10 A Physiologically Based Pharmacokinetic (PB-PK) Approach to Adjusting OELs

A sophisticated and more accurate approach than those previously proposed involves the use of a PB-PK model for determining adjustment factors for unusual exposure schedules (43, 56). The PB-PK model requires data on the blood-air and tissue-blood partition coefficients, the rate of metabolism of the chemical, organ volumes, organ blood flows, and ventilation rates in humans. Andersen et al. (56) illustrated the use of the approach on two industrially important chemicals—styrene and methylene chloride. Their analysis suggested that when pharmacokinetic data are not available, a simple inverse formula may be sufficient for adjustment in most instances; this makes application of complex kinetic models unnecessary. They noted that pharmacokinetic approaches alone should not be relied on for exposure periods greater than 16 hr/day or less than 4 hr/day because the mechanisms of toxicity for some chemicals may vary for very short or very long term exposure. For these altered schedules, biological information on recovery, rest periods and mechanisms of toxicity should be considered before any adjustment should be attempted.

Central to the development of an appropriate physiological model for nonconventional work shifts is the need to quantify the degree of risk associated with 8 hr exposure to the TLV so that the calculated limit for the long schedule poses an equivalent risk. For chemicals that possess a particular kind of toxic effect, it is a relatively straightforward task to adjust the exposure limits (Table 40.12). For most irritant gases, a ceiling TLV (C-TLV) has been identified that avoids air concentrations that produce direct irritant effects on the lungs or mucous membranes. In these cases, the parameter most likely to dictate the intensity of the adverse response, which will be defined as the risk index, is the maximum concentration in air. Consequently, for irritants, the ceiling value, or maximum air concentration, is the basis for TLV, irrespective of the duration of the work shift.

At the other extreme are chemicals with long biological half-lives (greater than 400 hr) where the total accumulated body burden is the parameter that will determine the severity of the response (i.e., the risk index) (56). For the persistent chemicals, many of which are heavy metals or very high molecular weight lipophilic organics, there is a daily or weekly limit implicit in the TLV and that amount (dose) becomes the basis by which the daily limit can be established. In these cases, adjustments to exposure limits can be based on a simple ratio and proportion approach. For example, if 1 mg/m^3 is the 8-hr TLV, then 0.67 mg/m^3 would be the 12-hr exposure limit.

At least half the chemicals for which TLVs have been established have systemic toxic effects (i.e., at tissues other than at the site of entry) and possess biologic half-lives that are not dissimilar to the duration of exposure in most workplaces (e.g., 6–12 hr). For these chemicals something must be known about the mechanism of toxicity before the most appropriate risk index and adjusted occupational limits can be calculated. For irritants, as shown by the dotted line in Figure 40.37, the concentration that causes discomfort will be the exposure limit regardless of exposure duration. For cumulative chemicals, such as PCBs or lead, shown in Figure 40.37 as the solid line, the relationship is a hyperbola where the TLV is related inversely to exposure duration. Most TLVs for industrial chemicals, however, will lie somewhere between lines A and B. The objective of any approach to adjusting limits, including the PB-PK approach, is to avoid recommendations that would require unnecessarily expensive controls, yet the limits ensure a safe level of exposure (Table 40.37). Potential indexing factors for adjusting OELs are presented in Table 40.13.

Table 40.12. Risk Indexes for Various Classes of Chemicals[a]

Class of Toxicant	Appropriate Risk Index
Irritant gases	Maximum air concentration
Cholinesterase inhibitor	Blood and acetylcholine esterase
Heavy metal	Total daily dose
Genotoxic carcinogen	Lifetime dose (with a weekly peak allowable concentration
Industrial solvents	Depends on mechanism of toxicity

[a]Ref. 56.

Table 40.13. Potential Indexing Factors for Adjusting Occupations Exposure Limits for Unusual Work Schedules[a]

Airborne concentration (exposure correlate)
Airborne concentration × time
Blood concentration (body burden)
Blood concentration × time
Peak metabolite concentration
Peak metabolite concentration × time
Target tissue concentration
Target tissue concentration × time

[a]Ref. 56.

A physiologically-based pharmacokinetic approach for examining the kinetic behavior of inhaled vapors and gases that are nonirritating to the respiratory tract has been developed (61). In this description (Fig. 40.36), the body is lumped tissue groups corresponding to the:

1. Highly perfused organs, excluding the liver
2. Muscle and skin
3. Fat
4. Organs with high capacity to metabolize the inhaled chemical. The physiological parameters of the metabolizing tissue groups are essentially those for the liver.

To describe the metabolism and fate of a chemical in humans or any living organism using a model, basic biological and physiological data on the species are used. The concentration of the inhaled contaminant in venous blood leaving each tissue can be determined by the tissue-blood partition coefficient. Blood flows and organ volumes are set consistent with literature values for these parameters. Organ partition coefficients and metabolic constraints for each chemical are determined by simple laboratory experimentation (24).

An essential element of the PB-PK models is a determination of the solubility of the test vapors in various biological fluids and tissues. For vapors, solubility is quantified by determining appropriate partition coefficients. Partition coefficients relate the relative amount of material in the liquid and gaseous phases at equilibrium. A blood-air partition coefficient of 10 means that there is 10 times as much substance in a unit volume on blood as in a corresponding unit volume of air at equilibrium. Partition coefficients can be determined for blood and tissues by a vial equilibration technique in which small amounts of test chemical vapors are added to the head space above the biological samples (24). After equilibration, the head space is sampled for test chemical. The partition coefficient is determined by a calculation based on the difference between the amount in the test vial and that in a control vial. Tissue-blood partition coefficients are determined by dividing the tissue-air by the blood-air values.

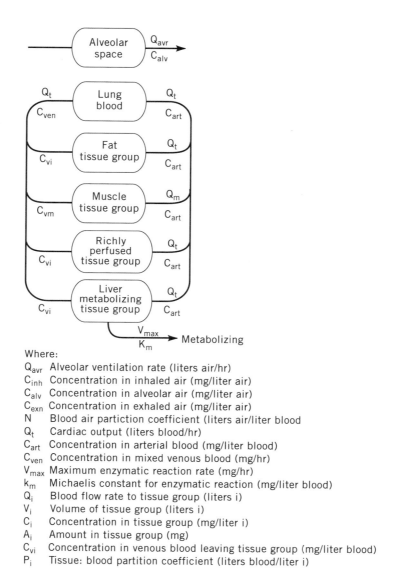

Figure 40.36. Illustration of how the body is described in physiologic pharmacokinetic modeling. From Ref. 95.

These various constraints are generally used in the four mass-balance differential equations that describe the time-dependent changes of tissue concentration in each of the compartments. Mixed venous blood concentration is determined as the weighted sum of effluent blood concentration from each tissue group, and the arterial concentration is determined from inhaled air concentration and venous blood concentration on the assumption that arterial blood leaving the lung is equilibrated with end alveolar air. Specifically, in the models built by Andersen et al. (92), cardiac output was assigned a value of 5.64 l blood/hr for a 0.30-kg rat, and to maintain a ventilation-perfusion ratio of 0.8, alveolar ventilation was set equal to 4.50 L air/hr. The liver, fat, muscle, and richly perfused tissue groups in the rat were assigned volumes, respectively, equal to 4, 7, 75, and 5 percent of body weight. Blood flow distribution to the liver, fat muscle, and richly perfused tissue groups was, respectively, 25, 9, 12, and 54 percent of cardiac output.

As discussed by Andersen et al. (92), to adjust the TLV for most gases and volatile liquids, toxicity will usually be related to the area-under-the-blood-curve (AUC) rather than to peak blood concentration (Figure 40.37). For the sake of simplicity, a simple inverse relationship (Fig. 40.37, line B) might be an acceptable way of adjusting many TLVs, since the risk index calculated when an AUC is used will be nearly identical to that derived from the inverse relationship. To adjust for shifts of 4–16 hr/day, the PB-PK approach is recommended whenever possible, or to use the direct ratio/proportion approach (Fig. 40.37, line B) if there is an insufficient amount of biologic data to develop a PB-PK model. For exposures of less than 4 hr or greater than 16 hr, the use of PB-PK modeling would be fairly difficult since information would be necessary on recovery and the mechanism of toxicity. An example of how a PB-PK model was used to adjust the OEL for styrene for unusual schedules is presented in Figure 40.38.

7 ADJUSTING LIMITS FOR CARCINOGENS

Unlike other toxic effects caused by xenobiotic substances, the carcinogenic response due to exposure to genotoxic agents is not currently believed to have a threshold (94). Although this hypothesis cannot be proven or disproved for most industrial chemicals since exposure to very low doses of the less potent carcinogens would require an exposed human population of enormous size before a response would be observed, this theory must be considered when setting and modifying exposure limits. The adjustment of exposure limits for carcinogenic materials has been addressed by Mason and Hughes (91) and their work is the basis for the following discussion.

7.1 Rationale

In general, toxic substances have been shown to exert their effect in proportion to the concentration in the body or within specific tissues. With most of these substances, it is common to find an all or none toxic response above a critical or threshold concentration (63, 98). In other words, at higher concentrations there will be a correlation between an increase in the response and increasing concentrations in tissue, and below this threshold a given effect will not be observed. For toxicants that act in this fashion, the peak con-

PHARMACOKINETICS AND UNUSUAL WORK SCHEDULES

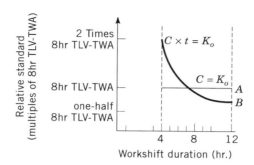

Figure 40.37. Strategies for adjusting 8-hr TLV-TWA to work shifts of shorter or longer duration. For irritants where a given airborne concentration has an effect, there would be no difference in proposed TLV regardless of work shift duration (solid line A between dotted lines). For chemicals that may accumulate with repeated exposure, the TLV is related inversely to the work shift duration (solid line B). It is not immediately obvious how adjustments would be made for most industrial vapors other than they should lie somewhere between lines A and B.

centration in tissue is generally thought to be the most important parameter to predict a toxic response (56). This group includes systemically acting substances such as chemical asphyxiants, narcotics and anesthetics, hemolytic agents, and probably some carcinogens or co-carcinogens. It *does not* include irritants that act with absorption, *nor does* it include allergens for which the severity of the response, once triggered, appears to be independent of the severity of exposure.

The lack of a totally safe concentration *may* apply to a small group of chemicals (e.g., carcinogens and mutagens) in which the biological response appears to result from chance molecular interactions that are independent of a tissue threshold. Response to these potent mutage substances is in some ways similar to the response observed following exposure to low levels of ionizing radiation (96, 97). For the most part, this latter group of substances act by forming covalent bonds with the genetic material of cells, causing an alteration of the genetic code, which in the case of somatic tissue may lead to cancer (100, 101). With many of these substances, the production of a diseased state appears to take place in separate phases. These begin with the initial chemical reaction or hits at a sensitive target (initiation) but end in a complicated series of interactions involving cell transformation, survival, and replication. Once the process of proliferation is initiated, progression to a diseased state may be independent of the concentration of the initiating substances, but may also be affected by the presence of promoting substances. The promoter may affect genetic repair, cell regulation, or the immune system and subsequently the survival of a transformed cell line (102, 103). As a result, the overall pattern of dose-response, especially at very low doses, is unclear.

Some chemicals can cause cancer in animals yet have no apparent genotoxicity or lack the ability to initiate cell transformation. These chemicals, which can be promoters, are often called epigenetic or nongenotoxic carcinogens. Although the exact mechanism of action is unclear for these substances, they apparently act in a manner much different from

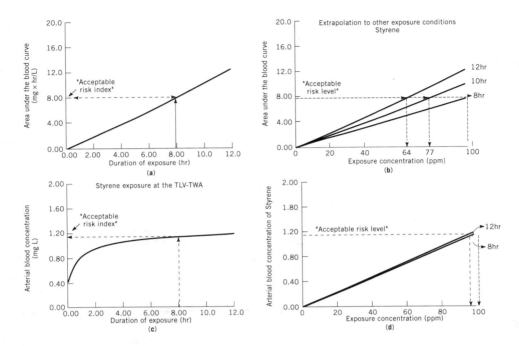

Figure 40.38. Shift adjustments using a physiological pharmacokinetic approach. Panel A: After the decision is made about the appropriate measure of tissue dose, the human PB-PK model is exercised at the TLV-TWA and the acceptable risk index determined for the 8-hr exposure. In this case you find 8 hr on the X axis and read off the accumulated area under the blood curve. Panel B: The curve in panel A is obtained by running a kinetic model at a single concentration. The curves in panel B are different. In this case the models are run for many concentrations with a specified exposure duration. The output of these so-called repetitive runs are used to construct a composite curve relating target tissue dose to various exposure concentrations. Three curves were generated for exposure durations of 8, 10, and 12 hr. In order to calculate the adjusted TLV, the risk index from panel A is used on the Y axis and a line drawn parallel to the X axis. The shift-adjusted TLV-TWA then is determined from the intersection with the particular work shift curves. Panel C: Shift adjustments based on peak blood concentrations. The PB-PK model is run to estimate the time course of styrene in blood and the value for the 8-hr exposure (i.e., the acceptable risk level) is read off the curve. Panel D: Adjusting the TLV to a nonstandard, a 12-hr shift. The model is exercised for exposures of 8 or 12 hr for a variety of exposure concentrations and end exposure peak blood styrene concentration plotted *vs.* exposure concentration. The new 12-hr TLV-TWA is determined as in panel B by taking the risk level from the Y axis and drawing the line parallel to the X axis to its intersection with the 12-hr curve. This is the new acceptable exposure level. From Ref. 56.

that of initiators (39, 102, 103). Since many experts believe that a threshold exists for nongenotoxic carcinogens, it has been suggested that any approach to adjusting the TLV for these substances should be similar to that used for systemic toxins (i.e., the pharmacokinetic approach) (104, 105, 107). A reasonable view of how to assess the risks posed by genotoxic and nongenotoxic carcinogens was recently proposed by the U.S. Environmental Protection Agency (108).

7.2 Method for Adjusting Exposure Limits for Carcinogens

The approach suggested by Mason and Hughes assumes that the most sensitive or critical step in the carcinogenic process is that of initiation, since it is predicated to result from the chance interaction of a single molecule of the substance, or its metabolic derivatives with an appropriate molecule within a tissue that is generally presumed to be DNA. It then follows that the chance of such an event occurring among a fixed population of receptors will be determined by the concentration of the substance that is available, and the time that it is available, that is,

$$P = f(d, t) \tag{43}$$

where P is the probability of initiation occurring and d and t are, respectively, the tissue concentration (dose) of the substance and duration of the concentration in tissue (109). In slightly different terms, the "effective dose" is the integral of the body burden:

$$D_{\text{eff}} = \int_{t_1}^{t_2} C\, dt \tag{44}$$

where t_1–t_2 is the span of interest, C the concentration in tissue, and D_{eff} is the effective dose associated with a given level of response in the exposed population (109). The assumption that it is the long-term average dose that is more important than day-to-day peak concentrations is supported in a study by Bolt et al. (110).

According to Mason and Hughes (91), if an acceptable level of response and a concomitant, albeit probably low, dose has been set for a substance to which exposure occurs over a 40-hr (5-day, 8-hr/day) workweek, that standard could be extrapolated to unusual shift work schedules by limiting the effective dose in the unusual shift to that predicted at the exposure limit for the 5-day, 40-hr workweek. Unlike the case of nonstochastic substances for which the peak concentration is of concern, the "dose" attributable to a series of integrated body burdens will be the simple sum of the contributions form the individual shifts. Using the indefinite integral, the effective lifetime contribution to the body burden (C_t) from any one exposure lasting time t_0 is thus

$$(C_t) = (k_i^*(M)/k_0)t_0 \tag{45}$$

where k_i^* and k_0 are effective mass transfer constants for the substances in humans and (M) its concentration in the environment (38).

Mathematically, the total dose over a series of exposure periods simply becomes the sum of the individual shift contributions, regardless of whether the work schedule is unusual or normal. Obviously, this influences the adjustment process since all exposure cycles of the same duration and concentration yield identical doses. Consequently, the authors have suggested that the TLV under unusual shift conditions should be

$$\text{TLV}_{(\text{unusual})} = \text{TLV}_{(\text{std})} \times \frac{\text{total exposure time (std shift sequence)}}{\text{total exposure time (unusual shift sequence)}} \quad (46)$$

Graphical depiction of the solution to this equation is presented in Figure 40.39.

The mechanics of shift arrangement for carcinogens then become a factor only if the novel work schedule results in a longer (or shorter) work week, work month, work year, or working lifetime. This is quite different from the case of nonstochastic agents for which the novel TLV is determined by iteration of the exponential function $(1 - e^{-k_0 t_0})$ to obtain the peak body burdens, which would result in the respective series (38). Consequently, biological kinetics become important only in making comparisons between substances and not in the extrapolation process.

In spite of its possible shortcomings and our lack of precise understanding of the carcinogenic processes, this approach seems reasonable for any chemical that has been shown to be positive in animal tests and also has demonstrated genotoxic potential.

7.3 Potential Shortcoming of the Proposed Approach for Adjusting TLVs for Carcinogens

As discussed by Mason and Hughes (91), the adjustment of TLVs for substances that produce a response in proportion to the time integral of the body burden is simple to

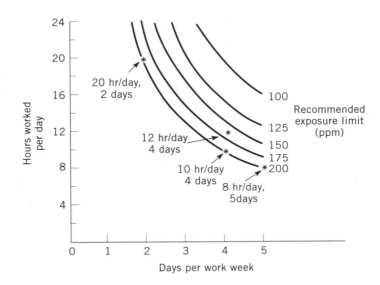

Figure 40.39. Graphical approach to adjusting exposure limits for genotoxic carcinogens based on the method of Mason and Hughes (91). Curves for adjusting limits for these substances are based on limiting the time integral of the tissue concentration for the work schedule rather than limiting the peak tissue concentration. Consequently, the three 40-hr work weeks illustrated have the same allowable concentrations. Exposure for longer periods, e.g., in four 12-hr shifts, requires lowering the degree of exposure. (Courtesy of Dr. J. Walter Mason.)

accomplish. However, the extent to which integrated body burdens adequately describe biological response for chemical carcinogens in humans is at present unknown (111). For some electrophilic substances, for example, ethylene oxide, this approach appears to be reasonable (112), while for others, such as benz(*a*)pyrene, it may not be appropriate because of the apparent role that enzyme induction, caused by repeated exposure, may play in modifying the biologic response. Also, for some chemicals, there may be an inverse relationship between dose and the onset of an observable response (time to response), and because this is not considered in the Mason and Hughes model, a different approach would be necessary.

8 ADJUSTING OCCUPATIONAL EXPOSURE LIMITS FOR MOONLIGHTING, OVERTIME, AND ENVIRONMENTAL EXPOSURES

In many cases, persons who work unusual shifts have much free time away from work. For example, these schedules often allow persons to farm 20–30 hr/wk. Free time for these shifts usually includes as few as three full days off work each week or as many as six continuous full days off work every two weeks. These long periods of time give persons an opportunity to work second jobs. In fact, in studies of 12-hr shift workers, it was reported that many persons hold a part-time job along with their regular job so as to gainfully occupy these large blocks of free time (4).

The adjustment of exposure limits for persons on long shifts has been a topic of interest since 1975, but if these persons are also exposed during their off hours, additional adjustment of the OEL may be needed. The approach to estimate the adjustment factor is the same for correcting limits to account for overtime as well as for environmental exposures. The following approach was developed and has been discussed by Hickey and Reist (65).

8.1 Adjustment of Limits for Overtime and Moonlighting

As before, the one-compartment biological model is used in this approach. The one-compartment model predicts that no adjustment is necessary to exposure limits for substances with very long (over 1000 hr) or very short (less than 1 hr) half-lives, but that adjustment is necessary for substances with intermediate half-lives (usually 6–100 hr).

For regular moonlighting or overtime on a 5-day/week basis, Eq. 27 devolves to

$$F_p = \frac{1 - e^{-8k}}{1 - e^{-t_s k}} \quad (47)$$

in which it is the daily exposure in hours. Adjustment factors derived from Eq. 47 for this situation are shown in Figure 40.40, which was developed by Hickey and Reist (65). Similar equations and curves can be developed from Eq. 27 for any schedule. For regular weekend moonlighting or overtime, working six or seven 8-hr days, Eq. 47 becomes

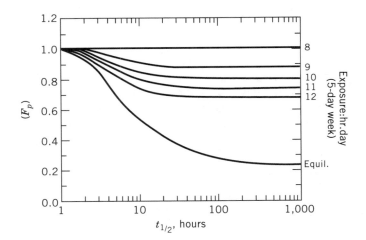

Figure 40.40. Adjustment factor (F_p) as function of substance half-life ($t_{1/2}$) for various daily exposures. The bottom line (equil.) represents the recommended adjustment factors for continuous exposure (24 hr/day). From Ref. 65.

$$F_p = \frac{1 - e^{-120k}}{1 - e^{-124km}} \qquad (48)$$

where m is the number of workdays per week. F_p values from Eq. 48 are shown in Figure 40.41.

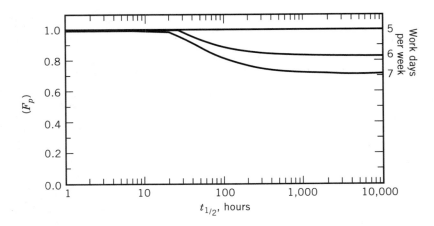

Figure 40.41. Adjustment factor (F_p) as a function of substance half-life ($t_{1/2}$) for workweeks of 5, 6, and 7 days. From Ref. 65.

Hickey has noted that the model is more complex when dealing with irregular or unplanned overtime or moonlighting added to an otherwise normal schedule. For example, suppose an employer decides Friday afternoon that workers must work overtime that same day or on Saturday. Recalling that in this model workers are normally presumed to have accumulated their allowable peak body burden of a substance by Friday afternoon, how does one adjust the exposure limit to prevent a predicted excess body burden accumulation?

If work (and presumed exposure) is to continue past normal quitting time Friday, the limit should be adjusted so that the body burden becomes no greater than the peak that would occur as a result of five 8-hr/day exposures per week at the TLV (PEL), which is the usual Friday-P.M. peak. In this case, Eq. 27 devolves to an equilibrium situation

$$F_p = \frac{(1 - e^{-8k})(1 - e^{-120k})}{(1 - e^{-168k})(1 - e^{-24k})} \tag{49}$$

For substances with short half-lives, no adjustment to exposure level is needed (the body is already at equilibrium), whereas for substances with long half-lives, the level should be reduced to 40/168 of normal. The problem does not end there because the worker has lost part of his or her weekend recovery time, and next week's exposure must be lowered to compensate. This involves complex manipulation of Eq. 27. Other irregular overtime and moonlighting situations must also be modeled individually to determine an appropriate F_p value. Since there are infinite variations, only the foregoing examples are given here.

The point to be emphasized is that these exposures do add to the body burden and should be taken into account in setting exposure limits. An illustrative example of the mathematical approach is shown in Example 21. Figure 40.42 and others have been developed by Hickey (65) for use during second shifts and are useful.

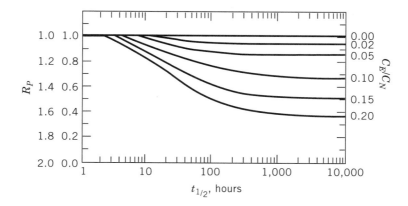

Figure 40.42. Predicted effect of off-the-job exposures to contaminants on adjustment factor (F_p) and peak body burden (R_p). From Ref. 65.

8.1.1 Illustrative Example 21 (Calculating an Adjustment Factor with Consideration Given to Overtime and Moonlighting)

Many persons work two jobs. If an employee were self-employed as a furniture stripper in his off hours, what modification to the normal daily TWA exposure limit would be necessary for methylene chloride given the following information?

A person works 10 hr/day, 4 days/week at a paint plant and is exposed to methylene chloride. He usually strips furniture as a second job for only 2 hr on the days he also works at the factory, but he usually strips furniture 8 hr/day during 2 of the 3 days off work each week. He is not exposed to methylene chloride on Sunday. The biologic half-life for methylene chloride in humans is 6 hr (assume TLV = 100 ppm).

Solution. The model considers only exposure time, not total work period, so F_p is based on four 2-hr daily exposures followed by two 8-hr daily exposures per week.

$$F_p = \frac{(1 - e^{-8k})(1 - e^{-120k})/(1 - e^{-24k})}{\begin{array}{c}1 - e^{-8k} + e^{-24k} - e^{-32k} + e^{-54k} - e^{-56k} + e^{-78k} \\ - e^{-80k} + e^{-102k} - e^{-104k} + e^{-126k} - e^{-148k}\end{array}}$$

$k = \ln 2/6 \text{ hr} = 0.1155.$

$$F_p = \frac{0.6434}{0.6413} = 1.0, \quad \text{TLV}_{\text{special}} = 1 \times 100 = 100 \text{ ppm}$$

The shorter exposure week (32 versus 40 hr) would allow an increase in exposure limit except that the $t_{1/2}$ is short (6 hr), the daily weekend exposure is the governing factor. Since weekend exposure is 8 hr/day, as in the normal schedule, no increase in TLV is predicted. This is illustrated in Figure 40.25.

Note that in the absence of weekend exposure, the model predicts an allowable increase in regular job exposure:

$$F_p = \frac{(1\ e^{-8k})(1 - e^{-120k})}{(1 - e^{-2k})(1 - e^{-96k})} = 2.92$$

The modified limit for 2 hr/day, 4 days/week exposures is 2.92 × 100 or 292 ppm.

8.2 Simultaneous Occupational and Environmental Exposures

As noted by Hickey, regulatory limits and recommended guidelines often assume zero off-the-job exposure to a substance. This is not always the case. The models discussed derive an expression for F_p, given a situation in which a person is exposed to an exposure limit of a substance during normal working hours, and to its environmental limit for the remainder of the time. The Hickey and Reist model was used to determine, first, how much the worker's peak body burden would be increased over that acquired without any off-the-job exposure, and second, how much the on-job limits should be reduced so that the

PHARMACOKINETICS AND UNUSUAL WORK SCHEDULES 1881

peak body burden would be no higher with both on-job and off-job exposure than it would be with normal on-job exposure and zero off-job exposure (65).

From Eq. 27 a value of F_p is found, representing, as before, the adjustment needed to the occupational limit to avoid a predicted higher-than-normal body burden accumulation because of the off-job exposure. If, on the other hand, it is assumed that no adjustment is made to the normal occupational limit, the relative increase in body burden due to the additional off-job exposure can be determined from the ratio of body burdens accumulated from normal and dual exposure. This approach is discussed in more detail in the original article (65).

It is clear (Fig. 40.42) that for agents with relatively long half-lives and with relatively high environmental limits (as compared to the normal exposure limits, C_n), the environmental exposure adds significantly to the body burden of a substance if no adjustment is made to the occupational limit. Under these conditions, significant reductions in occupational limits are necessary to avoid any increase in body burden as a result of the additional environmental exposure.

As noted by Hickey and Reist (65), the Environmental Protection Agency environmental limits for SO_2 and NO_2 are less than one percent of the OSHA occupational exposure limits, so dual exposure makes little difference to body burden accumulation. However, the EPA limit for carbon monoxide of 10 mg/m^3 (9 ppm), an 8-hr limit but in effect a ceiling limit, is 18 percent of the OSHA limit. This has some effect on F_p and R_p. For example, as shown in Figure 40.42 if a worker is exposed to 50 ppm CO on the job and 9 ppm the remainder of the time, his predicted peak body burden ($t_{1/2}$ in humans for carbon monoxide = 3–4 hr) with off-the-job exposure would be 1.06 times his on-the-job burden. If this environmental exposure occurs, the on-the-job exposure should be reduced to 0.94 of normal to avoid the predicted excessive peak body burden.

8.2.1 Illustrative Example 22 (Hickey and Reist Approach for Combined Environmental and Occupational Exposure)

In certain regions of the country, the ambient concentration of carbon monoxide averages 9.0 ppm, which is the current EPA limit for environmental exposure. Many persons are occupationally exposed to carbon monoxide at concentrations at or near the TLV (25 ppm in 1998). What modified occupational exposure limit would be suggested if a person worked 12 hr/day for 4 days each week if one also wanted to take into consideration the background concentration to which the person would be exposed when away from his job? Carbon monoxide has a biologic half-life of about 4 hr in humans.

Solution. Our objective is to find the allowable CO concentration at work (C_s) that will not result in a $B_{p/s}$ greater than $B_{p/n}$. In determining C_s, you account for both the work-related CO and the background CO. Of course, you can reduce only the work-related contribution of CO since the environmental levels are fixed. Consequently, the additional CO concentration during work is $C_s - C_e$.

If there were no additional environmental exposure, the TLV adjustment required by the special schedule would be

$$B_{p(\text{normal})} = C_N WK(f\!\!: t_n, k)$$

where

$$f\!\!: t_{n,k} = \frac{(1 - e^{-8k})(1 - e^{-120k})}{(1 - e^{-168k})(1 - e^{-24k})}$$

$$B_{p(\text{special})} = (C_s - C_e)WK(f\!\!: t_s, k) + C_e WK(f\!\!: t_e, k)$$

and

$$F_p = \frac{C_{\text{special}}}{C_{\text{normal}}}$$

The additional reduction from environmental exposure is

$$F_p = \frac{(f\!\!: t_n, k)}{(f\!\!: t_s, k)} - \frac{C_e}{C_n}\left[\frac{1}{f\!\!: t_p, k} - 1.0\right]$$

$t_{1/2} = 4$ hr

$$F_p = \frac{(1 - e^{-8k})(1 - e^{-120k})/(1 - e^{-168k})(1 - e^{-24k})}{(1 - e^{-120k})(1 - e^{-96k})/(1 - e^{-168k})(1 - e^{-24k})} - \frac{9}{50}\frac{1}{(1 - e^{-12k})(1 - e^{-96k})/(1 - e^{-168k})(1 - e^{-24k})}$$

Combined $F_p = 0.8571 - 0.0225 = 0.835$.
Modified TLV for the combined exposures: TLV $= 0.83 \times 25$ ppm $= 20.8$ ppm.
Since 9 ppm of carbon monoxide is already present in the ambient air, the work environment should contain no more than 11.8 ppm in order to maintain a peak body burden for this special situation ($B_{p/s}$) at the same level as a person occupationally exposed to 50 ppm ($B_{p/n}$).

9 ADJUSTMENT OF OCCUPATIONAL EXPOSURE LIMITS FOR SEASONAL OCCUPATIONS

Seasonal occupations are particularly associated with agriculture and related activities, including fertilizer and pesticide manufacture and use, seed treatment and distribution, cotton ginning, and food canning, as well as construction and innumerable other pursuits. Occupational Safety and Health Administration (OSHA) permissible exposure limits have not taken seasonal exposure patterns directly into account. In response to the lack of information in this area, Hickey (54) developed an approach to adjusting TLVs or PELs to suit particular seasonal exposures. The following discussion is based on his suggestions.

A one-compartment model was used to determine a "special" exposure limit that would predict equal protection to a worker in some special exposure situation that the TLVs (or PELs) provide in a "normal" exposure situation. This special limit is expressed as an adjustment factor (F_p), which is the ratio of the exposure limit for any special exposure schedule to the normal exposure limit (TLV or PEL). That is, the normal exposure limits times F_p equals the special exposure limit, or

$$(\text{TLV normal})(F_p) = (\text{TLV special}) \tag{50}$$

The subscript p indicates that the adjustment factor is based on "peak" body burden of a substance as the criterion for equal protection.

F_p is a function of the work schedule and the biological half-life ($T_{1/2}$) of a substance in the body. The general equation for F_p as modified to apply to seasonal exposure is

$$F_p = \frac{(1 - e^{-8k})(1 - e^{-120k})(1 - e^{-8400k})}{(1 - e^{kH})(1 - e^{-24kD})(1 - e^{-168kW})} \tag{51}$$

where H = length of daily work shift in the special schedule (hr/day)
D = length of the special workweek (days/week)
W = number of weeks in the special work season (weeks/year)
$-k$ = excretion rate of a particular substance in the body ($k = \ln 2/t_{1/2}$)
$t_{1/2}$ = biological half-life of the substance in the body

In this general equation, the special work schedule factors in the denominator are balanced against the "normal" work schedule factors in the numerator. These are a workday of 8 hr, 5 work days per week (120 hr), and fifty 168-hr weeks per year (8400 hr). This schedule assumes a 2-week vacation annually.

9.1 Application of the Hickey and Reist Model for Seasonal Shift Work

Hickey has offered the following example to illustrate the use of his model for seasonal shift work.

9.1.1 Illustrative Example 23 (Adjusting for Seasonal Work)

Take, for example, a seasonal job in which workers are exposed to a substance six 16-hr days/week for a 17-week season. Values of $H = 16$, $D = 6$, and $W = 17$ may be substituted in Eq. 51 and F_p found for any k value.

In this example, F_p goes from a value of 1.0 (no adjustment to TLVs) for exposure to substances with very short half-lives (high k values) to a low point of 0.43 for substances with a 400-hr half-life, and then rises to 1.2 for substances with very long half-lives (low k values) (54).

This phenomenon is explained as follows. If $t_{1/2} = 1$ hr, the peak body burden is reached in 5 or 6 hr exposure and gets no higher even if exposure continues for 16 hr/day. As the $t_{1/2}$ gets longer, the abnormally long work day and work week require a reduction in the

TLV to avoid a predicted higher-than-normal body burden. For a substance with a $t_{1/2}$ of 400 hr, F_p is 0.43, or nearly as low as the ratio of the normal 40-hr to the special 96-hr workweek (40/96 = 0.42).

To this point, there has been no effect on F_p from the short season. For substances with a half-life longer than 400 hr, as shown in Figure 40.43, F_p increases (curve B) to 1.2 at $t_{1/2} = 11$ years. This value approaches the ratio of the normal (2000 hr) to the special (1632) work year (2000/1632 = 1.22). F_p can exceed unity (i.e., TLVs can be adjusted upward) for substances with very long half-lives, because with such substances, the total intake/year is more important than whether the intake takes place over a 17- or a 50-week period. Since the seasonal work year has fewer work hours than a normal work year, the total intake and thus the predicted peak body burden will be no more than for a normal year even if the exposure level is increased by 20 or 22%.

Consider now the adjustment factor if the workday and workweek are normal and only the work year varies. In this case, Eq. 51 reduces to

$$F_p = \frac{1 - e^{-8400k}}{1 - e^{-168kW}} \tag{52}$$

F_p values from Eq. 52 are plotted in Figure 40.44 against $t_{1/2}$ for work seasons from 10 to 50 weeks. It can be seen that no "credit" (i.e., increase in TLV) can be taken for a short season unless the $t_{1/2}$ of the substance exceeds 30 hr (many particulate substances do have much longer half-lives). For substances with very long half-lives, the F_p again becomes the ratio of hours worked per normal year to hours worked per special season.

To avoid calculating each schedule separately, a quick approximation may be used to determine whether F_p is significantly different from 1.0 for a particular agent and work season. One may arbitrarily state that the effect of the shorter season on F_p is not worth considering if it permits raising the TLV by no more than 5%, or

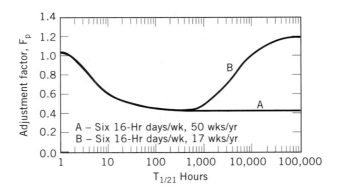

Figure 40.43. Adjustment factor (F_p) as a function of substance half-life ($t_{1/2}$) for the two seasonal exposure schedules indicated. From Ref. 54.

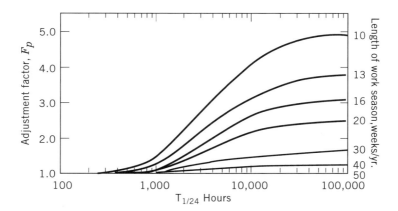

Figure 40.44. Adjustment factor (F_p) as a function of substance half-life ($t_{1/2}$) for various work seasons. From Ref. 54.

$$1/(1 - e^{-168kW}) < 1.05 \tag{53}$$

Solving this equation for $t_{1/2}$ ($k = \ln 2/t_{1/2}$):

$$t_{1/2} \text{ in hours} < 38\ W \tag{54}$$

That is, if the substance half-life in hours is less than 38 times the number of weeks (W) in the work season, the short season does not materially change the predicted peak body burden from that expected in a normal work year. Stated another way, the TLV may not be raised by virtue of the shorter work year unless the substance half-life in hours is greater than 38W. Example 23 illustrates the use of this approach.

9.1.2 Illustrative Example 24 (Approach to Adjusting Limits for Seasonal Occupations)

Seasonal jobs are common in many industries. Assuming that persons are exposed to styrene 16 hr/day for 6 days/week for an 18-week season, what modified TLV would be suggested if the biologic half-life in humans were 6 hr? (assumed TLV = 50 ppm.)

Solution
Part A. Since $t_{1/2}$ in hours <38 × (18 week), or <684 hr, the exposure limit may not be raised by virtue of a shorter work *year*.

Part B. Must it be lowered because of the longer workday?

$$F = \frac{(1 - e^{-8k})(1 - e^{-120k})/(1 - e^{-168k})(1 - e^{-24k})}{(1 - e^{-16k})(1 - e^{-144k})/(1 - e^{-168k})(1 - e^{-24k})}$$

$$= \frac{(1 - e^{-8k})(1 - e^{-120k})}{(1 - e^{-16k})(1 - e^{-144k})} = \frac{(0.6031)(1)}{(0.8425)(1)} = 0.72$$

This equation shows that only the length of the workday is important in determining TLV adjustment in this particular case.

This adjustment is the same as that predicted by Figure 40.43 for a 16 hr/day and a chemical with a 6-hr biologic half-life.

10 BIOLOGIC STUDIES

Very few studies have investigated the potential effects of unusual exposure regimens on the severity of toxic response. It appears that the health risks that might occur if an exposure limit were not lowered during an unusual work shift will, in all likelihood, be too subtle to be measured quantitatively in most animal studies. There is, however, some evidence that, in general, exposures that are intermittent or unusually long are likely to potentiate the response (113, 114). On a theoretical basis, measurable but perhaps clinically insignificant changes in the rates and routes of elimination of inhaled substances can be expected during long work shifts (22, 35, 42, 56, 95, 115).

Only a few biologic monitoring studies (116) have thus far been conducted to demonstrate whether a significant difference in effects between a normal and unusual schedule is likely to occur for any chemical at levels near its TLV. *It will, in all likelihood, be many years before clinical studies in humans, or toxicologic studies in animals will show whether there is the possibility of increased health risk for some chemicals if exposure limits are not lowered for most unusual workshifts.*

The rationale for such adjustments are therefore, in part, philosophical and involve a judgment about theoretical risk. Our knowledge of the pharmacokinetics of chemicals clearly tells us that if the airborne concentrations of a chemical are not lower during extra long periods of exposure, the peak levels of the toxicant in tissue *will be higher* during that period than would occur during "normal" 8-hr/day work schedules. Consequently, if health professionals believe that persons on both schedules should be at an equivalent level of risk, they can use one of the models available for determining the concentrations at which parity can be expected. We can, however, understand something about the magnitude of the hazard from the few studies that have been conducted.

Lehnert et al. (37) conducted a cross-sectional epidemiology study of workers on normal 8-hr/day shifts that involved the analysis of 6126 biological samples. These persons were exposed to either trichloroethylene, benzene, or toluene during their workday, and trichloroacetic acid, phenol, or hippuric acid, respectively, were measured in their urine as an indicator of exposure as well as body burden. They found that even during normal 8-hr/day schedules, exposures at or near the TLVs of these chemicals caused the concentration of the metabolites phenol and trichloroacetic acid in the urine to be higher on Friday afternoon than Monday afternoon. Lehnert's et al. data suggest that there may be some degree of accumulation, although of no apparent toxicological significance, occurring in these workers during normal schedules. From this, it might be expected that for those chemicals where day-to-day accumulation normally occurs, accumulation might be exaggerated when the exposure period is longer and the recovery period shorter.

At least one biological monitoring study was specifically conducted in an effort to determine whether persons exposed to dimethylformamide (DMF) for 12 hr/day accumulated the substance to a greater degree than persons who were exposed during a normal workweek (116). The conditions of the study were ideal in that a baseline set of urinary excretion data were collected on 80 employees who had been working the 8-hr/day shift for at least several months, and this was compared to the urinary data of the same group after they were placed on the 12-hr/day shift (only day shift). Consequently, through use of the paired-T test, the differences in the elimination for each person as well as the group could be determined. Their results indicate that *no difference* was observed in the concentration or the quantity of the urinary metabolites after they were placed on the 12-hr/day, 4-day/week shift. It should be noted that the implications of this study, like most other single studies, cannot be generally extrapolated to other chemicals because exposures to DMF averaged about 2 ppm, which was markedly below the TLV of 10 ppm. Second, the biologic half-life of DMF in humans is about 4-hr, therefore, it would not be expected to accumulate from day to day.

Few pharmacokinetic and toxicological studies have been conducted in animals exposed for 12 hr/day with the intent of determining any differences due to the longer exposure period. It appears that MacGregor conducted one of the first toxicity studies comparing the response due to 12 and 8 hr of exposure (117). She determined uptake factors for hexafluoroacetone using rats and predicted uptakes for periods of 6, 8, and 12 hr. MacGregor then developed a first-order model for determining uptake of inert gases in which she determined uptake constants for carbon monoxide using human volunteers. These predictions were compared to the data of Petersen (76) and good agreement was reported between the observed and predicted values of CO uptake. The Hickey and Reist model predictions of peak CO burdens have been compared to the predictions of the MacGregor model (51). In Hickey's validation procedure, exposure to 50 ppm CO was assumed for five 8-hr days per week, three 12-hr days per week, and 24 hr/day, 7 days per week. The comparisons were made in terms of F_p. The predictions of both models agreed quite closely. Hickey has noted that MacGregor also developed a predictive model for reactive vapors but that it has not been validated (51).

Paustenbach et al. (93, 115) conducted an experiment to evaluate differences in toxicity, distribution and pharmacokinetics due to exposures of 8 and 12 hr/day. One group of rats was exposed for 8 hr/day for 10 of 12 consecutive days (simulating 2 weeks on standard work schedule) and another for 11.5 hr/day for 7 of 12 consecutive days (simulated 12-hr/day schedule). Thus, each group received essentially the same dose (ppm-hr) of toxicant (carbon tetrachloride) during the 2-week test period. The results showed that the 11.5-hr/day exposure schedule produced minor changes in the distribution and concentration of CCl_4 in various tissues as compared to rats exposed 8 hr/day. There was no significant difference in hepatotoxicity between the groups following each week of exposure as measured by histopathology. However, exposure to the 11.5-hr/day dosing regimen consistently produced significantly higher levels of serum sorbitol dehydrogenase (SDH), an enzyme that indicates liver damage, than exposure to the 8-hr/day schedule (Table 40.14) (115). This is not surprising since the airborne concentrations were about ten-fold above the TLV and thus some toxicity would be expected via either dosing schedule.

It is noteworthy that the rates and routes of elimination were measurably different for the two schedules (Figs. 40.45 and 40.46) (95). Following 2 weeks of exposure to the

Table 40.14. Statistical Comparison of Serum Sorbitol Dehydrogenase (SDH) Activity in Selected Groups of Rats Following Exposure to 100 ppm of Carbon Tetrachloride (CCl_4) for Either 8 Hr/Day or 11.5 hr/day under a Number of Different Dosage Regimens[a,b]

Treatment Groups Compared	Dosage Regimen	SDH Activity Mean ± SE (IU/mL)
1	1 day, 8 hr	7.0 ± 1.5[c]
2	1 day, 11.5 hr	14.8 ± 3.7
3	2 days, 8 hr/day	11.5 ± 2.2
4	2 days, 11.5 hr/day	18.3 ± 4.0
5	3 days, 8 hr/day	21.0 ± 3.2
6	3 days, 11.5 hr/day	29.0 ± 6.2
7	5 days, 8 hr/day	22.5 ± 2.7[c]
8	4 days, 11.5 hr/day	68.3 ± 9.3
9	5 days, 8 hr/day[d]	20.3 ± 4.4[c]
10	4 days, 11.5 hr/day[c]	12.3 ± 0.6
11	5 + 5 days, 8 hr/day	39.0 ± 8.8[c]
12	4 + 3 days, 11.5 hr/day	65.0 ± 16.2
13	5 + 5 days, 8 hr/day[f]	4.9 ± 1.1[c]
14	4 + 3 days, 11.5 hr/day[c]	14.3 ± 1.3
7	5 days, 8 hr/day	22.5 ± 2.7[c]
11	5 + 5 days, 8 hr/day	39.0 ± 8.8
8	4 days, 11.5 hr/day	68.3 ± 9.3
12	4 + 3 days, 11.5 hr/day	65.0 ± 16.2
9	5 days, 8 hr/day[d]	20.3 ± 4.4[c]
13	5 + 5 days, 8 hr/day[f]	4.9 ± 1.1
10	4 days, 11.5 hr/day[e]	12.3 ± 0.6[c]
14	4 + 3 days, 11.5 hr/day[g]	14.3 ± 1.3
12	4 + 3 days, 11.5 hr/day	65.0 ± 16.2[c]
15	4 + 4 days, 11.5 hr/day	110.0 ± 26.7
12	4 + 3 days, 11.5 hr/day	65.0 ± 16.2[c]
16	4 + 5 days, 11.5 hr/day	102.3 ± 14.3

[a]Ref. 115.
[b]SDH activity was determined immediately after exposure except where indicated. Values shown are the mean of four rats per group.
[c]Significant difference in SDH activity between groups ($p < 0.05$).
[d]SDH activity was determined 64 hr after exposure.
[e]SDH activity was determined 84 hr after exposure.
[f]SDH activity was determined 64 hr after exposure.
[g]SDH activity was determined 108 hr after exposure.

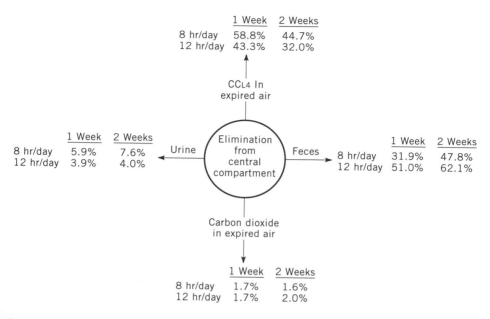

Figure 40.45. Differences in excretion of ^{14}C activity from rats following one and two weeks of exposure to 100 ppm of carbon tetrachloride during an 8-hr/day and 11.5-hr/day exposure schedule. From Ref. 93.

8-hr/day schedule, ^{14}C activity in the breath and feces comprised 52 and 41% of the total ^{14}C excreted. Following 2 weeks of exposure to the simulated 12-hr/day work schedule, the values were 32 and 62% indicating that the longer work shift altered both the rate and route of elimination of CCl_4. It was found that 97–98% of the ^{14}C activity in the expired air was $^{14}CCl_4$. The elimination of $^{14}CCl_4$ and $^{14}CO_2$ in the breath followed a two-compartment, first-order pharmacokinetic model ($r^2 = 0.98$). For rats exposed 8 hr/day, the average half-life for elimination of $^{14}CCl_4$ in the breath for the fast (α) and slow (β) phases for the 2-week schedule averaged 85 and 435 min, respectively. For rats exposed 11.5 hr/day, the average half-lives for the α and β phases over the two weeks averaged about 95 and 590 min, respectively. Differences in the rate of elimination of $^{14}CO_2$ and ^{14}C activity in the urine and feces were also observed (Table 40.15).

When a PB-PK model was developed to describe these data, the relevance to humans was quickly understood (33). As shown in Figure 40.47, it is clear that rats exposed to 5 ppm do not reach the same peak blood levels after repeated exposure as humans exposed to 5 ppm. Humans accumulated CCl_4 with repeated exposure while rats did not. The difference between the species is enhanced when exposed for 12 hr/day. This is, of course, the critical concern regarding unusually long work schedules; do workers achieve significantly higher body burdens or peak blood levels? This study was a good illustration of how the PB-PK model can successfully scale-up rat data to predict the human response.

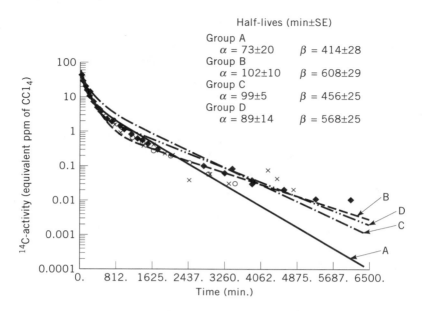

Figure 40.46. Elimination of ^{14}C activity (98% CCl$_4$) in the expired air of four groups of rats (4 per group) exposed to 100 ppm of carbon tetrachloride. Two groups were exposed for 8 hr/day for either 5 or 7 days (A) or 10 of 14 days (C). The other were exposed for 11.5 hr/day for either 4 or 7 days (B) or 7 of 10 days (D) and have markedly longer half-lives than those exposed 8 hr/day. From Ref. 93.

A comparative toxicity study involving 8- and 12-hr/day exposures was conducted by Kim and Carlson (118) for dichloromethane (methylene chloride). In this study, carboxyhemoglobin (COHb) formation and elimination in rats and mice exposed to an 8 hr/day, 5 day workweek or a 12 hr/day, 4 day simulated workweek at dichloromethane (DCM) concentrations of 200, 500, or 1000 ppm were compared. They showed that the effect of the 12-hr exposure period was insignificant as determined by the COHb level after first day's exposure, immediately prior to the second day's exposure, after the last workday's exposure and 2 or 3 days after the last exposure. They also measured the half-lives of COHb and DCM in blood. The relatively short half-lives of COHb and DCM in these two species indicated that neither COHb nor DCM would be present for prolonged periods after DCM exposure ceased. Treatment with SKF-525A did not affect the half-life of DCM, suggesting that DCM was rapidly exhaled. Even after correcting for physiological differences between the mice and rat and human, this study indicated that for compounds like DCM, with half-lives less than four hr in man, and where there are readily reversible biological effects, no increased toxicity would be expected in persons who were exposed 12-hr/day. Another related study of the same authors (119) evaluated aniline.

The Haskell Laboratories of DuPont Chemical Company have evaluated several chemicals to determine the influence of exposure duration on toxicological response (120). In

Table 40.15 Half-lives for Elimination of ^{14}C Activity in Rats Following Inhalation Exposure to 100 ppm of $^{14}CCl_4$ Following Several Different Dosing Regimens[a,b]

Dosing Regimen	$^{14}CCl_4$ in Expired Air		^{14}C Activity in Urine	$^{14}CO_2$ in Expired Air		^{14}C Activity in Feces
	$t_{1/2}(\alpha)$ ± SE	$t_{1/2}(\beta)$ ± SE	$t_{1/2}(\beta)$ ± SE	$t_{1/2}(\alpha)$ ± SE	$t_{1/2}(\beta)$ ± SE	$t_{1/2}(\beta)$ ± SE
8 hr/day, 5 days	73 ± 20[c]	414 ± 28[c,d]	1344 ± 149	123 ± 17[c-e]	1017 ± 95[d,e]	4900 ± 2100
11.5 hr/day, 4 days	102 ± 10	608 ± 29[d,e]	963 ± 107	221 ± 24[d,e]	1209 ± 124[d,e]	4300 ± 1400
8 hr/day, 5 + 5 days	96 ± 4	455 ± 24[e]	1066 ± 250	305 ± 33[d]	829 ± 81[d]	3700 ± 800[d]
11.5 hr/day 4 + 3 days	89 ± 14	568 ± 25[d]	944 ± 208	455 ± 42[d]	1824 ± 175	6700 ± 1400

[a]From Ref. 93.
[b]Results are an average of data obtained from four rats that made up each group exposed to a particular dosage regimen. Half-lives are expressed in minutes.
[c]Indicates that this pharmacokinetic parameter is statistically different ($p < 0.05$) from that calculated for the group exposed to the 11.5-hr/day, 4-day dosing regimen.
[d]Indicates that this pharmacokinetic parameter is statistically different ($p < 0.05$) from that calculated for the group exposed to the 8-hr/day, 5 + 5-day dosing regimen.
[e]Indicates that this pharmacokinetic parameter is statistically different ($p < 0.05$) from that calculated for the group exposed to the 11.5-hr/day, 4 + 3-day dosing regimen.

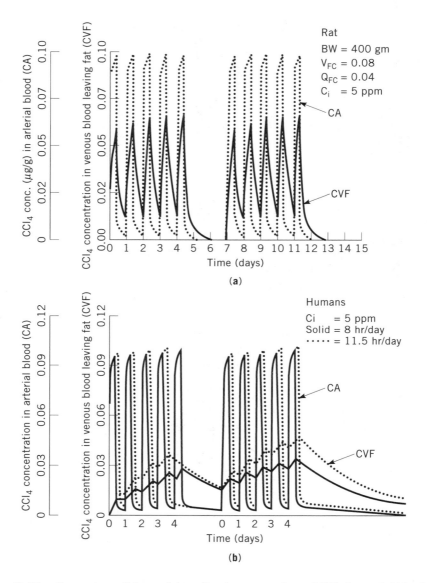

Figure 40.47. Comparison of the model-predicted concentrations of CCl_4 in arterial blood (CA) with those likely to be observed in venous fluid leaving the fat (CVF) for humans exposed to 5 ppm CCl_4. It is clear that there are significant differences between the rat and human even following just 8 hr of exposure at the 1993 TLV (5 ppm). As shown here, a PB-PK model can easily account for the pharmacokinetic differences between test animals and humans. From Ref. 33.

their study of aniline vapor, they exposed adult male rats to either 10, 30, or 90 ppm aniline vapors for either 3, 6, or 12 hr per day for two weeks. Daily indices of toxicological response included body weight and methemoglobin measurements. After the final exposure, red blood cell counts as well as spleen and liver weights were measured and these were examined histopathologically. Their results showed that aniline-induced hemolysis and consequent splenic enlargement and deposition of hemosiderin is slightly related to exposure duration, and strongly related to exposure concentration. Aniline-induced methemoglobin formation, however, was not related to exposure duration, but was linearly correlated with exposure concentration. Their results suggest that aniline concentration, rather than duration of exposure, predominantly influences toxicological response.

Although there is not a large number of careful studies that have evaluated the biologic response of humans exposed to unusual work schedules, our knowledge of the pharmacokinetics of chemicals is usually sufficient to allow toxicologists to make good predictions of the effects of extra long or continuous periods of exposure. The best method for approaching these problems is to use PB-PK models.

11 UNCERTAINTIES IN PREDICTING TOXICOLOGICAL RESPONSE

It has been noted that any model for adjusting exposure limits will have a number of limitations because models, by definition, are based on several assumptions. These limitations have been reviewed by Calabrese (57) and Hickey (51). As has been noted, one key assumption is that the pharmacokinetic models consider the body to function as one homogeneous compartment. The second, and possibly more important, limitation of the modeling approach is the toxicological assumption that neither repeated exposure, the length of the exposure, or the type of shift schedule (e.g., rapid rotation) will alter the way humans absorb, metabolize, and eliminate the substance. In most cases, this limitation will not be significant.

The major possible shortcoming of the modeling approach is the error involved in assuming that lengthening the exposure does not change the way the body handles the chemical. In an effort to predict the effects of changes in toxicity due exclusively to changes in dosage regimen (such as that involved in 12-hr shifts), toxicologists have frequently used Haber's law as a rule-of-thumb. Haber's law claims that the dose is of central interest rather than the time over which that dose is administered. A thorough study of the limitation of Haber's law was published by David et al. (70), and they showed that the severity of liver toxicity was markedly influenced by the time period over which a given dose (ppm-hr) was administered. Although the conclusions are similar to those obtained by other researchers (121) who have studied the limitations of extrapolating data from short-term inhalation exposures, this study is unique in that four different dosing regimens were used. Although the heavy emphasis on the total dose, rather than dose per unit time, has not been relied upon for nearly 40 years, industrial hygienists must all too often evaluate risks based on the amount of toxicant taken up per day rather than consider the pharmacokinetics. For example, this principle is used when hygienists extrapolate the results of standard 4- to 6-hr inhalation tests to estimate the adverse effects of exposure to 12-hr work schedules.

The affect of repeated exposure and long periods of exposure on a chemical's pharmacokinetic behavior are not accounted for in the models. This phenomena could have some affect on the toxicity of the material if the shift schedule were markedly different than an 8-hr/day, 5-day week. For example, MacGregor (117) has found that the rate of metabolism and elimination during 12-hr/day exposure periods is measurably different than that observed during 8-hr/day exposure periods. Paustenbach et al. (93) also showed that measurable differences in the rates and routes of elimination in the feces, breath, and urine between two different exposure schedules can occur.

There are several reasons why alterations in excretion rates and routes with repeated exposure have not been frequently reported. Colburn and Matthews (122) have noted that unless all of the inhalation exposures involve radiolabeled material, rather than using labeled material only during the first and last weeks, the 'last in-first out' phenomena may take place. When this occurs, potential effects on distribution, metabolism, and elimination due to repeated exposure may not be detected because nonlabeled chemical may not be uniformly distributed or excreted with the labeled material. In short, the first dose may be equilibrated more deeply in the fat than the second or subsequent doses.

The observation that repeated exposure to unusual work schedules may affect the distribution and excretion of some chemicals will not be limited to carbon tetrachloride, carbon monoxide, and 2-HFA. For example, similar effects would be expected for industrial solvents such as cyclopropane, cyclohexane, 1,1,1-trichloroethane, and perchloroethylene, which are similar to classic anesthetics (such as halothane) in that they are quite lipid soluble and not appreciably metabolized. For example, Fiserova-Bergerova and Holaday (42) have reported that the half times of uptake for halothane for the vessel-rich group (VRG), muscle group (MG), and fatty groups (FG) are in the magnitude of 2 min, 30 min, and 20 hr, respectively. They noted that most clinical anesthesia lasts 1–4 hr and that by the end of anesthesia the partial pressures of anesthetic agents in tissues of the FG compartment are far from equilibrium and that a steady-state has not been reached in that compartment. The result is a redistribution of vapor in the body after the offset of anesthesia. While clearance of the VRG and MG compartments starts instantly, the FG compartment continues uptake until the partial pressure in arterial blood declines to the partial pressure in the FG compartment. Consequently, desaturation curves are very much affected by the duration of exposure.

Another potential shortcoming of the modeling approach is the assumption that exposure to the toxicant does not inhibit or induce the microsomal enzyme system responsible for its metabolism (123). If this occurs, subsequent doses of the substance will alter the rate of toxication or detoxification. Along the same lines, models assume that one or more metabolic pathways will not be saturated. Although saturation is not very likely for exposure at or near the TLV, the likelihood that the metabolism of a compound will remain constant following repeated dosing is less likely. The induction or inhibition of its own metabolism by previous exposure has been demonstrated for nitrobenzene, acetone, and a few other xenobiotics (123).

All of the aforementioned factors, as well as those involving the likely affects of the circadian rhythm on toxic response, chronopharmacology and chronokinetics, make the modeling approach to the setting of modified exposure limits a crude approximation of the likely biologic processes that probably take place during unusual shifts. However, even though the current models may not account for the dozens of biologic phenomena that may be occurring, the available information clearly suggests that they are adequate to adjust OELs. Even though the existing biological data seem to suggest that, at levels near the TLV the increased risk of injury will usually not be appreciable, in light of the relative ease of adjustment, the procedure seems to be a worthwhile exercise.

It cannot be overstressed that modification of occupational exposure limits for situations other than 8-hr/day, 40-hr workweeks requires a clear understanding of the rationale for a particular limit. Blind use of any of the modeling approaches to modifying limits for either very short or very long periods of exposure can lead to either a lack of protection for those workers on unusual shifts or, as is more likely the case, a good deal of overprotection. Overprotection, although perhaps prudent, is not an optimal use of the limited resources allotted for minimizing occupational disease and, in some cases, could bring about undue economic hardships for both the employer and the employee.

BIBLIOGRAPHY

1. P. G. Rentos and R. D. Shepard, eds., *Shift Work and Health—A Symposium*, U.S. HEW PHS, National Institute Occupational Safety and Health, Washington, DC, 1976.
2. R. S. Brief and R. A. Scala, "Occupational Exposure Limits for Novel Work Schedules," *Am. Ind. Hyg. Assoc. J.* **36**, 467–471 (1975).
3. T. A. Yoder and G. D. Botzum, *The Long-Day Short-Week in Shift Work—A Human Factors Study*, 16th Annual Meeting of the Human Factors Society, Indianapolis, IN, 1971.
4. J. T. Wilson and K. M. Rose, *The Twelve-Hour Shift in the Petroleum and Chemical Industries of the United States and Canada: A Study of Current Experience*, Wharton Business School, University of Pennsylvania, Philadelphia, PA., 1978.
5. Z. Vokac, P. Magnus, E. Jebens, and N. Gundersen, "Apparent Phase-Shifts of Circadian Rhythms (Masking Effects) During Rapid Shift Rotation," *Int. Arch. Occup. Environ. Health* **49**, 53–65 (1981).
6. P. Knauth, B. Eichhorn, I. Lowenthal, K. H. Gartner, and J. Rutenfranz, "Reduction of Nightwork by Re-designing of Shift-Rotas," *Inter. Arch. Occup. Environ. Health* **51**, 371–379 (1983).
7. USBLS. U.S. Bureau of Labor Statistics, *News, Number of Days in the Workweek*, press release, March, 1979.
8. S. D. Nollen and V. H. Martin, "Alternative Work Schedules. Part 3: The Compressed Workweek," *AMACOM*, New York, 1978.
9. S. D. Nollen, "Work Schedules," in G. Salvendy, ed., *Handbook of Industrial Engineering*, Wiley-Interscience, New York, 1982.
10. E. Thiis-Evensen, "Shift Work and Health," *Ind. Med. Surg.* **27**, 493–513 (1958).
11. R. B. Dunham and D. I. Hawk, "The Four-Day/Forty-Hour week: Who Wants It?" *Academy of Management J.* **20**, 4–6 (1977).
12. P. J. Taylor, *The Problems of Shift Work*, Proceedings of an International Symposium on Night and Shiftwork, Oslo, Sweden, 1969.
13. K. Wheeler, R. Gurman, and D. Tarnowieski, "The Four-Day Week," *AMACOM*, New York, 1972.

14. M. Kenny, "Public Employee Attitudes Toward the Four-Day Workweek," *Publ. Personnel Management* **3**, 100–110 (March–April 1974).
15. R. F. Landry, "Off-Beat Rhythms and Biological Variables," *Occup. Health and Safety* **50**, 40–43 (1981).
16. M. H. Smolensky, "Human Biological Rhythms and Their Pertinence to Shift Work and Occupational Health," *Chronobiologia*, **7**, 378–390 (1980).
17. M. H. Smolensky, "Shiftwork," in Harris, Cralley and Cralley, eds., *Patty's Industrial Hygiene and Toxicology*, Volume III, Part B, John Wiley & Sons, New York, 1994, Ch. 6.
18. J. L. S. Hickey and P. C. Reist, "Application of Occupational Exposure Limits to Unusual Work Schedules," *Am. Ind. Hyg. Assoc. J.* **38**, 613–621 (1977).
19. M. Gibaldi and D. Perrier, *Pharmacokinetics*, 2nd Edition, Marcel Dekker, New York, 1982.
20. J. R. Withey, "Pharmacokinetic Principles," in G. Plaa, Ed., *First International Congress of Toxicology*, Academic Press, New York, 1979.
21. E. J. O'Flaherty, *Pharmacokinetics of Chemicals*, Wiley, New York, 1987.
22. A. G. Renwick, "Toxicokinetics—Pharmacokinetics in Toxicology," in A. W. Hays, ed., *Principles and Methods of Toxicology*, Third Edition, Raven Press, New York, 1994, Ch. 4, pp. 101–148.
23. H. W. Leung and D. J. Paustenbach, "Techniques for Estimating the Percutaneous Absorption of Chemicals Due to Occupational and Environmental Exposure," *Appl. Occup. Environ. Hyg.* **9**(3), 187–197 (1994).
24. M. L. Gargas, R. J. Burgess, D. E. Voisard, G. H. Cason, and M. E. Andersen, "Partition Coefficients of Low-Molecular-Weight Volatile Chemicals in Various Liquids and Tissues," *Toxicol. Appl. Pharmacol.* **98**(1), 87–99 (1989).
25. J. K. Piotrowski, *Exposure Tests for Organic Compounds in Industrial Toxicology*, Department of Health, Education and Welfare (NIOSH), Cincinnati, OH, pp. 77–144, 1977.
26. J. Piotrowski, *The Application of Metabolic and Excretion Kinetics to the Problems of Industrial Toxicology*, National Library of Medicine, U.S. Government Printing Office, Washington, DC, 1971.
27. R. E. Notari, *Biopharmaceutics and Pharmacokinetics*, 3rd ed., Marcel Dekker, New York, 1985.
28. J. Gabrielson and D. Weiner, *Pharmacokinetic and Pharmacodynamic Data Analysis, Concepts and Application*, Swedish Pharmaceutical Press, Stockholm, 1997.
29. J. V. G. Durnin and R. Passmore, *Energy, Work, and Leisure*, Heinemann Educational Books, LTD, London, 1967.
30. V. Fiserova-Bergerova, J. Vlach, and J. C. Cassady, "Predictable Individual Differences in Uptake and Excretion of Gases and Lipid Soluble Vapors Simulation Study," *Br. J. Ind. Med.* **37**, 42–49 (1980).
31. H. W. Leung and D. J. Paustenbach, "Physiologically Based Pharmacokinetic and Pharmacodynamic Modeling in Health Risk Assessment and Characterization of Hazardous Substances," *Toxicol. Letters* **79**, 55–65 (1995).
32. M. E. Andersen, H. J. Clewell, K. Krishnan, "Tissue Dosimetry, Pharmacokinetics Modeling, and Interspecies Scaling Factors," *Risk Anal.* **15**, 533–537 (1995).
33. D. J. Paustenbach, H. J. Clewell, M. L. Gargas, and M. E. Andersen, "A Physiologically Based Pharmacokinetic Model for Inhaled Carbon Tetrachloride," *Toxicol. Appl. Pharmacol.* **96**, 191–211 (1988).

34. E. Guberan and J. Fernandez, "Control of Industrial Exposure to Tetrachloroethylene by Measuring Alveolar Concentrations: Theoretical Approach Using a Mathematical Model," *Br. J. Ind. Med.* **31**, 159–167 (1974).

35. E. J. O'Flaherty, P. B. Hammond, and S. I. Lerner, "Dependence of Apparent Blood Lead Half-Life on the Length of Previous Lead Exposure in Humans," *Fund. Appl. Toxicol.* **2**, 49–54 (1982).

36. S. Riegelman, J. C. Loo, and M. Rowland, "Shortcomings in Pharmacokinetic Analysis by Conceiving the Body to Exhibit Properties of a Single Compartment," *J. Pharm. Sci.* **57**, 117–125 (1968).

37. G. Lehnert, R. D. Ladendorf, and D. Szadkowski, "The Relevance of the Accumulation of Organic Solvents for Organization of Screening Tests in Occupational Medicine. Results of Toxicological Analyses of More than 6000 Samples," *Inter. Arch. Occup. Environ. Health* **41**, 95–102 (1978).

38. J. W. Mason and H. Dershin, "Limits to Occupational Exposure in Chemical Environments under Novel Work Schedules," *J. Occup. Med.* **18**, 603–607 (1976).

39. P. J. Gehring and G. E. Blau, "Mechanisms of Carcinogenesis: Dose Response," *J. Environ. Path. Tox.*, **1**, 163–179 (1977).

40. J. C. Ramsey and P. J. Gehring, "Application of Pharmacokinetic Principles in Practice," *Fed. Proc.* **39**, 60–65 (1980).

41. D. G. Hoel, N. L. Kaplan, and M. W. Anderson, "Implications of Nonlinear Kinetics on Risk Estimation in Carcinogenesis," *Science* **219**, 1032–1037 (1983).

42. J. Fiserova-Bergerova and D. A. Holaday, "Uptake and Clearance of Inhalation Anesthetics in Man," *Drug Metab. Rev.* **9**(1), 43–60 (1979).

43. R. H. Reitz, A. L. Mendrala, R. A. Corley, J. F. Quast, M. L. Gargas, M. E. Andersen, D. Staats, and R. B. Connolly, "Estimating the Risk of Liver Cancer Associated with Human Exposures to Chloroform Using Physiologically Based Pharmacokinetics Modeling," *Toxicol. Appl. Pharmacol.* **105**, 443–459 (1990).

44. International Life Sciences Institute (ILSI), *Human Variability in Response to Chemical Exposures*, (Neumann and Kimmel, eds.), Washington, D.C., 1998.

45. C. C. Willhite, "Sensory Irritation in the Origin and Derivation of TLVs and AEGLs," *Amer. Ind. Hygiene Assn. J.* (in press).

46. *OSHA Compliance Officers Field Manual*, Occupational Safety and Health Administration, Department of Labor, Washington, DC, 1979.

47. *Criteria for a Recommended Standard—Occupational Exposure to Chloroform*, National Institute for Occupational Safety and Health, HEW Publication (NIOSH) 75-114, Rockville, MD, 1975.

48. J. S. Bus and J. E. Gibson, "Body Defense Mechanisms to Toxicant Exposure," in L. J. Cralley, L. V. Cralley, and J. Bus, eds., *Patty's Industrial Hygiene and Toxicology*, 3rd ed., Vol. 3B, John Wiley and Sons, New York, 1995, pp. 275–318.

49. D. J. Paustenbach, "Methods for Setting Limits for Acute and Chronic Toxic Ambient Air Contaminants," *Appl. Occ. Environ. Hyg.* **12**(6), 418–428 (1997).

50. *Threshold Limit Values for Chemical Substances and Physical Agents and Biological Exposure Indices for 1999*. American Conference of Governmental Industrial Hygienists, Cincinnati, OH, 1999.

51. J. L. S. Hickey, Application of Occupational Exposure Limits to Unusual Work Schedules and Excursions, Ph.D. Dissertation, University of North Carolina at Chapel Hill, Chapel Hill, NC, 1977.
52. P. Veng-Pedersen, D. J. Paustenbach, G. P. Carlson, and L. Suarez, "A Linear Systems Approach to Analyzing the Pharmacokinetics of Carbon Tetrachloride in the Rat Following Exposure to an 8-hour/day and 12-hour/day Simulated Workweek," *Arch. Toxicol.* **60**, 355–364 (1987).
53. S. A. Roach, "Threshold Limit Values for Extraordinary Work Schedules," *Am. Ind. Hyg. Assoc. J.* **39**, 345–364 (1978).
54. J. L. S. Hickey, "Adjustment of Occupational Exposure Limits for Seasonal Occupations," *Am. Ind. Hyg. Assoc. J.*, **41**, 261–263 (1980).
55. J. L. S. Hickey, "The 'TWAP' in the Lead Standard," *Am. Ind. Hyg. Assoc. J.* **44**(4), 310–311 (1983).
56. M. E. Andersen, M. G. MacNaughton, H. J. Clewell, and D. J. Paustenbach "Adjusting Exposure Limits for Long and Short Exposure Periods Using a Physiological Pharmacokinetic Model," *Am. Ind. Hyg. Assoc. J.* **48**, 335–343 (1987).
57. E. J. Calabrese, "Further Comments on Novel Schedule TLVs," *Am. Ind. Hyg. Assoc. J.* **38**, 443–446 (1977).
58. E. J. Calabrese, *Principles of Animal Extrapolation*, Wiley, New York, 1983.
59. Occupational Safety and Health Administration (OSHA), *The Field Operations Manual*, 6th Edition, Government Institutes, Inc., Washington, D.C., 1994.
60. M. Schaper, "Development of a Database for Sensory Irritants and Its Use in Establishing Occupation Exposure Limits," *Am. Ind. Hyg. Assoc. J.* **54**(9), 488–544 (1993).
61. M. E. Andersen, H. J. Clewell, M. L. Gargas, F. A. Smith, and R. H. Reitz, "Physiologically Based Pharmacokinetics and the Risk Assessment Process for Methylene Chloride," *Toxicol. Appl. Pharmacol.* **87**, 185–205 (1987).
62. Air Contaminants: Final Rule, Occupational Safety and Health Administration. *Federal Register* **54**, 2332–2983, 1989.
63. M. A. Ottobani, *The Dose Makes The Poison*, 2nd Edition, Van Nostrand Reinhold, New York, 1991.
64. R. L. Iuliucci, "12-Hour TLV's," *Pollution Eng.*, 25–27 (Nov. 1982).
65. J. L. S. Hickey and P. C. Reist, "Adjusting Occupational Exposure Limits for Moonlighting, Overtime, and Environmental Exposures," *Am. Ind. Hyg. Assoc. J.* **40**, 727–734 (1979).
66. S. A. Roach, "A More Rational Basis for Air Sampling Programs," *Am. Ind. Hyg. Assoc. J.* **27**, 1–19 (1966).
67. S. A. Roach, "A Most Rational Basis for Air Sampling Programs," *Ann. Occ. Hyg.* **20**, 65–84 (1977).
68. A. Ruzic, "Pharmacokinetic Modelling of Various Theoretical Systems," *J. Pharm. Sci.* **11**, 110–150 (1970).
69. Y. Henderson and H. H. Haggard. *Noxious Gases and the Principles of Respiration Influencing their Action*, Reinhold Publishing, New York, 1943.
70. A. David, E. Frantik, R. Holvsa, and O. Novakova, "Role of Time and Concentration on Carbon Tetrachloride Toxicity in Rats," *Inter. Arch. Occup. Environ. Health* **48**, 49–60 (1981).
71. J. R. Withey and B. T. Collins, "Chlorinated Aliphatic Hydrocarbons Used in the Foods Industry: The Comparative Pharmacokinetics of Methylene Chloride, 1,2-Dichloroethane,

Chloroform, and Trichloroethylene after I.V. Administration in the Rat," *J. Environ. Pathol. Toxicol.* **3**, 313–332 (1980).
72. R. Handy and A. Schindler, *Estimation of Permissible Concentrations of Pollutants for Continuous Exposure*, U.S. Environmental Protection Agency, Pub. No. EPA-600/2-76-155, Washington, DC, 1976.
73. M. Gibaldi and H. Weintraub, "Some Considerations as to the Determination and Significance of Biologic Half-life," *J. Pharm. Sci.* **60**, 624–626 (1971).
74. K. Nomiyama and H. Nomiyama, "Respiratory Retention, Uptake and Excretion of Organic Solvents in Man: Benzene, Toluene n-Hexane, Ethyl Acetate and Ethyl Alcohol," *Int. Arch. Arbeits Med.* **32**, 75–83 (1974).
75. C. G. Hunter and D. Blair, "Benzene: Pharmacokinetic Studies in Man," *Arch. Occup. Hyg.* **15**, 193–199 (1972).
76. J. E. Petersen, "Absorption and Elimination of Carbon Monoxide by Inactive Young Men," *Arch. Environ. Health* **21**, 165–171 (1962).
77. H. W. Leung and D. J. Paustenbach, "Setting Occupational Exposure Limits for Irritant Organic Acids and Bases based on Their Equilibrium Dissociation Constants," *Appl. Ind. Hyg.* **3**, 115–118 (1988).
78. H. W. Leung and D. J. Paustenbach, "Application of Pharmacokinetics to Derive Biological Exposure Indexes from Threshold Limit Values," *Am. Ind. Hyg. Assoc. J.* **49**(9), 445–450 (1988).
79. R. D. Stewart, H. H. Gay, D. S. Erley, C. L. Hake, and J. E. Peterson, "Human Exposure to Carbon Tetrachloride Vapor-Relationship of Expired Air Concentration to Exposure and Toxicity," *J. Occup. Med.* **3**, 586–590 (1961).
80. J. Adir, et al. "Pharmacokinetics of Fluorocarbon 11 and 12 in dogs and Humans," *J. Clin. Pharmacol.* **15**, 760–770 (1975).
81. N. Krivanek et al. "Monmethylformamide Levels in Human Urine after Repetitive Exposure to Dimethylformamide Vapor," *J. Occup. Med.* **20**, 179–187 (1978).
82. G. D. DiVincenzo, F. J. Yanno, and B. D. Astill, "Human and Canine Exposure to Methylene Chloride Vapor," *Am. Ind. Hyg. Assoc. J.* **33**, 125–135 (1972).
83. J. K. Piotrowski, "Evaluation of Exposure to Phenol: Absorption of Phenol Vapors in the Lungs and Through the Skin and Excretion in Urine," *Br. J. Ind. Med.* **28**, 172–178 (1971).
84. M. Ikeda, T. Immamura, M. Hayashi, T. Tabuchi, and I. Hara, "Biological Half-Life of Styrene in Human Subjects," *Int. Arch. Arbeitsmed.* **32**, 93–100 (1974).
85. R. D. Stewart, H. C. Dodd, E. D. Baretta, and A. W. Schaffer, "Human Exposure to Styrene Vapor," *Arch. Environ. Health* **16**, 656–662 (1968).
86. R. D. Stewart, A. Arbon, H. H. Gay, D. S. Erley, C. L. Hake, and A. W. Schaffer, "Human Exposure to Tetrachloroethylene Vapor," *Arch. Environ. Health* **2**, 516–522 (1961).
87. R. D. Stewart, E. D. Baretta, H. C. Dodd, and T. R. Torkelson, "Experimental Human Exposure to Tetrachloroethylene," *Arch. Environ. Health* **20**, 224–229 (1970).
88. A. C. Monster, "Difference in Uptake, Elimination and Metabolism in Exposure to Trichlorethylene, 1,1,1-Trichlorethane and Tetrachloroethylene," *Int. Arch. Occup. Environ. Health* **42**, 311–317 (1979).
89. R. D. Stewart, H. C. Dodd, H. H. Gay, and D. S. Erley, "Experimental Human Exposure to Trichloroethylene," *Arch. Environ. Health,* **20**, 64–71 (1970).
90. G. Kimmerle and A. Eben, "Metabolism, Excretion and Toxicology of Trichloroethylene After Inhalation. II. Experimental Human Exposure," *Arch. Toxicol.* **30**, 127–138 (1973).

91. J. W. Mason and J. Hughes, A Proposed Approach to Adjusting TLVs for Carcinogenic Chemicals (unpublished report), Univ. of Alabama, Birmingham, 1985.
92. M. E. Andersen, "Pharmacokinetics of Inhaled Gases and Vapors," *Neurolog. Toxicol. Teratol.* **3**, 383–389 (1981).
93. D. J. Paustenbach, G. P. Carlson, J. E. Christian, and G. S. Born, "A Comparative Study of the Pharmacokinetics of Carbon Tetrachloride in the Rat Following Repeated Inhalation Exposure of 8 hr/day and 11.5 hr/day," *Fund Appl. Toxicol.* **6**, 484–497 (1986).
94. NCRP. Maximum Permissible Body Burdens and Maximum Permissible Concentrations of Radionuclides in Air and in Water for Occupational Exposure, *NBS Handbook No. 69*, U.S. Government Printing Office, Washington, DC, 1959.
95. J. C. Ramsey and M. E. Andersen, "A Physiologically-Based Description of the Inhalation Pharmacokinetics of Styrene in Rats and Humans," *Toxicol. Appl. Pharm.* **73**, 159–175 (1984).
96. R. Peto, "Carcinogenic Effects of Chronic Exposure to Very Low Levels of Toxic Substances," *Environ. Health Perspect.* **22**, 155–159 (1978).
97. J. C. Bailer, E. A. C. Crouch, R. Shaikh, and D. Spiegelman, "One-Hit Models of Carcinogenesis: Conservative or Not?" *Risk Anal.* **8**, 485–490 (1988).
98. C. D. Klaasen, M. O. Amdur, and J. Doull, eds., *Cassaret and Doull's Toxicology: The Basic Science of Poisons*, 5th ed., Macmillan, New York, 1995.
99. R. H. Reitz, P. S. McCroskey, C. N. Park, M. E. Andersen, and M. L. Gargas, "Development of a Physiologically Based Pharmacokinetic Model for Risk Assessment With 1,4-dioxane." *Toxicol. Appl. Pharmacol.* **105**, 37–54 (1990).
100. C. J. Powell and C. L. Berry, "Non-genotoxic or Epigenetic Carcinogenesis," in B. Ballantyne, T. Marrs, and T. Syversen, eds., *General and Applied Toxicology*, 2nd Edition, Macmillian Reference Ltd., London, 1999, Ch. 51, pp. 1119–1137.
101. J. C. Barrett, Role of Mutagenesis and Mitogenesis in Carcinogenesis, *Environ. Mutagenesis* (1995).
102. G. M. Williams, and J. H. Weisburger, "Chemical Carcinogens," in C. D. Klaassen, M. O. Amdur and J. Doull, ed., *Cassaret and Doull's Toxicology: The Basic Science of Poisons*, 4th ed., Macmillan, New York, 1990, Ch. 5.
103. B. E. Butterworth and T. Slaga, *Nongenotoxic Mechanisms in Carcinogenesis: Banbury Report 25*, Cold Spring Harbor Laboratory, Cold Spring Harbor, New York, 1987.
104. A. M. Schumann, J. F. Quast, and P. G. Watanabe, "The Pharmacokinetics and Macromolecular Interactions of Perchloroethylene in Mice and Rats as Related to Oncogenicity," *Toxicol. Appl. Pharm.* **55**, 207–219 (1980).
105. P. G. Watanabe, R. H. Reitz, A. M. Schumann, M. J. McKenna, and P. J. Gehring, "Implications of the Mechanisms of Tumorigenicity for Risk Assessment," in M. Witschi, Ed., *The Scientific Basis of Toxicity Assessment*, Elsevier/North-Holland Press, Amsterdam, 1980, pp. 69–88.
106. B. N. Ames, "Six Common Errors Relating to Environmental Pollution," *Regul. Toxicol. Pharm.*, **7**, 379–346 (1987).
107. H. W. Leung, F. J. Murray, and D. J. Paustenbach, "A Proposed Occupational Exposure Limit for 2,3,7,8-TCDD," *Am. Ind. Hyg. Assn. J.* **49**(9), 466–474 (1988).
108. U.S.E.P.A. Proposed guidelines for Carcinogen Risk Assessment, Washington, DC, 1996.
109. G. C. Butler, "Estimation of Doses and Integrated Doses," in G. C. Butler, ed., *Principles of Ecotoxicology*, Scientific Committee on Problems of the Environment (SCOPE), Wiley, New York, 1979.

110. H. M. Bolt, J. G. Filser, and A. Buchter, "Inhalation Pharmacokinetics Based on Gas Uptake Studies, III: A Pharmacokinetic Assessment in Man of Peak Concentrations of Vinyl Chloride," *Arch. Toxicol.* **48**, 213–228 (1981).
111. S. H. Moolgavkar, A. Dwangi, and D. J. Venson, "A Stochastic Two-Stage Model for Cancer Risk Assessment: The Hazard Function and the Probability of Tumor," *Risk Anal.* **8**, 383–392 (1988).
112. L. Ehrenberg, K. D. Hieschke, S. Osteman-Golkar, and I. Wennberg, "Evaluation of Genetic Risks of Alkylating Agents: Tissue Doses in the Mouse from Air Contaminated with Ethylene Oxide," *Mutation Res.* **24**, 83–103 (1974).
113. D. L. Coffin, D. E. Gardner, G. I. Sidorenko, and M. A. Pinigin, "Role of Time as a Factor in the Toxicity of Chemical Compounds in Intermittent and Continuous Exposures. Part II. Effects of Intermittent Exposure," *J. Toxicol. Environ. Health* **3**, 821–828 (1977).
114. E. W. van Stee, G. A. Boorman, M. P. Moorman, and R. A. Sloane, "Time-varying Concentration Profile as a Determinant of the Inhalation Toxicity of Carbon Tetrachloride," *J. Toxicol. Environ. Health* **10**, 785–795 (1982).
115. D. J. Paustenbach, G. P. Carlson, J. E. Christian, and G. S. Born, "The Effect of the 11.5 hr/day Exposure Schedule on the Distribution and Toxicity of Inhaled Carbon Tetrachloride in the Rat." *Fund. Appl. Toxicol.* **6**, 472–483 (1986).
116. S. W. Dixon, G. J. Graepel, D. L. Leser, and L. F. Percival, "Effect of a Change from an 8-Hr to a 12-Hr Shift on the Levels of DMF Metabolites in the Urine," A Report by Haskell Labs, Dupont Corp, Wilmington, DE, 1984.
117. J. A. MacGregor, *Application of Pharmacokinetics to Occupational Health Problems*, Ph.D. Dissertation, University of California at San Francisco, 1973.
118. Y. Kim and G. P. Carlson, "The Effect of an Unusual Workshift on Chemical Toxicity: I. Studies on the Exposure of Rats and Mice to Dichloromethane," *Fund. Appl. Toxicol.* **6**, 162–171 (1986).
119. Y. C. Kim and G. P. Carlson, "The Effect of an Unusual Workshift on Chemical Toxicity: II. Studies on the Exposure of Rats to Aniline," *Fund. Appl. Toxicol.* **7**, 144–152 (1986).
120. T. P. Pastoor and B. A. Burgess, "Effect of Concentration and Duration of Exposure on the Inhalation Toxicity of Aniline for Periods of 3, 6, 9 and 12 Hours," presented at the 1983 Joint Conference on Occupational Health, 1983.
121. N. P. Kazmina, "Study of the Adaptation Processes of the Liver to Monotonous and Intermittent Exposures to Carbon Tetrachloride" (in Russian), *Gig. Tr. Prof. Zabol.* **3**, 39–45 (1976).
122. W. A. Colburn and H. B. Matthews, "Pharmacokinetics in the Interpretation of Chronic Toxicity Tests: The Last-In, First-Our Phenomena," *Toxicol. Appl. Pharm.* **48**, 387–395 (1979).
123. J. M. Wisniewska-Knyil, J. K. Jablonska, and J. K. Piotrowski, "Effect of Repeated Aniline, Nitrobenzene, and Benzene on Liver Microsomal Metabolism in the Rat," *Brit. J. Ind. Med.* **32**, 42–48 (1975).

CHAPTER FORTY-ONE

The History and Biological Basis of Occupational Exposure Limits For Chemical Agents

Dennis J. Paustenbach, Ph.D., CIH, DABT

1 HISTORY

Over the past 50 years, many organizations in numerous countries have proposed occupational exposure limits (OEL) for airborne contaminants. The limits or guidelines that have gradually become the most widely accepted both in the United States and in most other countries are those issued annually by the American Conference of Governmental Industrial Hygienists (ACGIH) and are termed Threshold Limit Values® (TLVs) (1–3).

The usefulness of establishing OELs for potentially harmful agents in the working environment has been demonstrated repeatedly since their inception (2, 4, 5). It has been claimed that whenever these limits have been implemented in a particular industry, no worker has been shown to have sustained serious adverse effects on his health as a result of exposure to these concentrations of an industrial chemical (6). Although this statement is arguable with respect to the acceptability of OELs for those chemicals established before 1980 and later found to be carcinogenic, there is little doubt that hundreds of thousands of persons have avoided serious effects of workplace exposure due to their existence.

The contribution of OELs to the prevention or minimization of disease is widely accepted, but for many years such limits did not exist, and even when they did, they were often not observed (1, 2, 6–8). It was, of course, well understood as long ago as the 15th century, that airborne dusts and chemicals could bring about illness and injury, but the concentrations and lengths of exposure at which this might be expected to occur were unclear (9).

As reported by Baetjer (10) "early in this century when Dr. Alice Hamilton began her distinguished career in occupational medicine, no air samples and no standards were available to her, nor indeed were they necessary. Simple observation of the working conditions and the illness and deaths of the workers readily proved that harmful exposures existed. Soon however, the need for determining standards for safe exposure became obvious."

Cook has reported that the earliest efforts to set an OEL were directed at carbon monoxide, the toxic gas to which more persons are occupationally exposed than any other. The work of Max Gruber at the Hygienic Institute at Munich was published in 1883. The paper described exposing two hens and 12 rabbits to known concentrations of carbon monoxide up to 47 hours over three days, he stated that "the boundary of injurious action of carbon monoxide lies at a concentration on all probability of 500 parts per million, but certainly (not less than) 200 parts per million." In spite of this conclusion, however, he reported no symptoms or uncomfortable sensations after three hours on each of two consecutive days at concentrations of 210 parts per million and 240 parts per million (2).

According to Cook, the earliest and most extensive series of animal experiments on exposure limits were those conducted by K.B. Lehmann and others under his direction at the same Hygienic Institute where Gruber had done his work with carbon monoxide. The first publication in the series, entitled *Experimental Studies on the Effect of Technically and Hygienically Important Gases and Vapors on the Organism*, was a report on ammonia and hydrogen chloride gas that was 126 pages in length within Volume 5 of *Archiv für Hygiene* (11). This series of reports on animal experimentation with a large number of chemical substances by Lehmann and associates continued through Part 35 in Volume 83 (1914), followed by a final comprehensive paper of 137 pages on chlorinated hydrocarbons in the September 2, 1986 Volume 116 (12), of the German Archiv. These reprints became the standard against which others would be compared for nearly 30 years.

Kobert (13) published one of the earlier tables of acute exposure limits. Concentrations for 20 substances were listed under the headings: (*1*) Rapidly fatal to man and animals, (*2*) Dangerous in 0.5 to one hour, (*3*) 0.5 to one hour without serious disturbances, and (*4*) Only minimal symptoms observed (2). In his paper on *Interpretations of Permissible Limits*, Schrenk (14) notes that the "values for hydrochloric acid, hydrogen cyanide, ammonia, chlorine and bromine as given under the heading "only minimal symptoms after several hours' in the foregoing Kobert paper agree with values as usually accepted in present-day tables of MACs for reported exposures." However, values for some of the more toxic organic solvents, such as benzene, carbon tetrachloride and carbon disulfide, far exceeded those currently in use (2).

One of the first tables of exposure limits to originate in the U.S. was that published by the U.S. Bureau of Mines (15). Although its title does not reflect the content, the 33 substances listed are those encountered in workplaces. Cook (2) also noted that most of the exposure limits through the 1930s, except for dusts, were based on rather short animal experimentation. A notable exception was the study of chronic benzene exposure by Greenburg of the U.S. Public Health Service conducted under the direction of a committee of the National Safety Council (16). From this, an acceptable OEL was derived based on long term animal experimentation.

According to Cook (2) for dust exposures, permissible limits established before 1920 were based on exposures of workers in the South African gold mines where the dust from drilling operations was high in crystalline free silica. The effects of the dust exposure were followed by periodic chest x-ray examination and the dust concentrations were monitored with an instrument known as a konimeter that collected a nearly instantaneous sample. In 1916, based on a correlation of these two sets of findings, an exposure limit of 8.5 million particles per cubic foot of air (mppcf) for the dust with an 80 to 90% quartz content was set (17). Later, the level was lowered to 5 mppcf. Cook (2) also reported that, in the U.S., standards for dust, also based on exposure of workers, were recommended by Higgins et al. following a study at the southwestern Missouri zinc and lead mines in 1917. The initial level established for high quartz dusts was 10 mppcf, appreciably higher than was established by later dust studies conducted by the U.S. Public Health Service.

The most comprehensive list of OELs up to 1926 was that for 27 substances published in Volume 2 of International Critical Tables (18). Sayers and Dalle Valle (19) published a table giving physiological response to five levels of concentrations of 37 substances. The first four refer to acute effects but the fifth is for the maximum allowable concentration for prolonged exposure. In 1930, the USSR Ministry of Labor issued a decree that included the first actual approval of workplace maximum allowable concentrations for the former USSR with a list of 12 industrial toxic substances. About this time, Lehmann and Flury (20) and Bowditch et al. (21) published papers that presented tables with a single value for repeated exposures to each substance.

As noted by Cook (2), many of the exposure limits developed by Lehmann were included in the Henderson and Haggard monograph (22) initially published in 1927 and a little later in Flury and Zernik's Schadilche Gase (23). According to Cook (2), this book acted as the bible on effects of injurious gases, vapors and dusts in industrial exposures until Volume II of Patty's Industrial Hygiene and Toxicology was published in 1949.

Baetjer (10) has reported that the first list of standards for chemical exposures in industry in the U.S. were called Maximum Allowable Concentrations (MAC) and these were prepared in 1939 and 1940. They represented a consensus of opinion of the American Standard Association (ASA) and a number of industrial hygienists who had formed the ACGIH in 1938. These "suggested standards" were published in 1943 by James Sterner.

A committee of the ACGIH met in early 1940 to begin the task of identifying safe levels of exposure to workplace chemicals. They began by assembling all the data they could locate that would relate the degree of exposure to a toxicant to the likelihood of producing an adverse effect (1, 6). This task, as might be expected, was a formidable one. After much painstaking research and labor intensive documentation, the first values issued by the ACGIH were released in 1941 by this committee, which was composed of Warren Cook, Manfred Boditch (reportedly America's first hygienist employed by industry), William Fredrick, Philip Drinker, Lawrence Fairhall and Alan Dooley (6).

In 1941, a committee, designated as Z-37, of the American National Standards Institute (ANSI), then known as the ASA, developed its first Standard—carbon monoxide, with an acceptable value of 100 parts per million (ppm). The committee issued separate bulletins through 1974 including exposure standards for 33 toxic dusts and gases.

At the Fifth Annual Meeting of the ACGIH in 1942, the newly appointed Subcommittee on Threshold Limits presented in its report a table of 63 toxic substances with the "maximum allowable concentrations of atmospheric contaminants" from lists furnished by the various state industrial hygiene units. The report contains the statement, "The table is not to be construed as recommended safe concentrations. The material is presented without comment" (2).

In 1945 a list of 132 industrial atmospheric contaminants with MACs was published by Cook. This is considered a landmark publication since it was thorough, included references on the original investigations, and provided the rationale leading to the values. The table included the then current values for the six states—California, Connecticut, Massachusetts, New York, Oregon, and Utah—values presented as a guide for occupational disease control by the U.S. Public Health Service, and 11 other standards. In addition, Cook included a list of MACs that appeared best supported by the references to original investigations (2).

At the 1946 Eighth Annual Meeting of ACGIH, the Subcommittee on Threshold Limits presented their second report with the values of 131 gases, vapors, dusts, fumes, and mists, and 13 mineral dusts. As stated in the report, the values were "compiled from the list reported by the subcommittee in 1942, from the list published by Warren Cook in *Industrial Medicine* (7), and from published values of the Z-37 Committee of the American National Standards Association (ANSI)." The committee emphasized that the "list of MAC values is presented . . . with the definite understanding that it be subject to annual revision."

The overall impact of these efforts to develop quantitative limits to protect humans from the adverse effects of workplace air contaminants and physical agents could not have been anticipated by the early TLV committees. To their credit, even though toxicology was then only a fledgling science, their approach to setting limits has generally been shown to be correct even by today's standards. For this reason, most of the techniques for setting limits established by this committee are still in use today (1, 4, 5, 7, 8, 24–26). The principles they used to set OELs were similar to those later used to identify safe doses of food additives and pharmaceuticals (27).

From the perspective of the hygienist, engineer and businessperson, there have been many benefits of setting OELs. The establishment of limits, by their very nature, implies that at some concentration or dose, exposure to a toxicant can be expected to be safe and pose no significant risk of harm to exposed persons. The key to the success of limits has not been only that they were established on solid scientific principles; rather, the setting of any goal gives a sense of purpose and direction to industrial, occupational, or medical programs which, prior to the TLVs, had been difficult to evaluate. The setting of goals, such as controlling workplace concentrations below an OEL, establishes an objective that can then be mutually pursued by the occupational health team, engineers and management. History has shown that by introducing the concept of "safe level of exposure" and by establishing a type of "management by objectives," occupational health professionals will establish and pursue a systematic program for reducing exposure (28).

1.1 Intended Use of OELs

The ACGIH TLVs and most other OELs used in the U.S. and some other countries are limits that refer to airborne concentrations of substances and represent conditions under which "it is believed that nearly all workers may be repeatedly exposed day-after-day without adverse health effects". In some countries, which will be discussed later, the OEL is set at a concentration that will protect virtually everyone. It is important to recognize that unlike some exposure limits for ambient air pollutants, contaminated water, or food additives set by other professional groups or regulatory agencies, exposure to the TLV will not necessarily prevent discomfort or injury for everyone who is exposed (29). The ACGIH recognized long ago that because of the wide range in individual susceptibility, a small percentage of workers may experience discomfort from some substances at concentrations at or below the threshold limit and that a smaller percentage may be affected more seriously by aggravation of a preexisting condition or by development of an occupational illness (30). This is clearly stated in the introduction to the ACGIH's annual booklet "Threshold Limit Values for Chemical Substances and Physical Agents and Biological Exposure Indices" (3).

This limitation, although perhaps less than ideal, has been considered a practical one since airborne concentrations so low as to protect hypersusceptibles have traditionally been judged infeasible due to either engineering or economic limitations. This shortcoming in the TLVs has, until about 1990, not been considered a serious one. In light of the dramatic improvements of the past 10 years in our analytical capabilities, personal monitoring/sampling devices, biological monitoring techniques, and the use of robots as a plausible engineering control, we are now technologically able to consider more stringent OELs.

The background information and rationale for each TLV are published periodically in the *Documentation of the Threshold Limit Values*. Some type of documentation is occasionally available for OELs set in other countries. The rationale or documentation for a particular OEL, as well as the specific data considered in establishing it, should always be consulted before interpreting or adjusting an exposure limit.

Threshold Limit Values, like OELs used in most other countries, are based on the best available information from industrial experience, experimental human and animal studies and, when possible, a combination of the three (31). The rationale for each of the values differs from substance to substance. For example, protection against impairment of health may be a guiding factor for some, whereas reasonable freedom from irritation, narcosis, nuisance or other forms of stress may form the basis for others. The age and completeness of the information available for establishing most OELs also varies from substance to substance; consequently, the precision of each particular TLV is not constant. The most recent TLV and its documentation should always be consulted in order to evaluate the quality of the data upon which that value was set.

Even though all of the publications that contain OELs emphasize that they were intended for use only in establishing safe levels of exposure for persons in the workplace, they have been used at times in other situations. It is for this reason that all exposure limits should be interpreted and applied only by someone knowledgeable of industrial hygiene and toxicology. The TLV Committee did not intend that they be used, or modified for use:

1. In the evaluation or control of community air pollution nuisances.
2. Estimating the toxic potential of continuous, uninterrupted exposures or other extended work periods.
3. As proof or disproof of an existing disease or physical condition.

4. By countries whose working conditions or cultures differ from those of the United States and where substances and processes differ.

It is noteworthy to remember that the ACGIH TLV Committee has repeatedly stated that "these limits are not fine lines between safe and dangerous contaminations." In short, without knowing the toxic endpoint to be avoided, they can't even be used as relative indices of toxicity.

The TLV Committee and other groups that set OELs warn that these values should not be "directly used" or extrapolated to predict safe levels of exposure for other exposure settings. However, if one understands the scientific rationale for the guideline and the appropriate approaches for extrapolating data, considering the pharmacokinetics and mechanism of action of the chemical, they can be used to predict acceptable levels of exposure for many different kinds of exposure scenarios and work schedules (32, 33).

The reason that the ACGIH has stated that the TLVs should not be used for other purposes and that they should be used only by professionals trained in the field is because of the history of misuse. For example, there are dozens of examples where lawyers, physicians and others have erroneously concluded that if a worker was exposed above a TLV concentration then the person was at "real" danger, or worse, that they may have been harmed. Often, such interpretations were self-serving and persons were adversely affected by such unfounded opinions. In addition, the TLVs have often been inappropriately used by regulatory agencies as a basis for establishing "temporary standards" for everything from ambient air guidelines to emergency evacuation criteria. Often, the group issuing the draft criteria knew that this was not a proper use of the TLVs but such action was considered justifiable because it was "science forcing." That is, it made professionals in the regulated community go about the task of doing the more detailed work necessary to develop proper standards since the regulatory agency was simply understaffed to handle such a large task. Because no one has been able to anticipate all of the various ways these values could be misused, the ACGIH decided to issue a "blanket disclaimer" many years ago. Obviously, the *Documentation of the TLVs* (32) provides a good deal of important information to the health scientist and, from this, a professional should be able to derive other criteria if all of the appropriate factors are considered.

1.2 Philosophical Underpinnings of TLVs and other OELs

TLVs were originally prepared to serve only for the use of industrial hygienists who could exercise their own judgement in their application. They were not to be used for legal purposes (10). However, in 1968 the Walsh-Healey Public Contract Act incorporated the 1968 TLV list, which covered about 400 chemicals. In the United States, when the Occupational Safety and Health Administration (OSHA) was formed, it allowed OSHA, for a period of two years, to adopt national consensus standards or established federal standards.

Exposure limits for workplace air contaminants are based on the premise that, although all chemical substances are toxic at some concentration when experienced for a period of time, a concentration (e.g., dose) does exist for all substances at which no injurious effect should result no matter how often the exposure is repeated. This premise applies to substances whose effects are limited to irritation, narcosis, nuisance or other forms of stress (6).

This philosophy thus differs from that applied to physical agents such as ionizing radiation, and for some chemical carcinogens, since it is possible that there may be no threshold or no dose at which zero risk would be expected (6). Even though many would say that this position is too conservative in light of our current understanding of the mechanisms by which cancer occurs, there are some data on some genotoxic chemicals that support this theory (34). On the other hand, a large number of toxicologists believe that a threshold dose exists for those chemicals that are carcinogenic in animals but that act through a non-genotoxic (sometimes called epigenetic) mechanism (35–38). Still others maintain that a practical threshold exists for even genotoxic chemicals, although they agree that the threshold may be at an extremely low dose (39–41).

With this in mind, some OELs for carcinogens proposed by regulatory agencies in the early 1980s were set at levels which, although not completely without risk, posed risks no greater than classic occupational hazards such as electrocution, falls, etc. (about one-in-one thousand) (42). Although a clear description of this risk level or the rationale for the criteria has rarely been presented or discussed, it is now apparent that it was used, in part, to justify these limits (42–44).

1.3 Occupational Exposure Limits in the United States

A comprehensive listing of the various OELs used throughout the world can be found in one of two references. One is the *Occupational Exposure Limits For Airborne Toxic Substances*, 4th Edition, published by International Labour Office of the World Health Organization (45), and the other is *Occupational Exposure Limits-Worldwide* published by the American Industrial Hygiene Association (AIHA) (2).

The philosophical underpinnings for the various OELs vary between the organizations and countries that develop them (2, 46, 47). For example, in the United States at least six groups recommend exposure limits for the workplace. These include the ACGIH TLVs, the Recommended Exposure Limits (RELs) suggested by the National Institute for Occupational Safety and Health (NIOSH) of the U.S. Department of Health and Human Services, the Workplace Environment Exposure Limits (WEEL) developed by the American Industrial Hygiene Association standards for workplace air contaminants suggested by the Z-37 Committee of ANSI, the proposed workplace guides of the American Public Health Association (48) and lastly, recommendations that have been made by local, state or regional government. In addition to these recommendations or guidelines, Permissible Exposure Limits (PEL), which are regulations that must be met in the workplace in the United States because they are law, have been promulgated by the Department of Labor and are enforced by OSHA (49).

In addition to the OELs established by professional societies and regulatory bodies, guidance has also been provided by many corporations who handle or manufacture specific chemicals (50). For example, beginning in the 1960s, Dow Chemical, DuPont, Celanese, Kodak, Union Carbide, and some other large firms began to establish internal OELs intended to protect their workers, as well as their customers who purchased the chemicals. Later, due to the fact that only their workers would be exposed to the chemicals that they manufactured (without a prescription), the pharmaceutical industry began to set limits for some of their intermediate and final products. Workers in the drug industry needed these guideline because the doses to which they could be exposed each day had the potential to be several fold greater than the therapeutic dose. Today, perhaps more than 40 firms in the United States set internal OELS.

Outside the United States, as many as 50 other countries or groups of countries have established OELs (2, 4, 46, 47). Many, if not most, of these limits are nearly or exactly the same as the ACGIH TLVs developed in the United States (2, 46, 51, 53). In some cases, such as in the East European countries, including former Soviet bloc countries, and in Japan, the limits can be dramatically different than those used in the United States. Differences among various limits recommended by other countries can be due to a number of factors:

1. Difference in the philosophical objective of the limit and the untoward effects they are meant to minimize or eliminate.
2. Difference in the predominant age and sex of the workers.
3. The duration of the average workweek.
4. The economic state of affairs in that country.
5. A lack of enforcement (therefore the OEL serves only as a guide).

For example, limits established in what is now the Organization of Russian States are often based on a premise that they will protect "everyone rather than nearly everyone, from any (rather than most) toxic or undesirable effects of exposure" (1, 4, 46, 47, 51–54).

In recent years, the ACGIH has worked closely with the European community (in particular, the German MAK Committee) in an attempt to harmonize the various approaches used to set OELs. For example, in 1997 a joint meeting between the TLV and MAK committees was held in Germany to move this initiative forward. The meeting went well and differences of opinion were shared freely. Due to a number of differences in their views of various scientific and social considerations, it is unlikely that a single method for dealing with each category of adverse effects will be adopted in the foreseeable future by these two organizations. In recent years, there has been an annual joint meeting of these two groups to promote a more frequent dialogue (55).

2 APPROACHES USED TO SET OEL

OELs established both in the United States and elsewhere are based on data from a wide number of studies and sources. As shown in Table 41.1, the 1968 TLVs (those adopted by OSHA in 1972 as federal regulations) were based largely on human experience (49). This may come as a surprise to many hygienists who have recently entered the profession since it indicates that, in most cases, the setting of an exposure limit was often after a chemical had been found to produce toxicity, irritation or other undesirable effects on humans. As might be anticipated, many of the more recent exposure limits for systemic toxins, espe-

Table 41.1. Distribution of Procedures Used to Develop ACGIH TLVs for 414 Substances Through 1968[a,b]

Procedure	Number	Percent Total
Industrial (human) experience	157	38
Human volunteer experiments	45	11
Animal, inhalation—chronic	83	20
Animal, inhalation—acute	8	2
Animal, oral—chronic	18	4.5
Animal, oral—acute	2	0.5
Analogy	101	24

[a]From Ref. 4.
[b]Exclusive of inert particulates and vapors.

cially those internal limits set by manufacturers, have been based primarily on toxicology tests conducted on animals, which is in contrast with waiting for observations of adverse effects in exposed workers (50, 56).

However, even as far back as 1945, animal tests were acknowledged by the TLV Committee to be very valuable and they do, in fact, constitute the second most common source of information upon which these guidelines have been established (4, 24, 57).

Several approaches for deriving OELs from animal data have been proposed and put into use over the past 40 years. The approach used by the TLV Committee and others is not markedly different from that which has been used by the U.S. Food and Drug Administration (FDA) in establishing acceptable daily intakes (ADI) for food additives as far back as the early 1950s (24, 27). An understanding of the FDA approach to setting exposure limits for food additives and contaminants can provide good insight to industrial hygienists who are involved in interpreting OELs (25, 42, 58).

Discussions of methodological approaches that can be used to establish workplace exposure limits based exclusively on animal data have also been presented (24, 25, 59–66). A review of the general procedures used by groups setting OELs has been published and it warrants evaluation (67). The OELs derived from the various approaches contain some degree of scientific uncertainty associated with them; that is, a particular OEL may also prevent illness or irritation at 2–3 times the selected OEL and conversely, in some cases, symptoms of overexposure may be seen in a few persons exposed at the recommended OEL. Many scientists believe that much of the uncertainty in identifying proper OELs can be reduced by the availability of better epidemiological information and through the use of physiologically based pharmacokinetic (PB–PK) models (a much better approach than the traditional qualitative extrapolation of animal test results to humans). These models are discussed in Chapter 40.

In 1968, approximately 50% of the TLVs were derived from the human data, and approximately 30% were based on animal data (Table 41.1). By 1998, about 50% of the 700 or so TLVs were derived primarily from animal data. The criteria used to develop the TLVs have been be classified into four groups: morphologic, functional, biochemical and miscellaneous (nuisance, cosmetic) (4). In recent years, these categories have been subdivided according to adverse effect (3).

Of those TLVs based on the human data, most are based on the effects observed in workers who were exposed to the substance for many years. In short, about half of the existing TLVs have been based on the results of workplace monitoring, coupled with qualitative and quantitative observations of the human response (4, 69). In recent times, TLVs for new chemicals have been based primarily on the results of animal studies rather than on human experience.

The rationale upon which most of the historical TLVs have been established is presented in Table 41.2. It is noteworthy that in 1968 only about 50% of the TLVs were intended primarily to prevent systemic toxic effects. Roughly 40% were based on irritation and about 2% were intended to prevent cancer. By 1998, about 50% of the TLVs were meant to prevent systemic effects, 30% to prevent irritation, 5–10% to prevent cancer and the remainder to prevent other adverse effects.

In the early years of the modern era of industrial hygiene, e.g., post-OSHA, very little information was available to the public regarding the precise methodology by which TLVs, MAKs, and other OELs were derived. To Warren Cook's credit, he wisely shared his rationale for selecting specific values in his documentation for the various limits (7) but it is clear from his writings that no set of uncertainty factors (UFs) or criteria were universally adopted in setting these limits of exposure. Furthermore, over the ensuing 50 years, the TLV Committee never adopted a standard approach and this has been troublesome to critics of the committee. For example, many professionals believed that a fairly generic approach to setting OELs could be applied to the various classes of chemicals. On only a few occasions has a prescribed approach to setting OELs been suggested in a published paper (56, 62–65, 83).

One reason that neither occupational nor environmental limits can be derived in a "cookbook" manner is because of the relatively "soft" nature of the data upon which these limits are based. For example, no animal bioassay can ever be large enough to conclusively assure us that we can identify a precise dose that will not pose some theoretical risk to

Table 41.2. Distribution of Criteria Used to Develop ACGIH TLVs for 414 Substances Through 1968[a,b]

Criteria	Number	Percent	Criteria	Number	Percent[a]
Organ or organ system affected	201	49	Biochemical changes	8	2
Irritation	165	40	Fever	2	0.5
Narcosis	21	5	Visual changes (halo)	2	0.5
Odor	9	2	Visibility	2	0.5
Organ function changes	8	2	Taste	1	0.25
Allergic sensitivity	6	1.5	Roentgenographic changes	1	0.25
Cancer	6	1.5	Cosmetic effect	1	0.25

[a]Ref. 4.
[b]Exclusive of inert particulates and vapors.
[c]Number of times a criterion was used of total number of substances examined × 100, rounded to nearest 0.25 percent. Total percentages exceed 100 because more than one criterion formed the basis of the TLV of some substances.

some individual, and no human epidemiology study can be so strong as to show that a chemical may not produce some type of adverse effect in some person. Nonetheless, in spite of our inability to prove the absence of risk, there is clear evidence that virtually safe doses for humans can be identified for all substances. The primary question that is difficult to answer is "What is the margin of safety between what we label as a 'safe' dose and the dose that might have some probability of producing some type of adverse effect in some individual?" (70).

The following sections describe, in a general way, some of the various approaches that could be or have been used to establish OELs. In the main, the UF approach can be used to establish OELs for nearly all adverse effects (except perhaps genotoxic carcinogens). The primary variable in setting these limits is the size of the UF, which will vary with the adverse effect to be avoided, as well as the amount of available data. It must be noted here that in most cases, the use of uncertainty or safety factors that have been used to establish the TLVs, for example, has rarely been explicitly described in the *Documentation of the TLVs* (32). This is a shortcoming in the process that deserves to be addressed.

2.1 Uncertainty Factors

Uncertainty factors, UFs, or safety factors (as they were called from 1950–1980), are used in health risk assessment to account for a lack of complete knowledge or uncertainties about the dose delivered, human variations in sensitivity, and other factors associated with extrapolating animal data to estimate human health effects (25). They are applied to animal or human data in an attempt to identify safe levels of exposure for most persons. The UF approach has been and continues to be used by FDA, EPA, OSHA, and virtually all agencies and scientific bodies who set acceptable levels of exposure to toxic substances.

Within the United States Environmental Protection Agency (EPA), for example, UFs are factors used to operationally derive environmental regulatory criteria. Specifically, the Reference Concentrations (RfCs) or Reference Doses (RfDs) are derived from experimental or epidemiological data (71, 73) involving animals or humans. The critical dose is usually defined as the No Observed Effect Level (NOEL) for the most sensitive adverse effect from the best animal study.

The RfC or RfD is calculated by dividing the critical dose by one or more UFs and sometimes a modifying factor (MF). Although the application of this number of UFs is not consistent with the history of setting occupational exposure limits, it is prudent for setting environmental limits if properly conducted.

UFs are intended to account for:

1. The variation in sensitivity among the members of the human population.
2. The uncertainty in extrapolating animal data to estimate human health effects.
3. The uncertainty in extrapolating from data obtained in a study that is of less than lifetime exposure.
4. The uncertainty in using Lowest Observable Adverse Effect level (LOAEL) data rather than the No Observed Adverse Effect Level (NOAEL) data.
5. An incomplete data base, generally with regards to reproductive or developmental toxicity.

Usually, each of these factors is set at 3 or 10 but a value as low as 1 may sometimes be appropriate (60). MFs are used the same way as UFs are used. However, MFs are applied when additional uncertainty exists after accounting for the uncertainty within the five listed categories. MFs would not generally be applied when setting an OEL because some human data are almost always available, thus avoiding the need for additional conservatism.

Additional information on EPA's justification and rationale for the use of UFs and MFs is presented in several documents, including *Interim Methods for Development of Inhalation Reference Doses, Methods for Derivation of Inhalation Reference Concentrations and Application of Inhalation Dosimetry, IRIS Supportive Documentation Volume 1* (72), Dourson and Stara (25) and Barnes and Dourson (73). These documents are worth consulting because when occupational exposure limits are set, the same factors are considered. The primary differences in setting an RfC vs. an OEL are the number and size of the UFs, the difference in the length of exposure (continuous vs. 40-hour workweek), the lack of a recovery period, the difference in the type of exposed population (old, young, women, and those with illness versus healthy workers) and differences in the definition of "acceptable risk" for the worker population and the public. In short, OELs do not attempt to protect everyone in the general population who could be exposed continually for a lifetime, while this is the intent of an RfD and an RfC.

2.1.1 Specific UFs Used To Set an RfC

As mentioned previously, the EPA issues guidance regarding safe airborne concentrations of various chemicals. This guidance is known as a reference concentration, or RfC. The formula for deriving the RfC is generally as follows:

$$RfC = \frac{\text{Critical dose}}{UF \times MF}$$

where *Critical dose* = Best estimate of the NOEL from the studies that have evaluated the most sensitive relative toxic effect.
MF = Modifying factor
UF = UF

When the EPA sets an RfC, documentation is presented in their data base, IRIS, for each chemical specifying the UFs applied and their rationale. This documentation is available to anyone via the World Wide Web. The EPA's application of UFs has evolved over time. Until recently they relied only on UFs representing an "order of magnitude" (factor of 10). Where appropriate, the EPA now uses an intermediate UF of 3, rather than 10. On a logarithmic scale or multiplicative scale, 3 is midway between 1 and 10. Since 3 is the approximate midpoint between 1 and 10 on a logarithmic scale, when 2 successive applications of UF 3 are performed, the combined UF is rounded to 10. Over the years, when setting OELs (rather than RfDs) the TLV Committee frequently applied an uncertainty or safety factor of 3 rather than 10 (although this was rarely explicitly stated in the docu-

HISTORY AND BIOLOGICAL BASIS OF OCCUPATIONAL EXPOSURE LIMITS

mentation) and only one or two, rather than three to five, UFs were applied to human or animal data to form the overall or aggregate UF.

UFs have often been applied by the EPA in the following manner when setting ambient air limits. It is noteworthy that only the EPA's conceptual approach, but not the specific size and number of UFs, is applicable to setting an OEL because a smaller population of healthy persons is the focus of workplace standards, they are exposed only 40 hours a week, and the levels of risk considered acceptable are much different for these two populations:

- When setting limits to protect the general population, a 10-fold UF is used to protect the sensitive individuals. However, as when they derived the RfC for carbon disulfide, the EPA used a three-fold UF because the critical dose was measured directly as an internal measure of dose.
- Where necessary, either a 10-fold or 3-fold UF is used to account for uncertainties in the extrapolation of data from animal studies to estimating effects on humans. When dosimetric adjustments are used, the smaller (3-fold) UF is used.
- The EPA uses a 10-fold or 3-fold UF to extrapolate from subchronic studies to predict the hazard due to exposure that may last a lifetime. The size of the UF depends on whether progression of the adverse condition is expected.
- In addition, EPA uses a 10-fold or 3-fold UF to extrapolate from a LOAEL to a NOAEL (in those studies where no clear NOAEL was observed). The smaller UF is used when the adverse effect is judged to be sufficiently mild.
- Where indicated, a 10-fold or 3-fold UF for database deficiencies can be applied. The most common deficiencies are lack of developmental or reproductive toxicity studies. This additional UF is most often used where there is reason to suspect the effect, but the necessary studies are lacking.

Finally, the MF is used by EPA only when there may be indications from previous studies or from supporting biological data that other effects may occur. The following example illustrates how this approach was used to derive a chronic RfC for methyl ethyl ketone for the general population.

2.1.1.1 Example #1: Setting an RfC for Methyl Ethyl Ketone.

Recently, the EPA recommended a chronic RfC for methyl ethyl ketone (MEK). It was based on decreased fetal birth weight. They reported that the NOAEL adjusted to a Human Equivalent Concentration (critical dose) from the best study was 2978 mg/m^3. From this beginning, they established an RfC in the following manner:

$$RfC = \frac{Critical\ Dose}{UF \times MF} = \frac{2978\ mg/m^3}{1{,}000 \times 3} \approx 1\ mg/m^3$$

A total UF of 1,000 was derived by multiplying the following factors:

- 10 to account for interspecies extrapolation.
- 10 to account for sensitive individuals.
- 10 to account for an incomplete database including the lack of chronic and reproductive toxicity studies.

An MF of 3 was used to address the lack of unequivocal data for respiratory tract (portal of entry) effects.

Critical dose = 2978 mg/m^3 [the NOAEL$_{(HEC)}$]

Note: This RfC was intended to protect virtually anyone in the population including the very young, the very old and those with one of a number of illnesses. The approach is unnecessarily conservative for establishing an occupational exposure limit intended to protect a much smaller and less diverse population that would *not* be continuously exposed to this concentration in ambient air. In the author's view, EPA should not have added an MF factor of three to account for possible respiratory effects since developmental effects were the most sensitive endpoint. Further, an aggregate 3,000-fold safety factor appears excessive for an RfC for this chemical given the large amount of toxicity data and the available information from worker studies.

2.1.2 UFs and OELs

The method by which one chooses the size of each of the specific UF has been debated for many years (25, 59, 75, 76). Not surprisingly, no single method has been embraced by all scientific bodies. For example, from about 1950–1985, the FDA often applied a UF of 10 to account of the possible increased susceptibility of humans vs. the animal tested and another factor of 10 to account for the differences in susceptibility across the human population (e.g., a total UF of 100) when establishing limits of exposure for food additives and pesticide residues. Often, if chronic animal data were available, this factor of 100 was considered adequate (76). On the other extreme, Weil (1972) once suggested that if only acute toxicity information was available, such as an LD$_{50}$, then an aggregate UF of 5,000 was reasonable if one is trying to prevent a chronic risk. Later research conducted at Harvard University generally supported his opinion.

When evaluating the various views about UFs, it is worthwhile to note that the size of the UF used to account for various "unknowns" has changed over the years. For example, it would not be fair to compare the views of Warren Cook in 1945 to those of Dr. Lisa Brosseau in 1999 (the current Chair of the TLV Committee). For example, one reason that the magnitude of the overall UF has changed over time is because the margin of safety thought to be necessary to protect "most workers" in 1945 is different than in 2000. In part, this is because society has begun to encourage scientists to prevent toxicological effects that are more subtle or less severe than the traditional endpoints. In addition, the margin of safety thought to be necessary to ensure protection of the diverse number of persons in the population has changed over the past 5–10 years (58, 70). Third, our knowledge of the degree of inter-individual variability in response to xenobiotics within the population has improved.

HISTORY AND BIOLOGICAL BASIS OF OCCUPATIONAL EXPOSURE LIMITS

The same four extrapolation factors used to adjust animal data to set an ambient air limit (RfC) can be used to derive an OEL although their magnitude will often be much smaller. A number of papers that have been published over the years discuss the rationale for selecting the various values (50, 71, 74, 75). The use of UFs to set OELs has been discussed by Zielhuis and VanDer Kreck (62, 63) and others. Unlike the approach used to set RfCs, aggregate or overall UFs much less than 100 have often been applied to lifetime animal studies to predict safe exposures for humans exposed in the workplace. This is because workers are only exposed as adults, they are generally more healthy than the general population, and the exposure is only for 40 hours per week and only for 30–40 years. Consequently, it is not unusual to see an overall UF as small as 10–100 being applied to chronic animal NOELs to establish an OEL for workers. This is not to say that larger safety factors are not often warranted. When the slope of the dose-response curve is known, it should influence the size of the uncertainty factors.

Regrettably, the ACGIH TLV Committee has not explicitly presented the quantitative basis for setting the vast majority of the TLVs. Thus, one must look at the value suggested for a specific chemical and the information from the specific study or studies that they seemed to rely upon in order to derive the size of "safety factor" that was used. Nevertheless, if one analyzes the historic TLVs for many of the systemic toxicants, a few generalizations can be offered. First, if a solid NOEL from a 6 month to 2 year animal study was available, it appears that a UF of 10–100 was usually applied to the NOEL to establish an OEL for a chronic toxicant that was not carcinogenic. Assuming there was no evidence of mutagenicity or carcinogenic potential, this margin of safety was considered adequate. Second, if there was a reasonable amount of human experience with the chemical and no adverse effects had been observed in carefully monitored workers, then UFs as small as 10 have often been used to establish TLVs when chronic animal data were available. Third, in recent years, there has been some interest by those involved in setting OELs to increase the margin of safety inherent in most OELs to accommodate the differences in susceptibility among workers. An interesting analysis of the various implied UFs in the TLVs has been published (78). The following section discusses various approaches one could adopt to set OELs based on the toxic effect of the agent.

2.2 Setting Limits for Systemic Toxicants

By far, the majority of chemicals for which OELs have been established are systemic toxicants. By definition, this class of chemicals brings about their adverse effects at a site or target organ distant from the site of contact (79). An example of a systemic toxicant is ethanol, which is usually ingested. Although the ethanol is absorbed in the GI, the adverse effects are on the central nervous system and the liver; thus, it is a systemic toxicant.

Before proceeding to interpret toxicity studies, it is important to evaluate the chemical structure of the substance in an attempt to see whether it has characteristics suggesting that it may act like another chemical for which the toxicology is well understood. This is called a structure-activity relationship (SAR) evaluation. For example, toxicity data on phenol are useful for understanding benzene (phenol is a major metabolite of benzene). Likewise, information about trichloroethylene is useful for understanding some aspects of perchloroethylene. Conversely, although one might expect 1,1,2 trichloroethane to be similar in toxicity to 1,1,1 trichloroethane, the two chemicals have markedly different potency. Understanding why there are differences or similarities among chemicals can be helpful to setting proper OELs.

Beyond evaluating the SAR, it is also useful to assess the available mutagenicity data. For those systemic toxicants that are slightly positive in the Ames test or other tests for genotoxicity, it is often useful to provide a slightly greater margin of safety than might otherwise be indicated if a lifetime bioassay has not been conducted.

The process used to set either an OEL for an industrial chemical or an acceptable daily intake (ADI) for a food additive is quite similar. After evaluating the SAR, the genotoxicity battery, the human experience, and all of the applicable toxicity studies, one attempts to identify the dose that is unlikely to pose any significant risk to humans. Generally, an emphasis is placed on those studies of the longest duration for which an adequate number of animals were used.

Because inhalation is generally the primary route of exposure to industrial chemicals, tests that use inhalation data are favored over those in which the animal was exposed via ingestion or gavage. If test results for both ingestion (dietary) and inhalation studies are available, then the results should be compared on a mg/kg basis (or other relevant dose metric) so that "first pass" effects can be evaluated. "First pass" effects are those produced prior to having the chemical metabolized by the liver.

Conceptually, the various groups that set OELs use the same approach as that used by EPA to set an RfC for a systemic toxicant. First, all relevant animal studies are evaluated and are placed in order of strength or quality. Second, the best study involving the most relevant species is selected. Third, the NOEL from this study is identified. One caution is that a NOEL should be identified for an endpoint with genuine biological significance. For example, it is widely held that transient changes in liver enzymes or slight liver hypertrophy due to the challenge of dealing with a xenobiotic are not usually considered a significant adverse effect so, in these cases, they should not be used as the basis for deriving an OEL. Fourth, the results of the study should be compared with those from other studies to determine if there is a consistent and clear message from all data sets. Fifth, if little else is known, the most simplistic approach to setting an OEL for a systemic chemical is to divide the NOEL from a high quality chronic animal study by a factor of 100. If a different route of exposure than inhalation is used, route-to-route conversions can be made. In all cases, the size of the safety factor should be influenced by the steepness of the dose-response curve.

2.2.1 Example #2: One Approach to Setting a Chronic OEL

Assuming that the NOEL for rats exposed for 2 years to 1,1,1-trichloroethane is 1,000 ppm (vapor) and that for mice the NOEL was 1,500 ppm; what is a reasonable OEL? Assume that the primary adverse effect observed in animals is liver toxicity (a systemic effect). Lastly, epidemiology data suggest that human exposure in the workplace to 100 ppm for 8-hours/day for a lifetime poses no adverse effects.

$$OEL = \frac{NOEL}{(UF_1)(UF_2)}$$

HISTORY AND BIOLOGICAL BASIS OF OCCUPATIONAL EXPOSURE LIMITS 1919

where $NOEL$ = no observed effect level in animals (most sensitive species).
UF_1 = 5 to 10 (to account for animal to human differences).
UF_2 = 3 to 10 (to account for differences in sensitivity among humans).

Therefore:

$$OEL = \frac{NOEL}{(UF_1)(UF_2)}$$

$$OEL = \frac{1000 \text{ ppm}}{(5)(4)}$$

$$OEL = 50 \text{ ppm}$$

If the epidemiology data suggested that human experience at 100 ppm had been favorable, then one would have even greater confidence in the validity of the OEL.

2.3 Setting OELs for Sensory Irritants

Sensory irritants are those chemicals that produce temporary and undesirable effects on the eyes, nose or throat. These include such chemicals as ammonia, hydrogen sulfide, formaldehyde, sulfuric acid mist and dozens of others. The general TLV policy on irritants can be found in the *Introduction to the Chemical Substances* in the TLV booklet, "The basis on which the values are established may differ from substance to substance, protection against impairment of health may be a guiding factor for some, whereas reasonable freedom from irritation, narcosis, nuisance or other forms of stress may form the basis for others (3)." Nearly 50% of the published TLVs are based on avoidance of objectionable eye and upper respiratory tract irritation (Table 41.2).

The approach typically used by the TLV Committee has been to assign ceiling values (CV) to rapidly-acting irritants and to assign short term exposure limits (STELs) where the weight of evidence from irritation, bioaccumulation and other endpoints (e.g., central nervous system depression, increased respiratory tract illness, decreased pulmonary function, impaired clearance) combine to warrant such a limit. As noted by Willhite (81), the MAK Commission has utilized a five-category system based on intensive odor, local irritation and elimination half-life but this system is being replaced to be consistent with the European Union Scientific Committee for Occupational Exposure Limit Values (SCOEL). It is now likely that the MAK will assign a 15 minute STEL when necessary and feasible and where the underlying data are sufficiently developed to justify a compound-specific excursion factor to control body burden (e.g., carbon monoxide) (80–82). These changes lead, among other things, to modification of eight hour time-weighted averages (TWAs) and STELs to accommodate unusual work schedules, where compliance often rests with availability of Biological Exposure Indices (BEI). In each instance, simplicity and practical application are considered along with the relationship between absorbed dose (e.g., area-under-the-curve) and biological/medical endpoint of concern.

A significant debate among health professionals has taken place in recent years because some believe that transient irritation does not constitute material impairment of health while others contend that the TLVs should protect against any irritation (80–82). It is noteworthy that in 1989, as part of OSHA's effort to promulgate PELs, 79 materials were proposed for regulation based on avoidance of sensory irritation (81):

> The recognition of sensory irritation as potentially being "material impairment of health" is consistent with the current scientific consensus related to health effects of environmental agents. Mucous membrane irritants can cause increased blink frequency and tearing, nasal discharge, congestion, and sneezing and cough, sputum production, chest discomfort, sneezing, chest tightness and dyspnea. Work environments often require levels of physical and mental performance considerably greater than encountered in daily living. Even in the absence of any permanent impairment, the symptoms listed can interfere with job performance and safety. Mucous membrane irritation is associated with respiratory illness, depending on the composition of specific exposure and on the dose, duration and frequency of exposure. No universally applicable conclusion can be drawn at this time regarding the association between irritative symptoms and permanent injury or dysfunction. Where certain individuals show no measurable impairment after an exposure, even when experiencing irritative symptoms, others may develop identifiable dysfunction (52).

OSHA concluded that exposure to sensory irritants can cause inflammation and increase susceptibility to other irritants and infectious agents, lead to permanent injury or dysfunction and permit greater absorption of hazardous substances (81). OSHA (52) also concluded that "Exposing workers repeatedly to irritants at levels that cause subjective irritant effects may cause workers to become inured to the irritant warning properties of these substances and thus increase the risk of overexposure." TLV treatment and interpretation of dose–response relationships for irritants are consistent with the OSHA description. While the TLV Committee assumes a sigmoid concentration-response relationship for common irritants and believes a NOEL (below which the risk of experiencing irritation is trivial) can be identified for "nearly all workers," the Committee seldom has available a data set sufficiently robust so as to assign a specific level of risk. Even though correlations have been drawn, there is no widely accepted method for extrapolation of animal irritation data to human beings (80–82).

Historically, OELs for irritants were based on observations of the response of workers to various airborne concentrations measured by industrial hygienists. Basically, concentrations that did not produce irritation were recorded and this information was sent to the TLV Committee so they could recommend an appropriate OEL. Armed with whatever information the committee was sent over the next two years following the "notice of intended change," individual hygienists were expected to submit information regarding whether workers were experiencing some irritation at the proposed TLV. Following receipt of comments, the committee could then choose to make adjustments to the value. It is because of the importance of the subjective information gathered in the workplace that the TLV Committee has placed such reliance on the advice offered by the "consulting members" of the committee; e.g., those toxicologists and industrial hygienists who work in industry.

More recently, methodologies for setting OELs for sensory irritants have been based on the results of eye and skin irritation tests conducted in rabbit or rodent studies. Since sensory irritation is generally considered an acute response, animal tests involving limited durations of exposure are now generally considered to be adequate to identify "interim" safe levels of exposure. Many acute toxicology tests of irritation, however, often don't provide a dose metric, which is helpful since they involve directly applying a liquid or solid to the eye or the skin. For example, how does one convert 0.1 mL of a chemical placed in the eye to an airborne concentration? Thus, unless the animals are exposed via inhalation, toxicologists and physicians have had to make rough estimates about the likely airborne concentration that might prevent irritation. Generally, if a number of persons might be exposed, animal studies need to be conducted with vapors to see if adverse effects are observed.

Historically, UFs applied to NOELs obtained in studies when animals are exposed to vapors to set an OEL for humans to prevent sensory irritation have been rather small. Often, the overall UF has been as low as 2–5. There are a number of reasons why this has been the case. First, rabbits have been considered more susceptible to eye irritants than humans, so if only animal data were available, these tests have been considered "worst case." Second, mild eye irritation due to vapors, if it occurs in humans, is generally transient. Third, mild or slight irritation is often accommodated and, by definition, the effects are reversible. Another reason the margins of safety have been small is that, until recent years, most TLVs and other OELs were established after, rather than before, workers were exposed to the chemical. Thus, when these committees met to establish OELs there was generally a significant amount of information about the airborne concentrations known to produce eye, nose, throat, or lung irritation in workers. Thus, having some information about human response and some animal data, the TLV Committee apparently believed that they could set TLVs at values which didn't require large UFs.

Today, there is a greater expectation that nearly everyone should be protected against even minor sensory irritation. Therefore, in recent years, committees establishing OELs tend to apply UFs of 5–10 or more to an animal NOEL to set an OEL if human data are not available. The UF should vary depending on the slope of the dose-response curve. Of course, for the period during which a new OEL (such as a TLV) is "out for comment," it is expected that those who use the chemical will collect industrial hygiene data and that they will submit comments regarding the reasonableness of the proposed guideline. In some cases, such as for chlorine, formaldehyde, and other widely used chemicals, data from controlled human studies have been available (83). In those situations where human studies have been conducted, it has not been uncommon for an OEL setting group to apply a UF as low as 1–2 to a well understood human NOEL. Whenever possible, it is recommended that expert panels identify all the studies considered to be of good quality and to build a comprehensive dose-response curve (83). As shown in Figure 41.1 these can be very informative and they represent a weight of evidence approach.

2.3.1 Setting OELs for Irritants Using Models

Unlike the period 1940–1975, when OELs for irritants were based primarily on the human experience or simple tests with rabbits, we now have a reasonably reliable capability for predicting safe levels of exposure using models. Two kinds of models are available. One is based on tests that consider the response of rats and/or mice (84–89) to irritants. The second is based on chemical properties (66). Going forward, it is recommended that these be used to identify OELs that can be used until experience has been gained with workers.

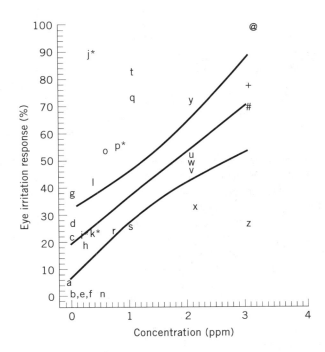

Figure 41.1. Linear concentration-response curve based on the data from various human studies regarding eye irration due to formaldehyde. Linear least-squares regression analysis of the data presented in Paustenbach et al. (Ref. 83), omitting the data for mobile home studies (points l^*, j^*, k^*, and p^*). The regression equation is %response = 19.6 + (17.4 × concentration in ppm): $n = 24$, r^2, = .45. The regression, that is positive slope, is significant ($p < .001$) and the 95% confidence interval for the regression lines are shown. The data points b, e, and f represent studies with zero response at zero concentration. The fit of the line does not vary appreciably if one fits the line with only the controlled human studies or all of the studies. From Ref. 83.

The first approach was developed at the University of Pittsburgh (84). It involves exposing rodents to various concentrations of contaminants and then measuring respiratory parameters. In this bioassay, mice are exposed to an airborne chemical and changes in their respiratory pattern are determined. For each chemical tested, the concentration capable of producing a 50% decrease in respiratory rate (RD_{50}) is obtained and its relative potency estimated. It is known that as the degree of irritation increases, the respiration rate decreases in rodents. Rodents, unlike other species, will decrease their metabolism to near death in order to avoid lung damage due to serious irritation.

In one of the more comprehensive papers, Schaper (89) described the success of their methodology, the RD_{50} approach, to accurately predict safe levels of human exposure for 89 chemicals. In this study, 295 such airborne materials, including single chemicals and mixtures, were found in the literature. A total of 154 RD_{50} values were obtained in male mice of various strains for the 89 chemicals in the database for which there were also

TLVs. An examination of the TLVs and RD_{50} values demonstrated, as previously with the smaller data set ($n = 40$), a high correlation ($r^2 = 0.78$) of the TLVs with $0.03 \times RD_{50}$. The authors concluded that these results supported the continued use of the animal bioassay for establishing exposure limits to prevent sensory irritation.

A second modeling approach, one based on chemical properties, has been used to set OELs for organic acids and bases, a class of chemicals that are well-known irritants (66). A generic method for understanding these chemicals was needed since only a few of the more than 40 chemicals in this class used in industry had OELs when the approach was developed. Although a great structural diversity exists among these chemicals, the primary biological effect produced by exposure to these materials is irritation. These researchers proposed that irritation should be related to their acidity or alkalinity. Since the strength of an organic acid or base is measured by its pK_a, it was shown that this term could be used to identify preliminary acceptable levels of human exposure.

As shown in Figures 41.2 and 41.3, the OELs for organic acids and bases correlate well ($r^2 \geq 0.80$) with pK. For organic acids and bases for which no OEL has been established, the following equations can be used to set a preliminary limit:

For organic acids: $\log OEL \ (\mu mol/m^3) = 0.43 \ pK_a + 0.53$

For organic bases: $OEL \ (\mu mol/m^3) = -200 \ pK_a + 2453$

Table 41.3 presents the OELs calculated with these formulae for a variety of organic acids and bases. A large number of corporate OELs have been adopted based on this approach.

In the coming years, it is quite likely that committees who set the TLVs and other OELs for the sensory irritants will recommend lower limits as society attempts to prevent even transient irritation from occurring in exposed workers (90).

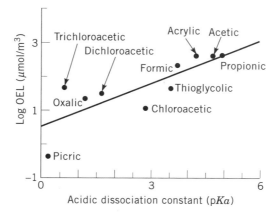

Figure 41.2. Correlation of OELs with equilibrium dissociation constants of organic acids. The correlation coefficient, $r = 0.80$. The regression equation is: $\log OEL \ (\mu mol/m^3) = 0.43 \ pK_a + 0.53$ (see Ref. 66).

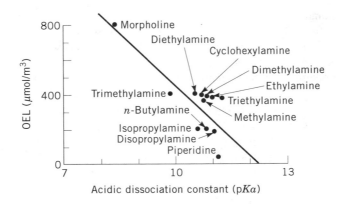

Figure 41.3. Correlation of OELs with equilibrium dissociation constants of organic bases. The correlation coefficient, $r = 0.81$. The regression equation is: OEL ($\mu mol/m^3$) = $-200\,pK_a + 2453$ (see Ref. 66).

Table 41.3. Occupational Exposure Limits for Selected High Volume Organic Acids and Bases Recommended by a Mathematical Formula Based on the Disassociation Constant[a,b]

Acid	mg/m³	ppm	Base	mg/m³	ppm
Acrylic	16	5	Allylamine	29	12
Butyric	35	10	Dialylamine	58	15
Caproic	49	10	Dibutylamine	43	8
Crotonic	30	8.5	Isobutylamine	21	7
Hepatanoic	55	10	Propylamine	21	7
Isobutyric	35	10	Trialylamine	109	20
Isocaproic	49	10			
Isovaleric	42	10			
Methacrylic	30	8.5			
Pentenoic	32	7.8			
Propiolic	1.5	0.5			
Valeric	42	10			

[a] From Ref. 66.
[b] Exposure limits were calculated by using the equations: acid: log OEL ($\mu mol/m^3$) = $0.43\,pK_a + 0.53$; base: OEL ($\mu mol/m^3$) = $-200\,pK_a + 2453$.

2.4 Setting Limits for Developmental Toxicants

Very few of the current OELs have been set by the ACGIH TLV Committee, the MAK, or the AIHA WEEL committee with the primary objective of preventing developmental effects. One reason is that relatively few of the 1,200 or so chemicals that have OELs have been tested in the standard Segment II test battery for assessing developmental toxicity and, of those chemicals tested, only a fraction have been found to be selectively toxic to the developing fetus (91–93). However, in recent years it has been recognized that OELs that specifically protect against developmental toxicity need to be established, and there has been an active dialogue within the toxicology community about the appropriate approach for estimating safe levels of human exposure based on animal data (94–97).

For many years, due to the technical difficulties involved in the safety evaluation/extrapolation process used to set OELs for developmental toxicants, toxicologists in industry typically recommended that women of child-bearing age not be placed in locations where exposure to these agents could occur. However, during the 1970s, in an attempt to satisfy federal labor laws in the United States and to give women greater access to higher paying jobs, it was no longer considered acceptable to restrict women from jobs where exposure to the truly significant developmental toxicants was possible. As a result, OELs must now address this hazard. The methodologies for setting OELs for these agents continues to evolve, as much due to changes in societal values about the required size of the margin of safety as to the changes in our scientific understanding of the developmental hazard (95–97).

By definition, a developmental toxicant is a chemical that can produce an adverse effect on the developing fetus (94). The range of possible adverse effects on development covers a very broad spectrum spanning small changes in birth weight to gross teratologic effects. The decision to classify a chemical as a developmental toxicant is clouded by the fact that at some dose, virtually all chemicals will produce an adverse effect on developing offspring (98). To complicate matters further, there are very significant differences in the susceptibility to various developmental effects among the various animal species and humans. There are also differences in the percentage of naturally occurring defects among rodent and non-rodent species, and humans (92). Lastly, when attempting to identify a reasonable OEL, one must assume that many of these effects are not reversible. For these and other reasons, it is not surprising that a great deal of deliberation needs to occur when attempting to identify safe levels of exposure to developmental toxicants (99).

In general, the current approach to setting an OEL for these agents is much like that used by FDA to identify acceptable exposure to certain drugs that might pose a developmental hazard. First, the critical NOEL observed in a well conducted Segment II developmental toxicity study is identified. A Segment II test involves exposing a rabbit and a relevant species to three or four doses of a toxicant. About two weeks prior to pregnancy, females are first exposed. Exposures continue throughout the pregnancy. About three days before delivery, the pups are removed via C-section and then examined for detrimental effects (93). Two species are always evaluated in these tests so two NOELs are produced. If both are similar, the lower of the two is used in the safety evaluation. If they are not similar, then a careful review of the specific adverse effects and the differences in metabolism between the species and human needs to be conducted. If known, the species thought to provide information most relevant to humans should be used. Second, one or more UFs should be applied to the animal NOEL. Historically, when setting tolerances for food additives and pesticide residues, an UC of 100 has been applied to the NOEL observed in the most sensitive species exposed in a Segment II test to identify the ADI. The size of the UF applied to developmental studies (like Segment II) used to set most OELs appears to have varied over the years from 10–50 (99). For some chemicals, the apparent UC incorporated into some TLVs has been smaller, depending upon the strength of the animal data, the mechanistic data, the warning properties of the chemical, and other factors.

Because developmental effects have a threshold, it is anticipated that the UF or benchmark dose approach will continue to be used to identify safe levels of exposure. The benchmark dose approach is a hybrid method (i.e., relies on both low-dose modeling and the safety factor approach) that has certain benefits versus either the modeling or UF approaches (100). Thus far, no OELs have been based on the benchmark dose approach but it should be one of the methods considered by groups who set limits in the coming years. As different techniques for identifying more sensitive developmental effects evolve, and as societal values change, it is likely that the OELs for these chemicals will become smaller.

2.5 Setting Limits for Reproductive Toxicants

"Reproductive dysfunction" can be broadly defined to include all effects resulting from paternal or maternal exposure that interfere with the conception, development, birth, and maturation of offspring to healthy adult life (93). For purposes of this discussion, reproductive effects are those that impair the ability of a male or female to produce offspring (102). The relation between exposure and reproductive dysfunction is highly complex because exposure of the mother, the father, or both may influence reproductive outcome. In addition, exposures may have occurred at some time in the past, immediately prior to conception, or during gestation. For some specific dysfunctions, the relevant period of exposure can be identified, and for others it cannot. For example, chromosomal abnormalities detected in the embryo can arise from mutations in the germ cells of either parent prior to conception or at fertilization, or from direct exposure of embryonic tissues during gestation. Major malformations, however, usually occur with exposure during a discrete period of pregnancy, extending from the third to the eighth week of human development (93).

While historically the bulk of interest has been on female reproductive function, the precise male contribution to reproductive failure and adverse pregnancy outcomes, although often unknown, is considered to be significant (102). When evaluating males, attention is focused primarily on toxic effects that involve testicular and postspermatogenic processes that are essential for reproductive success. Male reproductive failure resulting from germ cell mutation (i.e., genotoxicity), the role of the endocrine system in the support of reproductive function, and female reproductive toxicity are all important variables. Like the developmental toxicants, there have been few OELs established for chemicals whose primary adverse effect is reproductive toxicity. These are at least two reasons. First, not a great many industrial chemicals have been tested in the classic male or female reproductive toxicity batteries (93, 102). Second, of those chemicals tested, few have been shown to produce adverse effects selectively on reproduction at concentrations or doses lower than those known to produce significant adverse effects on other organs; thus, this is infrequently the "driving" health effect.

Similar to the historical approaches used to identify OELs that protect against other adverse effects, it appears that UFs in the range of 10–100 applied to a NOEL from well-conducted animal studies have been considered adequate to protect humans from reproductive effects. For example, the OSHA PEL for dibromochloropropane is 1 ppb and this appears to be based on applying an UF of less than 100 to the NOEL for adverse effects observed in rodents. Because there are a limited number of reproductive toxicants for which OELs have been set and there have been few long term follow-up studies of workers exposed to these agents, it is unclear whether UFs closer to 10, 30, 100 or slightly higher are the "best ones" for this class of chemicals. As with other adverse effects, the magnitude of the various UFs used to identify an OEL should be directly related to the strength of the animal and human data, the severity of the adverse effects, the reversibility of the effect, as well as the relationship to doses that cause other toxic effects.

2.6 Setting Limits for Neurotoxic Agents

Chemicals that can produce permanent neurological damage often present significant concern to toxicologists and physicians. This is because many neurotoxins can produce permanent damage at doses that produce no other adverse effects (116). For this reason, the FDA has traditionally regulated in a fairly aggressive manner neurotoxic agents that can enter the food chain. Many agencies, including the EPA and OSHA, often regulate neurotoxicants as stringently as the carcinogens. That is, they tend to apply large UFs to animal data.

It appears that UFs in the region of 5–100 have been typically applied to animal NOELs from high quality tests to set OELs. As in the setting of other OELs, a large degree of professional judgement is needed to identify the appropriate value. It is noteworthy that at times FDA has applied UFs as great as 2,000 to animal data to set acceptable levels of exposure to residues in foods of certain pesticides that cause permanent neurotoxicity, like dying-back neuropathies. Because most of the OELs for the neurotoxins are based on human experience and because of the fairly large database on these chemicals (like parathion), it appears that this is the justification for the relatively low UFs that have historically been used.

2.7 Setting Limits for Aesthetically Displeasing Agents and Odors

The process or approach to setting an OEL for a chemical that tends to have a low odor threshold or for those chemicals which are aesthetically displeasing has been fairly simple. One of two approaches is used. In the first, if the agent has an airborne odor threshold much lower than the concentration that produces even subtle toxic effects, the agent is categorized as one that has "self limiting" exposure characteristics. That is, because the odor is so objectionable, workers are not going to allow themselves to be placed at risk of injury due to exposure (unless they are exposed in a confined space without easy egress). These chemicals generally do not pose much of a concern to toxicologists or industrial hygienists as long as the agent doesn't cause rapid olfactory fatigue.

In the second approach, one usually identifies the airborne concentration at which most persons find the odor of the chemical objectionable and then divides that concentration by a factor of 2 or 3 to establish a preliminary OEL. As with other "preliminary OELs," during the two year period for receiving comments, the value can be raised or lowered based on the feedback from workers and industrial hygienists.

Historically, human experience in the workplace has been used to identify the concentration at which most persons recognized an odor (generally, one only focuses on the concentration that is found objectionable rather than simply detectable) and then the OEL was established. In recent times, some firms have used odor or irritation panels to identify the concentration at which detection occurs, as well as those concentrations where the odor is considered objectionable. Such panels include men and women of various ages since both sex and age are known to affect the threshold of smell. From the results of the odor panel, a concentration that is likely to be acceptable to most persons can be identified (82).

One aspect of setting an OEL for this class of chemicals that requires attention is the phenomenon of "accommodation." Accommodation is usually differentiated from olfactory fatigue in that accommodation means that with continuing exposure throughout the day, the objectionable nature of the odor diminishes. For example, many persons who work in factories that use chemicals like amyl alcohol initially find the odor objectionable but within 5–10 minutes, and for the remainder of the workday, the workers do not even recognize that it is present. Some have claimed that the fatigue provides an opportunity for chronic cellular irritation; thus, in their view accommodation is not a beneficial response. Although there is no commonly accepted approach for dealing with chemicals with this property, it is important that the concentration selected as the OEL is a fraction of that known to produce toxic effects (even if the odor is tolerable).

2.8 Setting Limits for Persistent Chemicals

In general, those groups involved in setting OELs have not attempted to quantitatively account for the differences in the pharmacokinetics of chemicals. That is, it has usually been assumed that chemicals with very long half-lives in animals will also have long half-lives in humans and that this is accounted for in the results of chronic animal studies.

In recent years, more attention has been focused on chemicals with very long half-lives and toxicologists now know that differences in the elimination between animals and humans can be substantial (even when relative life expectancies are considered). For example, the difference in the biologic half-life for dioxin between rodents and humans is sufficiently great that the steady-state blood concentrations at a given dose are quite different (103). For the so-called "long lived or persistent" chemicals, which in the late 1990s were called persistent organic pollutants (POPs), it is prudent to rely upon PB-PK models to account for interspecies differences when deriving OELs (especially those based on animal data). The basis for and the benefits of the PB-PK approach are discussed in the chapter on *Pharmacokinetics Schedules and Unusual Work Schedules* in Chapter 40 of this text.

An approach to setting OELs for persistent chemicals has been described by Leung et al. (103) and it is worthy of evaluation. Basically, the methodology incorporates information on the biological half-life of the chemical in humans (the pharmacokinetics), as well as the background concentration of the chemical in humans due to contamination of the food chain. The approach assumes that the biological half-life of a chemical in humans can be predicted based on animal data using a PB–PK model when human data are not available. These researchers reasoned that if the amount of chemical absorbed due to workplace exposure is about the amount every American ingests every day, or if the total uptake (occupational or dietary) is much lower than the NOEL, then the occupational contribution is unlikely to pose significant increased health risks.

This approach was applied to setting an OEL for tetrachlorodibenzo-*p*-dioxin [2,3,7,8-TCDD (dioxin)] because a good deal of toxicology information was available. Since TCDD is highly lipophilic and has a long biologic half-life in humans, it is expected to accumulate in adipose tissue with repeated daily exposure. TCDD levels in the adipose tissue of nonoccupationally exposed adults in the United States have been about 7 ppt.

The steady-state level of TCDD in adipose tissue resulting from exposure to an OEL of 200 pg/m³ (the value proposed in Ref. 103), for example, can be estimated by:

$$\text{Steady-state concentration} = \frac{1.44(D_t)(t_{1/2})}{10.5 + (59.5/10)}$$

where D_t = the daily intake
$t_{1/2}$ = biologic half-life (years)

This calculation assumes that the TCDD concentration in the liver and other tissues is about $1/10$ that in adipose tissue, and the average human weighs 70 kg with 10.5 kg (15%) body fat. If the half-life is assumed to be about 8 years, the steady-state TCDD concentration in adipose tissue resulting from occupational exposure at an airborne concentration of 200 pg/m³ will be 89–179 ppt. Thus, exposure to this OEL for 40 years could raise the steady-state body burden well above the 7 ppt background concentration (103).

Even though the increase in body burden for TCDD for an OEL of 200 pg/m³ is much greater than due to diet, the dose should not immediately be considered significant. The authors then began to evaluate other factors and they presented the following rationale for concluding that the OEL was reasonable. First, the concentration of TCDD measured in the adipose tissue of rats exposed for two years to the NOEL of 0.001 μg/kg-day was 540 ppt. Since humans sequester more TCDD in adipose tissue than lower species, which has been speculated by some scientists to lessen the toxic hazard, a comparable level in human fat should yield a lesser risk than that suggested in rodent studies. Second, humans exposed to 16 mg TCDD had a theoretical peak adipose tissue level of about 1,300 ppb (16 mg/12.25 kg), yet they only developed chloracne, which resolved within 6 months. None of those who received 8 μg TCDD developed chloracne, yet their peak adipose tissue levels were about 650 ppt (8 μg/12.25 kg).

The third factor these authors (103) considered was the relationship between the adipose tissue concentration for other persistent chemicals in the diet which also had OELs. Table 41.4 shows that the 26-fold increase over background (179 ppt/7 ppt) for TCDD, when compared with other industrial chemicals following workplace exposure at their corresponding TLVs, appears to be comparable. Fourth, and most importantly, the risk associated with the proposed OEL was 100-fold below the animal NOEL for carcinogenicity (i.e., 10 pg/kg-day) and this was thought to pose no significant human health hazard (103). Thus, from these data in 1988, the risk to humans appeared rather small (especially following a comparison with OELs for other persistent chemicals). Whether their conclusion about what constituted a safe dose of dioxin remains accurate today is irrelevant. The key

Table 41.4. Estimated Steady-State Adipose Tissue Concentration of Chemicals Following Chronic Exposure at the OEL Compared With the Levels Due to Background Exposure Alone[a]

Chemical	OEL	$t_{1/2}$ (yr)	Adipose Tissue Level		E/B^d
			Background[b]	Exposed[c]	
DDT[e]	1 mg/m³	1.5	6 ppm	480 ppm	80
Dieldrin	0.25 mg/m³	1	0.29 ppm	80 ppm	276
PCB[f]	1 mg/m³	2.5	1 ppm	800 ppm	800
TCDD[g]	0.2 ng/m[h]	8	7 ppt	180 ppt	26

[a]From Ref. 103.
[b]Background levels refer to those in nonoccupationally exposed general population.
[c]The levels in persons occupationally exposed to the OEL are calculated with the equation presented in Calabrese (1978).
[d]E/B = ratio of predicted steady-state adipose tissue level in persons occupationally exposed at the current TLV versus that measured in persons exposed to background levels.
[e]DDT-dichlorodiphenyl trichloroethane.
[f]PCB-polychlorinated biphenyl.
[g]TCDD-2,3,7,8-tetrachlorodibenzo-p-dioxin.
[h]OEL is the value suggested by Leung et al. (103).

point is that the methodology is a useful one for evaluating the hazard posed by persistent chemicals.

2.9 Setting Limits for Respiratory Sensitizers

Respiratory sensitization is an immune status whereas respiratory allergy is a clinical manifestation. Respiratory sensitization results from an immune response to antigen (usually, but not exclusively, exogenous antigen), which may result in clinical hypersensitivity upon subsequent inhalation exposure to the same or similar antigen. An allergic or sensitization response characteristically requires at least two encounters with antigen. Following first exposure, the susceptible individual mounts a primary immune response, which results in sensitization (the induction or sensitization phase). If the sensitized individual subsequently comes into contact with the same antigen, a clinical allergic reaction may be provoked (the elicitation phase). Allergic reactions may be attributable to either antibody or cell-mediated immune responses. Acute allergic reactions in the respiratory tract induced by exposure to exogenous antigens (e.g., some industrial chemicals) are almost invariably associated with specific antibody responses, frequently, but not always, of the IgE class (104).

Certain chemicals can produce an allergic response as a result of either dermal or inhalation exposure (105). The reason for the interest is that this class of chemicals can, after a sensitizing exposure (called induction) occurs, produce an adverse effect with subsequent exposures to very small quantities. Respiratory sensitizers can produce asthma in select persons and so-called "attacks or incidents" can be fatal if untreated.

In the late 1990s, respiratory sensitizers received perhaps the most attention of all categories of toxicants with respect to setting OELs. The concern about inhalation sensitizers or allergens came about because researchers reported that the incidence of asthma in children and adults appeared to be increasing dramatically in the United States and other Western countries. A few years ago, the German MAK began to identify with notation those chemicals that were known or suspected inhalation sensitizers. In 1996, the ACGIH TLV® committee chose to pursue the same approach and has recently attached a similar notation to chemicals for which they believe a sensitization hazard exists.

The toxicology community has made significant headway in developing methods for identifying likely dermal and respiratory sensitizers over the past 10 years. Since exposure to most dermal sensitizers is prevented by gloves and other personal protective equipment, the focus of the TLV and MAK committees has been on respiratory sensitizers. Fortunately, a model that relies on SAR has been developed for screening groups of chemicals to identify sensitizers (106) thus making the task a manageable one. The SAR model relies upon identifying chemical moiety in a substance that is known to increase the probability that it will be a respiratory sensitizer, like an aldehyde or cyano group. This model has been applied to nearly 100 different chemicals.

In addition to SAR models there are *in vitro* and *in vivo* test methods to identify sensitizers. The most common approach includes an *in vitro* assessment of protein binding potential, followed by *in vivo* evaluation using an animal model (106). Diverse species have been used including mice, rats, and guinea pigs, each possessing distinct advantages and disadvantages in representing human disease (106). The guinea pig model (106) assesses several hypersensitivity responses such as early and late airway reactions, airway hyperreactivity, production of allergen-specific IgE and IgG_1 antibodies, and eosinophilic inflammation. However, the model is costly and requires a high degree of technical skill (105). Mouse models have been described that associate an increase in total IgE with respiratory sensitizers or characterize the cytokines produced following exposure to chemical allergens. Each of the animal models has been tested with only a limited number of chemicals and requires further validation. An excellent paper that reviews current views on toxicology testing of respiratory sensitizers has been published (104).

Of all the tests, the guinea pig sensitization test has been used most frequently and refinements in the procedure have made it a much more powerful and reliable tool for identifying likely human sensitizers (104). Although a related test relies upon administration of an agent via dermal contact, the most reliable way to identify a likely occupational allergen is through inhalation testing. Using the results of the testing, it has been shown that "safe" levels of exposure can be identified by comparing the test results on a new chemical versus the results obtained with a known occupational sensitizer that has a TLV. Using a simple ratio approach, an OEL for the "new" chemical can be calculated directly from the animal data. This approach has been discussed by Graham et al. (106).

2.10 Setting Limits for Chemical Carcinogens

Carcinogen is the term applied to a chemical that has been shown to produce a significant increase in the occurrence of tumors (above background) in an appropriately designed and executed animal study or has been shown to produce an increase in the incidence of cancer in a human population. Chemical carcinogens have been the focus of many, if not most, environmental and occupational regulations for the past 25 years.

In the United States, the impetus to have the TLV Committee develop a classification scheme for occupational carcinogens began in 1970. At that time, lists were routinely published by numerous agencies and different groups who claimed that a large number of substances were likely to be occupational carcinogens. As noted by Stokinger (57), substances of purely laboratory curiosity, such as acetylaminofluorene (AAF) and dimethylaminobenzene, which were found to be tumorigenic in animals, were classed along with known human carcinogens of high potency and individual significance, such as bischloromethyl ether (BCME). In short, no distinction was made between an animal tumorigen and a likely human carcinogen. Union leaders, workers and the public would often become worried equally about the positive results of animal bioassays of different chemicals even though for a given dose the carcinogenic or mutagenic potency could vary by 1,000,000-fold (57).

During the 1970s and 1980s, the TLV Committee believed that the finding of a substance to be tumorigenic, often in a half-dead mouse or rat administered intolerable doses, as was the case for chloroform and trichloroethylene, was not suggestive evidence that it was likely to be carcinogenic in humans under controlled working conditions. It is for this reason that the ACGIH Chemical Substance TLV Committee, as early as 1972, made a clear distinction between animal and human carcinogens. As time has passed, the TLV Committee has stood firm that not all carcinogens pose an equal hazard, even when the potency of two chemicals may be equally great. One reason, among others, was that some chemicals are mutagenic or genotoxic while others produce tumors through epigenetic mechanisms (108, 109). By setting exposure limits, the ACGIH, as well as its sister organizations (like the MAK and the Health Executive of the United Kingdom) adopted the view that all chemical carcinogens should at least have a "practical threshold." This term simply means that some dose of these agents would not be expected to pose a significant cancer risk. The basic concept behind the term "practical threshold" is that humans, as they evolved mechanisms for handling naturally occurring carcinogens in the diet, these same mechanisms will detoxify small amounts of industrial chemicals (109).

In 1977, Herbert Stokinger, then chairman of the ACGIH TLV Committee, summarized the historical philosophy of the ACGIH with respect to TLVs for carcinogens:

> Experience and research findings still support the contention that TLVs make sense for carcinogens. First and foremost, the TLV Committee recognizes practical thresholds for chemical carcinogens in the workplace, and secondly, for those substances with a designated threshold, that the risk of cancer from a worker's occupation is negligible, provided exposure is below the stipulated limit. There is no evidence to date that cancer will develop from exposure during a working lifetime below the limit for any of those substances.
>
> Where did the TLV Committee get the idea that thresholds exist for carcinogens? We have been asked 'Where is the evidence?' . . . Well, the Committee thinks it has such evidence, and here it is.
>
> It takes three forms:
>
> 1. Evidence from epidemiologic studies of industrial plant experience, and from well-designed carcinogenic studies in animals,

2. Indisputable biochemical, pharmacokinetic, and toxicologic evidence demonstrating inherent, built-in anticarcinogenic processes in our bodies,
3. Accumulated biochemical knowledge makes the threshold concept the only plausible concept (57).

Although these comments were written about 20 years ago, a large fraction of industrial hygienists, industrial toxicologists, and occupational physicians generally continue to agree with Dr. Stokinger's position. This has been due, in part, to the work of Ames et al. (109) who have shown that man's diet is abundant with chemical carcinogens and that humans have clearly developed adequate mechanisms and this is generally validated by the continuing decrease in the incidence of cancer in the 20th century in spite of the tremendous increase in the amount of xenobiotics in our environment.

In recent times, the TLV and MAK Committees have attempted to keep pace with the increased understanding of the hazards posed by chemical carcinogens. For example, beginning in about 1985, they considered not only the results of mathematical models used to estimate response at low doses but also *in vitro* data, information on the mechanisms of action, case-reports, genotoxicity data, and other information on chemical carcinogens before setting a particular TLV. Evidence that such discussions occurred is presented in the documentation for the TLV for trichloroethylene, methylene chloride, 1,1,1-trichloroethane, and others that were revised between 1985 and 1995. Due to the variability in risk estimates between the various statistical or low dose models (e.g., Weibull, multistage) and their inability to incorporate biological repair mechanisms, the TLV Committee has been reluctant to place much emphasis on their results, and thus far, has not set a TLV based on the results of low dose models.

As noted in the current TLV booklet (3), when deciding on values for chemical carcinogens, the Chemical Substance Group within the TLV Committee gives greatest weight to epidemiologic studies based on good quantitative exposure data. When the weight of evidence is convincing, certain chemicals will receive an A1 categorization and these are called Confirmed Human Carcinogens. Next in importance, and more typically available, are positive bioassays involving rats or mice (but lacking human data). Such substances are given an A2 designation and are called "Suspected Human Carcinogens." In reviewing the key experimental toxicology studies, the Committee considers route of entry (greatest weight given to inhalation studies), dose–response gradient, potency, mechanism of action, cancer site, time-to-tumor, length of exposure, and underlying incidence rate for the type of cancer and species under study. Replication of results is important to the committee, especially if comparable results are obtained in different species. Other types of information, such as batteries of genetic toxicity studies, are useful in confirming that a substance is a carcinogen but are not usually helpful in setting a TLV.

Appendix A of the annual TLV booklet contains a description of categories into which chemical carcinogens have been placed (3). The goal of the Chemical Substances TLV Committee has been to synthesize the available information in a manner that will be useful to practicing industrial hygienists without overburdening them with needless details. The Committee reviewed current methods of classification used by other groups and in 1991 developed a new procedure for classification (111). This was generally accepted in 1992 and the following categories for occupational carcinogens are currently used by the TLV Committee (3):

- *A1—Confirmed Human Carcinogen*: The agent is carcinogenic to humans based on the weight of evidence from epidemiologic studies.
- *A2—Suspected Human Carcinogen*: Human data are accepted as adequate in quality but are conflicting or insufficient to classify the agent as a confirmed human carcinogen; OR, the agent is carcinogenic in experimental animals at dose(s), by route(s) of exposure, at site(s), of histologic type(s), or by mechanism(s) considered relevant to worker exposure. The A2 is used primarily when there is limited evidence of carcinogenicity in humans and sufficient evidence of carcinogenicity in experimental animals with relevance to humans.
- *A3—Confirmed Animal Carcinogen with Unknown Relevance to Humans*: The agent is carcinogenic in experimental animals at a relatively high dose, by route(s) of administration, at site(s), of histologic type(s), or by mechanism(s) that may not be relevant to worker exposure. Available epidemiologic studies do not confirm an increased risk of cancer in exposed humans. Available evidence does not suggest that the agent is likely to cause cancer in humans except under uncommon or unlikely routes or levels of exposure.
- *A4—Not Classifiable as a Human Carcinogen*: Agents about which there is concern that they could be carcinogenic for humans, but that cannot be assessed conclusively because of a lack of data. *In vitro* or animal studies do not provide indications of carcinogenicity sufficient to classify the agent into one of the other categories.
- *A5—Not Suspected as a Human Carcinogen*: The agent is not suspected to be a human carcinogen on the basis of properly conducted epidemiologic studies in humans. These studies have sufficiently long follow-up, reliable exposure histories, sufficiently high dose, and adequate statistical power to conclude that exposure to the agent does not convey a significant risk of cancer to humans; OR, the evidence suggesting a lack of carcinogenicity in experimental animals is supported by mechanistic data.
- Substances for which no human or experimental animal carcinogenic data have been reported are assigned no carcinogenicity designation.
- Exposures to carcinogens must be kept to a minimum. Workers exposed to A1 carcinogens without a TLV should be properly equipped to eliminate to the fullest extent possible all exposure to the carcinogen. For A1 carcinogens with a TLV and for A2 and A3 carcinogens, worker exposure by all routes should be carefully controlled to levels as low as possible below the TLV. Refer to the "Guidelines for the Classification of Occupational Carcinogens" in the Introduction to the *Documentation of the Threshold Limit Values and Biological Exposure Indices* for a more complete description and derivation of these designations.

The TLV Committee continues to evaluate the mechanisms through which various chemical carcinogens act and they are seeking improved methods for identifying more accurate guidelines (3, 5). In the future, for example, it is possible that the TLV Committee may place more emphasis on model derived cancer risk estimates for certain genotoxic agents (111, 112) rather than on the UF approach. As evidenced in their deliberations on benzene, formaldehyde, and vinyl chloride, the committee has tried to consider PB-PK models, controlled human studies, mechanisms of action data, pharmacokinetic data, and other relevant information when attempting to identify the appropriate TLV.

2.11 Two Approaches for Identifying OELs for Carcinogens

Even though the ACGIH TLV Committee, as well as many other groups that recommend OELs, may believe that there is likely to be a threshold for carcinogens at very low doses, another school of thought is that there is little or no evidence for the existence of thresholds for chemicals that are genotoxic (26, 110–113). In an attempt to take into account the philosophical postulate that chemical carcinogens do not have a threshold even though a NOEL can be identified in an animal experiment and because a test involving several hundred animals cannot describe the large differences among humans in the general population, modeling approaches to estimate the possible cancer risk to humans exposed to very low doses have been developed (28, 112–114).

The rationale for a modeling approach to identify safe levels of exposure is that it is impossible to conduct toxicity studies at doses near those measured in the environment because the number of animals necessary to elicit a response at such low doses would be too great (113). Consequently, results of animal studies conducted at high doses are extrapolated by these statistical models to those levels (e.g., doses) found in the workplace or the environment. By the early 1980s, mathematical modeling approaches for evaluating the risks of exposure to carcinogens were relied upon by various regulatory agencies who were attempting to protect the public; these models rapidly identified doses that almost certainly posed no health hazard. Interestingly, the limits derived by these models have rarely been the sole factor by which environmental regulatory limits have been established (42, 43, 115).

The most popular models for low dose extrapolation are the one-hit, multi-stage, Weibull, multi-hit, logit and probit. The pros and cons of these models have been discussed in many papers. Since it is usually presumed in these models that at any dose, no matter how small, a response could occur in a sufficiently large population, an arbitrary increased lifetime cancer risk level is usually selected (i.e., from 1 in 1,000 to 1 in 1,000,000) as presenting an insignificant or de-minimus level of risk. By identifying these de-minimus levels as virtually safe doses, regulatory agencies do not give the impression that there is an absolutely "safe" level of exposure or that there is a threshold below which no response would be expected. This has historically been considered prudent (110, 115).

Often the use of these statistical models to help assess risks of exposure to carcinogens has been erroneously called "risk assessment" (38, 115). In practice, modeling is only one part of the risk assessment process. A good dose-response assessment whose purpose is to help identify safe levels of occupational exposure requires exhaustive analysis of all of the information obtained from studies of mutagenicity, acute toxicity, subchronic toxicity, chronic studies in animals and metabolism data, human epidemiology data, and an understanding of the role of dermal uptake (115).

At this time in the evolution of our understanding of the cancer process, most scientists would support using quantitative risk modeling only as providing an additional piece of information to consider when setting an OEL. Because there are dozens of shortcomings associated with the models, especially their inability to consider complex biological events that undoubtedly occur at low doses, they have not been used as the sole basis for deriving occupational exposure limits.

Several papers have compared the model predicted upper bound cancer risk for workers exposed to TLV concentrations of several chemicals with risks often deemed acceptable by the EPA and FDA (43, 111). The results are presented in Table 41.5. As shown, the theoretical cancer risk for exposure to many, if not most, occupational carcinogens at the current OSHA PELs is about 1 in 1,000 rather than 1 in 1,000,000 (the goal of many environmental regulations). Some TLVs for carcinogens have model predicted risks as high as 1 in 100. The degree of acceptable exposure to carcinogens in the workplace considered "safe or acceptable" is even more interesting when one considers the estimated steady-state tissue concentration in humans following chronic exposure to the TLV vs. that due to background exposure to that chemical in our diet (Table 41.4).

The principal reason for the wide disparity between the ambient air guidelines recommended by the U.S. EPA (which attempts to limit the model-predicted cancer risk to 1 in 10,000 to 1 in 1,000,000) and the workplace values recommended by the TLV Committee can be explained primarily by the underlying philosophical principles governing the two organizations and the differences in the exposed populations, rather than the technical differences between the two methods for identifying "safe doses." The TLV Committee is governed by the precept that "Threshold Limit Values refer to airborne concentrations of substances and represent conditions under which it is believed that nearly all workers may be repeatedly exposed day after day without adverse effect" (3). Occupational exposure involves 40 hrs/week for about 40 years. In contrast, the U.S. EPA's Clean Air Act (CAA) and other environmental regulations are intended to ensure that virtually all members of the public are exposed to virtually insignificant risks. For example, the CAA states that air standards must protect the public health with an adequate margin of safety. The requirement for an "adequate margin of safety" is intended both to account for inconclusive scientific and technical information and to provide a reasonable degree of protection against hazards that research has not yet identified. The TLVs define "adequate margin of safety" differently from EPA since healthy workers allegedly make up the bulk of the workforce, e.g., those who report to work each day must be more healthy than the general population. Environmental exposure is assumed to occur continuously for 168 hrs/week for 70 years.

Table 41.5. Model Derived Estimates of Lifetime Risks of Death from Cancer per 1000 Exposed Persons Associated with Occupational Exposure at Pre-1986 and Post-1987 OSHA Permissible Exposure Limits (PELs) for Selected Substances[a]

Substance	Cases/1000 at Previous PEL	Cases/1000 at Revised PEL
Inorganic arsenic	148–767	8
Ethylene oxide	63–109	1–2
Ethylene dibromide (proposal)	70–110	0.2–6
Benzene (proposal)	44–152	5–16
Acrylonitrile	390	39
Dibromochloropropane (DBCP)	—	2
Asbestos	64	6.7

[a]Table reprinted from Rodricks et al. (42), reprinted with permission of *Regulatory Toxicology and Pharmacology*.

With respect to setting environmental standards, the use of conservative low-dose extrapolation models and the adoption of a 1 in 100,000 or 1 in 1,000,000 risk criterion have been justified because of a strong desire to protect virtually everyone in the public (e.g., the aged, young, and infirm), and to account for the fact that the public can be continually exposed for 70 years rather than a 40 year working lifetime. Due to the very different populations at risk and the fact that workers are sometimes compensated for accepting certain risks, it has been considered reasonable that the approaches used to set various limits (as well as the risk criteria) are different.

2.12 Setting Limits For Mixtures

The whole topic of setting OELs for mixtures began to be re-evaluated by the toxicology community in the middle 1990s and it can be expected that it will receive a good deal of discussion over the next ten years (117). Historically, the ACGIH TLV Committee approach has been to consider chemicals that act on the same target organ or act through the same mechanism of action as being additive with respect to their hazard. A number of meetings of experts were held in the 1970s through the 1990s to reassess this approach and, in the main, it was concluded that the methodology described in the TLV booklet was adequate (if not amply health protective) (118).

The approach currently recommended by the TLV Committee is as follows (taken from the recent TLV booklet) (3):

When two or more hazardous substances which act upon the same organ system are present, their combined effect, rather than that of either individually, should be given primary consideration. In the absence of information to the contrary, the effects of the different hazards should be considered as additive. That is, if the sum

$$\frac{C_1}{T_1} + \frac{C_2}{T_2} + \ldots \frac{C_n}{T_n}$$

exceeds unity, then the threshold limit of the mixture should be considered as being exceeded. C_1 is the observed atmospheric concentration and T_1 the corresponding threshold limit (see Example A.1 and B.1 in the TLV Booklet).

Exceptions to the above rule may be made when there is a good reason to believe that the chief effects of the different harmful substances are not in fact additive, but are independent as when purely local effects on different organs of the body are produced by the various components of the mixture. In such cases, the threshold limit ordinarily is exceeded only when at least one member of the series (C_1/T_1 + or + C_2/T_2, etc.) itself has a value exceeding unity.

Synergistic action or potentiation may occur with some combinations of atmospheric contaminants. Such cases, at present, must be determined individually since no model is available for predicting when this could occur. Potentiating or synergistic agents are not necessarily harmful by themselves and dose is a critical factor when one attempts to prohibit the health hazard. Potentiating effects of exposure to such agents by routes other than that of inhalation are also possible, e.g., imbibed alcohol and inhaled narcotic agents (trichloroethylene). Potentiation is characteristically exhibited at high concentrations, less probable at low concentrations.

As described in the ACGIH TLV booklet, the formulae apply only when the components in a mixture have similar toxicologic effects. The only exceptions involve sensory irritants or chemicals that are aesthetically displeasing since additivity of adverse effects should not occur. They should not be used for mixtures with widely differing reactivities or mechanisms of toxic action, e.g., hydrogen cyanide and sulfur dioxide. In such case, the formula for independent effects should be used. It is essential that the atmosphere be analyzed both qualitatively and quantitatively for each component present in order to evaluate compliance or noncompliance with this calculated TLV. Examples about how to perform the calculation are presented in Chapter 10 of Volume I of this series.

Recently, this approach to dealing with mixtures was questioned by the German MAK committee. Specifically, in 1997 the MAK committee reiterated their view that this simple approach may not be appropriate in some situations. They stated that it was advisable to conduct toxicology tests on the mixture of chemicals to which workers would be exposed rather than rely on equations that attempt to consider only the target organ. In particular, they believed it was essential to evaluate common commercial mixtures, like gasoline or certain other mixtures, in separate toxicology tests and not to rely on the above-mentioned formula. Although any group responsible for setting OELs would not take issue with such an approach, most would probably agree that until such data are available, the method recommended by the ACGIH appears to be reasonable. As noted by Perkins (56), there are several possible scientific shortcomings in the approach but it will take many years before these can be improved upon.

2.13 Dermal Exposure Limits

In recent years, there has been discussion within the industrial hygiene community that perhaps a new category of occupational exposure limit was needed to protect against excessive uptake of chemicals via the skin. A proposal to establish dermal occupational exposure limits (DOEL) was recently described and it warrants further evaluation (119, 120).

Conceptually, the proposal is not complex. The approach assumes that the dose absorbed following 8 hours of inhalation exposure to a particular OEL is acceptable. For example, if the OEL is 5 µg/m^3, it is inferred that a dose of 50 µg/day (5 µg/m^3 × 10 m^3/day) poses no significant risk. Clearly, the driver of the calculations is the dermal penetration rate, while surface area of exposed skin and time of contact are the secondary factors.

In the first paper (119), the authors present an example that involves a pharmaceutical agent. They assumed that during preparation of cyclophosphamide, it is expected that dermal exposure is limited to the hands and lower arms, an area of about 2,000 cm^2. Starting from an absorption percentage of 30%, a daily internal dose of 0.75 mg equals a D_A of (750 × 100/30)/2000· 1 µg/cm^2. However, the absorption percentage of 30% was estimated based on a dermal area dose (D_A) of 100 µg/cm^2. Considering the fact that absorption percentage may increase with decreasing D_A, an absorption percentage of 100% was assumed at a dermal dose/unit area of about 1 µg/cm^2. Therefore, the DOEL interpreted as D_A times area was set at 0.75 mg/day. For an estimated maximum value for A of 2000 cm^3 the D_A will be 750/2000· 0.4 µg/cm^3. Similarly, the cyclophosphamide dose of 7.5 µg associated with the lower reference value equals a D_A of 4 ng/cm^2.

In their second paper (120), the authors evaluated the two different techniques for assessing dermal exposure:

> The results were used to test the applicability of a recently proposed quantitative dermal occupational exposure limit (DOEL) for [4,4'-methylene dianiline] MDA in a workplace scenario. . . . For two consecutive weeks six workers were monitored for exposure to MDA in a factory that made glass fibre reinforced resin pipes. Dermal exposure of the hands and forearms was assessed during week 1 by a surrogate skin technique (cotton monitoring gloves) and during week 2 by a removal technique (hand wash). As well as the dermal exposure sampling, biological monitoring, measurement of MDA excretion in urine over 24 hours, occurred during week 2. Surface contamination of the workplace and equipment was monitored qualitatively by colorimetric wipe samples. . . . Geometric means of daily exposure ranged from 81–1762 micrograms MDA for glove monitoring and from 84–1783 micrograms MDA for hand washes. No significant differences, except for one worker, were found between exposure of the hands in weeks 1 and 2. Significant differences between the mean daily exposure of the hands (for both weeks and sampling methods) were found for all workers. . . . Excretion of MDA in 24-hour urine samples ranged from 8 to 249 micrograms MDA, whereas cumulative MDA excretion over a week ranged from 82 to 717 micrograms MDA. Cumulative hand wash and MDA excretion results over a week showed a high correlation ($R2 = 0.94$). The highest actual daily dermal exposure found seemed to be about 4 mg (handwash worker A on day 4), about 25% of the external DOEL."

The authors concluded that both dermal exposure monitoring methods were applicable, where the exposure relevant to dermal absorption is considered mainly restricted to hands. They concluded that setting a DOEL seemed to be relevant and applicable for compliance testing and health surveillance.

3 DO THE TLVS PROTECT ENOUGH WORKERS?

Beginning about 1988, concerns were raised by numerous persons regarding the adequacy or health protectiveness of the Threshold Limit Values® (78, 121, 122). The key question raised in these papers was "what percent of the working population is truly protected from adverse health effects when exposed to the TLV?"

In the first of their papers, Castleman and Ziem (121) claimed that the TLVs were excessively influenced by corporations and, as a result, suggested that they lacked adequate objectivity. In addition, they indicated that the scientific documentation for many, if not most, of the TLVs was woefully inadequate. They concluded by suggesting that "an ongoing international effort is needed to develop scientifically based guidelines to replace the TLVs in a climate of openness and without manipulation by vested interests."

In their second paper, Ziem and Castleman (122) further discussed their views about the inadequacies of the TLVs. To a large extent, this paper was a modification and expansion of their 1988 paper. They once again concluded that the TLVs were not derived with sufficient input from physicians and that many TLVs were simply not low enough to protect most workers. They believed that there was more than circumstantial evidence to show that there had been an excessive amount of industrial influence on the TLV Committee and that this resulted in TLVs that were not sufficiently low to protect workers.

The response to these two papers by occupational physicians and industrial hygienists was significant (29, 115, 123). Over the 12 months that followed, more than a dozen letters to the editor were published and editorials appeared in *Journal of Occupational Medicine*, *American Journal of Industrial Medicine*, and the *American Industrial Hygiene Association Journal* (AIHAJ). One editorial, written by Tarlau (124) of the New Jersey Department of Environmental Protection, suggested that industrial hygienists would be better off not relying on the TLVs. This prompted a rather lengthy response that discussed the historical benefits of the TLVs and suggested that the papers criticizing the TLVs had some merit but that the critics, to a large degree, were applying the social expectations and scientific standards of 1990 on risk decisions that were often performed more than 30 to 40 years ago (28).

During 1988–1990, the claims that the TLVs were not well based in science were, to a large extent, subjective or anecdotal. Although Castleman and Ziem (121) identified inconsistencies in the margin of safety inherent in various TLVs, alleged that companies had undue influence on the TLV Committee, and claimed that objective analysis had not been conducted, the significance of these claims with respect to whether employees were sufficiently protected at the TLV remained unclear. The situation changed when two professors, one from the University of California at Berkeley and the other from England, published a rather lengthy paper that analyzed the scientific basis for a large fraction of the TLVs (78). In this paper, they showed that for many of the irritants and systemic toxicants, the TLVs were at or near a concentration where 10 to 50% of the population could be expected to experience some adverse effect. Although for many chemicals the adverse effect might be transient or not very significant, e.g., temporary eye, nose or throat irritation, these authors did offer adequate evidence that there was only a small margin of safety between the TLV concentration for some chemicals and those concentrations that had been shown to cause some adverse effect in exposed persons.

Roach and Rappaport summarized their work in this manner:

Threshold Limit Values (TLVs) represent conditions under which the TLV Committee of the American Conference of Governmental Industrial Hygienists (ACGIH) believes that nearly all workers may be repeatedly exposed without adverse effect. A detailed research was made of the references in the 1976 *Documentation* to data on "industrial experience" and "experimental human studies." The references, sorted for those including both the incidence of adverse effects and the corresponding exposure, yielded 158 paired sets of data. Upon analysis it was found that, where the exposure was at or below the TLV, only a minority of studies showed no adverse effects (11 instances) and the remainder indicated that up to 100% of those exposed had been affected (eight instances of 100%). Although, the TLVs were poorly correlated with the incidence of adverse effects, a surprisingly strong correlation was found between the TLVs and the exposures reported in the corresponding studies cited in the *Documentation*. Upon repeating the search of references to human experience, at or below the TLVs, listed in the more recent 1986 edition of the *Documentation*, a very similar picture has emerged from the 72 sets of clear data which were found. Again, only a minority of studies showed no adverse effects and the TLVs were poorly correlated with the incidence of adverse effect and well correlated with the measured exposure. Finally, a careful analysis revealed that authors conclusions in the references (cited in the 1976 *Documentation*) regarding exposure-response relationships at or below the TLVs were generally found to be at odds with the conclusions of the TLV Committee. These findings suggest that those TLVs which are justified on the basis of "industrial experience" are not based purely upon health considerations. Rather, those TLVs appear to reflect the levels of exposure which were perceived at the time to be achievable in industry. Thus, ACGIH TLVs may represent guides or levels which have been achieved, but they are certainly not thresholds" (77).

The authors reported the following as their key findings:

Three striking results emerged from this work, namely, that the TLVs were poorly correlated with the incidence of adverse effects, that the TLVs were well correlated with the exposure levels which has been reported at the time limits were adopted and that interpretations of exposure-response relationships were inconsistent between the authors of studies cited in the 1976 *Documentation* and the TLV Committee. Taken together these observations suggest that the TLVs could not have been based purely on consideration of health.

While factors other than health appear to have influenced assignments of particular TLVs, the precise nature of such considerations is a matter of conjecture. However, we note that one interpretation is consistent with the above results, namely, that the TLVs represent levels of exposure which were perceived by the Committee to be realistic and attainable at the time (77).

A number of scientists published comments on the Roach and Rappaport analysis. One of the more thorough discussion papers was written by the past-chairs of the ACGIH (29). In their letter to the editor, they claimed that the Roach and Rappaport paper was flawed and that it did not assess the validity of the bulk of the TLVs. The essence of their criticism was that the:

... conclusions which they draw concerning the protection afforded by TLVs are based on incomplete consideration of all of the data relative to a given substance. The authors present information in their tables as though the effects and exposures are valid and generally accepted by the occupational health community. No single epidemiologic study normally stands by itself. Requirements for inferring a causal relationship between disease and exposure in epidemiological studies are well established and include criteria for temporality, biological gradient with exposure, strength of the association, consistency with other studies, and biological plausibility of the observed effect. Roach and Rappaport present an uncritical analysis of various reports which would lead the uninformed reader to conclude that these criteria have been satisfied. In developing exposure recommendations, the TLV Committee and most other scientific organizations consider all of the relevant data before drawing conclusions. This includes judgments as to the validity and quality of individual studies in addition to the overall weight of the scientific evidence.

Another set of comments on the Ziem and Castleman articles that contained a good deal of historical perspective was written by the ACGIH Board of Directors (125). In that paper, the Board stated that:

While some criticisms may be valid, these articles do not fairly present the facts concerning historical development of TLVs nor do they accurately portray procedures followed by the TLV Committee in developing and reviewing TLV recommendations. Both articles contain a substantial number of errors and omissions and freely exercise selective quotation and quotation out of context in an effort to make their points. The section of Ziem and Castleman's article which discusses "Origins of TLVs" is a masterpiece of selective quotation and quotation out of context. This begins with their quoting a statement made by L. T. Fairhall concerning the role of industrial hygienists in setting health standards: "He [industrial hygienist] is in contact with the individuals exposed and therefore soon learns whether the concentrations measured are causing any injury or complaint." The authors use this quote to imply that physicians were excluded from the process of developing exposure guidelines. Taken in context, Fairhall's statement is as follows: "The industrial hygienist is in contact with not one, but a number of plants, using a given toxic substance. He knows, as no one else knows, the actual aerial concentration of contaminant encountered in practice. And he is in contact with the individuals exposed and therefore soon learns whether the concentrations measured are causing any injury or complaint. His judgement and the combined judgement of this entire Conference group is therefore most valuable in helping formulate maximum allowable concentration values." Contrary to Ziem and Castleman's comments, Fairhall advocated a multi disciplinary approach, including physicians, to making exposure recommendations. This has continued to be the operating philosophy of the TLV Committee. The conference in its first ten meetings was chaired by five physicians in six of the ten years.

In 1993, Rappaport published a follow-up analysis regarding the adequacy of the TLVs. He noted that given the continuing importance of the ACGIH limits, it was useful to compare the basis of the TLVs with that employed by OSHA de novo in its 12 new PELs. Using benzene as an example, he showed that OSHA's new PELs had been established following a rigorous assessment of the inherent risks and the feasibility of instituting the limit. He concluded that the TLVs, on the other hand, had been developed by ad hoc procedures and appeared to have traditionally reflected levels thought to be achievable at the time. However, Rappaport noted that this might be changing. Specifically, he said that "Analysis of the historical reductions of TLVs, for 27 substances on the 1991–1992 list of intended changes, indicates smaller reductions in the past (median reduction of 2.0–2.5-fold between 1946 and 1988) compared to those currently being observed (median reduction of 7.5-fold between 1989 and 1991). Further analysis suggests a more aggressive policy of the ACGIH regarding TLVs for carcinogens but not for substances that produce effects other than cancer." He also noted that "Regardless of whether the basis of the TLVs has changed recently, it would take a relatively long time for the impact of any change to be felt, since the median age of the 1991–1992 TLVs is 16.5 years, and 75% of these limits are more than 10 years old" (126).

One of the more thought provoking proposals offered by Rappaport was whether the TLV Committee should consider redefining the definition of the protectiveness of these limits (126). Specifically, he suggested, among other things, that the ACGIH "define TLVs officially as levels that represent guides for purposes of control but that do not necessarily protect 'nearly all' workers. Such a move would be in keeping with what appears to be the traditional basis of TLVs. This direction could, in time, lead to explicit rules for establishing 'feasibility' and could allow for the direct participation of industry through the submission of data related to levels of exposure in facilities of various types and ages."

HISTORY AND BIOLOGICAL BASIS OF OCCUPATIONAL EXPOSURE LIMITS

It seems clear to most industrial hygienists and toxicologists who have reflected on this issue that, given our increased awareness of the differences in susceptibility of various persons in the workplace, there is a growing lack of confidence that "nearly all workers" are protected against some of the adverse effects at the current TLVs (such as irritation) unless "nearly all workers" is defined as 80–95% (127). Whether it is necessary for the ACGIH to ask the TLV Committee to reduce these values in an attempt to protect an even greater percentage of workers will be a topic of heated discussion as we enter the new millennium.

A related opinion paper, or commentary, that criticized Rappaport's 1993 analysis was published in 1994 by Castlemen and Ziem (128). The authors claimed that the TLV setting process continued to lack objectivity and that there was too much opportunity for conflicts of interest to occur. In support of their claims, they presented two tables that listed the names of certain chemicals where the primary authors were from various chemical companies or the chemicals for which their TLVs had been criticized. Since this paper, little additional debate has occurred and the ACGIH continues to work to improve the process to help satisfy its critics.

Although the merits of the Roach and Rappaport analysis, or for that matter, the opinions of Ziem and Castleman, have been debated over the past 10 years, it is clear that the process by which TLVs and other OELs will be set will never again be as it was between 1945 and 1990. In the future, it can be expected that the rationale, as well as the degree of "risk" inherent in a TLV, will be more explicitly described in its documentation. This degree of transparency in the documentation is necessary since the definition of "virtually safe" or "insignificant risk" with respect to workplace exposure will change as the values of society evolve regarding the definition of "safe" (129, 130).

3.1 Where the TLV Process is Going

The degree of reduction in TLVs or other OELs that will undoubtedly occur in the coming years will vary depending on the type of adverse health effect to be prevented, e.g., central nervous system depression, acute toxicity, odor, irritation, cancer, developmental effects, or others (130). It is unclear to what degree the TLV Committee will rely on various predictive toxicity models or the risk criteria (e.g., 1 in 1,000 cancer risk) they will adopt as we enter the next century. However, one thing is clear; transparency in the approach used to set any OEL will be expected by those who are going to apply these guidance values.

During the past 10 years, the Executive Director of the ACGIH TLV Committee, Bill Wagner, has worked diligently to make their work more transparent. The benefits of this effort are generally evident in the documentation for the various TLVs that have been published over the past 3–7 years. For example, the *1996 Documentation of the TLV* for formaldehyde is 24 pages in length and cites more than 200 references. The version published in 1990 was two pages long and cited about 20 papers. Regrettably, the fact that the documentations are much more thorough than in the past does not necessarily mean that all of the key papers have been carefully read and thoroughly understood. Often, the truly important or definitive papers have been weighed equally with those of lesser quality. Also, many documentations still lack a clear and concise explanation of how the limit was identified. That is, the critical study or studies upon which the NOEL was based are not always identified and the rationale for the size of the UF is rarely described. These shortcomings are simply a reflection of the time constraints placed on persons who serve on voluntary committees. Like the one that sets TLVs.

A question that might be raised is "Why hasn't the TLV Committee been able to carefully read each and every published paper on a chemical, understand the subtleties of the experimental method, document the entire process, then select the ideal occupational exposure limit?" Although in some or many cases the committee may have had the time to understand all published data regarding a chemical, there are a number of cases where it did not. Perhaps the primary reason the derivation for many of the TLVs is not well described and, in some cases, the ideal value was not selected, is that a marginally adequate number of human and financial resources has been dedicated to the task of setting these values. One must remember that those on the committee have always been unpaid volunteers. Each of these professionals allocates more than 200 hours each year to the business of setting TLVs simply because they believe these values have helped prevent disease in millions of workers. However, the time these professors or senior researchers can devote to conducting research-quality, detailed analyses is quite limited. In the past, at various times, the burden of finding the published papers, carefully comparing and contrasting the different experimental methods, and resolving differences in results, was done as a labor of love by government employees, professors, and industry consultants to the TLV committee. In some cases, many years ago, these professionals were given support by their employers and some spent as much as 50% of their workweek performing TLV Committee work. During the 1960s and 1970s, this was more commonplace because industrial hygiene groups in the government and in industry had much larger staffs and more generous budgets than have been seen in recent years. Based on discussions with those who have served, it appears that the industry consultants and the regular members of the TLV Committee now try to limit their time commitment to 3–5 days a month.

This level of historical employer support (both government and private sectors) lessened in the 1980s and 1990s as it was assumed that professionals within NIOSH and OSHA, using tax money, should begin displacing the TLVs with OSHA PELs and NIOSH RELs. In fact, the NIOSH criteria documents of the 1970s and 1980s were often more than 100 pages in length with dozens of references, and they rapidly became "the standard" by which future documentations of the TLVs would be compared. About 100 criteria documents were written between 1975 and 1985 and each provided a reasonably complete story of all toxicological and analytical chemistry issues related to a particular chemical. During the same time, the legal liabilities of setting and not meeting TLVs and other OELs (citable under Section 5A1 of OSHA) took on greater significance so corporations were less anxious to devote their resources to a program that might pose some incremental legal exposure and increase what was already considered a heavy regulatory burden. Thus, a serious chasm occurred about 1985 between the amount of effort expected to go into the development of the TLVs and the resources available to perform the work. To make matters worse, the consultants to the TLV Committee from industry who had for many years performed much of the literature searches, copying, and preliminary analyses were discouraged from being as active due to complaints that industry had had undue influence on the TLV setting process (122).

The availability of funds or professional resources is not the only reason that many current TLVs are not sufficiently well documented to satisfy every health professional. The other reason is that "the bar has been set much higher than its historical placement." In short, the quality of analysis that health professionals, lawyers, and the courts have come to expect since about 1985 has frequently gone beyond what can be provided by volunteers. The compilation of information and the analyses that were invested in the NIOSH criteria documents and the proposed OSHA standards of the 1980s and early 1990s were much more comprehensive than that envisioned as necessary by those who originally established the TLV setting process. One reason the TLV documentation fell short of these other efforts with respect to documentation was that it was not the original intent of the ACGIH to publish "permanent" values that would be "cast in stone." Rather, the goal was to disseminate information to health professionals to assist them in helping to protect workers from harm or discomfort. To perform this service, more than 50 years ago the ACGIH asked the most reputable and knowledgeable persons in the field to analyze what was known about a chemical and to suggest a value that was expected to be protective. After the value was proposed, then various parties could submit comments and documentation for the next two years so that a new TLV could be proposed, if appropriate. In short, the TLVs did not have to be scientifically "bulletproof" since they were to be dynamic values that could be changed in rather short periods of time.

The ACGIH anticipated that those who worked with the chemical each day would have an interest in providing the best possible guidance to fellow health professionals, so they expected any shortcomings in the documentation to be brought to their attention by their colleagues. One must remember that the TLVs came about as a direct result of a concern by health professionals that insufficient information was being shared within the occupational health community to help hygienists protect workers. They were never intended to be dissertation quality analyses. The committee has always encouraged those who were truly expert, including professionals in the firms who used various chemicals, to submit information that would help identify the best possible limits.

Does this discussion suggest that the TLV setting process is in need of change? The answer to this question resides within the ACGIH. If they feel obligated to issue "notice of intended changes" or new TLVs which are virtually bulletproof, then the current approach probably cannot provide that service. There is simply too much information, and the scientific expertise needed to interpret many of these studies is so specialized, that unassailable work products (e.g., Documentations of TLVs) cannot be reasonably provided by a group of volunteers who have a very limited support staff. If, however, the ACGIH continues to make their "Documentation Of The TLVs" more transparent, if it encourages interested parties to perform the vast amount of the tedious work needed to understand a chemical and to document the rationale for the suggested OEL, and if it is willing to provide modest stipends to the committee members to support graduate students who can do much of the background work, then the present system should be able to satisfy its critics. During this period of introspection about how to improve the TLV setting process, we should not forget that thousands of lives have been saved as a result of these values and that, to the best of our knowledge, as stated by Stokinger more than twenty years ago, few if any workers "have been shown to have sustained serious adverse effects on his health as a result of exposure to these concentrations of an industrial chemical" (6).

4 CORPORATE OELS

Although exposure limits or guides like TLVs or WEELs for most large volume chemicals have been established, and the vast majority of workers are exposed to processes for which these guidance values are applicable, the majority of the 3,000 chemicals to which workers are routinely exposed in industry do not have PELs, RELs, WEELs, or MAKs. As a result, about 50 companies in the United States have chosen to establish a number of internal or corporate limits to protect their employees, as well as the persons who purchase those chemicals.

The need for internal limits is generally identified by the manufacturing divisions within large companies, although the Health, Safety and Environmental Affairs Department may also initiate the process (50). A panel of toxicologists, industrial hygienists, physicians and epidemiologists usually gather the scientific data and make the technical assessment much like the ACGIH TLV Committee. The process used is depicted in Figure 41.4 (50). The data considered by the group are similar to those considered by other OEL setting bodies (Table 41.6). Their deliberations are often reviewed by an oversight group, which integrates the scientific input with information provided by the manufacturing, legal, law, regulatory and other groups in the company to establish, as appropriate, internal exposure levels, including in some instances maximum exposure levels and short term exposure levels.

Most firms who have established OELs believe that the management of occupational exposure requires limits or criteria much like a manufacturing group needs quality control criteria. Some companies have found that manufacturing groups use OELs to define acceptable versus unacceptable manufacturing conditions. Without these limits as guides, operations managers claim that they would not know when conditions are unhealthy, when personnel need to be protected and, if monitoring is performed, how the results should be interpreted. In short, the experience of the past 30 years indicates that corporate or internal OELs serve a useful purpose (50).

Many firms who have set their own OELs acknowledge that they have done so because they believe in several principles. Foremost is the concept that guidelines are needed whenever employees are being exposed. Second, firms should document the rationale for establishing their guides and share these with the TLV Committee and others. Third, if adequate toxicology data are not available, it is still valuable to set tentative exposure limits for a chemical based on exposure levels that have been measured and found to be acceptable. Since scientific data are generally lacking for establishing "exact" exposure levels, this kind of approach seems reasonable and more prudent than simply waiting to see if adverse effects are observed.

The question could be asked "If internal OELs are so beneficial, why don't all companies set them?" The answer to the question is complex. First, the cost of establishing limits through committees is substantial. Based on the author's experience, to establish and document a corporate OEL, about 80–240 professional hours are invested in:

1. Identifying the proper studies.
2. Reading and interpreting them.
3. Selecting a preliminary OEL.

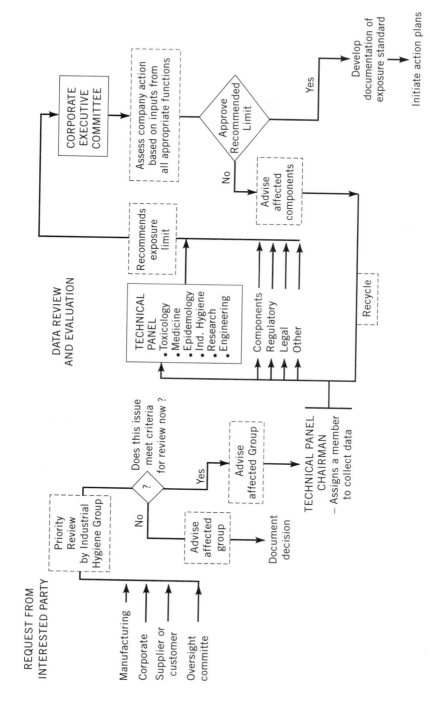

Figure 41.4. Organization and methods used to set corporate occupational exposure limits (50).

Table 41.6. Data often Used in Developing an Occupational Exposure Limit[a]

Physical properties
   Lipid solubility
   Water solubility
   Vapor pressure
   Odor threshold
Acute toxicity data
   Oral toxicity, LD_{50}
   Dermal toxicity, LD_{50}
   Dermal and eye irritation
   Inhalation toxicity, LC_{50}
Subacute and subchronic animal data (oral, dermal, or inhalation)
   14 day, NOEL
   90 day, NOEL
   6 month, NOEL
Other animal data
   Developmental (teratology and embyotoxicity)
   Mutagenicity (Arnes test, drosophilia, etc.)
   Fertility
   Reproductive (3-generation)
   Reversability study
   Dermal absorption test
   Pharmacokinetics
   Cancer bioassay (2 yr)
Epidemiologic data
   Morbidity
   Mortality
   Base reports
Controlled human studies
Industrial hygiene exposure data
   Area samples
   Personal samples

[a]Paustenbach and Langner (50).

4. Writing the documentation.
5. Having committee meetings.
6. Revising the documentation and OEL based on the committee's suggestions.
7. Obtaining reviews of corporate management and the legal department.

At a cost of about $175 per hour, this equates to an investment of about $25,000 to $60,000 per OEL. Second, many firms believe that setting an internal OEL establishes a legal responsibility to meet this limit at all times and that it generates a liability that is otherwise unnecessary. Although these are the two most important issues, others have been mentioned (50).

Nearly all the firms who set OELs have found that perhaps the most difficult and controversial aspects are the legal ramifications. For example, lawyers have noted that if a company develops internal standards on its chemicals or chooses to adopt values for a chemical that are more conservative than those of a regulatory agency, the firm had best plan to comply with them. On the other hand, many lawyers believe that perhaps an equal legal exposure exists with those firms who know a great deal about the potential hazards of a chemical yet do not set internal limits. Admittedly, such a scenario puts manufacturers between a rock and a hard place. For example, some firms may feel that the workmen's compensation immunity does not encourage them to set internal limits on their own chemicals. It is, however, worth bearing in mind that as the manufacturer of a chemical they could be sued by someone else's employee who, if injured, could claim that they did not supply enough data. Although internal limits have been set for nearly 40 years by various firms, there remains controversy about these complex legal issues.

5 MODELS FOR ADJUSTING OELS

Several researchers have proposed mathematical formulae or models for adjusting OELs (PELs, TLVs, etc.) for use during unusual work schedules and these have received a good deal of interest in the industrial and regulatory arenas (32, 62, 131–136). Although OSHA has not officially promulgated specific exposure limits applicable to unusual work shifts, at various times over the years they have published guidelines for use by OSHA compliance officers for adjusting exposure limits (137). These generally apply to work shifts longer than eight hours per day. The preceeding chapter in this volume of the *Patty's Industrial Hygiene and Toxicology* Series is devoted to describing how to adjust OELs. Only a couple of methods are described here.

5.1 Brief and Scala Model

In the early 1970s, due to the increasingly large number of workers who had begun working unusual schedules, the Exxon Corporation began investigating approaches to modifying the OELs for their employees on 12-hour shifts. In 1975, the first recommendations for modifying TLVs and OSHA PELs were published by Brief and Scala (138), wherein they suggested that TLVs and PELs should be modified for individuals exposed to chemicals during novel or unusual work schedules.

They called attention to the fact that, for example, in a 12-hour workday the period of exposure to toxicants was 50% greater than in the 8-hour workday, and that the period of recovery between exposures was shortened by 25%, from 16 to 12 hours. Brief and Scala noted that repeated exposure during longer workdays might, in some cases, stress the detoxication mechanisms to a point that a toxicant might accumulate in target tissues, and that alternate pathways of metabolism might be initiated. It has generally been held that given the margin of safety in most of the TLVs, there was little potential for frank toxicity to occur due to unusually long work schedules. Based on recent evaluations of the TLVs, the margins of safety are probably not as great as was believed prior to 1990 (78).

Brief and Scala's (138) approach was simple but important since it emphasized that unless worker exposure to systemic toxicants was lowered, the daily dose would be greater, and due to the lesser time for recovery between exposures, peak tissue levels might be higher during unusual shifts than during normal shifts. This concept is illustrated in Figure 41.5. The following formulas for adjusting limits are intended to ensure that this will not occur during unusual work shifts. The following equation was recommended for a five-day work week:

$$TLV\ Reduction\ Factor\ (RF) = \frac{8}{h} \times \frac{24-h}{16}$$

where h = hours worked per day

For a 7-day workweek, they suggested that the formula be driven by the 40-hour exposure period; consequently, they developed the following formula, which accounts for both the period of exposure and period of recovery:

$$TLV\ Reduction\ Factor\ (RF) = \frac{40}{h} \times \frac{168-h}{128}$$

where h = hours exposed per week

One advantage of these formulas is that the biologic half-life of the chemical and the mechanism of action are not needed in order to calculate a modified TLV. Such a simpli-

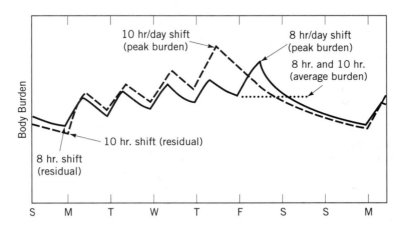

Figure 41.5. Comparison of the peak, average and residual body burdens of an air contaminant following exposure during a standard (8-hr/day) and unusual (10-hr/day) workweek. In this case the weekly average body burdens are the same for both schedules since each involved 40 hr/week. The residual (Monday morning) body burden of the 8-hr shift worker, however, is greater than the 10-hr shift worker and the peak body burden of the person who worked the 10-hr shift is higher than the 8-hr worker. Based on Hickey (132).

HISTORY AND BIOLOGICAL BASIS OF OCCUPATIONAL EXPOSURE LIMITS

fication has shortcomings since the reduction factor for a given work schedule is the same for all chemicals even though the biologic half-lives of different chemicals vary widely. Consequently, the Brief and Scala approach should overestimate the degree to which the limit should be lowered.

Brief and Scala (138) were cautious in describing the strength of their proposal and offered the following guidelines for its use. Their caveats should be considered when applying this model and also other approaches:

(1) Where the TLV is based on systemic effect (acute or chronic), the TLV reduction factor will be applied and the reduced TLV will be considered as a time-weighted average (TWA). Acute responses are viewed as falling into two categories: (a) rapid with immediate onset and (b) manifest with time during a single exposure. The former are guarded by the C notation and the latter are presumed time and concentration dependent, and hence, are amenable to the modifications proposed. Number of days worked per week is not considered, except for a 7-day workweek discussed later.

(2) Excursion factors for TWA limits (Appendix D of the 1974 TLV publication) will be reduced according to the following equation:

$$EF = (EF_8 - 1)RF + 1$$

where EF = Desired excursion
EF_8 = Value in Appendix D for 8-hour TWA
TW = TLV Reduction Factor

(3) Special case of 7-day workweek. Determine the TLV Reduction Factor based on exposure hours per week and exposure-free hours per week.

(4) When the novel work schedule involves 24-hour continuous exposure, such as in a submissive or other totally enclosed environment designed for living and working, the TLV reduction technique cannot be used. In such cases, the 90-day continuous exposure limits of the National Academy of Science should be considered, where applicable limits apply.

(5) The techniques are not applicable to work schedules less than seven to eight hours per day or ≤40 hours per week.

Brief and Scala (133) also correctly noted that:

The RF value should be applied (a) to TLVs expressed as time-weighted average with respect to the mean and permissible excursion and (b) to TLVs which have a C (ceiling) notation except where the C notation is based solely on sensory irritation. In this case the irritation response threshold is not likely to be altered downward by an increase in number of hours worked and modification of the TLV is not needed.

In short, the Brief and Scala approach is dependent solely on the number of hours worked per day and the period of time between exposures. For example, for any systemic toxicant, this approach recommends that persons who are employed on a 12 hours per day, three or four day workweek, should not be exposed to an air concentration of a toxicant greater than one-half that of workers who work on an eight hours per day, five-day schedule.

In their publication, Brief and Scala acknowledged the importance of a chemical's biologic half-life when adjusting exposure limits, but because they believed this information was rarely available, they were comfortable with their proposal. They noted that a reduction in an occupational exposure limit is probably not necessary for chemicals whose primary untoward effect is irritation since the threshold for irritation response is not likely to be altered downward by an increase in the number of hours worked each day (138); that is, irritation is concentration rather than time dependent. Although this appears to be a reasonable assumption, some researchers believe that it may not be entirely justified since duration of exposure could possibly be a factor in causing irritation in susceptible individuals who are not otherwise irritated during normal 8 hours per day exposure periods (89).

5.1.1 Illustrative Example #1 (Brief and Scala Model)

Refinery operators often work a 6-week schedule of three 12-hour workdays for three weeks, followed by four 12-hour workdays for three weeks. What is the adjusted TLV for methanol (1998 TLV = 200 ppm) for these workers? Note that the weekly average exposure is only slightly greater than that of a normal work schedule.

$$RF = \frac{8}{12} \times \frac{24 - 12}{16} = 0.5$$

Solution:

$$\begin{aligned}\text{Adjusted } TLV &= RF \times TLV \\ &= 0.5 \times 200 \text{ ppm} \\ &= 100 \text{ ppm}\end{aligned}$$

Note: The TLV Reduction Factor of 0.5 applies to the 12-hour workday, whether exposure is for three, four, or five days per week.

5.1.2 Illustrative Example #2 (Brief and Scala Model)

What is the modified TLV for tetrachloroethylene (1998 TLV = 25 ppm) for a 10 hours per day, four days per week work schedule if the biologic half-life in humans is 144 hours?

$$RF = \frac{8}{10} \times \frac{24 - 10}{16} = 0.7$$

Solution:

$$\text{Adjusted } TLV = 0.7 \times 25 = 15 \text{ ppm}$$

Note: This model and the one used by OSHA *do not consider* the pharmacokinetics (biologic half-life) of the chemical when deriving a modified TLV. Other models to be discussed later *do* take this into account.

5.2 Haber's Law Model

Most toxicologists believe that, in general, the intensity and likelihood of a toxic response is a function of the mass that reaches the site of action per unit time (68, 79, 139). This "delivered dose" is usually exposed as the concentration of the chemical in the blood for systemic agents. This principle is simplistic and, while it may not apply to irritants and sensitizers, it is clearly true for the systemic toxins. This assumption is the basis for the OSHA model for modifying PELs for unusual shifts (137), which is based on Haber's Law. The originators of the model assumed that for chemicals that cause an acute response, if the daily uptake (concentration × time) during a long workday was limited to the amount that would be absorbed during a standard workday, then the same degree of protection would be given to workers on the longer shifts. For chemicals with cumulative effects (i.e., those with a long half-life), the adjustment model was based on the dose imparted through exposure during the normal workweek (40 hours) rather than the normal workday (8 hours).

OSHA recognized that the rationale for the OELs for the various chemicals was based on different types of toxic effects. After OSHA adopted the 500 TLVs of 1968 (49) as PEL (29 CFR 1910.1000), and attempted to do the same in 1989 (52), they placed each of the chemicals into different toxicity categories to ensure that an appropriate adjustment model would be used by their hygienists. As can be seen in Examples 3 and 4, the degree to which an exposure limit is to be adjusted, if at all, is based to a large degree on the primary toxic effect of the chemical. The ACGIH TLV Handbook now lists the primary adverse effect associated with each chemical which is most helpful for knowing how to evaluate mixtures and unusually long work schedules (3).

Irrespective of the model that will be used to make the adjustments, including the pharmacokinetic models to be discussed, the table in the Federal Register (52) should be consulted before the hygienist begins the task of modifying an exposure limit. The use of the OSHA tables or the designations of primary adverse effect shown in the most recent ACGIH TLV booklet (3) combined with some professional judgment, will prevent hygienists from requiring control measures when they are unnecessarily restrictive, as well as minimize the risk of injury or discomfort from overexposure during an unusual exposure schedule.

The objective of OSHA's approach for acute toxicants is to modify the limit for the unusually long shift to a level that would produce a dose (mg) which would be no greater than that obtained during 8 hours of exposure at the PEL. Examples of chemicals with exclusively acute effects include carbon monoxide or phosphine. The following equation is recommended by OSHA for calculating an adjustment limit (Equivalent PEL) for these types of chemicals:

$$Equivalent\ PEL = 8\text{-}hour\ PEL \times \frac{8\ hours}{hours\ of\ exposure\ per\ day}$$

The other formula recommended by OSHA applies to chemicals for which the PEL is intended to prevent the cumulative effects of repeated exposure, e.g., the chronic toxicants. For example, PCBs, PBBs, mercury, lead and DDT are considered cumulative toxins because repeated exposure is usually required to cause an adverse effect and the overall biologic half-life is clearly in excess of 10 hours. The goal of PELs in this category is to prevent excessive accumulation in the body following many days or even years of exposure. Accordingly, the next equation is offered to OSHA compliance officers as a viable approach for calculating a modified limit for chemicals whose half-life would suggest that not all of the chemical will be eliminated before returning to work the following day. Its intent is to ensure that workers exposed more than 40 hours per week will not eventually develop a body burden of that substance in excess of persons who work on normal eight hours per day, 40 hours per week schedules.

$$\text{Equivalent PEL} = 8\text{-hour PEL} \times \frac{40\ hours}{hours\ of\ exposure\ per\ week}$$

The OSHA models, although less rigorous than the pharmacokinetic models which will be discussed, have certain advantages since they do account for the kind of toxic effect to be avoided, require no pharmacokinetic data, and tend to be more conservative than the pharmacokinetic models.

5.2.1 Illustrative Example #3 (OSHA Model)

An occupational exposure limit of 1 µg/m^3 has been suggested by NIOSH for polychlorinated biphenyls (PCBs). In studies of humans, it has been found that the biologic half-life of PCBs is several years. Animal studies have shown that some can cause cancer at certain doses. What adjustment to the occupational exposure limit might be suggested by NIOSH for workers on the standard 12-hour work shift involving four days of work per week if they adopted the simple OSHA formulae?

Solution: Recommended Limit = 8 hr *PEL* × 40 hr/48 hr
Recommended Limit = 1 µg/m^3 × 0.667 = 0.833 µg/m^3

5.2.2 Illustrative Example #4 (OSHA Model)

Many industries such as boat manufacturing are seasonal in their workload. During the months of January, February, March and April, the builders of boats work five days per week, 14 hours per day and could be exposed to concentrations of Toluene Diisocyanate (TDI) at the TLV of 0.005 ppm. What occupational exposure limit is recommended for TDI for a person who works 14 hours per day for five days per week but only works eight weeks per year?

Solution: No adjustment is necessary.

Note: TDI is categorized as a sensitizer and irritant. Substances in this category have limits that should never be exceeded and consequently the limits are independent of the length or frequency of exposure.

Exposure limits for chemical irritants such as these are currently thought not to require adjustment. Until more is known about human response to irritants during unusually long durations of exposure, the physician, nurse, and hygienist should make note of the employee tolerance to the presence of irritants at levels at or near the TLV. Eventually, human experience will provide information that will help identify those classes of chemicals for which irritation could be a time dependent phenomenon.

5.3 Pharmacokinetic Models

Pharmacokinetic models for adjusting occupational limits have been proposed by several researchers (33, 131–140). These models acknowledge that the maximum body burden arising from a particular work schedule is a function of the biological half-life of the substance. Pharmacokinetic models, like the other models, generate a correction factor which is based on the elimination half-life of the substance as well as the number of hours worked each day and week and this is applied to the standard limit in order to determine a modified limit. Unlike the OSHA, as well as the Brief and Scala approaches, by accounting for a chemical's pharmacokinetics, these models can also identify those exposure schedules where a reduction in the limit is not necessary.

The rationale for a pharmacokinetic approach to modifying limits is that during exposure to the TLV for a normal workweek, the body burden rises and falls by amounts governed by the biological half-life of the substance (see Fig. 41.6). A general formula provides a modified limit for exposure during unusual work shifts so that the peak body burden accumulated during the unusual schedule is no greater than the body burden accumulated during the normal schedule. This is the goal of all of the pharmacokinetic models that have thus far been developed.

It is worthwhile to note that the maximum body burden arising from continuous uniform exposure under the standard eight hours per day work schedule nearly always occurs at the end of the last work shift before the two-day weekend. The only exception is for some of the agents whose biologic half-life is quite long (greater than 16 hours). On the other hand, the maximum body burden under an extraordinary work schedule may not occur at the end of the last shift of that schedule (Fig. 41.6). This is especially true when the duration and spacing of work shifts that precede the last shift differ markedly from the standard week. Because unusual work schedules can be based on a 2-week, 3-week, 4-week, or even 11-week cycle and the work shift may be 10, 12, 16, or even 24 hours in duration, no generalization regarding the time of peak body burden can be offered. The time of peak tissue burden for unusual schedules must therefore be calculated for each specific schedule.

In 1977, Hickey and Reist (133) published a paper describing a general formula approach to modifying exposure limits that was fundamentally equivalent to that of Mason and Dershin (131). The benefits of their work were manifold. First, they validated their approach to some extent by comparing the results with published biological data. Second, they proposed broader uses of the pharmacokinetic approach to modifying limits and presented a number of graphs which could be used to adjust exposure limits for a wide number of exposure schedules. The graphs were based on (*1*) the biologic half-life of the material, (*2*) hours worked each day, and (*3*) hours worked per week.

Over the next three years they wrote publications that illustrated how their model could be used to set limits for persons on overtime (132) and for seasonal workers (135). Hickey's treatment of the topic of adjusting exposure limits is quite thorough and his publications

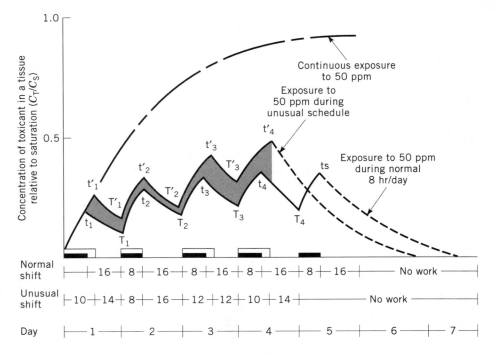

Figure 41.6. The accumulation of a substance during irregular periods of intermittent exposure that could occur during unusual work shifts and overtime. This plot illustrates that even when the total weekly dose (40 hr × 50 ppm) is unchanged, the peak body burden for two different shift schedules may not be equivalent. To help ensure that the peak tissue concentration for the unusual exposure schedule does not exceed the presumably "safe" level of the normal schedule, the air concentration of toxicant in the workplace may have to be reduced. K_{elim} has a value of 0.03 hr^{-1} in this illustration. (Courtesy of Dr. J. Walter Mason.)

are primarily responsible for most of the interest and research activity in this area. In recent years, other researchers have continued to assess the conditions under which OELs should be adjusted (140)

5.4 A Physiologically-based Pharmacokinetic (PB-PK) Approach to Adjusting OELs

As noted previously, the rationale for adjusting OELs for unusual work schedules is to ensure, as much as possible, that persons on these schedules are placed at no greater risk of injury or discomfort than persons who work a standard 8 hours per day, 40 hours per week schedule. For most systemic toxicants, the risk index upon which the adjustments are made will be either peak blood concentration or integrated tissue dose, depending on that chemical's presumed mechanism of toxicity.

HISTORY AND BIOLOGICAL BASIS OF OCCUPATIONAL EXPOSURE LIMITS 1957

The previous section described how one can adjust the OEL using mathematics to ensure that peak tissue levels are not exceeded during long shifts. Unfortunately, these models cannot account for biological factors like enzyme induction or approximate target tissue concentrations. The optimal method for adjusting exposure limits to account for peak blood levels for long workdays, or even continuous exposure, is to base it on a pharmacokinetic approach that accounts for the differences between animals and humans, as well as other biologic factors that aren't dealt with through the use of a single term like biologic half-life. The most sophisticated approach for incorporating these factors involves the use of a PB-PK model. At this time, about 40 papers describing evaluations of at least that many chemicals have been published (Table 41.7) (141).

At present, the PB-PK approach is recommended only for exposure periods of 4 to 16 hours per day. Pharmacokinetic approaches alone should not be relied on for exposure periods greater than 16 hours per day or less than 4 hours per day because the mechanisms of toxicity for some chemicals may vary for very short- or very long-term exposure. For these altered schedules, biological information on recovery, rest periods and mechanisms of toxicity should be considered before any adjustment should be attempted. As noted by Andersen et al. (139), when pharmacokinetic data are not available, a simple inverse formula may be sufficient for adjustment in most instances. Their paper illustrated the use of the PB-PK approach on two industrially important chemicals: styrene and methylene chloride.

6 OCCUPATIONAL EXPOSURE LIMITS (OELS) OUTSIDE THE UNITED STATES

OELs have been adopted or established in a number of countries outside the United States. Some information about these various limits is presented in this section. Most of the following information on various countries was obtained from Cook (2) and it represents only a fraction of the information on OELs used in other countries. A compilation of virtually all of the OELs used by various countries is routinely published by the World Health Organization (46).

Recently, a fairly comprehensive analyses of the process for setting OELs in other countries was published by Vincent (47). To prepare his assessment, he visited Australia, Britain, Norway, and Russia. Based on his study, Vincent offered the following general observations, which provide valuable insight that is necessary to properly interpret OELs set in other countries:

> Even for OELs set on the basis of the same scientific data, there are differences in how the OEL is set. The first may derive from divergent opinions about which health endpoint should drive the OEL or about what level or prevalence of ill-health is considered acceptable. Such differences are value-driven, dependent on variations in emotional response to certain types of ill-health, and variations in attitude to risk associated with certain substances. These in turn depend on local cultures and perceptions and cannot be quantified.

> If the scientific discussion leading to even a health-based OEL is subject to considerable differences at a number of levels, then opening up the process to consideration of questions beyond health effects leads to further amplification of the differences. Now a completely

Table 41.7. PB-PK Models for Toxic Substances[a]

Benzene	Methanol
Benzo(a)pyrene	Methoxyethanol
Butoxyethanol	Methyl ethyl ketone
Carbon tetrachloride	Nickel
Chlorfenvinphos	Nicotine
Chloralkanes	Parathion
Chloroform	Physostigmine
Chloropentafluorobenzene	PBB
cis-Dichlorodiamine platinum	PCBs
Dichloroethane	Styrene
Dichloroethylene	Toluene
Dichloromethane	TCDF
Dieldrin	TCDD
Diisopropylfluorophosphate	Tetrachloroethylene
Dimethyloxazolidine dione	Trichloroethane
Dioxane	Trichloroethylene
Ethylene oxide	Trichlorotribluoroethane
Glycol ethers	Vinylidene fluoride
Hexane	Xylene
Kepone	
Lead	

[a]This table was presented in a paper by Leung and Paustenbach (141).
At present, the PB-PK approach is recommended only for adjusting OELs for exposure periods of 4 to 16 hours per day.

different set of criteria and values comes into play. They are driven by the question of feasibility. Although feasibility relates to whether a given OEL is practically achievable using known or available technology, there is the underlying rationale related to the cost of achieving a given OEL versus the cost of loss of life or quality of life, and to the local and national need for a given industry to continue to operate and be profitable (and the resultant cost to society were it not to do so). At the societal or cultural level, it relates to individuals' and society's perceptions of the relative value of risk versus the need for employment and their attitudes toward the role of government in regulating industry and work, and to the ability of the government to set and enforce OELs within the machinery of the given regulatory framework and ethical climate. The issue then becomes part of the wider political discussion. Comparison of the British and U.S. approaches to occupational exposure standards illustrates how such issues can lead to differences. As noted in the SMA report, (6) the British process, based on consensus among industry, workers' unions, and independent experts and backed up by a strong, respected, and demonstrably impartial civil service, develops standards that are recommended to Parliament, which in turn can incorporate them into an existing regulatory framework. The goodwill engendered by this consensus-based process means that the possibility of legal actions to challenge occupational health standards is largely deflected. The same approach would not be possible in the United States because Congress does not have power of final approval over regulatory actions by an executive agency such as OSHA. In addition, culturally, the U.S. approach is greatly influenced by adversarial legal processes.

HISTORY AND BIOLOGICAL BASIS OF OCCUPATIONAL EXPOSURE LIMITS

"Although there are considerable differences in the types of OEL that are set by individual bodies, and the ways in which they are set, there does appear to be some common ground. Perhaps the most striking part is the fact that so many bodies depended strongly on the TLVs in the early years, and that the influence still remains strong today. The role of ACGIH was particularly important at the beginning in establishing a process by which (1) published—and sometimes unpublished—data on health effects and exposures, in humans and in animals, could be documented, evaluated, and discussed in a multidisciplinary peer group; (b) dose-response relationships identified; and (c) health-based OELs arrived at. Although many of the individual national bodies have striven over the years to reexamine the old data (and to include new data) to develop their own OELs independently, progress has been relatively slow. So progress in developing completely home-grown OELs has been slow in most countries and, in view of the effort required to carry out the full process for any given substance, it is likely to remain so. As a result, most of the OELs listed by many individual bodies remain numerically the same as the corresponding ones listed by ACGIH. The differences, where they occur, are for the relatively small number of "difficult" substances that have continued to generate interest by virtue of ongoing public concerns about associated occupational ill-health (e.g., crystalline silica, asbestos and other fibrous dusts, benzene and other solvents, etc.).

It is suggested that those who wish to better understand some of the differences in philosophy used to set OELs in other countries read not only Vincent (47), but several other papers on the topic (81). See also the General References in this chapter.

6.1 Argentina

Under ANNEX III, Article 60 of the regulation approved for Decree No. 351/79 lists OELs for Argentina that are essentially the same as those of the 1978 ACGIH TLVs. These OELs continue in force. The principal difference from the ACGIH list is that, for the 144 substances (of the total of 630) for which no STELs are listed by ACGIH, the values used for the Argentina TWAs are entered also under this heading.

6.2 Commonwealth of Australia

The National Health and Medical Research Council (NHMRC) of the Commonwealth of Australia adopted a revised edition of the "occupational health guide THRESHOLD LIMIT VALUES (1990–91)" in 1992. The OELs have no legal status in Australia, except where specifically incorporated into law by reference. The ACGIH TLVs are published in Australia as an appendix to the occupational health guides, revised with the ACGIH revisions in odd-numbered years. Vincent (47) provides a current analysis of the procedures used in this country.

6.3 Austria

The acronym MAC or maximal acceptable (or allowable) concentration was a term used in the USA during the years before the ACGIH introduced the expression TLV. The MAC was translated by both Austria and Germany to MAK for Maximale Arbeitsplatzkonzentration with the same pronunciation.

The Austrian list of values recommended by the Expert Committee of the Worker Protection Commission for Appraisal of MAC Values in cooperation with the General Accident Prevention Institute of the Chemical Workers Trade Union is used by the Federal Ministry for Social Administration. The MAC values were declared obligatory by the Federal Ministry on December 23, 1982, Number 61,710/24-4/82. They are applied by the Labor Inspectorate under the authority of the Labor Protection Law, Section 6, Abs. 2 (2).

6.4 Belgium

The Administration of Hygiene and Occupational Medicine of the Ministry of Employment and of Labor uses the TLVs of the ACGIH as a guideline in control of occupational health hazards. As of July 13, 1983, the periodically issued Belgian TLV publication was being brought up-to-date with the 1982 ACGIH TLVs (2).

6.5 Brazil

The TLVs of the ACGIH have been used as the basis for the occupational health legislation of Brazil since June 8, 1978, through the 3214/78 Edict published by the Ministry of Labor. As the Brazilian work week is usually 48 hours, the values of the ACGIH (1977 TLVs with changes recommended for 1978) were adjusted in conformity with a formula developed for this purpose. The ACGIH list was not adopted in its entirety but only for those air contaminants that at the time had nationwide application. The Ministry of Labor has brought the limits up-to-date with establishment of values for additional contaminants in accordance with recommendations from the Fundacentro Foundation of Occupational Safety and Medicine (2).

6.6 Canada

For each of the various Provinces in Canada, there can be different OELs. In the Province of Alberta, OELs are under the direction of Alberta Regulation 8/82 of the Occupational Health and Safety Act, Chemical Hazard Regulation with amendments up to and including Regulation 242/83. Section 2 of this Act requires the employer to ensure that workers are not exposed above the limits. They are under the direction of the Standards & Projects Section, Occupational Hygiene Branch, Occupational Health and Safety Division of the Workers' Health, Safety, and Compensation Department (2).

In the Province of British Columbia, the Industrial Hygiene Department, Industrial Health and Safety Division of the Workers' Compensation Board, administers the Industrial Health and Safety Regulations that are legal requirements with which most of British Columbia industry must comply. These requirements refer to the current schedule of threshold limit values for atmospheric contaminants published by the ACGIH (2).

In the province of Manitoba, the Department of Environment and Workplace Safety and Health is responsible for legislation and its administration concerning the OELs. Manitoba relies on the provisions of the Workplace Safety and Health Act (Chapter W210), ("assented to" June 11, 1976) for protection of workers. Section 4 provides that "every employer shall provide and maintain a workplace that is safe and without the risks of health, as far as is reasonably practicable" together with other sections on enforcement by the Department of Safety and Health Officer. The guidelines currently used to interpret risk to health are the ACGIH TLVs with the exception that carcinogens are given a zero exposure level "so far as is reasonably practicable" (2).

The New Brunswick Occupational Health and Safety Commission is the authority responsible for the health and safety of workers in this province. Under the Occupational Health and Safety Act, New Brunswick has adopted the ACGIH TLVs. The applicable standards are those published in the latest ACGIH issue and, in case of an infraction, it is the issue in publication at the time of infraction that dictates compliance.

The Northwest Territories Safety Division of the Justice and Service Department regulates workplace safety for non-federal employees in the N.W.T. The latest edition of the ACGIH TLVs has been adopted by the Division and these values have the force of law (2).

The Department of Labour and Manpower administers the Public Health Act of Nova Scotia, under which the list of OELs is a legal requirement. This list is the same as that of the ACGIH as published in 1976 and its subsequent amendments and revisions. These permissible exposure levels have been adopted and constituted as regulations under both the Industrial Safety Act and Regulations and the Construction Safety Act and Regulations (2).

In the Province of Ontario, regulations for a number of hazardous substances are enforced under the Occupational Health and Safety Act, Revised Statutes of Ontario, 1980, Chapter 321. These regulations are administered by the Occupational Hygiene Services, Occupational Health Branch, Ministry of Labour. Regulations are published each in a separate booklet that includes the permissible exposure level and codes for respiratory equipment, techniques for measuring airborne concentrations and medical surveillance approaches.

In the Province of Quebec, the Act is administered by the Commission of Occupational Health and Safety. Permissible exposure levels are similar to the ACGIH TLVs and compliance with the permissible exposure levels for workplace air contaminants is a legal requirement in Quebec (2).

6.7 Chile

The term "Concentraciones Ambientalis Maximas Permissibles" with the acronym "CAMP" is used for maximum permissible atmospheric concentrations in Chile. This refers to an average exposure during 8 hours daily with a weekly exposure of 48 hours which can be exceeded only momentarily. In the list of permissible exposure limits for Chile, the concentrations of eleven of the substances, indicated by an asterisk, cannot be exceeded for even a moment, these having the capacity of causing acute, severe or fatal effects.

The CAMP Standards are legally enforceable under Title III on environmental contamination of workplaces of Decreto No. 78 in conformity with the "Sanitary Code." The administration of the Regulation is by the Department of Occupational Health and Environmental Contamination, Institute of Public Health of Chile, Ministry of Health. The values in the Chile standard are those of the ACGIH TLVs, to which a factor of 0.8 is applied in view of the 48-hour week (2).

6.8 Denmark

The Danish OELs of April 1984 include values for 542 chemical substances and 20 particulates. It is legally required that these not be exceeded as time-weighted averages. Data from the ACGIH are used in the preparation of the Danish standards. About ¼ of the values are different from those of ACGIH, with nearly all of these being somewhat more stringent (2). A useful review of practices in the Nordic countries was published in 1993 (143).

6.9 Ecuador

The Division of Occupational Hazards of the Ecuadorian Institute of Social Security is responsible for control of occupational health hazards, but at this time Ecuador does not have a list of permissible exposure levels incorporated in its legislation. The TLVs of the ACGIH are used as a guide for good industrial hygiene practice (2).

6.10 Finland

The OELs for Finland are defined as concentrations that are deemed to be hazardous to at least some workers on long term exposure. In the establishment of these values, possible effects to especially sensitive persons such as those with allergies are not taken into consideration, nor are those exposures where the possibility of deleterious effect is very improbable.

It is stated that whereas the ACGIH has as their philosophy that nearly all workers may be reportedly exposed to substances below the threshold limit value without adverse effect, the viewpoint in Finland is just the opposite in considering that where exposures are above the limiting value, deleterious effects on health may occur.

6.11 Germany

The definition of the MAC value is "the maximum permissible concentration of a chemical compound present in the air within a working area (as gas, vapor, particulate matter) which, according to current knowledge, generally does not impair the health of the employee nor cause undue annoyance. Under these conditions, exposure can be repeated and of long duration over a daily period of eight hours, constituting an average work week of 40 hours (42 hours per week as averaged over four successive weeks for firms having four work shifts) . . . Scientifically based criteria for health protection, rather than their technical or economical feasibility, are employed."

6.12 Ireland

Workplace air contaminants are regulated by Section 20 of the Safety in Industry Act of 1980 under the direction of the Industrial Inspectorate of the Ministry of Labor. For the purpose of enforcing and interpreting this Section, the latest TLVs of the ACGIH are normally used. However, the ACGIH list is not incorporated in the national laws or regulations.

6.13 Japan

The process for setting limits in Japan has been described by Cook (2). It is not significantly different from that used in most other countries. In recent years, the process for setting limits for certain classes of chemicals has changed from the traditional approach that relied on safety factors to a greater emphasis on more complex approaches. For example, when setting OELs for carcinogens, low-dose extrapolation models and risk criterion like 1 in 1,000 have been adopted. Examples of how this approach has been applied were recently described by Kaneko et al. (142).

6.14 Netherlands

The OELs are used as a guide by the Labour Inspectorate administered by the Director General of Labour. The levels listed for the Netherlands in Table VII are those published as the "National MAC-list 1985 Arbeidsinspectie P no 145." Most desirably, the substances in this list include the Chemical Abstracts Service (CAS) numbers. These values are a revision of the 1981 and 1982/83 lists based on the advice of the National MAC Commission.

These MAC values are taken largely from the list of the ACGIH. A number of values are based on those of the Federal Republic of Germany's Senate Commission for the Investigation of Health Hazards of Work Materials; others are taken from the recommendations of the United States National Institute of Occupational Safety and Health.

The "Maximal Accepted Concentration" is the term used in Netherlands with emphasis on the word "accepted" rather than acceptable to point to the fact that the MAC value has been accepted by the authorities. The MAC of a gas, vapor, fume or dust of a substance is defined as "that concentration in the workplace air which, according to present knowledge, after repeated long-term exposure even up to a whole working life, in general does not harm the health of workers or their offspring." It is to be noted that the words "or their offspring" in this definition are not included in the definition of the ACGIH, on which the Dutch TLVs are based. In the MAC list, two types of values are used:

1. Maximum Aanaarde Concentratie—tijdgewogen gemiddelde (MAC-TGG). This is the maximal accepted concentration averaged over an exposure period up to 8 hours per day and 40 hours per week.
2. Maximale Aanvaarde Concentratie—Ceiling (MAC-C). This concentration may not be exceeded in any case. A MAC-C is based on short term toxic action.

6.15 Philippines

The Bureau of Working Conditions has adopted the entire list of the 1970 TLVs of the ACGIH. This list was incorporated in the Occupational Safety and Health Standards promulgated in 1978. In 1986, the only deviations from the 1970 list were 50 ppm for vinyl chloride and 0.15 mg/m^3 for lead, inorganic compounds, fume and dust (2).

6.16 Organization of Russian States (Former USSR)

The former USSR established many of its limits with the goal of eliminating any possibility for even reversible effects, such as those involving subtle changes in behavioral response, irritation or discomfort. The philosophical differences between limits set in the USSR and in the United States have been discussed by Letavet (144), a Russian toxicologist, who stated that:

> The method of conditioned reflexes, provided it is used with due care and patience, is highly sensitive and therefore it is a highly valuable method for the determination of threshold concentrations of toxic substances.
>
> At times disagreement is voiced with Soviet MAC's for toxic substances, and the argument is that these standards are founded on a method which is excessively sensitive, namely the method of conditioned reflexes. Unfortunately, science suffers not a surplus of excessively sensitive methods, but their lack. This is particularly true with regard to medicine and biology (144).
>
> Although the methods of examination of the higher nervous activity are very sensitive, they cannot be considered to always be the most sensitive indicator of an adverse response and to enable us always to discover the harmful after-effects of being exposed to a poison at the earliest time.

Such subclinical and fully reversible responses to workplace exposures have thus far been considered too restrictive to be useful in the United States and in most other countries. In fact, due to the economic and engineering difficulties in achieving such low levels of air contaminants in the workplace, there is little indication that these limits have actually been achieved in countries that have adopted them. Instead, the limits appear to serve more as idealized goals rather than limits manufacturers are legally bound or morally committed to achieve (51, 54, 55).

Vincent (47) provides a recent analysis of the process used in Russia. One of the most informative portions of his report summarizes the Russian MAC setting process:

> The MAC development process is entirely toxicological, without reference to occupational hygiene or epidemiology. It is carried out under the auspices of the Ministry of Health, acting through the Russian Federation Department of Sanitary and Epidemiological Surveillance (DSES).
>
> In preparing the MACs, the material considered is derived mostly from Russian sources. There is no discussion about prevailing exposure levels in industry, technical feasibility, or economical implications. The development of an MAC for a given substance is based entirely on its potential impact on the health of the worker, to the exclusion of all other considerations. But, even at the level of basic health effects, the criterion for setting an MAC is generally much more stringent than that adopted by other standards-setting bodies.

Of all countries, the OELs used by Russia or the other countries formerly part of the USSR are the most difficult to interpret due to the factors mentioned here. Their applicability to other countries is therefore limited.

7 CONCLUSIONS

Although it is not possible for any single book chapter to discuss how each of the various biological issues that need to be considered when establishing an OEL should be quantitatively accounted for, most of them have at least been generally addressed here. It should be clear from the discussion that the process for setting OELs remains remarkably similar to those that were used in the late 1940s but that the quality and quantity of data used to set these limits, as well as the methodology, has evolved with our increased level of scientific understanding. It is also clear that as occupational health professionals develop a better understanding of toxicology and medicine, techniques for quantitatively accounting for pharmacokinetic differences among chemicals, and better knowledge of the mechanisms of action of toxicants, more refined approaches for identifying safe levels of exposure will be developed. Hopefully, the end result will be that future occupational exposure limits will be based on the best scientific principles and, therefore, our confidence that workers will be protected at these limits will be even greater than it is today.

BIBLIOGRAPHY

1. M. E. LaNier. *"Threshold Limit Values: Discussion and 35 year index with recommendations (TLVs: 1946–81)"* American Conference of Governmental Industrial Hygienists, Cincinnati, OH, 1984.
2. W. A. Cook, *"Occupational Exposure Limits-Worldwide,"* Am. Ind. Hyg. Assoc. Akron, Ohio. 1986.
3. American Conference of Governmental Industrial Hygienists (ACGIH), *1999 Threshold Limit Values (TLVs)® for Chemical Substances and Physical Agents and Biological Exposure Indices (BEIs)*. American Conference of Governmental Industrial Hygienist, 1330 Kemper Meadow Drive, Cincinnati, OH 45240, 1999.
4. H. E. Stokinger, "Criteria and procedures for Assessing the Toxic Responses to Industrial Chemicals," in *Permissible Levels of Toxic Substances in the Working Environment*, ILO, World Health Organization, Geneva, 1970.
5. J. Doull, "The ACGIH Approach and Practice," *Appl. Occup. Environ. Hyg.* **9**(1), 23–24 (1994).
6. H. E. Stokinger. "Threshold Limit Values: Part I," in *Dangerous Properties of Industrial Materials Report*, 8–13. (May–June 1981).
7. W. A. Cook, "Maximum Allowable Concentrations of Industrial Contaminants," *Ind. Med.* **14**(11), 936–946 (1945).
8. H. F. Smyth. "Improved Communication; Hygienic Standard for Daily Inhalation," *Am. Ind. Hyg. Assoc. Qtrly.* **17**, 129–185 (1956).
9. B. Ramazinni. *De Morbis Atrificum Diatriba* (Diseases of Workers) (Translated by W. C. Wright in 1940). The University of Chicago Press, Chicago, IL, 1700.

10. A. M. Baetjer, "The Early Days of Industrial Hygiene. Their Contribution to Current Problems," *Am. Ind. Hyg. Assoc. J.* **41**, 773–777 (1980).
11. K. B. Lehmann, "Experimentelle Studien uber den Einfluss Technisch und Hygienisch Wichtiger Gase und Dampfe auf Organismus: Ammoniak und Salzsauregas," *Arch. Hyg.* **5**:1–12 (1886).
12. K. B. Lehmann and L. Schmidt-Kehl. "Die 13 Wichtigsten Chlorkohlenwasserstoffe der Fettreihe vom Standpunkt der Gewerbehygiene," *Arch. Hyg. Barkt.* **116**, 131–268 (1936).
13. R. Kobert. "The Smallest Amounts of Noxious Industrial Gases which are Toxic and the Amounts which may Perhaps be Endured," *Compendium of Practical Toxicology* **5** 45 (1912).
14. H. H. Schrenk. "Interpretation of Permissible Limits," *Am. Ind. Hyg. Assoc. Qtrly.* **8**, 55–60 (1947).
15. A. C. Fieldner, S. H. Katz, and S. P. Kenney. "Gas Masks for Gases Met in Fighting Fires," *USA Bureau of Mines Bulletin* 248. Pittsburgh, PA, 1921.
16. National Safety Council (NSC), *Final Report of the Committee of the Chemical and Rubber Sector on Benzene*, National Bureau of Casualty and Surety Underwriters. May, 1926.
17. Report of Miners'. *Phthisis Prevention Committee.* Johannesburg, Union of South Africa, 1916.
18. R. R. Sayers. Toxicology of Gases and Vapors. *International Critical Tables of Numerical Data, Physics, Chemistry and Toxicology*, Vol. 2, McGraw-Hill, New York, 1927, 318–321.
19. R. R. Sayers and J. M. Dalle Valle. "Prevention of Occupational Diseases other than those that are caused by Toxic Dust." *Mech. Eng.* **57**, 230–234 (1935).
20. K. B. Lehmann and F. Flury. *Toxikologie und Hygiene der technischen Losungsmittel*, Julius Springer Verlag, Berlin, 1938.
21. M. Bowditch, D. K. Drinker, P. Drinker, H. H. Haggard, and A. Hamilton, "Code for Safe Concentrations of Certain Common Toxic Substances Used in Industry," *J. Ind. Hyg. Tox.* **22**, 251 (1940).
22. Y. Henderson, and H. H. Haggard. *Noxious Gases and the Principles of Respiration Influencing their Action.* Reinhold Publishing Corp., New York, 1943.
23. F. Flury and F. Zernik. *Schadliche Gase, Dampfe, Nebel, Rauch-und Staubarten*, Julius Springer Verlag, Berlin, 1931.
24. World Health Organization. *"Methods Used in Establishing Permissible Levels in Occupational Exposure to Harmful Agents,"* Tech. Report 601, International Labor Office, WHO, Geneva, 1977.
25. M. L. Dourson and J. F. Stara, "Regulatory History and Experimental Support of Uncertainty (Safety) Factors," *Regulatory Toxicol. Pharm.* **3**, 224–238 (1983).
26. U.S. Environmental Protection Agency (EPA), *The Proposed Carcinogen Guidelines.* Washington, DC, 1996.
27. A. Lehman and O. G. Fitzhugh. "100-fold Margin of Safety," *O. Bull-Assoc. Food Drug Off.* **18**, 33–35 (1954).
28. D. J. Paustenbach. "Occupational Exposure Limits: Their Critical Role in Preventive Medicine and Risk Management (editorial)," *Am. Ind. Hyg. Assoc. J.* **51**, A332–A336 (1990).
29. L. E. Adkins, et al. "Letter to the Editor," *Appl. Occup. Environ. Hyg.* **5**(11), 748–750 (1990).
30. W. Clark Cooper, Indicators of Susceptibility to Industrial Chemicals. *J. Occ. Med.* **15**(4), 355–359 (1973).
31. R. G. Smith and J. B. Olishifski. Industrial Toxicology, in J. Olishifski, ed., *Fundamentals of Industrial Hygiene*, National Safety Council, Chicago, IL, 1988, Chap. 15, pp. 359–386.

32. American Conference of Governmental Industrial Hygienists (ACGIH). *Documentation of Threshold Limit Values* (TLVs)® American Conference of Governmental Industrial Hygienists, 1330 Kemper Meadow Drive, Cincinnati, OH 45240, 1999.
33. D. J. Paustenbach, "Occupational Exposure Limits, Pharmacokinetics, and Unusual Work Schedules," in (R. L. Harris, L. J. Cralley, and L. V. Cralley, eds.) *Patty's Industrial Hygiene and Toxicology.* Vol. 3(A), (3rd edition). Theory and Rationale of Industrial Hygiene Practice: The Work Environment, John Wiley & Sons, New York, 1994, Chapter 7, pp. 191–301.
34. J. C. Bailer, E. A. C. Crouch, R. Shaikh, and D. Spiegelman, "One-Hit Models of Carcinogenesis: Conservative or Not?" *Risk Anal.* **8**, 485–497 (1988).
35. P. G. Watanabe, R. H. Reitz, A. M. Schumann, M. J. McKenna, and P. J. Gehring. "Implications of the Mechanisms of Tumorigenicity for Risk Assessment," in M. Witschi, ed., *The Scientific Basis of Toxicity Assessment*, Elsevier/North-Holland Press, 1980 pp. 69–88.
36. B. Butterworth, R. B. Conolly, and I. J. Morgan, "A Strategy for Establishing Mode of Action of Chemical Carcinogens as a Guide for Approaches to Risk Assessments," *Cancer Letters* **93**, 129–146 (1995).
37. B. E. Butterworth and T. Slaga, *Nongenotoxic Mechanisms in Carcinogenesis: Banbury Report 25.* Cold Spring Harbor Laboratory, Cold Spring Harbor, NY, 1987.
38. C. N. Park and N. C. Hawkins, "Cancer Risk Assessment," in L. J. Cralley, L. V. Cralley, and J. Bus, ed. *Patty's Industrial Hygiene and Toxicology.* Vol. 3B, 3rd ed, John Wiley & Sons, New York, 1995, pp. 275–318.
39. J. P. Seiler. "Apparent and Real Thresholds: A Study of Two Mutagens," in (D. Scott, B. A. Bridges, and F. H. Sobels, eds.) *Progress in Genetic Toxicology* by, Elsevier Biomedical Press. New York, 1977.
40. C. F. Wilkinson, "Being More Realistic about Chemical Carcinogenesis," *Environ. Sci. Technol.* **9**, 843–848 (1988).
41. J. S. Bus and J. E. Gibson, "Body Defense Mechanisms to Toxicant Exposure" in L. J. Cralley, L. V. Cralley, and J. Bus, eds. *Patty's Industrial Hygiene and Toxicology.* 3rd ed., Vol. 3B, John Wiley & Sons, New York, 1995, pp. 159–192.
42. J. V. Rodricks, S. Brett, and G. Wrenn. "Significant Risk Decisions in Federal Regulatory Agencies," *Regul. Toxicol. Pharmacol.* **7**, 307–320 (1987)
43. C. C. Travis, S. A. Richter, E. A. Crouch, R. Wilson, and E. Wilson. "Cancer risk management: A review of 132 Federal Regulatory Decisions," *Env. Science Tech.* **21**(5), 415–420 (1987).
44. P. Hewitt. "Interpretation and Use of Occupational Exposure Limits for Chronic Disease Agents." in *Occupational Medicine: State-of-the-Art Reviews.* **11**(3), 561–590 (1996).
45. World Health Organization. *Occupational Exposure Limits for Airborne Toxic Substances*, 4th ed., Occ. Safety and Health Series, No. 37, International Labor Office, WHO, Geneva, 1995.
46. H. B. Elkins, "Maximum Acceptable Concentrations, A Comparison in Russia and the United States," *AMA Arch. Environ. Health* **2**, 45–50 (1961).
47. J. H. Vincent, "International Occupational Exposure Standards: A Review and Commentary," *Am. Ind. Hyg. Assoc. J.* **59**, 729–742 (1998).
48. American Public Health Association (APHA). *Health Based Exposure Limits and Lowest National Occupational Exposure Limits.* Draft #5. November 6, 1991. Washington, DC, 1991.
49. Occupational Safety and Health Administration (OSHA), U.S. Congress (91st) S.2193, Public Law 91-596, Washington, D.C. (1970).
50. D. J. Paustenbach and R. R. Langner. "Setting Corporate Exposure Limits: State of the Art," *Amr. Ind. Hyg. Assoc. J.* **47**, 809–818 (1986).

51. H. L. Magnusson. "Industrial Toxicology in the Soviet Union Theoretical and Applied," *Am. Ind. Hyg. Assoc. J.* **25**, 185–190 (1964).
52. Occupational Safety and Health Administration (OSHA). *Air Contaminants: Final Rule, Fed. Reg.* **54**(12), 2332–2983 (Jan. 19, 1989).
53. E. I. Lyublina, "Some Methods Used in Establishing the Maximum Allowable Concentrations," *MAC of Toxic Substances in Industry. IUPAC*, 1962 pp. 109–112.
54. H. E. Stokinger. "International Threshold Limits Values," *Amer. Ind. Hyg. Assoc. J.* **24**, 469 (1963).
55. C. C. Willhite and N. N. Oseas. "Reconciliation of American TLVs and German MAKs," *Occup. Hyg.* **4**, 1–16 (1997).
56. J. Perkins. *The Practice of Industrial Hygiene (a 3 volume text)*. Amer Conf of Gov't Industrial Hygienists. Cincinnati, OH, 2000.
57. H. E. Stokinger, "The Case for Carcinogen TLVs Continues Strong," *Occup. Health and Safety* **46**, 54–58 (March/April 1977).
58. B. Clayton, R. Kroes, J. C. Larsen, and G. Pascal, "Applicability of the ADI to Infants and Children," *Food Additives and Contaminants*, **15**(Supplement) 1–90 (1998).
59. C. S. Weil. "Statistics Versus Safety Factors and Scientific Judgment in the Evaluation of Safety for Man," *Toxicol. Appl. Pharmicol.* **21**, 454–463 (1972).
60. E. J. Calabrese, *Methodological Approaches to Deriving Environmental and Occupational Health Standards*, John Wiley & Sons, New York, 1978.
61. E. J. Calabrese, *Principles of Animal Extrapolation*. John Wiley & Sons, New York, 1983.
62. R. L. Zielhuis and F. Van Der Kreek. "Calculations of a Safety Factor in Setting Health Based Permissible Levels for Occupational Exposure. A Proposal I," *Int. Arch. Occup. Environ. Health* **42**, 191–201 (1979).
63. R. L. Zielhuis and F. W. Van Der Kreek. "Calculations of a Safety Factor in Setting Health Based Permissible Levels for Occupational Exposure. A Proposal II. Comparison of Extrapolated and Published Permissible Levels," *Int. Arch. Occup. Environ. Health* **42**, 203–215 (1979).
64. H. W. Leung and D. J. Paustenbach. "Application of Pharmacokinetics to Derive Biological Exposure Indexes from Threshold Limit Values," *Am. Ind. Hyg. Assoc. J.* **49**, 445–450 (1988).
65. H. P. Illing, "Extrapolating from Toxicity Data to Occupational Exposure Limits: Some Considerations," *Ann Occup Hyg.* **35**(6), 569–580 (1991).
66. H. W. Leung and D. J. Paustenbach, Setting Occupational Exposure Limits for Irritant Organic Acids and Bases Based on Their Equilibrium Dissociation Constants, *Applied Occup. Environ. Hyg.* **3**, 115–118 (1988).
67. American Industrial Hygiene Association (AIHA). *An International Review of Procedures for Establishing Occupational Exposure Limits*, AIHA, Prosperity Ave., Suite 250. Fairfax, VA 22031, 1996.
68. H. W. Leung and D. J. Paustenbach, "Physiologically-based Pharmacokinetic and Pharmacodynamic Modeling in Health Risk Assessment and Characterization of Hazardous Substances," *Toxicology Letters* **79**, 55–65 (1995).
69. J. M. Paull. "The Origin and Basis of Threshold Limit Values," *Amr. J. Ind. Med.* **5**, 227–238 (1984).
70. D. A. Neumann and C. A. Kimmel, "Human Variability in Response to Chemical Exposure," Inter Life Sci Institute, Washington, D.C. (1998).

71. A. M. Jarabek, M. G. Menache, et al. "The U.S. Environmental Protection Agency's inhalation RfD Methodology: Risk Assessment for Air Toxics," *Tox. Ind. Health* **6**(5), 279–301 (1990).
72. Environmental Protection Agency (EPA), IRIS: Integrated Risk Information System Database, Washington, D.C., 1998.
73. D. G. Barnes and M. Dourson, "Reference Dose (RfD): Description and Use in Health Risk Assessments," *Regul. Toxicol. Pharmacol.* **8**, 471–486 (1988).
74. M. L. Dourson, S. P. Felter, and D. Robinson, "Evolution of Science-based Uncertainty Factors for Noncancer Risk Assessment," *Regul. Toxicol. Pharmacol.* **24**, 108–120 (1996).
75. A. G. Renwick and N. R. Lazarus, "Human Variability and Non-Cancer Risk Assessment: An Analysis of Default Uncertainty Factor," *Regul. Toxicol. Pharm.* **27**, 3–120 (1998).
76. M. L. Dourson, "Methods for Establishing Oral Reference Doses (RfDs)," in *Risk Assessment of Essential Elements*, W. Mertz, C. O. Abernathy and S. S. Olin, eds. ILSI Press, Washington, D.C., 1994, pp. 51–61.
77. International Life Sciences Institute (ILSI) *Similarities and Differences Between Children and Adults*, (P. S. Guzelian, C. J. Itenry, and S. S. Olin, eds.), Washington, D.C., 1992.
78. S. A. Roach, and S. M. Rappaport. "But They Are Not Thresholds: A Critical Analysis of the Documentation of Threshold Limit Values," *Am. J. Ind. Med.* **17**, 727–753 (1990).
79. M. A. Ottobani. *The Dose Makes the Poison*, 2nd Edition, Van Nostrand, New York, 1991.
80. M. Meldrum, "Setting OELs for Sensory Irritants. The Approach in the European Union," *Amer. Ind. Hyg. Assn. J.* (in press).
81. C. C. Willhite, "Origin and Derivation of TLVs and AEGLs Intended to Prevent Sensory Irritation," *Amer. Ind. Hyg. Assn. J.* (in press).
82. P. Dalton, "Evaluating the Human Response to Sensory Irritation: Implications for Setting Occupational Exposure Limits (OELs)," *Amer. Ind. Hyg. Assn. J.* (in press).
83. D. J. Paustenbach, Y. Alarie, T. Kulle, N. Schachter, R. Smith, J. Swenberg, H. Witschi, and S. Horowitz. "A Recommended Occupational Exposure Limit For Formaldehyde Based On Irritation," *J Toxicol. Environ. Health* **50**, 217–263 (1997).
84. L. E. Kane and Y. Alarie. "Sensory Irritation to Formaldehyde and Acrolein during Single and Repeated Exposures in Mills," *Am. Ind. Hyg. Assoc. J.* **38**, 509–522 (1977).
85. Y. Alarie, "Dose Response Analysis in Animal Studies: Prediction of Human Responses," *Environ. Health Perspect.* **42**, 9–13 (1981).
86. M. H. Abraham, G. S. Whiting, Y. Alarie, et al. "Hydrogen Bonding 12. A New QSAR for Upper Respiratory Tract Irritation by Airborne Chemicals in Mice," *Quant. Struct. Relat.* **9**, 6–10 (1990).
87. G. D. Nielsen, "Mechanisms of Activation of the Sensory Irritant Receptor by Airborne Chemicals," *Crit. Reviews in Toxicol.* **21**, 183–208 (1991).
88. Y. Alarie and G. D. Nielsen, "Sensory Irritation, Pulmonary Irritation, and Respiratory Stimulation by Airborne Benzene and Alkylbenzenes: Prediction of Safe Industrial Exposure Levels and Correlation with their Thermodynamic Properties," *Toxicol. Appl. Pharmacol.* **65**, 459–477 (1982).
89. M. Schaper. "Development of a Database for Sensory Irritants and its use in Establishing Occupational Exposure Limits," *Am. Ind. Hyg. Assoc. J.* **54**(9), 488–544 (1993).
90. D. J. Paustenbach, "Approaches and Considerations for Setting Occupational Exposure Limits (OELs) for Sensory Irritants: Report of a Recent Symposia," *Appl. Occ. Env. Hygiene* (in press).
91. J. L. Schardein, *Chemically Induced Birth Defects*, 2nd ed., Marcel Dekker. New York, 1993.

92. E. M. Johnson. "Cross-Species Extrapolations and the Biologic Basis for Safety Factor Determinations in Developmental Toxicology," *Regul. Toxicol. Pharmacol.* **8**, 22–36 (1988).
93. J. M. Manson and Y. J. Kang. "Test Methods for Assessing Female Reproductive and Development Toxicology," in Hayes, ed., *Principles and Methods in Toxicology*. Raven Press, New York, Chapt. 29, 1994, pp. 989–1038.
94. U.S. Environmental Protection Agency (EPA). *Proposed Guidelines for Health Assessment of Suspect Developmental Toxicants*. Fed. Reg., Washington, DC, 1986.
95. D. W. Gaylor, "Quantitative Risk Analysis for Quantal Reproductive and Developmental Effects," *Environ. Health Perpect* **79**, 243–246 (1989).
96. E. M. Johnson. "A Case Study of Developmental Toxicity Risk Estimation Based on Animal Data. The Drug Bendectin," in D. J. Paustenbach, ed., *The Risk Assessment of Environmental Hazard: A Textbook of Case Studies*, John Wiley & Sons. New York, 1989, pp. 771–724.
97. C. A. Kimmel, R. J. Kavlock, B. C. Allen, and E. M. Faustman. "Benchmark Dose Concept Applied to Data from Conventional Development Toxicity Studies," *Toxicol. Lett.* **82/88**, 549–554 (1995).
98. D. A. Karnovsky, in J. G. Wilson and J. Warrany, eds. *Teratology: Principle and Techniques*, University of Chicago, Press. Chicago, IL, 1965, Chapt. 8.
99. D. J. Paustenbach. "Assessment of the Developmental Risks Resulting from Occupational Exposure to Select Glycol Ethers within the Semi-conductor Industry." *J. Toxicol. Env. Health.* **23**, 53–96 (1988).
100. K. Crump, "Calculation of Benchmark Doses from Continuous Data," *Risk Anal.* **15**, 79–85 (1995).
101. A. G. Renwick, "The Use of an Additional Safety or Uncertainty Factor for Nature of Toxicity in the Estimation of Acceptable Daily Intake and Tolerable Daily Intake Values," *Regul. Toxicol. Pharmacol.* **22**, 250–261 (1995).
102. H. Zenick, E. D. Clegg, S. D. Perreault, G. R. Klinefelder, and L. Earl Gray. "Assessment of Male Reproductive Toxicity: a Risk Assessment Approach," Chapter 27. in A. W. Hays, ed., *Principle and Methods of Toxicology*. Raven Press, New York, 1994, pp. 937–988.
103. H. W. Leung, D. J. Paustenbach, F. J. Murray. "A Proposed Occupational Exposure Limit for 2,3,7,8-TCDD," *Am. Ind. Hyg. Assn. J.* **49**(9), 466–474 (1988).
104. G. Briatico-Vangosa, C. L. Braun, G. Cookman, T. Hofmann, I. Kimber, S. E. Loveless, T. Morros, J. Pauluhn, T. Sorensen, and H. J. Niessen, "Review: Respiratory Allergy: hazard Identification and Risk Assessment," *Fundam. Appl. Toxicol.* **23**, 145–158 (1994).
105. K. Sarlo, and M. H. Karol. (1994). "Guinea Pig Productive Tests for Respiratory Allergy," Chapter 41 in Dean, Luster, Munson, and Kimber eds., *Immunotoxicology and Immunopharmacology,* Raven.
106. C. Graham, H. S. Rosenkranz, and M. H. Karol, "Structure–Activity Model of Chemicals that cause Human Respiratory Sensitization," *Regul. Toxicol. Pharmacol.* **26**(3), 296–306 (1997).
107. C. J. Powell and C. L. Berry, "Non-genotoxic or Epigenetic Carcinogenesis," Chapter 51, pp. 1119–1137, in B. Ballantyne, T. Marrs, and T. Syversen, eds., *General and Applied Toxicology*, 2nd Edition, Macmillian Reference Ltd., London, 1999.
108. H. C. Pitot and Y. P. Dragan, "Chemical Carcinogenesis," in C. Klaassen, J. Doull, and M. Amdur, eds., *Casarett and Doull's Toxicology: The Basic Science of Poisons* 5th *Edition*, Chapter 8, pp. 201–267, McMillan Publishers, New York, 1994.
109. B. N. Ames, L. S. Gold, and W. C. Willet, "The Causes and Prevention of Cancer," *Proc. Natl. Acad. Sci.* **92**(12), 5258–5265 (1995).

110. International Life Sciences Institute, "Low-dose Extrapolation of Cancer Risks," S. Olin, W. Farland, C. Park, L. Rhomberg, R. Scheuplein, T. Starr, and J. Wilson, eds., Washington, D.C., 1995.
111. M. C. R. Alavanja, C. Brown, R. Spirtas, and M. Gomez, "Risk Assessment of Carcinogens: A comparison of the ACGIH and the EPA." *Appl. Occup. Environ. Hyg.* **5**(8), 510–517 (1990).
112. D. Krewski, D. Murdoch, and J. Withey. "Recent Developments in Carcinogenic Risk Assessment," *Health Physics* **57**, 313–325 (1989).
113. K. S. Crump, D. G. Hoel, G. H. Langley, and R. Peto, Carcinogenic processes and their implications for low dose risk assessment *Cancer Res.* **36**, 2973–2979 (1976).
114. C. D. Holland and R. L. Sielken, Jr. *Quantitative Cancer Modeling and Risk Assessment.* Prentice Hall, Englewood Cliffs, NJ, 1993.
115. D. J. Paustenbach, "Health Risk Assessments: Opportunities and Pitfalls. *Columbia Journal of Environmental Law.*" **14**(2), 379–410 (1989).
116. J. P. J. Mavrissen and J. L. Mattsson, "Neurotoxicology: An Orientation," in L. J. Cralley, L. V. Cralley, and J. S. Bus, eds., *Patty's Industrial Hygiene and Toxicology*, 3rd ed., Vol. 3B, John Wiley and Sons, New York, 1995, Chapter 6, pp. 231–254.
117. R. S. H. Yang, *Toxicology of Chemical Mixtures: Case Studies, Mechanisms, and Novel Approaches*, Academic Press, New York, 1994.
118. U.S. Environmental Protection Agency. *Approaches to Risk Assessment for Multiple Chemical Exposures.* EPA 600/9-84-008. Washington, DC, 1984.
119. P. M. Bos, D. H. Brouwer, H. Stevenson, P. J. Boogaard, W. L. de Kort, and J. J. van Hemmen, "Proposal for the Assessment of Quantitative Dermal Exposure Limits in Occupational Environments: Part 1. Development of a Concept to Derive a Quantitative Dermal Occupational Exposure Limit," *Occup. Environ. Med.* **55**(12), 795–801 (1998).
120. D. H. Brouwer, L. Hoogendoorn, P. M. Bos, P. J. Boogaard, and J. J. van Hemmen, "Proposal for the Assessment to Quantitative Dermal Exposure Limits in Occupational Environments: Part 2. Feasibility Study for Application in an Exposure scenario for MDA by Two Different Dermal Exposure Sampling Methods," *Occup Environ. Med.* **55**(12), 805–811 (1995).
121. B. I. Castleman and G. E. Ziem, "Corporate Influence on Threshold Limit Values," *Am. J. Ind. Med.* **13**, 531–559 (1988).
122. B. I. Castleman and G. E. Ziem, "Toxic Pollutants, Science, and Corporate Influence," *Arch Environ Health.* **44**(2), 68, 127 (1989).
123. J. A. Finklea, "Threshold Limit Values: A Timely Look," *Am. J. Ind. Med.* **14**, 211–212 (1988).
124. E. S. Tarlau. "Industrial Hygiene with no Limits. A Guest Editorial," *Am. Ind. Hyg. Assoc. J.* **51**, A9–A10 (1980).
125. American Conference of Governmental Industrial Hygienists (ACGIH) "Threshold Limit Values: A More Balanced Appraisal," *Appl. Occup. Environ. Hyg.* **5**, 340–344 (1990).
126. S. M. Rappaport. "Threshold Limit Values, Permissible Exposure Limits, and Feasibility: the Bases for Exposure Limits in the United States," *Am. J. Ind. Med.* **23**(5), 683–694 (1993).
127. D. A. Neumann and C. A. Kimmel. *Human Variability in Response to Chemical Exposures: Measures, Modeling and Risk Assessment*, CRC Press, Boca Raton, 1998
128. B. I. Castleman and G. E. Zeim, "American Conference of Governmental Industrial Hygienists: Low Threshold Credibility," *Amer. J. Ind Med.* **26**, 133–143 (1994).
129. J. Graham and J. Hartwell, *The Greening of American Business*, Harvard Univ. Press, Cambridge, MA.

130. D. J. Paustenbach. "Putting Politics Aside, Updating the PELs." *Am. Ind. Hyg. Assoc. J.* **58**, 845–849 (1997).
131. J. W. Mason and H. Dershin. "Limits to Occupational Exposure in Chemical Environments Under Novel Work Schedules," *J. Occup. Med.* **18**, 603–607 (1976).
132. J. L. S. Hickey and P. C. Reist. "Adjusting Occupational Exposure Limits for Moonlighting, Overtime, and Environmental Exposures," *Am. Ind. Hyg. Assoc. J.* **40**, 727–734 (1979).
133. J. L. S. Hickey and P. C. Reist. "Application of Occupational Exposure Limits to Unusual Work-Schedules," *Am. Ind. Hyg. Assoc. J.* **38**, 613–621 (1977).
134. S. A. Roach. "Threshold Limit Values for Extraordinary Work Schedules" *Am. Ind. Hyg. Assoc. J.* **39**, 345–364 (1978).
135. J. L. S. Hickey. "Adjustment of Occupational Exposure Limits for Seasonal Occupations," *Am. Ind. Hyg. Assoc. J.* **41**, 261–263 (1980).
136. J. L. S. Hickey. "The "TWAP" in the Lead Standard," *Am. Ind. Hyg. Assoc. J.* **44**(4), 310–311 (1983).
137. Occupational Safety and Health Administration. *OSHA Compliance Officers Field Manual.* Department of Labor, Washington, DC 1976.
138. R. S. Brief and R. A. Scala, "Occupational Exposure Limits for Novel Work Schedules," *Am. Ind. Hyg. Assoc. J.* **36**, 467–471 (1975).
139. M. E. Andersen, M. G. MacNaughton, H. J. Clewell, and D. J. Paustenbach, "Adjusting Exposure Limits for Long and Short Exposure Periods Using a Physiological Pharmaokinetic Model," *Am. Ind. Hyg. Assoc. J.* **48**, 335–343 (1987).
140. R. Goyal, K. Krishnan, R. Tardif, S. Lapare, and J. Brodeur, "Assessment of Occupational Health Risk during Unusual Workshifts: Review of the Needs and Solutions for Modifying Environmental and Biological Limit Values for Volatile Organic Solvents," *Can. J. Public Health* **83**(2), 109–112 (1992).
141. H. W. Leung and D. J. Paustenbach, "Physiologically-based Pharmacokinetic and Pharmacodynamic Modeling in Health Risk Assessment and Characterization of Hazardous Substances," *Toxicol. Lett.* **79**, 55–65 (1995).
142. T. Kaneko, P. Y. Wang, and A. Sato. Development of Occupational Exposure Limits in Japan. *Int. J. Occup. Med. Environ. Health.* **11**(1), 81–98 (1998).
143. B. Beije and P. Lundberg, "Occupational Exposure Limits–Health Based Values or Administrative Norms?" *Proceedings of the First Intercourse on OELs*, April 19–23, 1993, Visby, Sweden, National Institute of Occupational Health, Washington, D.C., 1993.
144. A. A. Letavet. "Scientific Principles for the Establishment of the Maximum Allowable Concentrations of Toxic Substances in the USSR," in: *Proceedings of the 13th Annual Congress on Occupational Health July 25–29, 1960*, Book Craftsmen Assoc., New York, 1961.

GENERAL REFERENCES

M. H. Abraham, J. Andonian-Haftvan, J. E. Cometto-Muniz, and W. S. Cain (1996): An analysis of nasal irritation thresholds using a new solvation equation. *Fundam. Appl. Toxicol.* 31:71–76.

ACGIH, *Transactions of the Third Annual Meeting of the National Conference of Governmental Industrial Hygienists*. Bethesda, MD, April 30–May 2, 1940

ACGIH, *Transactions of the Fourth Annual Meeting of the National Conference of Governmental Industrial Hygienists*, Washington, DC, February 17–18, 1941

ACGIH, *Transactions of the Fifth Annual Meeting of the National Conference of Governmental Industrial Hygienists*, Washington, DC, April 9–10, 1942

ACGIH, *Transactions of the Seventh Annual Meeting of the National Conference of Governmental Industrial Hygienists*, St. Louis, Missouri, May 9, 1944

ACGIH, "The Toxicology of the Newer Metals," *Proceedings of the 8th Annual Meeting of the American Conference of Governmental Industrial Hygienists.* Chicago, IL, 1946 pp. 19–26.

ACGIH, *Proceedings of the 9th Annual Meeting of the American Conference of Governmental Industrial Hygienists*, Buffalo, N.Y., April, 26–29, 1947.

ACGIH, "Threshold Limit Values" (also, report of the committee on threshold limits). *Transactions of the 10th Annual Meeting of the American Conference of Governmental Industrial Hygienists.* Boston, MA, March 27–30, pp. 29–31, 1948

ACGIH, "Report of Committee on Threshold Limits," Transactions of the 11th Annual Meeting of the American Conference of Governmental Industrial Hygienists. Detroit, MI, April 2–5 1949, pp. 63–64.

ACGIH, "Threshold Limit Values of the American Conference of Governmental Industrial Hygienists," *Arch Ind. Hyg. Occup. Med.* **2**, 98–100 (1950)

ACGIH, "Interim report of Committee on Standard Labeling Procedures," *Transactions of the 16th Annual Meeting of the American Conference of Governmental Industrial Hygienists*, Chicago, IL, April 24–27, 1954 pp. 34–42.

ACGIH, Threshold Limit Values for 1955 (also, reports of the committees on threshold limits and standard labeling procedures). *Transactions of the 17th Annual Meeting of the American Conference of Governmental Industrial Hygienists.* Buffalo, April 23–26, 1955 pp. 44–50, 70–71.

ACGIH, "Threshold Limit Values for 1956. American Conference of Governmental Industrial Hygienists," *A.M.A. Arch. Ind. Health* **14**(2), 186–189 (1956).

ACGIH, Threshold Limit Values for 1957 (also, reports of the committees on threshold limits and standard labeling procedures). *Transactions of the 19th Annual Meeting of the American Conference of Governmental Industrial Hygienists.* St. Louis, Missouri, April 20–23, 1957 pp. 47–58.

ACGIH, Threshold Limit Values for 1958 (also, reports of the committees on threshold limits and standard labeling procedures). *Transactions of the 20th Annual Meeting of the American Conference of Governmental Industrial Hygienists.* Atlantic City, NJ, April 19–22, 1958.

ACGIH, Threshold Limit Values for 1959 (also, reports of the committees on threshold limits and standard labeling procedures). *Transactions of the 21st Annual Meeting of the American Conference of Governmental Industrial Hygienists.* Chicago, IL, April 18–21, 1959.

ACGIH, Threshold Limit Values for 1960. *Transactions of the 22th Annual Meeting of the American Conference of Governmental Industrial Hygienists.* Rochester, NY, April 23–26 1960, pp. 79–87.

ACGIH, Reports of the committees on threshold limits and standard labeling procedures. *Transactions of the 27th Annual Meeting of the American Conference of Governmental Industrial Hygienists.* Houston, TX, May 1–5 1965, pp. 121–125.

ACGIH, A Statement on the Use of Threshold Limit Values, *Transactions of the 28th Annual Meeting of the American Conference of Governmental Industrial Hygienists.* Pittsburgh, PA, May 15–17, 1966, pp. 179–180.

M. G. Allmark, "Observations on Acute and Chronic Toxicity Studies," *Canad. J. Pub. Health.* 45, 18–23 (1954)

M. G. Allmark, F. C. Lu, W. D. Graham, and J. R. MacDougal, "Assessment of Drug Toxicity in Canada," *Canad. Med. Assoc. J.* **77**, 1094–1097 (1957)

American Academy of Occupational Medicine, Symposium on "The Translation of Experimental Toxicology Data into Practical Criteria for Plant Exposure." *Proceedings of Fifth Annual Meeting.* February 12–13, Rochester, NY, 1953

American Conference of Governmental Industrial Hygienists (ACGIH), *Protection of the Sensitive Individual*, W. D. Kelly, ed., Cincinnati, Ohio, 1982.

American Industrial Hygiene Association, *Hygienic Guide Series*

Hygienic Guide Series Order Form, 1982, **43**(1); B93–94 (Jan. 1982).

Hygienic Guides, **36**(2): A15–16 (Feb. 1975).

Acetaldehyde, **16**(3), 237–238 (Sept. 1955).

Amyl Acetate, **16**(3), 241–242 (Sept. 1955).

Anhydrous Ammonia, **17**(3), 350–351, Sept. (1956)

Aniline, **16**(4), 331–332 (Dec. 1955).

Arsine, **17**(2), 231–232 (June 1956).

Benzene (Benzol), **16**(4), 333–335 (Dec. 1955).

Benzene (Benzol), **22** 504–506 (Dec. 1961).

Benzene (Benzol), **31**(3), 383–388 (May–June 1970).

Benzene (Benzol) (June 1978).

Beryllium and Compounds, **17**(3), 345–346 (Sept. 1956).

Bromochloromethane, **22**, 506–507 (Dec. 1961).

Butyl Alcohol, 1955, **16**(3), 240–241 (Sept. 1955).

Cadmium, **16**(3), 238–239 (Sept. 1955).

Carbon Disulfide, **17**(4), 446–447 (Dec. 1956).

Carbon Monoxide, **17**(3), 346–347 (Sept. 1956).

Carbon Monoxide, **26**(4), 431–434 (July–August 1965).

Carbon Monoxide (Nov. 1985).

Carbon Tetrachloride, **16**(3), 239–240 (Sept. 1955).

Carbon Tetrachloride **22**, 507–509 (Dec. 1961).

Carbon Tetrachloride (June 1978).

Chloroform **26**(6), 636–639 (Nov.–Dec. 1965).

Chloroform (Feb. 1978).

Chromic Acid, **17**(2), 233–234 (June 1956).

Chromic Acid (March 1980).

1,2-Dichloroethane, **17**(4), 447–448 (Dec. 1956).

1,2-Dichloroethane, **26**(4), 435–438 (July–Aug. 1965).

1,2-Dichloroethane (Feb. 1978).

Dichloromethane, **26**(6), 633–636 (Nov.–Dec. 1965).

Ethyl Alcohol, **17**(1), 94–95 (March 1956).

Ethyl Bromide, **26**(2), 192–195 (March–April 1965).

Fluoride-bearing Dusts and Fumes, **17**(2), 230–231 (June 1956).

Fluorine, **17**(2), 229–230, June 1956.

Formaldehyde, (1955), **16**(4), 336–337 (Dec. 1955).
Formaldehyde, **26**(2), 189–192 (March–April 1965).
Formaldehyde, (Dec. 1978).
Hydrazine, **17**(4), 445–446 (Dec. 1956).
Hydrogen Cyanide, **17**(3), 347–349 (Sept. 1956).
Hydrogen Cyanide, **31**(1), 116–119 (Jan.–Feb. 1970).
Hydrogen Fluoride, **17**(1), 98–99 (March 1956).
Hydrogen Sulfide, **16**(4), 335–336 (Dec. 1955).
Mercury, **17**(1), 97–98 (March 1956).
Methylene Dichloride, **17**(4), 448–449 (Dec. 1956).
Nickel Carbonyl, **17**(4): 449–450 (Dec. 1956).
Nitrogen Dioxide, **17**(2): 232–233 (June 1956).
Sulfur Dioxide, **16**(4): 332–333 (Dec. 1955).
Tetrachloroethylene, **21**(3): 270–271 (June 1960).
Tetrachloroethylene, **26**(6): 640–643 (Nov.–Dec. 1965).
1,1,1-Trichloroethane, **17**(1): 93–94 (March 1956).
Trichloroethylene, **17**(1): 95–96 (March 1956)
Trichloroethylene, **25**(1): 94–97 (Jan.–Feb. 1964)
Trichloroethylene, (Sept. 1986).
Zinc Oxide, **17**(3), 349–350 (Sept. 1956).

American Industrial Hygiene Association, *A Guide to Product Health and Safety and the Right to Know*. Ch. III "Product Liability: The Right to Know," Ch. V "Risk Assessment," and Ch. VI "Communicating Product Safety and Health Information," Akron, Ohio, 1986.

American Industrial Hygiene Association (AIHA). *Workplace Environment Exposure Limits.* (WEELS). Washington, DC, 1996.

American National Standards Institute, Inc., American National Standard for the Precautionary Labeling of Hazardous Industrial Chemicals, Jan. 15, Sponsored by the Manufacturing Chemists Association, Inc., 1976.

M. E. Andersen, (1991). Quantitative Risk Assessment and Chemical Carcinogens in Occupational Environments. *Appl. Occup. Env. Hyg.* 3, 267–273.

M. E. Andersen, (1995). Development of physiologically based pharmacokinetic and physiologically based pharmacodynamic models for applications in toxicology and risk assessment. *Toxicol Lett.* 79(1–3):35–44.

R. C. Anderson, F. G. Henderson, and K. K. Chen, "The Rat as a Suitable Animal for the Study of Prolonged Medication," *J. Am Pharm. A. (Scient. Ed.)* **32**(8), 204–208 (1943).

Association of Food & Drug Officials of the United States, 100-fold margin of safety. *Q. Bull. Assoc. Food & Drug Officials U.S.* January: 33–35, 1954.

Association of Food & Drug Officials of the United States, *A Procedure for Determining the Leachability of Uncertified Pigments for Printing Food Wraps and Coloring Plastic Food Containers.* 61st Annual Conference, Louisville, KY, May 6–11, 1957 pp. 113–115.

A. M. Baetjer, "The Early Days of Industrial Hygiene—Their Contribution to Current Problems." *Transactions of the 42th Annual Meeting of the American Conference of Governmental Industrial Hygienists*, 1981 pp. 10–17.

W. L. Ball, "Threshold Limits for Pesticides," *A.M.A. Arch. Indust. Health.* **14**(2), 178–185 (1956).

W. L. Ball, "The Toxicological Basis of Threshold Limit Values: 4. Theoretical Approach to Prediction of Toxicity of Mixtures," *Amer. Indust. Hyg. Assoc. J.* **20**(5), 357–363 (1959).

W. L. Ball, The toxicological basis of threshold limit values: Paper #6. Report of Prague symposium on international threshold limit values. *Amer. Ind. Hyg. Assn. J.* **20**, 370–373 (1959).

F. Bar, "The Toxicological Evaluation (Margin of Safety, Tolerances) for Food Legislation (Food Additives and Pesticide Residues)," *Arch. Toxicol.* **32**, 51–62 (1974).

Z. Bardodej, "Occupational Exposure Limits for Protection of Working People in Modern Society," *Ann. Am. Conf. Ind. Hyg.* **12**, 73–78 (1958).

J. F. Barkley, *"Accepted Limit Values of Air Pollutants"* Information Circular 7682, United States Dept. of the Interior, Bureau of Mines, 1954

D. G. Barnes, G. P. Daston, J. S. Evans, et al. (1995): Benchmark dose workshop: Criteria for use of a benchmark dose to estimate a reference dose. *Reg. Toxicol. Pharmacol.* 21:296–306.

E. C. Barnes, *Keeping Score on Toxic Materials and Processes in The Plant.* Industrial Hygiene Foundation of America, Inc., Seventh Annual Meeting of Members, Nov. 10–11, 1942, Pittsburgh, PA.

J. M. Barnes, The Use of Experimental Pathological Techniques in Assessing the Toxicity of Chemical Compounds, in *The Application of Scientific Methods to Industrial and Service Medicine.* Medical Research Council, His Mafesty's Stationery Office. London, 1951, pp. 49–55.

J. M. Barnes, "Toxic Hazards of Certain Pesticides to Man," *Bull. World Health Org.* **8**, 419–490 (1953).

J. M. Barnes and F. A. Denz, "Experimental Methods Used in Determining Chronic Toxicity," *Pharmacol. Rev.* **6**(2), 191–242 (1954).

J. M. Barnes, "The potential toxicity of chemicals used in food technology." *Proc. Nutrition Soc.* **15**(2): 148–154 (1956).

A. J. Bateman, "Testing Chemicals for Mutagenicity in a Mammal, *Nature.* **210**(5032), 205–206 (1966).

R. R. Bates, "Regulation of Carcinogenic Food Additives and Drugs in the US," in W. Davis and C. Rosenfeld, eds. *Carcinogenic Risks, Strategies for Intervention.* International Agency for Research on Cancer, Lyon, 1979.

B. D. Beck, R. B. Conolly, M. L. Dourson, D. Guth, D. Hattis, C. Kimmel, and S. C. Lewis (1993). Symposium Overview: Improvements in Quantitative Non-cancer Risk Assessments. *Fundam. Appl. Toxicol.* 20:1–14.

B. Behrend and E. Jaeger, "Drug Evaluation in Industrial Medicine—Purposes and Procedures," *Industr. Med. Surg.* **29**(7), 319–323 (1960).

E. F. Bellingham, J. J. Bloomfield, and W. C. Dreessen, Bibliography of Industrial Hygiene, 1900–1943. *Publ. Health Bull.* vol 289, United States Public Health Service, 1945, pp. 1–45.

V. R. Berliner, "U.S. Food and Drug Administration Requirements for Toxicity Testing of Contraceptive Products," in M. H. Briggs and E. Diczfalusy, eds., *Pharmacological Models in Contraceptive Development.* World Health Organization, Geneva, 1974.

C. M. Berry, "Safety in the Use of Economic Poisons," *Transactions of the 26th Annual Meeting of the American Conference of Governmental Industrial Hygienists.* Philadelphia, Penn, April 25–28, 1964, pp. 82–86.

F. C. Bing, Re: Perspective Versus Caprice in Evaluating Toxicity of Chemicals in Man. *JAMA*, **154**(11), 939 (1954).

E. Bingham and H. L. Falk, "Environmental Carcinogens, The Modifying Effect of Cocarcinogens on the Threshold Response," *Arch. Environ. Health.* **19**, 779–783 (1969).

C. I. Bliss, "The Determination of the Dosage–Mortality Curve from Small Numbers," *O. J. Pharmacy & Pharmacology.* **11**, 192–216 (1938).

J. J. Bloomfield, "What the ACGIH has done for Industrial Hygiene." *Am. Ind. Hyg. Assoc. J.*, **19**, 338–344 (1958).

J. J. Bloomfield, "Fragmentation in Industrial Hygiene, an Opportunity to Strengthen an Expanding field," *Transactions of the 30th Annual Meeting of the American Conference of Governmental Industrial Hygienists*, St. Louis, Missouri, May 12–14, 1968; pp. 16–19.

F. G. Bock, "Dose Response: Experimental Carcinogenesis," in E. L. Wynder, ed., *Toward a Less Harmful Cigarette, Nat. Cancer Inst.* Monograph 28, 1968; pp. 57–63.

H. M. Bolt, "Short-Term Exposure Limits," *Ann. Am. Conf. Ind. Hyg.* **12**, 85–87 (1985).

M. Borasi, "Standardization of Toxicity Tests," *Farmaco (fasc. straord.)*, **6**(7), 949–953 (1951).

L. L. Boughton, and O. O. Stoland, "The Effects of Drugs Administered Daily in Therapeutic Doses throughout the Life Cycle of Albino Rats," *Univ. Kansas Sci. Bull.* **27**(3), 27–60 (1941).

M. Bowditch, and H. B. Elkins, "Chronic Exposure to Benzene (Benzol). I. The Industrial Aspects," *J. Ind. Hyg. and Toxicol.* **21**(8), 321–330 (1939).

M. Bowditch, In "Setting Threshold Limits," *Transactions of the Seventh Annual Meeting of the National Conference of Governmental Industrial Hygienists*. St. Louis, Missouri, May 9, 1944 pp. 29–32.

E. M. Boyd, "Toxicological Studies," *J. New Drugs.* **1**(3), 104–109 (1961).

E. M. Boyd, "Predictive Drug Toxicity: Assessment of Drug Safety before Human Use," *Canad. Med. Assoc. J.* **98**, 278–293 (1968).

A. D. Brandt, *Industrial Hygiene Engineering*, John Wiley & Sons, New York, 1947.

A. D. Brandt, "Engineering and Chemical Application of Standards," *Am. Ind. Hyg. Assoc. O.* **17**, 286–292 (1956).

A. C. Bratton, "A Short-Term Chronic Toxicity Test Employing Mice," *J. Pharmocol. Exp. Therap.* **85** 111–118 (1945).

J. H. Brewer and H. H. Bryant, "The Toxicity and Safety Testing of Disposable Medical and Pharmaceutical Materials," *J. Amer. Pharm. Assoc.* **49**(10), 652–656 (1960).

P. C. Brinkley, "Industry's Responsibility in the Toxicity Testing, Manufacture, Compounding, and Use of Economic Poisons," *Am. Ind. Hyg. Assoc. J.* **26**(6), 611–614, (Nov.–Dec. 1965).

British Medical Association, "Labelling of Poisons," *Brit Med J.* (5204) (Oct. 1, 1960).

J. B. Brown, M. P. Fryer, P. Randall, and M. Lu, "Silicon Compounds in Plastic Surgery. Laboratory and Clinical Investigations, a Preliminary Report," *Plast. Reconst. Surg.* **12**(5), 374–376 (1953).

J. B. Brown, D. A. Ohlwiler, and M. P. Fryer, "Investigation of and Use of Dimethyl Siloxanes, Halogenated Carbons, and Poly(Vinyl Alcohol) as Subcutaneous Prostheses," *Ann. Surg.* **152**(3), 534–547 (1960).

J. B. Brown, "Studies of Silicones and Teflon as Subcutaneous Prostheses," *Plast. Reconstr. Surg.* **28**, 86–87 (1961).

J. B. Brown, M. P. Fryer, D. A. Ohlwiler, and P. Kollias, "Dimethylsiloxane and Halogenated Carbons as Subcutaneous Prosthesis," *Am. Surg.* **28**(3), 146–148 (1962).

E. Browning, *Toxicity of Industrial Organic Solvents*, Chemical Publishing Co., Inc., New York, 1953.

A. M. Brues, "Critique of the Linear Theory of Carcinogenesis," *Science.* **128**(3326), 693–699 (1958).

H. B. Burchell, "A Plea for the Labeling of Drugs," *Postgrad. Med.* **21**(6), 644–645 (1957).

E. J. Calabrese and E. M. Kenyon, *Air Toxics and Risk Assessment*, Lewis Publishers, Chelsea, Michigan, 1991.

H. O. Calvery, "Safeguarding Foods and Drugs in Wartime," *Am. Sci.* **32**(2), 103–119 (1944).

W. J. R. Camp, "The Toxicologist and Industrial Toxicology," *Indust. Med. Surg.* **19**(7), 321–322 (1950).

W. G. Campbell, "Progress in Food, Drug and Cosmetic control Under the New Federal Food, Drug and Cosmetic Act," *Assoc. Food Drug Officials of U.S.* **5**(1), 15–22 (1941).

Carbide & Carbon Chemicals Corp., *Manual of Hazards to Health from Chemicals Used at the Plants and Laboratories of Carbide and Carbon Chemicals Corp*, 1st ed. May 1, 1949.

C. P. Carpenter, H. F. Smyth, and U. C. Pozzani, "Assay of Acute Vapor Toxicity, and the Grading and Interpretation of Results on 96 Chemical Compounds," *J. Ind. Hyg. and Toxicol.* **31**(6), 343–346 (1949).

J. T. Carter, (1989). Indicative criteria for the new occupational exposure limits under COSHH. *Ann. Occup. Hyg.* 33:651–652.

V. L. Carter, "Development of Hygiene Guides," *Ann. Am. Conf. Govt. Ind. Hyg.* **3**, 113–115 (1982).

V. L. Carter, "Modus Operandi of Committee on Threshold Limit Values for Chemical Substances," *Ann. Am. Conf. Ind. Hyg.* **12**, 11–13 (1985).

B. I. Castleman and G. E. Ziem, (1994). American Conference of Governmental Industrial Hygienists: Low threshold credibility. *Amer. J. Ind Med.* 26:133–143.

Chemical Manufacturers Association, Legislative Action Alert, Pennsylvania Right to Know. State Affairs Report, Washington, D.C., 1984.

Chemical Manufacturers Association, An International Review of Procedures for Establishing Occupational Exposure Limits, Washington, D.C., 1992.

G. P. Child, H. O. Paquin, and W. B. Deichmann, "Chronic Toxicity of Methylpolysiloxane," *A.M.A. Arch. Indust. Hyg.* **3**(5), 479–482 (1951).

W. I. Clark and P. Drinker, *Industrial Medicine*, National Medical Book Co., New York, 1935.

Clayton Environmental Consultants, *OSHA's Hazard Communication Standard*, Issued. No. 15, January 1984.

A. L. Coleman, Report of Committee on Threshold Limits. *Transactions of the 16th Annual Meeting of the American Conference of Governmental Industrial Hygienists*. Chicago, IL, April 24–27, 1954 pp. 22–26.

A. L. Coleman, Threshold Limits of Organic Vapors. *Transactions of the 16th Annual Meeting of the American Conference of Governmental Industrial Hygienists*. Chicago, IL, April 24–27, 1954 pp. 50–52.

A. L. Coleman, The American Medical Association Proposed Act for Labeling Hazardous Substances; Legal Implications for Industry, *A.M.A. Arch. Indust. Health.* **19**(3), 271–273 (1959).

Commonwealth of Pennsylvania (November 1, 1971) Threshold Limit Values and Short-term Limits. Title 25, Part 1, Subpart D, Article IV, Chapter 201, Subchapter A, Threshold Limits. *Rules and Regulations, 1 Pa. B. 1985.*

B. E. Conley, "Principles for Precautionary Labeling of Hazardous Chemicals," *JAMA* **166**(17), 2154–2157 (1958).

W. A. Cook, "Maximum Allowable Concentrations of Industrial Atmospheric Contaminants," *Industrial Medicine*, **14**(11), 936–946 (1945).

W. A. Cook, "Present Trends in MACs," *Am. Ind. Hyg. Assoc. O.* **17**(3), 273–274 (1956).

W. A. Cook, "Problems of Setting Occupational Exposure Standards—Background," *Arch. Env. Health* **19**, 272–276 (1969).

W. A. Cook, "History of ACGIH TLVs," *Ann. Am. Conf. Ind. Hyg.* **12**, 3–9 (1985).

F. Cordle and A. C. Kolbye, Food safety and public health. *Interaction of science and law in the federal regulatory process. Cancer*, **43**, 2143–2150 (1979).

M. Corn and N. A. Esman, (1979) Workplace exposure zones for classification of employee exposures to physical and chemical agents. *Amer. Ind. Hyg. Asso. J.* 40:47–57.

C. E. Couchman, Echoes from the 1959 American Conference of Governmental Industrial Hygienists. *Transactions of the 30th Annual Meeting of the American Conference of Governmental Industrial Hygienists*. St. Louis, Missouri, May 12–14, 1968, pp. 23–25.

F. Coulston, "Some Principles Involved in Assessment of Drug Safety," *Canad. Med. Assoc. J.* **98**, 276–277 (1968).

L. J. Cralley, "Industrial Hygiene in the U.S. Public Health Service (1914–1968)," *Appl. Occup. Environ. Hyg.* **11**(3), 147–155 (1996).

G. M. Cramer, and R. A. Ford, "Estimation of Toxic Hazard—a Decision Tree Approach," *Fd. Cosmet. Toxicol.* **16**, 255–276 (1978).

C. W. Crawford, "Evaluation of Toxicity of Chemicals," *JAMA* **154**(11), 938–939 (1954).

A. S. Curry, "Toxicological analysis," *J. Pharm. Pharmacol.* **12**(6), 321–339 (1960).

W. C. Cutting, "Toxicity of Silicones," *Stanford Med. Bull.* **10**(1), 23–26 (1952).

T. D. Darby, "Pharmacologic Considerations in the Design of Toxicology Experiments," *Clin. Toxicol.* **12**(2), 229–238 (1978).

T. D. Darby, "Safety Evaluation of Polymer Materials," *Ann. Rev. Pharmacol. Toxicol.* **27**, 157–167 (1987).

G. T. Daughters, The Food Additive Amendment to the Food, Drug, & Cosmetic Act. Assn. of Food & Drug Officials of U.S. **23**(2), 73–76 (1959).

D. G. Davey, The Study of the Toxicity of a Potential Drug? Basic principles. Supplement to *Experimental Studies and Clinical Experience: The Assessment of Risk, Proceedings of the European Society for the Study of Drug Toxicity*, Vol. 6, Amsterdam: Excerpta Medica Foundation. 1965, pp. 3–13.

P. L. Day, "The Food and Drug Administration Faces New Responsibilities," *Nutr. Rev.* **18**(1), 1–5 (1960).

P. L. de la Cruz and D. G. Sarvadi, (1994). OSHA PELs: Where do we go from here? *Amer Ind Hyg Assoc J.* 55:894–900.

Deere & Co., "Labeling Code Category Classification Guide List," *Department of Industrial Health & Hygiene*, Moline, IL, 1952.

W. B. Deichmann and T. J. LeBlanc, "Determination of the Approximate Lethal Dose with about Six Animals," *J. Ind. Hyg. Toxicol.* **25**(9), 415–417 (1943).

W. B. Deichmann and E. G. Mergard, "Comparative Evaluation of Methods Employed to Express the Degree of Toxicity of a Compound," *J. Ind. Hyg. and Toxicol.* **30**(6), 373–378 (1948).

C. T. DeRosa, M. L. Dourson, and R. Osborne (1989). Risk assessment initiatives for noncancer endpoints: Implications for risk characterizations of chemical mixtures. *Tox. Ind. Health* 5(5):805–824.

H. J. Deuel Jr., "Toxicologic Determination of Suitability of Food Additives," *Food Res.* **20**(3), 215–220 (1955).

Deutsche Forschungsgemeinschaft Commission for the Investigation of Health Hazards of Chemical Compounds in the Work Area (1997): *List of MAK and BAT Values*. Weinheim, Federal Republic of Germany, Wiley-VCH, 185 pp.

R. Devignat, "Estimation of Virulent or Toxic Doses or their Antagonists by Means of Groups of Six Mice," *Rev. Immunol.* **17**(4–5), 211–223 (1953).

Doull, "The ACGIH TLVs: Past, Present, and Future," *Appl Occup Environ Hyg*, **6**(2), 89–90 (1991)

M. L. Dourson (1986). New approaches in the derivation of the acceptable daily intake (ADI). *Comments Toxicol.* 1:35–48.

M. L. Dourson and C. T. DeRosa (1991). The use of UFs in establishing safe levels of exposure. *In:* Statistics in Toxicology, D. Krewski and C. Franklin, Ed. Gordon and Breach Science Publisher, New York, NY.

M. L. Dourson (1994). Methods for establishing oral reference doses (RfDs). *In:* Risk Assessment of Essential elements, W. Mertz, C. O. Abernathy and S. S. Olin, Ed. ILSI Press, Washington, DC. Pages 51–61.

M. L. Dourson, S. P. Felter and D. Robinson (1996). Evolution of science-based UFs in noncancer risk assessment. *Reg. Tox. Pharmacol.*, 24:108–120.

H. N. Doyle, R. G. Smith, L. Silverman, V. H. Hill, S. J. Harris, and H. E. Stokinger, "Recent Industrial Hygiene Developments—A Symposium," *AIHA O*, **17**(3), 330–344 (1956).

H. N. Doyle, (1977) *The Federal Industrial Hygiene Agency. A History of the Division of Occupational Health*. United States Public Health Service, 1977.

J. H. Draize, G. Woodard, and H. O. Calvery, "Methods for the Study of Irritation and Toxicity of Substances Applied Topically to the Skin and Mucous Membranes," *J. Pharmacol. Exp. Ther.*, **82**, 377–390 (1944).

K. P. DuBois and E. M. K. Geiling, "General Principles of Toxicology," *Textbook of Toxicology* Oxford University Press, New York, Chapt. 2, 1959.

J. J. Duggan, "A Progress Report on In-Plant Hazard Identification Systems," *Arch of Environ Health*. **2**, 269–277 (Jan.–June. 1961).

ECETOC. (1994). *Strategy for Assigning a "Skin Notation"*. Revised ECETOC Document No. 31. European Centre for Ecotoxicology and Toxicology of Chemicals. Brussels.

R. E. Eckardt, "In Defense of TLVs," *J. Occup. Med.* **33**(9), 945–948 (1991).

E. F. Edson, "Pesticide Toxicology—Experimental, Applied and Regulatory," *World Rev. Pest. Control* **10**, 24–30 (1971).

G. E. Ehrlich, "Guidelines for Antiinflammatory Drug Research," *J. Clin. Pharmacol.* **17**(11–12), 697–703 (1977).

M. Eisler, "Pesticides: Toxicology and Safety Evaluation," *Trans. N.Y. Academy Sci., Ser. II* **31**(6): 720–730 (1969).

M. Eisenbud, *Environment, Technology and Health: Human Ecology in Historical Perspective*, New York Univ. Press, New York, 1978.

R. L. Elder, "The Cosmetic Ingredient Review—A Safety Evaluation Program," *Cosm. Ingred. Rev.* **11**(6), 1168–1174 (1984).

Electronic Industries Association, *Labeling of Hazardous Materials*, Safety and Health Committee, Industrial Relations Department, Data Sheet No. 101. Washington, DC Jan. 1986.

H. B. Elkins, "The Case for Maximum Allowable Concentrations," *Am. Ind. Hyg. Assoc. O.* **9**(1), 22–25 (1948).

H. B. Elkins, *The Chemistry of Industrial Toxicology*, John Wiley & Sons, New York, 1950.

H. B. Elkins, "Labeling Requirements for Toxic Substances," *AMA Arch of Ind Health* **18**, 451–456 (July–Dec. 1958).

H. B. Elkins, "Maximum Allowable Concentrations," *The Chemistry of Industrial Toxicology*, 2nd ed., John Wiley & Sons, New York, Chapt. 15, 1959.

H. B. Elkins, "Maximum Acceptable Concentrations," *Arch. Environ. Health* **2**, 45–49 (1961).

H. B. Elkins, "Maximum Allowable Concentrations of Mixtures," *Am. Ind. Hyg. Assoc. J.* **23**(2), 132–136 (1962).

H. B. Elkins, "Threshold Limit Values and their Significance," *Transactions of the 28th Annual Meeting of the American Conference of Governmental Industrial Hygienists*. Pittsburgh, PA, May 15–17, 1966, pp. 116–122.

H. B. Elkins, "Excretory and Biologic Threshold Limits," *Am. Ind. Hyg. Assoc. J.* **28**, 305–314 (1967).

H. B. Elkins, "The Real World or Science Fiction," *Ann. Am. Conf. Ind. Hyg.* **4**, 5–11 (1983).

"Exposure Limits for Occupational and Environmental Chemical Pollutants," Papers from an International Symposium, Prague, Czechoslovakia, 18–21 April 1989, *Sci. Total Environ.* **1**:101(1–2), 1–179 (1991).

E. J. Fairchild, "Occupational Exposure Limits and the Sensitive Worker: The Dilemma of International Standards," *Ann. Am. Conf. Govt. Ind. Hyg.* **3**, 83–89 (1982).

L. T. Fairhall, "The Relative Toxicity of Lead and Some of its Common Compounds." *Transactions of the Third Annual Meeting of the National Conference of Governmental Industrial Hygienists*. Bethesda, MD, April 30–May 2, 1940, pp. 155–164.

L. T. Fairhall, *Industrial Toxicology*, Williams and Wilkins, Baltimore, 1949.

L. T. Fairhall, "Inorganic Industrial Hazards," *Physiol. Rev.* **25**(1), 182–202 (1945).

S. Fairhurst, (1995). The UF in the setting of occupational exposure standards. *Ann. Occup. Hyg.* 39:375–385.

D. W. Fassett and R. L. Roudabush, "Short-term Intraperitoneal Toxicity Tests," *AMA Arch. Indust. Hyg. Occup. Med.*, **6**(6), 525–529 (1952).

D. W. Fassett and D. D. Irish, *Industrial Hygiene and Toxicology*, 2nd ed., Frank Patty, Interscience Publishers, New York, 1963.

D. W. Fassett, "Industrial Toxicology," in *Kirk-Othmer: Encyclopedia of Chemical Technology*, 2nd Edition, Wiley, New York, vol. 11, 1966, pp. 595, 610.

FDA, *Hazardous Substances: Definitions and Procedural and Interpretative Regulations*. U.S. Dept. of Health, Education, and Welfare, Food and Drug Administration, 1951.

FDA, *Read the Label on Foods, Drugs, Devices, and Cosmetics, Miscellaneous, Publication No. 3*, Food and Drug Administration. U.S. Dept. of Health, Education, and Welfare, 1957.

FDA, *Appraisal of the Safety of Chemicals in Foods, Drugs and Cosmetics*, The Association of Food and Drug Officials of the United States. Food and Drug Administration, 1959.

FDA, *Notices of Judgment under the Federal Hazardous Substances Labeling Act*, U.S. Dept. of Health, Education, and Welfare, Food and Drug Administration, 1963.

FDA, "Food and Drug Administration Advisory Committee on Protocols for Safety Evaluations: Panel on Reproduction Report on Reproduction Studies in the Safety Evaluation of Food Additives and Pesticide Residues," *Toxicol. Appl. Pharmacol.* **16**, 264–296 (1970).

Federal Focus (1996). *Epidemiology and Risk Assessment.* Washington, DC.

A. M. Finkel (1994). *A Quantitative Estimate of the Extent of Human Susceptibility to Cancer and Its Implications for Risk Management.* International Life Sciences Institute, Risk Science Institute, Washington, DC.

D. J. Finney, "The Statistical Treatment of Toxicological Data relating to more than One Dosage Factor," *Ann. Appl. Biol.* **30**(1), 71–79 (1943).

A. J. Fleming, C. A. D'Alonzo, and J. A. Zapp, *Modern Occupational Medicine*, Lea and Febiger, Philadelphia, 1954.

J. G. Fitzgerald, *The Practice of Preventive Medicine*, C. V. Mosby Co., St. Louis, MO, 1922.

F. Flury and W. Heubner. (1919). Uber Wirkung und Eingiftung eingeatmeter Blausaure. *Biochem. Z.* 95:249–256.

Food Safety Council, "Proposed System for Food Safety Assessment," *Food Cosmetics Toxicol.* **16**(Suppl. 2) (1978).

D. A. Fraser, "Re: Corporate Influence on Threshold Limit Values," *Am. J. Ind. Med*, **15**, 235–236 (1989).

J. P. Frawley, "Scientific Evidence and Common Sense as a Basis for Food-Packaging Regulations," *Food Cosmet. Toxicol.* **5**, 293–308 (1967).

A. C. Frazer, "Synthetic Chemicals and the Food Industry," *J. Sci. Food Agric.* **2**(1), 1–7 (1951).

A. C. Frazer, Pharmacological aspects of chemicals in food. Endeavour, **12**(45), 43–47 (1953).

D. E. H. Frear, *Pesticide Handbook*, College Science Publishers, State College, PA, 1960.

W. G. Fredrick, "The Birth of the ACGIH Threshold Limit Values Committee and its Influence on the Development of Industrial Hygiene," *Transactions of the 30th Annual Meeting of the American Conference of Governmental Industrial Hygienists*, St. Louis, Missouri, May 12–14, 1968, pp. 40–43.

L. Friedman, "Symposium on the Evaluation of the Safety of Food Additives and Chical Residues: II. The Role of the Laboratory Animal Study of Intermediate Duration for Evaluation of Safety," 1969.

D. G. Friend and R. G. Hoskins, "Reactions to Drugs," *Med. Clin. N. Am.* **44**(5), 1381–1392 (1960).

H. Frohberg, "Tasks and Possibilities of Toxicology in the Pharmaceutical Industry," *Drugs Made in Germany* **13**, 1–44 (1970).

J. T. Fuess, "The American Medical Association Proposed Act for Labeling Hazardous Substances," *AMA Arch of Ind Health*, **19**, 274–277 (Jan–June 1959).

S. C. Gad, Defining the Objective: Product Safety Assessment Program Design and Scheduling, in *Product Safety Evaluation Handbook*, Marcel Dekker, Inc. New York, 1988.

J. H. Gaddum, "The Estimation of Safe Dose," *Brit. J. Pharmacol.* **11**, 156–160 (1956).

W. M. Gafafer, *Manual of Industrial Hygiene (and Medical Services in War Industries)*, W. B. Saunders Co., Philadelphia, 1943.

D. M. Galer, H. W. Leung, R. G. Sussman, and R. J. Trzos, "Scientific and Practical Considerations for the Development of Occupational Exposure Limits (OELs) for Chemical Substances," *Regul. Toxicol. Pharmacol.* **15**(3), 291–306 (1992).

R. J. Gardner and P. J. Oldershaw. 1991. Development of pragmatic exposure-control concentrations based on packaging regulation risk phrases. *Ann Occup Hyg.* 35(1):59–59.

J. Gelzer, Governmental Toxicology Regulations: an Encumbrance to Drug Research? *Arch. Toxicol.* **43**, 19–26 (1979).

R. D. Gidel, "Providing Hazardous Substances Information," *Transactions of the 31th Annual Meeting of the American Conference of Governmental Industrial Hygienists*. Denver, CO, May 11–13, 1969, pp. 137–144.

R. P. Giovacchini, "Premarket Testing Procedures of a Cosmetics Manufacturer," *Toxicol. Appl. Pharmacol.* (Suppl. 3), 13–18 (1969).

R. P. Giovacchini, "Toxicological Evaluation of Product Safety," *Pediatric Clinics N. Am.* **17**(3), 645–652 (1970).

R. P. Giovacchini, "Old and New Issues in the Safety Evaluation of Cosmetics and Toiletries," *CRC Crit. Rev. in Toxicol.* **1**(4), 361–378 (1972).

E. M. Glaser, "Experiments on the Side Effects of Drugs," *Brit. J. Pharmacol.* **8**, 187–192 (1953).

M. L. Gleason, R. E. Gosselin, and H. C. Hodge, *Clinical Toxicology of Commercial Products: Actue Poisoning (Home and Farm)*, Williams and Wilkins Co., Baltimore, 1957.

G. Gobinet, "The Council of Europe Approach to Toxicity Testing and Toxicological Evaluation." *Arch. Toxicol* (Suppl. 5) 45–47 (1982).

L. Golberg, "The Assessment of Safety-in-use: Just How Much is Contributed by Feeding Studies in Animals," *J. Soc. Cosmetic Chem.* **15**, 177–194 (1964).

L. Goldberg, "Chemical and Biochemical Implications of Human and Animal Exposure to Toxic Substances in Food," *Pure Appl. Chem.* **21**, 309–330 (1970).

E. I. Goldenthal and D. Aguanno, W., "Evaluation of drugs," in *Appraisal of the Safety of Chemicals in Foods, Drugs and Cosmetics*, The Association of Food and Drug Officials of the U.S., 1959, pp. 60–67.

L. J. Goldwater, *Mercury: A History of Quicksilver*, York Press, Baltimore, 1972.

L. Goldwater, "Toxicology." in N. I. Sax, ed., *Dangerous Properties of Industrial Materials*, Reinhold Publishing Corp., New York.

L. G. Goodwin and F. L. Rose, "The Evaluation of New Drugs," *J. Pharmacy Pharmacol London*, **10**(Suppl.), 24–39 (1958).

R. Goyal, K. Krishnan, R. Tardif, S. Lapare, and J. Brodeur, "Assessment of Occupational Health Risk during Unusual Workshifts: Review of the Needs and Solutions for Modifying Environmental and Biological Limit Values for Volatile Organic Solvents," *Can. J. Public Health* **83**(2), 109–112 (1992).

P. Grandjean, A. Berlin, M. Gilbert, and W. Penning, (1988). Preventing percutaneous absorption of industrial chemicals: The "skin" denotation. *Amer. J. Ind. Med.* 14:97–107.

P. Grasso, "Bioactivation, Toxicity and Safety," *Fd. Cosmetics Toxicol.* **15**(4), 355–356 (1977).

S. Green, "Present and Future Uses of Mutagenicity Tests for Assessment of the Safety of Food Additives," *J. Environ. Path. Toxicol.* **1**, 49–54 (1977).

H. C. Grice, "The Changing Role of Pathology in Modern Safety Evaluation," *CRC Critical Rev. in Toxicol.* **1**(2), 119–152 (1972).

W. L. Guess, "Safety Evaluation of Medical Plastics," *Clin. Toxicol.* **12**(1), 77–95 (1978).

Gulf Oil Corporation, Summary, Your Rights to Hazard Communication Information as a result of the issuance by OSHA of the "hazard communications" standard: **29** *CFR* Part 1910.1200 (1984).

E. C. Hagan, "Acute toxicity," in: *Appraisal of the Safety of Chemicals in Foods, Drugs and Cosmetics*, The Association of Food and Drug Officials of the U.S., 1959, pp. 17–25.

A. Hamilton, *Industrial Poisons in the United States*, Macmillan, New York, 1929.

A. Hamilton, *Industrial Toxicology*, Harper and Brothers, New York, 1934.

A. Hamilton and H. L. Hardy, *Industrial Toxicology*, 3rd ed., Publishing Science Group, Inc., Acton, MA, 1974.

A. Hamilton, "The Toxicity of the Chlorinated Hydrocarbons," *Yale J. Biol. and Med.* **15**, 787–801 (1943).

A. Hamilton, "Forty Years in the Poisonous Trades," *Am. Ind. Hyg. Assoc. O*, **9**(1), 5–16 (March 1948).

R. Handy and A. Schindler (1976). *Estimation of Permissible Concentrations of Pollutants for Continuous Exposure.* U.S. Environmental Protection Agency, Pub. No. EPA-600/2-76-155, Washington, DC.

T. Hanley, V. Udall, and M. Weatherall, "An Industrial View of Current Practice in Predicting Drug Toxicity," *Brit. Med. Bull.* **26**(3), 203–211 (1970).

P. J. Hanzlik, H. W. Newman, W. Van Winkle, A. J. Lehman, and N. K. Kennedy, "Toxicity, Fats and Excretion of Propylene Glycol and some other Glycols." *J. Pharmacol. Exp. Therap.* **67**(1), 101–113 (1939).

F. L. Hart, "Controlling Labelling of Common Hazardous Substances," *Connecticut State Med. J.* **20**(12), 962–966 (1956).

A. L. Haskins, "Caveat emptor," *Bull. School Med. Univ. Maryland.* **43**(3), 51–53 (1958).

T. F. Hatch, "Significant Dimensions of the Dose–Response Relationship in Industrial Toxicology," *Transactions of the 29th Annual Meeting of the American Conference of Governmental Industrial Hygienists.* Chicago, IL, May 1–2, 1967 pp. 39–50.

W. J. Hayes, "Toxicological Evaluation of Some of the Newer Pesticides," *Transactions of the Fourteenth Annual Meeting of the American Conference of Governmental Industrial Hygienists.* Cincinnati, Ohio, April 19–22, 1952; pp. 17–21.

T. L. Hazlett, *Introduction to Industrial Medicine*, Industrial Medicine Pub. Co., Chicago, 1947.

D. Henschler (1984). Exposure Limits: History, Philosophy, Future Developments. *Ann. Occup. Hyg.* 28:79–92.

D. Henschler (1985). Development of Occupational Limits in Europe. *Ann. Amer. Conf. Govt. Ind. Hyg.* 12:37–40.

R. C. Hertzberg and J. Patterson (1987). Approaches to the Quantitative estimation of Health risk from exposure to Chemical mixtures. In: Health & Environmental Research on Complex Organic Mixtures, (R. H. Gray, E. K. Chess, P. J. Mellinger, R. G. Riley and D. L. Springer, Eds.) Twenty-Fourth Handford Life Sciences Symposium, October 20–24, 1985. P. 757–757.

P. Hewett, "Interpretation and Use of Occupational Exposure Limits for Chronic Disease Agents," *Occup. Med.* **11**(3), 561–590 (1996).

P. Hewett, Interpretation and use of occupational exposure limits for chronic disease agents. *Occup. Med: State of the Art Reviews* **11**(3), 561–590 (1996).

S. J. Hill, "The Manufacturing Chemists' Association Labeling Program," *AMA Arch of Ind Health.* **12**, 378–382 (1955).

C. H. Hine and N. W. Jacobsen, "Safe Handling Procedures for Compounds Developed by the Petro-Chemical Industry," *Am. Ind. Hyg. Assoc. J.* **15**, 141–144 (June 1954).

C. H. Hine, M. K. Dunlap, and J. K. Kodama, "Industrial toxicology. I. General principles and new developments," *AMA Arch. Intern. Med.* **104**(5), 816–826 (1959).

C. H. Hine, "Toxicology and Occupational Health," *J. Occup. Med.* **4**(9), 457–464 (1962).

H. C. Hodge and J. H. Sterner, "Tabulation of Toxicity Classes," *Am. Ind. Hyg. Assoc. Q.* **10**(4), 93–96 (1949).

H. C. Hodge and W. L. Downs, "The Approximate Oral Toxicity in Rats of Selected Household Products," *Toxicol. Appl. Pharmacol.* **3**, 689–695 (1961).

E. J. Hogan, "Basis Principles for Precautionary Labeling," *Industr. Med. Surg.* **29**(11), 530–533 (1960).

B. Holmberg and P. Lundberg, "Exposure Limits for Mixtures," *Ann. Am. Conf. Ind. Hyg.* **12**, 111–118 (1985).

S. Homrowski, "Current Problems in Safety Evaluation of Plastics (Introductory Lecture)," *Pol. J. Pharmacol. Pharm.* **32**, 65–75 (1980).

H. Hopkins, "Food Additives: Double Check on Safety," *FDA Consumer*, **11**(5), 8–13 (1977).

G. H. Horner, (ed.) "The Solubility of Silica," *Brit Med J.* **1**, 931–932 (1939).

G. H. Horner, (editor), "Industrial Solvents," *Brit Med J.* **2**, 177 (1939).

Hosey, "Labeling of Hazardous Substances Used in Industry," *Transactions of the 31th Annual Meeting of the American Conference of Governmental Industrial Hygienists.* Denver, CO, May 11–13, 1969, pp. 129–132.

O. B. Hunter, Jr. "Statement of the A.M.A. to the Subcommittee on Health and Safety, Committee on Interstate and Foreign Commerce, House of Representatives. Federal Hazardous Substances Labelling Act. March 14, 1960." *JAMA*, **173**, 263–265 (1960).

P. B. Hutt, "A History of Government Regulation of Adulteration and Misbranding of medical devices." *Fd. Drug Cosm. Law J.* **44**(2), 99–118 (1989).

Industrial Health Foundation, "The Voluntary Approach to Worker Health Since 1935," *Am. Ind. Hyg. Assoc. J.* **47**(11), 667–669 (1986).

Industrial Hygiene Foundation of America, Summary of Chemical-Toxicological Conference. 18th Annual Meeting Conferences, 1953.

Industrial Medical Association, "Report of an Investigation of Threshold Limit Values and their Usage," *J. Occup. Med.* **8**(5), 280–283 (1966).

"Industrial Medicine, Toxic logic (editorial)," *Ind. Med. Ind. Hyg. Sec.* **1**(4), 53–54 (1940).

H. P. Illing, "Extrapolating from Toxicity Data to Occupational Exposure Limits: Some Considerations," *Ann. Occp. Hyg.* **35**(6), 569–580 (1991).

D. D. Irish, "The Value to the Industrial Physician of Toxicological Information," *Ind. Hyg. Q.* **9**(1), 25–28 (March 1948).

R. L. Iuliucci (1982). 12-Hour TLVs. *Pollution Eng.* p. 25–57 (November).

R. H. Ivy, "Circumstances Leading to Organization of the American Board of Plastic Surgery," *Plast. & Reconstruct. Surg.* **16**(2), 77–87 (1955).

D. W. Jolly, "The History and Function of Toxicity Testing," in D. W. Jolly, ed., *How Safe is Safe? The Design of Policy on Drugs and Food Additives.* National Academy of Sciences, Washington, DC, 1974.

L. M. Kantner, "Drug Control in Maryland," *Assoc of Food & Drug Officials of the U.S.* **9**(4), 141–148 (1945).

D. Katz, "Psychology of Margin of Safety," *Nervenarzt.* **22**, 375–376 (1951).

W. T. Keane, "OSHA—Interpretation of the Industrial Hygienist," *Am. Ind. Hyg. Assn. J.* **33**, 547–557 (1972).

L. H. Keith and D. B. Walters, *The Compendium of Safety Data Sheets for Research and Industrial Chemicals*, VCH Publishers, 1986.

M. L. Keplinger, "Use of Humans to Evaluate Safety of Chemicals," *Arch. Environ. Health.* **6**, 342–349 (1963).

I. Kerlan and S. Molinas, "Current Status of the Federal Hazardous Substances Labeling Act," *New Physician* **11**, 72–76 (1962).

M. M. Key, A. F. Henschel, et al., *Occupational Diseases: A Guide to Their Recognition*, Revised edition, U.S. Dept. of Health, Education and Welfare, NIOSH, Washington, D.C., 1977.

C. H. Keysser, "Preclinical Safety Testing of New Drugs," *Ann. Clin. Lab. Sci.* **6**(2), 197–205 (1976).

Y. Kim and G. P. Carlson (1986a). The Effect of an Unusual Workshift on Chemical Toxicity: I. Studies on the Exposure of Rats and Mice to Dichloromethane. *Fund. Appl. Toxicol.* 6:162–171.

Y. C. Kim and G. P. Carlson (1986b). The Effect of an Unusual Workshift on Chemical Toxicity: II. Studies on the Exposure of Rates to Aniline. *Fund. Appl. Toxicol.* 7:144–152.

J. D. Kittelton, "Legal Considerations in Drafting Warning Labels," *Arch of Environ Health.* **2**, 263–268 (Jan.–June 1961).

T. G. Klumpp, "The Philosophy of the Administration of the Drug Sections of the Food, Drug, and Cosmetic Act," Presented at the Forty-Fifth Annual Conference of Food and Drug Officials. St. Paul, Minnesota, June 1941, pp. 83–89.

T. G. Klumpp, "The Food and Drug Administration and the Pharmaceutical Industry," *J. Indiana Med. Assoc.* **54**(11), 1680–1690 (1961).

K. G. Kohlstaedt, "Developing and Testing of New Drugs by the Pharmaceutical Industry;" *Clin. Pharmacol. Ther.* **1**(2), 192–201 (1960).

D. Krewski and Y. Zhu (1995). A simple data transformation for estimating benchmark doses in developing toxicity experiment. *Risk Anal.* 15:29–39.

D. Krewski, C. Brown, and D. Murdoch (1984). Determining "Safe" Levels of Exposure: Safety Factors or Mathematical Models. *Fund. Appl. Toxicol.* 4:383–394.

J. F. Lakey, "Food and Drug Laws; Interpretation and Enforcement," *Pediatrics.* **1**, 534–537 (1948).

K. Landsteiner and J. J. Jacobs, "Studies on the Sensitization of Animals with Simple Chemical Compounds," *J Experimental Med.* **61**, 643–657 (1935).

G. P. Larrick, "The Evolution of a Drug Control Program," Presented at the 49th Annual Conference. Buffalo, NY, June. 1945, pp. 42–48.

L. Lasagna, "Congress, the FDA, and New Drug Development: Before and After 1962," *Perspect. Biol. Med.* **32**(3), 322–343 (1989).

J. Lazarus and J. Cooper, Absorption, testing, and clinical evaluation of oral prolonged-action drugs. *J. Pharmaceutical Sci.*, **50**(9), 715–732 (1961).

P. Leeper, Finding the Bad Actors in a World of Chemicals. New Report, March 1984, pp. 4–10.

A. J. Lehman, The Toxicology of the Newer Agricultural Chemicals. *Assn. of Food & Drug Officials of U.S.* **12**(3), 82–89 (1948).

A. J. Lehman, Pharmacological Considerations of Insecticides. *Assoc. of Food & Drug Officials of the U.S.* **13**(2), 65–70 (1949).

A. J. Lehman, G. W. Laug, J. H. Draize, O. G. Fitzhugh, and A. A. Nelson, "Procedures for the Appraisal of the Toxicity of Chemicals in Foods," *Food Drug Cosm. Law Q.*, 412–434 (Sept. 1949).

A. J. Lehman, A. Hartzell, and J. C. Ward, "Effects on Beneficial Forms of Life, Crops and Soil and Residue Hazards," *JAMA*, 108 (Sept. 9, 1950).

A. J. Lehman, "Proof of Safety: Some Interpretations," *J. Am. Pharm. Assoc.* (Scient. Ed.). **40**(7), 305–308 (1951).

A. J. Lehman, "Chemicals in Foods: A Report to the Association of Food and Drug Officials on Current Developments," *Assoc. of Food & Drug Officials of U.S.* **15**(3), 82–89 (1951).

A. J. Lehman, "Chemicals in Foods: A Report to the Association of Food and Drug Officials on Current Developments," Part II. Pesticides. *Assoc. of Food & Drug Officials of U.S.* **15**(4), 122–133 (1951).

A. J. Lehman and W. I. Patterson, "F&DA Acceptance Criteria," *Modern Packaging*, 115–174 (1955).

A. J. Lehman, W. I. Patterson, B. Davidow, E. C. Hagan, G. Woodard, E. P. Laug, J. P. Frawley, G. Fitzhugh, A. R. Bourke, J. H. Draize, A. A. Nelson, and B. J. Vos, "Procedures for the Appraisal of the Toxicity of Chemicals in Foods, Drugs and Cosmetics," *Food Drug Cosm. Law J.*, 679–748 (Oct. 1955).

A. J. Lehman, "The Food and Drug Administration and Drug Safety," *Minnesota Med.* **41**(8), 574–576 (1958).

G. Lehnert, R. D. Ladendorf, and D. Szadkowski (1978). The Relevance of the Accumulation of Organic Solvents for Organization of Screening Tests in Occupational Medicine. Results of Toxicological Analyses of More than 6000 Samples. *Inter. Arch. Occup. Environ. Health* 41:95–102.

C. E. Lewis, A Method for Reporting Toxicological Information. *AMA Arch of Ind Health* **18**, 457–459 (July–Dec. 1958).

T. R. Lewis, "Identification of Sensitive Subjects not Adequately Protected by TLVs," *Ann. Am. Conf. Ind. Hyg.* **12**, 167–172 (1985).

E. W. Ligon, "Federal Hazardous Substances Labeling Act," *Arch Environ Health.* **10**(4), 596–598, (April. 1965).

A. Liljestrand, "Safety Aspects on Long-Term Medication," *Acta Pharm. Cuecica.* **10**, 371–380 (1973).

J. T. Litchfield and F. Wicoxon, "A Simplified Method of Evaluating Dose-effect Experiments," *J. Pharmacol. Exp. Ther.* **96**(2), 99–113 (1949).

J. T. Litchfield, Jr. "Evaluation of the Safety of New Drugs by Means of Tests in Animals," *Clinc. Pharmacol. Therap.* **3**(5), 665–672 (1962).

T. A. Loomis, *Essentials of Toxicology*, 3rd ed., Lea and Febiger, Philadelphia, 1978.

F. C. Lu, "International Activities in the Field of Food Additives, with Particular Reference to Carcinogenicity," *Ecotox. Environ. Safety.* **3**, 301–309 (1979).

F. C. Lu and M. L. Dourson (1992). Safety risk assessment of chemicals: Principles, procedures and examples. J. Occup. Med. Tox. 1(4): 321–335.

M. M. Luckens, "The Interpretation of Toxicity Data," *Safety Maintenance.* **123**(6), 44–47 (1962).

H. F. Ludwig, "Chemicals and Environmental Health," *Am. J. Pub. Health.* **45**(7), 874–879 (1955).

P. Lundberg (1994). National and International Approaches to Occupational Standard Setting Within Europe. *Appl. Occup. Environ. Hyg.* 9:25–27.

J. D. B. MacDougall, "Toxicity Studies on Silicone Rubber and Other Substances," *Nature* **172**(4368), 124–125 (1953).

H. J. Magnuson, Occupational Health in Official Agencies Retrospect and Prospect. *Transactions of the 30th Annual Meeting of the American Conference of Governmental Industrial Hygienists.* St. Louis, Missouri, May 12–14, 1968, pp. 44–49.

T. F. Mancuso, Forces and trends which developed the present state of the art. *Transactions of the 30th Annual Meeting of the American Conference of Governmental Industrial Hygienists*, St. Louis, Missouri, May 12–14, 1968, pp. 20–22.

N. Mantel, W. E. Heston, and J. M. Gurian, Thresholds in linear dose-response models for carcinogenesis. *J. Nat. Cancer Inst.* **27**(1), 203–215 (1961).

N. Mantel and W. R. Bryan, "Safety Testing of Carcinogenic agents." *J. Natl. Cancer Inst.*, **27**(2), 455–470 (1961).

N. Mantel, "Part IV. The Concept of Threshold in Carcinogenesis," *Clin. Pharm. Ther.* **4**(1), 104–109 (1963).

N. Mantel, N. R. Bohidar, C. C. Brown, J. L. Ciminera, and J. W. Tukey, "An Improved Mante-Bryan Procedure for "Safety" Testing of Carcinogens," *Cancer Res.* **35**, 865–872 (1975).

Manufacturing Chemists' Association, Inc., Washington, D.C., *Chemical Safety Data Sheets*
  Acetaldehyde, 1952, Manual Sheet SD-43.
  Acetylene, 1947, Manual Sheet SD-7.
  Ammonium Dichromate, 1952, Manual Sheet SD-45.
  Arsenic Trioxide, 1956, Manual Sheet SD-60.
  Benzene (third revision), 1960, Manual Sheet SD-2.
  Betanaphthylamine, 1949, Manual Sheet SD-32.
  Butyraldehydes, 1960, March, Manual Sheet SD-78.
  Carbon Tetrachloride (revised), 1963, Manual Sheet SD-3.
  Chloroform, 1974, Manual Sheet SD-89.
  Cresol, 1952, Manual Sheet SD-48.
  Chromic Acid (Chromium Trioxide), 1952, Manual Sheet SD-44.
  Diethylenetriamine, 1959, September, Manual Sheet SD-76.
  Dimethyl Sulfate, 1947, Manual Sheet SD-19.
  Ethyl Chloride, 1953, Manual Sheet SD-50.
  Formaldehyde (revised), 1960, April, Manual Sheet SD-1.
  Hydrocyanic Acid, 1961, Manual Sheet SD-67.
  Lead Oxides, 1956, August, Manual Sheet SD-64.
  Methyl Acrylate and Ethyl Acrylate, 1960, April, Manual Sheet SD-79.
  Methylamines, 1955, Manual Sheet SD-57.
  Methyl Ethyl Ketone, 1961, Manual Sheet SD-83.
  Naphthalene, 1956, Manual Sheet SD-58.
  Nitrobenzene (revised), 1967, August, Manual Sheet SD-21.
  Phthalic Anhydride (commercial), 1956, Manual Sheet SD-61.
  Sodium and Potassium Dichromates and Chromates, 1952, Manual Sheet SD-46.
  Styrene Monomer, 1950, Manual Sheet SD-37.
  Tetrachloroethane, 1949, Manual Sheet SD-34.
  Toluene, 1956, Manual Sheet SD-63.
  1,1,1-Trichloroethane, 1965, Manual Sheet SD-90.
  Trichloroethylene, 1947, Manual Sheet SD-14.
  Vinyl Chloride (revised), 1972, Manual Sheet SD-56.

Manufacturing Chemists' Association, *A Guide for the Preparation of Warning Labels for Hazardous Chemicals*. Manual L-1, 1945

Manufacturing Chemists' Association, *Product Caution Labels*. Manual L-2. May, 1945.

Manufacturing Chemists' Association, *Warning Labels, A Guide for the Preparation of Warning Labels for Hazardous Chemicals*. Manual L-1, 1949

Manufacturing Chemists' Association, *Warning Labels, A Guide for the Preparation of Warning Labels for Hazardous Chemicals*. Manual L-1, 1956.

Manufacturing Chemists' Association, *Guide to Precautionary Labeling of Hazardous Chemicals*, Manual L-1, 1961.

Manufacturing Chemists' Association, *Guide to Precautionary Labeling of Hazardous Chemicals* (7th ed). Manual L-1, 1970.

G. Markowitz and D. Rosner (1995). The limits of thresholds: silica and the politics of science, 1935 to 1990. *Am J Public Health.* 85(2):253–262.

K. E. Markuson, Plant Conditions. To what Extent Should Official Findings Regarding them be Made Available to Workers? *Transactions of the Seventh Annual Meeting of the National Conference of Governmental Industrial Hygienists.* St. Louis, Missouri, May 9, 1944 pp. 22–25.

F. A. Marzoni, S. E. Upchurch, and C. J. Lambert, "An Experimental Study of Silicone as a Soft Tissue Substitute," *Plast. Reconstr. Surg. & Transplantation Bull.* **24**(6), 600–608 (1959)

E. J. Masek, "Uniform Principles for Precautionary Labeling." *Transactions of the 31th Annual Meeting of the American Conference of Governmental Industrial Hygienists.* Denver, CO, May 11–13, 1969, pp. 133–136.

J. W. Mason and J. Hughes (1985). *A Proposed Approach to Adjusting TLVs for Carcinogenic Chemicals (unpublished report).* Univ. of Alabama, Birmingham.

E. Mastromatteo, Presented to the Mining Section, National Safety Council, October 26, 1971, Chicago, IL. In: *Ann. Am. Conf. Ind. Hyg.* **9**, 207–213 (1984).

E. Mastromatteo, "On the Concept of Threshold, Presented at the Am. Ind. Hyg. Conference, Portland, OR, May 24–29, in: *Ann. Am. Conf. Ind. Hyg.* **9**, 331–340 (1984).

B. E. Matter, "Problems of Testing Drugs for Potential Mutagenicity," *Mutation Res.* **38**, 243–258 (1976)

R. L. Mayer, "Shortcomings of the Experimental Methods for Detection of Sensitizing Properties of New Drugs." *J. Allergy* **26**(2), 133–140 (1955)

E. A. Maynard, "Toxicity Testing of Chemical Additives," *Food Technology* **6**(9), 351–353 (1952)

C. W. Mayo, M. Fishbein, and S. Covet. "A Plea for the Labeling of Drugs." *Postgraduate Med.,* 644–645 (June 1957).

G. W. McCarl, "Present Status of Labeling of Materials in Industry." *Am. Ind. Hyg. Assoc. O.* **17**(1), 510–513 (1956).

C. P. McCord, *Industrial Hygiene for Engineers*, Harper and Brothers, New York, 1931.

B. W. McCready, *On the Influence of Trades, Professions, and Occupations in the United States, in the Production of Disease*, ARNO Press, New York, 1972.

H. J. McDermott, *Handbook of Ventilation for Contaminant Control*, Ann Arbor Science, 1977.

L. C. McGee, "Chlorinated Insecticides," Toxicity for Man. *Indust. Med. Surg.* **24**(3), 101–109 (1955).

J. C. McGowan, "Physically Toxic Chemicals and Industrial Hygiene," *AMA Arch. Indust. Health.* **11**(4), 315–323 (1955).

A. E. M. McLean, "Testing of Industrial Chemicals," *The Lancet*, 1070–1071. (Nov. 19, 1977).

A. E. M. McLean. "Symposium on "Safety in Man's Food." Risk and Benefit in Food and Additives," *Proc. Nutr. Soc.* **36**, 85–90 (1977).

B. P. McNamara, "Toxicological Test Methods," *Assoc. Food Drug Off. U.S. Q. Bull.* **38**(1), 33–50 (1974).

B. P. McNamara, "Concepts in Health Evaluation of Commercial and Industrial Chemicals," *Advances in Modern Toxicology*, M. A. Mehlman, R. E. Shapiro and H. Blumenthal, eds., Vol. 1, Part 1: *New Concepts in Safety Evaluation.* Hemisphere Publishing Corp., Washington.

Medical Research Council, "Assessment of Toxicity," Memorandum of Toxicology Committee, *Month. Bull. Min. Health & Publ. Health Lab. Serv.* **16**, 2–5 (1957).

I. Mellan and E. Mellan, *Encyclopedia of Chemical Labeling*, Chemical Publishing Co., Inc., New York, 1961.

R. A. Merrill, "Regulation of Drugs and Devices: An Evolution," *Health Affairs* **13**(3), 47–69. (1994)
M. I. Mikheev, "Toward WHO-recommended Occupational Exposure Limits," *Toxicol. Lett.* **77**(1–3), 183–187 (1995).
W. Modell, "Problems in the Evaluation of Drugs in Man," *J. Pharm. Pharmacol.* **11**(10), 577–594. (1959)
R. J. Moolenaar (1992). Overhauling Carcinogen Classification. *Environ. Sci. Tech.* 8:70–75.
J. F. Morgan, "Summary of Chemical and Toxicological Conference," *Transactions of Thirteenth Annual Meeting of Industrial Hygiene Foundation of America Inc.*, Nov. 18, 1948, pp. 38–41.
K. M. Morse, "Industrial Hygiene—Progress and Inertia." *Transactions of the 30th Annual Meeting of the American Conference of Governmental Industrial Hygienists.* St. Louis, Missouri, May 12–14, 1968, pp. 12–15.
T. W. Nale, "Current Method of Determining Safety in the Application of New Chemicals," *Indust. Med.* **20**(11), 501–506 (1951)
T. W. Nale, "The Chemical Industry and Precautionary Labeling," *Transactions of the 20th Annual Meeting of the American Conference of Governmental Industrial Hygienists.* Atlantic City, NJ, April 19–22, 1958.
T. W. Nale, "The Federal Hazardous Substances Labeling Act," *Arch of Environ Health.* **4**, 239–246 (Jan.–June 1962).
National Academy of Sciences, *Principles and Procedures for Evaluating the Safety of Food Additives*, Publication 750. National Research Council, Washington, DC, 1959.
National Academy of Sciences, *Basis for Establishing Emergency Inhalation Exposure Limits Applicable to Military and Space Chemicals*, National Research Council, Washington, DC 1964.
National Academy of Sciences, *Principles and Procedures for Evaluating the Toxicity of Household Substances*, National Research Council, Washington, DC, 1964.
National Academy of Sciences, *Some Considerations in the Use of Human Subjects in Safety Evaluation of Pesticides and Food Chemicals.* Publication 1270. National Research Council, Washington, DC, 1965.
National Academy of Sciences, *Principles and Procedures for Evaluating the Toxicity of Household Substances*, National Research Council, Washington, DC, 1977.
National Academy of Sciences, *Principles of Toxicological Interactions Associated with Multiple Chemical Exposures*, National Academy Press, Washington, D.C., 1980.
National Fire Protection Association, *A Table of Common Hazardous Chemicals*, 7th ed., Committee on Hazardous chemicals and Explosives. Boston, 1944
National Institute of Occupational Safety and Health (NIOSH). *Recommended Exposure Limit (REL) Policy.* May 24, 1996. Cincinnati, Ohio.
National Inst. of Occupational Safety and Health (NIOSH), *Proceeding of a Workshop on Methodology for Assessing Reproductive Hazards in the Workplace*, Pub. No. 81-100, Washington, D.C., 1980.
National Safety Council, *Chemical Safety References*, Data Sheet 486, Revision A (extensive). Chicago, IL, 1968.
B. D. Naumann, E. V. Sargent, B. S. Starkman, W. J. Fraser, G. T. Becker, and G. D. Kirk (1996). Performance-based exposure control limits for pharmaceutical active ingredients. *Am Ind Hyg Assoc J.* 57(1):33–42.
NCRP (1959). Maximum Permissible Body Burdens and Maximum Permissible Concentrations of Radionuclides in Air and in Water for Occupational Exposure. *NBS Handbook No. 69.* U.S. Government Printing Office, Washington, DC.

E. E. Nelson, "Biological Assays and Enforcement Activities," *Presented at the Forty-Fourth Annual conference of Food and Drug Officials*. New Orleans, LA. 1940, pp. 22–28.

P. M. Newberne, "Pathology: Studies of Chronic Toxicity and Carcinogenicity," *J. the AOAC*. **58**(4), 650–656 (1975).

New Jersey State Department of Health, "Threshold Limit Values," *Occupational Health Bulletin*, **5**(3), 1–7 (1964).

New York State Department of Health, *Right to Know Bill—Resources for Training Materials*, Bureau of Toxic Substance Assessment, Jan. 1981

New York State Department of Health, *Chemical Fact Sheet, Trichloroethylene*, Bureau of Toxic Substance Assessment, July 1981.

New York State Department of Health, *Right to Know Bill—Implementation and Work Plan*. Bureau of Toxic Substance Assessment, May 1982.

NIOSH. A Recommended Standard for an Identification System for Occupationally Hazardous Materials. United States Department of Health, Education, and Welfare, Public Health Service, Centers for Disease Control, National Institute for Occupational Safety and Health, 1974

NIOSH, H. E. Christensen, T. T. Luginbyhl, and B. S. Carroll, eds., *The Toxic Substances List, 1974 Edition*. National Institute for Occupational Safety and Health, U.S. Dept. of Health, Education, and Welfare.

NIOSH, D. V. Sweet, ed., *Registry of Toxic Effects of Chemical Substances, 1985–86 edition, User's guide*. National Institute for Occupational Safety and Health, U.S. Dept. of Health, Education, and Welfare. April 1987.

T. Norseth (1994). Exposure limits: an ethical dilemma. In: *Advances in Modern Environmental Toxicology*, M. A. Mehlman and A. Upton (eds.). Princeton, NJ: Princeton Scientific Publishing Co., Inc.

R. D. Novak, M. H. Smolensky, E. J. Fairchild, and R. R. Reves (1990). Shiftwork and industrial injuries at a chemical plant in southeast Texas. *Chronobiol Int.* 7(2):155–164.

J. B. Olishifski, "The Elements of Industrial Toxicology," *National Safety News*, (Oct. 1967).

B. L. Oser, "Gaging the Toxicity of Chemicals in Food," *Chem. Eng. News.* **29**(28), 2808–2812 (1951).

B. L. Oser, "What are the Goals of a Safety Evaluation Program?" *Toxicol. Appl. Pharmacol.*, (Suppl. 3) 126–130 (1969).

OSHA, "Hazard Materials Labeling: Notice of Proposed Rulemaking and Public Hearings," *Fed. Reg.*, **42**(19), 5372–5374 (Jan. 28, 1977).

OSHA, "Hazard Identification; Notice of Proposed Rulemaking and Public Hearings." *Fed. Reg.* **46**(11), 4412–4453, (Jan. 16, 1982).

OSHA, "Hazard Communication; Notice of Proposed Rulemaking and Public Hearings," Fed. Reg., **47**(54), 12092–12101 (March 19, 1982).

OSHA, *Chemical Hazard Communication*, Occupational Safety and Health Administration, 1983.

OSHA, Technical Note # 19, OSHA Clarifies Warning Labels Required by Hazard Communication Standard. Occupational Safety and Health Administration, August 7, 1986.

G. E. Paget, "Toxicity tests: A Guide for Clinicians," *J. New Drugs.* **2**(2), 78–83 (1962).

G. E. Paget, "Standards for the Laboratory Evaluation of the Toxicity of a Drug? Viewpoint of the Expert Committee on Drug Toxicity of the Association of the British Pharmaceutical Industry." *Proc. Eur. Soc. Study Drug Toxicol.* **2**, 7–13 (1963).

W. J. Pangman, and R. M. Wallace, "Use of Plastic Prosthesis in Breast Plastic and Other Soft Tissue Surgery," *West. J. Surg. Obstetrics Gyn.* **63**, 503–512 (1955).

L. G. Parmer, "Toxicological Aspects of Common Food and Drugs," *Indust. Med. and Surg.* **27**(6), 285–286 (1958)

J. Patterson, R. Schoeny and P. Daunt (1991). The U.S. Environmental Protection Agency's Integrated Risk Information System (IRIS). In: *Regulating Drinking Water Quality.* C. Gilbert and E. Calabrese, ed. Lewis Publishers, Inc., Chelsea, MI. 273–281.

D. E. Patton, "Legal Aspects of Pesticides and Toxic Substances Testing Requirements," *J. Environ. Sci. Health*, **B15**(6), 645–663 (1980)

D. J. Paustenbach, G. P. Carlson, J. E. Christian, and G. S. Born (1986a). The Effect of the an 11.5 Hr./Day Exposure Schedule on the Distribution and Toxicity of Inhaled Carbon Tetrachloride in the Rat. *Fund. Appl. Toxicol.* 6:472–483.

D. J. Paustenbach, G. P. Carlson, J. E. Christian, and G. S. Born (1986b). A Comparative Study of the Pharmacokinetics of Carbon Tetrachloride in the Rat Following Repeated Inhalation Exposure of 8 hr/day and 11.5 hr/day. *Fund. Appl. Toxicol.* 6:484–497.

D. J. Paustenbach, "Methods for Setting Limits for Acute and Chronic Toxic Ambient Air Contaminants," *Appl. Occup Environ. Hyg.* **12**(6), 418–428 (1997)

H. M. Peck, "An Appraisal of Drug Safety Evaluation in Animals and the Extrapolation of Results to Man." In eds., D. H. Tedeschi and R. E. Tedeschi *Importance of Fundamental Principles of Drug Evaluation.* Raven Press, New York, 1968, pp. 449–471.

Pennsylvania Department of Health, *Short Term Limits for Exposure to Airborne Contaminants. A documentation*, Division of Occupational Health, 1969

The General Assembly of Pennsylvania, House Bill No. 1236, Section 6. Labeling, 1983, pp. 26–31.

The General Assembly of Pennsylvania, Worker and Community Right-to-Know Act. Act 1984–159, 1984 pp. 734–757.

S. M. Pier, R. C. Allison, E. M. Cunningham, and C. H. Ward, "Methods for Categorization of Hazardous Materials," in: *Proc. of the 1978 Natl Conf. on Control of Hazardous Materials Spills. April 11–13*, Miami Beach, FL, 1978.

C. J. Polson and R. N. Tattersall, "Advances in Clinical Toxicology," *The Practitioner* **187**, 549–556 (1961)

E. Poulsen, "Toxicological Aspects of Food Safety, Introduction to the Symposium," *Arch. Toxicol. Suppl.* **1**, 15–21 (1978).

U. C. Pozzani, C. S. Weil, and C. P. Carpenter, "The toxicological basis of TLVs: 5. The Experimental Inhalation of Vapor Mixtures by Rats, with Notes upon the Relationship between Single Dose Inhalation and Single Dose Oral Data," *Am. Ind. Hyg. Assoc. J.* **20**(5), 364–361 (1959)

P. E. Price, "Coming to Grips with the "Right to Know," *Metal Producing*, 36–37 (June 1986).

P. Price, R. Keenan, J. Swartout, C. Gillis, H. Carlson-Lynch and M. Dourson (1997). An Approach for Modeling Noncancer Dose responses with an Emphasis on Uncertainty. *Risk Anal.* 17(4):427–437.

Proceedings of Thirteenth International Congress on Occupational Health, July 25–29, 1961, Book Craftsman Associates, New York, 1961.

L. P. Prusak, "The Requirement for Proof of Utility in Patent Applications for Therapeutic Products," *Am. J. Pharm.* **125**(7), 240–249 (1953)

Public Health Service (PHS), *Survey of Compounds Which Have Been Tested for Carcinogenic Activity*, U.S. Govt. Printing Office, Washington, D.C., 1951.

W. A. Queen, Progress in Drug Control. Presented at the Conference of Pharmaceutical Law Enforcement Officials, Cleveland, OH, Sept. 7, 1944, pp. 49–64.

P. A. Raffle, W. R. Lee, R. I. McCallum, and R. Murray, *Hunter's Diseases of Occupations*, 6th ed., Little, Brown and Co., Boston, 1978.

T. G. Randolph, "Unlabeled Allergenic Constituents of Commercial Food and Drugs: A Critique of Food, Drug, and Cosmetic Act." *Annals of Allergy*, **9**, 151–165 (1951).

W. B. Rankin, "Criminal Evidence in Misbranded Drug and Device Cases," Presented at the 32nd Annual Conference of the Central Atlantic States Association, Baltimore, MD, May 26–27, 1948, pp. 105–111.

S. M. Rappaport (1981). The rules of the game: An analysis of OSHA's enforcement strategy. *Amer J Ind Med.* 6:291–303.

S. M. Rappaport, H. Kromhout, and E. Symanski (1993). Variation of exposure between workers in homogeneous exposure groups. *Amer Ind Hyg Assoc J.* 54:654–662, 1993.

R. S. Ratney, "The Development of Workplace Exposure Limits for Toxic Substances," *Appl. Indust. Hyg.*, Special Issue, 47–49 (Dec. 1989).

W. F. Reindollar, Laboratory Program for Drug Control. Presented at the Forty-fourth Annual Conference, New Orleans, LA, Oct. 1940, pp. 118–120.

P. Reznikoff, H. M. Rose, M. Stern, and L. R. Wasserman, "Toxic Effects of Therapeutic Agents; Transcription of a Panel Meeting on Therapeutics," *Bull. N. York Acad. M.* **32**(11), 796–818 (1956).

Rhode Island Department of Labor, *Hazardous Substance Right-to-Know Act. Chapter 28–21*, 1984.

W. E., Rinehart, M. Kaschak, and E. A. Pfitzer, Range-Finding Toxicity Data for 43 Compounds, *Industrial Hygiene Foundation of America, Chemical-Toxicological Series, Bulletin No. 6*, 1967.

S. R. Ripple (1992). The Use of Retrospective Health Risk Assessment: Looking back at nuclear weapons facilities. *Environ Sci Tech* 26: 1270–1276.

V. K. Rowe, H. C. Spencer, and S. L. Bass, "Toxicologic Studies on Certain Commercial Silicones and Hydrolyzable Silane Intermediates," *J. Indust. Hyg. Toxicol.* **30**(6), 332–352 (1948).

V. K. Rowe, H. C. Spencer, and S. L. Bass, "Toxicologic Studies on Certain Commercial Silicones," *Arch Ind. Hyg. Occup. Med.* **1**(5), 539–544 (1950).

V. K. Rowe, H. C. Spencer, D. D. McCollister, R. L. Hollingsworth, and E. M. Adams, Toxicity of Ethylene Dibromide Determined on Experimental Animals," *AMA Arch. Indust. Hyg. Occup. Med.* **6**(2), 158–173 (1952).

V. K. Rowe, D. D., McCollister, H. C. Spencer, F. Oyen, R. L. Hollingsworth, and V. A., Drill, "Toxicology of mono-, di-, and tri-propylene glycol methyl ethers," *AMA Arch. Indust. Hyg. Occup Med.* **9**(6), 509–525 (1954).

V. K. Rowe and E. M. Adams, "Problems of Health in the Marketing of Chemicals," *AMA Arch. Indust. Hyg. Occup. Med.* **10**(1), 50–53 (1954).

V. K. Rowe, M. A. Wolf, C. S. Weil, and H. F. Smyth, "The Toxicological Basis of Threshold Limit Values: 2. Pathological and Biochemical Criteria," *Am. Ind. Hyg. Assoc. J.* **20**(5), 346–349 (Oct. 1954).

V. K. Rowe, M. A. Wolf, C. S. Weil, and H. F. Smyth (1959). The toxicological basis of threshold limit values: 2. Pathological and biochemical criteria. *Amer. Ind. Hyg. Assn. J.* 20:350–356.

V. K. Rowe, "The Significance and Application of Threshold Limit Data," *Nat. Safety Cong.* **12**, 33–36 (1963).

E. M. Rupp, D. C. Parzychk, R. S. Booth, R. J. Ravidon, and B. L. Whitfield (1978). Composite Hazard Index for Assessing Limiting Exposures to Environmental Pollutants: Application through a Case Study. *Environ. Sci. Technol.* 12(7):802–807.

M. Sachs, "The Need for Threshold Limits," *Am. Ind. Hyg. Assoc. Q.* **17**(3), 274–281 (Sept. 1956).

C. O. Sappington, *Essentials of Industrial Health*, J. B. Lippincott Co., Philadelphia, 1943.

R. A. Scala (1992). Major issues in traditional workplace exposure limits. *Am Ind Hyg Assoc J.* 53(9):A438–439.

G. Scansetti, G. Piolatto, and G. Rubino (1988). Skin notation in the context of workplace exposure standards. *Am J Ind Med.* 14:725–32.

J. F. Schmutz, *Chronic Health Hazards, A National Challenge*, Presentation to the National Symposium on Chronic Hazards, Nov. 29, 1977.

M. A. Schneiderman and N. Mantel, "The Delaney Clause and a Scheme for Rewarding Good Experimentation," *Preventive Med.* **2**, 165–170 (1973).

H. H. Schrenk, "Toxic logic," *Ind. Med.* **1**(4), 53–54 (1940).

H. H. Schrenk, "Interpretations of Permissible Limits," *Am. Ind. Hyg. Assoc. Q.* **8**(3), 55–60 (1947).

H. H. Schrenk, *Interpretation of Permissible Limits in the Breathing of Toxic Substances in Air*, United States Department of the Interior, Bureau of Mines, May 1948.

H. H. Schrenk, "Development of New Products Includes Responsibility for Toxicological Data for Safe Manufacture and Use," *Ind. Eng. Chem.* **42**, 81A (Oct. 1950).

H. H. Schrenk, "Toxicity of Carbon Tetrachloride and Mercury Hazards in Laboratories are Subjects of Recent Reports," *Ind. Eng. Chem.* **44**, 131A (Oct. 1952).

H. H. Schrenk, "Pitfalls in Using Maximum Allowable Concentrations in Air Pollution," *Am. Ind. Hyg. Assoc. Q.* **16**(3), 230–234 (Sept. 1955).

H. H. Schrenk, "Growth and Progress in Industrial Hygiene," *Ind. Hyg. Q.* 113–118 (June 1957).

L. Schwartz, L. Tulipan, and S. M. Peck, *Occupational Diseases of The Skin*, Lea and Febiger, Philadelphia, 1947.

W. H. Sebrell, P. R. Cannon, C. S. Davidson, K. P. DuBois, D. W. Fassett, W. J. Hayes, G. Klatskin, A. J. Lehman and J. A. Miller, "Some considerations in the use of human subjects in safety evaluation of pesticides and food chemicals," A report of the ad hoc Subcommittee on Use of Human Subjects in Safety Evaluation of the Food Protection Committee, Food and Nutrition Board, National Academy of Sciences—National Research Council, 1965.

M. H. Seevers, "Perspective Versus Caprice in Evaluating Toxicity of Chemicals in Man," *JAMA* **153**(15), 1329–1333 (Dec. 12, 1953).

E. A. Sellers, "Unexpected Hazards Associated with Drugs and Chemicals," *Canad. Med Assoc J.* **86**(16), 721–724 (1962).

C. B. Shaffer, "Health Aspects of Production and Marketing of Chemicals," *AMA Arch. Indust. Health.* **19**(3), 298–301 (1959).

K. G. Shenoy, H. C. Grice, and J. A. Campbell, "Acute Toxicity as a Method of Assessing Sustained Release Preparations," *Toxicol. Appl. Pharmacol.* **2**, 100–110 (1960).

P. Shubik and J. Sice, "Chemical Carcinogenesis as a Chronic Toxicity Test, A Review," *Cancer Res.* **16**(8), 728–742 (1956).

P. Shubik, "Symposium on the evaluation of the safety of food additives and chemical residues: III. The role of the chronic study in the laboratory animal for evaluation of safety," *Toxicol. Appl. Pharmacol.* **16**, 507–512 (1970).

J. E. Silson, "The Significance of Maximum Allowable Concentrations," *Monthly Review of the Div. of Ind. Hyg. and Safety Standards. NYS Dept of Labor*, **28**(2), 5–8 (1949).

P. Silva (1986). TLVs to protect "nearly all workers." *Appl Occup Environ Hygiene.* 1:49–53.

L. Silverman, C. E. Billings, and M. W. First, *Particle Size Analysis in Industrial Hygiene*, Academic Press, New York, 1971.

I. Simon, "Suggestion of a Method for the Determination of the Minimal Lethal Dose of Drugs which Could Form the Basis of an International Convention on This Subject," *Sc. Med. Ital.* **2**(3–4), 653–661 (1951–1952).

W. Slikker, Jr., K. S. Crump, M. E. Andersen, and D. Bellinger (1996). Biologically based, quantitative risk assessment of neurotoxicants. *Fundam Appl Toxicol.* 29(1):18–30.

G. G. Slocum, "Pure Foods—Safe Drugs: The Food and Drug Administration's Role in Public Health," *Am. J. Pub. Health.* **46**(8), 973–977 (1956).

A. E. Smith, "Needed Medical Research Developments in Industry," *J. Am. Pharm. Assoc. Sci. Ed.* **34**(4), 123–126 (1945).

C. Smith, "A Short Term Chronic Toxicity Test," *J. Pharmacol. Exp. Therap.* **100**(4), 408–420 (1950).

R. G. Smith, Problems Raised by New Drugs. *Assn. of Food & Drug Officials of U.S.* **19**(4): 144–150; *Assn. of Food & Drug Officials of U.S.* **23**(1), 44–49 (1955).

R. G. Smith, "Evaluation of Safety of New Drugs by the Food and Drug Administration." *J. New Drugs* **1**(2), 59–64 (1961).

S. E. Smith, "The Safety of Drugs," *Nursing Times* 423–424 (March 31, 1967).

M. H. Smolensky, D. J. Paustenbach, and L. E. Schering (1985). Biological Rhythms, Shiftwork and Occupational Health. In *Patty's Industrial Hygiene and Toxicology* 2nd ed., Vol 3b, Biological Responses (Cralley and Cralley, ed.) John Wiley & Sons, NY; pgs. 175–312.

M. H. Smolensky and A. Reinberg, "Clinical Chronobiology: Relevance and Applications to the Practice of Occupational Medicine," *Occup. Med.* **5**(2), 239–272 (1990).

H. F. Smyth, Report of the Committee on Volatile Solvents to the Industrial Hygiene Section of the American Public Health Association, October 17, 1939.

H. F. Smyth, and C. P. Carpenter, "The Place of the Range Finding Test in the Industrial Toxicology Laboratory," *J. Ind. Hyg. Toxicol.* **26**(8), 269–273 (1944).

H. F. Smyth, "Solving the Problem of the Toxicity of New Chemicals in Industry," *The W. Virginia Med J.* **42**(7), 9–13 (1946).

H. F. Smyth and F. R. Holden, Summary of Conference on Chemistry and Toxicology. *Transactions of Eleventh Annual Meeting of Industrial Hygiene Foundation of America, Inc.*, Nov. 7, 1946.

H. F. Smyth and J. F. Morgan, Summary of Chemical and Toxicological Conference. *Transactions of Twelfth Annual Meeting of Industrial Hygiene Foundation of America, Inc.*, Nov. 20, 1947.

H. F. Smyth and C. P. Carpenter, "Further Experience with the Range Finding test in the Industrial Toxicology Laboratory," *J. Ind. Hyg. Toxicol.* **30**(1), 63–68 (1948).

H. F. Smyth, C. P. Carpenter, and C. S. Weil, "Range-Finding Toxicity Data, List III," *J. Ind. Hyg. Toxicol.* **31**(1), 60–62 (1949).

H. F. Smyth, C. P. Carpenter, and C. S. Weil, Range-finding toxicity data: List IV. *AMA Arch. Indust. Hyg. Occup. Med.* **4**(2), 119–122 (1951).

H. F. Smyth, C. S. Weil, E. M. Adams, and R. L. Hollingsowrth, "Efficiency of Criteria of Stress in Toxicological Tests," *AMA Arch. Ind. Hyg. Occup. Med.* **6**(1), 32–36 (1952).

H. F. Smyth, "Toxicological data—Sources of Information and Future Needs," *Am. Ind. Hyg. Assoc. Q.* **15**, 203–205 (1954).

H. F. Smyth, "Toxicology," in W. C. Cutting and H. W. Newman, eds., *Annual Review of Medicine*, Annual Review, Inc., Stanford, CA, 1954.

H. F. Smyth, C. P. Carpenter, C. S. Weil, and U. C. Pozzani, "Range-finding Toxicity Data. List V." *AMA Arch. Indust. Hyg. Occup. Med.* **10**(1), 61–68 (1954).

H. F. Smyth, "Improved Communication—Hygienic Standards for Daily Inhalation," *Am. Ind. Hyg. Assoc. Q.* **17**, 129–185 (1956).

H. F. Smyth, Jr. "The Toxicological Basis of Threshold Limit Values: 1. Experience with Threshold Limit Values Based on Animal Data," *Am. Ind. Hyg. Assoc.* **20**(5), 341–345 (1959).

H. F. Smyth, C. P. Carpenter, C. S. Weil, U. C. Pozzani, and J. A. Striegel, "Range-Finding Toxicity Data: List VI," *Am. Ind. Hyg. Assoc. J.* **23**(2), 95–107 (1962).

H. F. Smyth, "A Toxicologist's View of Threshold Limits," *Am. Ind. Hyg. Assoc. J.* **23**, 37–43 (1962).

H. F. Smyth, "Range-Finding Toxicity Data: List VII," *Am. Ind. Hyg. Assoc. J.* 470–476 (Sept.–Oct).

H. F. Smyth, "Current Confidence in Occupational Health. Presented at the Am. Ind. Hyg. Conf., Chicago, IL, May 27–June 1, 1979 in *Ann. Am. Conf. Ind. Hyg.* **9**, 323–329 (1984).

W. S. Spector, *Handbook of Toxicology*, Vol. 2, Wright-Patterson Air Force Base, Dayton, Ohio, 1955.

J. F. Stara, R. J. F. Bruins, M. L. Dourson, L. S. Erdreich, R. C. Hertzberg, P. R. Durkin and W. E. Pepelko (1987). Risk assessment is a developing science: Approaches to improve evaluation of siingle chemicals and chemical mixtures. In: *Methods for Assessing the Effects of Mixtures of Chemicals*, (V. B. Vouk, G. C. Butler, A. C. Upton, D. V. Parke and S. C. Asher, Ed.) 1987 SCOPE. Pages 719–743.

I. Starr, "The Testing of New Drugs and other Therapeutic Agents," *JAMA* **177**(1), 84–92 (1961).

M. Steinberg, "ACGIH TLVs and the Sensitive Worker," *Ann. Am. Conf. Govt. Ind. Hyg.*, **3**, 77–81 (1982).

J. H. Sterner, "A Program for Detecting Possible Toxic Responses to Varied Organic Chemical Exposures," *N.Y. State J. Med.* **41**(6), 594–599 (1941).

J. H. Sterner, "Determining Margins of Safety-Criteria for Defining a "harmful" Exposure," *Ind. Med.* **12**(8), 514–518 (1943).

J. H. Sterner, "Threshold Limits—A Panel Discussion," *Am. Ind. Hyg. Assoc. Q.* **16**(1), 27–39 (March 1955).

J. H. Sterner, "Methods of Establishing Threshold Limits," *Am. Ind. Hyg. Assoc. Q.* **17**(3), 280–286 (1956).

C. P. Stewart and A. Stolman, eds. "*Toxicology—Mechanisms and Analytical Methods*," Vol. 1, Academic Press, New York, 1960, Chapt. 1.

H. E. Stokinger, Present Status of M.A.C. of Gases. Transactions of the 16th Annual Meeting of the American Conference of Governmental Industrial Hygienists. Chicago, IL, April 24–27, 1954, pp. 53–55.

H. E. Stokinger, "Standards for Safeguarding the Health of the Industrial Worker," *Public Health Reports* **70**(1), 1–11 (1955).

H. E. Stokinger, "Advances in Industrial Toxicology for the Year 1955," *AMA Arch. of Ind. Health* **14**, 206–212 (1956).

H. E. Stokinger, "Toxicologic Aspects of Occupational Hazards," *Ann. Rev. Med.* **7**, 177–194 (1956).

H. E. Stokinger, "Threshold Limits and Maximal Acceptable Concentrations: Definition and Interpretation, 1961," *Arch. Env. Health* **4**, 121–123 (1962).

H. E. Stokinger (1962). Threshold limits and maximum acceptable concentrations; Their definition and interpretation. *Amer Ind Hyg Assoc J* 23:45–47.

H. E. Stokinger, Pharmacodynamic, biochemical, and toxicologic methods as bases for air quality standards. *Transactions of the 25th Annual Meeting of the American Conference of Governmental Industrial Hygienists*, Cincinnati, Ohio, May 6–10, 1963, pp. 25–40.

H. E. Stokinger (1964). Modus operandi of threshold limits committee of ACGIH. In *Transactions of Twenty-Sixth Annual Meeting of the American Conference of Governmental Industrial Hygienists* 26:23–2.

H. E. Stokinger, "Industrial Contribution to Threshold Limit Values," *Arch. Env. Health* **10**, 609–611 (1965).

H. E. Stokinger (1969). Current problems of setting occupational exposure standards. *Arch Environ Health* 19:277–280.

H. E. Stokinger (1971). Intended use and application of the TLVs. In *Transactions of the Thirty-third Annual Meeting of the American Conference of Governmental Industrial Hygienists*. 33:113–116.

H. E. Stokinger, "Criteria and Procedures for Assessing the Toxic Responses to Industrial Chemicals, presented at the first ILO/WHO meeting on international limits, June, in: *Ann. Am. Conf. Ind. Hyg.* **9**, 155–163 (1984).

H. E. Stokinger, Suggested Principles and Procedures for Developing Data for Threshold Limit Values for Air. *Industrial Hygiene Foundation of America, Chemical-Toxicological Series, Bulletin* No. 8–69, 1969.

H. E. Stokinger, "Sanity in Research and Evaluation of Environmental Health: How to Achieve a Realistic Evaluation (in seven commandments)," *Science* **174**, 662–665 (1971).

H. E. Stokinger, "Toxicity of Airborne Chemicals: Air Quality Standards—A National and International View," *Ann. Rev. Pharmacol. Toxicol.* **12**, 407–422 (1972).

H. E. Stokinger, "Concepts of Thresholds in Standards Setting: An Analysis of the Concept and its Application to Industrial Air Limits (TLVs)," *Arch. Env. Health* **25** 153–157 (1972).

H. E. Stokinger, "Industrial Air Standards—Theory and Practice," *J. Occup. Med.* **15**(5), 429–431 (1973).

H. E. Stokinger, "The Case for Carcinogen TLVs Continues Strong," Presented at the ACGIH symposium on Workplace Control of Carcinogens, October 25–26, 1976, Kansas City, MO, in *Ann. Am. Conf. Ind. Hyg.* **9**, 257–264 (1984).

H. E. Stokinger, Stokinger Lecture, Courtesy of Bill Wagner, American Conference of Governmental Industrial Hygienists, 1980.

H. E. Stokinger, "Historic Aspects of Occupational Health Standards and the Sensitive Worker." *Ann. Am. Conf. Gov. Ind. Hyg.* **3**, 65–69 (1982).

H. E. Stokinger, "Suggested Principles and Procedures for Developing Experimental Animal Data for Threshold Limit Values for Air," *Ann. Am. Conf. Ind. Hyg.* **9**, 177–186 (1984).

D. R. Stolz, L. A. Poirier, C. C. Irving, H. G. Stich, J. H. Weisburger, and H. C. Grice, "Evaluation of Short-Term Tests for Carcinogenicity," *Toxicol. Appl. Pharmacol.* **29**, 157–180 (1974).

G. B. Stone, "The New Product Challenge," *Am. J. Pharmacy.* **131**(9), 327–336 (1959).

R. T. Stormont, "New Program of Operation for Evaluation of Drugs," *JAMA*, **158**(13) 1170–1171 (1955).

T. O. Tengs, M. E. Adams, J. S. Pliskin, et al. (1995). Five-hundred life-saving interventions and their cost-effectiveness. *Risk Anal.* 15:369–390.

L. Teleky, "Toxic Limits," *Ind. Med., Ind. Hyg. Sec.* **4**, 68–71 (1940).

M. J. Thomas and P. A. Majors, "Animal, Human, and Microbiological Safety Testing of Cosmetic Products," *J. Soc. Cosmet. Chem.* **24**, 135–146 (1973).

L. F. Tice, "The Expansion of Drug Certification by the FDA," *Am. J. Pharmacy* **132**(12), 431–432 (1960).

L. F. Tice, "Package Inserts for Prescription Products," *Am. J. Pharmacy* **133**(7), 240–241 (1961).

M. D. Topping, C. R. Williams, and J. M. Devine (1997). *Industry's Perception and Use of Occupational Exposure Limits.* Paper presented at the 1997 Conference of the British Occupational Hygiene Society, Warwick, U.K., April 15–18, 1997.

R. Truhaut, "The Problem of Thresholds for Chemical Carcinogens—Its Importance in Industrial Hygiene, Especially in the Field of Permissible Limits for Occupational Exposure," *Am. Ind. Hyg. Assoc. J.* **41**, 685–692 (1980).

R. M. Tuggie (1981). The NIOSH decision scheme. *Amer Ind Hyg Assoc J.* 42:493–498.

G. E. Turfitt, "Recent Advances in Toxicologic Analysis," *J. Pharm. Pharmacol.* **3**(6), 321–337 (1951).

U.S. Environmental Protection Agency (EPA) (1992). *Dermal Exposure Assessments: Principals and Applications.* Exposure Assessment Group. Office of Health and Environmental Assessment, Washington, DC.

United States of America Standards Institute, *USA Standard Acceptable Concentrations of Formaldehyde*, sponsored by the American Industrial Hygiene Association, Sept. 12, 1967.

United States of America Standards Institute, *USA Standard Acceptable Concentrations of Carbon Tetrachloride*, sponsored by the American Industrial Hygiene Association, Oct. 9, 1967.

United States Congress, Federal Hazardous Substances Labeling Act. Pages 1–10. July 12, 1960.

United States Congress, Chemical Dangers in the Workplace, Thirty-fourth Report by the Committee on Government Operations. Washington, DC, 1976.

"Table of Common Hazardous Chemicals," *Chem. Eng. News* **23**(14), 1249–1256 (1945).

H. Vainio and L. Tomatis, "Exposure to Carcinogens: Scientific and Regulatory Aspects," *Ann. Am. Conf. Ind. Hyg.* **12**, 135–143 (1985).

P. M. Van Arsdell, "Health Hazards of Common Laboratory Reagents," *Chem. Eng. News* **26**(5), 304–309 (Feb. 2, 1948).

W. F. Von Oettingen, "Variations in the Toxicity of Chemical Compounds under Different Conditions," *Annual Safety Congress Transactions*, 1935.

W. F. Von Oettingen, "The Toxicity and Potential Dangers of Aliphatic and Aromatic Hydrocarbons," *Yale J. Biol. Med.* **15**(2), 167–184 (1942).

W. F. Von Oettingen, P. A. Neal, D. D. Donahue, J. L. Svirbely, H. D. Baernstein, A. R. Monaco, P. J. Valaer, and J. L. Mitchell, "The toxicity and potential dangers of toluene with special reference to its maximal permissible concentration," *Publ. Health Bull. 279*, United States Public Health Service, 1942, pp. 1–50.

W. F. Von Oettingen, "The aliphatic alcohols: Their toxicity and potential dangers in relation to their chemical constitution and their fate in metabolism," *Publ. Health Bull.* 281, U.S. Public Health Service, 1943.

W. F. Von Oettingen, *The Halogenated Aliphatic, Olefinic, Cyclic, Aromatic, and Aliphatic-Aromatic Hydrocarbons Including the Halogenated Insecticides, Their Toxicity and Potential Damages*, U.S. Dept. of Health, Education and Welfare, Washington, D.C., 1955.

F. A. Vorhes, "Requirements of Analytical Data," *Ag. Fd. Chem.* **4**(5), 415–416 (1956).

R. C. Wands, "Industrial health in 1978: a perspective," Presented at the Am. Ind. Hyg. Conf., Los Angeles, CA, May 7–12, in *Ann. Am. Conf. Ind. Hyg.* **9**, 317–321 (1984).

R. C. Wands, M. Steinber, E. K. Weisburger, and R. Spirtas, "Threshold Limit Values for Carcinogens—Current Status," *Ann. Am. Conf. Ind. Hyg.* **12**, 263–265 (1985).

University of Washington, "A Toxicity Evaluation of Several Trade-Name Solvents," in: *Occupational Health Newsletter*, Environmental Research Laboratory, Aug. 1954.

C. S. Weil, Tables for convenient calculations of median-effective dose (LD_{50} or ED_{50}) and instructions in their use, *Biometrics* **8**(3), 249–263 (1952).

C. S. Weil, C. P. Carpenter, and H. F. Smyth, "Specifications for Calculating Median Effective Dose," *Am. Indust. Hyg. A. Quart.* **14**(3) 200–206 (1953).

C. S. Weil, "Application of Methods of Statistical Analysis to Efficient Repeated-Dose Toxicological Tests. I. General Considerations and Problems Involved. Sex Differences in Rat Liver and Kidney Weights," *Toxic. Appl. Pharmacol.* **4**, 561–571 (1962).

C. S. Weil and D. D. McCollister, "Relationships between Short- and Long-Term Feeding Studies in Designing an Effective Toxicity Test," *J. Agr. Food Chem.* **11**, 486–491 (1963).

C. S. Weil, M. D. Woodside, J. R. Bernard, and C. P. Carpenter, "Relationship between Single-Peroral, One-Week, and Ninety-Day Rat Feeding Studies," *Toxicol. Appl. Pharmacol.* **14**, 426–431 (1969).

C. S. Weil and C. P. Carpenter, "Abnormal Values in Control Groups during Repeated Dose Toxicologic Studies," *Toxicol. Appl. Pharmacol.* **14**, 335–339 (1969).

C. S. Weil, "Editorial," *Toxicol. Appl. Pharmacol.* **17**(2), i–ii (1970).

C. S. Weil, "Guidelines for Experiments to Predict the Degree of Safety of a Material for Man," *Toxicol. Appl. Pharmacol.* **21**, 194–199 (1972).

C. S. Weil, "Statistics vs Safety Factors and Scientific Judgment in the Evaluation of Safety for Man," *Toxicol. Appl. Pharmacol.* **21**, 454–463 (1972).

J. H. Weisburger and E. K. Weisburger, "Food Additives and Chemical Carcinogens: on the Concept of Zero Tolerance," *Food Cosmet. Toxicol.* **6** 235–242 (1968).

G. M. Wheatly, "The Federal Hazardous Substances Labeling Act," *Pediatrics* **28**(3), 499–500 (1961).

WHO, "Principles for Pre-Clinical Testing of Drug Safety," *WHO Tech. Rep. Ser. No. 341*, World Health Organization, Geneva, 1966, pp. 3–22

WHO, "Principles for the Clinical Evaluation of Drugs," *Tech. Rep. Ser. No. 403*, World Health Organization, Geneva, 1968, pp. 9–10.

WHO, "WHO on Drug Safety," *Fd. Cosmetics Toxicol.* **7**(6), 674–675 (1969).

WHO, Permissible Levels of Occupational Exposure to Airborne Toxic Substances. Sixth Report of the Joint ILO/WHO Committee on Occupational Health. World Health Organization Geneva, 1969.

WHO, "Principles and Methods for Evaluating the Toxicity of Chemicals," Part 1, World Health Organization, Geneva, 1978.

WHO, "Evaluation of certain food additives and contaminants," *Technical Report Series 631*. Twenty-second Report of the Joint FAO/WHO Expert Committee on Food Additives. World Health Organization, 1978.

W. B. White, "The Addition of Chemicals to Food," *Fd. Drug Cosmetic Law Q.* **2**(4), 475–481 (1947).

G. M. Williams (1981). Epigenetic Mechanisms of Action of Carcinogenic Organochlorine Pesticides, ACS, Symp. Series: *ISS Pest. Chem. Mod. Toxicol.* 160:45–56.

A. Wilschut, W. F. Ten Berg, P. J. Robinson, and T. E. McKone (1995). Estimating skin permeation. The validation of five mathematical skin permeation models. *Chemosphere.* 30:1275–96.

H. K. Wilson (1997). Recent policy and technical developments in biological monitoring in the United Kingdom. *Sci Total Environ.* 199:101–5.

R. H. Wilson and F. DeEds, "Importance of Diet in Studies of Chronic Toxicity," *Arch. Ind. Hyg. Occup. Med.* 1(1), 73–80 (1950).

R. H. Wilson and W. E. McCormick, "Toxicology of Plastics and Rubber Plastomers and Monomers," *Indust. Med.* 23(11), 479–486 (1954).

M. Winnell, "An International Comparison of Hygienic Standards for Chemicals in the Work Environment," *AMBIO* 4(1), 34–36 (1975).

V. O. Wodicka, "Food Ingredient Safety Criteria," *Fd. Tech.* 31(1), 84–88 (1977).

V. O. Wodicka, "Risk and Responsibility," *Nutrition Rev.* 38(1), 45–52 (1980).

V. O. Wodicka, "Evaluating the Safety of Food Constituents," *J. Environ. Pathol. Toxicol.* 3, 139–147 (1980).

G. Woodard and H. O. Calvery, "Acute and Chronic Toxicity," *Ind. Med.* 12(1), 55–59 (1943).

R. Woolmer, "Clinical Tests of New Drugs," *Proc. Royal Soc. Med.* 52(2), 98–100 (1959).

A. N. Worden, "Toxicological Methods," *Toxicology* 2, 359–370 (1974).

W. P. Yant, "Industrial Hygiene codes and Regulations," Transactions of Thirteenth Annual Meeting of Industrial Hygiene Foundation of America, Inc., Nov. 18, 1948, pp. 48–61.

J. A. Zapp, "An Acceptable Level of Exposure," presented at the Am. Ind. Hyg. Conf., New Orleans, LA, May 22–27 1977 in *Ann. Am. Conf. Ind. Hyg.* 9, 309–315 (1984).

G. Zbinden, "The Problem of the Toxicological Examination of Drugs in Animals and their Safety in Man," *Clin. Pharmacol. Therap.* 5(5), 537–545 (1964).

G. Zbinden, Drug Safety: Experimental Programs. Problems and Solutions of the Past 10 Years are Critically Reviewed, *Science* 164(3880), 643–647 (1969).

C. Zenz, *Occupational Medicine: Principles and Practical Applications*, Yearbook Medical Pub., Chicago, 1975.

C. Zenz and B. A. Berg (1970). Influence of Submaximal Work on Solvent Uptake. *J. Occup. Med.* 12:367–369.

R. L. Zielhuis (1974). Permissible Limits for Occupational Exposure to Toxic Agents: A Discussion of Differences in Approach Between U.S. and USSR. *Int. Arch. Argbeitsmed.* 33:1.

R. L. Zielhuis (1994). Yant Memorial Award Lecture, 1992. A more informative list of occupational exposure limits with a supplements. *Am Ind Hyg Assoc J.* 55(2):102–111.

G. E. Ziem and B. I. Castleman (1989). Threshold Limit Values: Historical perspective and current practice. *J. Occup. Med.* 13:910–918.

CHAPTER FORTY-TWO

Biological Monitoring of Exposure to Industrial Chemicals

Vera Fiserova-Bergerova, Ph.D. and Jaroslav Mraz, Ph.D.

1 INTRODUCTION

Pollution of air, water, and soil by industrial chemicals presents a potential health risk to humans. Such chemicals can enter the human body by three routes, namely, by inhalation, dermal absorption, and ingestion. In the work place, pulmonary and dermal absorption are the main routes of entry, but poor personal hygiene and work habits can result in ingestion that contributes to the dose. Air monitoring provides reliable information on inhalation exposure, and patches can be used to estimate dermal exposure. Local adverse effects, such as skin and eye irritation, or nose and lung irritation, are closely related to the external exposure. Systemic adverse effects, on the other hand, are related to the absorbed amount (dose), or to the level of the pollutant or its metabolite in the target organ.

The actual dose—body burden—is determined by the difference between absorption and elimination rates of the pollutant. These rates, which are a function of time, are not easy to determine. The absorption rate (also called uptake rate) depends on the degree of external exposure. Both the absorption rate and the elimination rate depend on the chemical structure of the pollutant, on environmental factors such as temperature and humidity, and on physiological factors of the exposed subjects, such as pulmonary ventilation, cardiac output, tissue perfusion rate, and body surface. For example, physical activity, which affects pulmonary ventilation and cardiac output and its distribution, alters rates of pulmonary uptake and pulmonary wash-out of the volatile pollutants. The condition of the skin, its covering (e.g., clothes, gloves), and the delivery vehicle of the pollutant (e.g., grease) affect dermal absorption. The need for assessment of internal exposure—i.e., assessment

Patty's Industrial Hygiene, Fifth Edition, Volume 3. Edited by Robert L. Harris.
ISBN 0-471-29753-4 © 2000 John Wiley & Sons, Inc.

of actual exposure of the body tissues regardless of the route of absorption and of variables affecting the uptake and elimination—led to biological monitoring. Biological monitoring is based on the measurement of biological levels of the exposure indicators, also called biomarkers. They can be the pollutant itself (parent compound), its metabolite(s) and conjugate(s), adduct(s) with biological macromolecules, or induced reversible enzymatic and functional changes such as enzyme inhibition, or oxidation of oxyhemoglobin to methemoglobin. For the purposes of occupational exposure monitoring, epidemiological studies, and health risk assessment, the biomarkers are measured in urine, exhaled air, or blood. However, tissue bank samples can be used for retrospective studies of occupational and environmental exposures. In animals, biological levels of biomarkers in tissues are studied mainly for the purpose of elucidating the interaction of toxic agents with the body.

Exposure monitoring is an important tool in disease prevention in the work place. Thus, for example, the safety guidance values published for occupational exposure by the American Conference of Governmental Industrial Hygienists (ACGIH) for indicators of internal exposure (Biological Exposure Indices, BEIs) and the safety guidance values for airbornes (Threshold Limit Values for air contamination, TLVs) relate to each other on toxicokinetic bases. In practice, the dermal absorption and working conditions (such as workload and co-exposure to other chemicals) often upset the compatibility of inhalation exposure and internal exposure. If biological levels indicate an overexposure, compliance with the BEI should be the primary concern. The relationship between air monitoring and biological monitoring is pictured in Figure 42.1.

Timing of sample collection in biological monitoring is critical for estimation of the degree of exposure. Toxicokinetics provide guidance on the relation between external and internal exposure and on timing for collection of biological samples. The reference values define the sampling time that must be maintained when reference values such as BEIs (1) or BATs (2) are used.

Some biomarkers correlate better with adverse health effects than with the external exposure. Some adverse effects are related to the total absorbed amount (internal dose); other adverse effects are related to the actual concentrations of the biomarkers in tissues and body fluids. For example, nephrotoxicity of cadmium, which is induced by saturation of binding sites in the renal cortex, is closely related to the total absorbed amount of cadmium, regardless of delivery time. On the other hand, inhibition of CNS by organic solvents is related to their actual concentrations in the brain. Thus a high dose administered in a short time induces anaesthesia, but the same dose administered over a long period of time has no effect on CNS function.

Some biomarkers can be the direct cause of an adverse effect. For example, methemoglobin measurements in blood, used as a screening test for monitoring of exposure to some aromatic amines and nitrocompounds, is a measure of the potential risk of cyanosis. Biological levels of DNA adducts with electrophilic metabolites, investigated in association with carcinogenicity and genotoxicity, correlate poorly with the degree of exposure.

Irreversible biochemical and functional changes are not recommended as biomarkers for exposure monitoring because abnormal values already indicate a health injury which is supposed to be prevented by the monitoring. This chapter will discuss those biomarkers that are suitable for monitoring of occupational exposure.

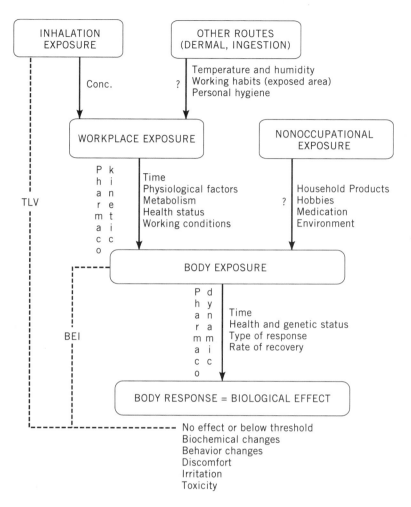

Figure 42.1. Sources of external exposure and their relation to internal dose and adverse effect. For explanation see text. Adapted from *Topics in Biological Monitoring* with the permission of the ACGIH, Cincinnati, OH, © 1995.

2 KINETICS OF UPTAKE, DISTRIBUTION, AND ELIMINATION OF BIOMARKERS

During the exposure, the biological levels of pollutants and their biomarkers rise, and, with few exceptions, start to decline immediately after the end of exposure. The physiologically based simulation model, developed 30 years ago for volatile organic solvents, describes the course of biological levels of inhaled vapors and gases and explains the role of physiological and physicochemical parameters in the uptake, distribution, and elimination of inhaled substances. The model is described in Chapter 40. A simplified one-compartment model is used below to identify some factors inducing variability of biological levels of biomarkers during and following exposure.

2.1 Linear Kinetics

At low exposures, permissible in the work place and in the environment, the uptake, distribution, and elimination are described by linear kinetics, characterized by rate constants k. The rising of the biological levels of the pollutants during exposure is described by the exponential function:

$$c_\tau = C(1 - e^{-k\tau}) \quad (1)$$

where e denotes the base of natural logarithm, τ is the time counted from the start of the exposure, c_τ is the biological level at a time τ, and C is the biological level which would be reached at steady state.

The declining of the biological levels after the exposure is described by the equation:

$$c_t = C_T e^{-kt} \quad (2)$$

where t is time counted from the end of the exposure, c_t denotes biological levels at a time t, T denotes the total exposure duration, and C_T is the biological level at the end of exposure. After substituting from Equation 1 in Equation 2

$$c_t = C(1 - e^{-kT})e^{-kt} \quad (3)$$

2.1.1 The Rate Constant

The rate constant k is a reciprocal value of the time constant, which indicates the time needed to reach 63% of the biological level at steady state. It is more common to relate the rate constant to the half-life $t_{1/2}$, which indicates the time needed to reach 50% of the biological level at steady state, or, following exposure, to reduce the biological level to half.

$$t_{1/2} = \frac{0.693}{k} \quad (4)$$

2.1.2 Determination of the Half-Life

The half-life can be determined experimentally by measuring the biological levels of biomarkers in samples collected at different times following the exposure, and plotting the measured values against the logarithm of the sampling time counted from the time the biological level reaches the peak value. The half-life is then determined as the time interval in which the level reduces to half. The half-life and rate constant can also be calculated if the alveolar ventilation, cardiac output, tissue perfusion, and clearance and biosolubility of the inhaled vapor or gas are known (see Equation 5).

2.1.3 Sample Collection Strategy and Half-Life

The best time to collect samples for biological monitoring is when the changes in biological levels of the biomarker are slow, so that the sampling time is not critical, and when the biological levels are high, so that the concentrations of the biomarkers are easy to measure (3). Biomarkers with half-lives shorter than one hour are not suitable for routine biological monitoring, because their biological levels change very rapidly, thus requiring precise sampling time, which is not practical to maintain in field conditions.

Biomarkers with a short half-life (<5 hr) do not significantly cumulate in the body; therefore, sampling at the end of the shift provides the best estimate of recent exposure. On the other hand, during intermittent exposures, such as occupational exposures, biomarkers with half-lives longer than five hours cumulate in the body over the work week, so that the biological levels are affected by the degree and timing of previous exposures. Biomarkers with long half-lives (>10 hr) cumulate in the body over weeks, months, and even years; thus their biological levels in samples collected at the end of the shift and at the end of the work week are indicators of recent as well as of past exposures. For biomarkers with a very long half-life, it takes months to reach the steady state, so that the lag time must be considered in the evaluation of data obtained in samples collected from newly employed workers. Biological levels are affected by the current exposure as well as by previous exposures. This applies for example to heavy metals in blood and urine. Table 42.1 contains summary guidelines on how to collect biological samples on the basis of the half-life of the biomarker.

2.1.4 Factors Determining the Half-Life

The half-lives of volatile pollutants can be predicted based on physiological parameters related to the body build and physical activity of the exposed subject, and on the biosolubility and bioreactivity of the pollutant and of its metabolite. The dependence of the half-life on these parameters is indicated in Equation 5, using a simplified one-compartment model:

Table 42.1. Recommended Sampling Time Based on the Half-Life of the Biomarker

Half-life	Shift[a]	Workweek[a]	Long-Term
<1 hour	Not suitable for routine monitoring		
<5 hours	End	N/C[a]	N/C[a]
<10 hours	End	End	N/C[a]
<4 days	N/C[a]	End	After 1 week[b]
<9 days	N/C[a]	N/C[a]	After 1 month[b]
<3 months	N/C[a]	N/C[a]	After 1 year[b]
>3 months	N/C[a]	N/C[a]	Each 6 months

[a]N/C means not critical.
[b]The biological levels approach steady state.

$$t_{1/2} = 0.693 V \lambda_{tis/air} \frac{(V_{alv} + F\lambda_{bl/air})}{V_{alv}F\lambda_{bl/air} + Cl(V_{alv} + F\lambda_{bl/air})} \quad (5)$$

Equation 5 shows that the half-life is directly related to the capacity of the body to retain the pollutant, i. e., to the body volume V and to the biosolubility of the pollutant in tissues, which is defined by the appropriate partition coefficient $\lambda_{tis/air}$. The half-life also depends on alveolar ventilation V_{alv}, blood flow F (both dependent on body surface and physical activity of the worker (4)), solubility of the pollutant in blood $\lambda_{bl/air}$, and total (intrinsic) clearance by excretion and metabolism Cl. Intrinsic clearance of volatile pollutants can be determined from the pulmonary uptake rate, when the exposure approaches the steady state. (The difference of concentrations of the pollutant in inhaled and exhaled air is multiplied by minute ventilation and the calculated tissue uptake rate is subtracted (5)).

2.1.5 Sources of Variability of Biological Levels

Alveolar ventilation, cardiac output, and tissue perfusion depend on physical activity of the exposed subject (4–6). Physical activity shortens the half-life. Thus the biological levels of biomarkers rise faster when the worker performs heavy work during the shift, and decline faster when the worker continues physical activity after the shift ends.

Dependence of the half-life on solubility in blood indicates that compounds well soluble in blood are distributed throughout the body faster than hydrophobic compounds. The half-life depends directly on body build. Body fat profoundly affects the kinetics of lipophilic solvents which have a large fat-gas partition coefficient and cumulate in relatively poorly perfused adipose tissues. Lipophilic compounds are very slowly washed-out from the body, especially in obese persons. The differences in the perfusion rate and in the solubility of compounds in different organs result in different elimination rates. Therefore the kinetics of uptake, distribution, and elimination are described by multicompartment models, usually by three or four exponential functions with distinguishable half-lives (5). Sampling time during the phase with the long half-life is not critical and therefore preferable in field conditions.

The effects of the above mentioned parameters are fully understood, and their role in producing interindividual differences in biological levels of biomarkers is predictable.

Metabolism and tissue binding play a significant role in the uptake and elimination of biomarkers. Polymorphism, and the availability of enzymes mediating the metabolism of xenobiotics, are the unpredictable sources of interindividual differences in the biological levels of biomarkers resulting from the exposure. Predictions of metabolic clearance based on chemical structure, and extrapolation from animal studies, can be deceptive.

Another factor with similar effects is binding of metals and of electrophiles to tissue components. Availability of binding sites is a source of variability of the biological levels of several biomarkers.

Ethnic differences in biological levels of biomarkers due to enzyme deficiency have been observed when monitoring occupational exposure to solvents (7).

Coexposure to two or more chemicals can profoundly affect the biological levels of those chemicals by inhibition or induction of the enzymes, or by competition for binding sites. As a result of inhibition, the biological levels of the parent compound increase, and the biological levels of the metabolites decrease, compared to the biological levels resulting from exposure only to the compound itself. Enzyme induction has opposite effects. While the effect of inhibition occurs immediately, induction takes time to develop.

2.1.6 Consequences of Variability

A single measurement of a specific biomarker in a properly collected sample can be sufficient for qualitative confirmation of exposure. Longitudinal evaluation over time is needed for quantitative evaluation of the exposure in the workplace. Outlying values may be caused by a special ability of the individual to handle the pollutant, by bad working habits, by poor personal hygiene, or by the individual's life style or personal habits (drinking alcoholic beverages, medication, diet). Therefore drinking habits, medication, coexposure to other chemicals, body build, and workload of the monitored workers should be recorded.

2.2 Saturable Kinetics

At high exposures, the capacity of the binding sites, as well as the capacity of enzymes that mediate the metabolism of xenobiotics, can be saturated. If either happens, linear kinetics is not valid. At such exposures, metabolic clearance (Cl_{met}) is not constant, but is defined by an analogue of the Michaelis-Menten constants K_m and V_{max}:

$$Cl_{met} = \frac{V_{max}}{K_m + c_t} \tag{6}$$

where c_t denotes the xenobiotic concentration at the metabolizing site at time t. As a result of saturation, the biological levels of the biomarkers cease to be a linear function of the degree of exposure and the half-lives no more define uptake and elimination. Nonlinear kinetic models are usually used in risk assessment when extrapolating safe degree of exposure from animal studies at high exposures (8).

At excessive exposures, the level of the biomarker remains approximately the same, thus making the biomarker an unsuitable indicator of the degree of exposure (Figure 42.2).

2.3 Interspecies Differences

Alveolar ventilation and cardiac output must be considered when extrapolating from animals to humans. Cardiac output and pulmonary ventilation, expressed per kilogram of body weight, are larger in small experimental animals than in a resting man. This means that biological levels in exposed laboratory animals rise faster than in a resting man. It has been calculated that the alveolar ventilation and cardiac output of rats are comparable with those of a man performing light work (about 50 W), and that these parameters for mice are comparable with those of a man performing heavy work (100 W) (4).

There are qualitative and quantitative interspecies differences in metabolism which must be considered when extrapolating from animals to humans. Controlled exposures at high concentrations are acceptable in experimental animals but are not acceptable in humans.

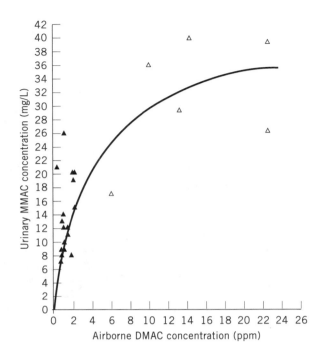

Figure 42.2. Excretion of monomethylacetamide (MMAC) in urine of workers exposed to dimethylacetamide (DMAC). The dependence of end-of-shift MMAC excretion on DMAC concentration in the air shows that MMAC can be used as a biomarker only for exposures lower than 5 ppm. At higher exposures, MMAC excretion levels off, and is no longer suitable for exposure monitoring. The triangles denote data from two different studies. Adapted from *Biological Monitoring of Exposure to Industrial Chemicals* with the permission of the ACGIH, Cincinnati, OH, © 1990.

Therefore kinetic constants derived from animal studies cannot be validated in humans, and the extrapolated values for humans can be deceptive.

3 BIOLOGICAL SPECIMENS FOR EXPOSURE MONITORING

Biomarkers can be measured in exhaled air, in blood and urine, in tissues (*in vivo* in adipose tissue), and in hairs, nails, perspiration, and saliva. In practice, only exhaled air, urine, and blood are used for monitoring of occupational exposure. The main advantages and disadvantages of each are discussed below. More detailed information can be found in the preface to the BEI documentation (10) and in special publications by ACGIH (11).

3.1 Exhaled Air

Biological monitoring based on analysis of exhaled air is limited to volatile chemicals—specifically, to stable, hydrophobic vapors and gases. Exhaled air is not suitable for biological monitoring of unstable pollutants that decompose upon contact with body fluid in the respiratory tract; nor is it suitable for monitoring of pollutants that dissolve in the mucus covering the walls of respiratory airways.

Interindividual differences in the metabolic rate introduce variability in the levels of parent compounds in blood, and consequently affect pulmonary excretion. Therefore, biological monitoring in exhaled air is more informative for poorly metabolized chemicals, such as perchloroethylene or methylchloroform, than for monitoring of extensively metabolized chemicals, such as trichloroethylene. The main advantage of monitoring in exhaled air is the simple analysis requiring no clean-up procedure.

Volatile chemicals also appear in exhaled air after dermal or gastrointestinal absorption (12, 13); this weakens the correlation between the concentration of pollutants in ambient air and in exhaled air.

3.1.1 Exhaled Air Composition and Sampling

Exhaled breath is a mixture of air coming from the lungs (alveolar air) and of air held in upper respiratory airways (dead space). Because stable, poorly soluble gases and vapors are not absorbed in the upper respiratory tract, their concentration at the beginning of exhalation (i.e., air from respiratory airways) resembles the concentration in the ambient air. In the alveoli, the concentrations of volatile pollutants equilibrate with their concentration in arterial blood. At rest, the air from the upper respiratory tract (dead space) accounts for about one-third of the mixed exhaled air. The dead space increases as the respiratory rate increases (4, 6, 14). When collecting multi-breath mixed-exhaled air samples in large glass or plastic containers, precautions must be taken that the sampling container has minimal resistance, so that the sample is taken under normal breathing conditions. End-exhaled air samples are representative of alveolar air. The donor of the sample must be trained to provide end-exhaled air properly, without force, since concentrations in forced-exhaled air differ from those in unforced end-exhaled air. The concentration changes during expiration are shown in Figure 42.3 (**b**).

Sampling via respirator or respirometer is not advisable unless the efficiency of the filter is tested.

Exhaled air samples collected from persons with abnormal pulmonary function are not suitable for exposure monitoring.

3.1.2 Time of Sampling

During exposure, the concentration of the pollutant in exhaled air fluctuates with the concentration of the pollutant in ambient air. At stable homogeneous exposures, the concentration of the pollutant in exhaled air rises and approaches the steady state level [Figure 42.3 (a)]. Biological monitoring during exposure is unnecessary, because measurements in breath samples provide approximately the same information as measurements in ambient air collected in the respiratory zone. Immediately after the exposure, the concentration of the pollutant in exhaled air declines very rapidly. At this time, the required accuracy of sampling time is unmanageable in field conditions and thus unsuitable for occupational exposure monitoring. Exhaled air samples collected 16 hours after exposure to solvents (prior to the next shift) are preferable. They contain the pollutants released from the fat

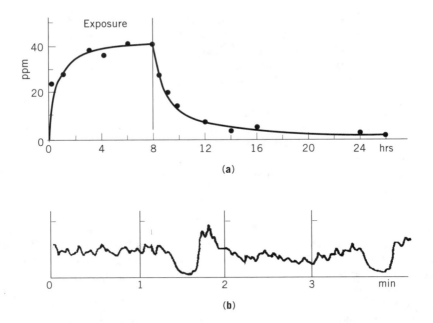

Figure 42.3. Exhalation of volatile compounds. (**a**) Benzene concentrations in mixed exhaled air of a volunteer during and following an 8-hr exposure to benzene. Note the rapid rise of the concentration at the beginning of exposure and its rapid decline after the end of exposure. (**b**) Continuous recording of mercury in exhaled air during exposure. UV-analyzer with small dead space was used to monitor mixed-exhaled air during the first and third minutes of the recording. During the second and fourth minutes, the volunteer was asked to exhale completely, so that the alveolar concentration could be recorded.

and tissue depot, and reflect the average concentration during the previous exposure (shift). These concentrations are about 100 times lower than the exposure concentration. To prevent contamination during sampling, the samples must be collected in an uncontaminated environment.

Thus when using exhaled air for biological monitoring of exposure to volatile pollutants, the worker must be instructed, and possibly trained, how to exhale in the sampling container. The timing of sampling with respect to exposure and elimination half-life is critical.

3.2 Urine

Urine is used for monitoring of hydrophilic biomarkers of low molecular weight. Blood transfers the biomarkers from the absorption or metabolizing sites to the kidneys, where they are excreted in the same way as endogenic metabolic waste. Urine collects in the bladder before it is voided; hence the concentration of biomarkers in a urine specimen represents the average excretion during the interval between voidings.

The advantages of urine for biological monitoring are noninvasiveness of sampling of sufficiently large samples and simple analysis that requires little clean-up procedure. The main disadvantage is that urine output, being influenced by water intake and loss (e.g., sweating, diarrhea), is highly variable.

3.2.1 Urine Composition and Sampling

Average daily urine output in adults is about 1.2 L (ranging between 0.6 to 2.5 L/day), which corresponds to an urinary flow rate of about 50 mL/hour. It contains about 50 g of solid metabolic end-products and electrolytes, the average specific gravity (density) being 1.020 (ranging between 1.001 to 1.030). Urine is slightly acidic, with an average pH of 6.0 (ranging between 4.6 and 8.0) (14). Urine samples with values beyond the ranges of the above parameters indicate abnormal stress on renal function and are not suitable for biological monitoring. Also, samples with density less than 0.005 are usually considered unsuitable.

3.2.1.1 Renal Clearance.
The measure of the efficiency of the kidneys to remove the compound from the body is called renal clearance (Cl_{ren}), which can be described mathematically as follows:

$$Cl_{ren} = \frac{V_u \cdot c_u}{c_p} \tag{7}$$

where V_u is urine output per time unit (usually per hour), and c_u and c_p denote concentrations of the compound in urine and plasma, respectively. Clearance is given in liters per time unit (L/hr or mL/min), and represents that volume of plasma, from which the compound would be completely removed in a given time unit.

3.2.1.2 Standardization of Biomarker Measurements in Urine.
To eliminate the effect of fluctuating urine flow rate, the urine volume and time between urine voiding is recorded, and the excretion rate of the biomarker is calculated. But such sampling is impractical in field conditions. A practical alternative is to relate the excretion of the biomarker to the excretion of solids. Such adjuster can be either specific gravity (density) or creatinine excretion.

The concentration of solids in urine (specific gravity) is indirectly related to the urine flow rate (15). The urine density of 1.020 is usually used for standardization of biomarker measurements. The adjustment factor f is calculated as follows:

$$f = \frac{1 - 1.020}{1 - sp.\ gr} \tag{8}$$

Measurement of density using a urinometer or refractometer is simple. The sediment occurring in the urine shortly after voiding makes the measurements in stored urine misleading.

The creatinine excretion rate in urine is more or less constant (about 1.0–1.6 g/day), and is independent of diet, hydration, and diuresis. Adjustment of biomarker excretion to gram of creatinine is suitable for biomarkers excreted by glomerular filtration (such as phenols, glucuronides, and organic acids), but is unsuitable for biomarkers excreted by tubular diffusion (such as organic solvents and gases). No concentration adjustment is required.

3.2.2 Time of Sampling

Twenty four-hour samples would be representative of daily excretion, but such sampling is impractical in field conditions. Urine collection two to four hours after the previous urine voiding is advisable. The concentration of biomarkers in urine rises during exposure, and is usually the highest in specimens collected at the end of exposure. Following exposure, the concentrations of biomarkers decline rapidly, usually with a half-life between three and twelve hours. Therefore the sampling time of urine must be carefully chosen. The reference values usually relate to urine voiding for the last two to four hours prior to the end of the exposure (shift) or voiding for the two to four hours prior to the next shift. The sampling time for biomarkers with long half-lives (e.g., metals) is discretionary, but the lag time has to be considered for newly employed workers. The reference values, such as American BEIs or German BATs, include information about the sampling time to which they apply. The urinary excretion of metabolites is mainly an indicator of recent exposure.

3.2.3 Effect of Binding

Large molecules (MW > 5,000) cannot be removed by kidneys. Therefore biomarkers bound to macromolecules, such as protein adducts or DNA adducts, cannot be monitored in urine.

The urinary excretion of heavy metals is affected by their binding to tissue components, namely to metallothionein and metallothionein-like proteins. As a result, the urinary excretion of metals is very slow until the binding sites are occupied. Afterwards, the urinary excretion of the metal suddenly increases. (Cadmium is an example.) Urinary excretion of metals is also affected by diet and by the pH of the urine, making their excretion rate unpredictable.

Thus, when using urine for biological monitoring, the timing with respect to the exposure and standardization of measurements of the biomarker must be carefully chosen. It should be verified that the parameters, such as urine density and pH, are in an acceptable range.

3.3 Blood

Blood levels of parent compounds are the best indicators of internal exposure. However, the sampling of blood is considered invasive, and can be done only by qualified personnel. Moreover, the analysis is usually complicated by a meticulous clean-up procedure. Blood is the only specimen available for measurement of macromolecular adducts, altered enzyme activity (cholinesterase), or some specific biomarkers resulting from interaction of the pollutant with hemoglobin (methemoglobin and carboxyhemoglobin). Blood may be preferable for monitoring of some metals and for monitoring of poorly metabolized solvents, such as perchloroethylene.

Blood distributes the pollutant from the site of entry (respiratory tract, skin, gastrointestinal tract) to other tissues, including to the excretory organ. The process reverses when the exposure ends and blood washes the pollutant and its metabolites from tissues and carries them to the excretory organs. A significant concentration gradient develops when blood passes through the absorption site or through the excretory organs. Thus the passage of volatile pollutants through the lungs causes a significant concentration gradient between arterial and mixed venous blood. During exposure, the absorption of the pollutant increases its concentration in arterial blood. On the other hand, as the blood passes through the other organs, the pollutant concentration in blood decreases due to retention of the pollutant in the tissues. As a result, the concentration of the pollutant in regional venous blood is lower than in arterial blood. Following the exposure, the pulmonary washout and tissue washout have the opposite effect, the concentrations in venous blood being higher than in arterial blood. This difference has to be kept in mind when sampling for biological monitoring of volatile pollutants. Thus, concentrations of volatile biomarkers in venous blood collected from the vena cubitalis can differ from concentrations in arterial blood collected from the finger.

Dermal absorption can induce local differences in biomarker concentration. For example, if one hand is immersed in solvent, the concentration in blood collected from the arm with the immersed hand can be higher than the concentration in blood collected from the other arm (16).

3.3.1 Blood Composition and Sampling

Determination of biomarkers can be done in whole blood, in cellular elements (red cells or lymphocytes), or in plasma or serum. The cellular fraction accounts for about 47% of blood volume in men and for about 42% in women (14). The cellular fraction is higher in a population that lives at a high altitude. To obtain plasma, an anticoagulant such as citrate, oxalate, EDTA, or heparin is added to the blood at sampling. The choice of anticoagulant depends on the biomarker and on the analytical method. Some biomarkers readily cross the cellular membranes and can be measured in plasma as well as in red cells. For example, solvents such as benzene, toluene, and trichloroethylene distribute evenly in blood, but their hydrophilic metabolites do not readily cross the cellular membranes and there is a concentration difference between plasma and red cells.

3.3.2 Time of Sampling

The sampling of blood is instantaneous. Therefore analysis of blood determines the momentary concentration of the biomarker at the time of the blood collection. Since biomarker concentration in blood rises during exposure and declines following the exposure, the sampling time for biomarkers with a short half-life is critical.

Thus, when collecting and analyzing blood for biological exposure monitoring, timing with respect to the exposure, sampling site, and blood specimen used for the analysis must be carefully chosen.

3.4 Hairs

Hairs seem to be suitable for retrospective monitoring of exposure to heavy metals. Arsenic is the the most studied (17). However, the difficult control of external contamination of hairs makes them unsuitable for routine monitoring of exposure in the workplace.

4 SELECTING A BIOMARKER

In selecting a biomarker for biological monitoring, the following factors should be considered:

1. Specificity of the biomarker.
2. Sensitivity and background level of the biomarker.
3. Purpose of the monitoring.
4. Availability of an analytical method.
5. Stability of the sample and of the biomarker during storage and transportation.

4.1 The Specificity of the Biomarker

Specificity is very important for identification of the toxic pollutant. Parent compounds are usually the most specific indicators of exposure, although there are exceptions. For example, carbon monoxide in exhaled air may appear to be a specific biomarker of carbon monoxide exposure. However, it has been shown that methylene chloride is metabolized to carbon monoxide, which is then exhaled. Thus, carbon monoxide in exhaled air and carboxyhemoglobin in blood are not specific biomarkers. Phenol in urine is another example of a parent compound being a nonspecific biomarker, since it is also a metabolite of benzene. Metabolites are usually less specific biomarkers than the parent compounds. For example, trichloroacetic acid occurs in urine following exposure to trichloroethylene, perchloroethylene, and methylchloroform.

Nonspecific biomarkers may have an advantage if the exposure to be evaluated is coexposure to several poorly identified pollutants with a biomarker related to the same adverse effect. For example, in the rubber industry, where workers are exposed to several unidentified pollutants that are metabolized to electrophiles which bind to glutathione, total thioethers are suitable for evaluation of exposure. In agriculture, where workers are exposed to mixtures of organophosphate pesticides, the nonspecific biomarker, acetylcholinesterase inhibition in erythrocytes, is favored as a screening test. Similarly, methemoglobin in blood is a suitable biomarker for exposure to several amines and nitrocompounds that are readily absorbed through the skin.

Metals measured in blood or urine are a specific biomarker for evaluating exposure to that metal. Metals and their compounds are usually present in the air as fumes, dust, or aerosol, and their biological levels depend on the size of the particles and the solubility of the inorganic or organic compounds in which they occur in the environment. The interaction of metals at the binding site significantly affects their biological level and excretion. When choosing a biomarker and its acceptable biological level to evaluate exposure to a

BIOLOGICAL MONITORING OF EXPOSURE TO INDUSTRIAL CHEMICALS

particular metal, the differences in toxicity of the different compounds of the metal must be kept in mind. Accordingly, the analytical method must be carefully chosen.

4.2 Sensitivity and Background Levels

The biomarker must be present in the biological specimen in a measurable amount. This depends not only on the availability of a sensitive analytical method; it also depends on whether the biological level of the biomarker induced by the pollutant differs from its normal level in an unexposed population. Deviation of the biological level of the biomarker induced by permissible exposure to the pollutant must be outside the range of its levels in the unexposed population.

The dependence of the selection of biomarkers on the permissible exposure in the work place is illustrated by benzene. The first TLV for benzene, published in 1946, was 100 ppm. To monitor benzene exposure at that level, sulfates and phenol in the urine were widely used as biomarkers. Over the years, the TLV was lowered several times, in the 1970s being 10 ppm. To monitor benzene exposure at that level, sulfates are not suitable biomarkers, because the levels of organic sulfates among workers occupationally exposed to benzene are not different from the levels of endogenic sulfates. In 1990, after benzene was confirmed to be a human carcinogen, its TLV was lowered to 0.5 ppm, making phenol in blood and urine an unsuitable biomarker because its increase in specimens collected from benzene-exposed workers is within the range of normal levels. In 1997, *S*-phenylmercapturic acid in urine was recommended as a suitable biomarker for monitoring of benzene exposures permissible in the workplace (1). Likewise, hippuric acid in urine as a biomarker for toluene exposure recently became ambiguous, because the increase of hippuric acid excretion caused by exposure to the new TLV of 50 ppm is in the range of excretion of endogenic hippuric acid (0.5–1.5 g/g of creatinine). Recently o-cresol has been recommended as a biomarker (1).

4.3 Purpose of Monitoring

The selection of biomarkers, biological specimens, and the timing of sample collection depend on the purpose of the monitoring. For confirmation of overexposure, or for medical screening, measurement in a single sample collected during or after the exposure is suitable, provided that the biomarker still persists in the body. Samples collected within the period of three half-lives following exposure are usually timely. For retrospective verification of past exposure(s), biomarkers with long half-lives, such as macromolecular adducts, are suitable.

For routine exposure monitoring and potential risk assessment, biomarkers that are excreted in urine are the first choice, because of convenience of urine sampling. For medical surveillance and health risk evaluation, biomarkers that are related to the toxicity are preferable (acetylcholinesterase inhibition, methemoglobin, DNA adduct). However, further steps should be taken to identify the pollutant causing the health problem.

4.4 Availability of Analytical Method

The importance of selecting a suitable analytical method should not be underestimated. A well-equipped laboratory should be engaged to perform the analysis. Gas chromatography or high pressure liquid chromatography with mass spectrographic detection are the best methods for determination and identification of organic biomarkers. Atomic absorption is usually the best choice for analysis of metals (18, 19).

Here are some questions to ask the laboratory:

1. How specific is the method? and what can interfere with the measurements? (Other pollutant, medication, diet).
2. Is the method sufficiently sensitive? What is the detection limit?
3. Is the precision of the method adequate?
4. What size sample is required?
5. Is the laboratory enrolled in a quality control program? How much does the analysis cost?

A laboratory where samples can be quickly and easily transported is preferable.

4.5 Stability of the Biomarker and Specimen

Biological samples are rarely analyzed in the field; they are usually transported to a laboratory and analyzed a few days later. Preservation of biological specimens, avoidance of contamination, and stability of the samples must be assured. To attain such assurance (assuming that the biomarker is stable and nonreactive), the following questions must be answered:

1. Whether the sample containers are made from material that will not contaminate the sample?
2. Whether the sample containers are made from material that does not absorb the biomarker, or does not react with the biomarker?
3. Whether the containers are adequately sealed to prevent loss of a volatile biomarker?
4. Whether the samples should be transported and stored anaerobically? That is, must the samples occupy the entire volume of the container, or is air space no problem?
5. What anticoagulant should be used for blood samples?
6. Whether the biomarker is photosensitive? If it is, the samples should be transported and stored in dark, nontransparent containers.
7. At what temperature should the samples be transported and stored?
8. Whether the samples need to be acted upon immediately after collection? (Centrifuging of blood or acidifying or alkalizing of urine.)

The answers to the above questions depend on the biomarker, on the biological specimen, and on the analytical method. Advice should be sought from the laboratory on how to collect and transport the samples.

5 REFERENCE VALUES FOR BIOLOGICAL MONITORING OF OCCUPATIONAL EXPOSURE

Already in the 19th century, the presence of solvents and their metabolites, as well as the presence of metals, was observed in the urine and blood of persons exposed to certain industrial pollutants. However, it was not until 1980 that biological monitoring was recognized as a powerful tool for industrial hygiene, and that reference values for biological monitoring of occupational exposure were established and used to protect workers' health.

5.1 History of Reference Values

Following is a brief history of biological monitoring and reference values of biomarkers for occupational exposure to industrial chemicals in the United States, in Germany and other European countries, and in Japan. These countries were selected because of the crucial differences in their concepts of reference values.

5.1.1 United States

The initiative in promulgating reference values for biomarkers was taken by the professional organization American Conference of Governmental Industrial Hygienists (ACGIH). The need of biological limit values for industrial exposure was first publicly mentioned by ACGIH in its TLV booklet for 1973 (20), but no action was taken to recommend any reference value until 1982, when ACGIH established the Biological Exposure Indices Committee (BEIC). That Committee recommended first reference values for six pollutants, which were published in the 1984–1985 TLV/BEI Booklet (21), and the BEI list has since been updated and new biomarkers added to it each year. The 1999 TLV/BEI Booklet (1) has a recommendation of BEIs for 71 biomarkers for 38 industrial pollutants (see Table 42.2). ACGIH also publishes a series of documentations which explain the basis for each BEI. The documentations are updated from time to time (10) and the BEIs are altered (usually lowered) as new information is available.

In 1986, the government agency NIOSH began promoting biological monitoring. NIOSH's objectives are implementation of biological monitoring in risk assessment, environmental control, and the validation of biological monitoring in epidemiological studies (22).

5.1.2 Germany

In 1979, the Commission for the Investigation of Health Hazards of Chemical Compounds in the Workplace established a working group for setting biological tolerance values, called BATs (Biologische Arbeitsstoff-Toleranz-Werte). The Commission published the first BATs in 1981 (23), and has since updated and added to them each year. In 1999, the Commission published BAT values for 58 biomarkers for 44 pollutants (Tables 42.2 and 42.3) (2). Table 42.2 shows the BATs for those biomarkers for which BEIs exist, so that their respective values can be easily compared; Table 42.3 lists the remaining BATs which are reference values for biomarkers for 16 other pollutants. The Commissions gives special consideration to carcinogens (Table 42.4). Documentations justifying BAT values are avail-

Table 42.2. Reference Values for Biomarkers Comparison of BEIs and BATs[a,b]

Chemical, Biomarker	Sampling Time	BEI	BAT
Acetone			
Acetone in urine	End of shift	50 mg/L	80 mg/L
Aniline			
Total *p*-aminophenol in urine	End of shift	50 mg/g cr.	
Methemoglobin in blood	During or end of shift	1.5% of hem.	1 mg/L
Free aniline in urine	End of shift after several shifts		100 µg/L
Aniline released from Hb in blood	End of shift after several shifts		Carcinogen
Arsenic and soluble compounds			
Inorganic As and methylated metabolites in urine	End of workweek	35 µg As/L	Carcinogen
Benzene			
S-Phenylmercapturic acid in urine	End of shift	25 µg/g cr.	
t,t-muconic acid in urine	End of shift	500 µg/g cr.	Carcinogen
Cadmium and inorganic compounds			
Cadmium in urine	Not critical	5 µg/g cr.	
Cadmium in blood	Not critical	5 µg/L	
Carbon disulfide			
2-Thiothiazolidine-4-carboxylic acid (TTCA) in urine	End of shift	5 mg/g cr.	4 mg/g cr.
Carbon monoxide			
Carboxyhemoglobin in blood	End of shift	3.5% of hem.	5% of hem.
Carbon monoxide in end-exh. air	End of shift	20 ppm	
Chlorobenzene			
Total 4-chlorocatechol in urine	End of shift	150 mg/g cr.	300 mg/g cr.
Total *p*-chlorophenol in urine	Prior to next shift		70 mg/g cr.
	End of shift	25 mg/g cr.	Carcinogen
Chromium (VI), water-soluble fume			
Total chromium in urine	Increase during shift	10 µg/g cr.	
	End of shift at end of workweek	30 µg/cr.	
Cobalt			
Cobalt in urine	End of shift at end of workweek	15 µg/L	Carcinogen
Cobalt in blood	End of shift at end of workweek	1 µg/L	
Dichloromethane (Methylene chloride)			
Dichloromethane in blood	During shift	0.5 mg/L	1 mg/L

Determinant	Sampling time	BEI
Dichloromethane in urine	End of shift	0.2 mg/L
Carboxyhemoglobin in blood	End of shift	5%
N,N-Dimethylacetamide		
N-Methylacetamide in urine	End of shift at end of workweek	30 mg/g cr.
N,N-Dimethylformamide (DMF)		
N-Methylformamide in urine	End of shift	15 mg/L
N-acetyl-S-(N-methylcarbamoyl) cysteine in urine	Prior to last shift of the week	40 mg/L
2-Ethoxyethanol (EGEE) and 2-Ethoxyethyl Acetate (EGEEA)		
2-Ethoxyacetic acid in urine	End of shift at end of workweek	100 mg/g cr.
Ethylbenzene		
Mandelic acid in urine	End of shift at end of workweek	1.5 g/g cr.
Mandelic acid plus phenylglyoxalic acid	End of shift	15 mg/L
Ethyl benzene in end-exh. air		Screening test
Ethyl benzene in blood	End of shift	100 mg/g cr.[c]
Fluorides		0.8 g/g cr.
Fluorides in urine	Prior to shift	2 g/g cr.
	End of shift	1.5 mg/L
Furfural		
Total furoic acid in urine	End of shift	200 mg/g cr.
n-Hexane		
2,5-Hexanedione in urine	End of shift	5 mg/g cr.
n-Hexane in end-exh. air		Screening test
Lead		
Lead in blood	Not critical	300 µg/L
δ-Aminolevulinic acid in urine	Discretionary	
Mercury		
Total inorganic Hg in urine	Preshift	35 µg/g cr.
Total inorganic Hg in blood	End of shift at end of workweek	15 µg/L
Methanol		
Methanol in urine	End of shift	15 mg/L
Methemoglobin inducers		
Methemoglobin in blood	During or end of shift	1.5% of hem.

Additional values shown in right column:
- 3 mg/g cr.
- 10 mg/g cr.
- 5 mg/L[d]
- 700 µg/L
- 300 µg/L[e]
- 15 mg/L
- 6 mg/L[e]
- 100 µg/L
- 25 µg/L
- 30 mg/L
- 4 mg/g cr.
- 8 mg/g cr.

Table 42.2. (continued)

Chemical, Biomarker	Sampling Time	BEI	BAT
2-Methoxyethanol (EGME) and 2-Methoxyethyl Acetate (EGMEA)			
2-Methoxyacetic acid in urine	End of shift at end of workweek	Screening test	
Methyl chloroform			
Methyl chloroform in end-exh. air	Prior to last shift at end of workweek	40 ppm	
Methylchloroform in blood	Prior to the shift after several shifts		0.55 mg/L
Trichloroacetic acid in urine	End of workweek	10 mg/L	
Total trichloroethanol in urine	End of shift at end of workweek	30 mg/L	
Total trichloroethanol in blood	End of shift at end of workweek	1 mg/L	
4,4-Methylene bis(2-chloroaniline) [MBOCA]			
Total MBOCA in urine	End of shift	Screening Test	
Methyl ethyl ketone (MEK)			
MEK in urine	End of shift	2 mg/L	5 mg/L
Methyl isobutyl ketone (MIBK)			
MIBK in urine	End of shift	2 mg/L	3.5 mg/L
Nitrobenzene			
Total p-nitrophenol in urine	End of shift at end of workweek	5 mg/g cr.	
Methemoglobin in blood	End of shift	1.5% of hem.	
Aniline released from Hb-conjugate in blood	After several shifts		100 µg/L
Organophosphorus cholinesterase inhibitors			
Cholinesterase activity in red cells	Discretionary	70% of individual's baseline	70% of ref. value
Parathion			
Total p-nitrophenol in urine	End of shift	0.5 mg/g cr.	0.5 mg/L
Cholinesterase activity in red cells	Discretionary	70% of individual's baseline	70% of ref. value
Pentachlorophenol (PCP)			Carcinogen
Total PCP in urine	Prior to last shift of workweek	2 mg/g cr.	
Free PCP in plasma	End of shift	5 mg/L	
Perchloroethylene (PCE)			
PCE in end-exh. air	Prior to last shift of workweek	5 ppm	
PCE in blood	Prior to last shift of workweek	0.5 mg/L	
Trichloroacetic acid in urine	End of shift at end of workweek	3.5 mg/L	

Chemical / Determinant	Sampling time	BEI
Phenol		
Total phenol in urine	End of shift	250 mg/g cr.	300 mg/L
Styrene		
Mandelic acid in urine	End of shift	800 mg/g cr.
	Prior to next shift	300 mg/g cr.
Phenylglyoxylic acid in urine	End of shift	240 mg/g cr.
	Prior to next shift	100 mg/g cr.
Mandelic and phenylglyoxylic acids in urine	End of shift after several exposures		600 mg/g cr.
Styrene in venous blood	End of shift	0.55 mg/L
	Prior to next shift	0.02 mg/L
Tetrahydrofuran		
Tetrahydrofuran in urine	End of shift	8 mg/g cr.	8 mg/L
Toluene		
Hippuric acid in urine	End of shift	1.6 g/g cr.
Toluene in venous blood	Prior to last shift of workweek	0.05 mg/L
o-Cresol in urine	End of shift		1 mg/L
Trichloroethylene		
Trichloroacetic acid in urine	End of shift	0.5 mg/L	3 mg/L
Trichloroacetic acid and trichloroethanol in urine	End of workweek	100 mg/g cr.
Free trichloroethanol in blood	End of shift at end of workweek	300 mg/g cr.
Trichloroethylene in blood	End of shift at end of workweek	4 mg/L
Trichloroethylene in end-exh. air		Screening test
		Screening test
Vanadium pentoxide		
Vanadium in urine	End of shift at end of workweeek	50 μg/g cr.	70 μg/g cr.
Xylenes (technical grade)		
Methylhippuric acids in urine	End of shift	1.5 g/g cr.	2 g/L
Xylene in blood	End of shift		1.5 mg/L

[a]Adapted from Ref. 1 with permission of American Conference of Governmental Industrial Hygienists (ACGIH®). The table includes changes proposed for 2000.
[b]cr. = creatinine; end-exh. = end-exhaled air; hem. = hemoglobin.
[c]BEI is given for sum of EGEE and EGEEA; BAT specifies 50 mg/g of creatinine for EGEE and 50 mg/g of creatinine for EGEEA.
[d]BAT includes 4,5-dihydroxy-2-hexanone.
[e]For women below 45 years of age.

Table 42.3. BATs For Biomarkers without BEI

Chemical, Biomarker	Sampling Time	BAT[a]
Aluminium		
Aluminium in urine	End of shift	200 μg/L
p-tert-Butylphenol (PTBP)		
PTBP in urine	End of shift	2 mg/L
Carbon tetrachloride		
Carbon tetrachloride in blood	End of shift after several shifts	70 μg/L
1,4-Dichlorobenzene		
Total 2,5-dichlorophenol in urine	End of shift	150 mg/g cr.[a]
	Prior to next shift	30 mg/g cr.[a]
Ethylene glycol dinitrate (EGDN)		
EGDN in blood	End of shift	0.3 μg/L
Ethylene glycol monobutyl ether		
Butoxyacetic acid in urine	After several shifts	100 mg/L
Ethylene glycol monobutyl ether acetate		
Butoxyacetic acid in urine	After several shifts	100 mg/L
Halothane		
Trifluoroacetic acid in blood	End of shift After several shifts	2.5 mg/L
Hexachlorobenzene		
Hexachlorobenzene in plasma or serum	Discretionary	150 μg/L
γ-Hexachlorocyclohexane (Lindane)		
Lindane in blood	End of shift	20 μg/L
Lindane in plasma or serum	End of shift	25 μg/L
2-Hexanone, (Methyl butyl ketone)		
Hexane-2,5-dione and 4,5-dihydroxy-2-hexanone in urine	End of shift	5 mg/L
Isopropyl Alcohol		
Acetone in blood	End of shift	50 mg/L
Acetone in urine	End of shift	50 mg/L
4,4'-Methylene diphenyl isocyanate		
4,4'-Diaminodiphenylmethane in urine	End of shift	10 μg/g cr.[a]
Nitroglycerin		
1,2-Glyceryl dinitrate in plasma or serum	End of shift	0.5 g/L
1,3-Glyceryl dinitrate in plasma or serum	End of shift	0.5 g/L
Organomercury		
Mercury in blood	Discretionary	100 μg/L
Tetraethyllead		
Diethyllead in urine	End of shift	25 μg/L (as Pb)
Total lead in urine	End of shift	50 μg/L (as Pb)

[a]cr. = creatinine.

Table 42.4. German Reference Values for Carcinogens

Chemical, Biomarker	Sampling Time	Air mg/m^3	Specimen, µg/L	
			Urine	Blood
Acrylonitrile				
Cyanoethylvaline (Ery)[a]	Discretionary	0.3		16
		0.5		35
		1.0		60
		7.0		420
Arsenic Trioxide				
Arsenic		0.01	50	
		0.05	90	
		0.10	130	
Benzene[b]			SPMA[c]	Benzene
	End of shift	1.0	0.010	0.9
		2.0	0.025	2.4
		3.0	0.040	4.4
		3.3	0.045	5.0
		6.5	0.090	14.0
		13.0	0.180	38.0
		19.5	0.270	—
Cobalt and its compounds				
Cobalt	Discretionary	0.05	30	2.5
		0.10	60	5.0
		0.50	300	25.0
Chromium trioxide				
Chromium (Ery)[a]	End of shift after several shifts	0.03	12	9
		0.05	20	17
		0.08	30	25
		0.10	40	35
Dimethyl sulfate				
N-Methylvaline (Ery)[a]	Discretionary	0.01		10
		0.03		13
		0.05		17
		0.20		40
Ethylene				
Hydroxyethylvaline (Ery)[a]	Discretionary	25		45
		50		90
		100		180
Ethylene oxide[d]				
Hydroxyethylvaline (Ery)[a]	Discretionary	0.92		45
		1.83		90
		3.66		180

Table 42.4. (continued)

Chemical, Biomarker	Sampling Time	Air mg/m^3	Specimen, µg/L	
			Urine	Blood
Hydrazine				
Hydrazine (plasma)	End of shift	0.013	35[e]	27
		0.026	70	55
		0.065	200	160
		0.104	300	270
		0.130	380	340
Nickel				
Nickel	After several shifts	0.10	15	
		0.30	30	
		0.50	45	
Pentachlorophenol (PCP)				
PCP (serum)	Discretionary	0.001	6	17
		0.05	300	1000
		0.10	600	1700
Toluene-2,4-diamine				
Toluene-2,4-diamine	End of shift	0.0025	6	
		0.01	13	
		0.017	20	
		0.035	37	
		0.100	100	
Vinyl chloride			mg/24hrs	
Thiodiglycolic acid	After several shifts	2.6	1.8	
		5.2	2.4	
		10	4.5	
		21	8.2	
		41	10.6	

[a](Ery) denotes that the measurements were done in erythrocytes.
[b]Similar data are suggested for *trans, trans*-muconic acid in urine.
[c]SPMA = *S*-phenyl mercapturic acid in urine in mg/g of creatinine.
[d]Similar data are suggested for ethylene oxide in alveolar air.
[e]The concentration is normalized to creatinine (µg/g of creatinine).

able from the Commission by request. The Commission established a Working Group for Analytical Chemistry which, since 1985, has published a series entitled "Analyses of Hazardous Substances in Biological Material" that describe analytical methods that have been tested by the Working Group (18).

5.1.3 Other European Countries

Biological monitoring of occupational exposure was practiced in Europe since World War II in a number of European countries, most actively in Czechoslovakia, Italy, Poland, and Sweden. International meetings were organized to share views and experiences on biological monitoring. The European Community Commission (ECC) began promoting biological monitoring of environmental and occupational exposure in 1983 with its initial publication of a series of documentations on the toxicity and biological monitoring of seven industrial pollutants (24). The series, originally entitled "Human Biological Monitoring of Industrial Chemicals Series," and renamed in 1984 "Biological Indicators For Assessment of Human Exposure to Industrial Chemicals," currently includes seven volumes containing information on the biological monitoring of 40 pollutants (Table 42.5) (25). The ECC does not recommend any reference values, but provides scientific bases on which each country can determine its own reference values (26).

In 1985, the ECCs Bureau of Reference Values began a voluntary laboratory certification program that has resulted in improved laboratory performance.

5.1.4 Japan

In 1986, a Society for Research of Biological Monitoring was established to develop reliable analytical methods, establish reference values, and implement and promote biological monitoring (27). In 1989, the Ministry of Labor began promoting biological monitoring, and accepted reference values recommended by the Society as guidelines for monitoring of lead and eight solvents.

Since 1993, the Japanese Society for Occupational Health has promoted OELs (Occupational Exposure Limits) for mercury, later adding lead, hexane, and 4,4'-methylene-

Table 42.5. Biological Indicators for the Assessment of Human Exposure to Industrial Chemicals

Documentations published by Commission of the European Communities

1983:	Benzene	1984:	Acrylonitrile
	Cadmium		Aluminum
	Chlorinated hydrocarbon solvents		Chromium
	Lead		Copper
	Manganese		Styrene
	Titanium		Xylene
	Toluene		Zinc
1986:	Alkyl lead compounds	1987:	Aldrin and dialdrin
	Dimethylformamide		Arsenic
	Mercury		Cobalt
	Oranophosphorus pesticides		Endrin
1988:	Aromatic amines		Vanadium
	Aromatic nitrocompounds	1989:	Beryllium
	Carbamate pesticides		Carbon monoxide
	Nickel		Ethylbenzene
1994:	Antimony		Methylstyrene
	Soluble barium compounds		Isopropylbenzene
	Hexane and methyl ethyl ketone		Inhalation anesthetics
	Thallium and tin		Selenium

bis-(2-chloroaniline) (MBOCA) (28). The conceptual basis of OELs, as well as the reference values selected, are similar to BEIs.

Japan has a well-developed quality control program for the analysis of biomarkers in biological specimens and for the implementation of biological monitoring on a regular basis (29).

5.2 Database for Reference Values

Reference values are based on human data that provide an understanding of the relationships among biological levels, external exposures, and adverse effects. Animal data are interpreted cautiously, with consideration given to species differences in metabolism of xenobiotics and in susceptibility to adverse effects. The database for determination of reference values for biological monitoring has four major sources of information:

1. Studies of the mechanism and pharmacodynamics of toxic endpoints, which include animal studies. Case reports on accidental overexposure provide valuable information. Special consideration is given to high risk populations.
2. Controlled laboratory studies in volunteers, which mainly provide information on toxicokinetic data at low exposures.
3. Simulation studies (modeling), which are used to elucidate the effects of circumstantial factors, such as exposure duration and physical activity, to determine the sampling time and to match biological levels to permissible external exposures.
4. Epidemiological field studies, in which the levels of biomarkers in specimens collected from exposed workers are related to the external exposure or to the adverse effects. These studies are also the ultimate tool for validating reference values.

The database is pictured in Figure 42.4

5.3 Concept of Reference Values

Although the concept of reference values varies from country to country, the values are in the same magnitude.

5.3.1 United States

The BEIs recommended by ACGIH are reference values for biomarkers that are indicators of the degree of integrated exposure to industrial chemicals. BEIs are intended to be guidelines for the evaluation of potential health hazards in the practice of industrial hygiene (1, 10). BEIs do not indicate a sharp distinction between hazardous and nonhazardous exposures. They are meant to be used for group evaluation rather than for evaluation of individuals, unless the individual is tested regularly. They are the biological levels of biomarkers that would most likely be reached during occupational inhalation exposure to the TLV, i.e., mean values that most likely result from inhalation exposures of eight hours a day five days a week.

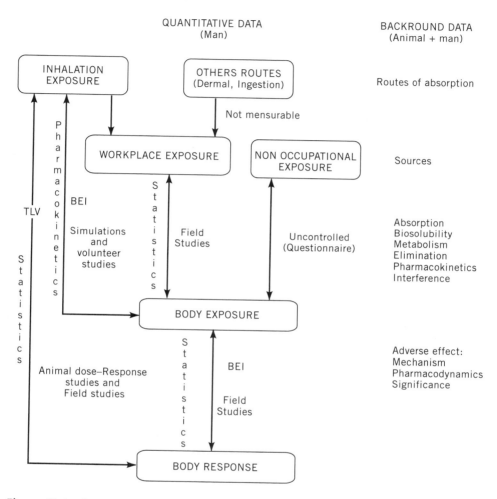

Figure 42.4. Data base for reference values for inhalation exposure (TLVs) and for biological levels (BEIs). For explanation see text. Adapted from *Topics in Biological Monitoring* with the permission of the ACGIH, Cincinnati, OH, © 1995.

The BEI and TLV Committees work together closely to make BEIs and TLVs compatible. The toxicokinetics provide a powerful tool to achieve this goal (30). This means that, assuming only inhalation exposure, the BEI and its corresponding TLV provide the same protection against systemic adverse effects. If dermal exposure occurs, biological monitoring provides earlier warning than air monitoring, and repetitive exceeding of the BEI should be taken seriously even if the TLV is not exceeded.

Irreversible biochemical and functional changes are not recommended by ACGIH as biomarkers, because the abnormal values associated with such changes indicate a health injury that should be prevented by the monitoring.

The government agencies OSHA and NIOSH do not provide reference values. Some states promote their own reference values. The leading state in this endeavor is California (31).

5.3.2 Germany

BAT values are defined as "the maximum permissible quantity of chemical substance or its metabolites, or the maximum permissible deviation from the norm of biological parameters induced by the substances in exposed humans" (2). Thus BATs can be interpreted as ceiling values. Therefore it is not surprising that BATs, in most instances, are higher than BEIs. It is surprising, however, that BEIs and BATs for *p*-nitrophenol (resulting from parathione exposure) and tetrahydrofuran are the same, and BATs for ethylbenzene and carbon disulfide are lower than BEIs (Table 42.2). BATs are based on scientific and medical data that indicate no adverse effect. They are intended mainly for medical surveillance. The BAT for lead indicates the recent concern for sex differences and reproductive effects (Table 42.2).

Special treatment is given to carcinogens. No BATs are given for biomarkers of carcinogens; instead, values corresponding to three to seven technical exposure limits are given for medical detection of the exposure (Table 42.4).

5.3.3 Other European Countries

For the most part, other European countries each have their own reference values and supporting documentations. The values are in the range of BEIs and BATs.

5.3.4 Japan

According to Ogata (27, 29), biological monitoring in Japan is part of a "specified medical examination" that is required by the Ministry of Labor every six months. Data obtained by biological monitoring are placed into one of the following three categories (Table 42.6): In Category I are biological levels below the permissible range; in Category II are biological levels within the permissible range; and in Category III are biological levels exceeding the permissible range. Values in Category II, which are based on the expected range of distribution of measurements in workers exposed to airborne concentrations permissible in the work place, are still in process of validation. If data from three consecutive six-month measurements are in Category I, future measurements are required only once a year. If data are in Category III, the measurements are promptly repeated, and if still in Category III, immediate steps are taken to remedy the situation.

5.4 Ethical Issues

There are two significant ethical issues related to biological monitoring, namely, the confidentiality of data, and the right of the worker to know results of measurements in his specimen. In the United States, there is great concern with infringement of confidentiality and possible abuse of information by employers when hiring and firing workers. Therefore obtaining written consent from the employee is advisable. In Japan, the Ministry of labor

Table 42.6. Reference Values for Biological Monitoring Approved by Ministry of Labor in Japan

Pollutant	Biomarker in Urine	Units	Category[a] 1	Category[a] 2	Category[a] 3
N,N-Dimethylformamide	N-Methylformamide	mg/L	≤10	10–40	>40
n-Hexane	2,5-Hexanedione	mg/L	≤2	2–5	>5
Lead	δ-Aminolevulinic acid	mg/L	≤5	5–10	>10
	Lead in blood	μg/100 mL	≤20	20–40	>40
	Protoporphyrin in erythrocytes	μg/100 mL	≤100	100–250	>250
Methylchloroform	Trichloroacetic acid	mg/L	≤3	3–10	>10
	Total trichloro-comp.	mg/L	≤10	10–40	>40
Perchloroethylene	Trichloroacetic acid	mg/L	≤3	3–10	>10
	Total trichloro-comp.	mg/L	≤3	3–10	>10
Styrene	Mandelic acid	g/L	≤0.3	0.3–1	>1
Toluene[b]	Hippuric acid	g/L	≤1	1–2.5	>2.5
Trichloroethylene	Trichloroacetic acid	mg/L	≤30	30–100	>100
	Total trichloro-comp.	mg/L	≤100	100–300	>300
Xylene	Methylhippuric acid	g/L	≤0.5	0.5–1.5	>1.5

[a]See section 4.3.4 for explanation of "Category".
[b]Lowering of reference values for toluene is under consideration.

has made it mandatory that workers be immediately informed of their personal results (27). In Europe, the ethical concerns vary from country to country.

6 ORGANIC COMPOUNDS

After entering systemic circulation, organic compounds are distributed by the blood to all organs, including the excretory organ(s). Hydrophilic compounds are excreted in urine; volatile compounds are also exhaled. The hydrophobic pollutants are metabolized to soluble compounds to facilitate the urinary excretion. The liver is the most efficient metabolizing organ, but other organs, such as the kidneys and bone marrow, can also mediate the metabolism of xenobiotics. The metabolic products are distributed by the blood throughout the body and are excreted via the kidneys. The metabolic intermediates of some compounds are electrophiles that covalently bind to glutathione and to macromolecules. These metabolic products have to be biodegraded before they can be excreted. The pathways of organic xenobiotics in the body are depicted in Figure 42.5.

The persistence of biomarkers in the body determines whether the measurements in samples collected after the end of exposure are suitable for evaluating recent exposure or are suitable for evaluating retrospective exposure. For example, the excretable metabolites of most organic compounds are excreted within two days after exposure, while metabolic products with macromolecules persist in the body for weeks (32).

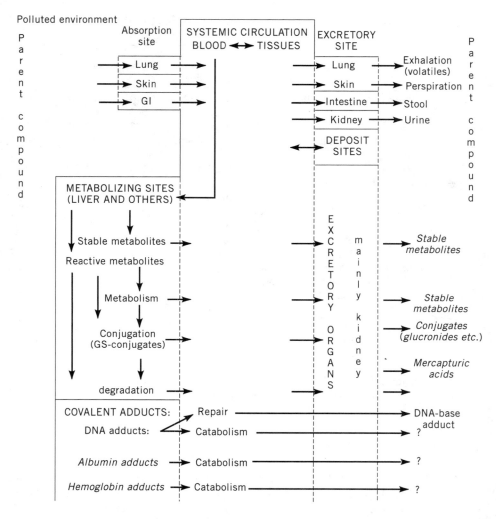

Figure 42.5. Pathways of organic pollutants in the body. GI stands for gastrointestinal tract; GS for glutathione; DNA for desoxyribonucleic acids. Metabolic products can be excreted in kidney, lungs, skin, and intestine. The potential biomarkers are in italics. They are measured in urine and/or blood.

6.1 Excretable Metabolites

Stable, hydrophilic metabolites are readily excreted in urine. Volatile metabolites are also exhaled. Excretion by other excretory pathways is usually insignificant. Unstable reactive metabolites are either converted into stable metabolites and excreted in urine (e.g., styrene oxide, an unstable metabolite of styrene, is oxidized to mandelic acid and excreted in urine); or, they can be detoxified by conjugation with glucuronic acid, sulfuric acid, or glycine (e.g., phenylglucuronide and phenylsulfate are excreted in the urine of workers exposed to benzene or phenol).

Concentrations of stable metabolites in blood and urine rise and decline in parallel, reflecting the exposure of the tissues. Figure 42.6 illustrates this parallelism, using the phenol levels in blood and the urinary excretion rate of phenol following benzene exposure. Excretable metabolites, because of their short elimination half-lives, are indicators of recent exposure, i.e., of the exposure on the day of the sampling.

The monitoring of excretable metabolites is a matter of common practice of hygienists and occupational health personnel. Documentation for specific pollutants should be consulted prior to conducting biological monitoring (10). The documentations provide information on sampling and interpretation of data, and indicate factors that can interfere with the interpretation of biological monitoring data. Valuable information on individual biomarkers can be also found in Refs. 24 and 25, and in the selected general references listed at the end of this chapter.

6.2 Deactivation of Electrophilic Metabolites

Many industrially important organic chemicals that enter the body are unstable or undergo biotransformation to highly reactive metabolic intermediates of electrophilic nature that covalently bind either to glutathione, which is rapidly biodegraded to mercapturic acids

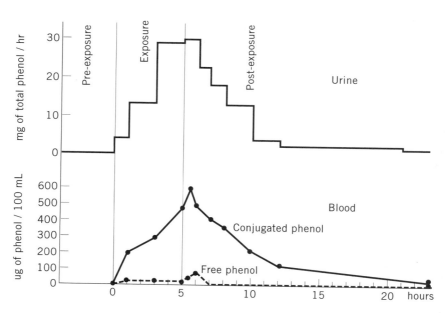

Figure 42.6. Phenol concentration in blood and urine of a volunteer exposed for five hours to benzene. Note the parallelism in blood levels and urinary excretion rate.

(N-acetylcysteine cojugates) and excreted in urine, or bind to macromolecules, such as albumin, hemoglobin, and DNA. The macromolecular products, being unable to cross cell membranes, cannot be excreted in urine, and persist in the body until destroyed by catabolism. Until recently, macromolecular adducts were rarely used as biomarkers of occupational exposure, because of lack of suitable analytical methods. Now that new technology makes identification and quantitation of these macromolecular metabolites possible, interest in their application to biological monitoring of occupational exposure has risen. To provide basic background information on these developing means in biological monitoring, the structures and principles of measurement of these potential biomarkers are reviewed below, and references are given to studies that lay the groundwork for their implementation in biological monitoring.

6.2.1 Glutathione Conjugates

Glutathione (GSH) is an endogenous tripeptide consisting of glutamine, cysteine, and glycine. It has a high affinity to electrophiles, and competes for covalent binding with macromolecules. Conjugation with GSH occurs mainly in the liver, where GSH is formed and where most of the enzymatic activation of xenobiotics occurs. However, some less reactive electrophiles can cross the cell membranes, be distributed in blood through the body, and be conjugated with GSH in remote tissues.

The conjugation of electrophiles with the SH-group of cysteine and glutamine cleavage are mediated by GSH-transferase and γ-glutamyl-transpeptidases, respectively. The conjugated cysteine-glycine dipeptide is further degraded by dipeptidases, leaving cysteinyl residues that are acetylated by N-acetyl-transferase to produce mercapturic acids. In humans, this mercapturic acid pathway takes place during hepato-biliary cycling. Polymorphism of the enzymes involved in GSH catabolism weakens the correlation between the degree of exposure and the urinary excretion of mercapturates.

The catabolism of GSH-conjugate is relatively rapid, and the urinary excretion of mercapturic acids is usually completed within 10 hours after the end of exposure. Mercapturic acids are therefore indicators of recent exposure.

Mercapturic acids in urine were first reported in 1879, in a person treated with bromobenzene. In the last two decades, mercapturic acids resulting from exposure to a number of electrophile-forming pollutants were determined in urine following alkaline hydrolysis as thioethers. Today, specific methods for determination of several mercapturic acid derivatives are available, and are beginning to be used for biological monitoring of exposure to pollutants biodegradated via electrophiles.

6.2.1.1 Principles of Mercapturic Acids Analysis.
A simple, nonselective method, using alkaline hydrolysis and Ellman reagent (5,5′-dithio-bis(2-nitrobenzoic acid)), determines urinary mercapturic acids as thioethers (33). This simple colorimetric method provides no information on the chemical structure of the electrophile attached to the N-acetylcystein. The method is used mainly for monitoring in the rubber industry and in workplaces where mixtures of solvents metabolized to electrophiles are in use (Table 42.7). The background levels measured by this unselective method are high and variable. Malonova and Bardodej used thioether measurement in urine for biological monitoring of occupational exposure

Table 42.7. Biological Monitoring using Degradation Products of GSH-Conjugates[a]

Monitored agent	Specificity	References	Ref. Values
	Single Exposure		
Acrylonitrile	S	34	84 μmol/L
Allyl chloride	S	35, 36[b]	0.35 mg/g cr.[c]
Benzene	S	1	BEI = 25 μg/g cr.[c]
1,3-Dichloropropene	NS	37	
	S	38[b]	14.4 mg/L Z-DCP
			3.2 mg/L E-DCP
N,N-Dimethylformamide	S	1	BEI = 40 mg/L
Epichlorohydrin	S	39, 36[b]	14 mg/L
Styrene	S	40	
	NS	41	
	Mixed Exposure		
Alkylating agents	NS	42	
Bitumen fumes	NS	43	
Chemical plants	NS	44	
Rubber industry	NS	45, 46	
Waste incinerator	NS	47	
Review		36, 48, 49	

[a]Nonspecific thioethers (NS) and specific mercapturic acids (S) in urine. Human studies pertinent to occupational exposure.
[b]The authors suggest a value acceptable for occupational exposure. If BEI is given see BEI-Documentation for references.
[c]cr. = creatinine.
Methods are also available for determination of mercapturic acids after exposure to acrolein, acrylamide, allyl amine, allyl halides, allyl phosphates, allyl sulfates, 1,3-butadiene, chlorobenzene, dibromopropanes, ethylene oxide, perchloroethylene, toluene, trichloroethylene, and *o*-xylene (36), but no application to monitoring of occupational exposure was found.

to several electrophile-forming pollutants (50). They suggested that the mercapturic acid excretion should not exceed 8 μmol/mmol of creatinine, which is the upper level they determined in the control group of unexposed persons. Thioether measurements are more suitable for potential health hazard evaluation than for exposure evaluation. The disadvantages of thioether measurements are high variability of background levels of thioethers in urine of unexposed persons and interference of smoking.

Selective methods are mainly based on gas chromatography or HPLC. If the former is used, mercapturic acids are extracted from acidified urine and derivatized to enhance their volatility. In some cases, selective detectors are used (36, 49, 51, 52). HPLC determination is suitable for those mercapturic acids which can be detected by fluorescence or UV detection.

In 1998, BEIs were proposed for N-acetyl-S-(N-methyl-carbamoyl)cysteine in urine for biological monitoring of exposure to N,N-dimethylformamide, and for S-phenylmercapturic acid for monitoring of benzene exposure (1). A number of papers suggests values for acceptable concentration of other mercapturic acids (Table 42.7). More information about biological monitoring of GSH-conjugates can be found in a recent reviews by de Rooij et al. (36) and by van Welie et al. (49).

6.3 Adducts with Biomacromolecules

Many industrially important organic compounds that enter the body are reactive, or undergo biotransformation to form highly reactive metabolic intermediates of electrophilic nature that bind to macromolecules such as proteins or DNA. The biomacromolecules possess abundant nucleophilic sites, such as amino-, thio-, and hydroxy-groups. The interaction between the electrophiles and the nucleophilic sites of macromolecules produces covalently bound macromolecular adducts. The binding takes place either at the site of the electrophiles formation (e.g., in hepatocytes) or at the site of their entry into the body (e.g., in respiratory tract). Reactive intermediates stable enough to enter systemic circulation may be bound to blood components, or to tissue components of remote organs (53).

Macromolecular adducts cannot be monitored in urine, because the kidneys are unable to excrete large molecules. They can, however, be monitored in blood. The macromolecular adducts persist in the body until the macromolecule is destroyed by catabolism. Because of their long dwelling in the body, macromolecular adducts are indicators of past exposure or of integrated exposure during the last months prior to sampling. The removal pathway of the electrophilic fragments from the body remains unknown.

DNA adducts, besides being removed by catabolism, are also removed by the repair system, which is an enzymatic mechanism to detect and remove the adducted nucleotides from the DNA. Therefore, the lifetimes of the DNA adducts in the body are shorter (usually less than 24 hours) than the turnover of human lymphocytes (more than 180 days). This process slows down the cumulation of DNA-adducts in the body. The DNA repair products, i.e., the DNA-base with the electrophile, are excreted in urine.

The following adducts with macromolecules have been extensively studied: DNA isolated from the lymphocytes and DNA repair products excreted in urine; globin isolated from erythrocytes; and albumin isolated from plasma.

6.3.1 Adducts with DNA

The use of DNA adducts as biomarkers for exposure monitoring is questionable, because their appearance signals a serious adverse effect that exposure monitoring is meant to prevent. DNA adducts are nevertheless useful in medical surveillance as indicators of mutagenic and carcinogenic potency of pollutants (49, 54, 55).

The level of the DNA adducts reflects not only the exposure to the pollutant, but also reflects the individual's ability to activate the pollutant and to remove the DNA products.

6.3.1.1 Binding Sites in DNA. The sites in DNA targeted by the electrophiles are predominantly nitrogen and oxygen atoms in guanine, and nitrogen atoms in adenine. The targeted sites preferred by individual electrophiles are not the same. The stability of the adducts depends on the position in the DNA to which the electrophile is attached. Furthermore, the spectrum of DNA adducts and their biological consequences are affected by interindividual differences in DNA repair capacity.

6.3.1.2 Principles of DNA Adducts Analysis.
(Reviewed in Ref. 55). The methods that have been used for detection of the xenobiotic-DNA adducts include immunoassay (56, 57), ^{32}P-postlabeling (58–61), fluorescence and phosphorescence spectroscopy (62, 63), and gas chromatography–mass spectrometry (GC–MS) (64). Currently, the ^{32}P-postlabeling method is the most commonly used in human studies. This method is highly sensitive (detects 1 adduct among 10^9 nucleotides) and requires small quantities of DNA (2–10 µg). The analysis starts with enzymatic DNA digestion and partial dephosphorylation. After enzymatic transfer of the ^{32}P-phosphate group from ^{32}P-ATP to the nucleosides, two-dimensional thin-layer chromatography or HPLC are used to separate the labeled adducts, which are then detected by autoradiography or radioactivity measurement. ^{32}P-postlabeling does not identify the structure of the adducts. The procedure has to be optimized for each type of adduct. The method is laborious and unsuitable for routine use.

6.3.1.3 DNA Repair Products in Urine.
The DNA repair products, excreted in urine as adducts of the electrophile with nucleoside, or with purine or pyrimidine bases, are potential biomarkers for biological monitoring of occupational exposure to carcinogens. For example, 7-methylguanine and 3-methyladenine derivatives can be used as biomarkers of exposure to methylating agents (64).

6.3.1.3.1 Analytical methods The nucleic acid bases with attached alkyls are extracted from urine, derivatized, and determined by GC–MS (64, 65). The method of choice for 8-hydroxy-2-deoxyguanosine, an indicator of oxidative damage by some pollutants, is HPLC with electrochemical detection (66).

6.3.1.4 Implementation of DNA Adducts to Biological Monitoring.
DNA adducts of polycyclic aromatic hydrocarbons (PAH) were used with mixed results for biological monitoring in industry where workers are exposed to mixtures of PAHs (review in Ref. 67). Thus, DNA products have been monitored in iron foundry workers (68–77), coke oven workers (78, 79), aluminum production plant workers (80, 81), fire fighters (82), garage workers and diesel exhaust workers (83–85). DNA adducts have also been monitored in workers exposed to styrene (86–89).

Monitoring of DNA adducts was mainly carried out as screening tests. The relation between external exposure and urinary excretion of DNA repair products has not as yet been determined, and no reference value has been suggested. Smoking is likely to interfere with biological monitoring of exposure (90).

6.3.2 Adducts with Proteins

Protein adducts as possible biomarkers for occupational exposure were first used by Osterman-Golkar et al. (91). Their primary interest was to assess the effective dose of alkyl groups at the critical site of DNA. They used proteins as a surrogate for DNA. It was later established in animals and in cell systems that all chemicals that react with DNA also react with hemoglobin. Several studies have shown that blood levels of protein adducts reflect the degree of exposure to a specific pollutant (reviewed in Refs. 32, 49, 53, 54, 92). Both hemoglobin and albumin can be easily isolated from blood.

The advantage of blood protein adducts as biomarkers of exposure is their long lifetime in the body. The turnover rate of hemoglobin in man is approximately 120 days, which is the lifetime of erythrocytes. Thus, it is conceivable that levels of hemoglobin adducts can be used to assess the degree of exposure well past the exposure. The slow declining of hemoglobin adducts is demonstrated in Figure 42.7 using N,N-dimethylformamide exposure as an example (93). Hemoglobin adducts found much wider application in biological monitoring than DNA or albumin adducts.

To bind to hemoglobin, the electrophiles have to penetrate the erythrocyte cell membranes which the highly reactive electrophiles may be unable to do. Albumin, on the other hand, is present in serum and readily available for binding with the reactive intermediates released from the metabolizing organs. Therefore albumin adducts may be more abundant and more sensitive biomarkers than hemoglobin adducts. Furthermore, albumin is synthesized in the liver, where are mainly located the enzymes mediating the metabolism of xenobiotics. The lifetime of albumin in humans is 20–25 days.

6.3.2.1 Binding Sites in Proteins. The nucleophilic sites in proteins that are most reactive towards electrophiles include the NH_2-group in terminal amino acid (e.g., valine in globin),

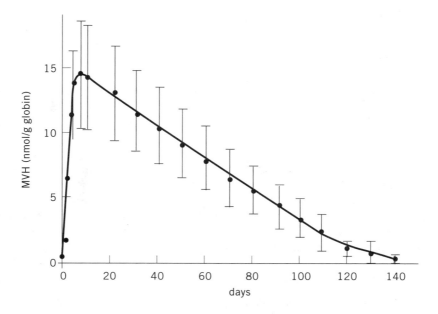

Figure 42.7. The slow declining of N-methylcarbamoyl adduct with globin (determined as 3-methyl-5-isopropylhydantoin, MVH) in blood of eight volunteers following 15 minutes of dermal exposure to N,N-dimethylformamide (DMF) (93).

the SH-group in cysteine, N-3 and N-1 in histidine and in some cases the NH_2-group in lysine, the OH-group in serine, and the carboxylic groups in aspartic or glutamic acids. All these groups are present in hemoglobin as well as in albumin molecules. Therefore procedures based on the same principles can be used for the analysis of adducts with both these proteins.

6.3.2.2 Analysis of Protein Adducts. The classical methods are based on isolation of protein adducts and determination of isolated protein adducts, either by an immunological or chromatographic method.

6.3.2.2.1 Isolation of protein adducts Prior to analysis, hemoglobin, globin, or albumin are isolated from the blood. Whole blood is centrifuged to separate erythrocytes from plasma. Plasma is removed and used for isolation of albumin adducts. Erythrocytes are washed with saline and lysed with distilled water. Hemoglobin is precipitated from the hemolysate by the excess of ethanol and further washed. Hemoglobin concentration is determined by spectrophotometry. Globin is precipitated from the hemolysate by the excess of acetone with 2% hydrochloric acid or by the addition of 2-propanol with 2% hydrochloric acid, and dried.

Albumin is isolated from plasma using fractional precipitation by means of ammonium sulfate, purified by dialysis or gel permeation chromatography on Sephadex, and dried. The albumin purification procedure is tedious, and may be harsh to labile adducts.

6.3.2.2.2 Determination of the protein adducts Two principal procedures, immunoassay and chromatography, are used to determine protein adducts:

Immunological methods have been used for determination of albumin adducts with PAH (reviewed in Ref. 94) or aflatoxin (95), but seldom for determination of hemoglobin adducts (96). The assays require polyclonal or monoclonal antibodies, production of which is tedious. The tests themselves are rapid and suitable for routine use. The immunological methods have been successfully employed for screening of albumin–carcinogen adducts.

Chromatographic methods consist of isolation of the protein (mainly globin) from blood, total acidic hydrolysis with 6 N hydrochloric acid, purification of the amino acid adducts by ion-exchange chromatography, and chemical derivatization consisting of methylation followed by perfluoroacylation prior to determination by GC–MS. The method is suitable for the analysis of alkylated adducts resulting from exposure to ethylene oxide and propylene oxide [detected as N-(2-hydroxyethyl)histidine, N-(2-hydroxyethyl)valine, S-(2-hydroxyethyl)cysteine and their (2-hydroxypropyl) analogues], acrylamide and acrylonitrile [detected as S-(2-carboxyethyl)cysteine], and methylating agents [detected as S-methylcysteine] (reviewed in Ref. 49). The tedious work-up of the samples has hampered application of this method for routine use.

6.3.2.3 Determination of Specific Adducts. In the 1980s, new gas chromatographic procedures for the determination of specific types of adducts were developed. These procedures do not require fragmentation of the protein and thus enable wider application.

6.3.2.3.1 Mild hydrolysis and implementation A mild acid or base hydrolysis method is suitable for determination of arylamine adducts with albumin or hemoglobin. Arylamines are activated in the body to nitrosarenes, which react with cysteinyl SH-groups in proteins to form arylsulfinamide adducts. These *in vivo* stable adducts can be easily hydrolyzed *in vitro* to form parent amines (97), which can be extracted, derivatized with perfluoroacyl agent, and determined by GC–EC or GC–MS (Ref. 18, vol. 4).

Implementation to biological monitoring. The mild hydrolysis method, applicable to determination of globin as well as to albumin adducts, was used for determination of adducts resulting from arylamines and PAH exposures (review Ref. 18, vol. 4).

The mild hydrolysis method was employed in biological monitoring of exposure to arylamines, which undergo either enzymatic acetylation, or are oxidized. The accrued electrophile binds to the protein. The yield of the arylamine adducts is, however, dependent on the acetylator status of the exposed subject. Since enzymatic acetylation of aromatic amines competes with the macromolecular adduct formation, slow acetylators produce significantly more globin adduct than the fast acetylators (reviewed in Ref. 97). The interindividual differences in adduct levels resulting from the competition interfere with exposure evaluation.

PAHs bind to proteins as esters following activation to epoxides or diolepoxides. The albumin and globin adducts have occasionally been used for biological monitoring of occupational exposure to PAH (98, 99).

A modified procedure was used for biological monitoring of styrene. The method simultaneously determines styrene oxide adducts with cysteine and carboxylic acid moiety in albumin or hemoglobin (100, 101).

Arylsulfinamide adducts are also produced by other pollutants yielding nitrosoarenes, such as nitroarenes, carbamates, urea derivatives, and isocyanates (102–107).

6.3.2.3.2 The modified Edman degradation method. This method is applicable to agents forming alkylated *N*-terminal valine in globin. The adducts are treated with the Edman reagent (pentafluorophenyl isothiocyanate) and mild base hydrolysis. The cleaved molecules, made up of alkyl, valine, and Edman reagent, are rearranged to cyclic thiohydantoins, which are extracted and determined by GC–MS or GC–EC (108). The method is selective and sensitive.

Isothiocyanates and isocyanates also bind to valine in globin molecules which, when treated with the Edman degradation procedure, form hydantoins consisting of valine and isocyanate that can be determined by GC–MS. Such adducts were observed in animals exposed to methylisocyanate (109) and 2,4-toluenediisocyanate (110).

Implementation to biological monitoring. This method has been used for biological monitoring of exposure to alkylating agents such as ethylene oxide, acrylonitrile, benzyl chloride, methylhalides, bis(chloromethyl)ether, dimethylsulfate (reviewed in Refs. 111, 112, and in Ref. 18, vol. 5), propylene oxide, styrene, acrylamide, and sulfur mustard (Table 42.8). Hemoglobin adducts of benzo[*a*]pyrene were also used to monitor the exposure of newspaper vendors to traffic exhausts (122).

The methylisocyanate adduct seems to be a potential biomarker for biological monitoring of exposure to *N,N*-dimethylformamide. It has been shown in volunteers as well as in workers that methylisocyanate, which is an unstable metabolite of *N,N*-dimethylformamide, binds to valine in globin molecules and can be determined as hydantoin (93, 121).

6.3.2.4 Reference Values. The German Commission for the Investigation of Health Hazards of Chemical Compounds in the Work Area recently suggested protein adducts as

Table 42.8. Selected Human Studies of Protein Adducts Pertinent to Biological Monitoring

Pollutant	Biomarker	Reference
Acrylonitrile	N-2-Cyanoethylvaline (Hb)	111, 114
Acrylamide	N-2-Cyanoethylvaline (Hb)	114
Ethylene oxide	N-2-Hydroxyethylvaline (Hb)	111, 116, 117
Propylene oxide	N-2-Hydroxypropylvaline (Hb)	111
Dimethyl sulfate	N-Methylvaline (Hb)	111
Sulfur mustard	N-(2-Hydroxyethylthioethyl)valine (Hb)	115
Styrene	Styrene oxide–cysteine adduct (Alb, Hb)	113
Styrene oxide	Styrene oxide–cysteine adduct (Alb)	88
Aniline and p-chloroaniline	Sulfinamide adducts (Hb)	118
4,4′-Methylenedianiline	Sulfinamide adduct (Hb)	105, 119
2,4,6-Trinitrotoluene	Sulfinamide adduct (Hb)	120
N,N-Dimethylformamide	N-Methylcarbamoylvaline (Hb)	93, 121

biomarkers for monitoring of occupational exposure to carcinogens. The Commission standardized the method for determination of N-2-cyanoethylvaline, N-2-hydroxyethylvaline, and N-methylvaline, and recommended it for evaluation of the exposure to acrylonitrile, ethylene oxide, and methylating agents. Since these pollutants are listed as human carcinogens no, BAT is recommended. Instead the Commission recommended "exposure equivalents for carcinogenic substances" (123).

7 METALS AND OTHER INORGANIC POLLUTANTS

Inorganic pollutants undergo little metabolism compared to organic pollutants. Their uptake, distribution, and excretion is primarily influenced by solubility of the metallocompound and by binding of the metal to tissue components. Heavy metals are the inorganic pollutants of principal concern.

7.1 Biological Monitoring of Metals

Metals and their compounds are inhaled as fumes, aerosols, or dust. The exceptions are arsine, nickel carbonyl, and metallic mercury, which are inhaled as gases or vapors. The absorption rate of particles from the respiratory tract determines the rise of the biological level of the absorbed metal. Excretion of metals is hindered by binding to extracellular and intracellular elements, mainly proteins.

Biological monitoring of metals is preferable to air monitoring, because air monitoring is relatively brief (recording actual exposure during sampling) compared to the cumulation and persistence of metals in the body. The presence of a metal in urine or blood is the most commonly used biomarker. For most metals, a time lag of several months is seen in newly exposed workers before the concentrations in blood and urine reflect the exposure.

7.1.1 Pulmonary Absorption and Clearance

Metallic dust is deposited throughout the respiratory tract. Tracheo-bronchial clearance usually takes about eight hours, depending on the size and solubility of the particles. Particles deposited in the upper respiratory airways are transferred by mucus to the larynx, from where they are swallowed, and either absorbed in the gastrointestinal tract and excreted via bile, or discharged in stool. Particles that reach the alveoli are either dissolved and absorbed into the blood circulation, or they are phagocytized and transported in macrophages to lymphatic vessels and nodes, from where they slowly penetrate into the blood circulation. The absorption rate in respiratory tract depends on the size of the particles, the solubility of the pollutant, and the valency of the metal. The digestion of metals depends on the physico-chemical properties of the metallic compound, on diet, and on the pH in the gastrointestinal tract. The slow absorption of metals in the lungs and gastrointestinal tract is reflected in the slow rise of their levels in blood, urine, and tissues.

7.1.2 Dermal Absorption

Significant dermal absorption has been documented for mercury, zinc, copper, gold, beryllium, thallium, cobalt, nickel, and tetraethyl lead.

7.1.3 Placental Penetration

Metals can penetrate the placenta. Considerable attention has been given to placental penetration of lead. Concern over protection of the fetus prompted the German Commission to lower the BAT value for lead for women in childbearing age (Table 42.2).

7.1.4 Distribution and Elimination

In the bloodstream, the distribution of a metal between the cellular fraction of the blood and the plasma depends on the ability of the metal to penetrate membranes or bind to the membranes. Penetration ability in turn depends on the form of the absorbed metal. For example, ingested salts of mercury were found mainly in erythrocytes (the erythrocytes-plasma ratio of mercury being about 2.5 to 1), while inhaled metallic mercury was found to be equally distributed between plasma and erythrocytes. However, about 98% of inhaled methylmercury was found in plasma. Distribution can also be affected by the valence of a metal. For example, trivalent chromium binds in plasma and is easily excreted. But hexavalent chromium enters erythrocytes, where it is reduced to trivalent chromium, binds to hemoglobin, and remains in erythrocytes until they are catabolized (124).

Metals in blood bind to metallothionein and to metallothionein-like proteins. These diffusible metallic compounds in plasma facilitate distribution and excretion of metals. At low exposures and at steady state, the concentration of a metal in plasma usually reflects its concentration in tissues.

7.1.5 Choice of Specimen and Analytical Method

Binding of metals to proteins can cause large differences between the concentration of the metal in the cellular fraction of the blood and its concentration in plasma or serum. Therefore urine is usually preferred for biological monitoring of metals such as mercury, chromium, cobalt, cadmium, and vanadium. Lead is the only metal for which measurement in blood is preferred.

The choice of analytical method may also affect the outcome of measurements. Combustion methods usually produce higher values than methods using hydrolysis or extraction treatment.

7.1.6 Factors Interfering with Biological Monitoring of Metals

7.1.6.1 Diet and Environment. The mineral composition of the diet and the source of protein can affect the biological level of metals. Thus the introduction of metals into the aquatic food chain can increase the levels of those metals in the local human population. This happened, for example, in Japan, when mercury and arsenic were discarded into Miamato Bay and were absorbed by fish in the bay. Consequently, the body burden of both metals was found to be higher in the Japanese population, depending on the local fish diet, than in the population of other countries. Cadmium cumulates in lobsters and shellfish. Consequently, eating these seafoods increases its biological levels.

The mineral composition of drinking water can also affect the biological level of metals. This is the situation in northern Chile, where the soil is rich in arsenic. Arsenic enters the sources of drinking water and increases the body burden of the local occupationally unexposed population.

Abnormal dietary and mineral water intake of essential metals, such as calcium, iron, zinc, and copper, can affect the biological levels of toxic metals and interfere with the balance between their intake and excretion.

Environmental air pollution affects the body burden of metals in the occupationally unexposed population. Thus city dwellers, as a result of air pollution from car emission, usually have higher blood levels of lead than the rural populations.

The problem of interference of these factors with biological monitoring of occupational exposure can be solved by simultaneous biological monitoring of the local, occupationally unexposed population. Safe biological levels, however, have to be maintained regardless of the source of exposure.

7.1.6.2 Other Factors Affecting Biological Levels of Metals. A person's age affects the biological levels of metals. As a person gets older, biological levels tend to increase because of the presence of metals in the diet and environment.

The binding of metals to tissue components usually is a saturable process. After binding sites are saturated, the plasma concentration and urinary excretion of the metal suddenly rise. An example is the urinary excretion of cadmium, which suddenly rises after the capacity of the binding sites in the renal cortex is saturated and renal dysfunction begins. The kidneys are the usual target organs of metals. Chronic as well as acute exposure to metals such as cadmium, lead, and mercury can cause kidney damage and alter the urinary excretion of metals and other biomarkers.

Coexposure to more than one metal may have a complex effect on their respective concentrations in blood and urine. Competition for binding sites results in changes in the distribution and excretion of any or all of the metals.

Treatment with complex-forming agents, such as EDTA, DMPS, and DMSA, or taking of essential metal supplements, can increase urinary excretion of toxic metals.

7.1.7 Reference Values for Biological Monitoring of Metals

In the United States, BEIs have been recommended for the biological monitoring of seven metals (Table 42.2). In Germany, BATs have been established for the biological monitoring of four metals, and another four metals are designated as carcinogens (Tables 42.2–42.4). The European Commission has published documentations for the biological monitoring of 10 other metals (Table 42.5) (24, 25). The specific documentation for the metal should be consulted when biological monitoring is planned.

7.1.8 Biological Monitoring for Medical Surveillance

Exposure to metals can also be monitored by measurement of biochemical changes in blood and urine. Such biochemical changes may provide valuable diagnostic information for medical surveillance, but they do not meet the preventive requirements of occupational hygiene, because their abnormal levels are indicators of adverse effects. For example, the interference of lead with heme synthesis can be monitored by measuring ALA-dehydratase activity in erythrocytes, by measuring protoporphyrin and zinc–protoporphyrin in erythrocytes, and by measuring the concentrations of ALA (δ-aminolevulinic acid) in urine. Another example is the interference of cadmium and mercury with renal function. The indicators of renal function impairment can be increased excretion of β-microglobulin and N-acetyl-β-D-glucosamidase (NAG) in urine.

Stimulation of urinary excretion by chelation of metals is used for therapy as well as for determination of the body burden of the metal. For example, DMPS administration raises excretion of mercury, and Ca-EDTA administration stimulates excretion of lead.

Detailed review of toxicity and biological levels of metals can be found in references 125–127.

7.2 Biological Monitoring of Inorganic Gases

Biological monitoring of inorganic gases is not recommended (except for carbon monoxide). This is because inorganic gases are either irritants, such as chlorine and sulfur dioxide, or their acute toxicity is critical, as with hydrogen cyanide, hydrogen sulfide, nickel carbonyl, and arsine. Air monitoring is preferred, because the acute adverse effect is usually related to peak concentration in the ambient air.

The exception is carbon monoxide, for which sensitive, relatively specific biomarkers are available (10, 25). BEIs are recommended for carboxyhemoglobin in blood and carbon monoxide in exhaled air. A BAT is recommended only for carboxyhemoglobin (see Table 42.2). The BEI documentation includes specific information on how to conduct biological monitoring of carbon monoxide exposure (10).

7.3 Biological Monitoring of Other Inorganic Pollutants

BEIs and BATs are currently available for fluorides and carbon disulfide. No other inorganic compounds have been considered for biological monitoring.

8 BIOCHEMICAL CHANGES AS BIOMARKERS OF POSSIBLE HEALTH RISK

Significant biochemical changes such as enzyme inhibition or interactions with vital molecules will not likely be observed if exposure is below the level permissible in the workplace. Some reversible biochemical changes, however, can be used as biomarkers of possible health risk. Any deviation from the norm of such reversible changes is a warning sign of possible incipient adverse effect from overexposure. For example, acetylcholinesterase inhibition is the warning sign of neurotoxicity resulting from overexposure to organophosphorus pesticides. Another example is an increased carboxyhemoglobin in blood resulting from overexposure to carbon monoxide and methylene chloride. This is a warning sign that the blood has a reduced capacity to transport oxygen. There currently exist a BEI and a BAT for both carboxyhemoglobin and for acetylcholinesterase. Several chemicals induce methemoglobinemia, which also reduces oxygen supply to the tissues. There is a BEI for methemoglobin in blood as an indicator of potential health risk from overexposure to methemoglobinemia inducers.

The reference values for these biomarkers are the respective highest or lowest levels observed in unexposed populations (Table 42.9). Deviations from normal levels are not suitable for quantitative evaluation of exposure. They should nevertheless prompt medical examination of exposed workers and further inspections of the workplace.

These biomarkers have two advantages: (*1*) They can be applied to pollutants with significant dermal absorption, for which air monitoring is insufficient. (*2*) Since they are nonspecific biomarkers of the pollutant, but specific biomarkers of the toxic end point, they are useful for monitoring workers who are exposed to mixtures of chemicals that have the same toxic end point (acetylcholinesterase inhibition and methemoglobinemia).

According to both the BEI and BAT, acetylcholinesterase activity in erythrocytes should not drop below 8% of the lowest normal value. Carboxyhemoglobin should not exceed the normal values measured in the urban commuting population. The BEI, as the "most

Table 42.9. Biochemical Changes as Biomarkers

Biomarker	Range of Normal Values	BEI	BAT
Acetylcholinesterase (Ery.)[a]	>78%	>70%	>70%
CO-hemoglobin			
Endogenic	0.4–0.7%		
Urban	1.0–2.0%	3.5%	5%
Commuters	<5%		
Smokers	<20%		
Methemoglobin	<2%	1.5%	
Lead			
Zn-protoporphyrin (Ery.)[a]	15–80 µg/100 mL		15 mg/L
δ-aminolevulinic acid (urine)	<5 mg/L		6 mg/L[b]

[a]Ery = measurements done in erythrocytes.
[b]Women in childbearing age.

likely expected level," is in the middle of the range of values measured in urban commuters. The BAT, as a "ceiling value," equals the highest value measured in urban commuters.

Products resulting from interference with heme synthesis are suitable indicators of overexposure to lead. There is a BAT for δ-aminolevulinic acid in urine. It specifies that the concentration in urine of women of childbearing age should not exceed the range of concentrations in the population unexposed to lead, but that the concentration in the urine of other workers can exceed that range three times. There is no BEI or BAT for zinc protoporphyrin, even though zinc protoporphyrin has been suggested as a biomarker by a number of investigators. A simple analytical procedure using a portable hematofluorimeter suitable for zinc protoporphyrin determination in field conditions is available. Aminolevulinic acid dehydratase in erythrocytes has also been suggested as a sensitive biomarker for monitoring the health risk induced by exposure to lead (24).

Proteinuria is an indicator of impaired renal function. As an indicator of cadmium exposure, the monitoring of low molecular weight proteins such as β-microglobulin has been promoted. Another promoted indicator of cadmium-induced renal impairment is urinary excretion of N-acetyl-β-D-glucosamidase (24).

9 BIOLOGICAL MONITORING OF EXPOSURES TO MIXTURES

Reference values such as BEIs and BATs apply to exposures to a single pollutant. In the workplace, workers are usually exposed to multiple pollutants; that is, to a mixture of pollutants. When this occurs, the reference values for exposure to mixture components and the values of biological levels of biomarkers must be adjusted. Such adjustment depends on the toxic endpoint of each pollutant comprising the mixture, and on the interaction of the pollutants and their metabolites with the body.

9.1 Toxicodynamic Effects

Mixture components may affect each other's toxicity. For exposure for mixtures of pollutants, the BEI committee recommends the same rules for adjustment of biological levels of biomarkers as the TLV committee recommends for adjustment of concentration in airborne mixtures (11). The effects of coexposure can be manifested in the following four ways:

1. Independent effects mean the toxicity of each mixture component is produced by a different mechanism, so that the adverse effect of one component is unaffected by the others. An example is a mixture in which one component is an irritant (e.g., sulfur dioxide) and another component produces a systemic effect (e.g., toluene).

2. Additive effects mean the mixture components have the same or similar endpoints and the toxicity of the mixture is the sum of the toxicities of its components. Examples are CNS depressants (e.g., organic solvents) or metals implicated in impairment of renal function (e.g., cadmium and lead). To adjust the biological levels of biomarkers of pollutants with additive effects, the hazard index k is calculated:

$$k = \sum_{i=1}^{i=n} \frac{c_i - C_i}{BEI_i - C_i} \quad (9)$$

where c_i is the measured biological level of the biomarkers of each pollutant, C_i is the background level of each biomarker (i.e., the level of biomarker in the unexposed population), and n is the number of pollutants comprising the mixture. The calculated hazard index k should equal 1 or less.

3. Potentiation or synergistic effect means the toxicity of one mixture component is enhanced in the presence of other mixture components. For example: Beryllium as a mixture component may have a synergistic effect, because it is known to initiate the autoimmune process. Coexposure to cobalt and tungsten carbide prompts the development of interstitial lung lesions which are not observed with cobalt exposure alone or tungsten carbide exposure alone. n-Hexane neurotoxicity and methyl-n-butylketone neurotoxicity are potentiated by methyl ethyl ketone. Hepatotoxins such as carbon tetrachloride enhance the toxicity of cadmium.

The reference values for exposures to mixtures of pollutants with synergistic effects cannot be set by applying any mechanistic formula or rule. Individual medical attention and significant exposure reduction are required. The simple solution to the problem is to avoid coexposure to pollutants with synergistic effects.

4. Antagonistic effects mean the toxicity of one mixture component is reduced in the presence of other mixture components. For examples: Selenium reduces the toxicity of mercury and of arsenic. And ethanol reduces the toxicity of methanol and of ethylene glycol.

A mixture with an antagonistic effect can be treated either as a mixture with additive effects (Eq. 9); or, it can be treated as a mixture with independent effects (meaning that no adjustment is needed).

BEIs and BATs for nonspecific biomarkers based on biochemical changes reflect the effect of the pollutants on the same toxic endpoint (Table 42.9). Reference values for nonspecific biomarkers apply to the mixture regardless of how many pollutants with the same effect are present in the mixture.

9.2 Metabolic Interactions

Mixture components may affect each other's metabolism, or affect each other's binding to tissue components. Such effects, which are mostly of a competitive nature, alter the uptake, distribution, and excretion of the mixture components, as well as alter the production, distribution, and excretion of their metabolites. Similar effects can be induced by other xenobiotics that workers may be exposed to, such as alcoholic beverages or medication. Coexposure to xenobiotics should therefore be watched for and considered in evaluating biological monitoring data.

9.2.1 *Metabolic Interactions among Pollutants*

Metabolic interactions of mixture components are usually caused by competition for metabolizing enzymes or by induction of enzyme activity. As a result of the interaction, the yield of the metabolite is either increased, which is a result of induction, or it is reduced by inhibition. Also, the rising of biological levels and the excretion of biomarkers is either accelerated or slowed down. The majority of organic pollutants is metabolized by the same microsomal enzyme system. The biotransformation of the component with the highest affinity to the enzyme is the least inhibited; the biotransformation of the component with the lowest affinity is the most inhibited. As a result, the biological level of the mixture component with low affinity to the enzyme is higher, and the biological level of its metabolite is lower, than if the worker had been exposed to that component alone. Competition has less effect on the biological levels of biomarkers of the mixture component with high affinity for the enzyme.

Competitive inhibition is common among mixtures of organic solvents, such as a mixture of benzene and toluene; of xylene and toluene; of xylene and butanol; of methyl ethyl ketone, methyl isobutyl ketone, xylene, and ethyl benzene; and of trichloroethylene and perchloroethylene.

The effects of inhibition are apparent from the start of exposure, while the induction effects occur after a few days of exposure. The induction is manifested by increased levels of metabolites and reduced levels of parent compounds.

The occurrence of inhibition or induction can be determined by simultaneous monitoring of a mixture component and its metabolite. If no interaction has occurred, the ratio of the biological level of the metabolite to the level of its parent compound will be the same as if the worker had been exposed to that pollutant alone.

9.2.2 Competition for Binding Sites

Metals and unstable pollutants bind to macromolecules and other tissue components. The effect of competitive binding on the biological levels of a metal depends on the affinity of the metal to metallothionein and to metallothionein-like molecules. The results of competition for binding sites are altered distribution of the metals in the body and their altered excretion. The altered availability of nucleotides for binding with organic electrophiles may affect the development of adverse effect.

The reviews of the effects of coexposure are discussed in References 11 and 128.

9.3 Effects of Alcohol Consumption

After consumption of one alcoholic drink, the ethanol concentration in blood can be about 1,000 times higher than the concentrations of pollutants resulting from exposure permissible in the workplace. Moreover, ethanol ingested by drinking is largely available for metabolism because of the first-liver-pass-effect. The experiencing of flashes by workers who consumed an alcoholic beverage during or shortly after exposure to trichloroethylene or N,N-dimethylformamide is an example of ethanol interaction. Drinking of alcoholic beverages significantly raises the blood levels of xylene, trichloroethylene, toluene, and methyl ethyl ketone during or shortly after exposure. Excretion of the metabolites of these pollutants, which can be used as a biomarker, is reduced during the time that ethanol concentration in the blood is high. In some cases, ethanol affects the excretion of one metabolite but not the excretion of other metabolites. For example, when a worker exposed to styrene drinks an alcoholic beverage, the excretion of mandelic acid is reduced, but the excretion of phenylglyoxylic acid is unaffected. Alcoholic beverages have a similar unilateral effect on the metabolites of N,N-dimethylformamide (for review see Ref. 129).

9.4 Effect of Medications

Medications may affect the biological levels of biomarkers in the following ways:

1. They may contain the biomarker, and thus increase the biological level of the biomarker of occupational exposure. Some commonly used medications that contain biomarkers used for biological monitoring of industrial pollutants are shown in Table 42.10.

2. They may inhibit or induce the activity of enzymes. This in turn can affect the metabolism of pollutants. Examples of enzyme inhibitors are sedatives (e.g., allobarbital, secobarbital) and aspirin; examples of enzyme inducers are phenobarbital, meprobamate, and antipyrine.

3. They may alter the physiological functions that control the uptake, distribution, and elimination of pollutants. For examples: anesthetics reduce pulmonary ventilation, thus reducing pulmonary uptake of pollutants; diuretics increase urine flow, thus diluting the concentration of biomarkers in urine; and cimetidine alters creatinine excretion, thus making creatinine excretion unsuitable for standardization of biomarkers.

4. They may affect binding of the pollutant or its metabolite to macromolecules. This effect can be found, for example, in workers who are occupationally exposed to toxic metals while taking vitamin supplements containing essential metals.

5. They may induce biochemical changes used as biomarkers of health risk. For examples: some antiglaucoma medications are cholinesterase inhibitors; nitroglycerine, sulfonamides, and phenacetin are methemoglobin inducers.

Table 42.10. Examples of Medication Possibly Affecting Biological Levels of Biomarkers

Affected Biomarker	Medication Containing Biomarker
Aluminum	Al-containing antacids
Aniline	Analgesic (phenylacetine)
Fluoride	Toothpaste
Mercury	Dental amalgam
Metals	Vitamins with essential metals
Phenol	Antiseptic lotions, lozenges
Trichloroacetic acid	Sedatives containing chloral hydrate
Enzyme inhibitors	Sedatives: allobarbital, secobarbital
	Aspirin
Enzyme inducers	Phenobarbital, Meprobamate
	Antipyrine

If a worker undergoing biological monitoring is taking medication, the pharmaceutical company's information sheet on the medication should be consulted. For more information on the effects of medication see References 11, 130, and 131.

10 IMPLEMENTATION OF BIOLOGICAL MONITORING (AIR MONITORING VERSUS BIOLOGICAL MONITORING)

The numerous factors affecting the biological levels of biomarkers may induce skepticism as to the usefulness of biological monitoring. Table 42.11 lists the sources of variability in both airborne measurements and in measurements of biological level. The source of variability makes either air monitoring or biological monitoring preferable, depending on the acquired information (Table 42.11).

10.1 Choice of Monitoring

Air monitoring is limited to evaluation of occupational inhalation exposures. It can locate and determine the peak concentrations of pollutants in the air, and thus help with the identification of the sources of air pollution in the workplace. It can also provide warning

Table 42.11. Causes of Variability—Choosing air Monitoring or Biological Monitoring

Variability Causes in

Air Monitoring (A.M.)	Biological Monitoring (B.M.)
1. Fluctuation of concentration	1. Physiological factors
2. Movements of workers	2. Interactions with the body
	3. Coexposure

Requested Information

	A.M.[a]	B.M.[a]
1. Identify peak air concentrations	+++	No
2. Identify the source of pollution	+++	No
3. Provide warning on acute health risk	++	+
4. Identify TWA concentration in the air	++	No
5. Identify only occupational exposure	+++	No
6. Identify total exposure	No	+++
7. Identify the internal dose	+	+++
8. Identify health risk of occupational exposure	+	+++
9. Identify health risk regardless of source of exposure	No	+++
10. Include dermal exposure and ingestion	No	+++
11. Testing of effectiveness of protective aids	No	+++
12. Suitable for medical surveillance	+	+++

[a] +++ good information; ++ reasonable information; + information dependent on circumstances; no information.

signals for acute health risks. Biological monitoring, on the other hand, provides information on workers' total exposure, regardless of the source of the pollutant (occupational as well as nonoccupational), the pollutant's entrance pathway into the body (inhalation, dermal, ingestion), or the circumstantial and physiological factors affecting the pollutant's absorption. Therefore biological monitoring data relate more closely to chronic health risks and systemic effects than does air monitoring data. And finally, biological monitoring is a major tool in evaluating the effectiveness of protective aids such as respirators and gloves.

10.2 Reference Values

Tables 42.2–42.4 and 42.6 list reference values for biological monitoring used in three industrial countries. U.S. BEIs are interpreted as average values based on health protection similar to the health protection provided by TLVs. German BATs, which are based on the same studies as U.S. BEIs, are interpreted as the highest biological levels (peaks) permissible in biological specimens collected from workers. Similarly, the reference values for airbornes, MAKs (meaning "Maximal Arbeitsplatz—Koncentrations"), are maximum airborne concentrations permissible in the work place. Japan ranks health risk into three categories, with the middle category (Category II) providing a range of warning levels (Table 42.6). For each biomarker, the ratio of values limiting Category II is about three (between 2.5 and 4).

A comparison of the values listed in Table 42.2 shows that BATs and BEIs ratios are between 0.53 and 2.8 (in average 1.39). BEIs and BATs for most pollutants fit into Japanese Category II (Table 42.6). On the basis of the respective concepts in the three countries, BATs should be at least twice as high as BEIs and about equal to the upper limit of Category II values, and BEIs should be in the middle range of Category II values.

These inconsistencies in concepts and numbers do not change the fact that the rulemaking organizations recognize the same range of biological levels associated with the potential health risk. Also, all of them consider the timing of biological specimen collection as a critical factor profoundly affecting the outcome of the measurements.

10.3 Interpretation of Data

Just as air monitoring data should be interpreted by professionals familiar with the factors affecting air pollution in the work place, biological monitoring data should be interpreted by professionals familiar with the external and internal factors affecting biological levels. External factors include coexposure to other pollutants and xenobiotics, non-occupational exposure to the pollutant or its biomarker, and exposure of different absorption sites. Internal factors are the interactions of pollutants with the body and the physiological parameters that affect uptake and absorption. These factors are usually the main cause of unexpected extreme biological levels of biomarkers.

Because of the variable and unpredictable factors affecting the biological levels, no hasty conclusions should be drawn from a single measurement or even from few measurements. Averaging the results of measurements in workers performing similar tasks in the same work place should be used for evaluation of the work place exposure. Averaging the results of repetitious measurements obtained in a single worker over a long period of time is a measure of the worker's personal exposure.

10.4 Conclusions

Air monitoring and biological monitoring are complementary tools used with the same objective: the protection of workers' health.

BIBLIOGRAPHY

1. *American Conference of Governmental Industrial Hygienists: 1999 TLVs® and BEIs® Threshold Limit Values for Chemical Substances and Physical Agents*, ACGIH, Cincinnati, OH, 1999, pp. 89–104 (new update each year).
2. *Commission for the Investigation of Health Hazards of Chemical Compounds in the Work Area: List of MAK and BAT Values 1999*. Report No. 34, Wiley-VCH, Weinheim, DFG, 1999, pp. 165–177 (new edition each year).
3. V. Fiserova-Bergerova and J. Vlach, "Timing of Sample Collection for Biological Monitoring of Occupational Exposure," *Ann. Occup. Hyg.* **41**, 345–353 (1997).
4. V. Fiserova-Bergerova, "Extrapolation of Physiological Parameters for Physiologically Based Simulation Models," *Tox. Let.* **79**, 77–86 (1995).
5. V. Fiserova-Bergerova, "Toxicokinetics of Organic Solvents," *Scand. J. Work Environ. Health* **11** (Suppl. 1) 7–21 (1985).
6. I. Astrand, "Effect of Physical Exercise on Uptake, Distribution and Elimination of Vapors in Man," in V. Fiserova-Bergerova, ed. *Modeling of Inhalation Exposure to Vapors: Uptake, Distribution and Elimination*, Vol. II, CRC Press Inc. Boca Raton, FL, 1983, pp. 107–130.
7. O. Inoue, K. Seiji, T. Watanabe, M. Kasahara, H. Nakatsuka, S. Yin, G. Li, S. Cat, C. Jin, and M. Ikeda, "Possible Ethnic Differences in Toluene Metabolism: A Comparative Study among Chinese, Turkish and Japanese Solvent Workers," *Tox. Lett.* **34**, 167–174 (1986).
8. M. E. Andersen, H. J. Clewell III, M. L. Gargas, F. A. Smith, and R. H. Reitz, "Physiologically Based Pharmacokinetics and Risk Assessment Process for Methylene Chloride," *Toxicol. Appl. Pharmacol.* **87**, 185–205 (1987).
9. G. L. Kennedy, Jr., "Biological Monitoring in the American Industry," In V. Fiserova-Bergerova and M. Ogata eds., *Biological Monitoring of Exposure to Industrial Chemicals*, ACGIH, Cincinnati, OH, 1990, pp. 63–67.
10. *American Conference of Governmental Industrial Hygienists: Documentation of Threshold Limit Values and Biological Exposure* Indices. Vol. III, ACGIH, Cincinnati, OH, 1991, suppl. 1996, 1997, 1998, 1999.
11. American Conference of Governmental Industrial Hygienists: *Topics in Biological Monitoring*, ACGIH, Cincinnati, OH, 1995.
12. R. D. Stewart and H. C. Dodd, "Absorption of Carbon Tetrachloride, Trichloroethylene, Tetrachloroethylene, Methylene Chloride and 1,1,1-Trichloroethane through the Human Skin," *Am. Ind. Hyg. Assoc. J.* **25**, 439–446 (1964).
13. V. Sedivec, M. Mraz, and J. Flek, "Biological Monitoring of Persons Exposed to Methanol Vapours," *Int. Arch. Occup. Environ. Health* **48**, 257–271 (1981).

14. *International Commission on Radiological Protection: Report of the Task Group on Reference Man. No. 23*, Pergamon Press Oxford, England, 1984.
15. H. S. Vij and S. Howell, "Improving the Specific Gravity Adjustment Method for Assessing Urinary Concentrations of Toxic Substances," *Am. Ind. Hyg. Assoc. J.* **59**, 375–380 (1998).
16. J. Teisinger and V. Fiserova-Bergerova, "Phenol in Blood of Workers using Benzene and Phenol," *Pracovni Lekarstvi* **7**, 156–160, (1955) (in Czech).
17. Y. Aizawa and T. Takata, "Biological Monitoring of Arsenic Exposure," in V. Fiserova-Bergerova and M. Ogata, eds. *Biological Monitoring of Exposure to Industrial Chemicals*, ACGIH, Cincinnati, OH, 1990, pp. 91–94.
18. J. Angerer and K. H. Schaller, eds., *Analyses of Hazardous Substances in Biological Materials*, Methods for Biological Monitoring, VCH Publisher, Deerfield Beach, FL., Vol. 1 (1985), 2 (1988), 3 (1991), 4 (1994), 5 (1997).
19. J. Angerer and K. H. Schaller, "Biological Materials, Hazardous Substances Analyses of., Biological Monitoring of Exposure to Chemical Agents," in R. A. Meyers, ed., *Encyclopedia of Environmental Analysis and Remediation*, John Wiley & Sons, Inc., New York, 1998, pp. 702–731.
20. American Conference of Governmental Industrial Hygienists: *TLVs® Threshold Limit Values for Chemical Substances and Physical Agents in the Workroom Environment with Intended Changes for 1973*, ACGIH, Cincinnati, OH, 1973, p. 8.
21. American Conference of Governmental Industrial Hygienists: *TLVs® Threshold Limit Values for Chemical Substances and Physical Agents in the Work Environment and Biological Exposure Indices with Intended Changes for 1984–1985*, ACGIH, Cincinnati, OH, 1984.
22. C. E. Adkins. "The Occupational Safety and Health Administration and Biological Monitoring," in V. Fiserova-Bergerova and M. Ogata, eds. *Biological Monitoring of Exposure to Industrial Chemicals*, ACGIH, Cincinnati, OH., pp. 55–57, 1990.
23. *Senatskommission zur Prufung gesundheitsschadlicher Arbeitsstoffe*, XVII Harald Boldt Verlag, DFG, 1981, pp. 71–76.
24. L. Alessio, A. Berlin, R. Roi, and M. Boni, eds. *Commission of the European Communities. Industrial Health and Safety: Human Biological Monitoring of Industrial Chemicals Series*. ECSC-EEC-EAEC, Luxembourg, 1983 (EUR 8476 EN).
25. L. Alessio, A. Berlin, M. Boni, and R. Roi, eds., *Commission of the European Communities. Industrial Health and Safety: Biological Indicators for the Assessment of Human Exposure to Industrial Chemicals*, ECSC-EEC-EAEC, Luxembourg, 1984 (EUR 8903 EN), 1986 (EUR 10704 EN), 1987 (EUR 11135 EN), 1988 (EUR 11478 EN), 1989 (EUR 12174 EN), 1994 (EUR 14815 EN).
26. K. Rasmussen, P. Lunde-Jensen, and O. Svane, "Biological Monitoring and Medical Screening at the Workplace in the EC Countries," *Int. Arch. Occup. Environ. Health* **63**, 347–352 (1991).
27. M. Ogata, "Goals and Activities of the Biological Monitoring Group in Japan," in V. Fiserova-Bergerova and M. Ogata, eds., Biological Monitoring of Exposure to Industrial Chemicals. ACGIH, Cincinnati, OH., 1990, pp. 3–7.
28. The Japan Society for Occupational Health "Recommendation of Occupational Exposure limits (1998–1999)," *J. Occup. Health*, **40**, 240–256 (1998).
29. M. Ogata, T. Numano, M. Hosokawa, and H. Michisuji, "Large-scale Monitoring in Japan," *Sci. Tot. Environ.* **199**, 197–204 (1997).
30. V. Fiserova-Bergerova, "Application of Toxicokinetic Models to Establish Biological Exposure Indicators," *Ann. Occup. Hyg.* **34**, 639–651 (1990).

31. J. Rosenberg, "Biological Monitoring Required by State Governments: Lead and Cholinesterase in California," in V. Fiserova-Bergerova and M. Ogata, eds. *Biological Monitoring of Exposure to Industrial Chemicals*, ACGIH, Cincinnati, OH, 1990, pp. 59–62.
32. R. F. Henderson, W. E. Bechtold, J. A. Bond, and J. D. Sun: "The Use of Biological Markers in Toxicology," *CRC Critical Reviews in Toxicology*, **20**, 65–82 (1989).
33. G. I. Ellman, "Tissue Sulphydryl Groups," *Arch. Biochem. Biophys.* **82**, 70–77 (1959).
34. M. Jakubowski, I. Linhart, G. Pielas, and J. Kopecky, "2-Cyanoethylmercapturic Acid (CEMA) in Urine as a Possible Indicator of Exposure to Acrylonitrile," *Br. J. Ind. Med.* **44**, 834–840 (1987).
35. B. M. de Rooij, P. J. Boogaard, J. N. M. Commandeur, N. J. van Sittert, and N. P. E. Vermeulen, "Allylmercapturic Acid as Urinary Biomarker of Human Exposure to Allyl Chloride," *Occup. Env. Med.* **54**, 653–661 (1997).
36. B. M. de Rooij, J. N. M. Commandeur, and N. P. E. Vermeulen, "Mercapturic Acids as Biomarkers of Exposure to Electrophilic Chemicals: Applications to Environmental and Industrial Chemicals," *Biomarkers*, **3**, 239–303 (1998).
37. R. T. H. van Welie, C. M. van Marrewijk, F. A. de Wolf, and N. P. E. Vermeulen, "Thioether Excretion in Urine of Applicators Exposed to 1,3-Dichloropropene: a Comparison with Urinary Mercapturic Acid Excretion," *Br. J. Ind. Med.* **48**, 492–498 (1991).
38. R. T. H. van Welie, P. van Duyn, D. H. Brouwer, J. J. van Hemmen, E. J. Brouver, and N. P. E. Vermeulen, "Inhalation Exposure to 1,3-Dichloropropene in the Dutch Flower-bulb Culture, Part II. Biological Monitoring by Measurement of Urinary Excretion of Two Mercapturic Acid Metabolites", *Arch. Environ. Contam. Toxicol.* **20**, 6–12 (1991).
39. B. M. de Rooij, P. J. Boogaard, J. N. M. Commandeur, and N. P. E. Vermeulen, "3-Chloro-2-hydroxypropylmercapturic Acid and α-Chlorohydrin as Biomarkers of Occupational Exposure to Epichlorohydrin," *Env. Tox. Pharm.* **3**, 175–178 (1997).
40. A. Norstrom, A. Low, L. Aringer, R. Samuelsson, B. Andersson, J. O. Levin, and P. Naslund, "Determination of *N*-acetyl-*S*-(2-phenyl-2-hydroxyethyl) cysteine in Human Urine after Experimental Exposure to Styrene," *Chemosphere* **24**, 1553–1561 (1992).
41. L. Aringer, A. Lof, and C. G. Elinder, "The Application of Measurement of Urinary Thioethers. A Study of Humans Exposed to Styrene during Diet Standardization," *Int. Arch. Occup. Environ. Health* **63**, 341–346 (1990).
42. F. Seutter-Berlage, H. L. van Dorp, H. G. J. Kosse, and P. Th. Henderson, "Urinary Mercapturic Acid Excretion as a Biological Parameter of Exposure to Alkylating Agents," *Inter. Arch. Occup. Env. Health* **39**, 45–52 (1977).
43. S. Burgaz, A. Bayhan, and A. E. Karakaya, "Thioether excretion of Workers Exposed to Bitumen Fumes," *Int. Arch. Occup. Health* **60**, 347–349 (1988).
44. H. Vaino, H. Savolainen, and I. Kilpikari, "Urinary Thioether of Employees of a Chemical Plant," *Br. J. Ind. Med.* **35**, 232–243 (1978).
45. I. Kilpikari, "Correlation of Urinary Thioethers with Chemical Exposure in a Rubber Plant," *Br. J. Ind. Med.* **38**, 98–100 (1981).
46. I. Kilpikari and H. Savolainen, "Increased Urinary Excretion of Thioethers in New Rubber Workers," *Br. J. Ind. Med.* **39**, 401–403 (1982).
47. R. van Doorn, C. M. Leijdekkers, R. P. Bos, R. M. E. Brouns, and P. Th. Henderson, "Enhanced Excretion of Thioethers in Urine of Operators of Chemical Waste Incinerators," *Br. J. Ind. Med.* **38**, 187–190 (1981).

48. N. P. E. Vermeulen, J. de Jong, E. J. C. van Bergen, and R. T. H. van Welie, "N-Acetyl-S-(2-hydroxyethyl)-L-cysteine as a Potential Tool in Biological Monitoring Studies? A Critical Evaluation of Possibilities and Limitation," *Arch. Tox.* **63**, 173–184 (1989).

49. R. T. H. van Welie, R. G. J. M. van Dijck, N. P. E. Vermeulen, and N. J. van Sitter, "Mercapturic Acids, Protein Adducts, and DNA Adducts as Biomarkers of Electrophilic Chemicals," *Crit. Rev. Tox.* **22**, 271–306 (1992).

50. H. Malonova and Z. Bardodej, "Urinary Excretion of Mercapturates as a Biological Indicator of Exposure to Electrophilic Agents," *J. Hyg. Epid. Microb. and Immunol.* **27**, 319–328 (1983).

51. P. Stommel, G. Muller, W. Stucker, C. Verkoyen, C. Schobel, and K. Norpoth, "Determination of S-Phenylmercapturic Acid in Urine—An Improvement in Biological Monitoring of Benzene," *Carcinogenesis* **10**, 279–282 (1989).

52. J. Mraz, "Gas Chromatographic Method for the Determination of N-Acetyl-S-(N-methylcarbamoyl)cysteine, a Metabolite of N,N-Dimethylformamide and N-Methylformamide, in Human Urine," *J. Chromatogr. Biomed. Appl.* **431**, 361–368 (1988).

53. M. J. Meyer and W. E. Bechtold, "Protein Adduct Biomarkers: State of the Art," *Environ. Health Perspect.* **104** (Suppl. 5), 879–882 (1996).

54. G. N. Wogan, "Markers of Exposure to Carcinogens," *Environ. Health Perspect.* **81**, 9–17 (1989).

55. M. C. Poirier and A. Weston, "Human DNA Adduct Measurements: State of Art," *Environ. Health Persp.* **104** (Suppl. 5), 883–893 (1996).

56. I. C. Hsu, M. C. Poirier, S. H. Yuspa, D. Grunberger, I. B. Winstein, R. H. Yolken, and C. C. Harris, "Measurement of Benzo(a)pyrene-DNA Adducts by Enzyme Immunoassays and Radioimmunoassay," *Cancer Res.* **41**, 1091–1095 (1981).

57. B. Schoket, D. H. Phillips, M. C. Poirier, and I. Vencze, "DNA Adducts in Peripheral Blood Lymphocytes from Aluminum Production Plant Workers Determined by ^{32}P-postlabeling and Enzyme-linked Immunosorbent Assay," *Environ. Health Perspect.* **99**, 307–309 (1993).

58. P. G. Shields, A. C. Povey, V. L. Wilson, A. Weston, C. C. Harris; "Combined High-Performance Liquid Chromatography ^{32}P-postlabeling Assay of N^7-methyldeoxyguanosine," *Cancer Res.* **50**, 6580–6584 (1990).

59. P. G. Shields, and C. C. Harris, S. Petruzzelli, E. D. Bowman, and A. Weston, "Standardization of the ^{32}P-postlabeling Assay for Polycyclic Aromatic Hydrocarbon DNA Adducts," *Mutagenesis* **8**, 121–126 (1993).

60. N. J. Gorelick, "Application of HPLC in the ^{32}P-postlabeling Assay," *Mutat. Res.* **288**, 5–18 (1993).

61. L. Moller, M. Zeisig, and P. Vodicka, "Optimization of an HPLC Method for Analyses of ^{32}P-postlabeled DNA Adducts," *Carcinogenesis* **14**, 1343–1348 (1993).

62. R. O. Rahn, S. S. Chang, J. M. Holland, and L. R. Shugart, "A Fluorometric-HPLC Assay for Quantitating the Binding of Benzo(a)-pyrene Metabolites to DNA," *Biochem. Biophys. Res. Commun.* **109**, 262–268 (1982).

63. K. Vahakangas, G. Trivers, M. Rowe, and C. C. Harris, "Benzo(a)pyrene Diolepoxide-DNA Adducts Detected by Synchronous Fluorescence Spectrophotometry," *Environ. Health Perspect.* **62**, 101–104 (1985).

64. E. Bailey, P. B. Farmer, and D. E. G. Shuker, "Estimation of Exposure to Alkylating Carcinogens by the GC-MS Determination of Adducts to Hemoglobin and Nucleic Acid Bases in Urine," *Arch. Toxicol.* **60**, 187–191 (1987).

65. W. G. Stillwell, H. X. Xu, J. A. Adkins, J. S. Wishnok, S. R. Tannenbaum, "Analysis of Methylated and Oxidized Purines in Urine by Capillary Gas Chromatography-Mass Spectrometry," *Chem. Res. Toxicol.* **2**, 94–99 (1989).
66. M. K. Shigenaga and B. N. Ames, "Assays for 8-Hydroxy-2'-deoxyguanosine, A Biomarker of in vivo Oxidative DNA Damage," *Free Radic. Biol. Med.* **10**, 211–216 (1991).
67. J. Angerer, C. Mannschreck, and J. Gundel, "Biological Monitoring and Biochemical Effect Monitoring of Exposure to Polycyclic Aromatic Hydrocarbons," *Int. Arch. Occup. Environ. Health* **70**, 365–377 (1997).
68. F. P. Perera, and D. L. Tang, J. P. O'Neill, W. L. Bigbee, R. J. Albertini, R. Santella, R. Ottman, W. Y. Tsai, C. Dickey, L. A. Mooney, K. Savela, and K. Hemminki, "HPRT and Glycophorin A Mutations in Foundry Workers: Relationship to PAH Exposure and to PAH-DNA Adducts," *Carcinogenesis* **14**, 969–973 (1993).
69. F. P. Perera, C. Dickey, R. Santella, J. P. O'Neill, R. J. Albertini, R. Ottman, W. Y. Tsai, L. A. Mooney, K. Savela, and K. Hemminki, "Carcinogen-DNA Adducts and Gene Mutation in Foundry Workers with Low-level Exposure to Polycyclic Aromatic Hydrocarbons," *Carcinogenesis* **15**, 2905–2910 (1994).
70. K. Hemminki, K. Randerath, M. V. Reddy, K. L. Putman, R. M. Santella, F. P. Perera, T. L. Young, D. H. Phillips, A. Hewer, and K. Savela, "Postlabeling and Immunoassay Analysis of Polycyclic Aromatic Hydrocarbons-adducts of Deoxyribonucleic Acid in White Blood Cells of Foundry Workers," *Scand. J. Work Environ. Health* **16**, 158–162 (1990).
71. R. M. Santella, K. Hemminki, D. L. Tang, M. Paik, R. Ottman, T. L. Young, K. Savela, L. Vodickova, C. Dickey, R. Whyatt, and F. P. Perera, "Polycyclic Aromatic Hydrocarbon-DNA Adducts in White Blood Cells and Urinary 1-Hydroxypyrene in Foundry Workers," *Cancer Epid. Biomark. Prev.* **2**, 59–62 (1993).
72. F. P. Perera, K. Hemminki, T. L. Yong, D. Brenner, G. Kelly, and R. M. Santella, "Detection of Polycyclic Aromatic Hydrocarbon-DNA Adducts in White Blood Cells of Foundry Workers," *Cancer. Res.* **48**, 2288–2291 (1988).
73. K. Hemminki, F. P. Perera, D. H. Phillips, K. Raderath, M. V. Reddy, and R. M. Santella, "Aromatic Deoxyribonucleic Acid Adducts in White Blood Cells of Foundry and Coke Oven Workers," *Scand. J. Work Environ. Health* **14** (Suppl. 1), 55–56 (1988).
74. D. Sherson, P. Sabro, T. Sigsgaard, F. Johansen, and H. Autrup: "Biological Monitoring of Foundry Workers Exposed to Polycyclic Aromatic Hydrocarbons," *Br. J. Ind. Med.* **47**, 448–453, (1990).
75. O. Omland, D. Sherson, A. M. Hansen, T. Sigsgaard, H. Autrup, E. Overgaard, "Exposure of Iron Foundry Workers to Polycyclic Aromatic Hydrocarbons: Benzo(a)pyrene-albumin Adducts and 1-Hydroxypyrene as biomarkers for Exposure," *Occup. Envin. Med.* **51**, 513–518 (1994).
76. D. H. Phillips, K. Hemminki, A. Alhonen, A. Hewer, and P. L. Grover, "Monitoring Occupational Exposure to Carcinogens: Detection by ^{32}P-postlabeling of Aromatic DNA Adducts in White Blood Cells from Iron Foundry Workers," *Mutat. Res.* **204**, 531–541 (1988).
77. M. V. Reddy, K. Hemminki, K. Randerath, "Postlabeling Analysis of Polycyclic Aromatic Hydrocarbon-DNA Adducts in White Blood Cells of Foundry Workers," *J. Toxicol. Environ. Health* **34**, 177–186 (1991).
78. K. Hemminki, E. Grzybowska, M. Chorazy, K. Twardowska Saucha, J. W. Sroczynski, K. L. Putman, K. Randerath, D. H. Phillips, and A. Hewer, "Aromatic DNA Adducts in White Blood Cells of Coke Workers," *Int. Arch. Occup. Environ. Health* **62**, 467–470 (1990).

79. C. C. Harris, K. Vahakangas, M. J. Newman, G. E. Trivers, A. Shamsuddin, N. Sinopoli, D. L. Mann, and W. E. Wright, "Detection of Benzo[a]pyrene Diol Epoxide-DNA Adducts in Peripheral Blood Lymphocytes and Antibodies to the Adducts in Serum from Coke Oven Workers," *Proc. Natl. Acad. Sci. USA* **82**, 6672–6676 (1985).
80. F. J. Van Schooten, F. J. Jongeneelen, M. J. Z. Hillebrand, F. E. Van Leeuwen, A. J. A. De Looff, A. P. G. Dijkmans, J. G. M. Van Rooij, L. Den Engelse, and E. Kriek, "Polycyclic Aromatic Hydrocarbon-DNA Adducts in White Blood Cell DNA and 1-Hydroxy-pyrene in the Urine from Aluminum Workers: Relation with Job Category and Synergistic Effect of Smoking," *Cancer. Epidemiol. Biomark. Prev.* **4**, 69–77 (1995).
81. S. Ovrebo, A. Haugen, K. Hemminki, K. Szyfter, P. A. Drablos, and M. S. Skogland, "Studies of Biomarkers in Aluminum Workers Occupationally Exposed to Polycyclic Aromatic Hydrocarbons," *Cancer Detect. Prev.* **19**, 258–267 (1995).
82. S. H. Liou, D. Jacobson-Kram, M. C. Poirier, D. Nguyen, P. T. Strickland, and M. S. Tockman, "Biological Monitoring of Fire Fighters: Sister Chromatid Exchange and Polycyclic Aromatic Hydrocarbon-DNA Adducts in Peripheral Blood Cells," *Cancer Res.* **49**, 4929–4935 (1989).
83. K. Hemminki, J. Soderling, P. Ericson, H. E. Norbeck, and D. Segerback, "DNA Adducts among Personel Servicing and Loading Diesel Vehicles," *Carcinogenesis* **15**, 767–769 (1994).
84. P. S. Nielsen and H. Autrup, "Diesel Exhaust-Related DNA Adducts in Garage Workers," *Clin. Chem.* **40**, 1456–1458 (1994).
85. P. S. Nielsen, A. Andreassen, P. B. Farmer, S. Ovrebo, and H. Autrup, "Biomonitoring of Diesel Exhaust-Exposed Workers. DNA and Hemoglobin Adducts and Urinary 1-Hydroxypyrene as Markers of Exposure," *Tox. Lett.* **86**, 27–37 (1996).
86. D. H. Phillips and P. B. Farmer, "Evidence for DNA and Protein Binding by Styrene and Styrene Oxide," *CRC Crit. Rev. Toxicol.* **24** (Suppl 1), S35–S46 (1994).
87. E. Horvath, K. Pongracz, S. M. Rappaport, and W. J. Bodell, "^{32}P-post-labeling Detection of DNA Adducts in Mononuclear Cells of Workers Occupationally Exposed to Styrene," *Carcinogenesis* **15**, 1309–1315 (1994).
88. S. M. Rappaport, K. Yeowell-O'Connel, W. Bodell, J. W. Yager, and E. Symanski, "An Investigation of Multiple Biomarkers among Workers Exposed to Styrene and Styrene-7,8-oxide," *Canc. Res.* **56**, 5410–5416 (1996).
89. P. Vodicka, L. Vodickova and K. Hemminki, "^{32}P-Postlabeling of DNA Adducts of Styrene-Exposed Lamination Workers," *Carcinogenesis* **14**, 2059–2061 (1993).
90. F. J. van Schooten, F. E. van Leeuwen, M. J. K. Hillebrand, M. E. de Rijke, A. A. M. Hart, H. G. van Veen, S. Oosterink, and E. Kriek: "Determination of benzo(a)pyrene Diol Epoxide-DNA Adducts in White Cell DNA from Coke-Oven Workers: The Impact of Smoking," *J. Natl. Cancer Inst.* **82**, 927–933 (1990).
91. S. Osterman-Golkar, D. Ehrenberg, D. Segerback, and I. Hallstrom, "Evaluation of Genetic Risks of Alkylating Agents. II. Haemoglobin as a Dose Monitor," *Mutat. Res.* **34**, 1–10 (1976).
92. F. J. Jongeneelen, "Methods for Routine Biological monitoring of Carcinogenic PAH-Mixtures," *Sci. Tot. Environ.* **199**, 141–149 (1997).
93. J. Mraz, E. Galova, H. Nohova, and S. Bouskova, "N-Methylcarbamoyl Adduct with N-terminal Valine of Globin: A New Biomarker of Exposure to N,N-Dimethylformamide (DMF) in Humans (Abstract)," *Proceedings of the 4th International Symposium on Biological Monitoring in Occupational and Environmental Health*, Seoul, Sept. 1998.
94. M. Dell'Omo and R. R. Lauwerys, "Adducts to Macromolecules in the Biological Monitoring of Workers Exposed to Polycyclic Aromatic Hydrocarbons," *Crit. Rev. Toxicol.* **23**, 111–126 (1993).

95. C. P. Wild, Y. Z. Jiang, G. Sabbioni, B. Chapot, and R. Montesano: "Evaluation of Methods for Quantitation of Aflatoxin-Albumin Adducts and their Application to Human Exposure Assessment," *Cancer Res.* **50**, 245–251 (1990).

96. M. J. Wraith, W. P. Watson, C. V. Eadsforth, N. J. Van Sittert, M. Tornquist, and A. S. Wright, "An Immunoassay for Monitoring Human Exposure to Ethylene Oxide," *IARC Sci. Publ.* **89**, 271–274 (1988).

97. M. Riffelmann, G. Muller, W. Schmieding, W. Popp, and K. Norpoth: "Biomonitoring of Urinary Aromatic Amines and Arylamine Hemoglobin Adducts in Exposed Workers and Non-exposed Control Persons," *Int. Arch. Occup. Environ. Health* **68**, 36–43 (1995).

98. B. W. Day, S. Naylor, L. S. Gan, Y. Sahali, T. T. Nguyen, P. L. Skipper, J. S. Wishnok, and S. R. Tannenbaum, "Molecular Dosimetry of Polycyclic Aromatic Hydrocarbon Epoxides and Diol Epoxides via Hemoglobin Adducts," *Cancer Res.* **50**, 4611–4618 (1990).

99. M. Ferreira Jr., S. Tas, M. Dell'Omo, G. Goormans, J. P. Buchet, and R. Lauwerys, "Determinants of Benzo(a)pyrenediol Epoxide Adducts to Haemoglobin in Workers Exposed to Polycyclic Aromatic Hydrocarbons," *Occup. Environ. Med.* **51**, 451–455 (1994).

100. S. M. Rappaport, D. Ting, Z. Jin, K. Yeowell-O'Connell, S. Waidyanatha, and T. McDonald, "Application of Raney Nickel to Measure Adducts of Styrene Oxide with Hemoglobin and Albumin," *Chem. Res. Toxicol.* **6**, 238–244 (1993).

101. K. Yeowell-O'Connell, W. Pauwels, M. Severi, Z. Jin, M. R. Walker, S. M. Rappaport, and H. Veulemans, "Comparison of Styrene-7,8-oxide Adducts formed via Reaction with Cysteine, N-Terminal Valine and Carboxylic Acid Residues in Human, Mouse and Rat Hemoglobin," *Chem. Biol. Interact.* **106**, 67–85 (1997).

102. H. G. Neumann, G. Birner, P. Kowallik, D. Schutze, and I. Zwirner-Baier, "Hemoglobin Adducts of N-Substituted Aryl Compounds in Exposure Control and Risk Assessment," *Environ. Health Perspect.* **99**, 65–69 (1993).

103. G. Sabbioni, "Hemoglobin Binding of Arylamines and Nitroarenes: Molecular Dosimetry and Quantitative Structure-Activity Relationships," *Environ. Health Perspect.* **102** (Suppl. 6), 61–67 (1994).

104. G. Sabbioni and H. G. Neumann, "Biomonitoring of Arylamines, Hemoglobin Adducts of Urea and Carbamate Pesticides," *Carcinogenesis* **11**, 111–115 (1990).

105. D. Schutze, O. Sepai, J. Lewalter, L. Miksche, D. Henschler, and G. Sabbioni, "Biomonitoring of Workers Exposed to 4,4'-Methylene-dianiline or 4,4'-Methylenediphenyl Diisocyanate," *Carcinogenesis* **16**, 573–582 (1995).

106. P. Lind, M. Dalene, G. Skarping, and L. Hagmar, "Toxicokinetics of 2,4- and 2,6-Toluenediamine in Hydrolysed Urine and Plasma after Occupational Exposure to 2,4- and 2,6-Toluene Diisocyanate," *Occup. Environ. Med.* **53**, 94–99 (1996).

107. P. Lind, M. Dalene, H. Tinnerberg, and G. Skarping, "Biomarkers in Hydrolysed Urine, Plasma and Erythrocytes among Workers Exposed to Thermal Degradation Products from Toluene Diisocyanate Foam," *Analyst* **122**, 51–56 (1997).

108. M. Tornqvist, J. Mowrer, S. Jensen, and L. Ehrenberg, "Monitoring of Environmental Cancer Initiators through Hemoglobin Adducts by a Modified Edman Degradation Method," *Anal. Biochem.* **154**, 255–266 (1986).

109. P. K. Ramachandran, B. R. Gandhe, K. S. Venkateswaran, M. P. Kaushik, R. Vijayaraghavan, G. S. Agarwal, N. Gopalan, M. V. S. Suryanarayana, S. K. Shinde, and S. Sriramachari, "Gas Chromatographic Studies of the Carbamoylation of Haemoglobin by Methyl Isocyanate in Rats and Rabbits," *J. Chromatogr. Biomed. Appl.* **426**, 239–247 (1988).

110. J. Mraz, E. Galova, and M. Hornychova, "Biological Monitoring of 2,4-Toluenediisocyanate (2,4-TDI). Formation and Persistence of 4-(2-amino)tolyl Adduct with N-terminal Valine of Hemoglobin in vivo in Rats (Abstract)," *Proceedings of International Symposium on Biological Monitoring in Occupational and Environmental Health*, Espoo, Sept. 1996.

111. J. Lewalter, "N-Alkylvaline Levels in Globin as a New Type of Biomarker in Risk Assessment of Alkylating Agents," *Int. Arch. Occup. Environ. Health* **68**, 519–530 (1996).

112. A. Kautiainen and M. Tornqvist, "Monitoring Exposure to Simple Epoxides and Alkenes through Gas Chromatographic Determination of Haemoglobin Adducts," *Int. Arch. Occup. Environ. Health* **63**, 27–31 (1991).

113. S. Fustinoni, C. Colosio, A. Colombi, L. Lastrucci, K. Yeowell-O'Connell, and S. M. Rappaport, "Albumin and Hemoglobin Adducts as Biomarkers of Exposure to Styrene in Fiberglass-Reinforced-Plastics Workers," *Int. Arch. Occup. Environ. Health* **71**, 35–41 (1998).

114. E. Bergmark, C. J. Calleman, F. He, and L. G. Costa, "Determination of Hemoglobin Adducts in Humans Occupationally Exposed to Acrylamide," *Toxicol. Appl. Pharmacol.* **120**, 45–54 (1993).

115. R. M. Black, R. J. Clarke, J. M. Harrison, and R. W. Read, "Biological Fate of Sulphur Mustard: Identification of Valine and Histidine Adducts in Haemoglobin from Casualties of Sulphur Mustard Poisoning," *Xenobiotica* **27**, 499–512 (1997).

116. P. B. Farmer, E. Bailey, S. M. Gorf, M. Tornqvist, S. Osterman-Golkar, A. Kautiainen, and D. P. Lewis-Enright, "Monitoring Human Exposure to Ethylene Oxide by the Determination of Haemoglobin Adducts Using Gas Chromatography-Mass Spectrometry," *Carcinogenesis* **7**, 637–640 (1986).

117. F. Sarto, M. A. Tornqvist, R. Tomanin, G. B. Bartolucci, S. M. Osterman-Golkar, and L. Ehrenberg, "Studies of Biological and Chemical Monitoring of Low-Level Exposure to Ethylene Oxide," *Scand. J. Work Environ. Health* **17**, 60–64 (1991).

118. J. Lewalter and U. Korallus, "Blood Protein Conjugates and Acetylation of Aromatic Amines: New Findings on Biological Monitoring," *Int. Arch. Occup. Environ. Health* **56**, 179–196 (1985).

119. E. Bailey, A. G. Brooks, I. Bird, P. Farmer, and B. Street: "Monitoring Exposure to 4,4'-Methylenediamiline by the Gas Chromatography-Mass Spectrometry Determination of Adducts to Hemoglobin," *Anal. Biochem.* **190**, 175–181 (1990).

120. G. Sabbioni, J. Wei, and Y. Y. Liu, "Determination of Hemoglobin Adducts in Workers Exposed to 2,4,6-Trinitrotoluene," *J. Chromatogr. Biomed. Appl.* **682**, 243–248 (1996).

121. J. Angerer, T. Goen, A. Kramer, and H. U. Kafferlein, "N-Methyl-Carbamoyl Adducts at the N-Terminal Valine of Globin in Workers Exposed to N,N-Dimethylformamide," *Arch. Toxicol.* **72**, 309–313 (1998).

122. R. Pastorelli, J. Restano, M. Guanci, M. Maramonte, C. Magagnotti, R. Allevi, D. Lauri, R. Fanelli, and L. Airoidi, "Hemoglobin Adducts of Benzo[a]pyrene Diolepoxide in Newspaper Vendors: Association with Traffic Exhaust," *Carcinogenesis* **17**, 2389–2394 (1996).

123. Ref. 2, pp. 159–162.

124. L. W. Miksche and J. Levalter, "Biological Monitoring of Exposure to Hexavalent Chromium in Isolated Erythrocytes," in M. L. Mendelsohn, J. P. Peeters, and M. J. Normandy, eds. *Biomarkers and Occupational Health, Progress and Perspective*, Joseph Henry Press, Washington DC, 1995, pp. 313–323.

125. B. L. Carson, H. V. Ellis III, and J. L. McCann, *"Toxicology and Biological Monitoring of Metals in Humans," Including Feasibility and Need*, Lewis Publishers Inc., Chelsa, MI, 1987.

126. T. W. Clarkson, L. Friberg, G. F. Nordberg, and P. R. Sager, eds. *Biological Monitoring of Toxic Metals*. Plenum Press, 1988.
127. "Proceedings of an International Workshop at Stockholm. Factors Influencing Metabolism and Toxicity of Metals: A Consensus Report." in G. Nordberg, B. A. Fowler, L. Friberg, A. Jernelov, N. Nelson, M. Piscator, H. H. Sandstead, J. Vostal, and V. B. Vouk, eds., *Environ. Health Perspec.* **25**, 1–171 (1978).
128. R. Tardif, R. Goyal, and J. Brodeur, "Assessment of Occupational Health Risk from Multiple Exposure: Review of Industrial Solvent Interaction and Implication for Biological Monitoring," *Toxicol. Ind. Health* **8**, 37–52 (1992).
129. Ref. 2, pp. 63–82.
130. P. J. A. Borm and B. de Barbanson, "Bias in Biological Monitoring Caused by Concomitant Medication," *J. Occup. Med.* **30**, 214–223 (1988).
131. C. Minoia, R. Pietra, E. Sabbioni, A. Ronchi, A. Gatti, A. Cavalleri, and L. Manzo, "Trace Element Reference Values in Tissues from Inhabitants of the European Community. III. The Control of Preanalytical Factors in the Biomonitoring of Trace Elements in Biological Fluids," *Sci. Total Environ.* **120**, 63–79 (1992).

GENERAL REFERENCES

Documentation of the Biological Exposure Indices, ACGIH publication #0301, 317 pages. The sixth edition (1993) was supplemented in 1996, 1997, 1998, and 1999.
 Explains the rationale for BEIs for 38 industrial chemicals (see Table 42.2) and provides guidelines for sampling of specimens and interpreting of data of biological monitoring. The documentations are updated when new data are available. The same documentation is included in the 3rd volume of Documentations of TLVs and BEIs (see Ref. 10).

Topics In Biological Monitoring: A Compendium of Essays by Members of the ACGIH Biological Exposure Indices Committee, American Conference of Governmental Industrial Hygienists, Inc., Cincinnati, OH, 1995, 99 pages.
 Nine essays, prepared by members of the BEI Committee, deal in more detail with the topics discussed in this chapter.

L. Alessio, A. Berlin, R. Roi, and M. Boni, ed., *Human Biological Monitoring of Industrial Chemicals Series*, the Commission of the European Communities, Luxembourg, 1983 (EUR 8476 EN), 175 pages.

L. Alessio, A. Berlin, M. Boni, and R. Roi, eds., *Biological Indicators for the Assessment of Human Exposure to Industrial Chemicals*, the Commission of the European Communities, Luxembourg, 6 volumes, respectively published in 1984 (EUR 8903 EN), 1986 (EUR 10704 EN), 1987 (EUR 11135 EN), 1988 (EUR 11478 EN), 1989 (EUR 12174 EN), and 1994 (EUR 14815 EN), each volume 80 to 110 pages.
 This series discusses toxicological data and health risks pertinent to biological monitoring of individual air pollutants (see Table 42.5).

R. R. Lauwerys and P. Hoet. *Industrial Chemical Exposure: Guidelines for Biological Monitoring*, 2nd ed., Lewis Publishers, CRC Press, Boca Raton, FL, 1993, 336 pages.
 General principles, and advantages and disadvantages of biological monitoring, are followed by description of the relation of biological level to external exposure and to biological effects, for 16 metals and a wide range of organic compounds.

S. Que Hee, ed., *Biological Monitoring: An Introduction*, ITP Van Nostrand Reinhold, a division of International Thomson Publishing, New York, 1993, 650 pages.

Written as a textbook for graduate courses, the book describes interdisciplinary factors, processes, and methodologies pertinent to biological monitoring of industrial chemicals and monitoring of microbiological elements in the workplace and in the environment. It deals specifically with biomarkers for 26 chemicals for which BEIs are available. Reference ranges of biological levels, based on international data, are listed. Laboratory testing of biomarkers of exposure as well as for tests detection of adverse effects are discussed.

T. W. Clarkson, L. Friberg, G. F. Nordberg, and P. R. Sager, eds., *Biological Monitoring of Toxic Metals*, Plenum Press, New York, 1988, 686 pages.

Lectures of invited speakers to the International Conference on Biological Monitoring of Metals (Rochester, NY, 1986) include: an overview of biological monitoring of toxic metals; their occurrence in the environment and in the workplace; their uptake, distribution, and excretion; and technicalities such as the feasibility and implementation of biological monitoring, analysis of biological specimens, and quality control. For 13 metals, specific information is provided on toxicity, uptake and excretion, and a variety of factors affecting their concentrations in body fluids. The information is well documented by ample references.

A. Aitio, V. Riihimaki, and H. Vainio, eds., *Biological Monitoring and Surveillance of Workers Exposed to Chemicals*, Hemisphere Publishing Corporation, 1984, 403 pages.

Lectures from International Course on "Biological Monitoring of Exposure to Industrial Chemicals" (Espoo, Finland, 1980) provide specific information on monitoring of six metals (Pb, Hg, Cd, Cr, Ni, As), seven aromatic and halogenated solvents, pesticides, and some additional compounds; deal with theoretical as well as practical issues, such as laboratory quality control and biological monitoring strategy; and application of biological monitoring to health surveillance and to the environment.

V. Foa, E. A. Emmett, M. Maroni, and A. Colombi, eds., *Occupational and Environmental Chemical Hazards: Cellular and Biochemical Indices for Monitoring Toxicity*, Ellis Horwood Limited, Chichester, England, 1987, 558 pages.

Proceedings of the International Symposium on Biochemical and Cellular Indices of Human Toxicity in Occupational and Environmental Medicine. (Milan, Italy, 1986) deal with general concepts of biological monitoring, emphasizing medical surveillance and specific biochemical and cellular indices related to hepatotoxicity, renal toxicity, genotoxicity, and neurotoxicity.

M. H. Ho and H. K. Dillon, eds., *Biological Monitoring of Exposure to Chemicals: Organic Compounds*, John Wiley & Sons, Inc., New York, 1987, 352 pages.

In the *Proceedings of the symposium on Biological Monitoring of Exposure to Organic Compounds* (St. Louis, Missouri, 1984), well known scientists discuss the state of art of biological monitoring: general issues, as well as specific biomarkers of organic solvents and pesticides, are discussed.

H. K. Dillon and M. H. Ho, eds., *Biological Monitoring of Exposure to Chemicals: Metals*, John Wiley & Sons, Inc., New York, 1991, 280 pages.

This book is a sequel to the proceedings cited directly above. It reports on assessment of chronic environmental and occupational exposure to four metals (Hg, Ni, Pb, Se) and on aluminum poisoning. In seven review chapters, topics such as monitoring of metals in hair and distribution of metals in the environment (soil, insects, other nonvertebrates) are discussed.

V. Fiserova-Bergerova and M. Ogata, eds., *Biological Monitoring of Exposure to Industrial Chemicals*, American Conference of Governmental Industrial Hygienists, Inc., Cincinnati, Ohio, 1990, 212 pages.

Proceedings from the United States—Japan Cooperative Seminar on Biological Monitoring of Exposure to Industrial Chemicals (Hawaii, 1989) provide insight into concepts and implementation of biological monitoring on three continents. Scientific grounds, as well as examples of field studies, are presented.

M. L. Mendelsohn, J. P. Peeters, and M. J. Normandy, eds., *Biomarkers and Occupational Health: Progress and Perspective.* Published by Joseph Henry Press, Washington DC, 1995, 335 pages.

Proceedings from International Workshop on Development and Applications of Biomarkers (Santa Fe, New Mexico, 1994) address recent developments in biomarkers related to adverse effect, such as macromolecular adducts. Also deal with general issues such as legal and ethical aspects, cost, and implementation.

B. L. Carson, H. V. Ellis III, and J. L. McCann, *Toxicology and Biological Monitoring of Metals in Humans: Including Feasibility and Need*, Lewis Publishers, Inc., Chelsea, MI, 1986, 328 pages.

Summarizes toxicological and toxicokinetic data on 52 metals and metaloids in humans and in other mammals. Can be helpful in interpreting data obtained by biological monitoring of metals.

C. Baselt, *Biological Monitoring Methods for Industrial Chemicals*, 2nd ed., PSG Publishing Company, Inc., Littleton, MA, 1988, 331 pages.

Discusses biological monitoring of about 100 industrial chemicals. For each chemical, a short summary on usage, metabolism, excretion, and toxicity is followed by description of classical methods for determination of the biomarker in biological specimens.

J. Angerer and K. H. Schaller, eds., *Analysis of Hazardous Substances in Biological Materials. Methods for Biological Monitoring*, Vols. 1–5, VCH Publisher, Deerfield Beach FL, 1985–1997.

This series provides basic information on toxic levels of biomarkers. The valuable contribution is description of analytical methods chosen and tested by members of the German Commission for the Investigation of Health Hazards of Chemical Compounds in the Workplace.

A. Aitio, ed., *Proceedings of the International Symposium on Biomonitoring in Occupational and Environmental Health, Sci. Tot. Environ.* **199** 1–2 (1997) 226 pages.

Selective papers presented at the International Symposium held in Espoo, Finland, in 1996, describe examples of more recent monitoring of exposure to organic and inorganic pollutants in industry and environment, mainly in Scandinavian countries, Finland, Italy, the United Kingdom, Germany, the Netherlands, China, and Japan.

CHAPTER FORTY-THREE

Cost-Effectiveness of a Multidisciplinary Approach to Loss Control and Prevention Programs

Judith F. Stockman, RN, C-ANP, COHN-S
and Ted C. Johnson, MSPH, CIH, CSP

1 INTRODUCTION

Internal occupational safety and health programs, managed by professionals in the field and employed by business, are in jeopardy. With corporate staff downsizing due to the competitiveness of the global market, occupational safety and health professionals must demonstrate cost-effectiveness or face the real possibility of elimination of their positions. It is no longer sufficient for such professionals to be retained solely to ensure regulatory compliance or as a value-added service.

Currently, the United States spends more than $1.3 trillion annually on health care. A recent study by the Health Care Financing Administration warns that this figure may nearly double to $2.1 trillion by 2007 (1). Because these expenses have a significant impact on the corporate balance sheet, management has begun to search for ways to contain and even reduce the spiraling growth of these expenditures. It is possible to limit expenses related to employee health and safety; however, it requires a new, coordinated, and multidisciplinary approach to loss prevention and control.

A truly effective loss control program must integrate the occupational health and safety team with that of risk management. The occupational health clinician, industrial hygienist,

---

Patty's Industrial Hygiene, Fifth Edition, Volume 3. Edited by Robert L. Harris.
ISBN 0-471-29753-4   © 2000 John Wiley & Sons, Inc.

and safety professional must coordinate their efforts with the risk manager and workers' compensation claims professional. As a team, goals must be established. With senior management's input and support, policies must be created and programs designed and implemented. Ultimately, to monitor program effectiveness, accurate data must be collected and organized so that reductions in costs related to health and safety are documented.

2 IMPACT OF HEALTH CARE AND RELATED COSTS

Members of the corporate occupational health and safety team must become increasingly involved in the economic viability of the company, not only to protect worker health and safety but also to preserve and enhance their careers. However, to accomplish this they must expand their roles dramatically from what was once perceived as a technical staff position to that of a cost-effective member of management. Historically, the perception of loss prevention and control has been limited to the prevention of industrial accidents and related costs; no longer is this limited approach viable.

Today and in the decades to come, skyrocketing costs of group medical benefits, expenses associated with tort litigation, impact of criminal litigation, and other expenses borne by industry add a completely new dimension to loss prevention and control. These costs place such a severe economic burden on industry that unless a proactive cost-containment program is initiated, health-related costs and, ultimately, the ability of corporations to compete will be adversely affected.

2.1 Direct Workers' Compensation Costs

Workplace injuries and illnesses continue to decrease while workers' compensation costs continue to escalate dramatically. Nationally, the number of workplace injuries (incidence rate) has decreased to an all-time low of 7.1 as of 1997 (number of injuries per 100 workers) (2).

Apparently, 87% of all workers were covered by the workers' compensation in 1993, with benefits varying from state to state (3). Conversely, workers' compensation program costs were $57.3 billion as of 1993, up from $25.1 billion in 1984 (4). Costs associated with workers' compensation continue to rise dramatically, even though workplace fatalities, illnesses, and injuries are at an all-time low.

The discrepancy between Occupational Safety and Health Administration (OSHA) injury and illness rates and workers' compensation costs is even more dramatic in California. The California OSHA incident rate in 1996 was 3.1, the lowest in the nation compared to the nationwide rate of 7.1 for the same period (5, 6). On the other hand, workers' compensation costs in California continue to be the highest in the nation, totaling $47,737,985 in 1994 (7). There are two major reasons for this discrepancy: (*1*) a higher litigation rate, and (*2*) increased medical costs in California, comparatively. For example, California employers paid $2.2 billion in litigation expenses in 1993, $700 more than in 1990. In California, litigation cost increases have averaged more than 50% every 2 years (8).

2.2 Indirect Workers' Compensation Costs

In addition to the direct (insured) cost of workers' compensation claims, indirect costs significantly escalate employers' injury expenditures. Available data indicates that for every dollar paid in workers' compensation, there is another dollar in indirect (uninsured) costs for which the employer is responsible. This figure is highly conservative, though indirect costs vary dramatically from company to company. Published estimates of indirect costs range from 1:1 to 5:1, illustrating the necessity of developing in-house data (9). On a nationwide basis, as of 1997, workers' compensation total direct costs ($65 billion) plus indirect costs ($106 billion) were estimated to be $171 billion (10).

The ratio of direct to indirect workers' compensation costs should reflect the specific experience of the individual company. Even ratios developed from data within the same general industry serve little useful purpose. Each business must develop its own data relative to the indirect costs based on the following:

1. Cost of productivity paid for time lost by workers witnessing an accident
2. Uninsured medical costs borne by the company
3. Cost of wages for time lost by the worker, other than workers' compensation benefits
4. Extra cost of overtime necessitated by the accident
5. Cost of wages paid to supervisors for time spent on activities relative to the accident
6. Costs associated with decreased productivity by an injured worker after returning to work
7. Costs inherent in training a replacement employee
8. Cost of damages to material and equipment
9. Cost of time spent by management and clerical personnel in processing forms and communicating with the employee, insurance company, health care providers, etc.
10. Other miscellaneous costs

2.3 Group Health and Related Costs

The cost of health care is rising at a rate that is as alarming as the increase in the workers' compensation system. As the costs escalate, the employer is paying an ever-increasing share of the bill. In 1986 it is estimated that U.S. firms paid $109 billion for their employees' health care. Additionally, $56.3 billion is estimated to have been paid in taxes that helped finance the nation's health care system (7). The occupational health team must become educated and in turn educate management to play an active role in reducing expenditures in this area, thereby increasing corporate profitability.

3 FACTORS PRECIPITATING INCREASED HEALTH CARE AND RELATED COSTS

The major reasons underlying the tremendous increases employers face relative to worker health care costs include the following:

1. Increased cost of the medical benefit system
2. Variability of workplace and environmental hazards
3. The complexity of workplace and environmental hazards
4. Unreliability of statistical data
5. Cost of a litigious society

3.1 Increased Cost of the Medical Benefit System

Increasing costs of the health care system have far exceeded the inflation rate. As of 1995, private employers contributed $206.44 billion in health care costs, the federal government contributed $239.8 billion, and state and local government contributed more than $241.3 billion, totaling $687.5 billion, or approximately 53% of the nation's health care expenditures (1, 11).

The aging worker population in America plays a significant role in the increased cost of medical benefits. As longevity increases and medical technology improves, the costs of complex treatment regimes such as organ transplants and kidney dialysis place a tremendous burden on the medical benefit system, which is funded largely by business. The cost of an aging worker population is also reflected in workers' compensation costs. Primarily, these costs are associated with older employees who develop cumulative trauma diseases that can be aggravated by and a result of the aging process. Additionally, the cost of providing medical benefits to retirees will continue to escalate as the baby boom generation ages.

Conversely, infectious diseases affecting younger workers are having a financial impact on business. Diseases such as acquired immune deficiency syndrome (AIDS), hepatitis, and tuberculosis, among others, require treatment regimes that are expensive and protracted. Some of these illnesses have no known cure, and the number of victims who will ultimately succumb can only be estimated; thus the financial impact cannot be projected. There are also complex legal issues regarding AIDS that have yet to be resolved, and these legal issues affect the cost that this disease represents to corporations (8).

As national boundaries become increasingly fluid, diseases that have not represented a significant cost in the United States are being imported as a result of immigration and increased importation of goods from other nations. As newly arrived immigrant workers become part of the labor force, American business is having to assume the financial responsibility for the diagnosis and treatment of these workers as well as the cost of prevention and medical surveillance for existing workers.

Owing to the rapidly escalating and unchecked rise in health care costs, the business community has attempted to contain costs utilizing a number of strategies, many of which have proven largely ineffective. The use of health maintenance organizations (HMOs) and preferred provider organizations has met with varying degrees of success. Unfortunately, the utilization of such plans has often resulted in price shifting. One example is that many HMOs, with prepaid contracts to provide medical benefits, report medical conditions that are nonindustrial as work-related in order to be reimbursed via the workers' compensation system. In essence, they are being paid twice.

Other obstacles to containing medical costs include the use of inappropriate and expensive diagnostic tests, unnecessary surgical intervention, and, often, treatment of dubious value. Additionally, health care practitioners frequently have financial agreements with, or outright ownership of, diagnostic treatment centers, hospitals, physical therapy centers, or other treatment modalities to which they refer. Obviously, such arrangements foster the potential for conflicts of interest.

3.2 Variability of Workplace and Environmental Hazards

The potential hazards encountered within the work environment are as diverse as the products being produced. The acute traumas associated with mechanical hazards have long been recognized. Unguarded belts, pulleys, gears, point-of-operation areas, and nip-points have accounted for untold injuries through the years. Similarly, exposures to a variety of chemicals and physical agents, including coal dust, crystalline silica, lead, noise, and others, have long been recognized as environmental hazards. Yet the technological improvements that enable us to identify increasingly smaller quantities of airborne contaminants, and the scientific advancements that enhance our capacity to detect subtle biological changes have greatly expanded the number of issues confronting the loss control team.

Rarely are exposures related to pure chemical or physical agents, rather; they generally involve a mixture of chemicals or multiple agents. Similarly, physical concerns seldom involve only noise and issues such as vibration, heat, improper lighting, or radiation from an adjacent operation must also be addressed. A large majority of employees work among a community of other workers; therefore, exposures also reflect adjacent activities. The shear variation in these operations makes the assessment and mitigation process very difficult. Lastly, the interface of humans and machines must also be viewed in light of both operational requirements and potential health concerns. Cumulative trauma disorders have become a serious component in the loss statistic, but rarely is an easy or financially viable solution readily apparent. Addressing these highly varied perimeters becomes a formidable task.

3.3 Complexity of Workplace and Environmental Hazards

Along with these technological and scientific advances, the sociopolitical response to worker health and safety issues has evolved. The acceptable risk level for harm to either the worker or the environment has steadily decreased over the years. Many occupational exposure standards have been decreasing, repetitive/cumulative trauma disorders have assumed greater importance, casual exposure of nonindustrial workers to chemicals and physical agents is increasingly problematic, and indoor air quality and fugitive emissions are but a few of the major concerns facing the employer.

Not only has science and technology revealed potential adverse health effects at exposures far lower than historically anticipated, but the industrial worker's and the community's perception of risk has also intensified. Countless newspaper, television, and magazine features have sensitized both workers and the general public to the potential hazards lurking within the industrial arena. The loss control and prevention team must contend with these highly complex and potentially volatile issues. No longer is it sufficient for the industrial hygienist to simply draw air through a sampling device solely to document compliance. The hygienist must be able and willing to discuss the results with those being monitored or affected. The hygienist must also recognize that his or her personal credibility may well come into question because the party paying for monitoring services will most likely be the employer.

More importantly, the health and safety team must make recommendations for change based on the monitoring data to ensure a healthful and safe work environment. To be effective, this must be accomplished through a joint effort between the health and safety team and appropriate members of the facilities and operations. Utilizing this team approach, the necessary changes can be accomplished in the least disruptive and most cost-effective manner.

3.4 Unreliability of Statistical Data

An area that affects employer health care cost but has gone largely unrecognized is the use of flawed workers' compensation data. Workers' compensation statistics are routinely used in research to substantiate the work-relatedness of occupational injury and illness. Unfortunately, much of these data are greatly distorted, and little or no effort is expended to identify the facts. The factors that may influence the validity of the statistical data include fraudulent claims, claims denied by the employer, and companion cases that have been included after a case has been adjudicated.

The impact on business of using flawed workers' compensation statistics can be extreme. Such data is often used to justify the passage of occupational health–related legislation and the formation of governmental standards. Compliance with these legally-mandated requirements is always costly. Equally damaging is that limited resources are sometimes diverted to support unwarranted programs or activities.

Because occupational health professionals are the major contributors to research related to health and safety issues, they have a particular responsibility to ensure that the supportive data is both accurate and valid so appropriate conclusions can be derived. Simply because workers' compensation data is published by a governmental agency or insurance carrier does not ensure its validity. It may be accurate in terms of the number of accepted workers' compensation claims, but not all accepted claims are legitimate.

It is incumbent on the researcher to review the data to ensure its validity. Unquestionably, large workers' compensation databases reflect a percentage of fraudulent claims, medical cost shifting from group health programs, malingering employees, poor claims management, and numerous other human factors that skew the data.

3.5 Cost of Litigious Society

The increasingly litigious nature of society in the United States has greatly contributed to the ever-escalating cost of health care. In no area are increased litigation costs more evident than in workers' compensation. In California, since 1988, when a workers' compensation claim becomes litigated, the cost of the claim increases approximately 52% (2). Many of these costs are related to medical and/or legal examinations, the duplication of diagnostic examinations, and unnecessary and often inappropriate medical treatment (9). For example, in a case involving alleged injury to several body parts, such as the back, neck, and pulmonary, the costs of forensic medical examinations frequently exceed $10,000 if the case is litigated. This does not include any medical treatment or other benefits.

4 STRUCTURE AND MANAGEMENT OF A COST-EFFECTIVE LOSS PREVENTION AND CONTROL PROGRAM

4.1 Program Design

To design, implement, and audit a cost-effective, integrated loss prevention and control program, it is imperative that all members of the occupational health team and the workers' compensation manager report to the same department head. Ideally, the manager should be a member of the occupational health profession, whether an industrial hygienist, an occupational health clinician, a safety professional, or a professional in a related field.

Numerous benefits are derived from having an occupational health professional manage the loss prevention and control program. First, the territorialization that can, and often does occur between occupational health and workers' compensation claims personnel is minimized. Second, by being part of a holistic approach to loss prevention and control management, the team learns more about one another's roles and develops greater appreciation and respect, thereby encouraging communication rather than conflict. Third, and most important, in areas where there is a great deal of overlap, the roles can be delineated. Only by utilizing such a team approach can a loss prevention program be presented to corporate management as part of the overall business plan that is able to document cost-effectiveness and promote employee health.

4.2 Program Implementation

Essential to the implementation of any program by a team of professionals from varying specialties is the need to identify specific areas of responsibility for each professional, and areas where overlap may occur. Within the field of occupational health and safety, the various elements of the loss control and prevention program should be addressed in a written policy detailing responsibilities and implementation procedures. These policies detail the special need for each program, such as hearing conservation, medical surveillance, and respirator programs, as well as the roles of the individual participants.

An effective communication mechanism between team members and participants must also be developed to achieve stated objectives. This may be in the form of weekly staff meetings, E-mails, progress notes, training programs, safety committee meetings, or a combination of these. However, it essential that the form of communication be both verbal and written to prevent possible misinterpretation of information. Audits, reviews of loss control data, or other monitoring procedures must also be established to ensure that the various program objectives are achieved and the program participants are held accountable.

The following discussion outlines the roles and functions of the various health and safety disciplines in an integrated, cost-effective loss prevention and control program. The development and implementation of a hearing conservation program will be discussed at the conclusion of this section. This will document how utilization of a comprehensive team approach will virtually eliminate noise-induced hearing loss (NIHL) and its associated costs. Because NIHL continues to be one of the top 10 leading occupational diseases (11), it will be particularly meaningful to this discussion.

4.2.1 Role of the Industrial Hygienist

Traditionally, the efforts of the industrial hygienist have been aimed at the prevention of work-related diseases through the recognition, evaluation, and control of health hazards. The development of safety and health policies and procedures, employee training, environmental monitoring programs, and the maintenance of health hazard evaluation data by the industrial hygienist have been instrumental in improving the quality of the work environment. However, the economic picture has changed dramatically during the past 10 years, so these roles, by necessity, must change. It is incumbent on the industrial hygienist to understand the financial impact of these programs and the impact industrial hygiene recommendations have on business. The industrial hygienist must learn to sell these programs to upper management based not only on their moral and compliance consideration but also on their economic viability.

The successful industrial hygienist must possess a host of technical skills. Chemistry, physics, engineering, acoustics, ergonomics, biology, and toxicology all come into play when an industrial hygiene program is implemented. The same skills provide the industrial hygienist with an opportunity to have a tremendous impact on the loss prevention and control program for the business. These opportunities interface directly with marketing, product research and development, and legal departments.

The assumption of such expanded responsibilities makes the role of the industrial hygienist an integral part of the loss prevention and control program, which in turn translates into improved support for the overall industrial hygiene effort. A partial listing of areas in which the industrial hygienist must take a leadership role follows:

1. Support new product development by assessing the potential exposure hazards associated with use of the product by the end-user and identifying the required protective measures. In many ways this is the classic role of the industrial hygienist, but an added function is now as a money-making member of the marketing team.
2. Assist in developing the premanufacturing notifications (PMN) required by the U.S. Environmental Protection Agency under the Toxic Substance Control Act. In the European Economic Community, a similar premanufacturing (premarketing notification) process defining the potential hazards is also required. Often the very approval of new products requires an estimate of the exposures expected within the customer's workplace. The role of the industrial hygienist in this process is essential.
3. Assess the hazards and potential liabilities inherent in implementing new processes and installing new equipment. The industrial hygienist is generally best qualified to assess the toxic exposure potential, the potential environmental impact, and the compliance cost inherent in such purchases.
4. Provide a detailed industrial hygiene survey when a facility is being shut down or sold. The study serves to document exposures, the effectiveness of engineering con-

trols, the use of personal protective equipment, and employee training as well as the operating procedures and their impact on employee exposure. Such information is invaluable when defending against future fraudulent workers' compensation and product liability claims.

5. Monitor facility-wide chemical usage and attempt to minimize the varieties of chemicals used within the organization. This can have a positive impact on procurement cost and significantly reduce hazardous waste–disposal costs.
6. Join maintenance engineering in the development and installation of engineering control measures. Direct involvement by an industrial hygienist can greatly enhance the engineering effort and can eliminate costly mistakes. The installation of local exhaust ventilation systems is an area where such involvement pays large dividends.
7. Advise corporate management of pending legislation and its potential impact on the organization. This provides management the opportunity to comment and greatly enhances the industrial hygienist's image as a member of the management team.
8. Utilize industrial hygiene expertise in the risk-management aspects of the business. In all too many cases, the workers' compensation insurance carrier or administrative agency has no knowledge of the available industrial hygiene reports that could eliminate or greatly mitigate the potential for a workers' compensation claim. For example, at one California company, numerous complaints and resulting claims arose from alleged exposure to chlorine. However, when workplace monitoring was initiated by a certified industrial hygienist, the results revealed ambient levels of chlorine well below the permissible limits. The health care provider then had appropriate objective data on which to make a decision rather than the employees' subjective history of exposure and physical findings that may or may not have been work related. In fact, the exposures were so low that the level of chlorine present was below that contained in potable water, thus negating even a potential claim for allergy to chlorine. These claims were successfully denied based on data provided by the industrial hygienist to the treating physician.

Many of the areas discussed in the preceding list generally go beyond the traditional industrial hygiene functions. The point is simple: expand the role of the industrial hygienist while maintaining those functions that are currently being performed. Through maintenance of an active employee exposure monitoring program, documentation of general working conditions within the facility (i.e., good housekeeping), subjective assessment of air quality (lack of nuisance dust, odors, etc.), and procedures requiring the use of personal protective equipment and employee training, the industrial hygienist will have a significant impact on not only workers' compensation costs but also other related costs.

It cannot be overemphasized that as industrial hygienists' roles expand, they must work even more closely with all other team members and upper management. In no other area is this interface more critical than between industrial hygiene and safety.

4.2.2 *Role of the Safety Professional*

The role of the safety professional encompasses much more than the traditionally perceived function of accident prevention and regulatory compliance. Such a limited role, although still very important, is not enough to ensure the economic viability of the function in today's corporate climate. Safety professionals must broaden their scope to become members of proactive, cost-effective loss control teams.

Obviously, safety program development incorporating facility inspection, safety training, and the like is critical in both accident prevention and compliance. However, for each program element implemented, there must be a mechanism to audit its effectiveness in terms of compliance, injury prevention, or quantifiable savings.

Prevention of accidents or illnesses is obviously the most effective way to limit health care costs. Unfortunately, all workplaces have some degree of hazard, and accidents can and will happen. When a mishap does occur, the safety professional should employ his or her skills to limit both the human and the economic severity of the incident (12).

Proper documentation is essential if a company is to defend itself successfully against spurious workers' compensation claims. Because workers' compensation in many states is subject to liberal construction, the employee's testimony is usually given more weight than that of the employer. Therefore, it is necessary for employers not only to act responsibly with regard to their employees' health and safety, but also to be able to prove it in a court of law.

Suppose an employee alleges a long history of exposure to toxic chemicals via inhalation, and as a result alleges occupationally related cancer. The employee claims that he or she was never provided with, or trained in the use of, a respirator. Regardless of whether instruction in the use of protective equipment was received, the case would most likely be decided against the employer, at great expense, unless proper documentation could be supplied. However, if the safety manager has maintained records documenting that the employee was provided the necessary respirator, was trained in its use and care, and was compliant in its use and care, the claim can be defended and the cost inherent either eliminated or greatly reduced.

One area that traditional accident investigation does not address is the legitimacy of an employee's claim that the injury sustained was work related. This should be one of the first issues addressed in any investigation. Many times the answer will be obvious, however some situations should raise questions. These include the following:

1. Unwitnessed injuries
2. A report of injury more than 24 hours after occurrence
3. Injuries where the mechanism of injury does not correlate with medical fact
4. Injuries alleged by employees with personnel problems, for example, absenteeism, probation, and lack of productivity
5. Injuries reported immediately before a layoff
6. Injuries to employees who have previously litigated suspect claims

By investigating such claims and developing a detailed history of the injury, the safety professional can provide the health care clinician with the information necessary to make an objective determination regarding the legitimacy of the injury. Thorough investigation of all accidents also serves to deter other employees from filing fictitious claims.

Documentation of all policies and procedures is an important aspect of the expanded role of the safety professional. Accurate data must be maintained regarding all written safety programs, safety committee meetings, and employee training in addition to required OSHA record keeping. Such documentation is necessary not only for compliance, but also as a key factor in reducing workers' compensation and related costs.

The data regarding the health and safety of the worker population, maintained by safety professionals, also enable them to perform trend analyses. Trend analyses allow the safety professional to pinpoint high-risk areas and the areas of highest cost relative to work-related injuries. Using data generated from the workers' compensation department, programs developed in-house, and OSHA records, an appropriate database can be developed. This information allows the safety professional to determine which department has the highest incidence of injuries, the incurred cost of injuries, the type of injuries, the shifts on which injuries most often occur, and the supervisor responsible for that shift. Only when a problem has been identified can a problem-solving approach be utilized, and the cost associated with it controlled.

It is incumbent on safety professionals to make the other members of the loss control team and management aware of the data and documentation they have compiled, thus enabling them to make use of it. By playing a more active role in loss control, the safety professional can alleviate the problem that has long plagued safety managers—lack of support.

Safety professionals have long been frustrated when management has failed to show the commitment necessary to achieve proposed goals. Development of a program that shows at least cost neutrality, and often cost benefit will enhance the role of the safety professional and make the safety program substantially more efficient.

4.2.3 Role of the Occupational Health Clinician

The drastic rise in the cost of health care has brought about a transformation of the role of the occupational health care clinician; what was once a limited technical position should emerge as a coequal member of business management. Regardless of what degrees they hold, occupational health specialists must utilize their technical and managerial skills to ensure that the employee population receives optimal health care at the lowest cost to the employer.

The treatment of minor injuries, cursory examinations, employee counseling, and medical records maintenance was the traditional province of the occupational health professional. In addition to having an excellent medical or nursing background and a comprehensive knowledge of occupational health, today's occupational health clinician must have an understanding of the rapidly escalating cost of health care to employers and its financial effect on the company. The clinician must then be zealous in implementing cost-containment measures which have an impact on health care as well as maintaining the occupational health of the employee population.

There are four occupational health programs known to be cost-effective in industry: preplacement health screening, appropriate management and treatment of injuries, employee assistance programs, and absence surveillance/disability management (14). These programs are discussed next.

4.2.3.1 Preplacement Health Screening. Many times injuries occur because a worker is physically unsuited to the task being performed. Additionally, employment of workers who are already injured or ill has a tremendous impact on the cost of health care. Preplacement health screening, performed mindful of the OSHA mandate to "provide a safe and healthful workplace" and in accordance with Federal Employment Guidelines, will serve to:

1. Place applicants in jobs for which they are physically qualified
2. Place workers in jobs that are both safe and healthful for them as well as for fellow workers
3. Provide baseline medical information that can be used in the event of future alleged work-related injury or illness
4. Identify individuals with previously unrecognized or inadequately managed health problems (15–17)

A preplacement health screening policy should be part of the company's overall occupational health policy. Standards utilizing job-related criteria should determine the fitness of the applicant. Employment criteria should be based on widely accepted standards that have been extensively researched or developed in-house, using data generated through a company's documented findings. Care should be taken to ensure that the standards are nondiscriminatory and that they are the same for all workers performing similar tasks.

Once a preplacement health screening policy has been written and the job-related criteria have been established, the occupational health clinician must utilize a comprehensive occupational health-medical history questionnaire and preplacement health screening examination format that documents negative as well as positive findings. Preplacement health screening exams preformed by outside providers may often be cursory and nondefinitive and, in the long term, of little value. To be effective, the preplacement health screening examination must be tailored to meet the needs of occupational health rather than the general population.

Basic diagnostic testing such as blood pressure, temperature, pulse, respiration, hemoglobin determination, urinalysis, and baseline audiogram should be performed on all employees in accordance with written policy. Specific diagnostic tests should be performed as needed, utilizing job-related criteria (18, 19). For example, persons working in some solvents or heavy metals should have blood chemistry panels before a determination is made if they are acceptable for the particular job. In many instances, it is neither prudent nor safe for employees to begin work in an area of potential exposure if they have preexisting abnormal blood chemistry findings that could be aggravated by potential workplace exposures.

Most injuries occur within the workers' first year of employment. Utilization of appropriate preplacement health screening can reduce injuries that are an aggravation of a preexisting injury or preexisting medical condition. The appropriate screening also reduces the number of cumulative trauma claims that arise over time.

However, for the preceding to be accomplished, the professional occupational health clinician, working with the human resources department and any outside health care providers, must design, implement, and audit the preplacement health screening process.

4.2.3.2 Management and Treatment of Work-Related Injuries. Inevitably, some workers do become injured. At this point, the technical skills of the occupational health professional come into play. Physicians and occupational health nurse specialists, working under standardized procedures, can provide quality primary health care in-house (20). This greatly reduces the number of work-related injuries that are referred to outside health care providers. The number of return doctors' visits and physical therapy sessions can also be greatly reduced by providing such care in-house. Lastly, the employees' progress can be more closely monitored when primary medical care is provided in-house, thus preventing complications, promoting early recovery, and reducing time lost from work, all elements of health care cost containment (21).

A key element in the medical management of injuries is the history of injury. Often, failure to elicit appropriate information and a lack of documentation have serious negative consequences, including misdiagnosis, assumption of responsibility for an injury that is not work related, and inappropriate medical care.

When it is necessary to use outside clinics or refer to specialists, diagnostic centers, emergency rooms, or hospitals, it is the responsibility of the clinician to obtain the best possible care for the injured employee at the lowest cost to the employer. It should be remembered that outside health care providers lack the incentive to take detailed histories or correlate medical findings, which may protect the employer. The occupational health specialist must work with other providers to make sure that the employer's needs, as well as those of the employee, are met.

The occupational health clinician plays a major role in the development of modified work programs. On a case-by-case basis, as part of a disability management program, it is advantageous to the injured employee and the employer to participate in a modified work program. In most instances, not only do workers who return to modified work have a speedier recovery, but also their progress can be monitored in-house, with basic physical therapy treatment provided, dressings changed, and minor complications treated before serious consequences result. This saves the employer not only the cost of temporary disability payments but also the cost of medical care that would otherwise be provided off-site (22).

At the time of an injured employee's discharge from medical treatment, the in-house occupational health professionals should provide their own assessment and documentation of case closure with specific notations regarding any positive physical findings, need for future treatment, and so on. The in-house clinician should also ensure that outside medical professionals provide appropriate documentation of case closure. Failure to provide such documentation can, and often will, lead to an employee's later claim that the injury never completely healed, or that there was an injury to another body part that was unreported at the time of the initial injury. Appropriate documentation will provide a defense against such fraudulent allegations. In the instance of true permanent disability, the same information provided to the workers' compensation carrier will speed the process of obtaining the injured worker's permanent disability benefit without need for legal counsel.

4.2.3.3 Employee Assistance Programs. Since the mid-1970s, the cost-effectiveness of employee assistance programs (EAPs) has been well documented by pioneers in the field, including the U.S. Navy and the United Parcel Service. The 1980s brought an increased awareness of the dramatic human cost, as well as tremendous financial burden, that has resulted from the use of drugs and alcohol in the workplace, making EAPs even more essential. The role of the occupational health professional in the development of EAPs is crucial because of the technical and confidentiality issues involved.

Even though substance abuse testing is a key element of the EAP, many types of testing are fraught with controversy. Preplacement testing for alcohol and drugs generally presents no legal problem for private industry as long as it is incorporated into an EAP policy, confidentiality is maintained, and the occupational health professional audits the collection mechanism, laboratory utilization, and documentation (23). Preplacement substance-abuse testing is, in and of itself, cost-effective. The range of positive tests varies from 12 to 25%.

Because it is estimated that 30 to 50% of all work-related injuries are the direct result of substance abuse in the workplace, it can be assumed that a substantial number of persons who are reported as positive on preplacement substance-abuse testing would have been otherwise injured on the job (24).

Other areas in which substance-abuse testing is used include for-cause testing, postinjury testing, and random testing. These have more potential for legal liability than preplacement testing; therefore the involvement of the occupational health professional is crucial in determining the appropriateness of utilization (23, 25).

In the development of a successful EAP, the occupational health clinician should (*1*) be involved in the selection of the employee assistance counselor, (*2*) provide education to management as to types of EAPs, (*3*) play a key role in writing the company employee assistance policy, and (*4*) be responsible for the implementation and audit of the program (26).

4.2.3.4 Disability Management. Unnecessary surgeries, unexplained medical costs, prolonged hospitalizations, overuse of inpatient hospitalization, and overuse of expensive diagnostic tests have all contributed to the dramatic increase in medical costs (27). Three-fourths of all health care costs can be attributed to fewer than 10 health conditions, but the proper management of these cases will reduce their cost (28).

Case management is defined as the process of organizing and mobilizing health services and resources to offer quality cost-effective care for an individual's health condition (28). It is an all-encompassing approach for development of individual plans of care for patients. In one study, 120 insurance claims were reviewed. A cost savings of $430,000 resulted from 29 of these cases, an approximate savings of $3,500 per case (29). Experts in the area of disability management suggest that case management works best when targeted at clients with very specific diagnoses, and when cases are referred and managed soon after the onset of the condition. It is the role of the occupational health professional to develop appropriate statistical data to determine the areas of greatest cost, and then present a plan to management for controlling such costs.

An increasing number of employers and insurance companies are using occupational health nurses as case managers to contain health care costs. It is the responsibility of the in-house professional to determine what methodologies of disability management are the most effective for their company, and to develop the program as well as an auditable method for determining quality assurance and cost-effectiveness.

In addition to the design and implementation of the programs known to cost-effective, the occupational health clinician must expand his or her role. There are numerous other areas in which the clinician's expertise will be of significant benefit to the employer.

Because of the tremendous escalation in the cost of medical benefits, it is incumbent on the occupational health professional to become knowledgeable and involved in the company's purchase of medical insurance for employees and their families. Often decisions as to benefit structure, choice of insurance carriers, and utilization of preferred provider organizations are left to either insurance brokers or risk managers who have little, if any knowledge of health care. Conversely, health care clinicians have little, if any knowledge regarding insurance. To become viable contributors in the mainstream of employee health, occupational health clinicians must assume responsibility for becoming knowledgeable and involved in the health care delivery system of the corporations in which they are employed.

All too often, wellness programs are seen simply as an employee benefit or are purchased as part of a package from a private firm with little thought given to quality assurance and cost-effectiveness. Even though wellness programs are shown to be beneficial and cost-effective, such benefits are limited to only those employees actually using the program. Unfortunately, most wellness programs have no incentives to ensure compliance or, as in the case of employee fitness centers, are used primarily by people who are already in good physical condition.

It is well documented that the health promotion programs most likely to be cost-effective have built-in incentives (27). Because wellness programs obviously involve numerous complex health issues, the occupational health clinician plays a key role in the design, implementation, and auditing of health promotion programs.

The role of the occupational health clinician has expanded tremendously. This has come about largely because of advances in technology and economic necessity. Unfortunately, the clinician is often underutilized within the corporate structure. As a key member of the loss control team, the occupational health clinician must use his or her technical skills to protect and promote worker health. Management skills must be employed to meet the challenges and the burden of rising employer health care costs. No longer can the health care professional be simply a reactive provider of aid to the injured (30).

4.3 Hearing Conservation: A Multidisciplinary Team Approach

Obviously, the first step in an effective hearing conservation program is the formation of a written policy and procedure. The policy should be written with the input of the entire occupational health and safety team and the workers' compensation manager. Health professionals often forget that the workers' compensation manager has little, if any, knowledge of the medical or safety issues related to hearing loss. Similarly, occupational health professionals are many times unaware of what constitutes a compensable hearing loss claim.

Only by working together and integrating their skills can they develop an effective policy. Certain components of the hearing conservation program will naturally be assumed by the health professional who maintains the appropriate technical skills to perform them.

In general, the industrial hygienist characterizes the industrial noise exposure to establish whether a hearing conservation program is warranted. Noise exposure data is essential to identify those personnel at risk, the level of risk (overall noise intensity), and personal protective equipment requirements. The feasibility of engineering controls, administrative controls, or both must be based on technology, costs, and business needs. Obviously, it is also essential to ensure compliance with federal and state occupational safety and health mandates.

Once the at-risk population has been identified, the industrial hygienist or safety professional must ensure that the affected employees are trained about the hazard and the need for hearing protection. The industrial hygienist or safety professional must also inform clinical members of the team of the exposed population to ensure that the required medical surveillance is initiated. The industrial hygienist must also ensure that the annual monitoring delineated in the federal regulations is conducted and remedial action taken as appropriate. Monitoring will also be conducted as processes are initiated or eliminated, as new equipment is integrated into the work environment, or in support of employee concerns.

The occupational health clinician is generally responsible for the medical surveillance aspects of the hearing conservation program. It is his or her responsibility to design and implement the audiometric testing program. Whether audiometric evaluations are conducted in-house or outsourced should be determined based on cost, quality, minimization of work environment disruption and other such factors.

Other critical components of the surveillance program include interpretation of results, retesting as indicated, medical referral when appropriate, and communication of results to the employee and other team members. Industrial threshold shift, as well as trends in any employee group, must be documented. The team can then present its findings to management using cost-versus-benefit data. This will ensure management's support in enforcing the various program elements. Because the workers' compensation manager is part of the team, he or she will be in a position to ensure that fraudulent workers' compensation claims involving hearing loss are successfully defended.

The industrial hygienist performs the environmental monitoring, the clinician is responsible for medical surveillance, and the safety professional assists with the design of the required engineering or administrative controls. However, other elements of the program, such as training in the use of hearing protection, fitting of hearing protection, and taking responsibility for compliance, may be overlapping functions. The responsibility regarding such functions must be delineated in the policy and procedures, and a system of accountability established by using a team approach. When a properly formulated policy with specific goals, and a viable course for reaching those goals are presented to management, along with statistical data showing potential cost savings, the probability of acceptance, and therefore success, is greatly enhanced.

The commitment of management is essential to the success of the hearing conservation program. It may be necessary for the loss prevention team to educate management as to its legal responsibilities and options. Only with the commitment of management can necessary compliance be achieved. A successfully implemented program will essentially eliminate all occupational hearing loss and enable the company to defend itself against future claims.

4.4 Program Management and Evaluation

For a multidisciplinary loss prevention and control program to be effective, a method of determining priorities, setting goals, and auditing program effectiveness must be developed. The loss prevention and control team must gain an understanding of the corporate philosophy, its financial situation, and its organizational structure. Such knowledge enables the occupational health and safety professionals to work within company policy and educate senior management as to the benefits of an integrated program. A loss analysis, using statistical data from all members of the loss prevention and control team, can be used to obtain management commitment, which is essential for comprehensive program implementation and success.

The loss analysis determines the areas of most serious concern and highest cost, thus enabling the loss prevention and control team, as well as corporate management, to establish priorities and set goals. The data used in the compilation of the loss analysis should be obtained for a minimum of the 3 years prior to program development. Basic data includes the following:

1. Number of new occupational injuries
2. Number of medical-only workers' compensation claims
3. Number of indemnity workers' compensation claims
4. Number of litigated claims in relationship to nonlitigated indemnity cases
5. Total number of OSHA-recordable injuries
6. Incidence rate (OSHA frequency)
7. Statewide incidence rate for industries of a similar nature
8. Manual adjusted workers' compensation insurance premium
9. Total incurred costs for workers' compensation (includes paid claims plus reserves)
10. Total incurred cost subcategorized as medical, indemnity, and expenses related to litigation
11. Payroll
12. Person-hours worked
13. Loss ratio (total incurred cost in relationship to the manual insurance premium)

The figures derived from the loss analysis also serve as baseline data against which future statistics can be compared to audit program effectiveness. The positive economic impact further enhances the necessity of a team approach to loss prevention and control.

4.5 Case Studies

The following examples provide data highlighting the effectiveness that a properly designed and implemented integrated occupational health program has on the number of injuries and the attendant workers' compensation costs.

4.5.1 Company A

This case study describes the cost advantage associated with using occupational health nurses in the medical management of work-related injuries. Company A is a major grocery retailer based in southern California and is self-insured and self-administered for workers' compensation. Beginning in January 1998, management made the decision to use the services of a medical case-management firm, staffed with registered nurses experienced in medical case management, to support and increase the effectiveness of their workers' compensation program. As a result of this decision, significant savings were achieved. In fact, for every dollar invested in the program, $17 in savings were realized during 1998.

The retailer has approximately 14,000 employees located in more than 450 markets, distribution centers, bakeries, and creameries in southern California. With such widespread, diverse operations, a decision was made to outsource medical case management with the goals of (*1*) providing optimum medical care in a timely manner to injured workers, and (*2*) identifying and controlling unwarranted costs. As part of the program design, it was determined that the program outcomes must be evaluated in terms of both benefits to the employee and savings to the company.

During 1998, 291 cases were referred for case management by the workers' compensation claims administrators. Of these cases, 136 were closed by year-end. There was a direct monetary savings to the company of $3,432,564 for the 136 cases closed. The total expenses for medical management services for these cases was $197,312. Thus a net savings was $3,235,252 (17:1 ratio) was achieved (see Fig. 43.1).

Using the data just described, average savings was calculated to be $25,239 per case. Average medical case management expenses for the 136 cases was $1,451 per case. Thus, net savings per closed case averaged $23,788 (see Fig. 43.2).

It is incumbent on the medical provider to document the cost-effectiveness of the case-management services by quantifying savings. Only by documenting savings will the employer recognize the true value of the professional nursing services provided to support the workers' compensation claims as well as the group health and general liability claims.

The following is a list of parameters that were applied when identifying the cost savings, as well as the benefit to the employee, by using professional nurses as case managers.

1. Assistance to administrators in determining the work-relatedness of injury/illness
2. Apportionment to preexisting or non-work-related disability as appropriate
3. Avoidance of litigation
4. Elimination of unwarranted or inappropriate medical treatment
5. Elimination of unwarranted or excessive medical expenses, such as physical therapy, durable medical goods, diagnostic tests, etc
6. Reduction and/or elimination of inappropriate temporary and permanent disability
7. Avoidance of unwarranted vocational rehabilitation costs

With the ability to track the various parameters affecting the cost of workers' compensation claims, it is possible to more accurately predict savings, which were historically considered soft. In the past, estimates of such savings, which included many of the items listed pre-

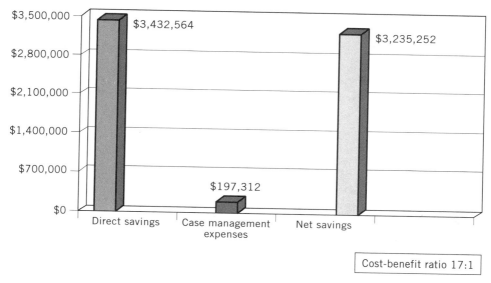

Figure 43.1. Case study, Company A. Monetary savings from 291 cases referred for case management in 1998.

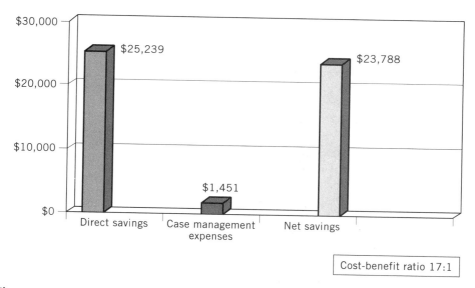

Figure 43.2. Case study, Company A. Monetary savings from 291 cases referred for case management in 1998, average values per closed case.

viously, could not be substantiated. With current tracking techniques and the availability of information referenced in, but not limited to *The Disability Advisor, American Health Care Policy Research*, various state labor codes, and industrial medical guidelines, actual savings can be quantified.

To determine the cost versus benefit of the overall medical case-management program, a close relationship between the provider and the claims adjuster is essential. Criteria for evaluating case-management performance must be mutually established at the outset. The claims administrator referring the cases for medical case management has significant latitude in determining cases to be referred as well as supportive information provided to the medical case manager; therefore, it is critical that the nurse have an ongoing dialog with the claims administrator to ensure the necessary exchange of information.

4.5.2 Company B

Company B is a southern California–based municipal agency employing approximately 1,500 full- and part-time workers. The municipality provides low-income subsidized housing and has employees at 32 locations. Occupations include office workers; a range of building trades, including carpenters, plumbers, electricians, and the like; police officers; and groundskeepers. The employees are represented by a collective bargaining unit. Those employees working in field activities have limited direct supervision and are also in an environment where health and safety controls are difficult to maintain.

In December 1998, the municipality made the decision to change the administrator responsible for its self-insured workers' compensation program. The firm selected was a full-service workers' compensation risk-management company, also California based. The municipality elected to use this firm's services in a multidisciplinary approach intended specifically to reduce the workers' compensation losses. The firm chosen uses an integrated approach involving workers' compensation claims administration, safety, industrial hygiene, and medical case management.

In the 1998 fiscal year there was a total incurred (paid plus reserved) workers' compensation cost of $1,879,995 (see Fig. 43.3). At fiscal year-end 1999, the total incurred cost decreased to $719,360, reflecting a savings of $1,160,365. Even with the inclusion of a 10% increase for incurred but not reported (IBNR) losses for fiscal year 1999, there remains a saving of $1,088,699, or 138%.

The reduction in cost is even more dramatic in that the new administrative firm had administrative responsibilities for the claims for only 9 months of fiscal year 1999. It is also important to note that the number of claims incurred between the two fiscal periods were similar.

This dramatic cost reduction was achieved using a variety of claim techniques and programs available from the administrative firm. As a result of the timely provision of benefits by the claims administrator and a close working relationship between the administrator and the medical case-management nurses, a significant reduction in medical costs were achieved (see Fig. 43.4). Lastly, by ensuring quality medical care, permanent disability was reduced, thus significantly decreasing indemnity costs (see Fig. 43.5).

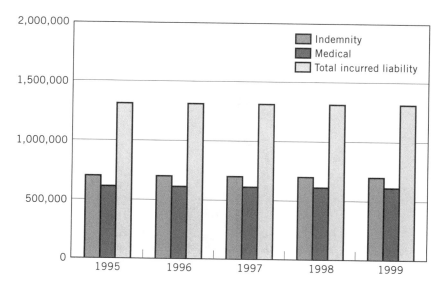

Figure 43.3. Case study, Company B. Total incurred cost, medical and indemnity for fiscal year-end, 1995 through 1999.

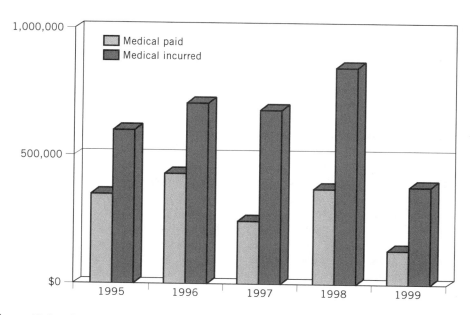

Figure 43.4. Case study, company B. Medical incurred and paid to date, fiscal year-end, 1995 through 1999.

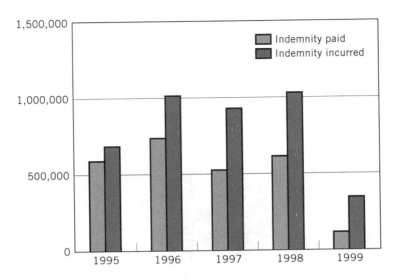

Figure 43.5. Case study, Company B. Indemnity incurred and paid to date, fiscal year-end, 1995 through 1999.

5 SUMMARY

The ballooning costs of occupational health care generally, and workers' compensation specifically, challenge occupational health and safety professionals to accept responsibility for preventing and controlling such expenditures. This approach is essential to creating a positive impact on the overall financial well-being of business and promoting optimum worker health. It is also critical to the survival of in-house occupational health and safety programs. The areas posing the most serious current financial and human liability can be identified by the use of a loss analysis. Once problem areas are identified, policies and procedures designed to rectify them can be implemented and future problems can be averted.

Policies developed by the loss prevention and control team should deal with both challenges that currently afflict the company, and those that provide a proactive approach to health and safety. Policies should be designed that are broad in scope and aimed at controlling and preventing losses. Limited programs tend to be reactive and less effective. A comprehensive loss prevention program requires a team approach. Integration of technical skills with a management perspective allows for an all-encompassing approach.

Industrial hygienists bring to the team a set of skills that enables them to monitor and assist in maintaining the quality of the environment of the workplace. Just as important, they should define and document potential hazards to prevent losses and ensure the credibility of reported injuries and illnesses. Safety professionals must establish practices that are designed to prevent injuries and illness and that also provide information, training, and documentation that reduces the cost of accidents both real and alleged. Occupational health clinicians possess a host of abilities that allow them to reduce the cost of employee health care while promoting employee health. Preplacement health screening, management of injuries, EAPs, and disability management are areas where the knowledge that occupational health care professionals possess can dramatically affect the cost of health care.

The effectiveness of an integrated loss prevention and control program is determined primarily by the occupational health and safety team members. To meet the challenges presented by a new and complex set of problems relating to loss prevention and control, they must be willing to expand their knowledge, interface with other staff, and be accountable. They must be able to demonstrate the positive financial impact of their roles, and they must make their employers aware of the potential benefits they offer in expanded management positions.

BIBLIOGRAPHY

1. "Leadership for a Healthy 21st Century," *Health Forum Journal*, January–February (Suppl.): 1–27 (1999).
2. Bureau of Labor Statistics, *Occupational Injury and Illness in the U.S. by Industry-1997*, 1998.
3. J. P. Leigh and J. Bernstein, "Public and Private Workers' Compensation Insurance," *J. Occup. Environ. Med.* **39**, (1997).
4. N. A. Ashford, "Workers' Compensation," in W. N. Rom, ed., *Environmental and Occupational Medicine*, 3d ed., Lippincott-Raven, Philadelphia, 1998.
5. California Workers' Compensation Institute, "California Employment Up, Work Injuries Down in 1996," *California Workers' Compensation Reporter* **26**, (February 1998).
6. Bureau of Labor Statistics, *Occupational Injuries and Illnesses in the U.S. by Industry for 1996*, 1997.
7. California Workers' Compensation Institute, "Workers' Compensation Benefits Decline Sharply in the U.S. Even More So in California," *California Workers' Compensation Reporter* **25**, (November 1997).
8. California Workers' Compensation Institute, "Litigation in California Workers' Compensation," *California Workers' Compensation Institute Research Notes* (November 1993).
9. D. H. Chenoweth, *Health Care Cost Management: Strategies for Employers*, Benchmark Press, Indianapolis, IN, 1988.
10. P. J. Leigh et al., "Occupational Injury and Illness in the United States," *Arch. Int. Med.* **157**, 1557–1568 (1997).
11. "Encouraging Action by Group Purchases," in *President's Advisory Commission on Consumer Protection and Quality in the Health Care Industry*, final report, July 19, 1998, chapter 6.
12. C. Lundeen, "Factors Affecting Workers' Compensation Claims Activity," *J. Occup. Med.* **31**, 653–656 (1989).
13. W. N. Rom, "Section 1: Environmental and Occupational Disease," in W. N. Rom, ed., *Environmental and Occupational Medicine*, Lippincott-Raven, Philadelphia, 1998.
14. National Safety Council, "Accident Investigation, Analysis and Cost," in *The Accident Prevention Manual for Industrial Operations: Administration and Programs*, 9th ed., Chicago, 1988.
15. P. Jacobs and A. Chovil, "Economic Evaluation of Corporate Medical Programs," *J. Occup. Med.* **25**, 273–278 (1983).

16. Industrial Relations, *Title 8 California Code of Regulations*, chapter 3, 1989.
17. *Bureau of Business Practice, Inc.*, Fair Employment Practices Guidelines No. 198(1), 1981.
18. Canadian Public Health Association, "Norms for Grip Strength," *Canada: Fitness and Amateur Sport*, 1977.
19. Joint National Committee, National Heart, Lung, & Blood Institute, *Detection, Evaluation and Treatment of High Blood Pressure*, Bethesda, MD, 1997.
20. S. Boydstun, "The Policy and Procedure Manual: Essential Component of an Employee Health Unit," *Occup. Health Nursing*, 334–337 (July 1985).
21. H. A. Zal, "The OHNs Influence on Employee Attitude and Ability to Return to Work," *Occup. Health Nursing*, 600–602, December 1985.
22. California Workers' Compensation Institute, "Vocational Rehabilitation Effectiveness in California," Bulletin no. 97-6, *California Workers' Compensation Institute Bulletin* (April 16, 1997).
23. C. Wright, "Alcoholism and Chemical Dependency in the Workplace," *Occupational Medicine: State of the Art Reviews* **4**, no. 2 (April–June 1989).
24. Department of Transportation, Research and Special Program Administration, "Control of Drug Use in Natural Gas, Liquefied Natural Gas, and Hazardous Liquid Pipeline Operations," *Federal Register* **53**, no. 224, (November 21, 1998).
25. *EAP Digest* **9**, 6 (Sept./Oct. 1989).
26. J. Nadolski and C. Sandonato, "Evaluation of an Employee Assistance Program," *J. Occup. Med.* **29**, 32–37 (1987).
27. B. J. Spain, "Evaluation of an Occupational Health Cost Containment Program," *Occup. Health Nursing*, 328–333 (July 1985).
28. W. E. Hembree, "Getting Involved: Employers with Case Managers," *Business and Health* (February 8–14, 1985).
29. K. C. Brown, "Containing Health Care Costs: The Occupational Health Nurse as Case Manager," *AAOHN Journal* **37**, 141–142 (1989).
30. J. Niland and C. Zenz, "Occupational Hearing Loss, Noise and Hearing Conservation," in *Occupational Medicine*, 3d ed., Mosby, St. Louis, MO, 1994.

CHAPTER FORTY-FOUR

Industrial Hygiene Survey, Records and Reports

Carolyn F. Phillips, CIH, CSP

1 INTRODUCTION

An integral and essential part of any good occupational health and industrial hygiene program is the administrative effort that supports and documents the work done by the hygienist. Often this is not perceived as one of the "fun" or challenging aspects of health and safety, but it is very necessary to maintain an effective program. The administrative activities include financial support for conducting the programs and suitable and useful records and reports, including program audits.

Industrial hygienists collect, maintain, and use data for recognizing, evaluating, and controlling health hazards. Evaluation of exposure data for trends can be a useful tool for anticipating potential problems. Hygiene records include, but are not limited to, assessment of hazards, exposure measurements, development and maintenance of controls, training and education, and auditing. The industrial hygienist will need information to document which employee has worked at which job for what time period, including job descriptions from an exposure viewpoint and representative exposure data for the jobs.

Record keeping in itself serves no useful purpose unless certain objectives have been defined and the records are translated into some form of report. The industrial hygienist must present data in a report format that is readily understandable, and in sufficient detail to permit the user to make adequate decisions. The report should reflect the special expertise of the industrial hygienist to interrelate all facets of the worker and the worker's environment in evaluating the potential impact on the worker's health. Effective reports can assist in future budget preparation and justification.

One key report is an audit. A thorough audit is crucial to maintaining a strong program. The main purpose of an audit report is to keep senior management up to date on both the strengths and the opportunities for improvement. The industrial hygiene audit can stand alone or be integrated with safety and environmental audits.

This chapter reviews the scope and contents of record keeping and reports and gives a few examples of workable forms. It also covers regulatory requirements and the role of the computer in record-keeping systems. Another key administrative factor for the industrial hygiene program is obtaining business support and financing for the programs. Historically, this was known as budgeting, but in today's environment it is often a more complex issue, and there is a growing need to show the business value added by the industrial hygiene programs.

2 INDUSTRIAL HYGIENE RECORDS

Record keeping serves as an essential tool for the industrial hygienist in managing, monitoring, and documenting efforts in evaluating and controlling employee exposure. This, however, is not the only function of the industrial hygienist's records. Physicians or epidemiologists will require employee exposure data if a causal relationship is suspected between a specific or mixed exposure and illness. Corporate legal and employee relations groups may consider the industrial hygienist's records essential for health-related grievance or arbitration cases and for compensation claims or litigation. Exposure records serve as the primary source for statistical evaluations and epidemiology studies and as a basis for further research. Properly documented exposure-illness records can assist in determining potentially unhealthy working conditions and developing new limits of exposure. In addition, many of today's regulations require that certain records be kept.

Records are only as good as the measurements or activity they document. It is of primary importance not only to record data accurately, but also to document the methods used to obtain the results. This allows the user of the records to evaluate and compare historical sample results. New technology applications may provide much more precise and accurate data than samples taken 20 or even 5 yrs ago. It is important to understand the methodology in order to make valid data comparisons.

Recognizing that record keeping is an essential part of the industrial hygienist's job, it is natural to ask what constitutes adequate records. The novice industrial hygienist soon discovers that there is no one "adequate" record-keeping system for all-purpose use. However, the various systems judged to be most effective all have certain characteristics in common. Records should be as detailed as possible for the data required (yet physically manageable), and they should be appropriately structured to relate to other available pertinent data (i.e., medical, personnel, weather). Care must be taken not to collect data for its own sake but with a specific program need in mind. Data may be collected for a major study or simply to respond to an employee exposure concern.

Any practical system for documenting industrial hygiene surveys and activities must be comprehensive, flexible, and simple. A system of storage and retrieval should be developed that will accomplish the following functions:

1. Allow the user to retrieve information in the form required
2. Cover all foreseeable areas of current and future interest
3. Exclude extraneous data not expected to be required
4. Minimize cost
5. Maximize efficiency

A major purpose of an exposure database is tracking people; chemicals; and other potential stresses, locations, and personal exposures.

Industrial hygiene studies often produce large amounts of data of various types. Consideration must be given to the use of the data for varying purposes. The data that constitute the original industrial hygiene records should be stored in a uniform or constant format. It must also be recognized that there will be many users of the data, and different outputs will be required. The permanent record from which the industrial hygiene report is developed should include notes logged in a field notebook or on a specialized survey form. The practice of jotting down random bits of information on pieces of scrap paper that may be lost is convenient in the field but poor practice in the long run. Details not recorded on the spot are often forgotten and so do not get into the report. Reports written from memory, though they may be sufficiently accurate in some instances, tend to be incomplete and are usually inadequate as a legal record. Some of the current technology, such as palm-size electronic data recorders, may provide a breakthrough in taking field notes and retaining them.

It is just as important to record all data pertinent to the sample as it is to collect the sample. The experienced industrial hygienist will later compare his or her field observations and assessments with analytic results and judge whether the data appear to be valid. Not every data record is as valid as some others are. Some are strictly factual; others are subjective judgments or approximations by the hygienist. A method is needed to indicate the validity of the data so that greater credibility is not given to the data than they warrant. Proper validation also reduces the likelihood of misuse of the data in the future.

Records should be sufficiently detailed to permit another individual to duplicate the previous survey without the assistance of the person who originally recorded the data. Last, the speed and accuracy of preparing a final report from survey records of an investigation depend largely on information recorded in the form of notes. Photographs tell stories with a minimum of explanation and serve as a permanent record of the specific conditions. Before-and-after pictures can be particularly effective. They may also provide information that was not recorded at the work site. Plot plans or engineering drawings of the facility showing location of equipment can also be useful, particularly for sound level surveys or noise dosimetry.

Evaluation of a potential health hazard may involve calibration of equipment, air sampling, laboratory analysis, evaluation of physical stresses (noise, light, radiation, heat), and biological monitoring (e.g., urinary phenol for benzene exposure). Each of these areas may require its own form or set of forms. When an industrial hygienist is called on to make many similar exposure-assessment surveys, it is advisable to design a specific form to be used at the time of sampling.

Emphasis is usually on time-weighted-average exposure sampling data, with consideration for area and short-term monitoring data. It is, however, important that all three types of data be obtained and recorded. Then, if adverse health effects are found in employees and personal time-weighted sampling data are found to be low, a review of short-term personal sample data and area data may point to the intermittent exposures as the causative factors.

In any event, recording data obtained in any survey activity requires the use of suitable forms and their orderly completion in a neat and efficient manner. This holds whether the data is recorded by hand or electronically. Forms with spaces labeled for essential data help avoid the common failure to ensure needed information has been obtained at the time of the survey. The use of a poorly designed or incomplete form will lead to incomplete reports or will require repetitive follow-up with field personnel or perhaps even a resurvey. In addition, the use of standardized forms within an organization assists in maintaining a consistent data-collection and documentation system for more than one user.

Recognizing potential health hazards involves an inventory of all materials and processes likely to create a health hazard by job or occupation or area. If the site is covered by the Occupational Safety and Health Administration (OSHA) Hazard Communication Rule (CFR 1920.1200), then the area chemical records developed for that purpose can be used as a checklist during a preliminary survey or during an audit as well as a record of potential exposure. Typically an initial walk-through appraisal is made, during which forms or at least a list of items can be reviewed and observations noted. Material safety data sheets (MSDS) from the suppliers should be available in the workplace for each material used in that workplace.

Table 44.1 lists the basic data elements that should be part of an industrial hygienist's record on exposure assessment. Examples of sample forms for industrial hygiene air samples are given in Figure 44.1(a), Figure 44.1(b), and Figure 44.2. There is no single standardized format. Many companies have designed their own forms to fit their own record-keeping systems. Records for the industrial hygiene laboratory should include adequate numbering and laboratory identification schemes for samples, as well as established calibration and quality-control standards. Calibration data for both field and laboratory equipment must be recorded and kept. This can be done on the air-sample form or the lab analysis form, or a separate record can be developed. The efficiency of the system depends on who does the calibrations and where they are done. The laboratory doing the analytical work should have documented records of its quality-assurance procedures and follow a strict quality-control program and good laboratory practice.

Physical stress data include noise, light, radiation, and heat. Separate forms may be developed or the basic air-sampling form can be amended to fit the specific needs. Figure 44.3 is an example of a noise exposure form. A biological monitoring form may also be appropriate. Records on inspection of ventilation systems may be recorded on forms such as in Figure 44.4 and Figure 44.5

3 REPORTS OF SURVEYS AND STUDIES

An effective report conveys, accurately and efficiently, both the data and pertinent observations and recommendations developed by the evaluation. These serve as a base for

Table 44.1. Industrial Hygiene Exposure Data

Employee name
Employee code number (Social Security and/or company no.)
Company name
Site and/or location name
Work area or unit name
Job title and/or job code
Operation condition (normal, shutdown, upset)

Sample date
Sample code number
Sample type (area, personal, source, bulk)
Substance(s) sampled
Sample length (time-weighted average, peak, task, grab)
Task description (work activity during sample)
Shift length
Shift time (morning, night, etc.)
Skin contact (potential or actual)
Personal protective equipment used (gloves, respirator, etc.)
Weather conditions (as applicable)

Sample results
Validity code (how good are the data?)
Sample and analytic method
Potentially interfering chemicals
Analytic lab (name and credentials)
Calibration data (lab and field)

solving immediate problems and also document information for future reference. Management needs carefully defined and timely information to initiate effective control action. Records and reports are also useful in meeting employee and community right-to-know information requests.

The report should present the facts, analyze and interpret the findings, and develop conclusions and recommendations as appropriate. The writer must try to anticipate and answer questions, because feedback from the written communication may not be received. The report must be organized and written for the needs and understanding of a specific reader or group of readers. The content, approach, style of writing, and choice of words depend on the varied backgrounds of the intended readers. The same data may be presented differently depending on whether the report is for management, engineering or medical personnel, or the employees. This means not that the content of the facts or conclusions and recommendations is different, but that it may be phrased differently. Reports should be written promptly after completion of the survey and a system must be established to retain these reports for future reference.

No standard format exists for all cases, but some technical writing books and company style manuals do prescribe various report structures. Books on writing style and on pre-

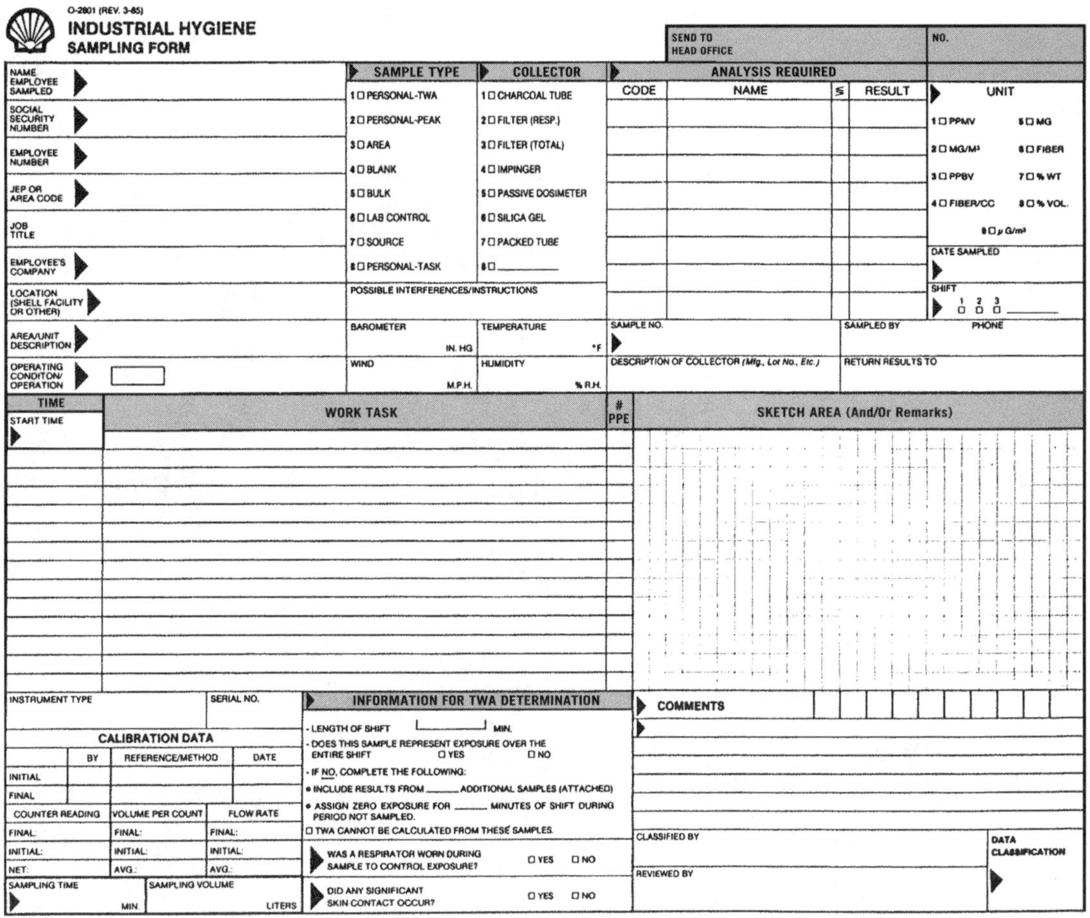

O-2801 (REV. 3-85)

INDUSTRIAL HYGIENE SAMPLING FORM

SEND TO HEAD OFFICE | NO.

NAME EMPLOYEE SAMPLED

SOCIAL SECURITY NUMBER

EMPLOYEE NUMBER

JEP OR AREA CODE

JOB TITLE

EMPLOYEE'S COMPANY

LOCATION (SHELL FACILITY OR OTHER)

AREA/UNIT DESCRIPTION

OPERATING CONDITON/ OPERATION

SAMPLE TYPE

1 ☐ PERSONAL-TWA
2 ☐ PERSONAL-PEAK
3 ☐ AREA
4 ☐ BLANK
5 ☐ BULK
6 ☐ LAB CONTROL
7 ☐ SOURCE
8 ☐ PERSONAL-TASK

COLLECTOR

1 ☐ CHARCOAL TUBE
2 ☐ FILTER (RESP.)
3 ☐ FILTER (TOTAL)
4 ☐ IMPINGER
5 ☐ PASSIVE DOSIMETER
6 ☐ SILICA GEL
7 ☐ PACKED TUBE
8 ☐ ______

ANALYSIS REQUIRED

CODE	NAME	≤	RESULT

UNIT

1 ☐ PPMV 5 ☐ MG
2 ☐ MG/M³ 6 ☐ FIBER
3 ☐ PPBV 7 ☐ % WT
4 ☐ FIBER/CC 8 ☐ % VOL.
9 ☐ μ G/m³

DATE SAMPLED

SHIFT 1 ☐ 2 ☐ 3 ☐ ______

POSSIBLE INTERFERENCES/INSTRUCTIONS

BAROMETER IN. HG | TEMPERATURE °F | SAMPLE NO. | SAMPLED BY | PHONE

WIND M.P.H. | HUMIDITY % R.H. | DESCRIPTION OF COLLECTOR (*Mfg., Lot No., Etc.*) | RETURN RESULTS TO

TIME	WORK TASK	# PPE	SKETCH AREA (And/Or Remarks)
START TIME			

INSTRUMENT TYPE | SERIAL NO.

CALIBRATION DATA

	BY	REFERENCE/METHOD	DATE
INITIAL			
FINAL			

COUNTER READING	VOLUME PER COUNT	FLOW RATE
FINAL:	FINAL:	FINAL:
INITIAL:	INITIAL:	INITIAL:
NET:	AVG.:	AVG.:

SAMPLING TIME MIN. | SAMPLING VOLUME LITERS

INFORMATION FOR TWA DETERMINATION

- LENGTH OF SHIFT ______ MIN.
- DOES THIS SAMPLE REPRESENT EXPOSURE OVER THE ENTIRE SHIFT ☐ YES ☐ NO
- IF NO, COMPLETE THE FOLLOWING:
- INCLUDE RESULTS FROM ______ ADDITIONAL SAMPLES (ATTACHED)
- ASSIGN ZERO EXPOSURE FOR ______ MINUTES OF SHIFT DURING PERIOD NOT SAMPLED.

☐ TWA CANNOT BE CALCULATED FROM THESE SAMPLES.

WAS A RESPIRATOR WORN DURING SAMPLE TO CONTROL EXPOSURE? ☐ YES ☐ NO

DID ANY SIGNIFICANT SKIN CONTACT OCCUR? ☐ YES ☐ NO

COMMENTS

CLASSIFIED BY

REVIEWED BY

DATA CLASSIFICATION

(a)

Figure 44.1. (**a**) Industrial hygiene sampling form. (**b**) Analytical report.

ANALYTICAL REPORT

CONDITION OF SAMPLE	DATE RECEIVED	DATE REPORTED	METHOD CODE	NO. OF ANALYSES	L.H. NO.
☐ OK ☐ SEE BELOW					
ANALYZED BY		DATE ANALYZED	LAB RECORD NO.	ACCOUNT NO.	

METHOD/REFERENCE

COMMENT

RESULTS	☐ PPMV ☐ PPBV	☐ MG/M³ ☐ μG/M³	☐ MG ☐ μG	USE < LIMIT OF DETECTION (NOT N.D.)			APPROVALS

OPERATIONAL CODES

OPERATING CONDITION CODE (01-15)
(For TWA Samples)

Normal	01
Startup	02
Upset/spill/leak	03
Shutdown/turnaround	05
Maintenance	06
Other (specify in written Comments section)	08

OPERATION CODE (16-60)
(For Peak or Task Samples)

Operations

Handling/disposal of product/process sample(s)	53
Working around spill/leak	34
Gauging tank(s)	31
Analyzing product/process sample(s)	24
Collecting product/process sample(s)	17
Other Operating Tasks (specify in Comments section)	36

Maintenance

Entry/work in confined space	16
Removing insulation	21
Applying insulation	54
Breaking into process line	23
Changing product/process filters	26

Maintenance (Continued)

Installing/removing blinds	55
Cleaning equipment (draining, purging, decontaminating)	37
Tank/equipment inspection (involving entry)	05
Degreasing/solvent washing	28
Welding/grinding/cutting	56
Painting	43
Abrasive blasting	44
Carpentry	22
Other mechanical work (specify in Comment section)	57

Loading/Shipping

Filling drum(s)	35
Emptying drum(s)	18
Filling container(s) (sacks, bins, etc.)	49
Emptying container(s) (sacks, bins, etc.)	50
Loading tank truck(s)	51
Unloading tank truck(s)	20
Loading tank car(s)	46
Unloading tank cars(s)	19
Loading barge(s)	45
Unloading barge(s)	42
Loading ship	48
Unloading ship	41
	47

COMMENT CODES

Monitoring Problems

Pump/dosimeter malfunctioned	02
Initial and final flow rates differ by > 10%	55
Pump may have not run for entire sampling period	14

Monitoring Conditions

Monitoring in conjunction with government agency	03
Local ventilation system in work area	04
Major engineering changes made	05
Random sample	08
Worst case situation	09
Duplicate sample	43

Analysis

Sample lost in shipment	28
Sample lost in analysis	20
Sample arrived contaminated/damaged	23
Incorrect sample collector used	35
Interference by other compound suspected	21
Sample shipped cold arrived ambient	61
Reported result is minimum value due to possible sample loss/breakthrough	56
Result corrected for recovery/desorption efficiency of < 75%	60

PPE (specify types in Comments section)

Respirator worn	57
Gloves worn	58
Protective clothing worn	59

(b)

Figure 44.1. Continued.

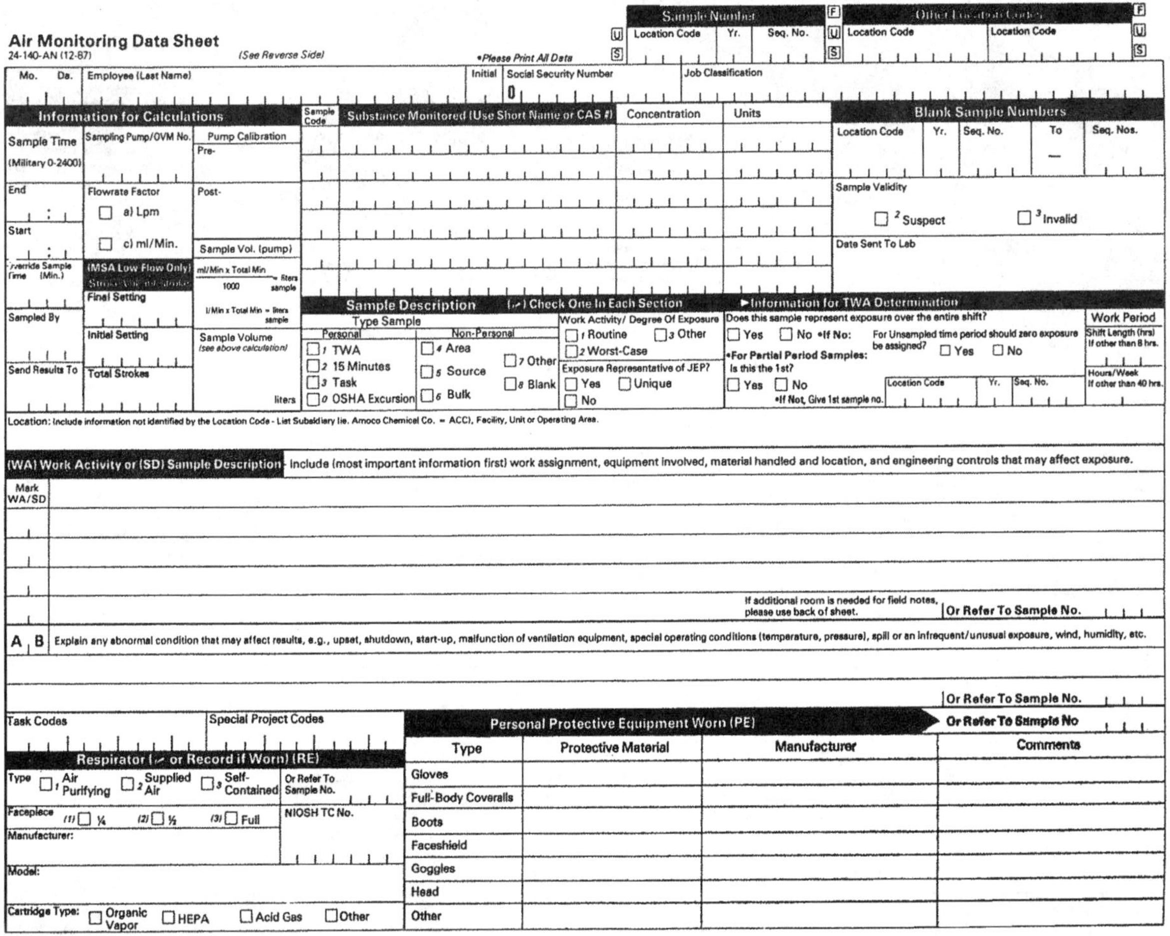

Air Monitoring Data Sheet
24-140-AN (12-87) (See Reverse Side)

*Please Print All Data

Sample Number		
Location Code	Yr.	Seq. No.

Other Location Codes	
Location Code	Location Code

Mo. | Da. | Employee (Last Name) | Initial | Social Security Number | Job Classification

Information for Calculations

Sample Time (Military 0-2400) | Sampling Pump/OVM No. | Pump Calibration: Pre- / Post-

End | Flowrate Factor: ☐ a) Lpm ☐ c) ml/Min.

Start | Sample Vol. (pump)

Override Sample Time (Min.) | (MSA Low Flow Only) | ml/Min x Total Min / 1000 = liters sample

Final Setting | l/Min x Total Min = liters sample

Sampled By | Initial Setting | Sample Volume (see above calculation)

Send Results To | Total Strokes | liters

Sample Code	Substance Monitored (Use Short Name or CAS #)	Concentration	Units

Blank Sample Numbers

Location Code	Yr.	Seq. No.	To	Seq. Nos.
			—	

Sample Validity: ☐ 2 Suspect ☐ 3 Invalid

Date Sent To Lab

Sample Description (✓) Check One In Each Section

Type Sample — Personal: ☐ 1 TWA ☐ 2 15 Minutes ☐ 3 Task ☐ 0 OSHA Excursion; Non-Personal: ☐ 4 Area ☐ 5 Source ☐ 6 Bulk ☐ 7 Other ☐ 8 Blank

Work Activity/ Degree Of Exposure: ☐ 1 Routine ☐ 3 Other ☐ 2 Worst-Case

Exposure Representative of JEP? ☐ Yes ☐ Unique ☐ No

▸Information for TWA Determination

Does this sample represent exposure over the entire shift? ☐ Yes ☐ No •If No: For Unsampled time period should zero exposure be assigned? ☐ Yes ☐ No

•For Partial Period Samples: Is this the 1st? ☐ Yes ☐ No •If Not, Give 1st sample no. Location Code | Yr. | Seq. No.

Work Period: Shift Length (hrs) If other than 8 hrs. | Hours/Week If other than 40 hrs.

Location: Include information not identified by the Location Code - List Subsidiary ie. Amoco Chemical Co. = ACC), Facility, Unit or Operating Area.

(WA) Work Activity or (SD) Sample Description - Include (most important information first) work assignment, equipment involved, material handled and location, and engineering controls that may affect exposure.

Mark WA/SD

If additional room is needed for field notes, please use back of sheet. | Or Refer To Sample No.

A | B Explain any abnormal condition that may affect results, e.g., upset, shutdown, start-up, malfunction of ventilation equipment, special operating conditions (temperature, pressure), spill or an infrequent/unusual exposure, wind, humidity, etc.

Or Refer To Sample No.

Task Codes | Special Project Codes

Respirator (✓ or Record if Worn) (RE)

Type ☐ 1 Air Purifying ☐ 2 Supplied Air ☐ 3 Self-Contained | Or Refer To Sample No.

Faceplece (1) ☐ ¼ (2) ☐ ½ (3) ☐ Full | NIOSH TC No.

Manufacturer:

Model:

Cartridge Type: ☐ Organic Vapor ☐ HEPA ☐ Acid Gas ☐ Other

Personal Protective Equipment Worn (PE) — Or Refer To Sample No

Type	Protective Material	Manufacturer	Comments
Gloves			
Full-Body Coveralls			
Boots			
Faceshield			
Goggles			
Head			
Other			

Figure 44.2. Air-monitoring data sheet.

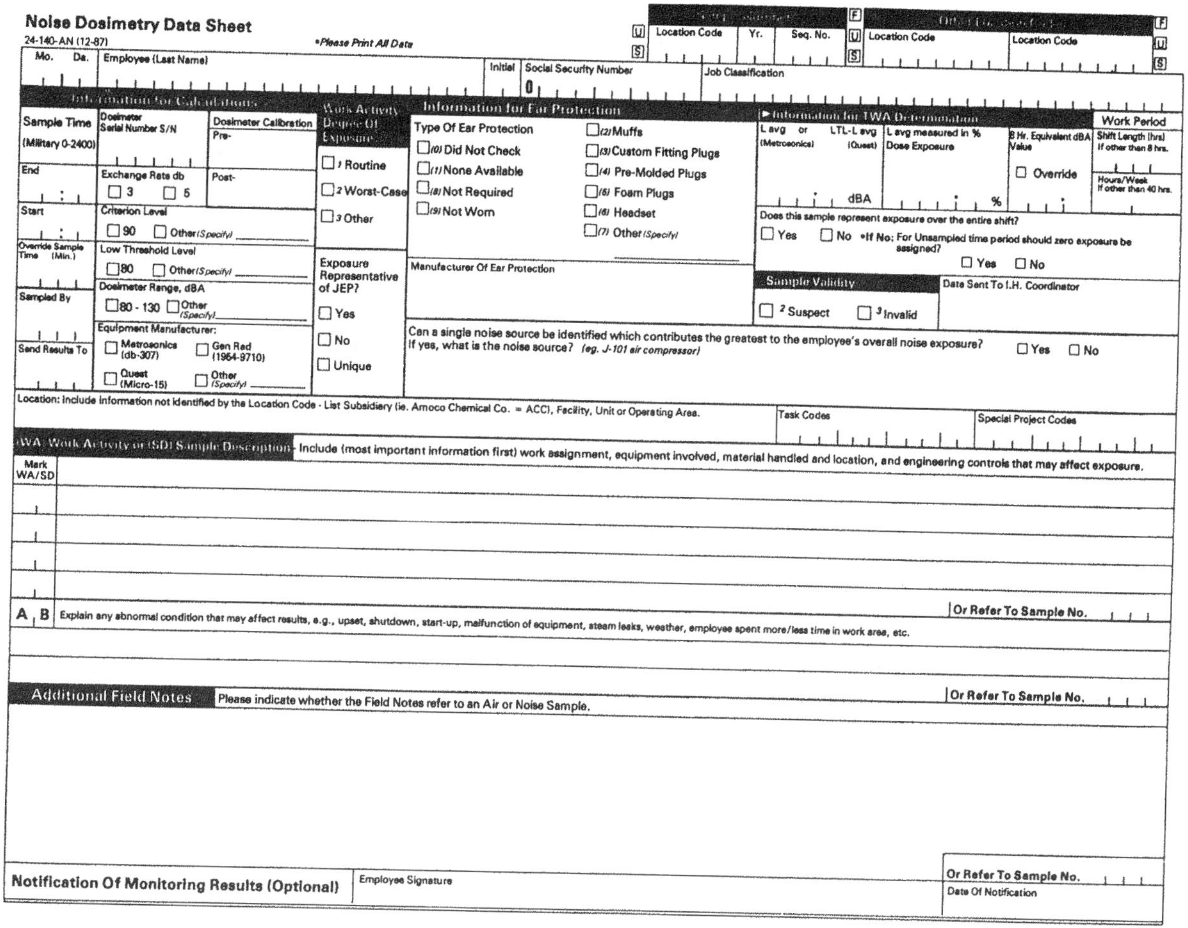
Noise Dosimetry Data Sheet
24-140-AN (12-87)
•Please Print All Data

U S Location Code | Yr. | Seq. No. F U S Location Code | Location Code F U S

Mo. | Da. | Employee (Last Name) | Initial | Social Security Number 0 | Job Classification

Information for Calculations
Sample Time (Military 0-2400)
End :
Start :
Override Sample Time (Min.)
Sampled By
Send Results To
Dosimeter Serial Number S/N
Dosimeter Calibration Pre- Post-
Exchange Rate db ☐ 3 ☐ 5
Criterion Level ☐ 90 ☐ Other (Specify)
Low Threshold Level ☐ 80 ☐ Other (Specify)
Dosimeter Range, dBA ☐ 80 - 130 ☐ Other (Specify)
Equipment Manufacturer: ☐ Metrosonics (db-307) ☐ Gen Rad (1964-9710) ☐ Quest (Micro-15) ☐ Other (Specify)

Work Activity Degree Of Exposure
☐ 1 Routine
☐ 2 Worst-Case
☐ 3 Other
Exposure Representative of JEP?
☐ Yes
☐ No
☐ Unique

Information for Ear Protection
Type Of Ear Protection
☐ (0) Did Not Check
☐ (1) None Available
☐ (8) Not Required
☐ (9) Not Worn
☐ (2) Muffs
☐ (3) Custom Fitting Plugs
☐ (4) Pre-Molded Plugs
☐ (5) Foam Plugs
☐ (6) Headset
☐ (7) Other (Specify)
Manufacturer Of Ear Protection

Information for TWA Determination
L avg or (Metrosonics) LTL-L avg (Quest) . dBA
L avg measured in % Dose Exposure . %
8 Hr. Equivalent dBA Value ☐ Override
Work Period
Shift Length (hrs) If other than 8 hrs.
Hours/Week If other than 40 hrs.
Does this sample represent exposure over the entire shift?
☐ Yes ☐ No •If No: For Unsampled time period should zero exposure be assigned? ☐ Yes ☐ No
Sample Validity ☐ 2 Suspect ☐ 3 Invalid
Date Sent To I.H. Coordinator

Can a single noise source be identified which contributes the greatest to the employee's overall noise exposure? ☐ Yes ☐ No
If yes, what is the noise source? (eg. J-101 air compressor)

Location: Include Information not identified by the Location Code - List Subsidiary (ie. Amoco Chemical Co. = ACC), Facility, Unit or Operating Area.
Task Codes
Special Project Codes

TWA Work Activity or (SD) Sample Description - Include (most important information first) work assignment, equipment involved, material handled and location, and engineering controls that may affect exposure.
Mark WA/SD

A B Explain any abnormal condition that may affect results, e.g., upset, shutdown, start-up, malfunction of equipment, steam leaks, weather, employee spent more/less time in work area, etc.
Or Refer To Sample No.

Additional Field Notes - Please indicate whether the Field Notes refer to an Air or Noise Sample.
Or Refer To Sample No.

Notification Of Monitoring Results (Optional)
Employee Signature
Or Refer To Sample No.
Date Of Notification

Figure 44.3. Noise dosimetry data sheet.

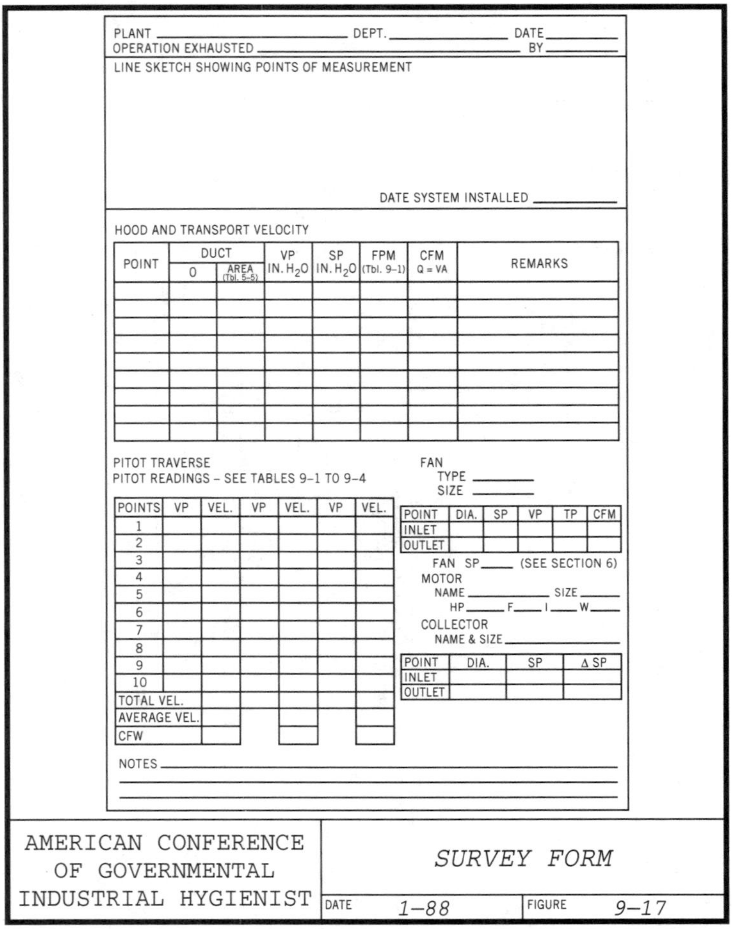
PLANT ______ DEPT. ______ DATE ______
OPERATION EXHAUSTED ______ BY ______
LINE SKETCH SHOWING POINTS OF MEASUREMENT

DATE SYSTEM INSTALLED ______

HOOD AND TRANSPORT VELOCITY

POINT	DUCT		VP	SP	FPM	CFM	REMARKS
	0	AREA (Tbl. 5-5)	IN. H_2O	IN. H_2O	(Tbl. 9-1)	Q = VA	

PITOT TRAVERSE
PITOT READINGS – SEE TABLES 9-1 TO 9-4

POINTS	VP	VEL.	VP	VEL.	VP	VEL.
1						
2						
3						
4						
5						
6						
7						
8						
9						
10						
TOTAL VEL.						
AVERAGE VEL.						
CFW						

FAN
TYPE ______
SIZE ______

POINT	DIA.	SP	VP	TP	CFM
INLET					
OUTLET					

FAN SP ______ (SEE SECTION 6)
MOTOR
NAME ______ SIZE ______
HP ______ F ______ I ______ W ______
COLLECTOR
NAME & SIZE ______

POINT	DIA.	SP	Δ SP
INLET			
OUTLET			

NOTES ______

AMERICAN CONFERENCE OF GOVERNMENTAL INDUSTRIAL HYGIENIST
SURVEY FORM
DATE 1–88 FIGURE 9–17

Figure 44.4. Survey form, American Conference of Governmental Industrial Hygienists.

paring technical reports are indispensable and should be part of a reference library. Writing skills improve through the use of such references, accompanied by practice. References (1–4) are examples of useful resources. In all cases, the final report format used by the writer should fit the needs of the organization as well as those of the reader and the writer. One commonly used report format contains the following sections:

1. A summary that concisely presents the work reported, including a statement of why the work was done, an abridgment of background information, conclusions, and recommendations
2. Recommendations that list all proposed changes, supported by brief comments on the reasons for the suggested courses of action

INDUSTRIAL HYGIENE SURVEY, RECORDS AND REPORTS

Laboratory Hood Ventilation Test Form

Area unit _____ Date tested _____

Bldg./Room _____ Tested by _____

Hood no. _____ Test instrument:
                                                                  Model no. _____

Hood door fully open[1]
                              Serial no. _____

	1	2	3
A	x	x	x
B	x	x	x
C	x	x	x

Temperature (°F) _____

Barometric pressure (in. Hg) _____

Hood use (check one:)
  Low toxicity materials (TLV > 10 PPM)
  _____

  Toxic materials (TLV < 10 PPM)
  _____

Velocity measurements

Location	Velocity (FPM)
A–1	_____
A–2	_____
A–3	_____
B–1	_____
B–2	_____
B–3	_____
C–1	_____
C–2	_____
C–3	_____
Sum measurements	_____

Hood dimensions:
  Height (in.) _____
  Width (in.) _____

Hood exhaust volume (Q):
  Q = area (ft.^{2}) × average
  Q = _____ CFM

Hood static pressure (in. H_2O) _____

Average velocity (ppm) _____

Minimum velocity (ppm) _____

Comments:

$$\text{Avg. velocity} = \frac{\text{Sum of measurements}}{\text{No. of measurements}}$$

Figure 44.5. Laboratory hood ventilation test form. Note: The hood door should be fully open or open to the designated operating level during measurements.

3. Discussion that presents findings at length and makes and evaluates conclusions
4. An attachment that contains result data and material too detailed for inclusion in the discussion, providing necessary support information

For a long report on a major study, a title page and a table of contents are useful. As an aid to the reader, headings and subdivisions should be used. Illustrations, tables, graphs, diagrams, and photographs reduce verbosity. The report should tell what was done, why it was done, what was seen, and what data were available to the industrial hygienist. Appropriate sections of the report should contain a brief description of the plant, department, or operation and any significant changes that may have taken place since any previous survey. It should also have a discussion of control measures already implemented; potential health effects resulting from exposure to the stresses surveyed; regulatory requirements, samples, measurements, and test results; and an interpretation of these results. Sampling, measuring, and analytical procedures; documentation of equipment calibration; and findings from other pertinent studies may be included or referenced as appropriate. The written report should be aimed at the intended readers, logically directing their attention to the facts in the shortest possible time. Writing that accomplishes this uses plain words and proper grammar for clarity and omits needless words and sentences. These characteristics will make the report more readable and will not distort the communication. In essence, the report should be as brief as possible but still contain all the relevant information.

Adequate follow-through is a prime factor in gaining acceptance of recommendations with a minimum of delay after a report has been issued. Follow-through methods include the following:

1. Offering assistance in the report transmittal letter
2. Presenting the report in person and reviewing contents with appropriate plant personnel, including management
3. Providing assistance or resources in carrying out recommendations
4. Offering to review designs for new control methods
5. Conducting a follow-up survey

4 LINKING INDUSTRIAL HYGIENE DATA TO HEALTH RECORDS

The health experience of workers in relation to exposure must be followed closely to achieve a complete occupational health program. Both the industrial hygienist and the physician are concerned with monitoring. The monitoring systems used are personal, environmental or area, biological, and medical. (Note: in this context, biological monitoring is restricted to measuring for the material or its metabolites in the body, usually via blood or urine analysis.) Personal and area monitoring provide the exposure information necessary for designing effective engineering controls and work practices. Biological and medical monitoring provide information on exposure only after absorption takes place. For adequate evaluation of the effect of the work environment, it may be necessary to use all four monitoring systems (5–7).

Industrial hygienists must keep in mind the value of reciprocal information. Medical findings may indicate areas for industrial hygiene study. Biological monitoring data may reveal exposure trends before illness symptoms. Although medical surveillance is not the primary means for evaluating employee exposure, it can be a supplementary tool to evaluate the effectiveness of a control program involving engineering or other control techniques, personal protective controls, or both. In cases where significant skin absorption of

INDUSTRIAL HYGIENE SURVEY, RECORDS AND REPORTS

a chemical may occur, personal and area monitoring may not provide sufficient information. Accordingly, there must be a close and ongoing relationship between industrial hygiene, toxicology, and medical to determine what exposure limits are needed for potentially hazardous materials when there are no guidelines or standards. Figure 44.6 indicates the data relationships involved in a combined industrial hygiene and medical assessment of the worker.

5 OTHER RECORDS

Coordinating record-keeping activities from many sources, including personnel records, medical data, environmental data, and chemical audits, is essential if linking of exposure data to individual employees is to be accomplished. Personnel records should contain the cumulative summary of jobs held, with dates and department. If personnel records are not kept for a sufficient length of time, the industrial hygienist will need to set up a method to retain the information for future use. Job lists will then identify potentially exposed workers. Identifying the chemical, biological, and physical agents is essential, as well as any process changes that could affect job activity or exposure levels.

Fundamental to the process is a review to identify the chemical, biological, and physical agents in the work environment. Any process changes that could affect exposure conditions

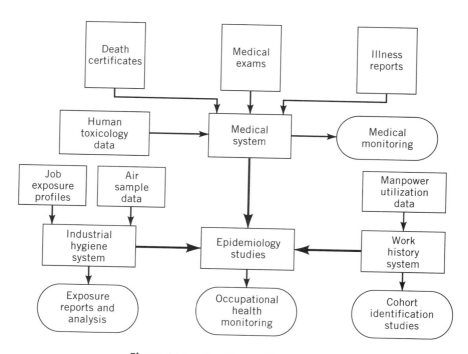

Figure 44.6. Health surveillance system.

should be reported promptly to the industrial hygienist for assessment. However, the real world often makes this a very difficult condition to meet. With the introduction of new processes, elimination of departments, and modifications and multitasking of jobs, operations are dynamic. Linking exposure data to employees through the personnel record system can be extremely complicated unless there is careful cross-referencing. Exposure records for a job may be considered representative for all employees who held that job during the given time interval. In such cases it may be easier to retrieve the needed information if exposure data are cross-referenced to job and department designations rather than to process and location in a department. Employees may frequently change jobs within a department, so information regarding an exposure assessment for all jobs should be available. Another reason to assess the exposure for all jobs is that the levels of exposure that are currently believed to be safe may later show a potential risk. This will often include monitoring, but professional judgment can be used to document estimated exposures based on objective data on the job and process. All this is just an indication that exposure records are more complex than they first seem. It is more than just the data for an air sample!

Many factors in addition to work exposure can contribute to the cause of diseases. Among these are individual habits such as smoking or the use of alcohol; dietary preferences; air pollution or polluted water supplies; exposure to noise, solvents, or other conditions or materials; or exposure through an avocation or second job. Unfortunately the employer generally does not always have access to this type of record, so there will be gaps in a total exposure record.

Obtaining appropriate industrial hygiene data is especially difficult in dealing with health effects that develop slowly or after a long latent period, yet acquiring such data is necessary. It may be necessary to obtain and retain such data for as long as 30 to 40 yrs for correlation with chronic health effects (see section 6). Cross-correlation of industrial hygiene data, medical history, and periodic medical surveillance examinations (e.g., audiometric examinations, pulmonary function tests) provides human exposure data of great value, particularly if large populations can be studied. Computerized data storage and retrieval is an asset for carrying out such epidemiological studies. Currently the weakest link in an epidemiology study is usually the historical exposure record.

Workers' compensation or claims records may provide information on employee exposures. Morbidity records may provide clues to developing trouble. For example, a high incidence of absenteeism for respiratory problems in a particular department or plant, as compared to other such locations, may be an indication of excessive environmental stress from a particular chemical or condition. Such situations should be investigated thoroughly by the physician to determine whether a common medical problem exists, and then, if these suspicions are confirmed, by the industrial hygienist. If workers' compensation claims costs are significant, this area may be a good place for health and safety staff to do some assessments: Are there real hazards causing claims, or are other issues leading to the claims? Is there an opportunity to reduce costs and be of benefit to the employees?

Personnel records can prove invaluable, especially in investigating causative factors in demonstrated or suspected occupational illness. For example, if medical examination reveals conditions such as silicosis or hearing threshold shift, a review of the worker's job assignment history may tell whether the worker has ever had related exposures. A check of personnel records might provide some assistance in determining whether the worker's condition may have resulted from exposure at a previous job.

6 OSHA RECORD-KEEPING REQUIREMENTS

This section is primarily concerned with the OSHA aspects of record-keeping requirements. Governmental requirements to maintain occupational records are standard in many countries. The OSH Act, Section (c)(1), states that

> Each employer shall make, keep and preserve, and make available to the Secretary or the Secretary of Health, Education and Welfare, such records regarding his activities relating to this Act as the Secretary may prescribe by regulation as necessary or appropriate to the enforcement of this Act or for developing information regarding the causes and prevention of occupational accidents and illnesses. In order to carry out the provisions of this paragraph such regulations may include provisions requiring employers to conduct periodic inspections. The Secretary shall also issue regulations requiring that employers, through posting of notices or other appropriate means, keep their employees informed of their protection and obligations under this Act, including the provisions of applicable standards (8).

OSHA has established numerous requirements for record keeping, including such items as training, certification, exposure data, programs, and injury/illness data. These requirements are contained in the *Code of Federal Regulations* Title 29, Labor Chapter XVII—Occupational Safety and Health Administration, Parts 1903 to 1920.1500 and 29 *CFR*, 1926 (9). Some examples of particular interest to industrial hygienists follow.

6.1 Part 1903—Inspections, Citations, and Proposed Penalties

OSHA is authorized to inspect, investigate, and review records required by the act and other records that are related to the purpose of the inspection.

6.2 Part 1904—Recording and Reporting of Occupational Injuries and Illnesses

Requires each employer to maintain a log and summary of all recordable injuries and illnesses no later than 6 days after receiving information of an occurrence. It also requires the posting of an annual summary of occupational injuries and illnesses for the previous calendar year (by February 1 for at least a 1-month period). Records shall be kept for 5 yr and be available to employees or their representative. Failure to maintain these records may result in a citation and penalties.

6.3 Part 1910—General Industry Standards

6.3.1 Subpart C—General Safety and Health Provisions

Section 1910.20—Access to Employee Exposure and Medical Records. This section gives OSHA, employees, and designated representatives the right of access to relevant exposure and medical records. These records include those taken both for compliance with other OSHA standards and for the companies' own purposes. Medical records are to be kept for length of employment plus 30 yr. Employee exposure records shall be maintained at least 30 yr. Any analyses using exposure or medical records shall be kept for at least 30 yr. The section also requires the employer to notify employees annually of their right to access these records.

6.3.2 Subpart G—Occupational Health and Environmental Controls

Section 1910.95—Occupational Noise Exposure. Employers shall have a hearing conservation program. The record keeping will involve exposure measurements and audiometric test data. Noise exposure data must be kept for 2 yr and audiometric data for the duration of employment.

Section 1910.96—Ionizing Radiation. The employer must maintain adequate past and current exposure records on exposure to ionizing radiation for individuals in restricted areas. The employer must advise the employee of the exposures on at least an annual basis.

6.3.3 Subpart I—Personal Protective Equipment

Section 1910.134—Respiratory Protection. When engineering controls are not feasible, respiratory protection may be used to protect employees from exposure. If respirators are used, the employer shall have a written respirator program covering safe use of respirators in normal and emergency situations. These written procedures shall include selection, fit, use, and maintenance and storage. Records shall be kept of inspections. Substance-specific standards may include more specific requirements, including fit testing and training records.

6.3.4 Subpart Z—Toxic and Hazardous Substances

In addition to the PEL list (1910.100), Subpart Z contains a number of substance-specific standards. The following ethylene oxide example is typical of record-keeping requirements in these standards.

Section 1910.1047—Ethylene Oxide. This section requires that records of objective data used to exempt certain operations be kept during the period the employer relies on the data. An accurate record of all measurements to monitor employee exposure to ethylene oxide must be kept at least 30 yr. Also to be retained are medical records for each employee required to be in the medical surveillance program.

Section 1910.1200—Hazard Communication. This section requires developing and using a hazard communication program. Manufacturers or importers of chemicals shall have a written hazard determination procedure. The written programs shall be available to employees and to OSHA. The standard requires area chemical lists that can be very useful to an industrial hygiene program and, when maintained efficiently, an asset to epidemiology studies.

OSHA is not unique in its record-keeping requirements. In 1988, a law was passed in the United Kingdom called Control of Substances Hazardous to Health Regulations (COSHH). The Health and Safety Executive administers this regulation. It introduced a new legal framework for controlling exposure to hazardous substances during work activities. The regulation requires both an assessment of health risk and measures to protect workers' health.

Part of the regulation requires records of the assessments. This requirement is to record why decisions about risks were made and the precautions taken. The records are to reflect the details with which the assessment is carried out, to be useful to those who need it now and in the future, and to indicate under what conditions the assessment must be reviewed. The regulations also indicate that the style and presentation of the assessment is to be influenced by the recipient. The records also have to be available to authorized inspectors. The regulations are similar to OSHA's requirement of maintaining monitoring records for at least 30 yrs if they are representative of personal exposure. It also requires that a record be kept for at least 5 yrs of tests and repairs to control measures.

7 CONFIDENTIALITY OF RECORDS

Between the employee and the company physician there is a patient–physician relationship dictating that all medical information be considered and retained in confidence by the physician. This confidentiality of medical records follows recognized standards of law and medical ethics. The role of a company-employed physician is further clarified by the provisions regarding proper conduct of an industrial physician, including codes of ethics of the American Medical Association, the American College of Occupational Medicine, and most state medical societies.

Management does not have a need to know specific diagnoses to fill the management function of assigning employees to specific tasks and operations. However, management does need to know what medically based restriction on work duties must be considered in assigning employees to jobs. Because management is not privy to an employee's medical history and specific disabilities, the physician must have an adequate knowledge of the various work environments and their potential adverse effects on employees and must advise management of any limitations in the placement of its employees. As an example, an employee with certain respiratory disabilities should not be assigned to operations presenting potential exposure to chemicals that are respiratory tract irritants. The industrial hygienist has a key role in working with the physician to review the work environments and the limitations for placement of employees.

Biological monitoring records of the employee, such as results of blood-lead analyses, are not quite as clear-cut. The results are important to both the physician and the industrial hygienist. Such data are needed by the industrial hygienist for a complete exposure assessment, for the results may indicate exposure not found during exposure monitoring of the employee. Management use of such data should be limited to implementing engineering or other control measures and to temporary assignment of the employee, if needed, to another area without that specific exposure.

When medical and exposure data are computerized, as discussed in the following section, it is important that the confidentiality of these restricted medical data be maintained. Only those privileged to see medical data should be able to obtain this information from the computer. The industrial hygienist will have a continuing need for exposure and biological monitoring data and should have access to them.

8 COMPUTERIZATION OF RECORDS

In the past, and even at present in many cases, numerous data on employee exposures were compiled in the form of narrative-style written reports. These are used primarily to document employee exposure evaluation as measured by some yardstick such as exposure limits and to recommend control measures where indicated. Medical records as confidential files, on the other hand, are largely kept on an individual employee basis. Under such circumstances, epidemiological studies and other correlations between industrial hygiene and medical data are often difficult and time-consuming. One need only consider the practice of attempting to correlate hearing tests (involving 14 measurements on each employee) with noise level measurements, each being repeated as often as yearly, to recognize the complexity of manually correlating industrial hygiene and medical data.

With the use of the computer for industrial hygiene data, there now exists a much better opportunity and likelihood that adequate evaluation of the various occupational health parameters will be accomplished. In the final analysis, this means better control of the work environment and early detection of any adverse health effects. A significant benefit of a computer database for industrial hygiene records is in usable output from the system (10,11).

Table 44.2 shows some of the activities that can be efficiently supported by a computerized health surveillance system. Table 44.3 lists the attributes required for such a system to be effective.

With an adequately designed computer program and input of pertinent data, the industrial hygienist can evaluate employee exposure to workplace contaminants more effectively than before. For example, the input data needed for benzene exposure would include those listed in Table 44.1. Urinary phenol measurements would also be included if available. In addition to personal monitoring data, area and source monitoring data would be included. Other data sets could include a record of all job assignments for an individual and a list of all chemicals used in the job over time. From such databases, a computer printout can be obtained of all employees potentially exposed to benzene, indicating which have measured exposure above or below the allowed limit.

Table 44.2. Activities to be Supported by Health Surveillance System

Epidemiology studies
Evaluation of workplace health conditions
Medical surveillance programs
Compliance with record-keeping and reporting mandates
Providing data for litigation purposes
Public understanding programs
Response to proposed legislation or regulatory activities

Table 44.3. Required Attributes of Health Surveillance System

Large storage capacity
Confidentiality
Operability (by noncomputer experts)
Quality assurance
Flexibility
Service levels
   Readily available
   Reliable
   Easily maintained
Cost-effective
Minimum impact on user locations
Easy data retrieval and display

Because the capability of computers is constantly evolving and because each company has different needs, this section can be no more than a general discussion of the considerations for developing a system. These include but are not limited to data to be stored, reports to be generated, statistical analysis, coordination with medical records, and work history for epidemiology studies. Other aspects to be considered include number of users, number of locations, size of database, financial resources, and staff capabilities and needs. Only after evaluation of this type of information can a decision be made as to use of a personal computer, mainframe, or network, or whatever the latest technology allows.

In a large company with numerous locations, the use of a companywide networked system has many benefits. These include the following:

1. Response to ad hoc queries that involve information on more than one facility
2. Standardizing data and record keeping
3. Ability to include data validity checks and error warnings in the base program
4. Potential for earlier acknowledgment and interpretation of trends related to exposure
5. Central repository of information for regulatory response
6. Reduction of clerical errors and data handling time

Finally, a summarized report of the employee's exposure potential and biological monitoring data could be made available to the examining physician at the time of the employee's periodic or special medical examination. This allows the physician to review the individual's medical findings in the light of environmental stresses.

In employing computer techniques, the industrial hygienist, working with the physician(s) and operations personnel, must first establish goals and define accurately what will be needed in the way of computer reports. Without these two factors carefully thought out in advance, the data recorded may not be retrievable, or if retrievable may not be usable. The actual users of the system should be included in the development process in order for the system to gain acceptance and to meet their needs. It is technically easy to enter data that will yield the usual statistical or other relatively simple information. What must be remembered is that the computer will not provide answers if the data input has not supplied the correct base for such answers, or if adequate formatting has not been provided to permit meaningful retrieval programming.

Another significant goal in computerizing a record-keeping system is assistance in complying with regulatory and legal requirements.

There are many commercial computer systems available that try to meet the industrial hygienists' needs as described here. A number of companies have chosen to design and develop their own systems tailored to interface with other corporate systems such as enterprisewide systems. Another fast-developing area is the direct interface of monitoring equipment with the computer database. New sampling and analysis devices do or will include smart samplers, data logging, and electronic dosimeters. In some cases, the data can be fed directly into the computer for processing. If a dosimeter can provide minute-by-minute exposure increments for the workday, the industrial hygienist will face the critical decision of what data are to be kept—some statistical summary of the time-weighted average and peak data or the entire record. For industrial hygienists to avail themselves of technologic advances, they must stay abreast of progress in microprocessor technology.

9 RELATED GOVERNMENTAL RECORD-KEEPING SYSTEMS

OSHA is implementing a computerized database for field inspection reports as a reference for compliance officers. In addition to being a resource for OSHA, it could be used to compare different facilities of the same company in various geographic locations so that citations in one location can be referenced for similar operations and added citations given in future visits if the same problem has not been corrected.

From a different perspective, the United Kingdom Health and Safety Executive (HSE) has initiated a national exposure database (NEDB) that aims to bring together exposure data on a wide range of workplaces in the United Kingdom (12). Input includes all airborne sampling data plus details on control measures gathered during visits to workplaces by HSE occupational hygiene inspectors. In the long term, the HSE also hopes to enter data from industry so that the database becomes a focal point for exposure information. The data will be used in developing new regulations and may offer a sounder base for future epidemiology studies.

Figure 44.7 is an example of the form used to collect the data for entry into the NEDB. The data includes substances monitored, industry, TWA values, control measures, job tasks, and other data. The database will be useful in identifying particular industries with problems or specific processes or tasks that may need improved controls.

Figure 44.8 shows a similar database concept for noise data developed by the Canadian Centre for Occupational Health and Safety (13). The contents have been compiled from data reported in journals, health and safety reports, and surveys by various industries and agencies. The objective is to share noise data by making it available either on compact disk or online.

Health and Safety Executive
Environmental monitoring results

FCG file reference		Date of visit		Substance											
Occupier				Units											
			Total number of people on site	Monitoring procedure											
				In-house atmos. monitoring											
				In-house biol. monitoring											
				Males exposed											
				Females exposed											
Reference number	Sample type	Sample description (eg name/process/job)	Sample period	Duration (minutes)	Result	TWA	Result	TWA	Result	TWA	Result	TWA
1												
2												
3												
4												
5												
6												
7												
8												
			Exposure limits	8 hour								
				10 min								

DEC1 (4/85)

(a)

Figure 44.7. Environmental monitoring results, (**a**) Part 1, (**b**) Part 2.

HSE Area no.			
Shield ID no.			
Industry			
Type of visit			
OHVR reference no.			
Exposure details		Control measures	
Type	Pattern	RPE	LEV
1			
2			
3			
4			
5			
6			
7			
8			

(b)

Figure 44.7. Continued.

10 AUDITING

There is a growing use of audits as a means to review management systems, gauge compliance with regulations and policies, and assess knowledge of employees. Systematic follow-up is also a critical factor. Audits are reports to senior management on the quality

NOISE LEVEL DATA BASE FIELDS (+ searchable)

*******ORIGIN OF DATA*******

Author+	The person(s) responsible for measurements.
Publication	The published source of measurements.
Title+	The title of the study (and language, if other than English).
Reference Data	Bibliographic details of the publication.
Year of Study+	The year of measurement.

*******WORKPLACE*******

Industry+	Standard industrial classification code and corresponding term, plus CIS facet code(s) and term(s) which describe a particular industry.
Operation+	Description of work-related human actions causing the noise.
Occupation+	Canadian classification and dictionary of occupations description of the worker group affected by the noise.
Source+	The source of the noise (eg. piece of equipment)
Type of Noise+	The duration characteristics of the noise measured (eg. continuous, intermittent, impulse).
Exposure Duration Per Day	The number of hours (per day) the workers are exposed to the measured noise levels.
Engineering Control	Devices or structures installed to reduce noise.
Protection	Ear protection worn by workers.

*******MEASUREMENT(S)*******

Location of Measurement	Detailed information concerning the position of measuring device from source of noise and/or worker.
Measuring Device Used	Brand name of the equipment and settings used to record the measurement.
Duration of Measurement	The amount of time the measuring devices were operating.
Noise Level	Sound pressure level (SPL) in dB(A) or peak SPL in dB. Minimum, maximum, and average sound pressure levels in dB(A). Equivalent continuous noise level; Time-weighted average (TWA) sound level in dB(A) and exchange rate used (5 dB or 3 dB). Octave band analysis (31.5–8000 Hz) in dB.

Figure 44.8. Noise level data base fields.

and effectiveness of health and safety programs that provide a much better picture of effectiveness than just the occupational injury and illness rates (i.e., OSHA 200 log). It is critical that audits are conducted in a user-friendly format with the goal of learning or developing opportunities for improvement versus a "we got you" approach.

The importance of health and safety audits to the well-being of employees is widely recognized. Effective audits are candid and self-critical. They should be perceived as a way to learn to be better. Effective audits and follow-up action are invaluable tools for identifying and correcting potential problems and conditions that could increase the likelihood of future accidents or overexposures. They are important for maintaining compliance with complex health, safety, and environment (HSE) rules and regulations.

In 1999 the Organization Resources Counselors (ORC) Occupational Health and Safety Group surveyed its member companies, and 100% of the respondent companies performed HSE audits. Some did multiple audits at plant, region/division, or corporate levels, others only at the corporate level. Because ORC member companies are among the major U.S. corporations, this is indicative of the value of auditing.

Assurance is a process that comprises both the execution of compliance activities as well as a technical assurance role of setting requirements and standards and then auditing for compliance. One approach for a plant site is establishing a sitewide assurance process that includes an auditing function. This can provide assurance to management that the plant is in compliance with all company policies and applicable laws and regulations as well as company standards.

Audits can be conducted with internal staff (i.e., using corporate staff or staff from other locations) or external auditors. They can be conducted and framed using internally generated checklists, external lists, or even computerized audit systems. The Chemical Manufacturers Association (CMA) and many partner organizations have implemented "Responsible Care" programs. Part of the program is annual self-reporting of member companies against the various codes of management practices, one of which is the Employee Health and Safety Code. CMA has now implemented a program of third-party verification of the company self-audits.

Some companies have developed a complete HSE management system to manage their HSE activities and integrate them with production operations and the business units. In some cases an International Environmental Management System from the International Standards Organization (ISO 14000) has been used as a template to develop an HSE system.

Audits of many different types and scopes are conducted. These include the following:

- Overall management of health and safety
- Full scope, very detailed
- Limited audit objectives (i.e., focus on a number of issues of concern)
- Special audits (narrow scope, such as Hazard Communication)
- Business based reviews
- Integrated HSE audits vs separate H, S and E audits

Audits provide many different benefits and serve different purposes:

- Verify compliance
- Identify and evaluate risk
- Inform senior management of status/needs
- Input to planning process
- Increase employee awareness
- Train staff (cross-fertilization)
- Identify emerging issues
- Correct deficiencies
- Evaluate employee awareness and attitudes
- Evaluate adequacy of available resources
- Identify business opportunities

Health and industrial hygiene audits may include but are not limited to the following:

1. Hazard communication implementation
2. Workplace exposure assessment
3. Hierarchy of control strategy implementation
4. Establishment and follow-through of an exposure-assessment strategy
5. Respiratory protection program effectiveness (and other personal protective equipment)
6. Risk evaluation and assessment

One excellent resource for developing an industrial hygiene audit program or the industrial hygiene section of a complex audit system is AIHA's (14) *Industrial Hygiene Auditing Manual*.

One current area of contention between the regulatory agencies and the companies is the confidentiality of the company audits when the regulator's compliance staff is conducting an inspection. There is concern that in a few cases company audits were used to support citations. This area is being reviewed at OSHA, and attempts are being made to resolve the conflicts.

10 BUSINESS VALUE OR BUDGETING

In these days of organizational restructuring, mergers, acquisitions, joint ventures, downsizing, and so on, it has become more important than ever for health and safety personnel to be able to show the benefit of their programs in concrete terms and their potential for contribution to the bottom line. Industrial hygienists may need to integrate their activities into other business processes. For a company to be successful it must maintain a competitive edge. There are several ways of doing this, including adding customer value or shareholder value. Companies need to look at the impact of health and safety on business values. Health and safety can play a role in such matters as the time it takes to get a new product to the marketplace, innovation in production processes, cost of production, and customer service, among others. The health and safety impact methods will vary depending on the nature of the business.

In many companies health, safety, and environmental activities are in a shared services organization with such areas as finance and accounting, legal, and other service groups. This type of structure means industrial hygienists will have to focus on demonstrating the value of their services to the business over and above the pure compliance aspects. It is also likely that they will be asked to determine the most cost-effective ways to implement compliance requirements. One way to approach this is to consider specific regulations and related items and look for ways to demonstrate cost avoidance rather than cost savings.

The various new organizational structures and cost reduction efforts also impact the industrial hygiene programs that have been implemented. Areas to think about include the following:

- Can they be done more cost-effectively?
- Will self-directed work teams impact the programs and the training needs?
- What is the impact on the structure? Has HSE become an integrated organization?
- Can best-practice sharing be improved, either in-house or external? In one company, one location had contracted for a very cost effective respirator fit-test program. When there was dialogue with other locations and the service supplier, it turned out it was possible for them to travel to other locations without a large cost increase.
- Will the health and safety organization structure change from corporate to local, or vice versa?
- Will outsourcing become the name of the game? If so, how can this be done most effectively within your structure? Is outsourcing really the cost saving it is perceived to be? More than hourly rates need to be considered; company, product, and people knowledge are critical in many situations.
- Is there a way, using operating staff, to leverage some of the short resources in industrial hygiene?
- Is the industrial hygienist willing and able to be that good team player?

11 CONCLUSIONS

It should be clear that for industrial hygiene to fulfill its obligation and goal of protecting the health of workers, excellent administrative processes are essential and must be employed to the fullest. The extent and sophistication of such records, reports, and business and financial support vary with the individual situation. The industrial hygiene report is the industrial hygienist's most forceful and enduring communication with management and employees. Done properly, in clear and concise language and understandable to those who are not industrial hygienists, it can effect the changes and improvements necessary to protect the health of the worker and demonstrate value to the company or organization.

If the report is done poorly, the effort that went into the original industrial hygiene survey record and exposure assessment may have been useless or the audit recommendations will never be followed.

Industrial hygiene records, though recognized as a necessary and valuable part of a report, can go far beyond the immediate intent of verifying environmental stresses that exist for the moment. Careful and full use of such records and data can assist the industrial hygienist and other health professionals to accomplish the following:

1. Evaluate the adequacy of current acceptable levels of exposure.
2. Aid in establishing acceptable levels of exposure for chemicals (or other hazards).
3. Provide a basis for input to the regulatory process.
4. Show trends in exposure and controls.
5. Conduct retrospective studies if occupational illnesses occur.
6. Confirm or refute the validity of any compensation claims.
7. Demonstrate value to business via customer service information.
8. In cooperation with the medical staff, detect any early evidence of developing occupational health problems and effect controls at an early stage.
9. Provide audit reports to management to enable them to track successful implementation and effectiveness of programs.
10. Provide support for legal staff or the industrial hygienist as an expert witness during litigation.

Finally, industrial hygiene records and reports must be retained long enough to ensure that they have fulfilled their intended purpose. Industrial hygiene engineering control recommendations, for example, may not be needed again after they have served their original purpose of effecting control of the work environment. However, the fact that controls were installed on a specific date (and the impact on the exposures) could be critical to epidemiology studies or future liability issues. Measurements of work stresses, on the other hand, may be useful 20, 30, or even 40 yr after serving their initial purpose. In some instances, retention of such records for a given period is required by regulation or law. In the final analysis, except for legal requirements, the industrial hygienist, with his or her background of training and experience, can best determine the point at which such records and data are no longer of value in protecting and promoting the health of the worker. The use of a carefully developed, flexible computer system can facilitate the handling of complex industrial hygiene data in an organized fashion and provide important information to the managers, medical staff, and employees. The use of carefully planned and executed useful audits can be very beneficial to involve management and increase their support of the program. Finally, to have the resources to do these activities, the hygienist needs to understand the business planning and budgeting process and provide appropriate input on value added by health and safety efforts.

BIBLIOGRAPHY

1. R. R. Rathbone and J. B. Stone, *A Writer's Guide for Engineers and Scientists*, Prentice-Hall, Englewood Cliffs, NJ, 1962.

2. W. Strunk, Jr., *The Elements of Style*, 3rd ed., revised by E. B. White, Macmillan, New York, 1979.
3. M. Stiertzer, *The Elements of Grammar*, Collier Books, Macmillan, 1986.
4. A. Plotnik, *The Elements of Editing*, Macmillan, 1982.
5. J. J. Bloomfield, "Industrial Health Records: The Industrial Hygiene Survey," *Am. J. Pub. Health* **35**, 559 (1945).
6. J. T. Siedlecki, "How Medicine and Industrial Hygiene Interact," *Int. J. Occup. Health Saf.* (Sept.–Oct. 1974).
7. M. G. Ott, H. R. Hoyle, R. R. Langner, and H. C. Scharnweber, "Linking Industrial Hygiene and Health Records," *Am. Ind. Hyg. Assoc. J.* **36**, 760–766 (1975).
8. Occupational Safety and Health Act of 1970. Public Law 91-596, Dec. 1970.
9. U.S. Dept. of Labor, "Occupational Safety and Health Standards for General Industry," (current version).
10. W. B. Austin and C. F. Phillips, "Development and Implementation of a Health Surveillance System," *Am. Ind. Hyg. Assoc. J.* **44**, 638–642 (1983).
11. L. T. Daigle and R. H. Cohen, "Computerized Occupational Health and Records System," *Appl. Ind. Hyg.* (May 1985, preview issue).
12. D. K. Burns and P. L. Beaumont, "The HSE National Exposure Database (NEDB)," *Ann. Occup. Hyg.* **33**(1), 1–14 (1989).
13. B. Pathak, K. Marha, and W. J. Louch, "An Industrial Noise Levels Database," *Ann. Occup. Hyg.* **33**(2), 269–274 (1989).
14. AIHA, Management Committee, A. J. Leibowitz, ed., *Industrial Hygiene Auditing, A Manual for Recommended Practice*, 1994.

CHAPTER FORTY-FIVE

Data Automation

Richard A. Patnoe, Ph.D.

1 THE NEED FOR DATA AUTOMATION

This chapter presents the basic concepts of electronic data processing (EDP) systems and describes the steps necessary to plan, design, and implement an effective computer application. Industrial health applications are also discussed and references are provided to assist those interested in pursuing additional aspects of data automation. A data processing glossary is included for reference.

The hygienist in executing his/her duties amasses lots of data. This data comes from various sources and includes calibration data, instrument manufacturers data, observations and comments while monitoring, the use of ventilation equipment and production rates. A hygienist can easily amass one hundred to two hundred data elements in a single sampling event. The data need to be consolidated and maintained in one place so that the integrity of the sampling process can be examined at later dates. Also, it is most useful to be able to compare the results from sampling across several years, with other similar samplings, and to accepted standards of exposure. To execute the aforementioned in an adequate manner can be an exhausting task and fraught with error due to translations and consolidations.

Fortunately, computers are designed to do this. They do NOT take the place of professional judgment, but assist the hygienist in the execution of professional judgment. This assistance is provided by quickly being able to compare, contrast, and translate the data gathered in sampling. Once the hygienist has entered the data in a well-conceived software package, the standard reports derived and the results presented, anomalies are quickly evident. The hygienist can then perform the in-depth analysis needed to examine the results and to assure worker protection.

Patty's Industrial Hygiene, Fifth Edition, Volume 3. Edited by Robert L. Harris.
ISBN 0-471-29753-4   © 2000 John Wiley & Sons, Inc.

These data, which have now been transformed into useful information, need to be shared. Besides industrial hygienists, other people or functional areas that are affected by the OSHA regulations (and data as relates to the regulations) are line management, employees, safety, transportation, medical, development engineering, manufacturing engineering, facilities engineering, chemical control and disposal, purchasing, shipping and receiving, personnel, laboratory, and similar corporate functions. Because of the large workload associated with record keeping, data analysis, and reporting, many industrial hygienists are turning to electronic data processing. An EDP system can greatly increase the productivity and quality of most information collecting, storage, and retrieval operations, while providing more timely and economical data analyses and reports. Automating a large industrial hygiene program is a complex task requiring careful planning and evaluation if costly mistakes are to be avoided.

The emphasis of industrial hygiene has gradually changed from discovering job-related causes of ill health to monitoring and controlling potentially harmful work environment situations before they result in injury to workers or the public. Associated with this modification has been a significant change in industrial hygiene methodology, namely, an increasing requirement for data collection, record keeping, statistical analysis, and reporting. Added to this professional responsibility for data management are the requirements of the OSHA for record keeping, reporting, and tracking a growing list of harmful or suspected substances used in modern industrial processes.

Discussions with industrial hygiene professionals about the demands of their jobs reveal that many are spending a significant amount of time arranging, refining, and manipulating data. Of the time spent working with data, many people estimate that 80% of the time is related to the above tasks and that 20% or less is concerned with interpretation of the data and in adding value to the results. This is a tragic waste of effort since the 80% discussed above are activities that can very easily, speedily, and conveniently be handled by automation. Computers and computing technology do a fantastic job of manipulating, collecting, and sorting of data. When industrial hygienists are working on these aspects of the problem, their talents and value to the enterprise are not being fully utilized. The industrial hygienist truly becomes valuable when efforts are spent in interpretation of data, deciding what additional data is needed, and presenting this data. The use of data automation frees up industrial hygienists to do the advanced and technical aspects of their jobs. Everyone is interested in making the maximum contribution to their job/employer and increasing their worth to the enterprise. Certainly, the employer shares that interest. Thus, the quest becomes one of how to optimize the effectiveness of the industrial hygienist. Automating data management can contribute to this. This chapter discusses how automating data can help to increase this value.

1.1 The Motivation for Automation

Many industrial hygienists pride themselves on the use of state-of-the-art instruments, expert technical advice, and a thorough engagement with the precision and accuracy of their work. Additionally, in planning their work, they are singularly devoted to enhancement of internal (to the industrial hygienist) department work flow. To improve efficiency, efforts are directed toward gathering more and more accurate data. When participating in professional conferences, the sessions attended are related to those that will enhance professional credentials and technical expertise. When assessing capital needs, the individual is most likely to budget for and spend money on a new tool or sampling instrument. If the tool is a software tool, the purpose is to enhance the internal productivity of the industrial hygienist.

The industrial hygienist is concerned with how the data is used and conveyed as information to the management, workers, or customers. The central focus is on the end user of the data. For example, how are the workers interpreting the message, how is management using the information, does management understand the information, and what is the value of the information to the users and to the enterprise? The emphasis is on how this information effects the economic health, reputation, or viability of the enterprise.

As an illustration of how hygienists effect business, consider the airborne monitoring results gathered for a particular chemical. The industrial hygienist focuses on gathering samples, the statistical treatment of the samples, and a means to more quickly gather the samples. In addition to these things, the hygienist is concerned with how management uses the information, how much gathering the information costs, whether fewer samples could be used, whether a less expensive analytical method could be used, how any sample exceeding the standards impacts personnel relations, how this would appear if found in the media, the impact of a high reading on the morale and performance of the person sampled, and how they can help the manager of the area get his or her job done.

Since the intent is to improve specifically the business processes, this very well could mean doing electronic logging of samples, and electronic data system for instrument calibration, a means of direct data acquisition from an instrument, or automating the hearing testing laboratory. Also, if one is considering how to best reach the people on the plant floor or get information from the plant floor, other things need to be considered. Often, plant floor systems can be used to give production figures, area samples, inventory of chemicals, and occasionally usage conditions such as ventilation flow rates or ambient temperatures.

By focusing on the business aspects of the enterprise the industrial hygienist professional contributes both to the overall enterprise and to the industrial hygiene area. Therefore the individual industrial hygienist's overall contribution is increased by considering the overall flow of data in and out of their area or department, provision of information to management in a format that is understandable to them, and with a minimum of resource expenditure. Automation can lower the overall costs if the process is expedited, the accuracy improved, and data and information quickly disseminated to the plant floor personnel.

1.2 Legal Requirements for Data

The purpose of OSHA is to ensure, as far as possible, safe and healthful working conditions for every industrial employee in the nation by providing mandatory occupational safety and health standards. These standards oblige the employer to maintain records of employee exposures, to give employees access to the records, to allow employees the opportunity to observe monitoring or measuring being conducted, to notify the employees of excessive exposures, and to inform them of corrective actions being taken. The government is also allowed access to all of the records. In some cases, record retention requirements are for 30 or more years. In addition to recording and reporting requirements, OSHA standards prescribe the training of the employee, suitable protective equipment, control procedures, type and frequency of medical exams, and posted warnings to ensure employee awareness of hazards, symptoms, emergency treatment, and safe use conditions. Thus the responsibilities of industrial hygiene management become exceedingly complex. In protecting the health of employees, the industrial hygienist must recognize potential health hazards, have them evaluated, assume that controls are in place, initiate exposure monitoring procedures, enter and delete employees from the record-keeping system, and comply with changing government requirements.

2 ELECTRONIC DATA PROCESSING CONCEPTS

2.1 Computer Equipment (Hardware)

Electronic data processing is the handling of data by an electronic computer and associated devices (peripherals) such as printers, scanners, and so forth in a planned sequence of operations to produce a desired result. The many types of EDP systems range in size from relatively simple desktop units to complex systems that fill several large rooms with interconnected devices. But regardless of the information to be processed or the complexity of equipment used, all EDP involves four basic functions:

1. Entering the source data into the system (input).
2. Storing the data in addressable locations (storage).
3. Processing the data in an orderly manner within the system (processing).
4. Providing the resulting information in a usable form (output).

2.1.1 Central Processing Unit

The central processing unit (CPU) is the main body or brain of a computer. The CPU is the controlling center of the entire EDP system. It is divided into two parts: the control section and the arithmetic/logical unit. The control section directs and coordinates all computer system functions. It is like a traffic cop that schedules and initiates the operation of input and output devices, arithmetic/logical unit tasks, and the movement of data from and to storage. The arithmetic/logical unit performs such operations as addition, subtraction, multiplication, division, shifting, moving, comparing, and storing. It also has a capability to test various conditions encountered during processing and to take action accordingly.

2.1.2 Input Devices

Input devices read or sense coded data and make this information available to the computer. Data for input can come from a keyboard, another computer, an instrument, a scanner, and so forth. Regardless of the origin of the data, it must be in a readable form by the receiving computer. The need for input, readable by the receiving machine, is much like person

DATA AUTOMATION

being able to call one of the Arabic countries on the telephone. Unless you speak Arabic (or they speak your language), although the connection is made, you cannot understand each other.

2.1.3 Output Devices

Output devices send information from the computer onto video display terminals, magnetic tape, printers, disk, or drums. They may print information on paper, generate signals from transmissions over teleprocessing networks, produce graphic displays or microfilm images, or take other specialized forms.

Frequently, the same physical device, such as a tape drive, is used for both input and output operations. Thus input and output (I/O) functions are generally treated together. The number and type of I/O devices that may be connected directly to a CPU depends on the design of the system and its application. Note that the functions of I/O devices and auxiliary storage units may overlap, thus, a tape drive or disk file may be used both for I/O operations and for data storage.

2.2 Computer Programs (Software)

2.2.1 Operating or Control Programs

To make possible the teleprocessing networks and the orderly operation of many types of I/O devices that may be on-line with a computer, control programs have been developed. Control programs, also known as monitor programs or supervisory programs, act as traffic directors for all the application programs (which solve a problem or carry out an operation or process data), then relinquish control of the computer to the control program. The control program may be constructed to allow the computer to handle random inquiries from remote terminals, to switch from one problem program within the computer to another, to control external equipment, or to do whatever the application requests.

Operating systems are the control programs that tell a computer how to function. There are a number of commercial products available for computer operating systems. For the PC there is from Microsoft Corporation© MS/DOS©, Windows 95©, Windows 98©, or Windows NT©. Other PC operating systems are OS/2© (IBM), MacOS© (Apple Computer Systems) and GNU/Linux (a freely available operating system that come in a variety of compatible distributions, RedHat, Caldera, Debian etc.) Generally associated with the workstations is UNIX© (American Telephone and Telegraph), AIX© (IBM Corporation), ULTRIX© (Digital Equipment Corporation), and HP_UX© (Hewlett-Packard Company), and Virtual System, VS© (Wang Computer Corporation). The larger IBM mainframes are Virtual Memory, VM©, and Multiple Virtual Memory System, MVS©. Although this list is not exclusive, it includes the vendors and names of the systems most commonly used.

The importance to the industrial hygienist is in understanding that a program compiled for one operating system will not function in another operating system. Thus, when selecting an application, the hygienist must consider the operating system that is used at his or her work site. Selecting a program that has the same operating system commonly employed at the work site can increase the effectiveness of a particular application. Porting is when a program that runs on one operating system is converted to run on another operating system.

2.2.2 Application Programs

Several examples of application programs are a laser inventory program, audiometric testing results, results of air monitoring, and so forth. A more complete list is found in Appendices A and B. Each program is designed to perform a specific number and type of operations. It is directed to perform each operation by an instruction. The instruction defines a basic operation to be performed and identifies the display device, or mechanism, needed to carry out the operation. The entire series of instructions required to complete a given procedure is known as an application (or problem) program.

The possible variations of a stored program afford the EDP system almost unlimited flexibility. A computer can be applied to a great number of different procedures simply by reading in, or loading, the proper program into storage. Any of the standard input devices can be used for this purpose because instructions can be coded into machine language just as data can.

The stored program is accessible to the computer, giving it the ability to alter the program in response to conditions encountered during an operation. Consequently the program selects alternatives within the framework of the anticipated conditions.

2.2.3 Storage

Storage is like an electronic filing cabinet, completely indexed and instantaneously accessible to the computer. All data must be placed in storage before it can be processed by the computer. Each storage element has a specific location, called an address, so that the stored data can be located by the computer as needed.

The size or capacity of storage determines the amount of information that can be held in the system at any one time. In some computers, storage capacity is measured in millions of digits or characters (bytes) that provide space to retain entire files of information. In other systems storage is smaller, and data are held only while being processed. Consequently the capacity and design of storage affect the method in which data are handled by the system.

The amount of computer storage available has increased dramatically and the cost per unit continues to drop. Storage is normally measured as kilobytes (1024 bytes), megabytes (1024 kilobytes), or gigabytes (1024 megabytes) which have been shortened to "meg" or "gig."

Rapid improvements and decreased costs have been made by using magnetic storage. Information is usually stored on mylar tape coated with a magnetic medium or a hard drive which is an aluminum alloy disk coated with magnetic medium. Or more recently, optical storage, where the metallic disk is coated with a material that is sensitive to laser light, allowing greater amounts of storage. With these refined technologies the amount of storage (from an industrial hygienists perspective) is essentially limitless.

Because storage is now cheap, the primary area of concern becomes the frequency and the convenience of access to the stored data rather than the amount of stored data. To illustrate a difficult situation, frequently the means of storing a large amount of data is a magnetic tape. The magnetic tape is stored in a cabinet. In order to retrieve the data, the user (or Information Systems group) must get the tape from the cabinet, load it on a tape drive, issue the command for retrieval, and wait for the data. Unless the tapes are well cataloged, this may not be a trivial exercise.

2.3 Networking

Networking is a means to join and communicate between more than one computer, including desktop computers. The types of networks vary considerably in complexity from the simple linking of two PCs together, to multinational networks linking business operations around the globe. It is necessary to discuss some of the simple terminology and meanings in this environment.

The joining of two PCs together is simple and the hardware and software can be purchased as a kit for less than $100 at this time. The joining allows the sharing of data across more than one machine and also may enable the use of a common printer, scanner, or other peripheral device. The requirements are simple, a card slot in the PC, some cables, and the software.

For a single site, the LAN, local area network, is the preferred solution. This will link all of the users of an application program on a site to a common database. The WAN, wide area network, is for use on an enterprise level. Although theoretically sound, the WANs to date have had some performance problems. The lack of speed of transmission, which most users have come to expect since they are working on an single stand alone PC, has not been met in the WAN environment. The lack of speed is particularly evident when transmitting scanned images (possibly MSDSs), which may have a packet size of 500K or larger. The transmission lines do not have enough band width (capacity) to accommodate all of the traffic on the lines. The result is slow transmission and degraded performance.

Largely, the hygienist will have to accept the networking system provided by the corporate enterprise or the site standard since the maintenance of the network is beyond the responsibility and knowledge level of the hygienist. However, a discussion with the Network Manager should be undertaken to understand what is available and the requirements that will be placed on the end user. Before entering this discussion, be prepared to tell the Network Manager what results you expect and how your system will be used including who and where people will be connecting. The importance of the network should not be overlooked since it frequently determines the speed of access to data, the ease of maintenance, and convenience.

3 METADATA, PUBLIC DOMAIN DATA, AND PRIVATE DATA

The data are the province and responsibility of the hygienist, and thus, deserve the most attention and understanding. The results and usability of the data depends upon the industrial hygienist. The hygienist must be able to elucidate very specifically how the data will be used, who the users are, the need for records retention and who will have access to the data.

Data can be broken down into three categories which are (*1*) data about data (metadata), (*2*) data that are generally available in the public domain (public domain data), and (*3*) data that are unique to a company (private data). Examples of metadata are information used to describe types of data such as numerical, integer, character, time, or decimal. A further discussion of metadata follows later in this chapter. Examples of public domain data are standards such as OSHA PELs, analytical chemistry standards, laws, legal case histories, etc. Examples of private data are sampling results, company/plant standards, location names, equipment names, and definitions within one's plant.

The category of metadata is information that describes the data and is best understood by providing an example. For example, time and date data (8:15 A.M., September 16, 1999) can be written in a number of different formats. The format is in itself a type of metadata. For illustration, the date may be written in a number of different ways such as MMDDYY (102898) or 10/26/98 for the date October 26, 1998. Some programs will keep the date and the time in two different fields, others will combine them into one field. If the time and date are combined into one field, it can be written as a fourteen character string, YYYYMMDDHHMMSS. If the dates in two sets of data are formatted differently, subtraction and addition of times and dates are not possible. They must be formatted in the same way before addition or subtraction can occur. Another example may be that of significant figures. If some of the numbers have two decimal accuracy and other less accuracy, then the results will assume the characteristics of the least accurate. The usual risks of significant figures then apply.

The hygienist is often called upon in professional practice to use both private and public domain data and compares personal data such as sampling data to known and recognized standards (public domain data). This creates the problem of merging the two different data types and in many instances creating additional data from the result. For example, if a hygienist samples for dust in the work environment, the results must be compared to historical data (of a private type) and to public domain data such as standard values found in a source, say the ACGIH Threshold Limit Values. In comparing to the standard values, a new set of data, derived data, is created which could be of the types "exceeds" or "does not exceed." The derived data is no less important than the private and public data, particularly when archiving data for long periods of time such as that required for sampling results.

3.1 Metadata and Data Types

The first class of data is metadata. Metadata is data (information) about data. There are a number of field types such as character, integer, decimal, time and date, and logical. For illustrative purposes, they are discussed.

Character. A character field contains those symbols found in the ASCII (American National Standard Code for Information Exchange) character set. Some of these are found on the standard English keyboard and others are not. They include the letters of the alphabet, symbols such as *, ?, #, @, ~, etc. In a given programmatic language, some of these symbols, as well as some words, are reserved for use by the programmatic instructions. For most nonprogrammer personnel, the word, "hygienist", which contains 9 characters from the alphabet would be an example. However, the hyphenated word, "user-

DATA AUTOMATION

friendly" contains a symbol not found in the 26 letters of the English alphabet. This hyphen is a perfectly allowable character as is a blank space or one of the symbols previously cited. It is important to know that since this is the case, the field length, 9 in the case of hygienist, or 13 in the case of user-friendly represent the field length. In the latter example, the hyphen is counted when determining the number of characters. If there were a blank between user and friendly, this would also be counted. Thus, this illustrates a type of data and the number of characters. Field length is discussed later.

Integer. Integer data is that data represented by the numerals 0, 1, 2, 3, 4, 5, 6, 7, 8, 9. The integers may have either positive or negative values.

Decimal. The decimal values have a decimal point in them. For example, 12.3 is a decimal. This is a straightforward example, but becomes complicated when doing calculations or manipulating data. For example, a field may allow for two numerals to the right of the decimal point. If one wishes to enter only one position to the right of the decimal or alternatively, 3 points to the right of the decimal point, there is a problem with the data entry. This seemingly innocuous example becomes very important when migrating data from one system to another or when doing calculations involving significant figures.

A further complication is introduced for those with operations in Europe, since the Europeans typically use a comma where Americans use a decimal, and vice versa. These factors can make the seemingly trivial task of data migration very complex and require a lot of decision making, much of which must be done by the hygienist rather than the programmer or data analyst.

Date and Time. The problem encountered with the time type of data is one of multiple formatting options. In some cases, the date and time are kept in the same field and written as in the following character string, YYYYMMDDHHMMSS, where Y is for year, M is for month, D is for day, H is for hour, M is for minute, and S is for second. In other programs, the time and date are kept as separate fields, and this is still does not include the recent problems with the Year 2000. For Americans, the date is written MM/DD/YY. Europeans would write the same date as DD/MM/YY. The problem becomes more complex when the slashes are not used. The reader can use his/her own imagination to identify date and format problems.

Logical Data. The logical field is a "yes" or "no" field. The field must read either one or the other. There is no maybe. The use of logical fields is much more common than some might expect. A few examples are answers to the questions: General Ventilation? Normal Operations? Reference Sample? Since the logical field can exist in only two ways, yes or no, the value is pre-set to one value or the other value. In short, the field has a default value. The field cannot be blank, it must be either yes or no. Thus, whereas character and numerical fields can be empty, a logical field cannot. This can give rise to some problems when analyzing data.

For example, if one were to analyze all data taken during normal operations (logical of yes, means normal operations), one outcome would be obtained. It is unlikely that the same result would be obtained if conditions were abnormal (default value of no). Knowing the value of the logical field is essential to understanding the results. How can this happen? If the default value is "yes" for the question of whether the operation was under normal conditions and someone fails to reset the value to no, meaning abnormal conditions, the composite results over a number of samples would produce misleading results.

Each of the above data types has its use and purpose and it is quite likely that all of these will be found in an application, so knowledge of at least these data types is necessary. There are others such as scientific notation, logarithms, certain symbols not found in the standard ASCII character fields, etc.

In summary, there is metadata, which is data about data. This includes not only the data just discussed, but such other information as the length of a field, the data format, a definition of the field, whether the field is an index in a table or whether the field is mandatory. In some cases the format, which includes the number of decimal places, can substantially effect the results.

Without going beyond the scope of this chapter, there are a couple more concepts of interest to the hygienist. If a field is an index, the rate at which data can be sorted (found), is increased significantly. An index also enforces uniqueness on a record, since no two records can then have the same index. A key field, usually also an index, is used to sequence stored records of a particular type within a file. An example of a key field in an employee record is the employee identifier/employee number/serial number. For example, when the employee number is an integer the records would be stored in sequential order.

3.2 Public Domain Data

Publicly held data is that data usually considered to be available to all and is not of a confidential or sensitive nature. Several examples were cited earlier but a more in-depth look at some items is appropriate.

Regulations. Typically, the regulations under consideration are legal ones of either a federal or state origin. The regulations are available to anyone. There are a number of different companies who vend these regulations. Among the better known vendors are I.H.S. Regulatory Products (*1*) and LEXIS-NEXIS (*2*). Although the regulations in electronic form can often be obtained from the agency, the commercially available packages have very powerful search engines and often include value added information such as case law, citations, and interpretations of the regulations. The commercial packages also can be a single source for both federal and state regulations, which is particularly useful where the industrial hygienist must consider multiple jurisdictions.

Another advantage of the commercially available packages is their ability to discriminate when doing searches. The problem usually occurs when a key word search is done and their are literally hundreds of "hits." The next step is to sort through this large number, a task that cannot be considered as trivial. The use of Boolean logic (and, or, and not, adjacent, etc.) in the search can refine the search considerably and get the hygienist closer to the appropriate materials.

Standards. Standards such as the ASHRAE, the American Society of Heating, Refrigerating and Air-Conditioning Engineers (3), are set by industry as standards of good practice. Standards are often incorporated into the regulations by reference but are available separately. Many analytical standards are maintained by the American Society of Testing and Materials (ASTM), and are used as analytical chemical procedures and practices. These organizations publish hard copy as well as electronic versions of their materials. Links to other sites are also frequently found on their Web sites. The Occupational Safety and Health

DATA AUTOMATION

Administration (4) maintains a web site as well as the Environmental Protection Agency. Material from the sites can be downloaded in a variety of formats.

3.3 Private Data

As stated above, there are a number of different types of privately owned data. A few important characteristics regarding this data are discussed below.

Monitoring Data. The number of data elements for this particular data is quite large. Also, many of the conclusions are based upon the notes and ancillary data that accompanies the sampling event(s). Some of the common data elements for personal air sampling are date, time, location, person's company identifier, analytes, calibration data, nature of activity, engineering controls, personal protective equipment worn, production rate, and whether the operation is operating normally or not. Many of these previously listed elements have been simplified since an element such as location can within itself have several aspects such as facility, building, production line, room, name, model and manufacturer of a particular type of equipment. Ideally, the location could be linked to a Geographic Information System, GIS, which is supported by members of the production force or facility engineering group.

The sampled party or person has much additional needed data that is linked to it. The additional data includes but is not limited to gender, ethnicity, first name, last name, department, internal mailing address or fax number, manager's name, in some cases age, job identifier, secondary job identifier, and employee status such as part time, contractor, or regular employee.

Since it is burdensome to collect and maintain all of this data for a sampling session, it is convenient and highly efficient to obtain data from other sources if possible. The personnel related data can be obtained from the personnel system. The problem then becomes one of availing oneself of the data from this source. With the cooperation of the Information Systems Manager, this should be possible. Assuming that the security issues regarding confidential employee information are surmounted, the real question becomes the ease of getting this data, time required, amount of information, and periodicity of refreshing the data. Employees would seldom maintain the same job identifier over a working lifetime, but it is important to capture that information at the time of the sampling and maintain it. The implication of the last statement is that there must be an employee history capability within an electronic system so that the job code *at the time of sampling* is maintained.

Material Safety Data Sheets. Although this data is quasipublic data, experience has shown that each facility or company has its own unique set of data. From the MSDS, the components of a substance as well as the percent composition can be obtained and maintaining this information is implied from the federal regulations. Frequently, the at-the-time values for OSHA PELs, statements regarding toxicity and precautions, and the evaporation rate are found on the sheet. A means must be found to maintain this information. There are a number of commercial vendors of material safety data sheets of which the Canadian Centre for Occupational Health and Safety (CCOHS) advertises to have several hundred thousand sheets. Unfortunately, the nature of industry is such that the sheets become outdated and new materials or even renamed or repackaged old materials may not show on these large files until some time after their introduction into commerce. Moreover, in most cases the datasheets are manufacturer or supplier specific, and thus, not useful if a different manufacturer or supplier is used even when the material has essentially the same composition. Since the record for exposure sampling must be kept and the identity of the material maintained, some MSDS data must be maintained.

An effort by the European Economic Community (Common Market) has resulted in hazard information on International Union of Pure and Applied Chemistry named materials for the 1500 most common substances in commerce in Europe. This is to be expanded to around 5000 materials in 1999. Both of these sources of information are available at modest cost, currently from about one hundred to several hundred dollars. There are numerous other sources of material safety data sheets.

The data sheets frequently refer to other standards, some of which may be obtained on-line. The example shown below is for butadiene and is taken from the *Federal Register*, 1910.1051 Appendix A.

> Disposal: This substance, when discarded or disposed of, is a hazardous waste according to Federal regulations (40 *CFR* part 261). It is listed as hazardous waste number D001 due to its ignitability. The transportation, storage, treatment, and disposal of this waste material must be conducted in compliance with 40 *CFR* parts 262, 263, 264, 268 and 270. Disposal can occur only in properly permitted facilities. Check state and local regulation of any additional requirements as these may be more restrictive than federal laws and regulation.

The maintenance of the personnel data and facility/site/location data is not the primary domain of the industrial hygienist. In the plant environment, others are responsible for the establishment and maintenance of this data. The industrial hygienist is the user of the data, and thus, has a vested interest. The issue is getting this data on a timely basis from the owners and maintainers and in a form that is usable to the industrial hygienist. Each hygienists, or at least at each facility or within each company, this poses different problems. Obtaining the personnel data on a regular basis is necessary because of changes that occur in the data, new job, new employee, etc. Thus, a refreshing of the data on a period basis, perhaps weekly, is needed. These repetitive tasks can be accomplished electronically and for efficiency can be set up to run as batch jobs. This data needs to be maintained by the industrial hygienist.

The question becomes, what personnel information must be brought over for use by the industrial hygienist? Is the data being used for Occupational Illness and Injury reporting? The answer depends on how the data are to be used. Minimally, it is reasonable that the data must include employees first, middle, and last names, employee number or social security number, gender, age, job code and description at the time, and a secondary job code, which may refer to ancillary duties such as emergency response team, etc. The implication of the last items, job code and description, is that the job codes will change over time. If reporting is done from a relational database, the job code "at the time of the personnel sampling" must be used, not some previous job code or successive job code.

Consistency of data entry is another problem that occurs within a business or enterprise. If data is to be summarized, it must be summarized around common terms. As an example, if the data refers to sampling within a particular building, e.g., building two. This building designator can be written in a number of different ways such as B002, B2, Building2, Bldg002, etc. All of these may be recognizable to a knowledgeable person, however, to a computer, they would all refer to a different building. Thus in aggregating data, sorting data, or filtering data, different results may be obtained dependent upon the designator. The author recommends using the designator applied by the financial services part of the business since this is apt to remain the most consistent over time. With a properly engineered software program, the building designator can be taken from a pick list (list of agreed upon designators) and then the problems are eliminated.

3.4 Combining and Comparing Personal Data with Public Data

After obtaining the personal data and the public data, the next step is to use the two sources together. For example, a hygienist performs an air sample and calculates a TWA for an eight-hour period based upon the sampling results. The next step is to compare this data to that of legal standards. Figure 45.1 is a graph used to illustrate the point.

The sampling data, shown on the x-axis is personal data. The line across the graph designates the TLV taken from an outside source, ACGIH (American Conference of Governmental Industrial Hygienists). The preparation of this report utilizes data from the two separate sources, one personal and one public, and in effect could be used for an exception report, i.e., those TWAs that are greater than the ACGIH Threshold Limit Values. In the Figure shown, sample number 4 represents a case where the TLV is exceeded. This new piece of information, those samples that exceed the TLV, is a derived piece of data based upon the two different data sources.

3.5 Long Term Data Storage Problems

The practicing hygienist is well aware of the need for data retention for employee exposure records of thirty years required by the OSHAct (5). This presents some particularly vexing

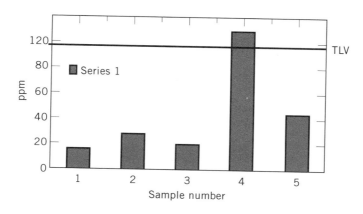

Figure 45.1. Threshold limit values for dichloroethane.

problems. These problems stem from the rapid and expected continuing advances in computing systems. If one looks back about 30 years, the PC was unknown. The author sees no reason that the pace of change will be any different in the next 30 years. To envision what will exist is beyond the capabilities of the author. But, there are still some steps that can be taken to assure that records can be recovered.

These records are a personal opinion based upon 25 years experience in the health and safety field. During this period of time, it has been necessary to go back and reconstruct records various times. But first, one must consider some assumptions. It is assumed that: (*1*) any software running today will not be running in thirty years; (*2*) the hardware operating today will not be in use; (*3*) the operating systems will be obsolete.

So, what are the options? There are several. The hygienist can:

1. Save hard copies of the monitoring data. If needed, these may be placed on microfiche or similar method of capturing the records such as scanning the documents.

2. Save a database containing the data. This is different from the data and includes the data structures, table relationships, indices, etc. The advantage here is to be able to manipulate the data. The author is reasonably comfortable that if the data is ANSI SQL compliant, there will be a querying tool that will allow searching of the database, related tables, and fields, and the generation of reports. To facilitate this, it is extremely helpful to have the metadata (described above) to assist in the generation of the queries. The recommendation is to retain this data and data diagrams in hard copy since this will not be that much data and can easily be stored.

The Application Program and the database are separate parts of the programming package. Although programs vary, it is frequently possible to access the database without the presence of the program. The access may be controlled by the operating system security programs and this issue needs to be taken up with the system administrator.

3. Save the data in the database and the program. Most programs are upwardly compatible, i.e., a more recent version can read an older version. Thus, data may be used by newer and more recent versions of the program. The bold presumption is that the same program is still in use and available. For reasons of confidence, it is recommended that if this method is used, it is in addition to those above. This is a precaution because of the inability to predict what will be the situation in thirty years. For a few years, one can have comfort, but thirty years is a long time span for the active life of any software program and spans the career of both hygienists and programmers.

One final precaution should be taken. It is best to use standard rather than proprietary types of data storage. When an ANSI, ASCII or other standard exists, the standard is the recommended option. For example, for documents that are text documents, the extension of 'txt' should be used instead of proprietary extensions such as "doc" or "wpd". Because there is proprietary formatting embedded in the files, it may not be readable by a reader other than the original proprietary reader. The downside, of course, is that some of the imbedded formatting may be lost and the new reader will not have the exact formatting of the original.

3.6 Data Conversion Problems

Among the more simple problems conceptually, but more vexing technically is the problem of the conversion of data from one program to another. This occurs when legacy data is to be put into a new software program. This legacy data may be from a previous relational database program, from spreadsheet files such as Excel™ or Lotus Notes™ programs or from a number of simple relational database programs like Access™ or DBIII™.

There are a number of problems. These problems are related to the previous section on data types and metadata. If the reader has not read that section, he/she is advised to do so before reading this section. Some of the problems are

- No corresponding field—the table for the new application may not have the same fields as the old data. Also, the new field may have a mandatory field that is not found in the original data.
- Field lengths are different—the new field may not have as may spaces or characters as the old one and the data may need to be truncated if the new field is too short or padded if it is too long.
- Field type is not the same—for example, one field may be an integer, whereas the other is a decimal field.
- Field format is not the same—for example, the date field may be formatted differently such as frequently occurs with date fields.
- Format is different—for example, one may have the hyphens in a CAS, Chemical Abstracts Service number, and the other may just have the digits.
- Tables are different—the two programs have different fields in their respective tables.

There are a number of ways to cope with each of the problems stated above. This section is included to apprise the hygienist of some of the problems that may occur. There are ways to resolve each of the above problems, but the solutions are likely to be beyond the capabilities of the hygienist, who is advised to seek expert advise before starting the data conversion. Remember, scrambled data may be worse than no data at all.

4 BUSINESS CONSIDERATIONS

4.1 Cost to Automate

The cost to put in place a data management system varies considerably. A simplistic view is to consider only the cost of the software. This is a mistake! This single item may be a small part of the overall cost of implementation. As a base, the hardware costs, software costs (both operating system and application), integration services to connect the application program to other programs, personnel time including both the time to select and specify as well as to train to use, and travel to visit other sites constitute the system cost.

There are numerous trade-offs that will occur in this process. The cost of a solution may be cheaper if it is a PC solution, but it may not have the functionality, capacity, or speed of a more sophisticated solution. It is tempting to buy little pieces of several different functions from different vendors. Then the systems would not operate in the same way, training costs go up and the utility of the solution suffers. A third example is keyboard entry of material safety data sheets. If the number of sheets is small, keyboard entry may be cheaper than the use of image, but accuracy and speed of entry is lower.

The costs of software alone can be almost nothing if some programming is copied out of a book to do simple things such as calculating exposure limits. Some examples of this are found in *Computers in Health and Safety* by the American Industrial Hygiene Association. There is also the opportunity to get these already on a diskette by paying a small fee, a bit of commerce known as shareware. Single-application-only software typically costs from $500–5000 if purchased from a vendor. Some vendors also offer a time share function where you can put your data on their hardware and then pay a charge for the amount of processing and the duration of the connect time. The connection is made through a modem installed in the computer and connected to the vendor over a telephone line.

4.2 Size of a System

Since it is assumed that the reader is familiar with the personal computer, it is well to discuss some of the other computing solutions available. Computer analysts break down the types of computers into four classes: personal computer, workstation, midrange system, and mainframe. There is considerable overlap in price and function between these classes. Also, these different computer sizes are linked together and processing of data occurs on all of the systems, the advantages of each size of system being optimized. In general as one moves from PC to mainframe, the power, speed, and capability increase along with the total price for a complete system. These computing systems can be single user/single task, single user/multiple task, multiple user/multiple tasks. In general the PC using MS/DOS© is a single user/single function system. This means that only one person can be using the one function at a time. IBM's OS/2©, or Operating System 2, is a single-user environment but a multiple-tasking environment. With OS/2© on a single PC, the PC may be collecting data from an audiometric measuring device while at the same time printing out a chart of the data. In the workstation environment, usually associated with reduced instruction set computing (RISC), or with a large mainframe, many users may be connected to a system performing multiple tasks at the same time such as accounting, word processing, database searching, graphics processing, industrial hygiene applications, and process control.

4.3 Steps in EDP Application Selection Development

In designing and implementing any new computer application, there are logical steps of analysis, planning, development, testing, and installation that must be carried out. A means of addressing some of the problems is now described.

4.3.1 Getting Help

In the same way that a person from the factory floor looks to the industrial hygienist to provide specialized expertise, the industrial hygienist should look to the computer specialist to provide special skills and insight. Too frequently, the industrial hygienist begins data automation by seeking out software providers. Although this step will provide some valuable input into what functions are available and some approximate costs, it will NOT provide the necessary internal foundation to make wise investments in data processing.

One of the first steps necessary is to get someone with the needed systems background to aid with the decision. Before making a decision on the size of a computing system, what database design is optimal, what are the data transfer and networking requirements, it is well to consult an expert. Most industrial hygienists do not possess the requisite background to answer all the questions successfully particularly in an efficient and timely manner. The first place to start is with a systems analyst. The job of the systems analyst is to work with the end user in defining requirements and system selection. All large corporations and most locations and divisions of corporations have someone with either this title or with these skills and duties. Far too many industrial hygienists, in an effort to go it alone, overlook the first step—seek an expert. In all probability, the industrial hygienist can use the extra help also since they are extremely busy.

If a company systems analyst is not available, all of the large computer manufacturing companies have staff available for this purpose. The service of a systems analyst is also available through companies whose primary business is systems integration. The specialists are available by consulting a telephone directory or asking for advice from professional associates.

Once the specialist is on board, he or she will be able to guide the industrial hygienist through the entire process of defining requirements, software selection, providing a justification, giving opinion of provider expertise, providing some reality testing, and assurances that the decisions are the correct ones.

Deciding what, how, when, costs, timing, and so forth for automation are big decisions. It is well worth the time and effort in planning the approach to the problem with all the expertise available.

4.3.2 What Is Currently Being Used at Your Location?

Operating systems are the control programs that tell a computer how to function. The importance to the industrial hygienist is in understanding that a program written for one operating system will not function in another operating system. Thus, when selecting an application, the industrial hygienist must consider the operating system that is used at the work site. Selecting a program that has the same operating system commonly employed at the work site will increase the effectiveness of a particular application and maximize efficiency of the department.

4.3.3 Defining and Sizing the Scope of the Automation

When a new computer application is desired, the first task is to clearly define the problem. A report should be prepared that states clearly and concisely just what is to be performed by the proposed system and the scope (size) of the effort. This report describes the problem in terms of subject, scope, objectives, and recommendations. This report will establish the basis for communicating needs to the system analyst, to management, and will form the foundation for a proposal for purchase or coding of a program.

4.3.4 Analyzing the Problem

The analysis phase of a design effort is not an isolated step. The system analyst, who views a new computer application in terms of its scope and objectives, works closely with the users to determine their needs, the information in use in the present system, and the information needed in the new system. The analyst must also consider what equipment would be the most cost-effective for the necessary functions of the new system, determine the basic functional specifications, approximate price, duration of project, personnel requirements, and contractual performance requirements.

4.3.5 System Selection

Some broad policy level decisions need to be made as a necessary first step. First, the decision must be made as to how many persons the system is to serve. If the answer is a simple application such as recording the building and type of laser at a location for use by one industrial hygienist, the task is straightforward. If this is the only requirement, there are number of very simple databases that could accommodate the needs, and a simple PC can accommodate this task. Frequently, the need is much greater, even for an individual with a simple concept. As in the example above regarding lasers, the industrial hygienist would like a clock or calendar function so if a new laser is registered, the data and time can automatically be noted. If the industrial hygienist wishes to sort by laser type, another variable is needed. Tracking of the users and their education or retraining could be another addition to the system as could the location, last date of inspection, and so forth. Further, it is desirable to have the education record added to the personnel file, so some integration of the data is useful. Another consideration is whether other hygienists or associated personnel such as physicians need to access the information during the normal course of their work or during absences of the industrial hygienist. Thus, it is advantageous if others have knowledge of the way the data management system works so that access can be quick, easy, and reliable. As this simple example illustrates, the problem grows quickly.

The primary question facing the industrial hygienist is "Do I need an automated system for my data?" It depends, is the answer. Although this is not a very satisfying answer, it does state the truth. There are many conditions to be reviewed. Some of the questions are summarized below:

- Quantity of data to be managed?
- How will the data be used? Federal or state reports? Management reports?
- How will the data be accessed? Terminal? Batch reports? Other applications?
- Who needs the data besides the end user? Corporate? Medical? Division?
- How often, when, and under what circumstances do I need the data?

If the decision to automate is affirmative, then several other questions need to be answered:

- Which applications will be automated and in what priority?
- Do the applications stand alone or do they require data from another source such as material safety data sheets?

- Is the project going to be combined with another such as the automation of environmental, medical, or safety data.
- How is the investment justified?
- Is a customized system needed? Semicustomized? Off the shelf?

It is advisable to make every effort in answering all of these questions correctly and thinking through how the data management system will be used. This will save much time and labor later when it is time to bring the system into production.

4.3.6 Combining Needs with Other Areas

Each person wishing to automate their data needs has to consider many things besides functionality. To the industrial hygienist, the paramount issue may be air sampling data, to the environmental engineer it may be EPA's Superfund Amendment and Reauthorization Act report (also known as SARA, Title 3, form R, or SARA 313), or to the safety engineer, it may be accident data tracking. One legitimate question seems to be whether to combine the project(s) for each of the above four organizations or to go it alone. Experience has shown that compromises are hard to reach since each party wants to have his or her functions done first or has some special need that the others do not have.

How do you resolve this problem? One means is to have a structured discussion format with moderator and analyst. In the study of information science this method is known by various names but the words joint application design (JAD) or joint application requirements (JAR) are often used to describe the techniques. The JAD or JAR has five fundamental pieces: (*1*) A top-down system design process with an executive sponsor who can resolve issues if not resolvable at a lower level; (*2*) participants work together in a group session generally of about 8–15 people; (*3*) synergy is developed between management, users, and information systems; (*4*) the participants jointly produce requirements and (if a JAD) design specifications; (*5*) the project is facilitated by experienced people, typically one facilitator and one analyst. The outcome of the effort is a prioritized list of requirements for the project and a design if carried through to this point. Appendix C gives a more detailed description of a JAD.

4.3.7 Justification of the Investment

Cost justification for automation is a difficult but necessary task. There are three fundamental ways that savings can be accomplished by the use of computerized data management system: (*1*) the reduction of manpower used to perform specific tasks such as the preparation of reports, summaries, and planning documents; (*2*) better analysis of data for accurate assessment of trends, quicker analysis of data, more complete analysis such as with the use of statistical tools and patterns analysis; and (*3*) rapid transmission of this data to those who use the data such as production managers, corporate functions, and so forth.

Regardless of the business model used, most justifications will include reduction in clerical time, shifting of responsibilities from more expensive personnel, such as industrial hygienists, to clerical or data entry personnel, ability to respond quickly to requests for information such as an OSHA visitation, making the data available to others either in the department, within a division, to an overseas affiliate, or to the corporate offices. Although hard to accurately quantify the benefit, the ability to respond quickly and accurately to questions posed by production floor managers, is certainly valuable. With a computerized database, the industrial hygienists can respond to these requests quickly and accurately, can prepare a presentation for employees in a reasonable time, and then go on to do other things. Better use of existing data, such as on statistical comparisons, epidemiology, etc., are greatly aided by computerization.

4.3.8 Breaking Down the Costs

The costs of a system may be broken down as follows:

- Costs to purchase
  Hardware
  Software
  Personnel
- Costs to produce
  Initial costs for the functions sought
  Systems analysis costs
  Programming costs
  Testing costs
  Training costs
  Implementation costs
  Documentation costs
- Costs to operate
  Day-to-day costs
  Repair costs (hardware and software)
  Maintenance costs (hardware and software)

4.3.9 Spreading the Costs

Prudence is needed in estimating the costs. For example, a particular application may require a graphically enabled monitor. It is more expensive than a monitor with less capability, but since a monitor is being purchased for some other reason, only the incrementally larger purchase price should be used. If a workstation is included in the purchase order for this application, is this the only application for which the workstation will be used? The same logic holds for larger hardware items such as an upgrade to the size of the mainframe because of additional load. It is tempting to add this upgrade cost onto the price of this one application even though there are likely many other applications running on the same system. The rhetorical question is "Is it the first application put onto the computer system causing the upgrade or the most recently added application?" Larger firms will have a methodology for coping with this seemingly intractable problem. Unfortunately, too often the justification model calls for putting all of the costs onto the most recent project (yours). There is no easy solution to this problem, but recognition that this justification methodology may be used will help to prepare your business case. One must become familiar with the business case used in justification of investment in EDP. Ask the information systems personnel for model proposals.

There are similar problems to consider when purchasing software. There is a range of tools that are used by various applications. They may be graphical enablers (those software packages that allow you to use a graphic system), imaging systems that may be used for many applications in addition to the industrial hygiene applications, specific workstation software that may compress and decompress images for storing or viewing, statistical packages, instrument connects, and many more. These special tools packages are usually called by an application program when needed. Some are built into a particular software application, but others are called when needed. A good example of the latter is a printer driver program. Most applications need to print out results, thus there is no point in the printer driver being in every application, so it becomes one that is called up by the application as needed. This saves valuable memory space and is more efficient then writing the driver for each application. Also, if the printer is changed, you do not want to change each application to accommodate the new printer.

If a software package is a commercial off-the-shelf system (COTS), the base cost of the application package needs to be determined. Different vendors of software use different pricing structures. Some are based on the size of the CPU processor, others use as a basis the number of users, still others price by the number of employees at the location, and still others use a fixed cost. Regardless of the software providers pricing structure, consideration needs to be given as to whether and where there is training, is the training billed separately or a part of the initial cost, is it on location or at the vendor's location, etc. Also, is there any customization included in the cost? The base cost of the system may NOT be the ultimate cost if customization is required. If this is a customized package, the internal costs of a systems analyst, programmer, etc must be considered. These are costs to the company or site. Sometimes these costs are charged directly to the internal customer (the industrial hygienist), sometimes they are fixed fee, and sometimes they are covered in a site overhead-account. Each alternative has a different impact on your budget. Understand this impact before committing!

4.3.10 Common Look and Feel (Usability and Familiarity)

If a software package is purchased, it is still likely that there is a need for some systems analysis time and some customization of the software. The screen design (this is the way a computer screen appears) typically replicates some form that is in current use. Significant decisions need to be made at this point since a computer screen has a different number of lines than an 8½ by 11-in. sheet of paper (there are large screens that can accommodate the whole page, but this will increase costs). The users of the system will be much more accepting if it looks familiar. Thus, if the function key (or F key), $F = 12$ is usually used for filing, it is wise to use this same key on your program. If the help key is $F = 1$, then this ought to be the one used in your program.

Custom front ends, that screen that is visible to the user, may be created. This adds costs, but also increase acceptability. Some applications are now being made with a Web front end. These use a web-browser front end that links to a database in the background. The use of a web front end has the advantage of a familiar screen and the advantage of a commercial database in the background. It is some advantages of both the "build" solution and the "buy" solution. It is expected that this will become an increasingly popular way of deploying the application program.

In addition to functional characteristics, buy or build decisions, there are some general considerations that are of great importance. The first of these is the usability of the system. Regardless of the power of the system, if it is difficult to learn and use, chances are pretty good that the user will not be satisfied with it. If the decision is made to use several smaller packages, then a number of problems are incurred. One of the real advantages of a common system for a company is the ability for all to use it easily and readily. Another way of saying the same thing is to minimize training costs. If a common system is used, normally one supplied by one provider or vendor, the following advantages will occur. They are:

Minimized training costs

Minimized development or customization costs

Purchasing leverage

Ease of integration with other applications

Remote locations with equal access

The author would be remiss if he did not discuss a down side to the use of a common system. At the time of this writing, it is believed that the only vendor offering a single system for environmental, safety, and industrial hygiene is Quantum Compliance Systems (6). A listing of software programs is typically found in the July issue of the *Journal of Occupational Health and Safety* (7).

4.3.11 Buy or Build

The most critical question to be answered is "What are the requirements for data automation?" If this question is not answered, then it is foolish to even consider the buy or build decision. The hygienist must first determine exactly the need that is being filled. Once this determination is done, then one can begin shopping for what is available. Unfortunately, many people start out by looking at software packages rather than analyzing their own data processing needs. This is truly the cart in front of the horse. While looking around for ideas is certainly an acceptable way to gain insights into what is needed, the base decision comes back to the industrial hygienists' needs.

The list of choices available in making the buy or build decision is as follows:

1. Buy a software package and integrate it into an existing system in the company.
2. Buy a hardware and software package that can be integrated into another system being developed in house.
3. Buy a hardware and software package that will meet the complete needs of the project with no major modification.

DATA AUTOMATION

4. Buy a hardware and software package that meets part of the needs (50%) of the project and contract with the vendor to develop the remaining requirements.
5. Buy all hardware and operating software and contract with a vendor to develop the application software according to your functional and technical requirements.
6. Buy all hardware and operating software and develop the application software completely in house.

None of these options is exclusive. Normally, some additional hardware must be purchased, some additional software must be purchased. Additional hardware is usually needed if facsimile devices are used, scan images of material safety data sheets are used, more people are using computers, etc. Even if application software is purchased, there may be some additional report generation features, scanning software, graphics programs, etc., that are needed. Because different means and ways of doing business are used, the methodology will change. For example, if the calibration of the instruments is part of the software, there is no need for calibration for the equipment.

In the seventies and early eighties, there was a dearth of software programs available for the industrial hygienist. Today, with the popularity of workstations, midrange systems, and mainframes, the industrial hygienist has much more from which to choose.

A few additional comments are appropriate in making the buy or build decision:

- Cost effectiveness is a major factor in the decision process. But remember that if the system is not used by the right people for whatever reason, the cost per user rises proportionately. The lowest total price is not necessarily the right choice. The cost should be prorated over the entire lifetime of the product, the number of users, the costs to run the system (computer people like to say the number of cycles that are burned), and the cost maintenance as well as convenience of use.
- The buy or build decision process does not reduce the need to assess what it is one wants to do with the information. It does not allow skipping any of the important phases of the project. A feasibility study (a pilot project is recommended) and a functional system design are still required to determine the project requirements.
- The buy or build decision will carry over and greatly impact future project tasks such as testing, installation, documentation, training, implementation, and maintenance requirements.
- The buy or build decision should take into account how soon the product is needed. Build will usually take much longer to get to production than when buying a system ready to go.
- What are the long-term ownership requirements and future maintenance requirements? Does the hygienist want to be able to do the work by him/herself.
- Does the hygienist want to retain full ownership and possibly market the product at a later date? If a marketing organization is not geared to this product and customer set, funded for the effort, and imbued with the skills, prospects for success are dismal. Regardless, the business case should be built on one's needs and any fortunate outcomes considered to be gravy.

4.3.12 Selecting an Area for Automation (A-Pilot Effort?)

There are numerous application areas that could be selected for automation. These include air sampling data, training, lasers, and so forth. The reader is referred to Appendices A and B for a more complete list of the possibilities. By using a weighting system for the applications some initial assessment can be made. However, caution is needed in this selection since an item with lower priority may be a stand-alone application and can be done easily, thus the pay back is quick and the solution simple. If a simple application is selected first, experience is gained in the automation process that may be applied to future automation efforts.

4.3.13 Implementation

Too often, industrial hygienists believe that the job is finished when a selection process has taken place or a functional specification has been written. Nothing could be further from the truth. Now the real work begins. The implementation phase is absolutely critical to success of the project. No matter how careful requirements are gathered, no matter how much probing and examining is done in the selection or programming of the software, no matter how sophisticated one or two users are, the project is going to have a very rocky start if the implementation plan is flawed, understaffed, or poorly organized. It is of paramount importance to get off to a good start, experience some early successes, and get the user community friendly toward the project. There are bound to be some detractors among the users since some will not want to change, some have a vested interest in the old system, some are threatened, and some do not see anything in it for them. Thus, the project organizer has two aspects of the project to address, the technical and organization plus the public relations and communicative aspects; both are addressed below.

4.3.14 Organizational and Technical Project Aspects for the Users

Although the following list is not complete, it outlines the major tasks that must be completed in order to be successful. The outline is for use by an end user or industrial hygienist/manager. The systems analyst will have a different list of actions to do although there is a relationship between them.

- Define current environment
  Hardware configuration
  Networking
  Hardware available
- Define new hardware requirements
  Hardware, printers, monitors, cables, computers
  System requirements, additional burden to network or mainframe
  Data storage requirements, how much data storage
  Define transaction rate, optimum, mean, low
- Define new software requirements
  New systems software or version of the software

New workstation or PC software required
New application software
- Define test project/environment
  Develop test plan
  Define test time frame
  Define beginning/end of test
- Order/receive products
  What is minimum needed for test?
  Lease/buy/borrow?
  Configure
- Define personnel requirements for test
  Who will test?
  Is the test environment representative of working environment?
  What will be left out/included?
  What are the criteria for success? What concludes the test period? Measurable results?
- Preparation for full installation
  Roll out plan
  Define personnel requirements
  Define training plan
  Create user's manual
  Establish "hotline" or Question & Answer board or electronic bulletin board
- Install full system
  Acquire remaining hardware items
  Acquire remaining software items
  Assign personnel
  Develop security plan—who has access, when?
  Develop project schedule
  Develop transition plan—critical decisions must be made about when and what data is migrated to the new system. Does the users start with day 0 and put only new data in the system?
  Does one migrate only last year's data? Two years? All data? What subsets of the data?
  How does one manage when elements of the old system are still in place and a new system is being brought up?
- Exploitation of the system
  The new system has capabilities that did not exist before, how can those be discovered and taken advantage of?
  What changes could be made to take even greater advantage of the system?
  Generation of standard reports

What are these reports?
What is the periodicity?
How do we distribute the reports?
- Plan a celebration.
Recognition
Summarization

5 RELATED TECHNOLOGIES

Other developments in related optical technology make the storage of information more direct. One such development is the WORM (write one-read many) optical drive. This machine allows the user to enter data directly into an optical disc. The information then becomes permanently archived on the disc and can be accessed in a similar manner to mass-produced CD-ROM discs. Another technology is called DRAW (direct read after write). Both WORM and DRAW provide direct means for the generator of the data to write directly onto the disc, thus, bypassing the expensive mastering step. These technologies are becoming increasingly common, however, and are not directly compatible with CD-ROM.

Videodiscs, which use an analog recording technique, are related to CD-ROM, and record visual images and sound. While the main market for video discs has been in the entertainment industry, they offer much potential for health and safety training. Videodiscs can be programmed with interactive software, to provide customized training and testing capabilities.

5.1 Hardware and Software Requirements

CD-ROM discs require a special player called a CD-ROM drive, which is somewhat different from the audio CD players is music systems. There are numerous manufacturers of the CD-ROM drives, but the most widely used models are made by SONY, Hitachi, Toshiba, and LMSI (formerly Philips). There are various ways to accommodate more than one disc when several different CD-ROM systems are installed on the same computer, or when a single system is on more than one disc. A CD-ROM drive may be internal to the PC. There are also CD-ROM drives that can be attached to the central processor as a single unit or as a "juke box". A multiple reader many be needed if there are lots of MSDSs for instance.

5.2 CD-ROM Delivery Systems

In general, information from a CD-ROM can be distributed in the same ways as any other electronic information. Besides the freestanding PC for a single user, most CD-ROM products can run on a LAN (local area network) for multi-user access. The theoretical upper limit to the number of simultaneous CD-ROM users on a network is a complex function of the hardware and network software.

DATA AUTOMATION

The development of portable PCs has allowed CD-ROM information systems to be truly portable as well. At the present time, the HazMat response trucks of the Fairfax County, Virginia, Fire Department and others have portable computers on the trucks with CD-ROM databases installed. Thus the HazMat response team can have full and rapid access to critical information at the scene of incidents involving hazardous materials.

Portable or laptop PCs can also be used with modems and cellular phones in the field to call a host base where the CD-ROM system resides. This true remote access capability requires additional software programs, including communications software and a data-linking program such as Carbon Copy. Remote access to health and safety data is especially helpful in high-risk field operations, such as logging, construction, petroleum exploration and drilling, mining, and pipeline maritime operations.

5.3 Suitability of Information for CD-ROM

CD-ROM is especially useful when a large amount of information needs to be accessed or stored, when the information does not change rapidly, and when rapid access to any or all of the information is required. All of these conditions need to be fulfilled in order to make CD-ROM storage the most cost-effective technology in comparison to paper, microfiche, or magnetic media.

Some examples of information that are well-suited for CD-ROM in the health and safety field are access and storing of industrial hygiene, medical and safety surveillance records, analytical laboratory methods, collections of reference visible, infrared, and ultraviolet spectrophotometric, nuclear magnetic resonance, and mass spectral patterns, and analytical laboratory data; computer-based training for hazard communication and other OSHA requirements; collections of material safety data sheets (MSDS); bibliographic databases containing references and abstracts of published studies; compiled databases containing reviews and summaries of health, safety, and environmental data; recommendations for medical treatment of occupational and environmental overexposures and for HazMat emergency response; selection of protective clothing and equipment; statutes and regulations. An expanded discussion follows on each of these uses, along with information on commercial CD-ROM products in these areas.

Industrial hygiene, medical, and safety surveillance records are especially well-suited for CD-ROM. Typically these involve large amounts of data that may legally need to be archived for many years (see previous section on data storage for long term). Archiving these records in paper form creates problems of space, accessibility, durability, and vulnerability. Often the paper on which these records exist deteriorates rapidly, especially heat- or pressure-sensitive recording paper used for audiograms, pulmonary function testing, electrocardiograms, and clinical laboratory data. Even magnetic media is prone to deterioration when undisturbed for long periods of time. Single-paper copies of critical health and safety records are vulnerable to loss by fire or natural disaster. It is important to remember, however, that CD-ROM (or any other form of EDP) cannot totally eliminate the need to archive original medical records in the United States.

Large collections of MSDSs can be put on CD-ROM. It should be possible to control access to particular MSDSs at the level of a company, location, department, or employee. MSDSs on CD-ROM can be on a network for accessibility in the workplace.

Compiled databases containing information on toxicology, medical treatment recommendations, HazMat emergency response, and selection of protective clothing and equipment, are generally suitable for CD-ROM. It may be desirable, or even necessary, for this kind of information to be portable or accessible from remote locations, or at the very least as part of a network with access directly from the workplace.

Statutes and regulations on CD-ROM are useful for many purposes, especially where rapid access to legal information and savings relative to on-line access are required. One disadvantage of regulations on CD-ROM, or any other electronic media with periodic rather than continuous updating, is that some of the information may be out of date.

5.4 Summary and Conclusions

CD-ROM is the method of choice for information storage and retrieval when large amounts of data need to be accessed rapidly, and when the data do not change often. CD-ROM can make very large amounts of information available through a personal computer, thus, saving the expense of acquiring and maintaining this information in a mainframe environment. It is relatively expensive to produce one or a few copies of a CD-ROM disc, but its excellent archiving properties and accuracy in comparison with magnetic media may justify this cost for companies to convert information to CD-ROM for internal use. Appendix D lists some sources for CD-ROMS.

6 DATA AUTOMATION SUMMARY

The intent of this chapter is to provide a useful guide for practicing industrial hygienists in the development of a system for managing the data records with which they work. Considerable gains in productivity are expected by more efficient gathering, searching, comparing, and reporting of this data. It is expected that the automation of the data management process will continue for some time to come, with ever more sophisticated and cost-effective solutions being developed. The leading edge industrial hygienists will want to use these new solutions to enhance their job enjoyment as well as enable them to contribute in even greater ways to their employers or clients.

APPENDIX A. REPORTING REQUIREMENTS FOR OCCUPATIONAL HEALTH INFORMATION SYSTEMS

Medical	Evaluation	Hazardous Communication	Evaluation
Employee Medical Report	_____	Training Status	_____
Summary Medical Report	_____	Protective Equipment	_____
Work Restrictions	_____	First Aid Information	_____
Health Risk Appraisal	_____	Spill Information	_____
Hearing Threshold Shift	_____	Carcinogen List	_____
Analysis of Abnormal Employee Groups	_____	Agents Inventory	_____
Schedules	_____	Workplace History of Environmental Agents	_____
Others	_____	MSDS	_____
		MSDS Revisions	_____
Industrial Hygiene		Labels	_____
Sample Schedules	_____	MSDS Updates	_____
Sample Report	_____	Customer Letter	_____
Sample Summaries	_____	Customer Report	_____
Sample Follow-up	_____	Product Report	_____
Incomplete Samples	_____	Materials Report	_____
Lab Audit	_____	Supplier Report	_____
Workplace Inspection	_____	Right-to-know	_____
Problem Summary	_____	SARA Tier I	_____
Agents Inventory	_____	SARA Tier II	_____
Workplace History of Environmental Agents	_____	Others	_____
Carcinogen List	_____		
Material Safety Data Sheet	_____		
MSDS Revision Review Sheet	_____		
Labels	_____		
Others	_____		

APPENDIX B. DATA REQUIREMENTS FOR OCCUPATIONAL HEALTH INFORMATION SYSTEMS

Medical	Evaluation	Work Place	Evaluation
Audiometry	____	Description	____
Chemistry	____	Inventories	____
Cytology	____	Medical Requirements	____
Demographics	____		
Diagnosis	____	**Claims**	
Drug Screen	____	Case Descriptions	____
ECG	____	Medical Costs	____
Exposures	____	Legal Costs	____
Hematology	____	Salary Costs	____
Immunizations	____	Award Costs	____
Medical History	____	Other Costs	____
Medications	____	Recovered Costs	____
Mortality	____	Status Costs	____
Office Visits	____		
Physical Exams	____	**Workplace**	
Physiological Problems	____	Description	____
Recommendations	____	Inventory	____
Scheduling	____	IH Estimates	____
Serology	____	Engineering Controls	____
Spirometry	____	Inspections	____
Urinalysis	____	Incidents	____
Vision	____	Incident Costs	____
X-ray	____		
		IH Sampling	
Wellness Program		Description	____
Demographics	____	Scheduling	____
Follow-up	____	Method	____
Objectives	____	Details	____
Physiology	____	Data	____
Stress Tests	____	Results	____
Visits	____	Recommendations	____
Workshops	____	Exception Reports	____
Employee Assistance		**Equipment**	
Demographics	____	Description	____
HRA	____	Scheduling	____
		Calibration	____
Medical Inventory		Inspections	____
Description	____		
Suppliers	____	**Provider**	
Purchases	____	Demographics	____
		Certification	____
		Continuing Education	____
		ICD-9 Codes	

APPENDIX C. JOINT APPLICATION DESIGN (JAD) PROCESS

The purpose of a JAD is to define the objectives, scope, requirements, and external design of a system, by external design, is meant as part of the application that the end user sees, but not the coding behind it.

A JAD plan session is used to define system objectives and scope. The purpose of a JAD plan is to define the scope for large applications or groups of applications. A JAD plan provides an organized method for gathering and documenting the data required to estimate, plan, and schedule JAD sessions.

For a JAD plan, individuals with a solid understanding of the functional areas affected by the application are brought together for a brief, but intensive-cooperative effort. These individuals, led by a JAD leader, define the scope of the project by developing a list and description of business functions to be included in the system. They also establish a preliminary list of required screens, reports, and connection points to other systems (system I/Os). This list serves as a basis for estimating the length and complexity of the design process. Any significant issues regarding the application are documented as well.

The product of a JAD plan is a documented and agreed-upon statement of what is to be done. It contains the system objectives, functions, and scope. Such a statement, and the commitment it represents, is an excellent basis for estimating, planning, and obtaining management approval for the projected system.

After what is to be done has been defined, the "how" can be addressed. A JAD design is intended for this purpose.

Once again, individuals directly involved with the proposed system are brought together for brief, but intensive, sessions under the guidance of a JAD leader. The result is a document containing a definition of the system requirements and the external design. This document is the foundation for the implementation and installation of the system.

It is felt that a JAD is an efficient and economical technique for obtaining user and management commitment for the remainder of the application development effort.

The same basic types of activities take place in both a JAD plan and a JAD design. These are

1. Preparation and kick-off
   a. Establishment of the executive objectives
   b. Orientation to the application
   c. Preparation of materials to be used in the JAD
   d. Kick-off meeting of the executive sponsor and JAD participants to assure a common understanding of the objectives of the project.
2. AD sessions, whether for planning or design purposes, are intensive meetings with a carefully selected user, management, and data processing representatives. These sessions are chaired by the JAD leader and documented by JAD analysts.

Activities during a JAD design session are different from those in a JAD plan. Because the JAD plan session activities have already been described, the activities in a JAD design session are concentrated on here. These activities deal with the system first on a general level and then go on to specify the system in detail.

Called system definition, the general definition activities obtain participant commitment to the following:

- Statement of purpose of the system
  System functions
  Assumptions on which the system is based
  Constraints on the system

These topics form the basis of the JAD documentation. Special forms are used to document the system functions during the sessions. These include:

- Work flow diagrams of business process and their relationships
- Work flow descriptions of inputs, reports, and users of the system
- Function flow diagrams of data groups and inputs/outputs of each business function.

When the general overview of the system is complete, the more detailed specification of the system can begin. This is called external design. During external design, JAD participants commit to, and document the following:

- Definition of data elements (date, time, CAS, etc.)
- Screen and report layouts (if a query language is to be used, the query search requirement are identified but screen layouts will be omitted)
- Edit and validation requirement for screens and reports
- Function descriptions including security and architecture requirements, volumes of transactions, and frequencies

3. Wrap-up. During wrap-up, the conclusions documented during the JAD sessions are edited and checked for consistency and completeness.
4. Executive presentation. At the executive presentation, the JAD participants summarize the results of their JAD sessions and present the completed documentation to the executive sponsor.

APPENDIX D. CD-ROM APPLICATIONS
EXAMPLES OF COMMERCIALLY AVAILABLE CD-ROM PRODUCTS

Type of Information	Publishers[a]
Bibliographic databases	
Administrative (Health)	6
CANCERLIT	2, 6
Ecological Abstracts	4
Environment Abstracts	3
Life Sciences Collection	4
Health and Safety Science Abstracts	4
MEDLINE	2, 4, 6, 7, 8, 9, 19
NIOSHTIC	4, 5
NTIS Database	7, 18, 19
Nursing and Allied Health (CINAHL)	6
Pollution Abstracts	4
Toxicology Abstracts	4
TOXLINE	3, 4
Books	
Concise Encyclopedia of Science and Technology	13
Dictionary of Science and Technical Terms	13
Physician's Desk Reference	10, 14
Shepard's Catalog of Teratogenic Agents	15
Directories of CD-ROM Products	21, 22
Emergency Medical Response and Treatment	
EMERGINDEX System	15
POISINDEX System	15
TOMES Plus System	15
Environmental	
Acid Rain (Canadian Government Documents)	42
Ecological Fact Sheets	20
OHM/TADS	15, 19
TOMES Plus System	15
Toxic Release Chemical Inventory	20
First Aid	
CHRIS	15, 19
TOMES Plus System	15
HazMat Emergency Response	
CHRIS	15, 19
DOT Emergency Response Guides	15
OHM/TADS	15, 16
TOMES Plus, System	15
Material Safety Data Sheets	
Collections	1, 5, 17, 20
Patents and Trademarks	

APPENDIX D. (continued)

Type of Information	Publishers[a]
Regulatory Information	
Federal Register	10, 7
Statutes and Regulations	5, 11, 12
TOMES Plus System	15
Toxic Release Chemical Inventory	20
Right-to-Know	
New Jersey Fact Sheet	15, 20
Risk Assessment	
IRIS	15
Storage, Shipping, Waste Disposal	
CHRIS	15, 19
HSDB	15
TOMES Plus System	15
Toxicology, Health Effects	
CISDOC	19
HSDB	15
HSELINE	19
OHM/TADS	15, 19
POISINDEXÓSystem	15
REPRORISK, System	15
RTECS	15, 19
Shepard's Catalog of Tetragenic Agents	15
TERIS (Teratogen Information System)	15
TOMES Plus, System	15

[a]Information was believed to be accurate at the time of writing (January, 1999). Publishers listed in the References section may be contacted for current information.

CD-ROM Publishing Corporations

1. Aldrich Chemical Company, Inc., 1001 W. St. Paul Ave., Milwaukee, WI 53233, (800) 558-9160, (414)273-3850.
2. Aries Systems Corp., One Dundee Park, Andover, MA 01810, (508) 475-7200.
3. Biosis, 2100 Arch St., Philadelphia, PA 19103-1399, (800) 523-4806, (215) 587-4800.
4. Cambridge Scientific Abstracts, 7200 Wisconsin Ave., Bethesda, MD 30814, (800) 843-7751.
5. Canadian Centre for Occupational Health & Safety, 250 Main Street East, Hamilton, Ontario L8N 1H6, Canada (800) 263-8340, (416) 572-4444.
6. CD Plus, 333 7th Ave., 6th Floor, New York, NY 10001, (212) 563-3006.
7. Dialog Information Services, Inc., 3460 Hillview Ave., Palo Alto, CA 94304, (800) 334-2564, (415) 858-3785.

8. Healthcare Information Services, Inc., 2335 American River Dr., Suite 307, Sacramento, CA 95825, (916) 648-8075.
9. DYNASTAR, Inc., 1228 S. Park St., Madison, WI 53715, (608) 258-7420.
10. EBSCO Electronic Information, P.O. Box 13787, Torrance, CA 90503, (800) 888-3272.
11. ERM Computer Services, Inc., 855 Springdale Dr., Exton PA 19341, (800) 544-3118, (215) 524-3600.
12. IHS Regulatory Products, 15 Inverness Way East, Englewood, CO 80150, (303) 790-0600.
13. McGraw-Hill Book Company, 11 West 19th St., New York, NY 10011, (212) 512-2000.
14. Medical Economics Co., 680 Kinderkamack Rd., Oradell, NJ 07649, (201) 262-3030.
15. Micromedex, Inc.; 600 Grant St., Denver, CO 80203-3527, (800) 525-9083, (303) 831-1400.
16. NASA/Ames Research Center, c/o Steven Hipskind, MS 245-5, Moffett Field, CA 94035-1000; (415) 604-5076, FTS 464-5076.
17. Occupational Health Services, Inc., 11 W. 42nd St., 12th Floor, New York, NY 10036, (800) 445-MSDS, (212) 789-3535—NY.
18. Online Products Corp., 20251 Century Blvd.; Germantown, MD 20874, (800) 922-9204.
19. SilverPlatter Information, Inc., One Newton Executive Park, Newton Lower Falls, MA 02162-1449; (800) 343-0064, (617) 969-2332.
20. United States Government Printing Office; Superintendent of Documents; Washington, DC 20402-9325, (202) 783-3238. (Department of Defense Hazardous Materials Information Services (HMIS), 008-000-00567-2, EPA Regulatory and Community Right to Know Toxic Release Chemical Inventory 055-000-00356-4.)
21. United States Department of Health and Human Services, Public Health Service, Centers for Disease Control, Center for Environmental Health and Injury Control, Information Resources Management Group, Atlanta, GA 30333.
22. The H. W. Wilson Co., 950 University Ave., Bronx, NY 10452, (800) 622-4002.

GLOSSARY

Only the most usual and frequently encountered definitions are given here. The reader is advised to consult a more comprehensive list for additional material.

Baud. (1) A unit of signaling speed equal to the number of discrete conditions or signal events per second; for example, one baud equals one-half dot cycle per second in Morse code, one bit per second in a train of binary signals, and one 3-bit value per second in a train of signals each of which can assume one of eight different states.

Cache Memory. A special buffer storage, smaller and faster than main storage, that is used to hold a copy of instructions and data in main storage that are likely to be needed next by the processor, and that have been obtained automatically from main storage. (T)

Case. Computer-assisted software engineering. A set of tools or programs to help develop complex applications.

Data Dictionary. (1) A database that, for data of a certain set of applications, contains metadata that deal with individual data objects and their various occurrences in data structures. (T) (2) A centralized repository of information about data such as meaning, relationships to other data, origin, usage, and format. It assists management, database administrators, system analyst, and application programmers in planning, controlling, and evaluating the collection, storage, and the use of data. (3) In the System/36 interactive data definition utility, a folder that contains field, format, and file definitions. (4) In IDDU, an object for storing filed, record format, and file definitions.

Database. (1) A collection of data with a given structure for accepting, storing, and providing, on demand, data for multiple users. (2) A collection of interrelated data organized according to a database schema to serve one or more applications. (3) A collection of data fundamental to a system.

EDI. Electronic data interchange. A standard format for the transmission of data and the use of that data.

Fourth-Generation Language (4GL). A high-level language used in programming that has additional tools for screen generation, interfaces, and system design.

Information Engineering. The application of rigorous discipline to the management of data. Usually involves goals, data design, and data relationships among other things.

LAN. Local area network. A LAN is a cabling configuration that is used to link several computing systems at one location.

Object Code. Output from a compiler or assembler that is itself executable machine code or is suitable for processing to produce executable machine code.

Operating System. Software that controls the execution of programs and that may provide services such as resource allocation, scheduling, input/output control, and data management. Although operating systems are predominantly software, partial hardware implementations are possible.

RAM. Random-access memory. The memory found on a computer chip that is in a computing system.

Record. (1) In programming languages, an aggregate that consists of data objects, possibly with different attributes, that usually have identifiers attached to them. In some programming languages, records are called structures. (2) A set of data treated as a unit. (3) A set of one or more related data items grouped for processing.

Relational Database. A database in which the data are organized and accessed according to relations.

RISC. Reduced instruction set computer. A type of operating system that is frequently used in scientific computing. Many manufacturers have their own versions.

ROM. Read-only memory. The memory of a computer that allows reading only. It cannot be written over without being destroyed.

Shared Data. Data that are used for more than one applications or purpose of giving efficiency to the total system.

Source Code. The input to a compiler or assembler, written in a source language. Contrast with object code.

SOL. Structured query language. This allows the querying or seeking out of information in a database in a standard form.

Third-Generation Language (3GL). A high-level computer language that uses ordinary English words.

Time Sharing. (1) A method of using a computing system that allows a number of users to execute programs concurrently and to interact with the programs during execution.

UNIX. UNIX (trademark of UNIX Systems Laboratories, Inc.) operating system. An operating system developed by Bell Laboratories that features multiprogramming in a multi-user environment. The UNIX operating system was originally developed for use on minicomputers but has been adapted for mainframes and microcomputers.

Upwardly Compatible. The capability of a computer to execute programs written for another computer without major alteration, not vice versa.

Utility Program. (1) A computer program in general support of computer processes; for example, a diagnostic program, a trace program, a sort program; synonymous with service program. (2) A program designed to perform an everyday task such as copying data from one storage device to another.

WAN. Wide area network, a network that extends over many locations.

ACKNOWLEDGMENTS

Many people and companies have contributed to the thinking that is contained in this chapter. Principal among those is the IBM Corporation, which has provided material and input for the technical aspects of the chapter. Also, Micromedex in Denver, Colorado, provided useful material on the use of CD-ROMs, General Research Corporation in Vienna, Virginia, has graciously provided the material found in Appendices A and B to be used in selecting software.

BIBLIOGRAPHY

1. American Society for Testing and Materials, 100 Barr Harbor Drive, West Conshohocken, Pennsylvania USA 19428-2959, http://www.astm.org/.
2. LEXIS-NEXIS, P. O. Box 933, Dayton, Ohio 45401-0933, USA, (937) 865-6800, (800) 227-9597 http://lexis-nexis.com/.
3. American Society for Heating and Refrigeration Engineers, ASHRAE, 1791 Tullie Circle, N.E., Atlanta, GA 30329, Phone: (404)636-8400, Fax: (404)321-5478, http://www.ashrae.org/.
4. OSHA, http://www.osha.gov/

5. CFR1910.1020(d)(1)(ii)(A)
6. Quantum Compliance Systems, 4251 Plymouth Rd., Suite 1200, Ann Arbor, MI 48105, (734)761-3058, http://www/qcs-facts.com
7. *Journal of Occupational Health and Safety*, Stevens Publishing, 5151 Beltline Road, Suite 1010, Dallas, TX 75240, (972)687-6700.

CHAPTER FORTY-SIX

Risk Analysis for the Workplace

Robert G. Tardiff, Ph.D., ATS

1 INTRODUCTION

Occupational exposures to chemicals, often inevitable, are generally viewed as without consequence to the health of the workers, provided that the doses are no greater than limits set by some recognized authority.

Risk analysis (also referred to as risk estimation) is often vital to decisions on controlling risks and to communicating about the significance of risks before and after abatement of emissions. By its nature, the risk analysis process enables the systematic evaluation of data and the quantitative presentation of complex information. It facilitates both comparisons among alternatives and incorporation of seemingly disparate information (e.g., technological choices, economics, and policy goals) into complex decision-making processes.

Applied to emissions of chemicals in the work environment, risk analysis (*1*) characterizes injurious effects that might be associated with an activity, (*2*) enables comparison of health risks of existing technologies with health risks or benefits of proposed replacement or supplemental technologies, (*3*) helps to distinguish between major and minor sources of risks and their relative impact on health for resource allocations, and (*4*) permits establishment of priorities in cases presenting multiple potential health problems with limited resources to address them. With an understanding of the extent and seriousness of potential health effects, reasoned choices can be made about allocation of resources and about engagement in specific activities.

Risk analysis provides an orderly method to communicate risks associated with complex processes. For instance, workers may fear chemical injury from continuous or accidental releases of numerous chemicals into their work environment by industrial practices. Risk analysis can help managers and skilled workers understand the nature and magnitude of

Patty's Industrial Hygiene, Fifth Edition, Volume 3. Edited by Robert L. Harris.
ISBN 0-471-29753-4 © 2000 John Wiley & Sons, Inc.

those risks in a context to which they can relate. When used to foster understanding of the magnitude of risks, such understanding can serve either to dispel unfounded fears or to indicate a genuine basis for concern and hence for risk-avoidance behavior.

The estimation of the degree of risk (and conversely, safety) to human health from substances in the air—whether workplace or ambient—has been used to achieve a variety of objectives.

The most prominent use of risk assessment for substances in air breathed by humans is the setting of tolerable levels of exposure (usually as concentrations in air). By taking into account the toxic properties of a substance and applying predetermined norms of safety (such as the magnitude of some margin of safety), one or more concentrations deemed to be tolerable for humans can be derived. To become a guideline or limit, such a concentration needs to be tempered by technological and economic feasibility. Using this approach in part, standards have been set by the Occupational Safety and Health Administration (OSHA) not only to limit concentrations of chemicals (e.g., benzene) in workplace atmospheres but also to prescribe rules for the use of protective equipment by workers. Likewise, limits have been set for pollutants (e.g., ozone) in ambient air and in other media with which humans come into contact. The American Congress of Government and Industrial Hygienists (ACGIH), a private organization, recommends concentrations of chemicals in the workplace that provide for tolerable levels of risk to workers. Internationally, the World Health Organization (WHO) through its International Programme of Chemical Safety (IPCS) has promulgated a highly comparable approach to distinguish safe from unsafe levels of exposure to chemicals (1).

Selections of one technological approach over another to reduce emissions of chemicals to the atmosphere can be greatly influenced by an understanding of the degree of health safety or health risk associated with each option. Furthermore, knowledge of the health benefits from a specific selection can be shared with those exposed to ease the concern for their well-being.

This chapter first provides an overview of risk analysis and describes some detailed procedures applied to the interpretation of physical, chemical, and toxicity data to estimate risks (section 2). Next, an illustration of the estimation of risks from exposure to a carcinogen in the workplace is presented (section 3). That is followed by a comparable illustration to demonstrate the derivation of the degree of safety from workplace exposures to reproductive toxicants (section 4). A variety of resource documents are available to the reader that provide more detailed information about risk analysis procedures and their applicability; some include References 2–12.

2 THE SUM AND SUBSTANCE OF RISK ANALYSIS

Risk analysis is the process of determining quantitatively the likelihood of adverse effects resulting from exposures, and alternatives for managing them. The risk analysis paradigm widely acknowledged by the U.S. scientific community to be the current standard used in industrial risk management and public policy settings is that first articulated by the U.S. National Academy of Sciences in its report, *Risk Assessment in the Federal Government: Managing the Process* (3). In a sequel entitled *Science and Judgment in Risk Assessment*

(13), the National Research Council defined major elements in the estimation of risks and placed the process within contexts of managing such risks. Among the elements elaborated in detail are (*1*) the values and limitations of default assumptions and the circumstances under which they should be replaced with empirical findings; (*2*) the importance of understanding the mode of toxic action of a compound to provide increased accuracy in defining the presence of hazards to human health; (*3*) the role of variability in human response as a basis for deciding the degree of health protection to specific population groups; and (*4*) aggregation of exposures and risks from all sources of exposure so as to provide realistic estimates of risk to specific groups such as workers.

Health risks are estimated for chemicals known to have a potential for systemic effects in the body after repeated and prolonged exposure (referred to as chronic exposure). Further assessments can be conducted for noncarcinogenic effects, such as respiratory irritation, that can be manifested after relatively brief (i.e., acute) exposure to elevated concentrations of chemicals.

The four basic elements of a risk analysis, as presented in Figure 46.1, are hazard identification, exposure assessment, dose-response evaluation, and risk characterization. Each is described in the following subsections. A more detailed depiction of the complexity and integration within these elements is presented in Figure 46.2.

2.1 Hazard Identification

In hazard identification a determination is made as to whether a causal relationship exists between a chemical and an injurious effect on health. It involves gathering and evaluating toxicity data on the types of health injury or disease that may be produced by a chemical and on the conditions of exposure under which injury or disease is produced. Hazards other than toxicity (e.g., flammability, explosivity) are not considered in this type of review. It may also involve characterization of the behavior of a chemical within the body and the interactions it undergoes with organs, cells, or even parts of cells. Data of the latter types may be of value in answering the ultimate question of whether the forms of toxicity known

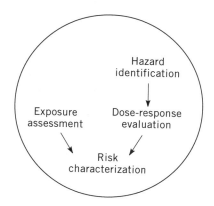

Figure 46.1. Components of a risk analysis.

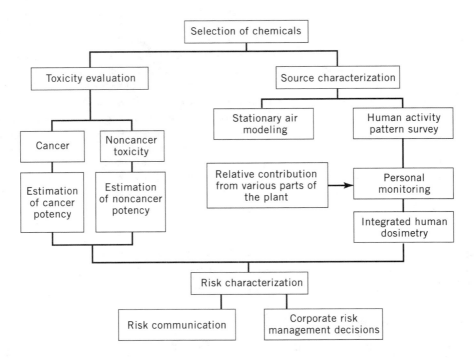

Figure 46.2. Process in the analysis of risks for chemicals of interest.

to be produced by a chemical agent in one population group or in laboratory animals are also likely to be produced in the human population group of interest. Risk is not assessed at this stage. Hazard identification is conducted to determine whether and to what degree it is scientifically correct to infer that toxic effects observed in one setting will occur in other settings (e.g., are chemicals found to be carcinogenic or teratogenic at high doses in experimental animals also likely to be so in exposed workers?).

Within the framework of risk analysis, hazard identification is conducted to fulfill two objectives:

- To ascertain from empirical scientific literature the critical health effects associated with specified chemical(s), including the particular location(s) in, and functions of, the human body that are targets of potential injury. This step includes a determination of health effects associated with both long-term (years to decades) and short-term (24 h or less) exposures.
- To determine, from all that is known about chemical(s) of interest, the weight of evidence for causal relationship(s) between specified chemicals and human injury, and to avoid mischaracterizing as causal influences that are within the causal chain of events.

The hazardous properties of compounds are ascertained from human epidemiological and toxicological data derived from a variety of scientific studies. Epidemiology studies are reports of human disease associated with workplace or other environmental chemical exposure; in the workplace, such investigations generally take the form of not only retrospective but also prospective surveys and case control studies. Toxicity studies are controlled experiments examining the effects of compounds on experimental subjects (generally laboratory animals). The results of these studies are judged in total to assess the hazardous properties of the compounds and the weight of evidence of the results.

Epidemiology studies have an advantage in that they reflect the effects of chemicals directly on human health. These studies are particularly valuable to determine causation in either acute or rare chronic diseases. However, the power of epidemiology to assign causation in more common chronic disease is less, due to its relative insensitivity. In addition, individuals are seldom exposed to only one chemical but rather are exposed to a number of potentially hazardous substances virtually at the same time, both in the workplace and in the home. An additional difficulty in interpreting the findings of epidemiology studies is that precise determinations of worker exposure levels are generally very difficult.

Laboratory animal studies of toxicity have the advantage that an investigation's experimental design can control many of the variables that cannot be controlled in the real setting. In particular, exposure concentration and duration can be determined and maintained with considerable precision. Furthermore, many subtle pathological conditions can be examined in experimental animals. Results from toxicity studies can be used to verify hypotheses derived from epidemiologic associations; however, in the absence of information about the kinetics of the compound, they are limited in being able to provide extrapolations from test animals to the human population of interest.

The health effects from exposure to a chemical are typically classified as acute or chronic. Acute health effects can range from subtle and reversible changes in the body (e.g., a temporary rise in an enzyme level) to debilitating, long-term irreversible effects (e.g., stripping of cells lining the lung after a rapid, large exposure to a caustic respiratory irritant). Acute effects occur relatively quickly, vary in severity, and may be reversible. By contrast, chronic injuries generally follow repeated and prolonged exposures (years) to relatively low doses. Injurious effects appear gradually and may be partially or wholly reversible over time once exposure ceases, although less so than effects from most acute exposures. Cancers and chronic obstructive lung disease are illustrations of chronic effects.

Toxicity data, whether from humans or laboratory animals, need to be analyzed comprehensively to determine the strength of associations between exposures and specific pathologies. A number of schemes exist to systematically weight the evidence for causation. Such approaches are generally specific to a pathology or injury. The toxicity for which weight-of-evidence schemes are most prevalent are cancer and reproductive injury.

The procedures for interpreting these data are those that have been developed over several decades and used to make numerous types of decisions on the safe use of chemicals in food, air, and water and to differentiate between significant and insignificant risks to human health. For the most part, these procedures and techniques have been either developed or sanctioned by the National Academy of Sciences in advising various branches of the federal government about the health risks associated with diverse chemical activities (3–5).

For chemicals suspected of being able to cause cancer, weight-of-evidence criteria have been promulgated by the International Agency for Research on Cancer and by USEPA (6,

14). In each instance, three sets of data (epidemiology, toxicity, and ancillary but relevant findings) are first evaluated separately. Next, conclusions as to the strength of evidence from human observations and from laboratory animal studies are combined in a numerical rank that reflects the extent to which a substance is likely to cause cancer in humans. The IARC and USEPA schemes are conceptually similar; however, they differ in some of the details of ranking.

For chemicals suspected of being able to cause reproductive injury, Brent and USEPA have each proposed a weight-of-evidence scheme to deal with one specific form of toxicity: birth defects (15, 16). Brent proposed a five-step procedure for identifying substances or processes that may pose a threat of birth defects and other childhood maladies as a result of chemical exposures by one or both parents. Other elements in the estimation of risk of reproductive impairment have been reported by Christian, Francis and Kimmel, and ILSI (17–19).

1. Several well-controlled epidemiologic studies consistently demonstrating an increased incidence of a particular congenital malformation in an exposed human population
2. Secular trends demonstrating a relationship between the incidence of a particular malformation and exposures in the human population
3. An animal model that mimics the human malformation at "clinically comparable exposures," including
    a. No evidence of maternal toxicity
    b. No reduction in food or water intake
    c. Careful interpretation of malformations that occur in isolation
4. A dose-response relationship for teratogenic effects
5. An understanding of the mode(s) of action of birth defects (teratogenesis)

Although this set of guidelines is useful, several practical difficulties exist in meeting some of the criteria. For instance, if exposure to a teratogen is rare and the defect caused by the teratogen is common, then it would be very difficult to detect a secular trend between exposure and incidence. Furthermore, the general population is exposed to so many chemicals that for an epidemiology study to isolate exposure to only one substance would be rare indeed. Thus, unless a compound is exquisitely toxic, these criteria are generally considered too restrictive to be of practical value.

On the other hand, with USEPA's weight-of-evidence approach for qualitatively assessing teratogenic risk of chemical exposure, findings from human and animal studies are separately evaluated for their scientific quality. Based on USEPA guidelines (20), the quality of the studies is evaluated, and the subject chemical receives an evaluation of the likelihood that it is a reproductive toxicant in animals or humans. The results of these evaluations are combined, and the chemical receives a rating of the qualitative evidence for teratogenicity.

With either approach, great attention must be given to the methods used in reproductive toxicity studies to ensure that the appropriate scientific conclusions can be drawn with confidence based on the data. In particular, greater confidence is given to studies in which

RISK ANALYSIS FOR THE WORKPLACE

(*1*) an adequate number of doses of the test substance are used, (*2*) a sufficient number of animals is in each dose group to ensure adequate power to detect the ability to cause injury, and (*3*) evaluation of the data is complete (to ensure that a potential effect is not missed due to failure to examine a target organ).

The determination of the degree to which exposures in the workplace are likely to be protective of reproductive injury is a function, in part, of the extent to which a chemical's toxicity has been investigated and the degree to which one understands the circumstances under which toxicity is likely to be modified (19, 21). The confidence in protective values is dependent on factors such as those listed in Table 46.1. Another major consideration is the extent to which the workplace may impart exposures not only to a single compound but also to mixtures of substances, some of which might act together to increase the level of risk. Increasingly, the setting of safe levels of exposure in all circumstances calls for aggregation of all doses from all routes of contact and for consideration of uniquely susceptible groups of individuals.

All relevant studies need to be carefully reviewed to ascertain the presence, or absence, of injurious effects at doses in humans lower than those used to treat experimental animals. For instance, if a compound produces no reproductive toxicity at a specified dose in humans or animals, the absence of effect may be due either to its inability to damage the function of the organs or to the administration of too low a dose of the chemical. Finally, if the subcellular mode(s) of action by which a chemical causes injury in an experimental animal is unique to that species and is not present in the human population, then justification exists for not using such data as sentinels of human risk.

Several factors should always be considered when selecting one of the many analytic approaches available for evaluating the total database. The critical ones are replication of

Table 46.1. Guidelines for Confidence in Toxicity Database

High Confidence

Chemical tested in more than one species and sex
Adequate design of studies
Consistent response within species
Consistent response among species
Fetotoxicity present at maternally nontoxic doses

Medium Confidence

Adequate design of studies
Chemical tested in only one species or one sex
Fetotoxicity present at maternally nontoxic doses

Low Confidence

Inconsistent response within species
Inconsistent response among species
Inadequate study design
Minor or marginal effects that could have been due to other causes
Fetotoxicity present only at maternally toxic doses

results, reproducibility of results, and concordance of results. It is also necessary to consider the general body of scientific knowledge concerning the particular toxic end point under evaluation and the extent of its applicability to those exposed in the workplace.

These concepts are defined here as follows:

- An experimental result is said to be replicated if it is found in experiments of identical design.
- An experimental or epidemiological finding is said to be reproducible if it is produced in the same species under different conditions, such as different genders, strains, dose groups, and routes of exposure.
- Experimental and epidemiological findings are said to be concordant if they are consistent across species and by a variety of study designs.

For a given set of data pertinent to a given toxic end point (assuming each study has been critically evaluated), it is advisable to describe the degree of replication, reproducibility, and concordance. In general, as the degree of data replication, reproducibility, and concordance increases, it becomes more certain that a substance possesses the capacity to cause the specific toxic effect under review.

Moreover, it is necessary to consider the degree of correspondence between observations in experimental animals and expected responses in humans for a given form of toxicity, particularly when the only data available derive from animal studies and a judgment must be made about expected effects in humans. If data indicate that certain types of animal carcinogens are likely to be similarly active in humans, inferences from animal data can be confidently drawn in specific cases. In the absence of data, of course, the reliance on inferences can greatly reduce confidence in the risk estimates for humans. Consequently, risk assessors must maintain their awareness of emerging scientific knowledge so they can make well-informed inferences from data on animals and other types of experimental information.

It is always important to separate true from apparent lack of data on replicability, reproducibility, and concordance. For example, several data sets may, on superficial examination, appear to lack reproducibility, but more careful examination may reveal that this lack is not substantive. An attempt should be made to learn whether the apparent lack is due to differences in study design and conduct or to true differences in response. It is generally inappropriate to characterize toxicity tests as "positive" or "negative" and then to conclude that the degree of reproducibility of a given effect is low or absent simply because some tests yielded "positive" results and others yielded "negative" results. This oversimplification of test results can be avoided if each study has been critically evaluated so that the differences in various tests (e.g., in extent of examination of test animals, sample size, duration, and magnitude of exposure) are fully known. Once this examination has been completed, it becomes possible to assess the degree of reproducibility (or lack thereof) for a given effect of a compound. This type of careful examination should also be applied when assessing degree of replication and concordance. Judging the degree of a response's concordance among various species is probably the most difficult aspect of data evaluation, so that the critical evaluation becomes especially important in order to avoid the oversimplification just described.

A final factor to be remembered when judging total data sets is the difference between absence of data and absence of replication, reproducibility, and concordance. A distinction should be made between the presence of positive evidence showing that an effect is not reproducible and the absence of negative evidence because no attempt has been made to reproduce an effect. Clearly, the failure to reproduce an effect (assuming that it is a true failure) may raise serious doubts about the toxic capacity of a substance, whereas the absence of data concerning the reproducibility of an effect should not raise such suspicions. This same principle of evaluation applies not only to reproducibility but also to judgments about replication and concordance. The actual steps in this process are described elsewhere (2). Consistent with this principle is the increased emphasis toward a clear understanding of the behavior of a chemical within the body, a process referred to as toxicokinetics, particularly to determine whether injury might occur at doses much lower than those found in laboratory animal studies (1, 8, 13).

2.2 Dose-Response Assessment

Dose-response assessment describes toxic potency or the quantitative relationship between the amount of exposure (e.g., the dose received by an individual each work day for a specified duration such as 20 or 45 yr) to a chemical and the extent of toxic injury or disease.

Data are derived from animal studies or, less frequently, from studies in exposed human populations. Many dose-response relationships may exist for a chemical agent, depending on conditions of exposure (e.g., single versus repeated and prolonged exposures) and the variety of response (e.g., central nervous system [CNS] dysfunction, cancer, birth defects) being considered. This process is highly complex, taking into account diverse information about the body's ability to create metabolites more toxic than the parent compound, its ability to detoxify potentially toxic compounds or metabolites, variations in sensitivity to doses of toxic substances, and differences between the mechanisms of toxicity in test organisms (e.g., laboratory rodents) and in human target organs. In many cases, the features of a dose (e.g., duration, frequency, and route) have a great impact on the degree of toxic potency. Specialized procedures are used to assure that later characterization of toxic risk is scientifically defensible.

The presence of a chemical in workplace air, even at concentrations slightly above a health-protective limit for a brief time, is no guarantee that an adverse effect will result. Among the many factors that influence the ability of a chemical to cause harm, the magnitude and duration of the dose determine to the largest extent whether any adverse reaction is possible. For convenience, exposure is defined as the opportunity for a dose, such as concentrations in food, air, or water, and is generally reported in units such as parts per million, parts per billion, milligrams per liter, and so on. Dose is defined as the amount received by the body of the target organism (e.g., humans) or a target organ (e.g., the liver or kidneys); it is generally reported in units of weight of the substance (e.g., milligrams or micrograms) per body weight of an individual per unit of time (e.g., hours or days). Inhalation doses are generally described as concentrations in air either as parts per million or as milligrams per cubic meter—particularly for pulmonary irritants and sensitizers. Doses resulting from skin contact are generally described as concentrations per unit of surface area (e.g., square meters); however, when the toxicity from skin contact is systemic, the dose is also described in the traditional unit of milligrams per kilogram per day. Other factors include the nature of the chemical species, the route of exposure, and modifying factors such as inborn (i.e., genetic) predispositions and preexisting disease. Yet, during the course of a day, every individual is exposed to a variety of chemicals with no resulting harm, largely because the doses are well within the range of the body's ability to avoid injury to vital functions, through either effective detoxification or efficient and rapid repair of molecular alterations that if unheeded could lead to functional impairment.

Dose-response assessment characterizes the relationship between the dose of an agent administered and the incidence and severity of adverse health effects in an exposed population. The process leads to a determination of toxic potency, which is defined as the dose of a substance needed to cause either a specific incidence of injury (e.g., 50% of group manifesting an effect, such as eye irritation) or a specific incidence of severity (e.g., 50% of a group manifesting mild or severe gastrointestinal upset). The slope of the dose-response curve is highly valuable for it provides insights as to the seriousness of adverse reactions within a population group. A steep curve (such a quadratic form) usually indicates that the number of individuals affected may increase sharply over a relatively small dose range, whereas a shallow slope usually suggests that effects might be minor over a large range of doses.

Dose-response relationships can be developed from data obtained in epidemiological studies or experimental toxicology studies. In some cases, epidemiological and human experimental data contain insight into dose-response relationships. These studies are especially useful to determine causation in acute and rare chronic disorders, particularly since toxicity is assessed directly on human health. Their disadvantages to establish dose-response relationships relate to (*1*) exposures to mixtures of chemicals rather than single substances, thereby complicating the identification of the causative agent(s) in the workplace and elsewhere; (*2*) determination of the dose of a substance that an individual receives over an extended time frame; and (*3*) the presence of external factors (e.g., alcohol consumption, smoking cigarettes, high-fat diet) that can greatly influence the manifestations of toxicity. Even when relevant epidemiological data are available, their findings generally require extrapolation from observed exposures to substantially different ones encountered elsewhere to estimate risks to populations of interest (*5*).

In the absence of, or in addition to, reliable epidemiological data, experimental toxicity data (usually in rodents) are used for dose-response assessments. The inference that results from animal bioassays are generally applicable to humans is fundamental to toxicological research and risk assessment. This premise has been extended from experimental biology and medicine into the experimental observation of carcinogenic effects. Situations exist, however, where observations in animals may not be of relevance in humans, but generally laboratory animal studies have proved to be a reliable indicator of the carcinogenic potential of many chemicals (*5*).

Many variables, not readily controllable in epidemiological studies, are controlled by design in animal studies. By regulating the route and duration of exposure to a specific chemical, the dose given to experimental animals can be determined quite precisely. Animals can be observed during the course of studies; on completion of the study, they can be examined for pathological conditions. Animals can also be used to define the toxicokinetic behavior of a compound such that reasonably precise estimates of human systemic doses can be made for specific routes of exposure.

Most animal bioassays have been designed primarily to determine which organ is injured by a substance rather than to determine dose-response relationships. Frequently, animal bioassays (with the possible exception of toxicokinetic studies) generally provide little data about responses in animals at doses encountered in the workplace or the general environment. Studies with dose-response structure require extrapolation from high to low doses using mathematical modeling that incorporates to varying degrees information about physiologic processes in the body (5).

Dose-response functions are often grouped into two classes based on two distinct groups of modes of toxicity: (*1*) those adverse effects expected to have a nonlinear dose-response relationship (sometimes referred to as *biological threshold*) and (*2*) those likely to have a linear (i.e., no threshold) dose-response structure. A threshold for a particular toxic response is defined as the dose rate below which the response attributable to the specific agent is virtually impossible (22). Acute toxic responses, for instance, have long been associated with thresholds (23). In many cases, the biological basis for thresholds for acute and chronic responses can be demonstrated based on mode of action information, that is, how a substance elicits injury at the molecular level (9, 13, 20, 24).

The existence of thresholds for cancer-causing chemicals (at least those called *initiators*) is, at least hypothetically, unlikely (23); hence, risk is imputed at all doses, no matter how small. The reason for the presence of risk of cancer even at very low doses of a carcinogen stems from observations that initiators irreversibly damage genetic material (DNA) of a cell, causing uncontrolled cell division; that uncontrolled cell replication is viewed as a self-sustaining process, in theory no longer requiring the presence of the toxicant that started the process. Biologically, the absence of a threshold is plausible, but generally it cannot be confirmed experimentally (5). The choice of an appropriate mathematical model to extrapolate to low doses is based partially on knowledge of whether a chemical is likely or not to have a linear or nonlinear response curve for the form of pathology of interest.

Each issue plays a significant role in determining how useful the data obtained from epidemiological and/or experimental toxicity studies are to estimate the likelihood that the general human population is likely to experience any form of toxicity observed under very different circumstances. Because animal bioassay data are available more often than epidemiological data, the issues of extrapolating animal data from high to low doses and from animals to humans are two of the most crucial decisions made concerning the estimation of risk from exposures to a particular chemical.

If a substance causes cancer, the slope of the cancer dose-response curve is used as the unit to describe potency and is called the cancer potency factor (CPF; also technically designated as $q1^*$). This approach is used because chemical carcinogens are often viewed as having no demonstrable biological threshold, although some empirical evidence exists to suggest that biological thresholds may well be present (i.e., in humans and laboratory animals, such defense mechanisms as detoxification of carcinogens and efficient repair of molecular lesions).

If a substance causes any form of chronic toxicity other than cancer, the highest dose level that causes no observed adverse effects (NOAEL) serves as an index of toxic potency by delineating doses above which some injury might occur and below which none is likely because of the existence of efficient defense mechanisms in the body. When a NOAEL is obtained from laboratory animal data, the findings are then extrapolated to humans by mathematically taking into account the physiologic and pharmacokinetic differences between the two species (20). The result of that process produces either a *reference dose* or a *reference concentration* (RfD or RfC; terms used by some U.S. regulatory agencies) or an *acceptable daily intake* (ADI; a term used mostly by the World Health Organization and western European countries).

2.2.1 Quantitative Extrapolation of Dose-Response Curves for Cancer and Other Forms of Toxicity Based on the Hypothesis of Linear Toxic Potency

Possible incidences in humans of cancer can be estimated for doses far below those in the range of observations only by the use of appropriate mathematical models that extend dose-response curves taking into account biological considerations (5, 9, 13). Because such dose-response functions cannot be determined empirically, the actual shapes of such dose-response curves at the lowest ends of the spectrum are unknown.

Since nonlinear dose responses (i.e., biological thresholds) are thought not to exist for most carcinogens, only certain mathematical models are used to predict carcinogenic responses at low doses (nonthreshold models). If a chemical has no known threshold for adverse effects, the dose-response curve will necessarily pass through the origin. In the case of nonthreshold chemicals, it is assumed that there is no dose (except zero dose) that corresponds to zero risk of injury. In practical terms, any dose of a carcinogen results in an incremental increase in the risk of cancer (5). Conversely, if a chemical is known to have a threshold for adverse effects, the dose-response curve will not be restricted to passing through the origin. In practical terms, this means that below some exposure level, risk of chronic injury is not anticipated.

Many types of models have been developed to assess the effects of low doses of carcinogens. They may be broadly classified into three classes: statistical, mechanistic, and enhancements. The statistical models assume that each individual in a population has a threshold below which cancer will not occur, and the distribution of thresholds in a population is a probability function. In contrast, the mechanistic models assume that no threshold exists for carcinogenic effects and that any exposure to a carcinogen results in an incremental risk of cancer. The enhancement models modify the mechanistic ones by incorporating experimental data on the behavior of the chemical in the body and data on the modes of action of specific carcinogenesis.

The most frequently used cancer risk model is a mechanistic type, the linearized multistage model. The key tenet of this model is that it assumes that the production of a malignant cell is a multistep process. Furthermore, at low doses, the risk of cancer is directly proportional to the dose of the carcinogen. At the time that this model was chosen by some regulatory agencies (6), its theoretical basis most closely approximated current thought on the modes of cancer induction (25). Since then considerable interest has been placed on a two-stage model that seems to describe well some forms of chemically induced cancers (26). Furthermore, several important refinements have been implemented, such as the application of life-table analyses combined with consideration of the modes of cancer causation (e.g., initiation vs. promotion) in relation to age at onset of first exposure. As an illustration, the LTL model permits a more comprehensive evaluation of less-than-lifetime exposure to a carcinogen by allowing consideration of the mechanism of carcinogenic action of the agent and the effect of exposure during different periods of a person's lifetime (27).

Such models generate the slope of a dose-response curve and its upper 95% confidence limit (UCL), which are often referred to as cancer potency factors. Such estimates assume exposure to be constant for a working lifetime of 45 yr (birth to 70 yr of age for environmental exposures). By its nature, UCL overestimates risk and is therefore suited more to standard settings than to defining risks to a specific population. Quantitative calculations of risk are unlikely to be higher than those derived by this approach and may be lower or zero.

A major limitation with low-dose extrapolation models is that they all often fit the data from animal bioassays equally well, and it is not possible to determine their validity based on goodness of fit. Each model may fit experimental data equally well, but they are not all equally plausible biologically. The dose-response curves derived from different models diverge substantially in the dose range of interest (5). Therefore, low-dose extrapolation is more than a curve-fitting process, and considerations of biological plausibility of the models must be taken into account before choosing the best model for a particular set of data.

For several decades, regulatory agencies in the United States have applied risk-estimation techniques to establish tolerable levels of exposure in all media (air, water, soil), in industrial and commercial settings and for foods, drugs, and other consumer products. The application of this tool for regulatory standard setting is different from its use in a nonregulatory setting, where the objective is to define the risks to a specific target population for a defined chemical exposure. In the regulatory setting, conservative "upper-bound" conditions are used for standard setting as a way of indicating that, even in the extreme, risks are unlikely to be greater than the selected value. In the nonregulatory setting, "most-likely" conditions are used to estimate the most probable risk to a target population (e.g., workers in a chemical manufacturing plan or in an oil refinery). Thus, the regulatory process strives for worst-case exposures and risk; the nonregulatory process strives for likely case exposures of risk.

This difference in application of risk-assessment methods can often lead to confusion because of a failure to accurately describe the basis for the results. For instance, the use of upper-bound conditions should lead to a conclusion that the risk "may be as high as" some magnitude of risk (e.g., one in a million) with no indication of how large the risk actually is, whereas the use of most-likely conditions should lead to a conclusion that the risk "is" a range of the magnitude (e.g., five in 10 million), with an indication of the degree of variance.

2.2.2 Extrapolating Animal Data to Humans

The physiological differences between rodents (rats and mice) and humans are at times responsible for differences in toxic potency of the same compound. The dose unit often governs the degree of correspondence in toxic potency between species. Several methods are used to make this adjustment and assume that animal and human risks are equivalent when doses are measured as milligrams per kilogram per day; as milligrams per square meter of body surface area; as parts per million in air, diet, or water; or as milligrams per kilogram per lifetime (5).

Scaling factors are used to correct the differences between species when making comparisons among species. At the level of the mode of action, scaling factors consider two independent physiological processes: (1) differences in toxicokinetics, which determine the dose delivered to target tissues; and (2) differences in tissue sensitivity between species to an identical delivered dose. On a practical level, scaling factors use dose adjustments across species that are based on a normalizing factor such as body weight or total body surface area. The most commonly used scaling factor is body weight, but the most appropriate scaling factor to use in interspecies extrapolations for carcinogenicity has been debated. Surface area, which is approximately equivalent to (body weight)$^{2/3}$ is an alternative to using body weight as a scaling factor between species (23); a widely used factor is body weight to the 3/4 power (8). Surface area may be more accurate than body weight used directly (28), although some data supports the use of 3/4 as the exponent for a better fit of the data (29, 30). No single scaling factor is best used in all circumstances. The inclusion of toxicokinetics in the creation of case-specific factors provides the most scientifically defensible cross-species extrapolations (31).

2.2.3 Extrapolating Data from One Route of Exposure to Another Route

The route by which an individual is exposed to a chemical may influence its toxic potency. For systemically toxic compounds, such differences in toxic potency are generally due to degrees or rates of absorption, to the sequence of organs that are exposed to the substance, or to the nature and rate of the metabolism of the compound. Compounds that injure the tissues at the portals of entry into the body (e.g., ozone and chlorine gas) are dependent substantially on the concentration of the compound that reaches and reacts with these tissues rather than the kinetics of the compound. Hence, when risk is to be estimated for a substance by a particular route of administration, experimental data by that same route are preferred to that from another route. However, when experimental data from the actual route of administration are absent, then data from another route must be extrapolated to the one route of interest if risk is to be estimated.

In route-to-route extrapolation, equivalent exposure concentrations from two routes (e.g., ingestion and inhalation) that yield the same absorbed dose are calculated by (1) determining the amount of a chemical absorbed via the experimental route (e.g., ingestion), and then (2) calculating the applied dose that would yield the same absorbed dose by the route of interest (e.g., inhalation). This procedure uses information about the physical-chemical properties of a substance and about the physiological characteristics of membranes.

To be confident that a specific route-to-route extrapolation is scientifically justified, several criteria should be satisfied: (1) a chemical injures the same organ regardless of route—but to varying degrees (exceptions include corrosives and irritants, which primarily affect local tissue at the portal(s) of entry and are not systemically absorbed to any appreciable extent); (2) the toxicity occurs at a site distant from the portal of entry; and (3) after absorption, the behavior (e.g., distribution, storage, metabolism, and excretion) of a substance in the body is similar by alternative routes of contact.

2.2.4 Risk Extrapolation for Threshold (Noncancer) Chronic Toxicity

Many compounds are capable of causing injury to human health provided that sufficiently large daily doses are obtained repeatedly over long periods of time. However, not all chronic toxicants can cause cancer. This section deals solely with substances that are capable of causing chronic toxicity other than benign or malignant tumors.

The ability to cause chronic, noncancer toxicity is generally the result of either the accumulation of damage in susceptible tissues or the accumulation of chemical that ultimately reaches a critical concentration and precipitates injury. Affecting the dose at sensitive tissue sites are processes in the body such as rates of absorption and distribution, redistribution to and from storage sites, metabolism to more or less toxic substances, and rates of excretion.

Historically, the accepted approach to determine the degree of chronic toxic risk of a substance is to identify that dose range below which no adverse effects are likely to occur. This inflection point in a dose-response curve is known as the no-observed-adverse-effect level (NOAEL)—referred to by some a threshold. This term underscores the realization that the studies from which the observations are made have inherent limitations in their ability to detect adverse effects; that is, the studies are known to have a finite ability to detect toxic responses. NOAELs may be obtained from studies of humans; however, more commonly, they are derived from studies with laboratory animals. When a NOAEL is obtained from laboratory animals, it usually must be converted to a no-adverse-effect level (NAEL) for humans by taking into account differences between the experimental conditions and actual human exposures and physiological differences between test and target (i.e., human) species.

When a NOAEL is based on data in humans whose circumstances are appreciably different from those in the human setting of interest, some uncertainty also exists about the location of the range of doses below which injury is likely to occur. The NOAEL derived from human observations must also be adjusted to produce a corresponding NAEL.

In either case, uncertainties stem from several factors, each of which can be used to quantitatively modify a NOAEL and estimate a NAEL for humans. The degree of confidence in the NAEL influences directly the degree of confidence in margins of safety (MOS) that are estimated in the risk-characterization step described next. The major adjustments of modifications address the following considerations, some or all of which may apply to a specific chemical exposure:

1. *Interspecies variation in susceptibility.* A factor of between 1 and 10 is used when data from laboratory animals are used to derive a NAEL in humans; if the toxic potency of a compound is similar in humans and experimental animals, then a factor less than 10 may be justified.

2. *Intraspecies variation in susceptibility.* Tests conducted in a homogeneous laboratory animal population (or even a small human volunteer group) often do not account fully for the heterogeneous human population. Individuals may vary considerably in their susceptibility to chemical insult due to genetic makeup, lifestyle factors (e.g., smoking cigarettes), age, hormonal status (e.g., pregnancy), immune system integrity, and preexisting illness. To take into account the diversity of human populations, a factor between 1 and 10 is used;

when sufficient experience with a compound in humans indicates the existence of a narrow range of susceptibilities, a factor less than 10 may be justified.

3. *Interspecies differences in size and other physiological characteristics.* When differences in toxic potency between species represent less the innate toxicity of the chemical and more the differences in the species themselves (e.g., surface area or rate of blood flow), then physiological or allometric scaling factors (e.g., body weight to the 3/4 power) are justified to provide accurate comparability of doses; one such procedure restates the dose (expressed in milligrams per kilogram of body weight) in laboratory rodents to one of milligrams per square meter surface area in humans.

4. *Route-to-route variability in toxic potency.* When the toxicologic information is obtained from one route of administration (e.g., ingestion) yet workers are exposed by another route of contact (e.g., inhalation), resulting differences in toxic potency are taken into account by applying a factor between 1 and 10; if evidence indicates little or no such variability, a factor less than 10 may be justified.

5. *Weight of evidence.* The quality of studies on which risk and safety conclusions are to be based for humans may vary considerably and thus influence confidence in the conclusions. To take into account the reliability of the experimental results, the range of species tested for toxicity, differences in durations of exposure, the inability to identify a NOAEL (in which case a lowest-observed-adverse-effect level, or LOAEL, is used), and the degree of reproducibility of findings, an uncertainty factor between 1 and 10 may also be applied (23).

Although methods for quantitative extrapolation of laboratory animal data to estimated human exposures have focused primarily on cancer risks, efforts have been made to develop dose-response models for extrapolation of noncarcinogenic health effects such as developmental toxicity (32–35). These models need to be characterized more fully before they can be used in risk assessment (23).

2.3 Exposure Assessment

This process is used to describe the nature (i.e., distribution of age, sex, and unique conditions such as pregnancy, preexisting illness, and lifestyle) and size of the various populations exposed to a chemical agent and the magnitude and duration of their exposures. The assessment might include past, current, and anticipated future exposures.

As conceptualized, exposure assessment should include two major steps: (*1*) determining concentrations at the breathing zone and the membrane interface (skin and gastrointestinal tract), and (2) ascertaining the dose to the target tissue, usually at some distance from the point of entry into the body. The first step relies on data obtained from industrial hygiene monitoring activities, including data gathered with stationary and personal monitors. At times, these can be supplemented by the use of air-dispersion modeling using computer simulations or dose-reconstruction modeling using physical models. The second step records work activities that influence rates and degree of absorption into the body and the physical properties of the chemical that influence the rate of absorption into the bloodstream. The outcome of this step provides reliable and defensible quantitative expressions of the relationship between chemical concentrations in the body and associated injury to tissues, organs, and cells. This two-step approach permits the delineation of the contribution of each work area to total dose, and the amalgamation of total dose through a workday in which a worker may move from one location to another. The details of each step are presented in sections 2.3.1 and 2.3.2.

Estimates of human doses in the workplace are generally derived for (*1*) prolonged and continuous exposures over a protracted time period (i.e., years, particularly for exposure to carcinogens) and (*2*) single peak short-term exposures of 8 h or less.

The determination of inhaled doses requires the use of appropriate physiological factors, such as body weight, skin surface area, and breathing rates (Table 46.2). Exposure factors are derived from empirical data describing the distributions of human weight, skin surface areas, breathing rates, food and water ingestion rates, activity patterns, and life expectancy as a function of gender and age (37). Because dose is expressed in terms of body weight (occasionally as surface area), appropriate body weights and surface areas for adults must be selected. When estimating cancer risks applicable to adults (because of long latency, particularly at doses much lower than those used in experimental studies), 70 kg is a standardized body weight for an adult male and 62 kg is that for an adult female. Surface areas of skin for adults along with inhalation rates for men and women are found in Anderson et al. (36) and USEPA (37).

2.3.1 *Atmospheric Concentrations as a Screening Tool*

Most often only the first step (measurement of air concentrations in breathing zones) is accomplished, and these values are treated as surrogates for actual dose to the body and to the target organs. This abbreviated approach is of pragmatic value when health-protective standards exist and are expressed as concentrations in air (mg/m^3 or ppm). Comparisons between the two can be used to demonstrate degrees of compliance; however, such comparisons cannot be used to ascertain reliably the degree of health risk or safety associated with actual air concentrations.

Table 46.2. Summary of Human Inhalation Rates by Activity Level (m^3/h)[a]

Subject	Resting[b]	Light[c]	Moderate	Heavy[d]
Adult, male	0.7	0.8	2.5	4.8
Adult, female	0.3	0.5	1.6	2.9
Adult, average	0.5	0.6	2.1	3.9

[a]Exposure factors are derived from empirical data describing the distributions of human weight, skin surface areas, breathing rates, food and water ingestion rates, activity patterns, and life expectancy as a function of gender and age (37).
[b]Includes sitting or standing at an assembly line or sitting at a desk.
[c]Includes activities such as level walking at 2–3 mph, stacking firewood, simple construction, and pushing a wheelbarrow with a 15-kg load.
[d]Includes vigorous physical exercise such as cross-country skiing, stair climbing with a load, and chopping wood with an axe.

2.3.2 Estimation of Human External Dose

Many chemicals exert their chronic toxic effects in organs distant from the portals of entry (i.e., the lungs, skin, and gastrointestinal tract). The body burden, or actual human dose, for systemic toxicants is not solely dependent on the concentration in the air but depends also on the amount that enters the body. Expressing dose on a milligram of substance per mass of body weight per unit of time (e.g., mg/kg-day) is often reflective of the body burden to the individual exposed. Conversion of air concentrations to actual doses entails multiplying the air concentration for a specified time by the breathing rate of the individual, and dividing the product by the individual's body weight. If groups of individuals are involved, then group averages and ranges can be used. If workers spend parts of the day in work locations whose concentrations of the substance vary considerably, then the contribution of each work location to the body burden would be added together to obtain an expression of the body burden on a daily basis.

External doses are calculated for either of two situations: (*1*) average daily dose over a year for chronic exposure to noncarcinogens, and (*2*) lifetime average daily dose for repeated and prolonged exposure to carcinogens. In either case, the biological modes of action for either type of toxicant dictate a somewhat different approach to the expression of dose that is most directly related to toxic potency. The steps are as follows.

Yearly Average Daily Dose (ADD$_y$). The yearly average daily dose is computed in three or four steps. Average or maximum daily air concentrations based on duration-specific samples are each aggregated into yearly values to develop weighted annual averages (one for mean concentrations and one for maximum concentrations).

Step 1. Average daily concentration (in mg/m^3) for each individual (ADC$_i$) is calculated by the following equation:

$$\text{ADC}_i = \frac{C_1 + C_2 + \cdots + C_n}{D} \tag{1}$$

where C_n = measured daily concentration in mg/m^3 as 8- or 10- or 12-hr average for an individual on the nth day (Averages can be replaced by the maximum concentration to obtain an indication of the upper bound.)
D = number of days on which measurements were made

Step 1a. ADC$_i$ can be adjusted for the number of months (or weeks) an individual spends at various locations of a workplace by the following equation to obtain the A-ADC$_i$ (in mg/m^3):

$$\text{A-ADC}_i = \text{ADC}_{i1}(T_1) + \text{ADC}_{i2}(T_2) + \cdots + \text{ADC}_{in}(T_n) \tag{2}$$

where T_n = fraction of time a worker spends in location S_n
S_n = location within a facility represented by ADC$_{in}$

RISK ANALYSIS FOR THE WORKPLACE

Step 2. The average daily dose per person (ADD_i; in mg/kg of body weight/day for a year) is calculated by the following equation:

$$ADD_i = \frac{(\text{A-ADC}_i)(m^3 \text{ inhaled air per workday})}{BW} \quad (3)$$

where BW = body weight (kg)

Step 3. The average daily dose for the workforce (ADD_f; in mg/kg-day for a year is calculated by the following equation:

$$ADD_f = \frac{ADD_{i1} + ADD_{i2} + \cdots + ADD_{in}}{N} \quad (4)$$

where ADD_{in} = average daily dose of *n*th worker
N = total number of workers

Lifetime Average Daily Dose (LADD). The LADD can be calculated in one or two additional steps, depending on whether the dose is for an individual or a group of workers. Average or maximum daily air concentrations based on duration-specific samples are each aggregated into lifetime values to develop weighted lifetime averages (one for mean concentrations and one for maximum concentrations).

Step 1. To calculate the LADD for an individual ($LADD_i$), the following equation is used:

$$LADD_i = \text{A-ADC}_i\left(\frac{y}{Y}\right) \quad (5)$$

where y = years worked
Y = years of total (usually 70) or partial lifetime

Step 2. To calculate the LADD for the workforce ($LADD_f$), the following equation is used:

$$LADD_f = \frac{LADD_{i1} + LADD_{i2} + \cdots + LADD_{in}}{N} \quad (6)$$

where $LADD_{in}$ = lifetime average daily dose for *n*th worker
N = total number of workers

2.3.3 Estimation of Human Internal Doses

In some circumstances, dose estimates could be refined by estimating the bodily dose that could reach susceptible organs. Such estimates are obtained by modifying the ADD or

LADD estimated in section 2.3.2 by the degree of absorption via the portal of entry. For instance, if a systemic dose of 10 mg/kg-day had been calculated for a chemical of interest, and the degree of absorption over 1 day through the lungs was known to be 50% and through the skin to be 10%, then the absorbed dose would be 6 mg/kg-day [= (10 × 50%) + (10 × 10%)].

Determination of internal human doses at specific sites in the body where chemicals are likely to exert their effects relies on techniques based on concepts of

- Distribution of exposure assumptions (e.g., inhalation rates associated with various activities and the time spent on them)
- Dosimetry (e.g., the rate of a chemical's absorption, once inhaled, from the lung into the bloodstream)
- Toxicokinetics (e.g., the relative distribution of a chemical throughout the body and its rates of metabolism and excretion)

In combination, such approaches can be used to estimate internal human dose and present an opportunity to understand risks to human health with increased accuracy. This refinement is applicable only when absorption data are available for both humans and the species in which the toxicity data were obtained for use to judge safety or risk from workplace exposure.

When exposure is estimated for a group of workers (e.g., a plant shift or site or the entire workforce), the distribution of doses among those individuals provides a much more informative depiction of the overall degree of danger or safety at a work site. Examples of successful distributional analyses in the workplace have been reported (38–40).

In conclusion, although air concentrations may be used as an index of workplace exposure to chemicals, the accuracy of human doses can be enhanced greatly through detailed consideration of factors that influence systemic doses, such as activities that modify breathing rates and those that control the degree and rate of absorption through the lungs and other portals into the bloodstream.

2.4 Risk Characterization

The final step of risk characterization involves integration of the data and analyses from the other three steps of risk assessment to determine the likelihood that specified workers may or may not experience any of the various forms of toxicity associated with a chemical under its known or anticipated conditions of exposure. This step includes estimations of risk for individuals and population groups and a full exposition of the uncertainties associated with the conclusions. Scientific knowledge is usually incomplete so that inferences about risk are inevitable. A well-constructed risk assessment relies on inferences that are most strongly supported by general scientific understanding and, to the extent feasible, do not include assumptions derived solely from risk management or public policy directives.

2.4.1 Cancer Risks

The approach used to estimate risks from exposure to carcinogens involves extrapolation of observations of cancer at relatively high doses to much lower doses anticipated or measured in the workplace. The risks for a specified chemical in a defined set of circumstances are estimated by combining the cancer potency factor and the various measures of dose. The risk is expressed as a probability, for instance, as the number of individuals, among all individuals exposed to a cancer-causing agent, that might contract (or die from) the disease attributable to that agent over a specified time, usually a lifetime of 70 yr. A specific example would be a lifetime risk of one in a million, meaning one person may get cancer from among a million identically exposed persons. Risk is frequently expressed in a notational format of 10^{-3}, 10^{-4}, 10^{-5}, or 10^{-6}, meaning risks of one in 1,000, one in 10,000, one in 100,000, or one in a million, respectively. Such cancer estimates reflect the chance that an event may occur and not that they must inevitably occur; however, because of limitations in knowledge about the processes of cancer causation, it is also possible that the risk may be zero and that no cancer would ensue from a specified exposure. Depending on the quality of the data that underlie an estimated risk, the uncertainties may be small or large; such uncertainties are to be described explicitly and comprehensively at this stage.

2.4.2 Noncancer Risks

For adverse health effects other than cancer, a margin-of-safety (MOS) approach is used to establish whether a potential human dose is lower than a theoretical limit on exposure to provide an acceptable safety margin to indicate the unlikelihood of adverse effects. The MOS can be determined for a variety of exposure scenarios and for different toxic health effects, such as alterations of some functions of the nervous system or of the kidney. Using this procedure, first the human NAEL is obtained either from epidemiologic observations or laboratory animal findings. In some cases, all examined exposures produce an effect, and only a LOAEL is available. In such cases, the effects occurring at the lowest doses in the most sensitive species/strain/sex are generally used as the basis for estimating a NOAEL. When the NAEL is derived from human data, the measured NOAEL for one or more studies in humans is adjusted for variability in individual susceptibilities, which generally appears to be one order of magnitude (i.e., between 1 and 10-fold) or less. For instance, a human NOAEL of 10 mg/kg-day for a hypothetical compound can be converted to a human NAEL by dividing the NOAEL by 1 or 10 or some intermediate value to account for greater variability in sensitivity in an exposed population than in the study group; the result would be a human NAEL of 5, or 1 mg/kg-day, respectively.

When a human NAEL is derived from results of animal studies, the human NAEL must account for not only the range of individual susceptibilities (i.e., 1 to 10-fold) but also the possible differences in sensitivity to toxic substances between humans and the laboratory species in which the toxicity had been measured (also ranging between 1 and 10-fold). For example, an animal NOAEL of 100 mg/kg-day for a hypothetical compound could be converted to a human NAEL by dividing the animal NOAEL by 100 to account for differences in sensitivity between test animals and humans and for the degree of variability in sensitivity in an exposed human population; the result would be a human NAEL of 1 mg/kg-day, respectively.

The human NAEL is divided by the daily dose obtained by the population of interest; the result is a MOS. If the margin of safety is greater than 1, a dose unlikely to cause human injury is obtained and may serve as a basis for setting tolerable levels of exposure in the workplace or elsewhere. The larger the MOS, the greater the certainty that no injury will occur.

Human MOS can be estimated for either acute or chronic toxicity by relying on the relevant toxicity and exposure data. For instance, to determine a MOS for acute exposure, peak doses are to be compared to acute NAELs for humans. Depending on the quality of the data that underlie an estimated risk, the uncertainties may be small or large; such uncertainties are to be described explicitly and comprehensively at this stage.

The concept of MOSs and their application to the determination of acceptable daily intakes (ADIs) was established by the World Health Organization (WHO). Recently, WHO, through the International Programme of Chemical Safety, restated and refined this approach to establish safe levels of exposures for workers and the general public (1). Because the terms "safety" and "acceptable" are value-laden and inherent in the ADI/safety factor approach to determine regulatory values for threshold responses, USEPA has modified this procedure by standardizing "uncertainty factors" and a "modifying factor" as an additional uncertainty factor that allows for professional judgment on the level of confidence about a NOAEL. The terminology of "acceptable" level has been replaced with the terms reference dose (RfD) and reference concentration (RfC). An RfD or RfC is calculated by dividing the NOAEL obtained from an animal bioassay by the product of the uncertainty factors and the modifying factor (23).

2.4.3 Mixtures

Because workers may be exposed to more than one compound simultaneously, the possibility of adverse interactions among these chemicals can be assessed. Guidance for conducting health risk assessment for chemical mixtures has been developed (41). One compound may affect the toxicity of another compound in three possible ways:

1. *Additivity* results when the effect of combined exposure to a combination of compounds equals the sum of the toxicity that would be expected from each of the compounds acting independently.
2. *Synergism* results when the effect of combined exposure to a combination of compounds is greater than the sum of the toxicity that would be expected from each of the compounds acting independently.
3. *Antagonism* results when the effect of combined exposure to a combination of compounds is less than the sum of the toxicity that would be expected from each of the compounds acting independently.

For compounds affecting the same target organ and for which no mechanistic data reject the concept of additive toxicity, their toxicities are assumed to be additive. The effect of additivity may be assessed by calculating a hazard index:

$$\text{HI} = \sum_{n=1}^{i} \frac{\text{dose}_n}{\text{NAEL(H)}_n} \tag{7}$$

where NAEL(H) = the human NAEL. If the hazard index is less than 1, reasonable certainty exists that no adverse effects will occur.

To illustrate the application of risk assessment to the workplace, two case studies are presented, one for a cancer (benzene) and another for reproductive toxicity (benzene, p-chlorotoluene, and toluene).

3 ASSESSING THE CANCER RISKS OF BENZENE: A CASE STUDY

For this illustration, benzene is present in the workplace atmosphere at a hypothetical petrochemical facility. In this illustration, the data from industrial hygiene monitoring indicate that workers are exposed to benzene at levels that are below the OSHA permissible exposure limits (PEL) of 3.2 mg/m^3 (1 ppm) as a result of application of state-of-the-art practices in industrial hygiene. Nevertheless, both workers and medical department personnel are asking whether a risk to human health exists and, if so, its magnitude. Verma and des Tombe reviewed the history of setting the TLV for benzene by ACGIH (42).

3.1 Hazard Identification

Benzene is a volatile, colorless, and flammable liquid aromatic hydrocarbon with a characteristic odor. It is used primarily as a raw material in the synthesis of styrene, phenol, cyclohexanes (nylon), aniline, maleic anhydride (polyester resins), alkyl benzenes (detergents), chlorobenzenes, and other products used in the production of drugs, dyes, insecticides, and plastics. Benzene is also a component of crude oil and gasoline.

3.1.1 Chronic Toxicity

Several reviews provide detailed descriptions of the experimental evidence used to characterize the chronic toxicity of benzene (43–49). The only two documented forms of chronic toxicity of benzene are cancer (usually acute myelogenous leukemia in adults) and injury to the hematopoietic system (questions have been raised that benzene can cause reproductive injury; that is dealt with in the illustration in section 4). Because cancer is viewed as a toxic phenomenon unlikely to have a threshold, some risk can be estimated for every chronic dose. For hematopoietic injury, a biological threshold is expected to be present below which no injury to this organ system is likely to occur. Some investigators have put forth the proposition that damage to the hematopoietic system is a necessary antecedent to the production of leukemia; however, this has not been proven.

Benzene, which can be readily absorbed through skin, lungs, and gastrointestinal tract, is known to be converted in the body to one or more metabolites that are likely to cause cellular damage in the bone marrow. Thus, although benzene is rapidly excreted and does not accumulate in the body, some forms of subcellular damage may accumulate, as reflected by a sense that, at least hypothetically, cancer risk is cumulative with repeated exposures. As a result, the risk of cancer is often expected to persist even at relatively low doses. Only the carcinogenic property of benzene is addressed in this illustration.

Benzene seems to produce few forms of other toxicity, except at doses so high as to be encountered only in unusual circumstances. This is the case for reproductive toxicity in laboratory animals in which benzene is not known to injure a developing fetus except when mothers are exposed to atmospheric concentrations much higher than those permitted in workplaces (addressed in section 4).

The information on which these conclusions are based is summarized as follows.

3.1.1.1 Observations in Humans.

The chronic toxicity and carcinogenicity of benzene have been investigated extensively (50, 51). Chronic exposure to concentrations of benzene greater than 200 ppm (640 mg/m^3) in air can seriously damage the bone marrow, causing pancytopenia and aplastic anemia. Benzene is a confirmed human carcinogen and is associated, in most cases, with acute myelogenous leukemia at workplace concentrations as low as 35 ppm (112 mg/m^3). Benzene may be capable of causing cancer at lower concentrations; however, epidemiology studies are unable to detect an increased risk of leukemia due to their relative insensitivity (14, 51, 52). Recently, Rothman et al. (52) reported findings of an epidemiologic study of Chinese workers heavily exposed (31 ppm as an 8-hour time-weighted average) to benzene. They found a dose-response relationship between benzene exposure and white blood cell count, absolute lymphocyte count (most sensitive indicator of exposure), and mean corpuscular volume; and an excess of red blood cell abnormalities among the more highly exposed individuals.

An association between benzene and leukemia was suggested as early as 1928 (54). Since then, a number of case reports have suggested that benzene can cause leukemia, most notably a series of leukemias reported by Aksoy and coworkers among Turkish shoemakers (55–57). These case reports have been confirmed by epidemiology studies, most notably in a cohort of rubber workers (52, 58, 59) and chemical workers (60, 61). Based on the weight of evidence, IARC has classified benzene as a confirmed human leukemogen (14, 51).

The weight of evidence that benzene is a known carcinogen to humans is summarized as follows. Aksoy (50) examined a series of 34 cases of leukemia among shoe workers in Istanbul. Based on a government estimate of the number of shoe workers in Istanbul, the investigator estimated the incidence of acute leukemia, or preleukemia, among these workers was 13 per 100,000 compared with an estimated incidence of 6 per 100,000 in the general population (relative risk = 2). Because the shoe workers labored in small shops, obtaining reliable estimates of benzene exposure for the population as a whole was not possible. The general exposure was probably high; peak exposures to benzene were reported to be in the range of 210–650 ppm (679–2075 mg/m^3) in some poorly ventilated shops. The duration of exposure was estimated to be 1–15 yr with a mean of 9.7 yr. Due to the uncertainties in the population at risk and levels and durations of exposure, neither reliable estimates of the risk of cancer in this group of exposed individuals nor an estimate of individual exposure could be obtained (62). Although these studies provide little quantitative information on the relationship between benzene exposure and leukemia, they do support a positive qualitative relationship between benzene exposure and leukemia.

Infante et al. (58) and Rinsky et al. (52) made a retrospective cohort analysis of 748 workers occupationally exposed to benzene between 1940 and 1959 in two factories producing rubber hydrochloride. Vital status was determined for workers up to 1975. Expected rates of death due to leukemia were obtained from U.S. white male mortality statistics. A statistically significant increase in mortality from all leukemias was observed (observed/expected = 7/1.25; SMR = 560). In a follow-up study, Rinsky et al. (59) extended this cohort to employees hired between 1940 and 1964. In addition, vital status was determined up to 1981. A statistically significant increase in mortality from all leukemias was observed (observed/expected = 9/2.27; SMR = 337). Both studies by Rinsky et al., (52, 59) attempted to characterize the exposure of the workers. Workers were classified by job title and, based on industrial hygiene records, the exposures of the individual workers were estimated. A significant correlation between the intensity and duration of benzene exposure and the risk of leukemia was observed, further supporting the assumption that benzene is the etiologic agent.

Wong et al. (60) examined the mortality of 4,062 workers exposed to benzene in the chemical industry in Canada. Workers were classified by job description as to the duration and intensity of benzene exposure. Control subjects were not exposed to benzene in the workplace. A dose-dependent increase in the incidence of leukemia, lymphatic cancer, and hematopoietic cancer was observed in benzene-exposed individuals. The observed increase in incidence of leukemia may have been due in part to a lower-than-expected incidence of leukemia among control workers, making this study of questionable significance.

Ott et al. (61) examined the mortality of 594 workers at a Dow Chemical plant. Three leukemias were observed. The increase in cancer was not statistically significant, and no dose-response was observed. On the other hand, the small number of workers exposed makes this study relatively insensitive statistically. In addition, the benzene exposures were relatively low (generally less than 15 ppm), which decreased the probability of detecting a carcinogenic effect. Leukemia cases in this study may have been underreported in that one death certificate listed leukemia as a secondary rather than a primary cause of death. Listing leukemia as a primary cause of death would have added the case to the three cases previously identified, which would have made this study's increased incidence of cancer statistically significant. Because of these factors, this study is considered to be of borderline value.

In addition to causing neoplasms, benzene can alter the function of the bone marrow. Exposure to high concentrations of benzene can result in the development of aplastic anemia. This lesion is characterized by a progressive decrease in the number of circulating formed elements in the blood. The bone marrow in these cases is necrotic, with fatty replacement of the functional bone marrow (63). The lowest benzene concentration associated with the development of aplastic anemia is 60 ppm (192 mg/m^3) (64).

Acute exposure to benzene in humans can result in headache, lassitude, and weariness at concentrations greater than 50 ppm (160 mg/m^3 = 3.8 mg/kg) for 5 hr. At higher concentrations, benzene can depress the central nervous system and even lead to death.

3.1.1.2 Observations in Animals. In a series of studies, Cronkite and colleagues examined the carcinogenic and hematotoxic effects of benzene. Cronkite (65) exposed CBA/Ca male mice to benzene at 100 and 300 ppm (320 and 960 mg/m^3) for 6 h/day, 5 days/week, for 16 weeks, followed by observation for the remainder of their lives. These doses of benzene caused a significant increase in the incidence of leukemia in exposed mice. Cronkite also observed anemia, a decrease in stem cell content of bone marrow, and a decrease in marrow cellularity at doses of 100 ppm (320 mg/m^3) for 6 h/day, 5 days/week, for 2 weeks. Cronkite et al. (66) observed increased incidence of leukemia in another strain of mouse (female C57B1/6) following doses of 300 ppm benzene for 6 h/day, 5 days/week, for 16 weeks. In addition, Cronkite et al. (67) exposed CBA/Ca mice to a range of inhaled concentrations of benzene. A NOAEL was observed at 10 ppm (32 mg/m^3) after benzene exposure 6 h/day, 5 day/week, for 2 weeks. A dose-dependent decrease in blood lymphocytes, bone marrow cellularity, and marrow content of spleen colony-forming units was observed for concentrations above 25 ppm (80 mg/m^3).

Maltoni et al. (68) exposed Sprague-Dawley rats for 15 weeks to 200 ppm (640 mg/m^3) benzene for 4 h/day, 5 days/week, for 7 weeks and then for 7 h/day, 5 days/week, for 85 weeks. Exposure at this concentration was associated with an increase in total malignant tumors when compared to controls. Malignant and benign tumors were found in the zymbal glands, the oral cavity, the mammary glands, and the liver. Because humans have no zymbal glands, the relevance of the zymbal gland tumors observed in this study is of questionable significance to human health and generates considerable controversy.

In addition to causing neoplasms, benzene is also hemotoxic in experimental animals. Snyder et al. (69) exposed male AKR/J mice exposed to 100 ppm (320 mg/m^3) via inhalation after 6 weeks of age for 6 h/day, 5 days/week, for life. A significant increase in the incidence of anemia, lymphocytopenia, and bone marrow hypoplasia was observed. Rozen and Snyder (70) exposed mice to benzene concentrations of 300 ppm (960 mg/m^3) for 6 h/day, 5 days/week, for 23 weeks. Mice were examined for lymphoid parameters at 1, 6, and 23 weeks. A progressive reduction in the response of T- and B-cells to mitogenic stimuli, a reduction in the numbers of B-lymphocytes in the bone marrow and spleen, and a decrease in the number of T-lymphocytes in the thymus and spleen was observed. At 23 weeks, mitogen-induced proliferation of bone marrow and splenic B-lymphocytes was eliminated. The authors concluded that 300 ppm benzene exposure results in severe depression of B- and T-lymphocyte numbers and response to mitogens. These alterations may be a factor in the carcinogenic response to benzene.

3.1.2 Genotoxicity

Significant increases in chromosomal aberrations of bone marrow cells and peripheral lymphocytes from workers with high exposure to benzene have been cited by numerous investigators (51). High doses of benzene have also induced chromosomal aberrations in bone marrow cell from mice (71) and rats (72). Positive results in the mouse micronucleus have been reported (71). Benzene was not mutagenic in several bacterial and yeast systems, in the mouse lymphoma cell forward mutation assay, or the sex-linked recessive lethal mutation assay with Drosophila melanogaster (51). These findings suggest that benzene's mode of carcinogenesis is likely to be mediated via damage to the genome, thereby increasing the probability that the dose-response is linear.

3.1.3 Toxicokinetics

The toxicokinetics of benzene have been actively studied (73, 74), and several physiologically based pharmacokinetic (PB-PK) models have been proposed recently for benzene metabolism (75–77). Following absorption into the body, benzene is widely distributed to tissues (73, 78). Because of its lipid solubility, benzene is preferentially distributed to fat (78). In rats, the half-life of benzene in fat is longer than the half-life in other tissues (1.6 h versus half-lives between 0.4 and 0.8 h for other tissues) (78). The bone marrow, lung, and kidney are also major sites of benzene deposition in experimental animals (as indicated by higher benzene concentrations in these tissues than in the blood). The half-life of benzene in humans is approximately 0.75 h at inhaled doses of less than 10 ppm for 8 h and approximately 1.2 h at doses of 100 ppm for 1 h (79, 80).

It is generally accepted that benzene must be metabolized before it can cause injury (81). Metabolism of benzene occurs principally in the liver; however, extrahepatic metabolism appears to contribute significantly to benzene toxicity, especially in the bone marrow (73). In the liver, benzene is metabolized via cytochrome P450 to form benzene epoxide. Benzene epoxide undergoes further metabolism to form phenol, catechol, and hydroquinone as primary metabolites (82). These ring metabolites may be conjugated with sulfate or glucuronic acid and excreted in the urine (73). Open-ringed metabolites are also formed and may eventually become the potentially toxic metabolites *trans, trans*-muconaldehyde and *trans, trans*-muconic acid (83, 84). Variations in the route of exposure do not significantly alter the production of benzene metabolites (85).

The toxicity of benzene is likely due to the action of one or more of its metabolites rather than of benzene itself (81, 83). In several studies, the modification of benzene metabolism has led to altered benzene toxicity. For example, toluene, a competitive inhibitor of benzene metabolism, decreased benzene hematotoxicity in mice (86). In another study, Longacre et al. showed that benzene metabolites were found at higher concentrations in benzene-sensitive DBA/2 mice than in benzene-resistant C57B1/6 mice (87, 88).

3.1.3.1 Absorption: Inhalation. For continuous doses of 50–100 ppm benzene for several hours, the absorption of benzene by humans is approximately 50% (51, 81). Nomiyama and Nomiyama (89, 90) reported that men and women who were exposed to 52–60 ppm benzene for 4 h had an estimated respiratory retention (the difference between respiratory uptake and excretion) of 30% of the dose, with little difference between the genders.

Inhalation studies in animals confirm that benzene is rapidly absorbed through the lungs. Sabourin et al. (91) reported that uptake of benzene vapor in rats and mice is inversely proportional to the air concentration. The percentage of inhaled benzene that was absorbed and retained during a 6-h exposure period decreased from 33 to 15% in rats and from 50 to 10% in mice as the exposure increased from 10 to 100 ppm (31.9–390 mg/m^3).

3.1.3.2 Absorption: Skin. Benzene can be absorbed through the skin but at a greatly decreased rate when compared to inhalation exposure. Blank and McAuliffe (92) calculated that an adult working in an ambient air containing 10 ppm (31.9 mg/m^3) benzene would absorb 6.6 mg/h from inhalation and 1.32 mg/h from whole-body (2 m^2) exposure. The authors propose that because of the water solubility of benzene, diffusion through the stratum corneum is the rate-limiting step for dermal absorption.

3.1.4 Summary

Exposure to benzene is known to cause leukemia after workers have been exposed to high concentrations of the solvent over prolonged periods. Benzene or its metabolites is also carcinogenic in experimental animals. Based on results obtained from epidemiology studies, IARC has classified benzene as a known human carcinogen (51). Benzene or its metabolites is genotoxic, causing chromosomal aberrations that in some cases might be precursors of lymphopoietic tumors. Limited information precludes any conclusion about the potential for alterations of the immune and central nervous systems resulting from long-term benzene exposure.

3.2 Dose-Response Characterization

The quantitative descriptor of benzene's cancer potency is the cancer potency factor (CPF). The critical chronic toxic end point for benzene exposure is ordinarily acute myelogenous leukemia in adult humans. Because the studies associating cancer with benzene exposure contain data on worker exposure to benzene, USEPA based its CPF on these epidemiologic data. Three studies were used to model the dose-response relationship (52, 61, 60). A total of 21 estimates were made, using six models and various combinations of the epidemiologic data, with a mathematical correction for the findings of Wong et al. (60). Evidence exists that benzene is incompletely absorbed by inhalation (81, 89, 90); consequently, USEPA assumed in calculating an inhaled (absorbed) CPF of 2.9×10^{-2} (mg/kg/day)$^{-1}$ that 50% of inhaled benzene is absorbed (93).

For acute toxicity, the critical toxic end point is CNS depression (headaches, lassitude, and weakness). These symptoms are associated with benzene concentrations between 160 and 480 mg/m^3. The acute NOAEL in humans for these effects was identified as 160 mg/m^3 (94). The human NAEL is obtained by dividing the human NOAEL of 3.7 mg/kg by an uncertainty factor of 2 to account for the intraspecies differences in sensitivity; the result is a dose of 1.9 mg/kg.

3.3 Exposure Assessment

At this hypothetical facility, 100 individuals are employed to perform assorted tasks in either of two chemical production areas (A or B). From employment records, workers generally perform tasks approximately 40% of the workday in area A and approximately 60% of the workday in area B. Length of service at the facility ranges from 2 to 40 yr, with 26 yr as a median value. Daily work schedules frequently last 10 h.

Information about levels of benzene in each work area has been collected by two means: (1) six area monitors (3 in area A and 3 in area B), collecting data hourly during the workday; and (2) personal monitors affixed to 20 plant workers during four 10-h shifts. After 2 yr of employment, workers were at the plant 48 weeks of the year.

To enable the calculation of doses of benzene for each employee, physiological factors have been obtained. To determine approximate breathing rates of these employees that would in turn influence the magnitude of each employee's body burden of benzene, a questionnaire has been administered to 40 employees. This information provides a basis for determining the absorbed dose of benzene. The body weight of each employee has been obtained from recent records of periodic medical examinations.

3.3.1 Chronic Doses

Chronic doses (in the unit of lifetime average daily dose, or LADD) have been calculated to estimate the chance of developing cancer from benzene at this facility. Personal moni-

RISK ANALYSIS FOR THE WORKPLACE

toring data have indicated that the concentration in the breathing zones of the workers averages 0.6 mg/m^3 (± 0.08 mg/m^3) integrated across areas A and B.

According to survey responses, while in area A workers performed moderate to heavy activity, whereas relatively light activity was performed in area B. The average body weight of the employees was 66 kg.

The LADD can be calculated for the workers using Eq. (5) and Eq. (6). The results ranged from 2 mg/kg-day for an individual working at the facility for 2 yr to 53 mg/kg-day for the 40-yr veteran; for the median 26-yr duration of employment, the estimated dose was calculated to be 35 mg/kg-day (Table 46.3). Because these data have been collected solely during 1 yr of operation, the findings are extrapolated to 26 and 40 yrs, assuming that conditions for the median and maximum of employee work lives have remained constant during the working lifetime.

The area monitoring has yielded annual average concentrations of 0.9 mg/m^3 (± 0.01 mg/m^3) in area A and 0.2 mg/m^3 (± 0.01 mg/m^3) in area B. The records of yearlong monitoring indicate relatively few incidences of excursions greater than one standard deviation from the mean. Area monitoring data during the time of the personal monitoring are compared to those for the balance of the yearlong data collection. The results indicate no statistically significant differences (Table 46.3); consequently, the results of the personal monitoring are deemed to be reasonably representative of repeated and prolonged exposure, at least for the 12 months during which data have been collected. The doses obtained from these results were 3, 43, and 66 mg/kg/day for the 2, 26, and 40 yr of employment, respectively.

3.3.2 Acute Doses

To estimate acute doses for which the MOS can be estimated, data from either personal or area monitors can be used. Specifically, the highest measured concentrations of 1 exposure durations are selected. As such, the exposures calculated by this method are plausible worst case and are consistent with the conditions that are capable of causing acute toxicity. The results indicate that the peak concentrations measured in area A range from 1.6 to 2.4 mg/m^3 slightly less than 1% of the time. Peak concentrations in area B were considerably less. The highest concentration measured in area A is equivalent to a dose of 0.1 mg/kg.

3.4 Risk Characterization

The cancer risk associated with exposures to benzene in this hypothetical workplace were estimated by combining the CPF with the calculated doses. The results are presented in

Table 46.3. Chronic Doses of Benzene

Years of Work	LADD (µg/kg-day)	
	Personal	Fixed
2	2.6	3.3
26	3.5	4.3
40	5.3	6.6

Table 46.4. The estimated risks steadily increased from approximately 7 in 10,000 to 2 in 1,000 as a function of years of employment. The risks estimated from either fixed or personal monitoring data were not significantly different from one another; however, exposures calculated from the personal monitoring data are deemed to be somewhat more accurate than those obtained from the fixed or area monitors.

The MOS for the potential for acute toxicity was derived by dividing the human NAEL for acute toxicity by the peak dose in work areas A and B. The results indicate that the MOS is 19. Because the NAEL was derived with ample uncertainty factors, the magnitude of the MOS is sufficient to assure that acute toxicity is unlikely to occur.

3.4.1 Analysis of Degrees of Uncertainty

Exposure assessments employ a series of assumptions, each with its own range of uncertainty that, when combined, can produce an exposure estimate with widely separated numerical bounds. A number of sources of variability and uncertainty impinge on any given exposure assessment:

- Measurement error, which arises from random and systematic error in a given measurement technique
- Sampling error, which arises from uncertainties relating to the representativeness of the actual population being measured
- Variability, which arises from the natural differences in exposure-related parameters among individuals, such as inhalation rate
- Limitations in applicability of generic or indirect data used as surrogate data for actual exposure levels when these are unknown
- Professional judgment, which, if used to assign values to exposure parameters that may not be accurately known, may produce misleading results, or widely varying results if different values are assigned by different assessors

Ideally, exposure assessments can be based on measurements of actual concentrations; residues in biological fluids and tissues; data on emissions sources; exposure duration; exposure frequency; and knowledge of environmental transport, transformation, fate, and toxicokinetics. However, it is rare that an exposure can be precisely defined for a given chemical and exposure scenario, because some key data, ranging from detailed knowledge

Table 46.4. Estimated Cancer Risk from Hypothetical Exposure to Benzene in the Workplace

Years of Work	Estimated Cancer Risk	
	Personal	Fixed
2	7.5×10^{-5}	9.2×10^{-5}
26	9.7×10^{-4}	1.2×10^{-3}
40	1.5×10^{-3}	1.9×10^{-3}

of source terms to characteristics of the exposed population (which would influence the choice of exposure parameters), may be unknown. Thus, any value for an input parameter for estimating individual or population exposures via an exposure model or algorithm has an associated level of uncertainty.

3.4.1.1 Uncertainties Associated with Measured Air Concentration Data.
Many of the current air-monitoring methods have inherent variability. For example, a high-quality measured data set may have measured concentrations that typically are accurate to within $\pm 25\%$. A fixed monitoring program with less-rigorous control may be accurate to within a factor of 2 or more. Thus, all measured concentrations contain some degree of sampling and analysis errors. These uncertainties should be considered when using measured air quality data for exposure or risk assessment.

3.4.1.2 Uncertainties Associated with Personal Monitoring and Time-Activity Surveys.
Attempts have been made to increase certainties in the exposure and risk estimates by applying the result of the personal monitoring studies and time-activity surveys. Personal monitoring increases certainties by accounting for sources of exposure that can effectively be accounted for in no other way; time-activity surveys increase certainties by providing information on activities that can be used indirectly to define more precisely the daily inhalation rates of individual members of the population. Time-activity data also enhance the accuracy of estimates regarding the time spent in different microenvironments; from these findings, it is known that the average individual spends approximately 40% of the time in area A and approximately 60% in area B.

3.4.1.3 Uncertainties Associated with Exposure Parameters.
The use of accurate and appropriate exposure parameter values provides an important step in increasing certainties in exposure assessment. The results of a study conducted for the Chemical Manufacturers Association (95) underscores that the use of accurate and representative exposure factors (such as were obtained through the time-activity survey) increase the certainties in the resulting exposure estimate.

4 ASSESSING THE RISKS OF REPRODUCTIVE TOXICITY: A CASE STUDY

At a hypothetical chemical-manufacturing facility, both female and male employees are working under conditions in which three compounds (benzene, toluene, and p-chlorotoluene) are present in large quantities in the workplace atmosphere. Despite the demonstration that the facility meets federal standards of industrial hygiene practice and of atmospheric concentrations (i.e., PELs), workers and management have questioned whether these standards of operation are sufficient to protect against impairment of the reproductive performance in women and men. Particular concern surrounds the possibility that developing fetuses of female workers may be at risk of injury from chemical exposures to the mothers. In this instance, the workplace standards are silent on whether they protect against reproductive impairment; hence, risk assessment, relying on industrial hygiene monitoring data, provides some means to address this matter in a systematic manner.

A risk assessment was performed to evaluate the significance to reproductive health of the occupational exposures. The analysis first examined each compound individually and then collectively. The results of the evaluation are summarized in the subsections that follow. For each compound, the workers' levels of exposure were compared to the derived tolerable exposure concentration and to the current OSHA and ACGIH standards.

As background, the individuals work 8 h/day, 5 days/week, 50 weeks/yr. All of the activity on the job is considered light to moderate for purposes of estimating breathing rates. The air concentrations of benzene were derived from fixed monitoring data collected by the firm's industrial hygiene specialists for the period 1981 to 1993. Nine job codes were identified for 30 workers.

4.1 Benzene

The data for benzene were analyzed according to the four steps in risk assessment. Reproductive toxicity is often divided into development toxicity (injury to the developing fetus, with particular emphasis on birth defects, also called terata; the term teratogenesis is used to refer to the process by which birth defects (skeletal or soft tissue) are produced; a teratogen is one that causes birth anomalies) and fertility dysfunction (injury to the reproductive function of either male or female that results in impairment of the ability to reproduce). This distinction is important because laboratory animal studies can be designed to address either one, the other, or both; hence the existence of a teratogenicity study may shed little light on the ability of a substance to alter the fertility of test animals or humans.

4.1.1 Hazard Identification

Evidence exists that benzene can cause injury at levels of exposure below 200 ppm; however, none of these are related to reproductive toxicity. ACGIH (96) cites several reports of workplace exposures to benzene at air concentrations around 40 ppm having an adverse effect on blood components. This position is supported by results from Deichmann et al. (97), which suggest that inhalation exposure to 44–47 ppm (131–150 mg/m^3) can affect the bone marrow function in rats.

Nine studies examined the possibility that benzene may adversely alter reproductive outcome in humans (50). In a study by Holmberg (98), a statistically significant association was found between exposure to various organic solvents (benzene, toluene, etc.) and CNS defects in newborns. However, only one case reported exposure to benzene, and the number of controls with benzene exposure was not reported. A serious flaw with this study is that no control was made for other factors that may influence pregnancy outcome. Thus, this study provides insufficient information to draw any conclusions on the potential adverse effects of benzene on pregnancy outcome in humans.

In the eight other studies of pregnancy outcome in benzene-exposed women, none found a significant increase in the frequency of birth defects in the offspring of women occupationally or environmentally exposed to benzene. The benzene doses were not specified, although two of the women were suffering from benzene-induced anemia, suggesting that the repeated doses were likely to have been quite high. According to ACGIH (96), no blood dyscrasias—believed to be the most sensitive index of noncancer toxicity—occur below an 8-h time-weighted average (TWA) of 25 ppm (80 mg/m^3).

Several reproductive toxicity studies using laboratory animals have been reported. Kuna and Kapp (99) exposed pregnant rats to 0, 10, 50, and 500 ppm (0, 32, 160, and 1600 mg/m^3) benzene in air for 7 h/day on gestation days 6–15. Dams exposed to benzene had a dose-dependent decrease in weight gain between days 5 and 15, which was significant only at 50 and 500 ppm. No overt signs of benzene-related toxicity were observed in exposed rats. Benzene treatment had no effect on pup viability, but there was significantly decreased pup weight at 50 and 500 ppm. At both 50 and 500 ppm benzene, pups had delayed skeletal development as evidenced by lagging ossification; but current scientific thinking holds that this manifestation is not considered teratogenic (100). However, several developmental abnormalities were noted in the group exposed to 500 ppm, including exencephaly, angulated ribs, dilated lateral and third ventricles of the brain, and forefoot ossification out of sequence. Only a few of the pups had these defects (no more than 4 for each category out of 98 fetuses); however, these anomalies are extremely rare in historical controls, suggesting a causal relationship for birth defects. Thus some evidence exists to indicate that benzene is fetotoxic and, at relatively high doses (i.e., 500 ppm and above), may be teratogenic to laboratory rodents.

Green et al. (101) exposed rats to 100, 300, and 2,200 ppm (320, 960, and 7040 mg/m^3) benzene in air for 6 h daily on days 6–15 of gestation. Rats exposed to 2,200 ppm benzene had decreased weight gain and showed signs of narcosis during exposure, suggesting the presence of acute toxicity. No effect on weight gain was noted in rats exposed to 100 or 300 ppm benzene. At 2,200 ppm benzene, decreased litter weight and significant increase in skeletal anomalies (primarily missing sternabrae and delayed ossification of the sternabrae) were observed. At the lower benzene doses, no effect was observed on fetal weight; however, increases in skeletal abnormalities were found, but with no clear dose-response in the incidence of the anomalies: The incidence of missing sternebra was highest in the 100- and 2,200-ppm benzene exposed group, but the 300-ppm benzene group had no significant increase in missing sternebra. These data indicate that benzene is fetotoxic at doses at which no overt effects on maternal well-being are present.

Murray et al. (102) exposed pregnant mice and rabbits to 500 ppm (1600 mg/m^3) benzene in air for 7 h/day on gestation days 6–15 and 6–18, respectively. No compound-related toxicity was observed in dams. Benzene was slightly embryotoxic as evidenced by decreased fetal weight in the benzene-exposed mice compared to control. In addition, benzene caused delayed development of the skeletal system, suggesting an embryotoxic effect.

Tatrai et al. (103) continuously exposed rats to 0 and 124 ppm (0 and 300 mg/m^3) benzene in air on days 7–14 of gestation. Exposed rats had decreased weight gain, indicative of maternal toxicity. The pups had decreased mean fetal weight with retarded bone development compared to controls. In a similar study, rats were exposed to 0 and 313 ppm (0 and 1000 mg/m^3) benzene in air continuously on days 9–14 of gestation (104). In this study, the dams manifested decreased weight gain, and the pups had decreased birth weight with retarded bone development.

Watanabe and Yoshida (105) injected pregnant mice with 2,600 mg/kg benzene subcutaneously on days 11–15 of gestation and sacrificed them on day 19. No controls were used in this study. They observed an increase in the incidence of fetuses with cleft palate and decreased jaw size in mice injected on day 13 of pregnancy compared to other days of injection. This dose of benzene caused a significant decrease in the white blood cell count of mice.

4.1.2.1 Confidence in Database.
Consistently among nine epidemiologic studies, benzene has not been reported to cause birth defects in the offspring of exposed women, even when the women themselves were suffering from signs of benzene-induced chronic damage to the blood-forming organs (e.g., pancytopenia). This observation suggests that benzene has only a weak ability to cause birth defects in humans. The inability of benzene to cause developmental toxicity in experimental animals has been reported in numerous studies; these findings are consistent with the observations in humans. In these studies, high doses of benzene were embryotoxic (without being teratogenic) in three species of laboratory animals (rats, mice, and rabbits) at doses that would have adversely affected or would have been expected to adversely affect maternal health.

Because the results of these studies in humans and various species of laboratory animals have been consistent, strong assurance is provided to conclude that benzene is unlikely to be teratogenic. However, benzene appears to have the potential to be fetotoxic at high doses (>160 mg/m^3). From the collective evidence (see Table 46.5), one can conclude that exposure to benzene at levels that do not affect maternal well-being would have no adverse effects on the offspring.

4.1.2 Dose-Response Assessment

Epidemiological evidence from workers exposed to benzene suggest that the threshold for all noncancer adverse effects is around 40 ppm (128 mg/m^3 = 19.7 mg/kg/8-h day) and that no noncancer toxicity is likely to occur at or below 25 ppm (80 mg/m^3 = 12.3 mg/kg/8-h day) (96). To obtain a human NAEL that accounts for the possibility of the existence of a wider range of susceptibility to the potential developmental toxicity of benzene than was observed among those in the published studies, the dose was divided by 3 to obtain a human NAEL of 8.3 ppm (26.6 mg/m^3 = 4.1 mg/kg/8-h day). In light of the absence of reproductive toxicity in human studies of benzene (despite limitations in those investigations), the NAEL may be conservatively low, implying that somewhat higher doses could well be tolerated without the possibility of reproductive injury.

Table 46.5. Summary of Results[a]

Compound	Developmental Toxicity		Fertility Dysfunction	
	Conducted	Result	Conducted	Result
Benzene	+	$T_\pm$; F_+	−	0
p-Chlorotoluene[b]	+	$T_\pm$; F_+	−	0
Toluene	+	$T_\pm$; F_+	+	−

[a] + = positive; ± = equivocal: − = negative; 0 = no data; T = tetratogenicity, F = fetotoxicity secondary to maternal toxicity.
[b] Based on o-chlorotoluene.

4.1.3 Workplace Exposure

Before reproductive risk can be characterized, the magnitude and nature of human exposure and dose must be estimated.

Time-weighted averages (TWAs) were plotted by job code and year against the TWA concentration. The scatter plots were examined for temporal trends. Because the data ranged over several orders of magnitude, the method of presentation of choice was in the form of semilog graphs. Estimates of worker exposures were made using cumulative frequency distributions. Due to the small number of workers in certain job categories, it was decided to pool the industrial hygiene data for the cumulative frequency distributions. Because workers were exposed to a wide range of benzene concentrations, the 80th percentile (i.e., the air concentration at or below which 80% of the workers were exposed) was selected at a realistic upper-bound exposure scenario. This value is realistic in as much as reproductive toxicity is generally related to high concentrations for relatively brief periods (i.e., a year) rather than to decade-long averages.

Information provided by industrial hygienists suggest that for most processes, the worker activity would be of light to moderate exertion. The average lightly working male breathes 0.8 m^3/h or 6.4 m^3/workday; the average female breathes 0.5 m^3/h or 4 m^3/workday. The average male weighs 78 kg and the average female weighs 65 kg. Pulmonary absorption was obtained from published values. Because by definition the manufacturing process is closed, dermal exposure to the workers is discounted.

The TWA benzene concentrations for the various job codes are depicted in Figure 46.3, along with the current OSHA PEL of 1 ppm (3.2 mg/m^3) (106). During the 8-yr monitoring period, only three samples exceeded the current OSHA PEL. Inspection of the plot reveals no obvious trends in exposure level with time for any job category. The 80th percentile exposure level was determined to be 0.5 ppm or 1.47 mg/m^3 (dose = 230 mg/kg/day).

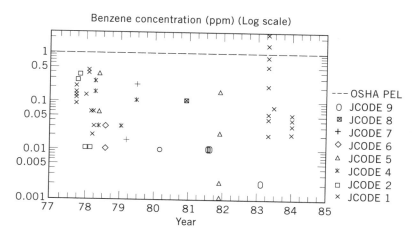

Figure 46.3. Scatter plot of benzene TWA concentration by job code and year.

4.1.4 Risk Characterization

The evidence from several observations in humans suggests that even at maternally toxic doses the development of the fetus may not be altered. However, observations in laboratory animals are compelling that excessive exposure to benzene may have adverse effect on the fetus. Based on human observations, such adverse effects are unlikely to occur below 8.3 ppm (26.6 mg/m^3; dose = 4.1 mg/kg day). The 80th percentile workplace exposure to benzene is about 0.5 ppm (1.5 mg/m^3 = 230 mg/kg/day) at this facility. The MOS for the fetuses of pregnant women exposed to benzene is, therefore, 17.7 (26.6 ÷ 1.5 mg/m^3. Because OSHA's current PEL is 1 ppm (3.2 mg/m^3) (106), the MOS for benzene developmental toxicity at the PEL is at least 2 (3.2 ÷ 1.47 mg/m^3). Therefore, meeting the OSHA standard as a maximum (and not an average) should be protective against developmental toxicity.

No data were available on whether benzene can alter fertility. Therefore, no conclusion can be reached on whether the PEL of 1 ppm is protective against fertility dysfunction in workers.

4.2 Toluene

The data for toluene were analyzed according to the four steps in risk assessment.

4.2.1 Hazard Identification

The toxicity of toluene has been well studied over the years (107). The primary effects of toluene are on the CNS. Short-term exposure to toluene (750 mg/m^3) can cause minor signs of CNS toxicity (e.g., headache). Higher exposure levels can result in narcosis (drowsiness). Finally, repeated chronic deliberate inhalation of very high concentrations of toluene (glue sniffing) can result in brain damage.

Chronic toluene abuse associated with glue sniffing has been shown to injure the fetus. Infants born to mothers who sniffed glue during pregnancy have had low birth weights, and some have had a condition resembling fetal alcohol syndrome (108, 109), but no systematic reports of pregnancy outcomes in glue sniffers have been made. At present, no studies have been reported on pregnancy outcome to toluene vapors at levels encountered in the workplace.

Toluene has been examined for its effects on pregnant rats, mice, and rabbits. When rats were exposed to 6000 mg/m^3 toluene for 24 h/day on days 9–14 or 9–21 of pregnancy, no birth defects were noted. However, toluene was toxic to the embryos, as evidenced by increased embryo death and decreased birth weight of the pups (107). Additional studies in mice and rabbits have confirmed that toluene can cause low birth weights and delayed development of the bones, although it has not been found to cause birth defects (i.e., not teratogenic). In addition, spontaneous abortions were reported in rabbits exposed to 1000 mg/m^3 toluene, 24 h/day on days 6–20 of gestation (the rabbit gestation period is 33 days) (111). By contrast, rats exposed to lower concentrations of toluene (375–1500 mg/m^3) for 6 h/day on days 6–15 of pregnancy showed no signs of developmental toxicity (112).

Toluene has also been studied for its effects on reproductive organs. CIIT (113) reported no adverse pathology in the ovaries and testes of rats exposed to 1100 mg/m^3 for 24 months. In addition, mice exposed to 375 and 1100 mg/m^3 had no altered reproductive behavior, although mice exposed to 7000 mg/m^3 had decreased growth but no effect on reproductive parameters (114).

The actual dose received by the mice in the API study (114) can be calculated by a similar set of assumptions as were used for calculation of human dose. In this case, the exposure is assumed to have lasted for 6 h/day, and a mouse is assumed to breathe 0.0018 m^3/h (0.011 m^3/6 h) (112) and weigh 30 g. Without data on pulmonary absorption, pulmonary absorption is assumed to be 50%. The dose is calculated to be 200 mg/kg/day {[(1100 mg/m^3) (0.011 m^3)]/(0.03 kg)}.

4.2.1.1 Confidence in Database. The effects of toluene on pregnant animals are generally consistent across species, with one exception (i.e., the spontaneous abortion in rabbits exposed to toluene is inconsistent with observations in the majority of the reports from other species). Nevertheless, the results consistently demonstrate that the adverse effects of toluene on pregnancy appeared only at doses that were maternally toxic. Teratogenic effects have been observed in women who chronically abuse toluene; however, in these cases, these women had a history of malnutrition and often alcohol abuse, each of which is known to injure the developing fetus. Thus, these cases are not considered indicative of workplace exposures. Based on these considerations, toluene is expected to have low teratogenic potential. Likewise, the evidence in experimental animals suggests that toluene is fetotoxic, but only at doses sufficient to injure maternal health.

Collectively, the evidence indicates that exposure to toluene at levels that do not affect maternal well-being would have no adverse effects on the developing offspring. In two species of laboratory animals and in appropriately designed studies, toluene consistently did not alter fertility.

4.2.2 Dose-Response Assessment

The available evidence in experimental animals suggests that toluene is fetotoxic only at doses that are greater than those injurious to maternal health. Therefore, exposure to toluene at levels that do not affect maternal well-being would have no adverse effects on the offspring. The highest dose of toluene that was found to have no adverse effect on pregnancy outcome was 1500 mg/m^3 (112); however, because of the toxicity to rabbits at 1,000 mg/m^3, the most confident NOAEL is 375 mg/m^3 (dose = 68 mg/kg/day). Because this value is lower than NOAEL in other studies, this may be considered a worst-case scenario.

The human NAEL of 2.7 mg/kg day is obtained by dividing the NOAEL in laboratory animals (68 mg/kg/day) by 25 (i.e., 5 × 5) to account for interspecies and intraspecies variability in response. If the results in rabbits overstate the toxic potency of toluene (which may well be the case), the human NAEL could justifiably be four times higher.

4.2.3 Workplace Exposure

A scatter plot of the TWA toluene concentrations for the various job codes are depicted in Figure 46.4. The ACGIH TLV is 100 ppm. Inspection of the plot reveals no obvious trends in exposure level with time for any job category. The frequency distribution of worker exposure (Figure 46.5) shows that the 80th percentile exposure level was 0.32

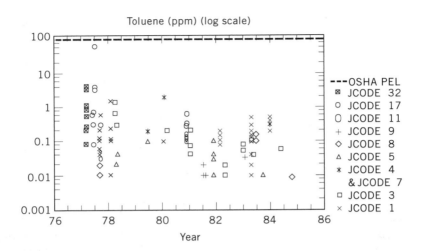

Figure 46.4. Scatter plot of toluene TWA concentration by job code and year.

ppm, or 1 mg/m^3. The estimated dose is 0.041 mg/kg-day $\{[(1\ \text{mg/m}^3)\ (6.4\ \text{m}^3/\text{day})\ (0.5)]/7.81\ \text{kg}\}$ for males and 0.031 mg/kg-day $\{[(1\ \text{mg/m}^3)\ (4.0\ \text{m}^3/\text{day})\ (0.5)]/65.4\ \text{kg}\}$ for females.

4.2.4 Risk Characterization

The evidence is compelling that exposure to toluene may have an adverse effect on the fetus, but only at very high toluene concentrations that are toxic to the pregnant animal (Table 46.5). Dividing the human NAEL of 2.7 mg/kg-day by the dose reported at the facility (0.31 mg/kg-day) produces a MOS of 8.7, which appears sufficiently large to protect against fetotoxicity. The actual MOS may be four times larger if the data from rabbits are not applicable to humans.

In two species of laboratory animals and in appropriately designed studies, toluene consistently did not alter fertility. Therefore, it is inappropriate to calculate a NOAEL, NAEL, or MOS for toluene. Consequently, the TLV of 100 ppm should provide ample protection against dysfunction in fertility in workers.

4.3 p-Chlorotoluene

Compound p-chlorotoluene (PCT) represents a special case in which the safety or risk to the reproductive functions can be estimated solely from the data for a chemical congener because no appropriate data are available for the compound of interest. The data for this compound were analyzed according to the four steps in risk assessment.

4.3.1 Hazard Identification

Because no data were available on the reproductive toxicity of PCT, the reproductive toxicity data for chemical congener of PCT, o-chlorotoluene (OCT), were evaluated.

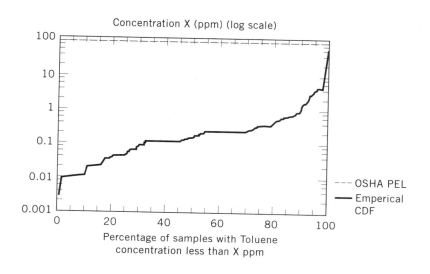

Figure 46.5. Cumulative distribution of toluene TWA concentrations.

In a study conducted by Huntingdon Research Centre (115), rabbits were exposed to 1.5, 4, and 10 g/m³ OCT in air for 6 h/day on days 6–29. Toxicity was noted in the highest-exposure group (ataxia and lachrymation and/or salivation); 5 of the 15 pregnant rabbits died or were sacrificed. In the middle exposure group (4 g/m³), slight ataxia occurred during exposure; no mortality was observed among the 13 rabbits during treatment. Two out of 12 rabbits in the control and the 1.5 g/m³ groups died during the study. The authors attributed the mortality to factors other than OCT. No significant effect on mean values for litter size, pre- and postimplantation loss, or litter and mean fetal weight were noted. Thus, OCT produced no reproductive toxicity in pregnant rabbits.

In another study performed at the Huntingdon Research Centre (116), rats were exposed to air concentrations of 1, 3, and 9 g OCT/m³ for 6 h/day on days 6–19 of pregnancy. At the highest level of exposure, rats were ataxic with lachrymation, salivation, or both in some animals. At 3 g/m³, some ataxia was noted in exposed rats. No overt signs of toxicity were noted in the rats exposed to 1 g/m³ OCT. There was a dose-dependent decrease in food consumption and body weight gain in rats exposed to 3 and 9 g/m³. At 9 g/m³ the mean litter and mean fetal weights were significantly reduced. In addition, an increase was noted in the incidence of fetal malformations (brachydactyly of a single forepaw or hind paw). Also, an increase was observed in sternebrae variants that was attributed to decreased fetal weight. On the other hand, no increase in visceral anomalies were found in rats exposed to 9 g/m³. No adverse effects were noted at 1 or 3 g/m³.

The actual dose at a concentration of 3 g/m³ received by the pregnant rats can be calculated by a set of assumptions similar to that used for calculation of human dose. Average physiological conditions were assumed to be that a rat breathes 0.006 m³/h (0.036 m³/6 h) (112) and weighs 278 g (taken from gestation day 14 body weights) (116). Without data on pulmonary absorption, absorption is assumed to be 50%. The dose is calculated as 1.9 g/kg/day {[(3 g/m^3) (0.36 m^3) (0.5)]/0.278 kg}.

4.3.1.1 Confidence in Database. The studies followed standard design for developmental toxicity studies. The doses of OCT were high because significant maternal toxicity was present at the two highest doses. The deleterious effects of OCT on the offspring can be attributed to the toxic effect of the chemical on maternal well-being. The absence of signs of teratogenic effects at 1 and 3 g/m^3 supports the proposition that OCT would not be expected to be teratogenic in humans (Table 46.5).

4.3.2 Dose-Response Assessment

The study in rats is best suited for determining the toxic potency of the OCT (116) because of the absence of mortality among treated animals. Because 9 g/m^3 OCT adversely affected fetal development, possibly due to effects on maternal health, this dose is a LOAEL. The NOEL for laboratory animals is 3 g/m^3, which is equivalent to a dose of 1900 mg/kg-day. The human NAEL is calculated by dividing the animal NOAEL of 1900 mg/kg-day by an uncertainty factor of 200 (i.e., 10 × 10 × 2) to account for inter- and intraindividual variability in sensitivity and for the possibility that PCT may have somewhat greater toxic potency than OCT. The result is a human NAEL of 9.5 mg/kg-day.

4.3.3 Workplace Exposure

A scatter plot of the TWA PCT concentrations for the various job codes are depicted in Figure 46.6. The ACGIH TLV for OCT of 50 ppm is also noted. Inspection of the plot reveals no obvious trends in exposure level with time for any job category. The cumulative frequency distribution of PCT TWA concentration is shown in Figure 46.7. It was deter-

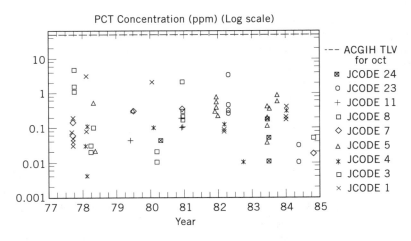

Figure 46.6. Scatter plot of *p*-chlorotoluene by job code and year.

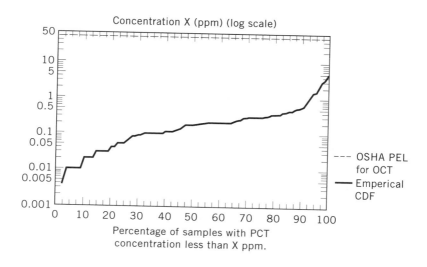

Figure 46.7. Cumulative frequency distribution of *p*-chlorotoluene TWA concentrations.

mined that the 80th percentile exposure level was 0.35 ppm, or 1.8 mg/m^3. Because reproductive data were available only for females, no dose derivation was made for male workers. Because the workers were working at light to moderate activity when exposed to PCT, the volume of air the female workers breathed was 4.0 m^3/day. The estimated dose for female workers is 0.055 mg/kg/day {[(1.8 mg/m^3) (4.0 m^3/day)(0.5)]/65.4 kg}.

4.3.4 Risk Characterization

The studies suggest that the congener OCT has a low order of toxicity with no apparent effects on fetal development at doses up to 3 g/m^3 for 6 h/day, which is equivalent to a dose of 1,900 mg/kg-day. The human NAEL is 9.5 mg/kg-day.

The MOS for OCT exposure by females is approximately 170 (9.5/0.055). This finding indicates that is unlikely that PCT at this or lower doses would adversely affect fetal development, even if PCT were 2 or 3 times more potent toxicologically than OCT.

Neither PCT nor OCT has been tested for effects on fertility in males or females; consequently, no conclusion about their potential for injury to such functions can be reached.

4.4 Assessment of Exposure to Mixtures

Because of the possibility that workers are likely to encounter two or three of these compounds during the workday, it is advisable to determine whether the combined exposures might be of some consequence to the health of the fetus (there are no corresponding data on fertility dysfunction to undertake such as analysis for that toxic end point). Such an analysis was performed using Eq. 7. The result was a hazard index of 0.025 when the higher human NAEL for benzene was used and 0.933 when the lower human NAEL was used. In either case, the index is less than 1 and hence indicative that fetal injury is unlikely to occur at the concentrations measured in the workplace.

5 SUMMARY AND CONCLUSIONS

Risk analysis is a systematic means of interpreting complex information to estimate risks to human health from chemical exposures. Applied initially to exposures to substances in the general environment, risk analysis has a role in estimating the impact of chemicals in the workplace, even in the presence of health-based standards.

As currently fashioned, risk analysis examines information in four discrete yet interrelated steps. First, the toxic properties of substances are evaluated to determine what types of injury they are capable of producing in humans and under what circumstances—a process designed hazard identification. Next the amount or dose of substances—that is, their toxic potency—that is needed to cause varying rates of injury or degrees of severity among exposed individuals is determined—the dose-response assessment. Exposure assessment is the third step, which characterizes the doses of specified chemicals obtained by target populations (e.g., individuals in a specific facility). Risk characterization amalgamates the information into estimates of possible cancer incidence and of the likely absence of noncancer injury.

The process contains numerous uncertainties that influence confidence in the conclusions. Some of the predominant ones include (*1*) the degree to which laboratory animal data are predictive of human responses, (*2*) the variations among humans in their susceptibility to chemical toxicity, and (*3*) the amount of concordance between the findings in a database and the actual situation for which the risks are being estimated. To the extent that understanding of the influence of such factors for specific substances in the workplace is increasing continually, confidence is enhanced incremental in the validity of conclusions derived from the risk-assessment process. However, it is important to recognize that, for some chemicals and some forms of exposure, the data are sufficiently incomplete that they necessarily contain wide error bands surrounding the risk estimates. Such ranges in data quality are apparent in the two illustrations in this chapter. For example, little doubt remains that benzene causes an adult form of leukemia at high doses; less certainty exists about its ability to cause birth defects; yet, without further study, no comparable claim can be made about the possibility, or lack thereof, that benzene might impair reproductive success.

Despite its limitations, risk analysis is a tool finding increasing use of the workplace to inform workers and management of the extent to which they may be at risk or relatively safe in the implementation of control technologies and in the setting of policies by private industry and government. Used wisely, risk analysis can indeed be a suitable guide to that safe use of chemicals in the workplace.

BIBLIOGRAPHY

1. WHO/IPCS. (1999). *Principles for the Assessment of Risks to Human Health from Exposure to Chemicals*, Environmental Health Criteria 210, World Health Organization/International Programme of Chemical Safety, Geneva, Switzerland, 1999.

2. R. G. Tardiff and J. V. Rodricks, eds., *Toxic Substances and Human Risk. Principles of Data Interpretation*, Plenum Press, New York, 1988.
3. NRC, "Chemical Contaminants: Safety and Risk Assessment," in *Drinking Water and Health*, National Research Council, National Academy of Sciences, Washington, DC, 1977, pp. 19–62.
4. NRC, "Problems of Risk Estimation," in *Drinking Water and Health*, vol. 3, National Research Council, National Academy Press, Washington, DC, 1980, pp. 25–65.
5. NRC, *Risk Assessment in the Federal Government: Managing the Process*, National Research Council, National Academy Press, Washington, DC, 1983.
6. USEPA, "Guidelines for Carcinogen Risk Assessment," *Federal Register* **51**, 33992–34003 (September 24, 1986).
7. USEPA, "Guidelines for Exposure Assessment," *Federal Register* **57**(104), 22888–22938 (May 29, 1992).
8. USEPA, *Proposed Guidelines for Carcinogen Risk Assessment*, Office of Research and Development, U.S. Environmental Protection Agency, Washington, DC, September 1996.
9. USEPA, *Draft Revisions to the Proposed Guidelines for Carcinogen Risk Assessment*, Office of Research and Development, U.S. Environmental Protection Agency, Washington, DC, December 1998.
10. WHO, *Principles and Methods for Evaluating the Toxicity of Chemicals. Part I*, Environmental Health Criteria 6, World Health Organization, Geneva, Switzerland, 1978.
11. WHO, *Principles for Evaluating Health Risks to Progeny Associated with Exposure to Chemicals During Pregnancy*, Environmental Health Criteria 30, International Programme on Chemical Safety, World Health Organization, Geneva, Switzerland, 1984.
12. WHO, *Principles and Methods for the Assessment of Neurotoxicity Associated with Exposure to Chemicals*, IPCS Environmental Health Criteria 60, International Programme on Chemical Safety, World Health Organization, Geneva, Switzerland, 1986.
13. NRC, *Science and Judgment in Risk Assessment*, National Research Council, National Academy Press, Washington, DC, 1994.
14. IARC, *IARC Monographs on the Evaluation of Carcinogenic Risks to Humans*, suppl. 7, *Overall Evaluations of Carcinogenicity: An Updating of IARC Monographs Volumes 1 to 42*, World Health Organization, International Agency for Research on Cancer, Lyons, France, 1987.
15. R. L. Brent, "Etiology of Human Birth Defects: What Are the Causes of the Large Group of Birth Defects of Unknown Etiology?" in J. A. McLachlan, R. M. Pratt, and C. L. Markert, eds., *Developmental Toxicology*: Mechanisms and Risk, Banbury Report 26, Cold Spring Harbor Laboratory, Cold Spring Harbor, NY, 1987, pp. 287–303.
16. USEPA, "Proposed Amendments to the Guidelines for the Health Assessment of Suspect Developmental Toxicants," *Federal Register* **54**(42), 9386–9403 (March 6, 1989).
17. M. Christian, A critical review of multi-generation studies. *J. Am. Coll. Toxicol.* **5**; 161–180 (1986).
18. E. Francis and G. Kimmel "Proceedings of the Workshop on One- vs Two-Generation Reproductive Effects Studies," *J. Am. Coll. Toxicol.* **7**, 911–925 (1988).
19. ILSI, *An Evaluation and Interpretation of Reproductive Endpoints for Human Health Risk Assessment*, International Life Sciences Institute, Washington, DC, 1998.
20. USEPA *Draft USEPA Science Advisory Board Report on the Revisions to the Proposed Guidelines for Carcinogen Risk Assessment*, Office of the Science Advisory Board, U.S. Environmental Protection Agency, Washington, DC, May 1999.

21. N. Kaplan et al., "An Evaluation of the Safety Factor Approach in Risk Assessment," in J. A. McLachlan, R. M. Pratt, and C. L. Markert, eds., *Developmental Toxicology: Mechanisms and Risk*, Banbury Report, Cold Spring Harbor Laboratory, Cold Spring Harbor, NY, 1987, pp. 335–346.
22. C. C. Brown, "Approaches to Interspecies Dose Extrapolation," in R. G. Tardiff and J. V. Rodricks, eds., *Toxic Substances and Human Risk: Principles of Data Interpretation*, Plenum Press, New York, 1987, pp. 237–268.
23. C. D. Klaassen and D. L. Eaton, "Principles of Toxicology," in C. D. Klaassen, M. O. Amdur, and J. Doull, eds., *Casarett and Doull's Toxicology: The Basic Science of Poisons*, 4th ed., Macmillan, New York, 1991, pp. 12–49.
24. W. N. Aldridge, "The Biological Basis and Measurement of Thresholds," *Annu. Rev. Pharmacol. Toxicol.* **26**, 39–58 (1986).
25. P. Armitage and R. Doll, "Stochastic Models for Carcinogenesis," in *Proceedings of the Fourth Berkeley Symposium on Mathematical Statistics and Probability*, vol. 4, University of California Press, Berkeley, 1961, pp. 19–38.
26. S. H. Moolgavkar and A. G. Knudson, Jr., "Mutation and Cancer: A Model for Human Carcinogenesis," *J. Natl. Cancer Inst.* **66**, 1037–1052 (1981).
27. M. Ginevan, "Appendix A. Explanation and Sensitivity Analysis of the Modified Armitage and Doll Model (LesLife)*f*," in N. L. Nagda, R. C. Fortmann, M. D. Koontz, S. R. Baker, and M. E. Ginevan, eds., *Airliner Cabin Environment: Contaminant Measurements, Health Risks, and Mitigation Options*, Report to U.S. Department of Transportation, Washington, DC, by Geomet Technologies, Germantown, MD, 1989, pp. A1–A10.
28. I. W. F. Davidson, J. C. Parker, and R. P. Beliles, "Biological Basis for Extrapolation Across Mammalian Species," *Regul. Toxicol. Pharmacol.* **6**, 211–237 (1986).
29. C. C. Travis and R. K. White, "Interspecific Scaling of Toxicity Data," *Risk Anal.* **8**, 119–125 (1988).
30. USEPA, "Draft Report: A Cross-species Scaling Factor for Carcinogen Risk Assessment Based on Equivalence of mg/kg3/4/day; Notice," *Federal Register* **57**(109), 24152–24173 (June 5, 1992).
31. S. L. Brown et al., "Review of Interspecies in Risk Comparisons," *Regul. Toxicol. Pharmacol.* **8**, 191–206 (1988).
32. D. Gaylor, "Quantitative Risk Analysis for Quantal Reproductive and Developmental Effects," *Environ. Health Perspect.* **79**, 243–246 (1991).
33. R. Kodell et al., "Mathematical Modeling of Reproductive and Developmental Toxic Effects for Quantitative Risk Assessment," *Risk Anal.* **11**, 583–590 (1991).
34. D. Gaylor and M. Razzaghi, "Process of Building Biologically Based Dose-Response Models for Developmental Defects," *Teratology* **46**, 573–581 (1992).
35. D. Gaylor and J. Chen, "Dose-Response Models for Developmental Malformations," *Teratology* **47**, 291–297 (1993).
36. E. Anderson et al., *Development of Statistical Distribution or Ranges of Standard Factors Used in Exposure Assessments*. Final Report, U.S. Environmental Protection Agency, Office of Health and Environmental Assessment, Washington, DC, 1985. Available from National Technical Information Service, Springfield, VA 22161, as PB85-242667.
37. USEPA, *Exposure Factors Handbook*, Report no. EPA/600/8-89/043, Exposure Assessment Group, Office of Health and Environmental Assessment, U.S. Environmental Protection Agency, Washington, DC, 1989.

38. LaKind et al. 1999a.
39. LaKind et al. 1999b.
40. LaKind et al. 1999c.
41. USEPA, *Guidance for Conducting Health Risk Assessment of Chemical Mixtures*, external scientific peer review draft, Office of Research and Development, U.S. Environmental Protection Agency, Washington, DC, April 1999.
42. D. K. Verma, and K. des Tombe, Measurement of benzene in the workplace and its evolutionary process, Part I: Overview, history, and past methods. *Am. Ind. Hyg. Assoc. J.* **60**, 38–47 (1999).
43. WHO/IPCS, *Benzene*, Environmental Health Criteria 150. World Health Organization/International Programme of Chemical Safety, Geneva, Switzerland, 1993.
44. G. L. Gist and J. R. Berg, "Benzene—a Review of the Literature from a Health Effects Perspective," *Toxicol. Ind. Health* **13**, 661–714 (1997).
45. D. A. Savitz and K. W. Andrews, "Review of the Epidemiological Evidence on Benzene and Lymphatic and Hematopoietic Cancers," *Am. J. Indust. Med.* **31**, 287–295 (1997).
46. M. A. Medinsky, P. M. Schlosser, and J. A. Bond, "Critical Issues in Benzene Toxicity and Metabolism: The Effect of Interactions with Other Organic Chemicals on Risk Assessment," *Environ. Health Perspect.* **102** (suppl 9); 119–124 (1994).
47. K. Hughes, M. E. Meek, and S. Bartlett, "Benzene Evaluation of Risks to Health from Environmental Exposure in Canada," *Carcinogenesis and Ecotoxicology Reviews* **12**, 161–171 (1994).
48. R. Snyder, G. Witz, and B. Goldstein, "The Toxicology of Benzene," *Environ. Health Perspect.* **100**, 293–306 (1993).
49. R. Snyder and G. F. Kalf, "A Perspective on Benzene Leukemogenesis," *Crit. Rev. Toxicol.* **24**, 177–209 (1994).
50. ATSDR, *Toxicological Profile for Benzene*, Report no. ATSDR/TP-88/03. Prepared by Clement Associates, Inc., Fairfax, VA. Agency for Toxic Substances and Disease Registry, Atlanta, GA, 1989. Available from National Technical Information Service, Springfield, VA 22161, as PB89-209464.
51. IARC, "Benzene," in *IARC Monographs on the Evaluation of the Carcinogenic Risk of Chemicals to Humans*, vol. 29, *Some Industrial Chemicals and Dyestuffs*, World Health Organization, International Agency for Research on Cancer, Lyons, France 1982, pp. 93–148.
52. R. A. Rinsky, R. J. Young, and A. B. Smith, "Leukemia in Benzene Workers," *Am. J. Ind. Med.* **2**, 217–245 (1981).
53. K. N. Rothman et al., "Hemototoxicity among Chinese Workers Heavily Exposed to Benzene," *Am. J. Ind. Med.* **29**, 236–246 (1996).
54. P. Delore and C. Borgomano, "Acute Leukaemia Following Benzene Poisoning. On the Toxic Origin of Certain Acute Leukaemias and Their Relation to Serious Anaemias," *J. Med. Lyon (Fr.)* **9**, 227–233 (1928).
55. M. Aksoy, "Leukemia in Workers Due to Occupational Exposure to Benzene," *New Istanbul Contrib. Clin. Sci.* **12**, 3–14 (1977).
56. M. Aksoy, "Different Types of Malignancies Due to Occupational Exposure to Benzene: A Review of Recent Observations in Turkey," *Environ. Res.* **23**, 181–190 (1980).
57. M. Aksoy, S. Erdem, and G. DinCol, "Leukemia in Shoe-workers Exposed Chronically to Benzene," *Blood* **44**, 837–841 (1974).
58. P. F. Infante et al., "Leukaemia in Benzene Workers," *Lancet* **2**, 76–78 (1977).

59. R. A. Rinsky et al., "Benzene and Leukemia: An Epidemiologic Risk Assessment," *N. Engl. J. Med.* **316**, 1044–1050 (1987).
60. O. Wong, R. W. Morgan, and M. D. Whorton, *Comments on the NIOSH Study of Leukemia in Benzene Workers*, technical report submitted to Gulf Canada, Ltd., by Environmental Health Associates, 1983.
61. M. G. Ott et al., "Mortality among Workers Occupationally Exposed to Benzene," *Arch. Environ. Health* **33**, 3–10 (1978).
62. IRIS, "Benzene," Toxicology Data Network (TOXNETE), National Library of Medicine, Bethesda, MD, 1989 (retrieved September 1991).
63. R. Snyder and J. J. Kocsis, "Current Concepts of Chronic Benzene Toxicity," *CRC Crit. Rev. Toxicol.* **3**, 265–288 (1975).
64. ACGIH, *1991–1992 Threshold Limit Values for Chemical Substances and Physical Agents and Biological Exposure Indices*, American Conference of Governmental Industrial Hygienists, Cincinnati, OH, 1991.
65. E. P. Cronkite "Benzene Hematotoxicity and Leukemogenesis," Blood Cells **12**, 129–137 (1986).
66. E. P. Cronkite et al., "Benzene Inhalation Produces Leukemia in Mice," *Toxicol. Appl. Pharmacol.* **75**, 358–361 (1984).
67. E. P. Cronkite et al., "Hematotoxicity and Carcinogenicity of Inhaled Benzene," *Environ. Health Perspect.* **82**, 97–108 (1989).
68. C. Maltoni et al., "Zymbal Gland Carcinomas in Rats Following Exposure to Benzene by Inhalation," *Am. J. Ind. Med.* **3**, 11–16 (1982).
69. R. Snyder et al., "The Inhalation Toxicology of Benzene: Incidence of Hematopoietic Neoplasms and Hematotoxicity in ADR/J and C57BL/6J Mice," *Toxicol. Appl. Pharmacol.* **54**, 323–331 (1980).
70. M. G. Rozen and C. A. Snyder, "Protracted Exposure of C57BL/6 Mice to 300 ppm Benzene Depresses B- and T-lymphocyte Numbers and Mitogen Responses; Evidence for Thymic and Bone Marrow Proliferation in Response to the Exposures," *Toxicology* **37**, 13–26 (1985).
71. J. Meyne and M. S. Legator, "Sex-related Differences in Cytogenetic Effects of Benzene in the Bone Marrow of Swiss Mice," *Environ. Mutat.* **2**, 43–50 (1980).
72. D. Anderson and C. R. Richardson, "Chromosome Gaps Are Associated with Chemical Mutagenesis," *Environ. Mutat.* **1**, 179, Abstr. no. Ec-9 (1979).
73. C. C. Travis, J. L. Quillen, and A. D. Arms, "Pharmacokinetics of Benzene," *Toxicol. Appl. Pharmacol.* **102**, 400–420 (1990).
74. R. Snyder et al., "Studies on the Mechanism of Benzene Toxicity," *Environ. Health Perspect.*, **82**, 31–35 (1989).
75. M. A. Medinsky et al., "A Physiological Model for Simulation of Benzene Metabolism by Rats and Mice," *Toxicol. Appl. Pharmacol.* **99**, 193–206 (1989).
76. F. Y. Bois, M. T. Smith, and R. C. Spear, "Mechanisms of Benzene Carcinogenesis: Application of a Physiological Model of Benzene Pharmacokinetics and Metabolism," *Toxicol. Lett.* **56**, 283–298 (1991).
77. R. C. Spear et al., "Modeling Benzene Pharmacokinetics across Three Sets of Animal Data: Parametric Sensitivity and Risk Implications," *Risk Anal.* **11**, 641–654 (1991).
78. D. E. Rickert et al., "Benzene Disposition in the Rat after Exposure by Inhalation," *Toxicol. Appl. Pharmacol.* **49**, 417–423 (1979).

79. J. Srbova, J. Teissinger, and S. Skramovsky, "Absorption and Elimination of Inhaled Benzene in Man," *Arch. Ind. Hyg. Occup. Med.* **2**, 1–8 (1950).
80. R. J. Sherwood, "Pharmacokinetics of Benzene in a Human after Exposure at about the Permissible Limit," *Ann. N.Y. Acad. Sci.* **534**, 635–647 (1988).
81. R. Snyder et al., "Biochemical Toxicology of Benzene," *Rev. Biochem. Toxicol.* **3**, 123–153 (1981).
82. G. F. Kalf, "Recent Advances in the Metabolism and Toxicity of Benzene," *Crit. Rev. Toxicol.* **18**, 141–159 (1987).
83. D. V. Parke and R. T. Williams, "Studies in Detoxification. 49. The Metabolism of Benzene Containing []14C]Benzene," *Biochem. J.* **54**, 231–238 (1953).
84. G. Witz et al., "The Metabolism of Benzene to Muconic Acid, a Potential Biological Marker of Benzene Exposure," in C. M. Witmer et al., eds., *Biological Reactive Intermediates*, vol. 4, Plenum Press, New York, 1990, pp. 613–618.
85. R. F. Henderson et al., *Environ. Health Perspect.* **82**, 9–17 (1989).
86. L. S. Andrews et al., "Effects of Toluene on the Metabolism, Disposition and Hemopoietic Toxicity of [3H]benzene," *Biochem. Pharmacol.* **26**, 293–300 (1977).
87. S. L. Longacre et al., "Toxicological and Biochemical Effects of Repeated Administration of Benzene in Mice," *J. Toxicol. Environ. Health* **7**, 223–237 (1981).
88. S. L. Longacre, J. J. Kocsis, and R. Snyder, "Influence of Strain Differences in Mice on the Metabolism and Toxicity of Benzene," *Toxicol. Appl. Pharmacol.* **60**, 398–409 (1981).
89. K. Nomiyama and H. Nomiyama, "Respiratory Retention, Uptake, and Excretion of Organic Solvents in Man. Benzene, Toluene, n-hexane, Trichloroethylene, Acetone, Ethyl Acetate, and Ethyl Alcohol," *Int. Arch. Arbeitsmed.* **32**, 75–83 (1974).
90. K. Nomiyama and H. Nomiyama, "Respiratory Elimination of Organic Solvents in Man. Benzene, Toluene, n-hexane, Trichloroethylene, Acetone, Ethyl Acetate and Ethyl Alcohol," *Int. Arch. Arbeitsmed.* **32**, 85–91 (1974).
91. P. Sabourin et al., "Effect of Dose on Absorption and Excretion of]14C Benzene Administered Orally or by Inhalation," *Toxicologist* **6**, 163 (1986).
92. I. H. Blank and D. J. McAuliffe, "Penetration of Benzene through Human Skin," *J. Invest. Dermatol.* **85**, 522–526 (1985).
93. USEPA, *Health Effects Assessment for Benzene*, Report no. EPA/600/8-89/086, Environmental Criteria and Assessment Office, Office and Research and Development, U.S. Environmental Protection Agency, Cincinnati, OH, 1989.
94. E. E. Sandmeyer, "Aromatic Hydrocarbons," in G. D. Clayton, and F. E. Clayton, eds., *Patty's Industrial Hygiene and Toxicology*, vol. 2, 3d ed., Interscience, New York, 1981, pp. 3253–3283.
95. RiskFocus, *Analysis of the Impact of Exposure Assumptions on Risk Assessment of Chemicals in the Environment. Phase I. Evaluation of Existing Exposure Assessment Assumptions*, Report to Chemical Manufacturers Association by RiskFocus*f* Division, Versar, Inc., Springfield, VA, 1990.
96. ACGIH, *Documentation of the Threshold Limit Values and Biological Exposure Indices*, 6th ed., American Conference of Governmental Industrial Hygienists, Cincinnati, OH, 1991.
97. W. B. Deichmann, W. E. MacDonald, and E. Bernal, "The Hemopoietic Tissue Toxicity of Benzene Vapors," *Toxicol. Appl. Pharmacol.* **5**, 201–224 (1963).
98. B. Holmberg, "Central Nervous System Defects in Children Born to Mothers Exposed to Organic Solvents," *Lancet* **2**, 177–179 (1979).

99. R. A. Kuna and R. W. Kapp, "Embrotoxic/Teratogenic Potential of Benzene Vapor in Rats," *Toxicol. Appl. Pharmacol.* **57**, 1–7 (1981).
100. C. A. Kimmel and J. G. Wilson "Skeletal Deviations in Rats: Malformations or Variations?" Teratology **8**, 309–316 (1973).
101. J. D. Green, B. K. J. Leong, and S. Laskin, "Inhaled Benzene Fetotoxicity in Rats," *Toxicol. Appl. Pharmacol.* **46**, 9–18 (1978).
102. F. J. Murray et al., "Embryotoxicity of Inhaled Benzene in Mice and Rabbits," *Am. Ind. Hyg. Assoc. J.* **40**, 993–998 (1979).
103. E. Tatrai, K. Rodics, and G. Y. Ungvary, "Embryotoxic Effects of Simultaneously Applied Exposure of Benzene and Toluene," *Folia Morphol.* **28**, 286–289 (1980).
104. A. Hudak and G. Ungvary, "Embryotoxic Effects of Benzene and Its Methyl Derivatives: Toluene, Xylene," *Toxicology* **11**, 55–63 (1978).
105. G.-I. Watanabe and S. Yoshida, "The Teratogenic Effect of Benzene in Pregnant Mice," *Acta Med. Biol.* **17**, 285–291 (1970).
106. OSHA, "Occupational Exposure to Benzene," *Federal Register* **52**, 34460–34578 (September 11, 1987).
107. WHO, *Toluene*, Environmental Health Criteria 52, International Programme on Chemical Safety, World Health Organization, Geneva, Switzerland, 1985.
108. J. H. Hersh et al., "Toluene Embryopathy," *J. Pediatr.* **106**, 922–927 (1985).
109. T. M. Goodwin, "Toluene Abuse and Renal Tubular Acidosis in Pregnancy," *Obstet. Gynecol.* **71**, 715–718 (1988).
110. A. Hudak et al., "Effects of Toluene Inhalation on Pregnant CFY Rats and Their Offspring," *Munkavedelem* **23**(1–3)(Suppl.), 25–30 (1977).
111. G. Ungvary and E. Tatrai, "On the Embryotoxic Effects of Benzene and Its Alkyl Derivatives in Mice, Rats, and Rabbits," *Arch. Toxicol.* **8** (suppl.), 425–430 (1985).
112. L. S. Gold et al., "A Carcinogenic Potency Database of the Standardized Results of Animal Bioassays," *Environ. Health Perspect.* **58**, 9–319 (1984).
112. Litton Bionetics, *Teratology Study in Rats. Toluene*, LBI Project no. 20847, final report to American Petroleum Institute by Litton Bionetics Inc., Kensington, MD, American Petroleum Institute, Washington, DC, 1978.
113. CIIT, *A Twenty-Four Month Inhalation Toxicology Study in Fischer-344 Rats Exposed to Atmospheric Toluene*, Chemical Industry Institute of Toxicology, Research Triangle Park, NC, 1980.
114. API, *Two-Generation Reproduction/Fertility Study on a Petroleum-Derived Hydrocarbon* [i.e., Toluene], vol. 1, American Petroleum Institute, Washington, DC, 1985.
115. Huntingdon Research Centre, *Effect of 2-chlorotoluene Vapour on Pregnancy of the New Zealand White Rabbit*, Report to Occidental Chemical Corporation, Huntingdon Research Centre, Huntingdon, England, August 3, 1983.
116. Huntingdon Research Centre, *Effect of 2-chlorotoluene Vapour on Pregnancy of the Rat*, Report to Occidental Chemical Corporation, Huntingdon Research Centre, Huntingdon, England, August 6, 1983.
117. R. L. Brent and M. I. Harris, *Prevention of Embryonic, Fetal and Perinatal Disease*, DHEW Publication (NIH)76-853, U.S. Department of Health and Human Services, National Institutes of Health, Bethesda, MD, 1976.

CHAPTER FORTY-SEVEN

Health Surveillance Programs in Industry

Mitchell R. Zavon, MD

1 INTRODUCTION

There is no universal agreement on what is included or what should be included in the term *health surveillance in industry*. Some apply a narrow interpretation, for example, making it synonymous with "health effects monitoring" or "the periodic medical–physiological examinations of exposed workers with the objective of protecting health and preventing occupationally related disease. The detection of established diseases is outside the scope of the definition" (1). It was defined as distinct from *biological monitoring*, "the measurement and assessment of workplace agents or their metabolites either in tissues, secreta, excreta, expired air or any combinations of these to evaluate exposure and health risk compared to an appropriate measure." (1). This discussion, however, is not limited to such a narrow definition of health surveillance; the maintenance and protection of worker health must include evaluations of exposure as well as evidence of biologic effects, if any (2).

Any discussion of health surveillance in industry must take into account the reasons for such surveillance. First and foremost in the rationale for such a program should be concern with health maintenance and health protection of those in the work force. This should not exclude mental health, an area of concern which has rarely been addressed when looking at health surveillance in industry (2, 3). Any discussion of health surveillance in industry must, however, take into account the important influence of government regulations. These regulations will differ greatly around the globe. This discussion will be directed primarily toward the American experience and regulations. These include mandates by the Occupational Safety and Health Administration (OSHA), the Mine Safety and Health Administration (MSHA), the Environmental Protection Agency (EPA), the Equal Employment Opportunities Commission (EEOC), and others. There are also influential guidelines by nonregulatory agencies, such as the National Institute for Occupational Safety and Health (NIOSH). The specific requirements embedded in regulations promulgated by these agencies, and some of the fallacies therein, will be addressed after a more general review of current medical and scientific concepts regarding the role of such surveillance in worker health protection.

There have been several excellent and comprehensive reviews of the topic that should be read by all who are interested (4, 5). A major conference on medical screening and biological monitoring, sponsored by NIOSH, EPA, and the National Cancer Institute (NCI) held in 1984 had over 70 presentations. The conference proceedings were published in 1986 (6).

Although the emphasis in this chapter will be on hazard-oriented health surveillance aimed at protection of workers, it is still important to regard this as part of a broader program of health maintenance. The effective incorporation of special elements into a more general health program is the preferred approach. For this reason it is not feasible to discuss hazard-oriented examinations out of the context of examinations to detect preexisting conditions or the development of abnormalities unrelated to the work environment. Reference 7 pointed out similar overlapping when considering biomarkers, that is, indicators of exposure, effect, and susceptibility.

2 OBJECTIVES OF HEALTH SURVEILLANCE

The objective must always be to avoid an adverse effect on the worker of any physical or chemical exposure in the workplace and to detect as early as possible any nonwork related health problems. This requires that the environment of the workplace be monitored often enough to know the actual conditions in the workplace. The medical examinations are basic, but must incorporate any additional elements to a general medical examination that is dictated by knowledge of the workplace exposures. An occupational health and safety program has many elements. Physical examinations and laboratory testing are only parts of a comprehensive program. Perhaps the most valuable and important part of the examination is the history. Self-history questionnaires should be used. Many are available but the one used should have been validated in actual use and structured to facilitate responses. The examiner can readily pursue leads opened up by the self-history. But the physical examination and the laboratory testing no more prevent illness or injury than a film badge prevents radiation exposure. All parts of the examination should be evaluated individually. Each part of the examination should only be included if it can be justified by the information it provides. The entire examination should be designed and performed to secure maximum preventive benefits at minimal risk to the worker and minimal cost and inconvenience to the employer. As noted earlier, the mental health of the workforce should also be taken into consideration. Not that we have any specifics to offer at this time but because depression, stress and other mental health problems exact such a huge toll in our society. It would be advisable to include mental health in any surveillance program and plan for possible corrective action if necessary. The following sections outline the types of examinations that are commonly performed.

2.1 Preplacement Examinations

Preplacement examinations are those performed on otherwise qualified applicants, persons who have already been offered employment. There should be no implication that an examination is a "preemployment" examination or that an employment offer is contingent on "passing an examination." In the past this examination was done before an offer of employment was made, and it was often stated categorically that an offer of employment would be made only if the applicant "passed" the examination. The objective was to obtain the most physically fit work force that was available and to exclude individuals with any suggestion of medical or psychological problems. Such practices were defended by pointing out the cost of hiring and training individuals who could not perform their jobs, who would work but a brief time, and who could create expensive medical care problems.

For many years this has been recognized as ethically and socially unacceptable (8). The Americans with Disabilities Act (9), now places severe restrictions on an employer's response. It stipulates that if the examination detects a condition that would jeopardize the health or well-being of the employee or of fellow employees, it is now the responsibility of the employer to modify the work situation so that it is safe for the employee and for fellow employees, if technologically and economically feasible. Deciding what is technologically and economically feasible can be a difficult exercise.

During the past two or three decades, there has also been increasing ability to detect genetic factors that are associated with higher risks for many diseases (genetic predisposition) as well as some that are indicative of hypersusceptibility to specific hazardous agents. In 1983, the Office of Technology Assessment of the Congress of the United States (10) published a booklet on genetic testing. They differentiated genetic screening, "a one-time test to determine the presence of particular genetic traits in individuals," from *genetic monitoring*, the periodic testing of workers to assess damage to their DNA or chromosomes from exposure to hazardous materials. The user of any genetic testing procedure should be well aware of the need for validation of the test procedure and the great distance between the original report of the gene location and its relation to a particular disease and the clinical use of such test procedure (11–14).

The ethical and legal implications of genetic screening and indeed of all screening procedures have been the subject of many very spirited discussions. (7, 15–18). Certainly medical or genetic screening can only be done with a full understanding of the need for consideration of the psychological effects of what may be learned, availability of counseling for individuals and their families and full awareness of the requirements of the Americans with Disabilities Act. These requirements can, in turn, have a negative impact on the performance of any medical examinations at all. They may present problems, not prevent them. The foregoing issues may modify but do not justify the elimination of preplacement examinations. As will be pointed out, they serve an essential function by providing: (*1*) a record of previous work experiences; (*2*) information on the state of health; prior to joining the work force; (*3*) in a few situations useful tests for hyper-susceptibility; and (*4*) a baseline for comparison with later observations and measurements. The core preplacement examination, will vary somewhat with the industry. It will usually include: history, physical examination, and a limited clinical laboratory appraisal. This examination serves as a baseline against which future evaluations can be measured and will often include such procedures as a chest X-ray, pulmonary function, specifically FEV 1 and FVC, EKG, and audiometry. In addition there should be added elements tailored to the special potential hazards of the location or job under consideration.

2.2 Preassignment Examinations

When a worker is being transferred from one operation to another and the new job has known potential hazards different from the former operation, it may be necessary to carry out a special examination. This may include inquiries about the present state of health and activities off the worksite that might, in some manner unsuspected by the worker, interact with a substance to which there might be exposure at the new job. This examination might also include a special physical examination, clinical laboratory tests, or, in some cases, biological monitoring to provide a baseline against which future comparisons might be made. Similar considerations and special examinations may also be justified in case of significant changes in processes or working conditions. And, in instances where some persons are known to be peculiarly susceptible to a chemical agent, if available, test for susceptibility may be justified. Special change of job examinations of this type also offer opportunity for additional worker education about possible health hazards associated with the new job or process, with emphasis on personal protection.

2.3 Periodic Examinations

Periodic examinations, an essential part of health monitoring, are performed to detect changes in physical condition or any signs or symptoms of incipient disease. When appropriate, they may include biological monitoring for absorption of harmful agents and the use of other biomarkers. Although most regulations for specific chemical exposures require that examinations be made available at least annually, when exposures are controlled, such frequent examinations are rarely productive. In most industrial situations an examination every three years between the ages of 18 and 45, every two years between 46 and 55 years and annually thereafter, should be adequate to detect changes in a timely fashion.

The periodic examination should update the history including job assignments, medical attention in the interim since the last examination and any current health problems. The examination itself may be a repeat of the baseline examination or a repeat of specific portions of that examination. Unless there is a specific medical indication or it is required by governmental regulation, chest x-ray and EKG need not be repeated. Though the chest X-ray is particularly useful for comparison, question has been raised as to the value of the EKG for future comparison.

Periodic examination provides a splendid opportunity for reinforcing education regarding good health habits and noting the potentially disastrous effects of bad health habits such as smoking cigarettes, eating a high fat diet, or failure to get a reasonable amount of exercise. The periodic surveillance of a group of workers having similar exposures may identify changes that are not of sufficient magnitude to be notable in an individual, but that suggest an abnormal situation and perhaps an increased risk when analyzed statistically for the group (19). The absence of positive findings is equally useful as part of the monitoring needed in a preventive program. Individual workers benefit by knowing the status of their health and by the potential for detecting early indications of increased blood pressure, diabetes, and other potentially life threatening conditions not necessarily related to the work environment. The employer benefits from the potential for earlier detection of occupationally related health problems and for the maintenance of a healthier more productive work force as a result of the detection of nonoccupational health problems which may be amenable to corrective action.

2.4 Termination Examinations

The health status of the individual worker at the termination of employment should be documented. This can be helpful for both the worker and the employer. Evidence of any changes that have occurred during the employment period can be evaluated at the time when exposure data are more readily acquired than at a later date. Changes in the audiogram of significant degree can be evaluated in an attempt to determine if such changes are job related. If corrective action is indicated it can be recommended immediately. If a compensation claim should be filed, that too can be done immediately.

There is of course, no way to rule out the possibility that effects may show up long after employment has terminated. The long latency before cancer caused by asbestos, cancer caused by smoking cigarettes, radiation-induced leukemia, or numerous other long-delayed disease can make etiologic diagnoses very difficult in some situations.

2.5 Special-Purpose Examinations

Other examinations, which may not fit into the preceding four categories, may be needed for special purposes such as the following:

1. Requirements of regulatory agencies for evaluating the health status of specific types of employment, such as vehicle drivers in interstate commerce or airline pilots.
2. Evaluation of the effects, if any, of accidental exposure to a specific chemical or physical agent or a mixture of chemical and/or physical agents.
3. Evaluation of health status before return to work after absence for disease, injury, or a prolonged layoff.
4. Evaluation of the health status of an employee who has difficulty in performing work satisfactorily, in the absence of any known hazard.
5. Determination of impairment of function or disability after specific worker complaints.
6. Undue complaint of stress associated with employment.
7. Certification of fitness to wear a respirator (discussed later).

3 SURVEILLANCE FOR GENERAL HEALTH MAINTENANCE

3.1 Content and Scope

The presence or absence of long-term benefits and the cost effectiveness of periodic examinations have been subjects for investigation and discussion for many years (20–23). The American Medical Association in a council report published in 1983 (24) provides general guidelines that includes recommendations for medical evaluations at intervals of 5 years until age 40, with shorter intervals until age 65. At which time annual examinations are suggested. Most physicians who have given thought to the subject tend to agree that multiphasic screening, if used judiciously, can provide the occupational physician with a valuable tool. Inasmuch as the workplace is the complement of the school where children gather in one place, adults generally gather at their place of work. The workplace serves as a focus where surveillance of blood pressure, detection of diabetes, and numerous other life promoting and life extending health procedures can be implemented with minimum inconvenience to the individual and minimum need for self-initiation. It is important that all screening be subject to rigorous quality control. Each test or testing procedure has its own range of normal values. An understanding is needed of the range of normal values. A relaxed attitude toward minor deviations from the average range and vigorous follow-up of significant findings should be standard operating procedure. Screening should never be undertaken without a written procedure in place specifying precisely how the results will be conveyed to the individual, what will be done about significant findings as far as medical follow-up, and how the cost for follow-up will be borne. This latter issue can cause significant employee relations problems if the results, on recheck with the personal physician, are deemed to be insignificant yet the employee is presented with a bill for services not covered by insurance.

4 HAZARD-ORIENTED MEDICAL EXAMINATIONS

It is necessary to consider hazard-oriented medical examinations in terms of (*1*) what is legally required, (*2*) what has been recommended by official agencies such as NIOSH, and (*3*) what has persuasive medical and scientific justification. The resulting programs are not necessarily the same. Primarily, the surveillance program should be dictated by "good practice" and then reviewed to make sure that it meets all legal and regulatory requirements. Regulations seldom clearly define the objectives of medical surveillance nor consider risks as well as benefits of such programs. Most people would question the wisdom of mandating complex and expensive examinations of thousands of workers whose exposures are at levels so low that hazard-derived abnormalities would be extremely rare (6). Nevertheless, a mandatory examination, if incorporated in a regulatory standard, must be met faithfully and with careful attention to the quality of information and its relationship to work exposures.

Before leaving this subject, a word about "newly discovered illnesses." Cumulative trauma disorder can be used as an example of such an illness. Does it exist to the extent that is implied by the number of reports of its presence and the number of people compensated for disability? Health surveillance for such newly discovered, or newly popularized entities must be very carefully constructed and implemented (25).

4.1 OSHA Requirements

4.1.1 General Requirements

All permanent standards of the *Occupational Safety and Health Administration (OSHA)* contain *requirements for medical surveillance*. As of 1998 there were 20 such permanent standards, 5 of which apply to work situations rather than specific chemical agents. Table 47.1 shows the particular substance or work situation and the standard, with the specific pertinent paragraph in the standard.

4.1.2 Examination for Ability to Wear a Respirator

In almost all instances any person fit to work will be sufficiently fit to wear a respirator. However the ability to wear a respirator has to be determined by the ability to have a proper fit of the respirator to be used, as well as the two problems that may be absolute contraindications to use of the respirator: allergy to the face piece and claustrophobia. The use of a respirator in extremes of heat or the use of self-contained breathing apparatus (SCBA) which requires carrying a heavy tank of air pose a need for extra effort which must be considered in evaluating the ability to wear a respirator. OSHA Regulations (*29 CFR* 1910, 19,26) require that "every employee must be medically evaluated prior to fit testing and initial use of a respirator". This regulation requires the employer to select a physician or other "licensed health care provider (PLHCP)" who then must conduct an examination consisting of a mandatory medical questionnaire and an initial medical examination before the employee can be fit tested or perform the work and use the equipment. The respirator user's medical status need only be reviewed thereafter when there is a significant change in job or job assignment (*29 CFR* 1910-134 (e)). The reader is advised to consult the Regulation for details. At this writing parts of the regulation are being litigated and much of the medical examination requirements remains to be clarified. In as much as no medical examination procedure has yet been validated for determining a person's ability to wear a respirator, there are numerous approaches to the problem of complying with this part of the regulation (26, 27). The questionnaire mandated by the OSHA Regulation has many questions about matters which, on the face of it, would seem to have little relevance to an ability to wear a respirator.

4.2 Mine Safety and Health Administration

The Federal Coal Mine Health and Safety Act of 1969 (28) was landmark legislation in its requirement for medical examination of coal miners. It set schedules for chest roentgenograms, required the Secretary of Health, Education, and Welfare to prescribe classification schemes for radiographic interpretation and tied compensability to film interpretations. Regulations developed under the act added provisions relating to the training and proficiency of physicians who interpreted films and specified methods of mea-

Table 47.1. OSHA Standards Requiring Medical Surveillance

Substance	Standard
Hazardous waste operations	1910.120(f)
Asbestos	1910.1001(1),(m)(3) Appendix D,E,H
	1910.1011(j)
4-Nitrobiphenyl and 13 carcinogens	1910.1003(g)(1)(i)
α-Naphthylamine	1910.1004(g)
Methyl chloromethyl ether	1910.1006(g)
3,3′-Dichlorobenzidine and its salts	1910.1007(g)
bis-Chloromethyl ether	1910.1008(g)
β-Naphthylamine	1910.1009(g)
Benzidine	1910.1010(g)
4-Aminodiphenyl	1910.1011(d),(g)
Ethyleneimine	1910.1012(g)
β-Propiolactone	1910.1013(g)
2-Acetylaminofluorene	1910.1014(g)
4-Dimethylamino azobenzene	1910.1015(g)
N-Nitrosodimethylamine	1910.1016(g)
Vinyl chloride	1910.1017(k)
Inorganic *arsenic*	1910.1018(n) Appendix C
Lead	1910.1025 (j), Appendix C
Cadmium	1910.1027(l)
Benzene	1910.1028 (i), Appendix C
Coke oven emissions	1910.1029(j)
Blood-borne pathogens	1910.1030(f)
Cotton dust	1910.1043 (h), Appendix B-I,C,D
1,2-Dibromo-3-chloropropane	1910.1044 (i)(5), Appendix C
Acrylonitrile	1910.1045 (n),
Ethylene oxide	1910.1047(i),
Formaldehyde	1910.1048(l) (1). Appendix C
Laboratory chemicals	1910.1450(g)
4,4′-Methylenedianiline	1910.1050 (proposed)
1,3-Butadiene	1910.1051(k)(8) Appendix C
Methylene chloride	1910.1052 (j) App. B

suring pulmonary function, among other detailed requirements. Thus was born the "B" Reader.

The legislative mandate, PL 95-164 (29), for the Mine Safety and Health Administration included the provision that its standards shall "where appropriate, prescribe the type and frequency of medical examinations or other tests which shall be made available by the operator." Current MSHA regulations pertaining to medical examinations and the specific requirements are listed in Table 47.2.

Table 47.2. MSHA Regulations and Their Medical Requirements

MSHA Regulation	Requirement
30 *CFR* Pail 70.510 30 *CFR* Pail 71.805	Audiometric examinations for underground and surface coal miners
30 *CFR* Pail 49.7	Physical requirements for mine rescue teams
42 *CFR* Pail 37	X-ray program for underground coal miners

4.3 Medical Surveillance Recommendations by the National Institute for Occupational Safety and Health (NIOSH)

The NIOSH criteria documents, which recommend standards for occupational exposures, always include recommendations to OSHA for medical surveillance. In nearly all cases these prescribe mandatory surveillance. The objectives (e.g., for epidemiologic studies, for early detection and treatment, for detection of group risk factors, or merely for good occupational medical practice) are rarely stated. In comparatively few instances have these recommendations been incorporated in the OSHA regulations. Although they do not have the force of regulations, they constitute a powerful coercive influence, suggesting as they do a level of good practice. They are a useful guide but must be viewed critically as to justification and whether implementation is justified by the benefit to be derived.

In 1974 NIOSH commissioned the preparation of so-called mini-criteria documents for nearly 400 chemicals. The NIOSH/OSHA Standards Completion Program, Draft Technical Standards, included recommendations for biologic monitoring and medical examinations. These were included in the Proctor and Hughes handbook *Chemical Hazards in the Workplace* (30), which provided a valuable summary to aid the occupational physician in examinations of workers. In their 1988 update (31) for 198 of 398 substances the authors did not see a need for periodic physical examinations solely on the basis of exposures to a specific agent. They stressed that a brief interim history is usually sufficient for such surveillance.

4.4 Other Sources for Recommendations

It is no longer sufficient to assume that the principal objective of health surveillance is to observe changes in cardiovascular, pulmonary, hepatic or hematopoietic functions. Reproductive function (32), nervous system function, indeed, any organ or organ system may prove to be critical and therefore must be observed periodically to ensure that adverse effects are not occurring. At least one of the medical specialty societies, the American Thoracic Society, has developed detailed recommendations for surveillance of the respiratory tract when there is a potential for exposure to respiratory hazards (33).

4.5 Justification for Hazard-Oriented Medical Surveillance

Monitoring of the environment provides an ongoing indication of exposure to chemical and physical agents of use for engineering control and evaluating medical findings. Medical surveillance of the individuals working or living in that environment provides another level of monitoring while biologic indicators of absorption of chemical agents based on analysis of the agent or a metabolite in expired air, urine, or blood is a monitoring method increasingly used by the industrial hygienist and the medical community (34). The physician can use such measurements as part of the hazard-oriented surveillance program in addition to monitoring minor physiologic changes, reductions of function or pathologic changes that may be early manifestations of toxicity in individual workers or groups of workers. In further justification of periodic medical surveillance it should be noted that every contact with the health establishment provides an opportunity for education of the worker, education that can create a healthy awareness of risk without creating unnecessary fears.

4.5.1 Indicators of Absorption

Biologic monitoring has become increasingly important in medical surveillance (35, 36). A substance present in the workplace may also be present in consumer products and in the general environment. Only by measuring the concentration of the substance or a metabolite in body fluid, expired air, or other bodily substance can an evaluation of total absorption be made. Sampling should be done at a specific time for some substances but may be done at any time for long biologic half-lived materials, such as lead. But the conditions of collection of the sample, how it is stored and transported, and how it is to be analyzed should all be determined before collection. Abnormally high or low values, depending on the biological marker being used, if confirmed by additional biologic sampling, may show the need for careful environmental monitoring and increased controls or better hygienic education. A more detailed discussion of bio-monitoring is provided in Chapter 45.

The American Conference of Governmental Industrial Hygienists (ACGIH) Biological Exposure Indices Committee publishes an annual update of information on selected biomarkers (37). These biological exposure indices (BEIs) are defined as reference values or guidelines, representing the level of determinants that are most likely to be observed in specimens collected from a healthy worker who has been exposed to chemicals to the same extent as a worker with inhalation exposure to the threshold limit value (TLV). Documentation is published periodically, with the most recent having appeared in 1991 (38). Space does not permit a detailed discussion of this subject but a list of the 36 substances included in the 1998 annual report appears on Chapter 45. The complete report includes sampling times and all of the BEIs that have been adopted as well as new or revised BEIs that have been proposed but not yet adopted.

4.5.2 Indicators of Early Effects

Biological monitoring is an indicator of absorption of the substance being monitored. Finding the substance being looked for or monitored is not an indication of an effect. The use of the term *subclinical poisoning* or *subclinical effect* has served to confuse the question of what is an early adverse effect versus what is simply evidence of absorption (35). There are many recommendations regarding when a biological marker for health effects should be a trigger for action, but few of these recommendations are based on specific criteria

HEALTH SURVEILLANCE PROGRAMS IN INDUSTRY

and there is little consistency among the recommendations for various substances. Absorption of organophosphorus cholinesterase inhibitors will result in inhibition of red blood cell and plasma cholinesterase. At what point should the worker be removed from further exposure when there is reduction in cholinesterase? (39–41). Similar questions can be asked about biologic markers for lead, cadmium, cotton dust, benzene, and a host of other materials. Care must be exercised to differentiate between change that is part of the normal homeostatic response and change that must be considered pathological.

A major problem in interpreting early changes is that normal variations occur in virtually all tests. A preexposure baseline and serial follow-up testing is required for many of these biological markers. If there is indication of a change from baseline-determined values, there is then the problem of deciding not only whether removal from exposure is justified but also the decision as to whether a suspected abnormality is sufficient to be reported as an occupational illness on the OSHA log 200.

Problems associated with the early detection of cancer are discussed in Section 4.5.5.

4.5.3 Indicators of Hypersusceptibility

All individuals exposed to a chemical do not respond alike. The differences in response are not necessarily the result of different levels of exposure. Although variation in response is the rule rather than the exception, there is usually a normal range of response that the clinician can come to expect. Occasionally, a person will prove to be responsive to such a degree that the individual can be properly labeled "hypersusceptible." This classification is usually one of exclusion. Before it can be made a careful investigation should be made to determine a possible explanation: Is there a synergistic or additive effect of occupational exposure plus nonoccupational exposure? Examples are cigarette smoking and asbestos exposure or ethyl alcohol and chlorinated hydrocarbon solvents. Preexisting disease may be the cause of hypersusceptibility. There may be inborn errors of metabolism that interfere with the detoxification of a chemical or augment its effects. Inherited traits that may predispose to hypersusceptibility were discussed in Section 2.1.1. Although the reality of such inherited factors is indisputable, it has been difficult to establish firmly their importance in occupational medicine. Cooper reviewed the status of the most promising indicators of hypersusceptibility and could not recommend their application in routine screening. The tests he considered were those for sickle cell trait, glucose-6-phosphate dehydrogenase (G6PD), α-1-antitrypsin, and cholinesterase deviants. All appeared to be appropriate subjects for controlled research but not for mandatory inclusion in regulations or as positive indicators for exclusion from particular jobs (42). In the intervening years, there has been little practical application of genetic screening, although it has been the subject of a great deal of debate (see Section 2.1.2). It certainly cannot be applied to discriminate against individuals or ethnic groups.

4.5.4 Tests for Effects on Fertility

The discovery in the late 1970s that dibromochloropropane (DBCP), a nematocide that had been widely used for more than 20 years, could cause reduction in sperm or sterility in human males at very low doses caused great discussion and consternation (32). The impact of exposures to lead on the ability of both males and females to reproduce has been a cause for inquiry and research for many years. The concern has not necessarily been only because of its impact on fertility but also because of the possibility that *in utero* lead exposure might impact adversely on the fetus (43). Investigations related to reproduction are fraught with considerable difficulty because of the sensitivity of the subject in our culture and the lack of generally acceptable monitoring methods (44). Questionnaires of past reproductive experience can be considered only an investigational tool after the fact. Hormone studies of various types have thus far not become routine monitoring methods. It is likely that for the immediate future we will continue to be limited to retrospective studies to determine whether a chemical or physical agent does have an impact on fertility. But as something to be included in routine surveillance, present technology and lack of any evidence of a widespread problem does not justify inclusion in routine medical surveillance. Where any suspicion of an effect on fertility exists, if the population at potential risk is of sufficient size, specific attention to this parameter can be included in the routine surveillance in order to determine whether or not the subject population is demonstrating a lower birth rate than might be expected. Where the suspect population is too small for such an investigation to be meaningful, a detailed clinical investigation may be indicated.

4.5.5 Early Detection of Cancer

Cancer has become a major preoccupation of the public as life expectancy has increased. Where workers are exposed to suspected or proven carcinogens, medical surveillance to detect premalignant changes or early cancer has become a fundamental part of health surveillance (45). Unfortunately it has so far resulted in very limited benefits in reducing occupational cancer mortality. For example, lung cancer and mesothelioma are major concerns in asbestos-exposed workers. However, periodic examinations of current and former asbestos workers rarely detect evidence of tumors that result in curative intervention (46). This is consistent with other major studies where large populations of cigarette smokers were observed by three medical centers over 5-year periods, with tests including periodic sputum cytology and chest roentgenograms. No statistically significant effects on lung cancer mortality were observed, even in one of the studies where tests were done every 4 months (47).

Bladder cancers have been recognized for many years as a type of occupational or environmental tumor that might respond well to early detection by screening tests, such as urine cytology. However, there are still many uncertainties as to the best procedures and the actual benefits, as pointed out in an excellent international conference on "Bladder Cancer Screening in High Risk Groups" sponsored by NIOSH in 1989 (48).

The current permanent General Industry asbestos standard (49) does not address itself to lung cancer or mesothelioma in its medical surveillance requirements, which are limited to annual chest films, measurements of pulmonary ventilatory function, and history. None of these monitoring procedures has been shown to result in any specific benefit to the worker at risk, yet can result in additional anxiety.

The standard for vinyl chloride (50) includes provisions for a battery of liver function tests, presumably because vinyl chloride is a known hepatotoxin, and it has been suggested that such hepatoxic effects would precede angiosarcomas caused by exposure to vinyl chloride. There is no evidence that this is the case. In any event, if exposures are kept below 1 ppm, it is probable that thousands of workers would be repeatedly examined before any vinyl chloride-related disorder were discovered. It is now recognized that the United States and, indeed, the entire world, is at the beginning of an epidemic of hepatitis C. The requirements for surveillance of vinyl chloride workers was enacted before hepatitis C and its long term effects were discovered and before the effect or lack of effect from exposures to vinyl chloride of 1 ppm or less was fully appreciated. Positive liver function studies may well result in unnecessary removal from exposure or advice to terminate employment. There is provision in the OSHA regulations for the attending physician to make the final determination as to what must be included in the examination. However, in part because of the legal pit falls, only rarely will the examining physician deviate from the standard examination.

The standards for coke oven emissions and for arsenic include provisions for sputum cytology. The former also requires urine cytology (51). Frequent sputum examinations in those who have worked in high-risk areas for many years may detect an occasional operable lung cancer. This is probably an unproductive exercise in relation to individuals working under controlled conditions, even though it is required.

Urine cytology is also unlikely to be a useful screening procedure in coke oven workers, in view of the relatively low incidence of urinary tract cancers in the group. It can be a useful test in those heavily exposed in the past to proved bladder carcinogens such as β-naphthylamine and benzidine. But even with these compounds, urine cytology has generally proved of limited benefit (52).

Specifying medical surveillance for cancer, particularly for chemicals whose effects in humans are speculative and based solely on findings in experimental animals, has produced difficulties for regulatory agencies. The published standards include requirements for preplacement and annual physical examinations. Many stipulate a "personal history of the employee, family and occupational background, including genetic and environmental factors." Some also stipulate that in all physical examinations the examining physician should consider whether there exist conditions of increased risk including reduced immunological competence, of those undergoing treatment with steroids or cytotoxic agents, pregnancy and cigarette smoking.

It is unclear how the examining physician would evaluate all the information obtained or how it might affect employability. It is unclear whether anyone is presently capable of relating family history, for example, to the potential for development of cancer from exposure to most of the known or suspected occupational carcinogens. As can be seen, to follow some of these rules blindly, and without careful thought as to the consequences, could lead one into a maze of problems.

4.5.6 Biochemical Markers of Cancer

Such biochemical markers of cancer or incipient cancer as carcinoembryonic antigen (CEA), α-fetoprotein (AFP), and prostate specific antigen (PSA), deserve mention. The value of these markers remains controversial. PSA, the most recently used of these antigens, is well established and is used routinely in the investigation and confirmation of prostatic cancer. Whether it should be used as a screening tool is controversial only because of the large number of false positives and false negatives. The cost–benefit ratio has to be evaluated both as regards the economics of the screening and the psychologic cost of both false negatives and particularly of false positives (53). As pointed out by Brandt-Rauf (54) studies of possible biochemical tools for effective monitoring of occupational cancer are making rapid strides, but their utility still has to be proven.

5 RELATIONSHIP OF MEDICAL SURVEILLANCE AND EXPOSURE DATA

The availability of computer power has made possible the relatively easy coordination of exposure and medical data. Numerous examples of such coordination in both large and small plant populations are now available. Commercially available software for use with desktop personal computers has reduced the technical and financial constraints on such programs. Computerization of the findings on medical examination makes possible the analysis of single point items or of the employee health status as a group. Thus blood pressure, hearing, cholesterol level, or a specific test for occupational exposure such as lead can be surveyed in order to detect an adverse trend before health is affected and a major problem develops. Further, this can be correlated with the results of air sampling to determine the effect, if any, on groups of employees at given levels, the effectiveness of control measures or to determine safe levels of exposure.

6 U.S. EQUAL EMPLOYMENT COMMISSION (EEOC)

Another agency that has acquired legal sanctions that impact on health surveillance programs is the EEOC, which has responsibility for enforcing the Americans with Disabilities Act (ADA) of 1990 (9). The employment provisions of this act, covers all employers with 15 or more employees.

Included in its many provisions involving public transportation, building accommodations, and other provision for those with disabilities are many stipulations regarding the employment of those with disabilities. What may and may not be included in the medical examination and appraisal of potential employees and many of the requirements for accommodating these employees and potential employees are stipulated in the regulations to implement this act. The requirements involve consideration of hardship to employees and employers (55). Whether accommodation of an individual with a disability is economically and technically feasible is one consideration. How much risk is entailed to the worker and co-workers as a result of placing the worker with a disability in a particular job? As can be easily appreciated, decisions on many of these questions will be highly subjective and often arbitrary. How much expenditure is economically feasible to place a worker with a particular disability? What is technically feasible? But for the health program, the question needs to be posed once again, when should the preplacement examination be given; what can be justified for inclusion in the examination; what are the markers for an unsafe situation in which either the worker or co-workers would be at greater than acceptable risk? Good practice and good will must prevail if the examining physician, the employee, and the employer are all to avoid an attack of acute anxiety as well as legal penalties. The reader is referred for additional reading (56, 57).

7 RECORDING

7.1 Maintenance of Records

Because of the long latent period between exposure and the appearance of chronic effects such as asbestosis or occupationally related cancers, there is need for preservation of medical records for what is essentially the lifetime of the employee. OSHA requires that occupational medical records be retained for the duration of employment plus 20, 30, or 40 years, depending on the particular regulation. Many of the early OSHA regulations specify that, if the employee terminates employment, retires, or dies or if the employer ceases business without a successor, the records must be forwarded to the director of OSHA. It is recommended that all medical records be kept for a minimum of 40 years and that duplicate records be kept in storage on computer-accessible film or microfiche. On occasion it has been recommended that employees be invited to return and destroy their own medical records on their 100th birthday.

The use of these many records for epidemiological studies of morbidity and mortality is envisioned as a by-product of the OSHA requirement. How useful they will be in this regard remains to be seen.

7.2 Confidentiality of Records

Records containing personal information on workers must be protected from transmission to, or perusal by, those not responsible for medical services or care. Without assurance of confidentiality, it is difficult to elicit vital medical information and impossible to maintain the professional relationship so important in performing the responsibilities of the physician. The ADA regulations require that medical records must be kept confidential and that the guardian of these confidential records need not be a physician but simply a person so designated and properly instructed (55). Management can be informed of a person's limitations without providing confidential medical information. For example, the manager can be informed that the employee should not work around moving machinery, at heights, or driving a vehicle. The manager need not be told that the person has a seizure disorder that is not under complete control. The activities of a diabetic on insulin might be limited by simply stating that the person should not work alone for any extended period.

Confidentiality of records has been difficult to achieve in small as well as large companies. Curiosity has killed the cat, but it has not necessarily harmed the fellow employee who tries to get a look at the medical record or the human resources manager who "just has to know." These are difficult situations. The physician or a designated health person or persons should be in charge of the health records and must be given full authority by top management to protect the confidentiality of the records. Release of records to anyone, except on court order, should only be with the express written consent of the employee. Access to medical records by OSHA, NIOSH, or EEOC personnel should be looked at very carefully and provided only with proper legal authorization. Whether a medical or nonmedical person is responsible for the medical records, there should be careful indoctrination of the responsible individual as to how confidentiality must be maintained and what legal requirements must be met, for example a legal subpoena, before records can be released to someone other than the individual or a health care provider designated by that individual.

Provisions in OSHA, ADA, or other regulations regarding accessibility and dissemination of medical records vary. The record guardian should consult the specific regulation before agreeing to release any record without a court order. Whereas the standard for asbestos requires that medical reports be sent to employers, more recent standards have been less explicit (16).

8 PROBLEM AREAS

8.1 Compulsory versus Voluntary Examinations

All regulations so far promulgated provide that medical surveillance be made available by employers, only the regulation regarding respirators requires the employee to take an examination to determine ability to wear a respirator. Nowhere else is it stated that employees shall be required to take the examinations. Individual employers may, however, make examinations a condition of employment. The issues here are obviously complex and the Americans with Disabilities Act must only serve to increase the complexity. If a regulatory agency believes that a given examination or test is so important to workers' health that every employer must be prepared to provide it, how can it not require employees to take the examination if they are to work with a specified chemical agent? Present regulations leave the question of such medical surveillance at the level of negotiation between the workers or their representatives and the industry and necessitate assurances relating to protection of job rights. For that part of the work force that is not organized, the vast majority in the United States, such protection would have to be guaranteed by law.

8.2 The Problem of Small Companies

The majority of new jobs, and in a few years, the majority of jobs, will be in small companies. Medical surveillance in small industry may be aided by the extensive development of free-standing occupational medical clinics and similar hospital-based clinics. Major corporations with medical departments gradually adjust to regulations and recommendations requiring medical surveillance. The small employer must meet these requirements without the staff support found in some large organizations. These new clinics and the many consultants now available may help fill the needs of the smaller companies. The small company requirements may actually be merging with those of larger companies that have jettisoned their own in-house medical departments and are relying on contract and consulting services.

8.3 Handling Abnormal Findings

A major problem for the occupational physician in the current medical/legal climate is that of borderline abnormal findings. Regulations stipulate that a physician must certify that a worker has no condition that might be adversely affected by exposures on the job.

It can be very difficult for a cautious physician to make such a certification if there are any abnormal findings. The presence of metaplastic cells in the sputum of an asbestos worker or a coke oven worker would make certification difficult, even though current exposures were low and controlled. Similarly, some physicians may find certification of ability to wear a respirator difficult for an individual who years earlier had had a coronary occlusion. The risk, however slight, might give pause. Situations that can be handled easily in a normal doctor–patient relationship become matters with serious medico-legal implications because they have been the subject of certification under a federal regulation. Reviews of these many medico-legal problems as well as the accompanying ethical problems are available (5, 16).

8.4 Consequences of Overregulation

Inherent in much of the foregoing discussion has been concern over the inclusion of detailed requirements for medical surveillance in governmental regulations. Many practices that are highly desirable for physicians to follow in selected occupational groups, or for plants to carry out with full understanding of their implications, are not necessarily right for mass application mandated by law. When applied to individuals with very low exposures as required by present workplace standards, the usual physical examination is likely to yield few abnormal findings. This becomes a stultifying overuse of physician time, needless expense, and a waste of national resources. Often it is former employees, no longer covered by regulations, who could most benefit by periodic medical examinations. But it is the new employees, with relatively little exposure, who receive the attention. Industrial hygienists and occupational physicians need to be listened to and heard when these regulations are eventually revised.

9 LEGAL CONSIDERATIONS

As will be evident from reading earlier sections in this chapter, there are many legal burdens placed on the physician responsible for the medical surveillance of the worker (16, 58). A comprehensive review of such legal strictures is not presented here, but a few areas are discussed briefly.

9.1 Informing the Worker

There is general agreement and a regulatory requirement (59) that the worker must be given information about the hazards of the workplace. This is management's responsibility, but the physician must and will have a role. For many hazards the explanation is easily and readily understood by the worker. For carcinogens, particularly for those substances designated as "suspect" carcinogens, the message is more difficult to impart. Does the physician even have available all the data necessary to evaluate whether the evidence for carcinogenicity is solid or weak? The best that can be said is that the physician should be honest in giving an appraisal of the evidence, should create sufficient concern and anxiety to encourage observance of rules for containment and protection against known carcinogens, and should indicate the relative probability of effects from low exposures and for weak or suspect carcinogens. To do this properly the physician should seek the assistance of the industrial hygienist, the toxicologist and any other resources that may be available, to come to reasonable conclusions about the risk under the specific circumstances in which the worker is laboring.

9.2 Certification for Continued Employment

Regulations that require the examining physician to certify that workers will not be adversely affected by continued employment are in many instances asking to certify as to the unknown or the unknowable. There will be a tendency by some physicians to take no chances. There is relatively little information to say whether an individual with an elevated serum glutamic oxaloacetic transaminase (SGOT) or other elevated liver enzyme would be harmed by exposure to 1 ppm of vinyl chloride. The probability is extremely high that it would make no difference. If, however, a worker should develop liver disease after certification, regardless of whether it was due to vinyl chloride, the certifying physician could conceivably be sued. A person certified to work with asbestos, even at low levels, after the finding of moderate metaplastic changes in the sputum, could present a similar problem. The examining physician is best advised to use the best clinical judgment of which he or she is capable, learn what is known about the hazard, attempt to learn something of the actual exposures their patients are experiencing, and realize that depriving a person of a job unnecessarily is a very serious matter.

9.3 Wearing of Respirators

Physicians were previously designated as the responsible professional for deciding if someone could safely wear a respirator. That is no longer the case. Objective criteria of ability to wear a respirator are limited, as noted earlier in this chapter (26, 27). The translation of pulmonary function test results to respirator use has thus far not been validated. Actual trial of individuals using the respirator that they will be called on to use at work is the best test available. Differentiation must be made between situations when a respirator is worn briefly for emergency situations for a worker's own protection, when it is required for the worker to perform duties essential to the safety and health of others, and where the worker may be required to wear it over long periods of time.

10 SUMMARY

The health surveillance of workers requires a baseline examination usually the preplacement examination, and periodic examinations aimed both at general health maintenance and the prevention or early detection of effects from specific job hazards. Good practice calls for a comprehensive preplacement evaluation, with an occupational and medical history and review of all systems, baseline laboratory studies (including blood chemistry and urinalysis as well as a hemogram), electrocardiogram and study of visual and hearing acuity, simple tests of pulmonary function, a chest roentgenogram, and a general physical examination. The content of periodic examinations should be based on regulatory requirements with overall consideration of what is good practice. There should be special hazard-oriented studies when indicated, such as biologic monitoring for chemicals to which the worker is known to be exposed, early indicators of toxic or other biological effects, and audiometry when indicated by noise exposures of 80 dB or greater.

Scheduling of examination frequency must consider age, nature of hazards, and regulatory requirements, but it is essential that hazard-oriented surveillance, environmental, and general medical data be closely interlocked.

ACKNOWLEDGMENTS

W. Clark Cooper, co-author in the 3rd edition of this work, laid the foundation for this chapter. He has the author's greatest appreciation. For their help and helpful criticism the author should like to thank Don Hillman and Peter Zavon.

BIBLIOGRAPHY

1. R. L. Zielhuis, "Biological Monitoring: Confusion in Terminology," *Am. J. Ind. Med.* **8**, 515–516 (1985).
2. Editorial, "Prevention and Treatment of Occupational Mental Disorders," *The Lancet* **352**, 999 (Sept. 26, 1998).
3. I. L. D. Houtman, A. Goudswaard, S. Dhondt, M. P. van der Grinten, V. H. Hildebrandt, and E. G. T. van der Poel, "Dutch Monitor On Stress and Physical Load: Risk Factors, Consequences, and Preventive Action," *Occup. Environ. Med.* 55, 73–83 (1998).
4. World Health Organization, *Early Detection of Health Impairment in Occupational Exposure to Health Hazards*, Technical Report Series 571, WHO, Geneva, Switzerland, 1975.
5. M. A. Rothstein, *Medical Screening of Workers*, The Bureau of National Affairs, Inc., Washington, DC, 1984.
6. W. F. Halperin, P. A. Schulte, and D. G. Greathouse, Guest Eds., "Conference on Medical Screening and Biological Monitoring for the Effects of Exposure in the Workplace, Part 1," *J. Occup. Med.* **28**(8), 543–788 (1986).
7. P. A. Schulte, "Contribution of Biological Markers to Occupational Health," *Am. J. Ind. Med.* **20**, 435–446 (1991).
8. C. R. Goerth, "Physical Standards: Discrimination Risk," *Occ. Health Saf.* **52**(6), 33–34 (1983).
9. *Americans with Disabilities Act*, PL 101-336, 42 *USC* 12101 et seq. (1990).
10. OTA, *The Role of Genetic Testing in the Prevention of Occupational Disease*, prepared by Advisory Panel on Occupational Genetic Testing, Office of Technology Assessment, Congress of the United States, Washington, DC, 1983.
11. N. A. Holtzman, P. D. Murphy, M. S. Watson, and P. A. Barr, "Predictive Genetic Testing: From Basic Research to Clinical Practice," *Science* **278**, 602–605 (1997).
12. R. E. Pyeritz, "Family History and Genetic Risk Factors," *JAMA* **278**(15), 1284–1285 (1997).
13. D. Seligman, "Outlawing DNA," *Forbes*, 110–115 (July 6, 1998).

14. W. C. McKinnon, et al. "Predisposition Genetic Testing for Late-Onset Disorders in Adults." *JAMA*, **278**(15), 1217–1220 (1997).
15. S. W. Samuels, "Medical Surveillance: Biological, Social and Ethical Concerns," *J. Occup. Med.* **28**(8), 572–577 (1986).
16. N. A. Ashford, D. B. Hattis, C. J. Spadafor, and C. C. Caldert, *Monitoring the Worker for Exposure and Disease, Scientific, Legal and Ethical Considerations in the Use of Biomarkers*, Johns Hopkins University Press, Baltimore, MD, 1991.
17. E. Draper, *Risky Business. Genetic Testing and Exclusionary Practices in the Hazardous Workplace, Cambridge Studies in Philosophy and Public Policy*, Cambridge University Press, Cambridge, England, 1991.
18. AMA, Council on Ethical and Judicial Affairs, "Council Report: Use of Genetic Testing by Employers," *JAMA* **266**, 1827–1830 (Oct. 2, 1991).
19. M. R. Zavon, "Methyl Cellosolve Intoxication," *Am. Ind. Hyg. Assoc. J.* **24**, 36–41 (1963).
20. N. J. Robert, "The Values and Limitations of Periodic Health Examinations," *J. Chronic Dis.* **9**(2), 95–116 (1959).
21. G. Siegel, "An American Dilemma—The Periodic Health Examination," *Arch. Env. Health* **13**(9), 292–295 (1966).
22. W. K. C. Morgan, "The Annual Fiasco (American Style)," *Med. J. Australia* **2**(11), 923 (1969).
23. P. Jacobs and A. Chovil, "Economic Evaluation of Corporate Medical Programs," *J. Occup. Med.* **25**(4), 273–281 (1983).
24. AMA, Council on Scientific Affairs, "Medical Evaluations of Healthy Persons," *JAMA*, **249**, 1626–1633 (1983).
25. N. M. Hadler, "Cumulative Trauma Disorders, An Iatrogenic Concept," *J. Occup. Med.* **32**(1), 38–41 (1990).
26. P. Harber, "Medical Evaluation for Respirator Use," *J. Occup. Med.* **26**(7), 496–502 (1984).
27. T. K. Hodous, "Screening Prospective Workers for the Ability to Use Respirators," *J. Occup. Med.* **28**(10), 1074–1080 (1986).
28. *Federal Coal Mine Health and Safety Act of 1969*, PL 91-173, December 30, 1969.
29. *Federal Mine Safety and Health Amendments Act of 1977*, PL 95-164. November 9, 1977.
30. N. H. Proctor and J. P. Hughes, *Chemical Hazards of the Workplace*, J.B. Lippincott, Philadelphia, PA, 1978.
31. N. H. Proctor, J. P. Hughes, and M. L. Fischman, *Chemical Hazards of the Workplace*, 2nd ed., J. B. Lippincott, Philadelphia, PA, 1988.
32. D. Whorton, R. A. Krauss, S. Marshall, and T. H. Milby, "Infertility in Male Pesticide Workers," *Lancet* **17**, 1259–1261 (1977).
33. American Thoracic Society, "Surveillance for Respiratory Hazards in the Occupational Setting," *Am. Rev. Respir. Dis.* **126**, 952–956 (1982).
34. R. R. Lauwerys and P. Hoet, "Industrial Chemical Exposure", *Guidelines for Biological Monitoring*, 2nd ed., Biomedical Publications, Davis, CA, 1993.
35. L. F. Lowry, "Biological Exposure Index as a Complement to the TLV," *J. Occup. Med.* **26**(8), 578–589 (1986).
36. A. Bernard and R. Lauwerys, "Present Status and Trends in Biological Monitoring of Exposure to Industrial Chemicals," *J. Occup. Med.* **28**(8), 558–562 (1986).

37. *Documentation of the Threshold Limit Values and Biological Exposure Indices*, 5th ed., American Conference of Governmental Industrial Hygienists. Cincinnati, OH, 1986 (6th ed., 1991.).
38. *Biological Exposure Indices*, American Conference of Governmental Industrial Hygienists, Cincinnati, OH, pp. 93–107, 1998.
39. M. J. Coye, J. A. Lowe, and K. Maddy, "Biological Monitoring of Agricultural Workers Exposed to Pesticides: 1. Cholinesterase Activity Determinations," *J. Occup. Med.* **28**(8), 619–627 (1986).
40. M. A. Gallo and N. J. Lawryk, "Organic Phosphorus Pesticides," in W. J. hayes. Jr., and E. R. Laws, Jr., eds., *Handbook of Pesticide Toxicology*, Vol. 2, Academic Press, New York, 1991, pp. 917–1123.
41. R. McConnell, L. Cedillo, M. Keifer, and M. Palomo, "Monitoring Organophosphate Insecticide-Exposed Workers for Cholinesterase Depression," *J. Occup. Med.* **34**(1), 34–37 (1992).
42. W. C. Cooper, "Indicators of Susceptibility to Industrial Chemicals," *J. Occup. Med.* **15**(4), 355–359 (1973).
43. W. N. Rom, "The Effect of Lead on the Female and Reproduction: A Review," *Mount Sinai J. Med.* **43**, 542–552 (1976).
44. R. J. Levine, M. J. Symons, S. A. Balogh, D. M. Amdt, N. T. Kaswandik, and J. W. Gentile, "A Method for Monitoring the Fertility of Workers: 1. Method and Pilot Studies," *J. Occup. Med.* **22**(12), 781–791 (1980).
45. B. S. Hulka, "Screening for Cancer: Lessons Learned," *J. Occup. Med.*, **28**(8), 687–691 (1986).
46. W. C. Cooper, "Asbestos Protection in the Workplace: Problems of Medical Surveillance," presented at World Symposium on Asbestos, Montreal, Quebec, Canada, May 25, 1982.
47. N. I. Berlin, C. R. Buncher, R. F. Fontana, J. K. Frost, and M. R. Melaned "Screening for Lung Cancer," *Am. Rev. Resp. Dis.* **130**, 565–570 (1984).
48. P. A. Schulte, W. E. Halperin, E. M. Ward, and A. M. Ruder, Guest Eds., "Bladder Cancer Screening in High-Risk Groups," *J. Occup. Med.* **32**(9), 787–945 (1990).
49. U.S. Department of Labor rule **29** *CPL* 1915, 1001 (l) and 1910.1001(m) (1983).
50. U.S. Department of Labor, **29** *CPR* 1910.1017(k) (1983).
51. U.S. Department of Labor, **29** *CFR* 1910, 1029(j) and 1910, 1018(n) (1983).
52. G. M. Farrow, "Cytology in the Detection of Bladder Cancer: A Critical Approach," *J. Occup. Med.* **32**(9), 817–821 (1990).
53. J. R. Drago, "The Role of New Modalities in the Early Detection and Diagnosis of Prostate Cancer," *Ca-A Can. J. Clinicians* **39**(6), 326–336 (1989).
54. P. W. Brandt-Rauf, "New Markers for Monitoring Occupational Cancer: The Example of Oncogene Proteins," *J. Occup. Med.* **30**(5), 399–404 (1988).
55. Americans with Disabilities Act (ADA) Regulations, **29** *CFR* 1630, July 26 (1991).
56. R. N. Anfield. "Americans with Disabilities Act of 1990: A Primer of Title 1 Provisions for Occupational Health Care Professionals," *J. Occup. Med.* **34**(5), 503–509 (1992).
57. S. St. Clair and T. Shuts, "Americans with Disabilities Act: Considerations for the Practice of Occupational Medicine," *J. Occup. Med.* **34**(5), 510–517 (1992).
58. Felton, J. 5. (1978). "Legal Implications of Physical Examinations," *West. J. Med.*, 128, 266–273.
59. U.S. Department of Labor. Occupational Safety and Health Administration (1987). OSHA Safety and Health Standards (29 CFR 1910:1200) Hazard Communication Standard. August 24.

CHAPTER FORTY-EIGHT

Health Promotion in the Workplace

B. Toeppen-Sprigg, MD, MPH, K. Schmidt, RN, MS, COHN-S and A. Hart, Ph.D., CIH, CSP

1 INTRODUCTION

Worksite health promotion programs first appeared in industry during the 1970s. Interest in these programs has grown exponentially since then. In the late 1970s and early 1980s less than 5% of employers had employee health promotion programs (1). The 1992 *National Survey of Worksite Health Promotion Program Activities* revealed that 81% of workplaces offered at least one health promotion initiative, compared to 66% at the time of the previous survey in 1985 (2). The majority of the companies with health promotion programs stated that health care cost containment and improved worker health and safety were the primary goals of their health promotion programs. It is unclear how many of these programs attained their stated goals. Indeed, only 12% of the companies surveyed reported that they conducted a formal program evaluation (3).

The purpose of this chapter is to present a broad overview of the concept and application of outcome-based worksite health promotion. The theory behind worksite health promotion is reviewed, including the definitions of health and health promotion and the theoretical frameworks that attempt to explain how people decide whether or not to participate in these programs. Worksite health promotion is defined and several program-planning models discussed.

Following this review of the basics, there is a discussion of outcome analysis as a means to evaluate the effectiveness of worksite programs. Several examples of successful health

---

Patty's Industrial Hygiene, Fifth Edition, Volume 3. Edited by Robert L. Harris.
ISBN 0-471-29753-4 © 2000 John Wiley & Sons, Inc.

promotion programs within industry are reviewed. The chapter concludes with an overview of the trends that will affect the delivery of health promotion services in the future and suggested approaches to address these challenges.

2 THEORY OF HEALTH PROMOTION

The best way for a company to institute an employee health promotion program is to develop a clear concept of health at the workplace and then to operationalize this concept throughout the program planning, implementation and evaluation (4). Interestingly, the concept of health has changed a great deal through the years.

2.1 Definition of Health

Prior to the 1970s, the sciences of public health and medicine described health as the absence of morbidity and mortality. In 1974, the World Health Organization (WHO) proposed a different definition of health. WHO conceptualized health as a state of complete physical, mental and social well being. Ardell coined a new term for this concept of health in his landmark work, *High Level Wellness*. Ardell described wellness as "a way of life which you design to enjoy the highest level of health and well-being possible during the years you have in life" (5). Pender further refined this conceptualization of health as the actualization of human potential through goal directed behavior, competent self-care and satisfying relationships (6).

2.2 Definition of Health Promotion

Within the context of this changing definition of health, the concept of health promotion has also evolved. Before the 1970s, programs were designed to prevent or halt disease through different levels of preventive actions (7). Health promotion is no longer thought of as simply taking actions to prevent illness. Rather, it is the science and art of helping people choose a lifestyle that moves them toward a state of optimal well being (8). Pender pointed out that prevention is reactive but health promotion is proactive (6). However, there is a danger in divorcing the concepts of wellness and disease avoidance. In reality, wellness and disease avoidance exist together, so that health promotion can be seen as any intervention that attempts to move people toward a state of optimal health, including traditional disease prevention efforts (9). O'Donnell conceptualized health as a continuum with premature death on one end and wellness, or optimal health on the other end. Individuals move toward wellness by using many strategies, including traditional medicine, health protection, knowledge, health behaviors, positive attitudes, and health promotion activities (10). Given that, effective health promotion programs must include a wide variety of strategies targeted to increase awareness, change lifestyle behaviors and provide supportive environments (11).

Unfortunately, even the most sophisticated and comprehensive workplace or community sponsored health promotion programs may have limited success. To be most effective, programs must take into account the body of knowledge that describes how humans decide whether or not to engage in health protecting and health promoting behaviors.

2.3 Theoretical Frameworks

For many decades researchers have developed and studied theoretical models to explain why some people engage in health promotion and preventive behaviors and others do not. Many theories have been derived from psychological decision-making theories, which attempt to explain action taken in situations of choice.

One psychological decision-making theory, behavior-motivation, was pioneered by Lewin who visualized the individual as living in a life space composed of positively and negatively charged regions (12). Illness was viewed as negatively valenced, exerting a force to move individuals away from its region and toward the more positively valenced region of health.

Further, Lewin hypothesized that behavior depended on two factors: the value placed by an individual on a particular outcome and his estimation of the likelihood that a given action will result in the outcome. Although the individual aspires to attain movement to positively valenced regions, he/she will choose the route that he/she perceives will be the most likely to succeed. Thus, the individual's perceptions, and the many variables that affect those perceptions, play an important role in the decision to act. This theory forms the basis of The Health Belief Model, a popular model used to explain health-protecting behaviors.

The Health Belief Model was originally developed to explain preventive health behavior (13, 14). It has since been amended by Becker and has been used to explain compliance with medical treatment regimens (15). Consistent with Lewin's theory, the Health Belief Model postulates that the likelihood that an individual will perform a health-related behavior depends on the individual's perceptions, factors that modify those perceptions, and variables, that, when taken together affect the likelihood of taking action. In addition, a cue or stimulus is needed to prompt the individual to take action.

The individual perceptions identified in this model include estimates of susceptibility to the illness/condition, perceived threat, the estimated benefit of the health action and the expected barriers of taking action. The actual likelihood of taking action, then, is the result of the perceived benefits weighed against the perceived barriers. It is hypothesized that these individual perceptions are modified by a variety of demographic, sociopsychological, and structural variables.

Pender developed the Health Promotion Model as a complement to the Health Belief Model (6). The Health Promotion Model shares the same structure and some of the variables of the Health Belief Model. However, it differs from the model in some important respects. The Health Promotion Model focuses on the individual's movement toward a state of well being, rather than focusing on behavior that decreases the probability of encountering illness. Given this, the model includes variables regarding the individual's perception of health, which are not included in the Health Belief Model.

Like the Health Belief Model, the Health Promotion Model suggests that determinants of health-promoting behavior are categorized into cognitive/perceptual factors, modifying factors and variables affecting the likelihood to act. Within this framework, modifying factors affect health-promoting behaviors by their influence on the cognitive-perceptual factors. These include demographics, biological characteristics, interpersonal influences, situational factors and behavioral factors. In addition, the likelihood to take action depends on internal or external cues.

Within the context of these decision-making theories, health promotion program planners can influence an individual's decision to change lifestyle behaviors. First of all, planners can promote lifestyle changes by influencing the cognitive-perceptual factors, such as providing information aimed at increasing the awareness of the threat of illness or the benefit of wellness and decreasing barriers to success. Secondly, program planners can provide cues to action in the form of incentives. This involves finding ways to motivate individuals to act. Different incentives motivate individuals in a variety of ways. Examples of incentives include cash rewards, benefits, rebates, and personal recognition (16).

There are several other social-learning theories that propose that peoples' thoughts have a strong effect on their health behaviors. The Model of Health Promotion Behavior states that self-efficacy beliefs play a central role regarding health beliefs and behavior (8). Self-efficacy is defined as a person's conviction that he can successfully execute the behavior that is required to produce the outcome. Self-efficacy is affected by four sources of information: performance accomplishments, vicarious experiences, verbal persuasion and emotional arousal (17). It is easy to understand the importance of strategies to increase perceived self-efficacy in smoking-cessation programs, where participants may have experienced one or more previous failures at attempts to stop smoking.

The work by Bandura on self-efficacy has been utilized by Prochaska and others as part of the Transtheoretical Model. There are several major dimensions to this model, including the stages of change. These stages are ordered categories along a continuum of motivational readiness to change a problem behavior—precontemplation, contemplation, preparation, action and maintenance. Intervening or outcome variables in this model include self-efficacy, decisional balance (the pros and cons of change), situational temptations, and behaviors specific to the problem area. Transitions between the stages of change are affected by a set of ten independent variables known as the processes of change (18). These include such variables as consciousness raising, helping relationships, and reinforcement management.

The Theory of Reasoned Action proposes that intentions are the immediate determinants of behavior (19). The individual's attitude and subjective norms control intentions. Attitudes are the result of one's evaluation of the consequences of an action or inaction. Subjective norms are the individual's perceptions of what others think he/she should or should not do. According to this theory, health promotion program planners can affect a change in the individual's attitudes through awareness programming, but must also attempt to address subjective norms through a supportive environment.

No single theoretical model successfully predicts human behavior in situations of choice. However, knowledge and understanding of behavioral theories contributes to the program planner's worldview and guides the development of program goals as well as the interventions used to reach those goals. Theory-based health promotion programs add to the body of knowledge and in turn guide future program development.

3 WORK-SITE HEALTH PROMOTION PROGRAMS

Health promotion within the context of a workplace can best be understood as an organized program intended to assist employees and their families in making voluntary behavior changes to reduce their health risks and enhance their productivity (16). It is the maintenance and enhancement of existing levels of health through the implementation of effective programs, services and policies (20).

Effective programs take into consideration the idea that worker health is affected not only by his own individual attitudes and actions regarding health, but is also affected by the characteristics of the work environment itself and the community in which the work environment exists. Researchers have called this idea an ecological framework of worksite health promotion (21).

3.1 Ecology of Health Promotion

The ecology of health promotion looks at the issue from a larger, broader perspective. Social ecology studies the interaction between people and their environment (22). In the case of health promotion, the interaction between people in the workplace and the various social and environmental factors involved in health promotion is studied. There may also be factors outside the workplace that may influence the interactions. Among the personal factors that affect the individual's health and the interaction are genetic makeup, psychology, and behaviors. The workplace factors are all the things that affect the workplace health and safety climate as well as the effects that outside social contacts have on the workplace health and safety climate.

The ecological approach emphasizes a multidisciplinary and multilevel approach to workplace health and safety involving physiology, psychology, interpersonal interactions, organizational effects, and community effects. This approach takes into account the whole range of interactions, which leads to a comprehensive health promotion program addressing the variability of the workforce. The ecological approach is also more likely to lead to closer integration with community health programs and support services.

A health promotion program must take into consideration the culture of a workplace and social context of the employees. In 1993 Glasgow and co-workers made recommendations about program design, organizational size and climate, worker activities, making programs fit the company, and program promotion (Table 48.1) (23). These recommendations were based upon a review of the literature on health promotion programs. Other reviews provide a framework to begin the discussions necessary to design an effective health promotion program (9, 24).

A study of low long-term participation in employee fitness programs in a group of 236 employees was designed to look at what influenced participation in health promotion activities (25). Determinants of adherence such as attitude toward health factors, support from partners, co-workers, supervisors and issues such as shift, time, muscle soreness were tested and found to influence long term commitment to changing health behaviors. It was suggested that a program needs to focus on overcoming barriers to participation, as well as reinforcing positive attitudes and social supports. The ecological approach recognizes the commonality of themes or interests between health promotion programs and occupa-

Table 48.1. Recommendations on Program Design[a]

Variable	Actions
Program design	1. Offer employees options 2. Offer incentives to participate 3. Consider continuing, multi-risk factor programs 4. Make programs accessible (on-site) and convenient (work hours)
Company/Employee variables	1. Place greater emphasis on recruiting small businesses 2. Identify and work with organizational climate aspects that facilitate or impede participation 3. Include screenings, health risk appraisals to increase employee awareness of their health risks
Company/Program fit	1. Establish employee steering committees to guide and tailor interventions 2. Use surveys, focus groups, peer counselors to maximize employee input and involvement 3. Implement programs in ways consistent with management philosophy
Promotion and recruitment	1. Publicize program using top management, supervisors, and union leaders 2. Use repeated advertising and multiple channels 3. Target barriers to participation among low participation employee subgroups

[a]Reprinted with permission of R. E. Glascow, K. B. McCaul and K. J. Fisher

tional health and safety programs in the workplace (26). It encourages the working together of practitioners in different disciplines and integration of programs into a comprehensive effort to define and achieve common goals and increase the effectiveness of the programs (21). The ecological approach requires strategic planning that includes plans for evaluating the effectiveness of the health promotion activities (27). A team composed of occupational health nurses, human resources, health, safety and environment, management, physicians and external resources, should complete this strategic planning. A team representing these varying areas of expertise leads to a program reflecting the inputs of human behavior, organizational values and the experience of differing approaches to health promotion programs. The following section discusses several approaches for comprehensive strategic planning.

3.2 Program Designs

It is apparent from the review of the definitions of health and health promotion and the theory explaining health-promoting behaviors that no single approach to worksite health promotion will be effective for all individuals or groups of individuals. It is imperative that program planners follow a system of assessment, planning, implementation and evaluation to design a program to meet the unique needs of their workforces.

Several researchers and health promotion specialists have proposed program models to aid in the systematic planning of effective programs. Green and Kreuter suggest that lifestyle changes require a multifaceted approach that pays attention to cultural, socioeconomic, environmental, regulatory and organizational issues (28). They proposed the PRECEDE-PROCEED model to planning programs within this ecological framework. In the PRECEDE phases, the planners assess the multiple factors that shape health and focus them into specific program objectives and plans for subsequent program evaluation. The PROCEED phases involve the actual implementation of the program, followed by evaluation of process, impact and outcome.

The Wellness Councils of America propose that the success of a worksite health promotion program is directly related to how it fits within the corporate culture (16). They suggest an integrated business approach to program planning. This approach involves forming a business plan that begins with the formation of program goals that support the business goals of the company. The next step is the development of an implementation plan and timeline, budget, promotional strategy, and formal evaluation plan.

Rogers takes this business approach one step further in the Health Promotion Program Plan (Fig. 48.1) (4). This program design incorporates four phases; assessment, planning, implementation, and evaluation. In the assessment phase, program planners work to obtain management commitment, use a variety of tools to understand the health needs and interests of the workforce and identify the existing health promotion resources.

In the planning phase, the planners from a committee or work group of interested managers and employees. This group begins by defining program goals and objectives. The most commonly cited goals of worksite health promotion programs are to improve the health and safety of employees and to decrease the cost of providing health care to employees (4). Potential benefits to employers include improved worker productivity, a reduction in medical benefits costs, a reduction in workers' compensation costs and enhanced company image (29). Unfortunately, these benefits may be difficult to measure. However, participation rates, reductions in health risks, rates of recordable injuries, and changes in medical service utilization can be convincing statistics if tied to program goals (9). Therefore, it is imperative that the planning committee identify realistic goals and define how the attainment of these goals can and will be measured. During the planning phase, the work group also identifies further needed resources, develops the program strategies, marketing strategies, budget, and timetable.

In the implementation phase, the plans are put into action. Program content is aimed at increasing employee awareness of health, encouraging behavior change, and developing a supportive work and community environment. In the fourth and final phase, evaluation of how the program met the designated objectives is completed. This includes an evaluation of the process, its immediate impact and its long-term effects on the health of the workforce. Data obtained during program evaluation, in turn, are used to revise the program.

Evaluation is such an important part of the program that it must begin in the planning stage. Program planners must identify how the program outcomes will be measured and evaluated at the outset of the program. Unfortunately, program evaluation is often forgotten or completed in only a rudimentary way. The next section of the chapter explores various methods of outcome analysis as a means of program evaluation.

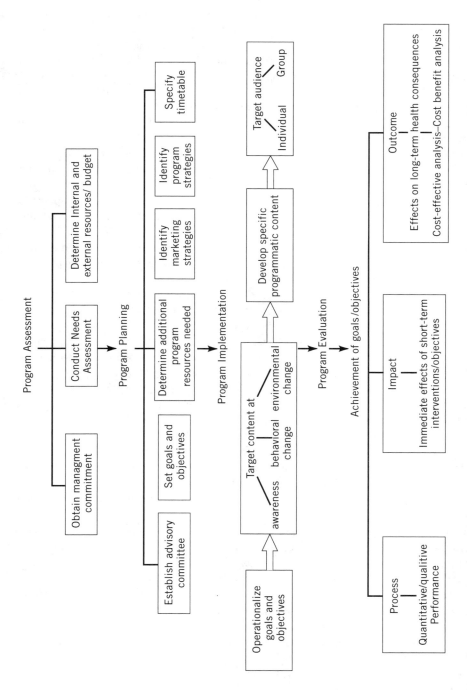

Figure 48.1. Health Promotion Program Model, reprinted with permission from B. Rogers, Ref. 4. and the W.B. Saunders Company.

4 OUTCOME ANALYSIS

In the initial period of development of worksite based health promotion activities, the indicators used to measure success were those that demonstrated change in the health measurement of the individuals who participated in the program over relatively brief periods of time such as one or two years. With increasing experience and the development of models such as those outlined earlier, demonstration of effectiveness and dissemination has also become much more complex. Programs themselves have tended to become more comprehensive and broadly based, rather than narrowly focussed on a single risk factor or measured result. In addition, it has become clearer that a reduction in health risks can produce a reduction in benefit costs. It remains easier to demonstrate changes in health benefits costs, however, than to demonstrate changes in health outcomes. At present, there is no single study that addresses all the issues. The primary focus of much of the research has been to evaluate the short-term efficacy of worksite programs among active participants.

In internal evaluation of programs, structural, process and outcome indicators are used. Important structural elements include adequacy of staff skills for the program elements, demographics of participants as contrasted with nonparticipants, the match of the program with the mission and goals of the organization and the commitment of management to wellness and support of the program (29). Current research continues to demonstrate that the demographics of participation are the most difficult to affect. Some programs, such as that of the City of Birmingham, require participation in a health appraisal as a condition of the receipt of health benefits, so at least the baseline is known in some regards, and measurement of the whole population can occur (30). When characteristics of both the numerator and denominator are known, then program penetration by risk group classification can be assessed. This issue is particularly important in the design of programs for smoking cessation and nutritional impact.

The term "process" directs attention to the way in which services are offered and delivered. Attention to process involves evidence of a method of assessment to assure that the specific activities in the programs are appropriate for the employee groups that are being targeted, and methods of monitoring for needed changes in the program are in place.

The major contribution of health promotion programs comes from evidence that health risks or costs have been reduced by the program elements; i.e., that the outcome or result has changed. The program must demonstrate that the outcome of an intervention has changed commensurate with the cost in time and resources that have been required. A single list of outcomes on health promotion by Lusk from a recent review article can be categorized into six groups (Table 48.2) (24).

The first three categories—risk awareness, reduction, and identification—are indicators of the personal health status and beliefs of individuals, and the latter three—work productivity, cost control, and cost reduction—are reflections of that behavioral change in the organization. O'Donnell has characterized the two general categories as (*1*) health promotion and (*2*) demand management (31). Programs may evaluate just the alteration in personal indicators, and infer that alterations have occurred in economic parameters, but the greatest strength for continued program support comes with outcome assessment that combines both.

Table 48.2. Outcomes Used in Health Promotion Programs

Risk Awareness	Work Productivity
Infectious disease, especially HIV/AIDS	Absenteeism
Alcohol	Productivity
Cognitive changes in regard to nutrition	Tardiness
Fitness facility membership	Turnover
Behavior scores	
	Cost Control
Risk Reduction	Procedure utilization
Blood pressure control	Frequency of medical care visits
Cholesterol levels	Hospitalizations
Morbidity/mortality rates	Program costs
Seat belt use	Workers' Compensation claims
Smoke free environments	
	Cost Reduction
Risk Identification	Benefit to cost ratio
Clinical breast exam	Disability/illness
Breast feeding duration and rates	Medical/health care claims/costs
Dietary intake	
Exercise/aerobic capacity	
Injuries	
Job satisfaction	
Mammography intention rates	
Smoking status	
Speeding	
Stress levels/psychological state	
Body weight/fat	
Membership in a high risk group Health risk	

Pelletier has regularly reviewed comprehensive health and cost-effective outcome studies, most recently in 1997 (32). This review identified only 12 studies that met the defined criteria of a randomized, control trial and examined the clinical and/or cost impact multifactorial programs. All of the programs included an initial medical screen or a variety of Health Risk Analyses (HRA), and often included physiological measures. These items were then repeated at intervals to evaluate the clinical and/or cost impact of the program. Most programs provided only generic information on risk behaviors, but did offer opportunities to learn and practice new skills. Most programs also incorporated aspects of organizational policy, e.g., smokeless workplace, on-site exercise facilities. They varied in the extent to which they focused on the entire workplace or offered intensive interventions for high-risk individuals. An additional 77 studies that appeared in a series of reviews in *the American Journal of Health Promotion* were also utilized. Conclusions from this review included the following:

1. Worksite health promotion interventions that are more intensive and include individualized follow-up with behavioral counseling are likely to be most effective in reducing cardiovascular risks.
2. The administration of Health Risk Appraisal (HRA) alone results in little, if any, sustained risk reduction.
3. Sustained and repeated strategies are required to assist employees in sustaining their initial risk factor reductions.
4. Populations to which the findings are generalizable are limited. None of the cardiovascular interventions, and few of the general worksite programs have been implemented with dependents or retirees, and none considered differential responses from the "working poor" or racial or ethnic subpopulations in the workplace.
5. There is a profound lack of standardization of what constitutes cost or benefits in interventions and their subsequent evaluations.
6. Programs must be of sufficient duration to demonstrate clinical and/or cost outcomes. Results strongly suggest that a program must be sustained for a minimum of one year to bring about risk reductions among employees and three to five years to demonstrate cost effectiveness.

Not all programs have strong statistical analysis available, but all must include evaluation. O'Donnell, a researcher in this field who has also designed and administered programs, suggests the following realistic minimums:

1. Participation and attendance records to assess penetration to selected work groups.
2. Annual health assessment of the organization.
3. Periodic employee surveys concerning:
    a. Practices
    b. Desires
    c. Attitudes
4. Program specific pre and post health assessment for each behavior change program.
5. In-depth assessments of special programs. (33)

Over the thirty years during which health promotion has been in development, the thoughtful work of many has led to a variety of designs that are now available and from which much has been learned. The cycles of planning, implementation and evaluation which have led to successful programs that result in changes in the lives of individuals have led in turn to broader social benefits such as increasing numbers of smoke-free worksites and greater social support for positive health behaviors.

4.1 Comprehensive Models

4.1.1 Live for Life

One of the now classic programs is the Johnson & Johnson LIVE FOR LIFE Program®, a comprehensive, multifaceted worksite health promotion/disease prevention program that has been very successful in the home corporation and is still extensively utilized. Six main program components, primarily offered on company time, illustrate current principles of health promotion:

1. Health screening including questionnaires, biologic measurement, and risk assessment.
2. Seminars to discuss testing results and promote awareness of company programs for change.
3. Educational programs offered in a variety of formats concerning management of the basic risks: tobacco cessation, weight control, stress management, physical activity, and high blood pressure.
4. Shorter educational and health promotion programs including the above plus breast self-exam, biofeedback, and carbon monoxide analysis for smokers.
5. Participation incentives in a variety of modes: promotion by company leaders, incentives and prizes, permanent program administrators, organized information feedback to company sponsors.
6. Attention to environmental support: a wide variety of services including appropriate exercise facilities, cafeteria nutritional information, referral systems for troubled employees, flextime, carpooling, health fairs, etc.

A variety of publications have reported on the evaluation of effectiveness for this comprehensive program (34–37). Although appropriate critiques have been made concerning some of the study designs, positive effects on smoking cessation, organizational commitment, physical fitness, weight reduction, blood pressure, absenteeism, seat belt use, and reductions in health care costs and utilization have been documented (2, 4). When Duke University contracted with Johnson & Johnson Health Management for the LIVE FOR LIFE® Program, a significant decrease in absenteeism was noted among hourly employees who volunteered to participate in the program at any time during the three year introductory period, with greater reduction as the program progressed (38).

A critique of the data analysis used by the LIVE FOR LIFE® program led Kingery's group at Texas A&M University to combine a health promotion program with analysis of the risk of high claims cost within four gender-specific groups (39). Five risk factors were examined: cholesterol, blood pressure, cardiovascular fitness, body fat, and smoking status. Screened employees who released their claims ($n = 367$) were examined against a random sample of employees ($n = 587$). Linear regression analysis was then used to determine the risk of having high claims costs across the four groups. A formula was then applied to determine that more than 43% of the cost of medical claims was associated with elevated risk. This and other studies have directed attention to the application of intensive programs for high-risk populations.

4.1.2 Take Heart

The Take Heart program illustrates a continuous low-cost, low-intensity health promotion intervention following a public health model. This intervention relied heavily on one of the recommended intervention strategies, that of the formation of an employee steering committee in each of 26 relatively diverse worksites. This steering committee played an active role in the tailoring of the program to their worksite (40). A second primary component was a "menu" approach to the components offered. Worksites were assessed on a number of worksite organizational characteristics and employee behaviors. Participants attended assessments on work time, which included questionnaires and a fingerstick total cholesterol, and later a HDL cholesterol. Immediate feedback was provided. Questions from the assessment included:

1. Items regarding perceived support from supervisors and coworkers for tobacco and dietary-related behavior change
2. Stage of change in these areas.
3. Attempts to reduce fat intake or quit smoking in the previous year.
4. Intent to limit fat or quit smoking.
5. Current level of tobacco use.
6. Exercise frequency, and
7. Demographic information.

Additional questionnaires estimated the mean grams of fat per day and high fat eating patterns. Participation rates varied from 23 to 83% (mean = 58% in the 1993 study initiation)

Additional elements included:

1. Employee steering committees to help tailor the interventions to the culture of the particular worksite. Members of the steering committees were trained and provided with written guidelines.
2. Activities were developed by a menu matrix. For both tobacco and nutrition, four classes of activity (motivational/incentive, educational/skills training, policy/workplace environment, and maintenance) were offered.
3. Emphasis was placed on the use of community health organizations such as the American Heart Association and the American Cancer Society and on networking among participating worksites.
4. Biometric assessments and questionnaires were repeated after two years.

Cohort outcome analysis showed a somewhat greater decrease in smoking among intervention worksites (31 vs. 19% cessation), but this was not statistically significant because of the small sample size per worksite and the associated variability across worksites. The strongest results were found in dietary outcomes. All dietary measures revealed greater reductions among intervention than control worksites, but the measures were those of self-report. Unfortunately serum cholesterol, exercise, and overall cardiovascular heart disease risk did not reveal significant changes. The process variable of stage of change and perceived support for healthy behaviors revealed greater improvement among intervention than control worksites. The results suggest that continuous, low-cost, low intensity health promotion interventions delivered at the worksite following a public health model can be effective in producing beneficial changes in measures of employee dietary fat intake and dietary habits.

4.1.3 WellWorks

A different approach was used in the WellWorks worksite cancer prevention project, a randomized controlled study of an integrated health promotion/health protection intervention. This project was part of the Working Well Cooperative Agreement, a five year project funded by the National Cancer Institute, which tested the effectiveness of worksite health promotion interventions by achieving individual and organizational changes to reduce cancer risk. Working Well was a randomized, prospective field experiment with 114 worksites and over 360,000 surveyed workers. The WellWorks component assessed the effectiveness of a model integrating the two components of integrated health promotion and health protection for exposure to workplace hazards. As in the overall project, the WellWorks intervention model applied three key elements:

1. Joint worker and management participation in program planning and implementation.
2. Consultation on worksite changes.
3. Coordinated programs targeting health behavior change (41).

Despite extensive efforts, blue-collar workers were less likely to participate than white-collar workers. In the project, however, controlling for job status and gender, workers who participated in classes or presentations about reducing exposure to harmful substances were significantly more likely also to have participated in nutrition education activities than workers who did not participate in activities related to workplace exposures. In addition, workers who reported that their employers made changes to reduce exposures perceived as harmful in the workplace were significantly more likely to have participated in both smoking control and nutrition activities, compared to workers not reporting that management had made such changes. These data were cross sectional and correlational in nature, and there was no design that randomly assigned workers to health promotion alone versus the addition of health protection programs. However, the WellWorks study suggests that integrated programs can affect participation in health-related activities.

4.1.4 Health Check

Health Check, a workplace health promotion program in The Procter and Gamble Company (P&G), was able to demonstrate that high-risk screening and intensive one-on-one counseling results in lower total and lifestyle related health care costs (42). A retrospective cross-sectional study compared two groups of active P&G employees who were eligible for the company's medical benefits plan. The size of the total cohort was 8,334 individuals. The first group voluntarily completed a Health Check health profile questionnaire and participated in follow-up high risk interventions while the second group completed the Health Check but did not participate in any aspect of the worksite health program.

After completion of questionnaires and biometric measures, participants received individualized reports outlining their health status in relation to specific risk areas. Individuals were then identified to be high risk if they displayed any of the following risk factors: elevated blood pressure, elevated cholesterol, cigarette smoking, lack of activity, poor diet, stress, and multiple risks. Participants deemed to be at high risk were provided one-to-one counseling and behavior change support, with quarterly follow-ups and design of a health improvement action plan. Ancillary health support programs for all participants included (a) fitness flex time, (b) after-work on-site aerobics, (c) noon aerobics, (d) diet/weight management programs, (e) cholesterol and blood pressure education, (f) smoking cessation programs, (g) brown-bag educational programs, (h) annual mammography screening, and (i) exercise incentives as participation prizes.

Medical costs and participation were tracked for three years. Participants and nonparticipants began at about the same baseline costs, but over time, costs for participants grew at a much slower pace than costs of nonparticipants. A subset of costs, those specifically linked to potential lifestyle-related diagnoses, were even lower for participants in year three when compared to total costs. Although it was not clear how much of the effect was due to specific intervention with high-risk individuals, the data suggests support for graded intervention offerings.

The four programs cited are samples of the thoughtful work being done in health promotion and demand management. Theoretical and practical models are becoming clearer, and support in general for worksite based programs remains strong. O'Donnell, et al, in a recent survey of 26 health promotion programs, identified the following as characteristics of the "Best Practice" programs:

1. Program plans linked to organizational business objectives.
2. Effective communication programs.
3. Effective incentive programs.
4. Presence of an evaluation component:
   a. Systematic evaluation.
   b. Evaluation results shared with top management and employees.
   c. Top management values evaluation results.
5. Strong efforts to create a supportive environment.
6. Strong top management support (33).

The four programs illustrate, in different ways and with different histories, those major characteristics.

5 FUTURE TRENDS

Thus far, this chapter has discussed what is known about worksite health promotion programs today. The ever-changing workplace will demand new approaches to meet the needs of employers, employees and their families in the future.

Within the next decade the demographics of the American workforce will be very different than today. The number of women in the workforce will increase as well as the number of aging workers. There will be continued growth in the hiring of workers from various ethnic groups. The percentage of disabled, part-time, and contingent workers will increase. Workers will be more mobile and live farther away from work, resulting in longer commutes, increased fatigue and overall stress (29).

The workplace itself will change in the future. The number of manufacturing jobs will decrease with a concomitant increase in service-sector jobs. An increase in technology will result in a more geographically dispersed workforce as well as a workforce that expects and demands high-tech, multi media presentations of information. The globalization of the economy will result in continued work redesign, corporate downsizing and the use of temporary service personnel. The spillover of community problems such as substance abuse and violence will continue to affect the worksite. There will be continued shifting of health care costs to employees and changes in third party reimbursement for health care services. Specifically, the trend toward capitated managed care and away from fee-for-service reimbursement will continue (21, 29, 32). These and other trends will challenge program planners to develop innovative approaches to the delivery of health promotion services.

5.1 Diversity in the Workplace

The workplace is increasingly diverse in terms of gender, age, ethnic background, race, economic status, blue-collar vs. white-collar and other factors. This diversity adds to the richness of the workplace as people from different cultures and social backgrounds work together to achieve company production or service goals. However, this richness presents a challenge to health promotion planners. Educational, linguistic, social and cultural barriers may exist for individual workers or groups of workers. These barriers need to be identified and addressed in order to design health promotion programs that will reach out to all workers.

Understanding workplace diversity is critical in designing programs that will be effective in not only communicating to the targeted work group but also in influencing permanent changes in life-styles. A cultural assessment should be included as part of future health promotion program planning. Lowenstein and Glanville presented a model for assessing and intervening in intercultural and racial conflict (43). The model suggests that both individual and organizational characteristics need to be assessed. The culture of the workplace and the social climate of employees strongly influences their participation and long term commitment to changing their lifestyle and health behaviors.

A number of organizational factors affecting employee participation in health promotion activities at Federal worksites have been analyzed (44). The variables were evaluated using focus groups and individual interviews. It was found that employees were more likely to participate when there was strong co-worker support. A more comprehensive program structure, increased marketing, and time off to encourage participation or on-site facilities led to more participation by lower rank employees and by minority employees. These results are supported by the work of Sorenson et al. in looking at participation of blue-collar workers as a part of the WellWorks worksite cancer prevention project as was discussed in section 4.1.3 (41). It was found that blue-collar workers were less likely than white-collar workers to report participation in health promotion activities. When management made changes to reduce the exposure of workers to occupational hazards, blue-collar workers were more likely to participate in nutrition and smoking control programs. There needs to be visible management support for a comprehensive program of health risk reduction. A health and safety climate supporting changed behavior must exist before improved health behaviors can be successfully initiated and sustained.

5.1.1 Gender

Gender is a significant diversity issue that must be considered in health promotion. A study of women in a variety of manufacturing environments focused on the participation of women in workplace health promotion (45). The results were based on the analysis of focus group discussions of the reasons women in low, medium, or high health risk categories participated or did not participate. The low-risk women were younger, in good health, and took responsibility for their health. The workplace health promotion programs had little influence on their lifestyles. Women in the medium-risk group were most susceptible to negative workplace influence on their health behaviors. They were most easily discouraged and less likely to continue participation or make permanent lifestyle changes. Women in the high-risk group were most dependent on workplace health promotion to support changes in lifestyle and healthy behaviors. It would be most desirable to have all the groups participate. The movement of individuals from one risk group to another with time is unknown. Ideally, there would be a movement toward lower risk health behaviors, with the low-risk group having existing healthy behaviors supported by the workplace health and safety climate. In general, employees with higher levels of education, younger employees and women correlate with higher health promotion participation, while men are more likely to participate in physical fitness activities.

The generalities mentioned above were examined in a study considering effects of gender, race and level of education on definitions of health, value of health, health promoting behaviors, and appraisal of problem solving in stressful situations in a group of 331 employees (46). Gender was found to have the major effect on health promoting behaviors, such as relaxation and exercise. Men exercised more, but women engaged in more relaxation and health promoting behaviors. Women were observed to value health more than men were, however, no differences in the definition of health were noted. Women and black men felt less personal control in solving problems in stressful situations. No effect of the level of formal education was noted on problem solving, although all the participants were, at minimum, high school graduates.

5.1.2 Ethnicity

Every culture has a system of beliefs and practices that reflect its worldview, including its view of health and illness (47). In addition, socioeconomic status and the major causes of morbidity and mortality differ among cultural groups. African-Americans, for example, are disproportionately represented in the lower socioeconomic groups in the United States. High risk health behaviors, such as smoking, poor nutrition, and lack of physical activity are over-represented in this ethnic group (48). A major concern of this group is stress. Black men in particular report that they feel less control in stressful situations (8).

African-Americans have a strong sense of community and dislike the "blame the individual" approach of some health promotion programs (48). This must be considered when planning health promotion programs for this group. Gary et al. studied the participation of African-American women in workplace health promotion programs and suggested that programmers look beyond the workplace to reach and influence the health of this group (49). The suggestion was made that program planners collaborate with community organizations and churches to influence African-American women to overcome the economic and social barriers to healthy behaviors. In those situations in which a health and safety climate is not supportive or a lack of trust exists, innovative ways of developing and sustaining support must be considered. The boundaries of health promotion programs will be an area for discussion as these programs are developed and community alliances may be necessary for success.

Program planners must also address the needs of Hispanic-Americans. It is estimated that by the year 2013, Hispanics will replace African-Americans as the largest minority group in the American workplace (29). Significant challenges to promoting healthy life behaviors for this group of workers exist. First of all, great diversity exists within this group. Hispanics may be urban Chicanos, rural Chicanos, Puerto Ricans or Cuban-Americans (47). Each ethnic subgroup may have its own beliefs regarding health and health promotion. There may also be language and educational barriers to program participation. A study of health promotion among Hispanic workers resulted in the recommendation to develop culturally and linguistically sensitive programs, resisting the easy adaptation of existing interventions (50). Programs must have goals and objectives targeting the specific needs of the workers and need to address knowledge, attitude and skills.

The National Heart, Lung, and Blood Institute (NHLBI), a division of the, National Institutes of Health, Department of Health and Human Services, recently developed a community-based health promotion program for Hispanics that can serve as a model for workplace programs. The NHLBI implemented a comprehensive Latino outreach program called "Salud para su Corazon" (51). Realizing that cardiovascular disease is the leading cause of death for Latinos, the NHLBI decided it was time to take action. The program uses Latino traditions to offer televised heart health messages, bilingual written educational materials, and strategies for implementing the program in various communities. The program has been so well received that the NLHBI is developing a similar culturally based program for African-Americans.

Many other ethnic groups are represented in workplaces. The variability of minority groups must be remembered and appreciated. People can be in multiple diverse groups and flow from one group to another depending upon the social climate of the workplace and the community in which it exists. The interaction or contact between the individual and program is an important determinant in the individual's long term commitment. Perhaps the best strategy to ensure that programs that will positively affect the health of all workers is to include members of ethnic groups on program planning committees and advisory boards.

5.1.3 *Other Workforce Variables*

In addition to the ethnicity of the workforce, health promotion program planners need to consider other workforce variables, such as the rising average age of workers, the increasing number of disabled workers and the presence of workers with varying educational levels. As the baby boom generation ages, the average age of the workforce will increase. Aging workers bring a different set of challenges to worksite health promotion providers. The incidence and prevalence of musculoskeletal problems and chronic diseases increase with age. Therefore, program planners need to focus their programs in these areas. Aging workers also experience a decrease in vision, hearing and agility. Effective programs must include strategies to assure that older workers can process information and handle the technologies. Suggested strategies include self-paced learning, training materials with high-contrast colors and bold typeface and the use of older trainers to teach older workers (29).

There have been examples of successful health promotion programs among work groups with low levels of formal education in the literature, particularly if the programs are carefully tailored to the target population. Fouad and co-workers identified barriers to a program for the control of hypertension in a group of 4,000 city workers (30). The barriers identified were low literacy, lack of treatment understanding, variability in health beliefs and priorities. An educational program was developed with sessions on a variety of health topics. These sessions were offered to hypertensive employees, 130 enrolled and 81 completed the program, 81 controls were also selected. The sessions were 30 to 40 minutes at lunchtime or immediately after work to enable convenient attendance. Group size was limited to 10 participants to maintain mutual support within small workgroups. Participants in the program showed a significant blood pressure decrease. The study showed that a program carefully tailored to a target audience could be successful and influence life-style behaviors.

Workforce diversity will continue to be a major challenge in health promotion activities and may be increasing as movement from less developed countries to the developed countries continues or increases (52). Health promotion programs will be involved with the challenge of communicating and influencing healthy behaviors while dealing with workplace diversity and the issues of interaction of minorities in the workplace (3, 53).

5.2 The Changing Work Environment

The workplace of today is in a constant state of change. The changing workplace can negatively impact the health of workers and make old models of health promotion programs obsolete. Specific issues that health promotion planners will need to address include the globalization of the economy, the increased use of technology in the workplace and the spillover of community mental health issues into the workplace.

5.2.1 *Globalization*

It is anticipated that the trend toward globalization of the world economy will continue to change the American workplace. The concern is raised that the current trends in industry toward globalization and resultant downsizing and the use of outsourced or contingent workers will reduce worker control or empowerment and negatively impact worker health. Johnson argues that social role or power is a major factor affecting health (54). Those in the upper class have the best health profiles while those in the working class or marginally employed the worst. The access to social resources such as power and control over aspects of daily life makes a significant difference. A group of Swedish workers with low social support and low workplace control studied over a 25-year period had a 170% excess risk of death from cardiovascular disease after adjustments were made for other factors. It has also been asserted that there is sufficient evidence to cause concern that working over 50 hours a week over long periods adversely affects health and safety of employees (55).

Outsourcing is currently an economic driver in industry (56). The primary reason cited is to improve efficiency and save money by obtaining the necessary expertise in specific areas. The majority of organizations surveyed in 1997 have outsourced business and support areas. The resulting increase in contingent workers in the workplace presents a special challenge to health promotion program planners. These employees share the same health concerns as company employees but may not be offered the same level of benefits, including health promotion programs.

5.2.2 Technology

The increased use of technology is changing the design of the workplace. Advanced telecommunications tools are making virtual offices, or office without walls more commonplace. This, combined with the global nature of many businesses, will result in the geographic dispersion of workers. Health promotion planners will no longer have large captive audiences of workers under one roof. Therefore, an important feature of a health promotion program will be the design of the communication system or the media component of the program (57).

Among the options that are available is the traditional model of one on one, patient to physician encounters. This approach has been used when the health promotion is an outgrowth or continuation of the periodic physical examination which employees receive. This does not take advantage of peer and social supports for behavior changes, and for a large workplace, could be prohibited by resource limitations.

Innovative use of telephone technologies is another communication method which has been successful. It has been used in combination with written communication tools as a resource to provide answers to employee questions and to provide regular support for lifestyle changes. The telephone has also been used as a means to survey the target work group (6, 13, 58).

Among the alternative communication tools are health risk appraisals. There are now over 50 different health risk appraisals available (59). These are available to be given in groups, one on one, or interactively by computer. The tradeoffs remain whether to administer in groups to reduce cost and increase peer group support or one on one to preserve privacy. The use of interactive computer systems can be a step forward by ensuring privacy, guiding individual health education and, if well designed, aiding in patient-physician education (60).

Health risk appraisals are also available on the World Wide Web enabling individuals to privately evaluate their health risk. These interactive tools are available at little or no cost to individuals who are motivated to review and possibly change their health behaviors. This individual approach has attraction for some individuals because it is private, there is no pressure to participate and it can be done according to the individual's schedule. The individual has total control. However, products available on the Internet should be used with caution since they are not regulated and the quality of programs may vary considerably.

A combination of techniques may be the most prudent course of action. Bell Atlantic designed a program to give a dispersed population access to the features of health promotion as if they were in a single locale (58). The features included a toll free phone number to receive a health risk appraisal, record exercise data, and receive on-line information from a nurse. A film library, faxable information, an employee assistance program and a network of fitness facilities were also available. A network of volunteers forms a core group that keeps the program going and provides employee support. The multiple media used for communication allows the diverse groups to participate and support by volunteers encourages co-worker commitment.

Despite the availability of electronic technologies, paper questionnaires and written information remains the method most used in health promotion programs described in the literature. The challenge for health promotion in the workplace will be to take advantage of all available tools, to make them part of workplace health promotion, and to empower, encourage and educate individuals to make wise use of their availability.

5.2.3 Preventive Mental Health Programs

Society as a whole continues to struggle with community health issues such as addictive behaviors, the increasing incidence of personal violence and prevalence of depressive illness. These problems spillover into the workplace and will continue to do so in the future.

Among the addictive behaviors, smoking looms as a significant issue for the workplace. As a result, smoking cessation programs have been a major contributor to studies and a major component of health promotion programs. The elimination of smoking would reduce many of the health concerns and costs companies encounter. A report of a follow-up study of smoking control programs in 11 communities had only a moderately hopeful result (61). A program of smoking control consultations and support services was provided to employers in 11 communities. Four years after the intervention, the communities were re-examined and the results compared with matching communities at which no intervention had taken place. The level of worksite smoking cessation programs was higher in the communities in which intervention activities occurred. However, the percentage of companies which had smoke-free policies was the same between the matched communities. This indicates the necessity of a continuous effort and the continuing support required by health promotion programs. Although smoking cessation may be the easiest to measure and yields the largest return on the health promotion investment, it is also the most difficult to accomplish and maintain over extended periods.

Despite these difficulties, there have been programs with documented results. For example, the Indiana Chamber of Commerce has encouraged the development and implementation of smoking control and smoking cessation programs in its member companies (62). These programs have contributed to significant decreases in the number of smokers in Indiana workplaces and decreased risk to the health of nonsmokers. In a study of over 20,000 predominantly blue-collar workers, educational level was found to be a major influencing factor in dietary and smoking habits and a strong predictor of willingness to give up smoking (63). A major deterrent to smoking cessation and healthy food selection was lack of support for healthy habits by co-workers and more importantly supervisors. One conclusion was that top management must visibly support healthy behaviors to have an effective program.

A major predictor of companies that may start active smoking programs is having an active health promotion program (64). It appears that if the health promotion program does not address smoking control initially, smoking will be addressed as the program progresses. Smoking remains a visible sign of high risk health behavior.

A comprehensive Employee Assistance Program (EAP) may best address smoking and other addictive behaviors, as well as other mental health concerns. Employee Assistance Programs address employee personal problems to reduce their impact on employee performance at work (65). The goals of employee assistance are to reduce employee stress, absenteeism, improve morale, decrease accidents, and illness. These are some of the same goals of a comprehensive health promotion program. As discussed in section 3.1 on ecology of health promotion, expanding the boundaries and integrating program goals can result in improvement in the narrower focus of the different areas within the comprehensive program (27). As companies become self-insured, econometric studies indicate that workplace alcohol treatment and prevention programs become more economically attractive (66). Unionized and larger workplaces are also more likely to offer worksite alcohol treatment programs.

Well designed employee assistance programs, wellness program, or comprehensive health promotion programs also address depressive disorders (67). Depressive disorders are a common form of mental illness. The length of disability, relapse rate and medical costs were greater for this disease than for any other disease used for comparison. At the First Chicago Corporation, a study was conducted comparing short-term disability data, medical plan costs, and Employee Assistance Program costs for depressive disorders with selected common chronic medical conditions. The average length of disability and the disability relapse rate was greater for depressive disorders than for the selected medical groups. Depressive disorders were also found to have the largest medical costs of all behavioral health diagnoses. The employee assistance programs, wellness program, or comprehensive health promotion programs addressing these diseases have been shown to significantly reduce the associated costs of treatment.

5.3 Future Direction of Health Promotion

As with other activities in nearly all organizations there will be continuing and possibly increasing pressure to justify, and if possible, reduce the costs of worksite health promotion programs. These pressures will increase the efforts to improve targeting and evaluate program effectiveness.

One way to increase efficiency may be to outsource part of or the entire program. This may be an effective strategy in organizations with very small health and safety staffs. However, steps must be taken to ensure that programs offer high quality service and useful program evaluation, remain credible among the workforce, provide cost containment and maintain state of the art technology. Without careful attention to the written agreements with outside vendors, loss of long term control of the program may occur.

With the ever increasing medical costs and possibly aided by the outsourcing movement, workplace health promotion programs will become broader, encompassing areas previously considered separate or part of employee assistance programs. Health promotion programs will also begin to reach beyond the boundaries of the workplace. These "outreach" efforts will first involve offering programs to employee families to reduce spiraling medical costs and to create an environment of positive reinforcement of healthy lifestyle behaviors for everyone tied to the organization or covered by the organization's medical insurance umbrella. Another driver of this process will be occupational health issues such as ergonomics and stress, which can have nonoccupational components in addition to occupational.

A second "outreach" effort will be to community organizations. In order to effectively communicate with and positively influence some groups, the support of community organizations will be necessary. This approach will be justified as another way to reduce the escalating medical costs.

6 CONCLUSION

From the foregoing discussion of health promotion programs it is apparent that no single strategy or approach to workplace health promotion will be successful in all settings. However, based upon the evidence presented in this review, development of a health promotion program requires the following steps and components. First of all, a preliminary plan based upon an understanding of the theoretical models, composition of the workforce, available resources and organizational goals must be developed. The preliminary plan must be presented to management to achieve a commitment to visible support and participation. Based upon this plan, and with management support, a cross-functional team representing the targeted workforce needs to design the wellness factors to be addressed, communication methods to be used and support mechanisms to be made available. An implementation plan must be prepared and evaluation tools developed. Finally, it is necessary to accept and analyze the information received from the evaluation tools and use it as feedback to adjust the program and continuously support and improve it. The on-going program should be multidisciplinary, multimedia and flexible enough to address the challenges of the future. Comprehensive health promotion programs that follow these strategies will be valued as part of the total health, safety, medical and environmental program of an organization. This will be understood to be good business and necessary to the survival of the company.

BIBLIOGRAPHY

1. M. P. O'Donnell and J. Harris, *Health Promotion in the Workplace*, Delmar Publishers, New York, 1994.
2. USDHHS, *National Survey of Worksite Health Promotion Activities*, Government Printing Office, Washington DC, 1992.

3. USDHHS, *Health Promotion Goes to Work: Programs with Impact*. Government Printing Office, Washington DC, 1993.
4. B. Rogers, *Occupational Health Nursing*, W. B. Saunders Company, Philadelphia, 1994.
5. D. B. Ardell, *High Level Wellness*, Rodale Press, Pennsylvania, 1977, p. 93.
6. N. J. Pender, *Health Promotion in Nursing Practice*, 2nd ed., Appleton & Lange, Norwalk, Connecticut, 1987.
7. H. Leavell and E. Clark, *Preventive Medicine for the Doctor in the Community*, McGraw Hill, New York, 1965.
8. M. P. O'Donnell, *Am. J. Health Promot.*, **3**(3), 5 (1989).
9. L. S. Saphire, *AAOHN J.* **43**(11), 570 (1995).
10. M. P. O'Donnell, *Am. J. Health Promot.* **1**, 4 (1986).
11. M. P. O'Donnell, *Am. J. Health Promot.* **1**, 6 (1986).
12. K. Lewin, *A Dynamic Theory of Personality: Selected Papers*, McGraw-Hill, New York, 1933.
13. I. M. Rosenstock, *Milbank Memorial Fund Quarterly*, **44**, 94 (1966).
14. I. M. Rosenstock, "Historical Origins of the Health Belief Model," in M. H. Becker, ed., *The Health Belief Model and Personal Health Behavior*, Charles B. Slack, New Jersey, 1974.
15. M. H. Becker, ed., *The Health Belief Model and Personal Health Behavior*, Charles B. Slack, New Jersey, 1974.
16. S. Wendel, ed., *Healthy, Wealthy, and Wise. Fundamental of Workplace Health Promotion*, Wellness Councils of America, Nebraska, 1995.
17. A. Bandura, *Psychological Rev.* **84**(2) 101 (1977).
18. J. O. Prochaska, C. C. DiClemente, and J. C. Norcross, *Am. Psychol.* **47**, 1102 (1992).
19. I. Ajzen, *Attitudes, Personality and Behavior*, Open University Press, Milton Keynes, 1988.
20. M. S. Goodstadt, R. I. Simpson, and P. O. Loranger, *Am. J. Health Promot.* **11**, 58 (1987).
21. E. Baker, B. A. Israel, and S. Schuman, *Health Educ. Q.* **23**, 175 (1996).
22. D. Stokols, K. R. Pelletier, and J. E. Fielding, *Health Educ. Q.* **23**, 137 (1996).
23. R. E. Glascow, K. B. McCaul, and K. J. Fisher, *Health Educ. Q.* **20**, 391 (1993).
24. S. L. Lusk, *Annu. Rev. Nurs. Res.* **15**, 187 (1987).
25. L. Lechner and H. De Vries, *JOEM* **37**, 429 (1995).
26. C. A. Heaney, and L. M. Goldenhar, *Health Educ. Q.* **23**, 133 (1996).
27. N. Smith, *Health Promotion* **2**, 48 (1997).
28. L. Green and M. Krueter, *Health Promotion Planning: An Educational and Environmental Approach*, Mayfield Publishing Company, Mountain View, CA, 1991.
29. M. K. Salazar, ed., *AAOHN Core Curriculum for Occupational Health Nursing*, W. B. Saunders Company, Philadelphia, PA, 1997.
30. M. N. Fouad, C. I. Kiefe, A. Bartolucci, N. M. Burst, V. Ulene, and M. R. Harvey, *Ethn. Dis.* **7**, 191 (1997).
31. M. P. O'Donnell, C. A. Bishop, and K. L. Kaplan, *Am. J. Health Promot.* **1**(1), 1 (1997).
32. K. R. Pelletier, *JOEM*, **39**(12), 1154 (1997).
33. M. P. O'Donnell, *Design of Workplace Health Promotion, Am. J. Health Promot.* Rochester Hills, MI, 1992.

34. J. Bly, R. C. Jones, and J. E. Richardson, *JAMA* **256**, 3255 (1986).
35. L. Breslow, J. Fielding, A. A. Herrman, and C. Wilbur, *Prev. Med.* **19**, 13 (1990).
36. J. Fielding, "Assessing Direct and Indirect Economic Benefits of Worksite Health Promotion: The LIVE FOR LIFE Experience," in J. Opatz, ed., *The Economic Impact of Worksite Health Promotion*, Human Kinetics Publishers, Champaign, IL, 1993.
37. C. S. Wilbur, "The Johnson & Johnson LIVE FOR LIFE® Program: Its Organization and Evaluation Plan," in M. Cataldo and T. Coates, eds., *Health and Industry: A Behavioral Medicine Perspective*, John Wiley & Sons, New York, 1986.
38. K. K. Knight, R. Z. Goetzel, J. E. Fielding, M. Eisen, G. W. Jackson, T. Y. Kahr, G. M. Kenny, S. W. Wade, and S. Duann, *JOM* **36**, 533 (1994).
39. P. M. Kingery, C. G. Ellsworth, B. S. Corbett, R. G. Bowden, and J. A. Brizzolana, *JOM* **36**, 1341 (1994).
40. R. E. Glasgow, J. R. Terborg, L. A. Strycker, S. M. Boles, and J. F. Hollis, *J. Behav. Med.* **20**(2), 143 (1997).
41. G. Sorensen, A. Stoddard, J. K. Ockene, M. K. Hunt, and R. Youngstrom, *Health Educ. Q.* **23**, 191 (1996).
42. R. Z. Goetzel, B. H. Jacobson, S. G. Aldana, K. Vardell, and L. Yee, *JOEM* **40**(4), 341 (1998).
43. A. Lowenstein and C. Glanville, *Nurs. Econ.* **13**, 203 (1995).
44. C. E. Crump, J. A. Earp, C. M. Kozma, and I. Hertz-Picciotto, *Health Educ. Q.* **23**, 204 (1996).
45. K. M. Emmons, L. Linnon, and D. Abrams, *Women Health Issues* **6**, 74 (1996).
46. G. M. Felton, M. A. Parsons, and M. G. Bartoas, *Public Health Nurs.* **16**, 361 (1997).
47. K. H. Kavanaugh and P. H. Kennedy, *Promoting Cultural Diversity Strategies for Health Care Providers*. Sage Publications, Newbury Park, CA, 1992.
48. USDHHS, *Strategies for Promoting Health for Specific Populations*, Government Printing Office, Washington, DC, 1981.
49. F. Gary, D. Campbell, and C. Serlin, *J. Florida Med. Assoc.* **83**, 489 (1996).
50. F. S. Mas, R. L. Papenfuss, and J. J. Guerrero, *J. Community Health* **22**, 361 (1997).
51. USDHHS, *Heartmemo*, (Summer 1998).
52. A. J. Marsella, *Scan. J. Work Environ Health*, **23**(suppl.4), 28 (1997).
53. L. Grensing-Pophal, *Nursing* **97**, 78 (1997).
54. J. V. Johnson, *Scan. J. Work Environ Health*, **23**(suppl.4), 23 (1997).
55. A. Spurgeon, J. M. Harrington, and G. L. Cooper, *Occup. Environ. Med.* **54**, 367 (1997).
56. R. Sunseri, *Hosp. Health Network* **71**, 54 (1997).
57. D. C. Helmer, L. D. Dunn, K. Eaton, C. Macedonio, and L. Lubritz, *AAOHN J.* **43**, 558 (1995).
58. M. Foran and L. C. Campanelli, *AAOHN J.* **43**, 564 (1995).
59. C. J. Turner, *AAOHN J.* **43**, 357 (1995).
60. S. Krishna, E. A. Balas, D. C. Spencer, J. Z. Griffin, and S. A. Boren, *J. Family Practice* **45**, 25 (1997).
61. R. E. Glasgow, G. S. Sorensen, C. Giffen, R. H. Shipley, K. Corbett, and W. Lynn, *Prev. Med.* **25**, 186 (1996).
62. G. Perry Jr., *Indiana Med.* **89**(2), 157 (1996).

63. J. Heimendinger, Z. Feng, K. Emmons, A. Stoddard, S. Kinne, L. Biener, G. Sorensen, D. Abrams, J. Varnes, and B. Boutwell, *Prev. Med.* **24**, 180 (1995).
64. G. Sorensen, R. E. Glasgow, M. Toper, and K. Corbett, *JOEM* **39**, 520 (1997).
65. M. K. Monfils, *AAOHN J.* **43**, 263 (1995).
66. D. S. Kenkel, *Journal Studies on Alcohol* **2**, 211 (1995).
67. J. Conti and W. N. Burton, *JOM* **36**, 983–988 (1994).

CHAPTER FORTY-NINE

Occupational Health and Safety Management Systems

Charles F. Redinger, CIH, MPA, Ph.D.
and Steven P. Levine, Ph.D., CIH

1 INTRODUCTION

As organizations become globally integrated, occupational health and safety (OHS) professionals are forced to focus on the strategic management of their organizations. In this environment, new approaches to OHS integration with business functions and performance measurement become critical. Compliance with varying national regulations is a challenging starting point. International consensus standards add an increasingly interesting dimension to this challenge. For OHS professionals to play a key role in their organizations, they need to understand existing management system approaches, know how to develop and implement such systems, and understand the conformity-assessment structures that affect these systems.

The objective of this Chapter is to provide an overview of necessary information to understanding OHS management system issues. Various aspects of formal occupational health and safety management systems (OHSMS) and conformity assessment structures are addressed. The topics contained here are important to industrial hygienists as increased emphasis is being placed on strategic management and "beyond compliance" strategies to reduce OHS injuries, illness, and fatalities.

1.1 The Search for Beyond-Compliance Strategies

Since the passage of the Occupational Safety and Health Act of 1970 (P.L. 91-596), the incidence rate of occupational fatalities and total injury/illness case rates have been reduced. In 1997 these reductions were 76 and 27% respectively (1). Even with these positive changes, the frequencies of occupational health and safety (OHS) fatality and injury/illness incidents, coupled with a stubbornly high and unchanging total lost-work-day case rate, continue to affect adversely the lives of millions of workers and present a substantial burden on the cost of health care in the United States. This was confirmed in a comprehensive study which, among other things, found that approximately 6,500 job-related deaths from injuries, 60,300 deaths from disease, and 862,200 illnesses are estimated to occur annually in the American work force. The total direct and indirect costs are estimated to be $171 billion, of which, injuries cost $145 billion and illnesses $26 billion (2).

With the goal of reducing these rates and associated costs, strategies for augmenting traditional command-and-control regulatory approaches are explored on an ongoing basis. One such approach is the application of systems models to OHS management. The attention given to OHS management system (OHSMS) approaches in the late 1990s derived from three events. The first was the development by the International Organization for Standardization (ISO) of a quality-assurance-system model (ISO 9001) and an environmental-management-system model, ISO 14001 (3, 4). The second was the Occupational Safety and Health Administration's (OSHA) development and refinement of the Voluntary Protection Programs (VPP) and related programs based on the agency's Safety and Health Program Management Guidelines (5, 6). Finally, public and private organizations reported that the implementation of OHSMSs resulted in improvements in workplace health and safety while simultaneously achieving cost reductions (7, 8).

Numerous governmental and nongovernmental bodies throughout the world have of developed OHSMS standards or are considering such development. Within ISO the development of an OHSMS standard was considered in late 1996, but not pursued. At the nation-state level several countries, including the United Kingdom, Australia, Japan, Spain, Norway, Jamaica, and others are developing or have developed OHSMS national standards. In the United States, rule making for a comprehensive OHS program standard was initiated by OSHA in 1996 (9). Furthermore, in 1995, researchers at the University of Michigan (UM) developed an ISO 9001-based OHSMS that has been published by the American Industrial Hygiene Association (10). The UM/AIHA OHSMS has received significant attention from various stakeholders and standards-making organizations (11, 12). In addition to these activities, numerous custom management system approaches can be identified in private and public organizations.

Interest in OHSMSs has increased as the need for a global approach to OHS management was recognized as a logical and necessary response to the growth of the "global economy." Most major companies in the industrially developed world are multinational and favor a standardized approach to OHS. Japan, for example, has been manufacturing products and dealing with safety concerns around the world for a considerable period of time. Most companies recognize the need and benefits of meeting world standards or best practices for OHS while striving to meet local requirements of the host country. Current management science theories suggest that performance is better in all areas of business, including OHS, if it is measured and continuous improvement sought in an organized fashion as is done with formal management systems. Central to the ISO approach is to harmonize existing standards or create new ones that promote free trade. Two of ISO's recent standards, ISO 9001 and 14001, developed by the world community, address areas

OCCUPATIONAL HEALTH AND SAFETY MANAGEMENT SYSTEMS 2249

analogous to OHS. Both standards integrate these functions within a business (management) framework.

One of the advantages that has been identified to an OHSMS approach is resolution of the common criticism that OHS is rarely integrated into business systems, but rather is typically a stand alone adjunct in most companies. Additional value realized through the use of OHSMSs include:

- Alignment of OHS objectives with business objectives.
- Integration of OHS programs/systems into business systems.
- Establishment of a logical framework upon which to establish an OHS program.
- Establishment of a universal set of more effectively communicated, policies, procedures, programs, and goals.
- Applicability to, and inclusive of cultural and country differences.
- Establishment of a continuous improvement framework.
- Provide an auditable baseline for performance worldwide.

In addition to the benefits and open issues presented in Table 49.1 (13), the development of an international OHSMS standard would indeed need to be considered in light of existing international trade agreements and organizations such as the GATT, NAFTA, and WTO. An analysis conducted in 1996 by Levine and Markey shows that OHSMS standards such as those discussed in this chapter, do not present a non-tariff trade barrier to GATT, NAFTA, and WTO (14).

1.2 Theoretical Roots, Concepts, and Considerations

Occupational health and safety is an interdisciplinary field which encompasses, among others, the disciplines of industrial hygiene, occupational medicine, occupational nursing, engineering, epidemiology, and toxicology. Traditionally, OHS is considered a subset of the broader discipline of public health. Some public health scholars suggest that the field of public health, and for that matter occupational health, can be considered a social science (15). As such, while there are numerous theories of public health, no one encompasses the entire reach of the field. To the extent that a theory can be identified, it is expressed in terms of prevention and intervention.

In 1920, Charles Winslow defined public health as "the Science and Art of (*1*) preventing disease, (*2*) prolonging life, and (*3*) promoting health and efficiency through organized community efforts . . . (15)." Broadening this definition, and reflecting a social science perspective, Pickett proposes a definition of public health as "the organization and application of public resources to prevent dependency, which would otherwise result from disease or injury (15)." Based on these definitions, much of the focus in public health has first been on the prevention of disease or injury, and second, on intervention to reduce the frequency of incidents of impairment, disability, or dependency.

As the concepts expressed in these definitions have been applied to OHS, the focus has been on the anticipation, recognition, evaluation, and control of occupational hazards (16).

Table 49.1. Potential Benefits and Open Issues of an Environment, Safety and Health Management Standard[a]

Potential Benefits

National/International
1. Benefit to environment and workers
2. Based on ISO 9000 series standards
3. Integratable with ISO 9000 and 14000
4. Benefit to multinational entities
5. International "reach" through procurement language
6. Probably GATT-legal

Industrial/Governmental
7. Flexibility and incentives for innovation
8. Helps small/medium entities into ES&H management systems
9. Leverage of scarce resources for governmental agencies
10. Existing knowledge to effect institutional change
11. Prevention system approach
12. Opportunities for ES&H professionals

Open Issues

Applications
1. Need for policies and procedures
2. Industry sector-specific practices
3. Liability of the registrar
4. Ensuring against a drift to mediocre performance
5. Relationship between government and Registrars
6. Evaluation and updating of ISO system requirements
7. Governmental enforcement authority

Ethics
8. Potential for fraud and abuse by Registrars
9. Political, socioeconomic and public pressures

Cost
10. Added expense for companies with responsible programs
11. Financial impact on small and medium entities

International
12. Difficulty in reaching consensus
13. One vote for each country in the ISO process
14. Potential as a trade barrier and GATT-illegal practices
15. Alternate use of trading blocs and unilateralism
16. Internationally Recognized Worker Rights

[a]From Ref. 13

The International Labour Organization and the World Health Organization (WHO) have defined the objectives of OHS as:

> The promotion and maintenance of the highest degree of physical, mental, and social well-being of workers in all occupations; the prevention among workers of departures from health caused by their working conditions; the protection of workers in their employment from risks resulting from factors adverse to health; the placing and maintenance of the worker in an occupational environment adapted to his physiologic and psychological condition (17).

The traditional means of achieving the goals expressed by WHO have been focused primarily on the assessment of exposures to occupational hazards and risks, and the setting of standards (18). This practice is reflected in the activities of the American Conference of Governmental Industrial Hygienists (ACGIH) Threshold Limit Value (TLV) committee and subsequently the structure of the OSHA Act of 1970.

The traditional approach in OHS practice is reflected in boxes A and B in Figure 49.1. Necessary attention has been given to understanding the relationship between occupational stressors and human impact (18). The OHS literature has been dedicated to the stated general principles of anticipation, recognition, evaluation, and control of occupational hazards. As a better understanding of occupational hazards has been achieved, the question of the best way to manage these hazards has received increased attention. The primary focus of this effort has been to establish exposure limits for chemical and physical stressors present in the workplace and the means to achieve these limits. Building on the OHS principles presented above, Jacqueline Corn states that "The concept of occupational standards, as we understand it today, is the philosophical basis for creating a healthy workplace" (18). Written in 1992, this focus on OHS standards provides insights about the way OHS programs have developed and are managed.

The traditional approach to OHS management can be described as:

- Driven by OSHA regulations which are based on command-and-control principles (19).
- Focused on individual programs and not on the manner in which these programs may interact, or on the ways that the programs affect the organization as a whole (20).

Some scholars have argued that because OHS management has been driven by a rigid OSHA command-and-control regulatory structure, it has not been the most effective approach to fulfilling the OSHA Act mandate (19, 21). This is aptly described by J. Corn.

> From the onset, establishing numerical values for occupational exposure limits acceptable to society was both a political and scientific activity. It was, and still is value laden, highly controversial, and rife with conflicting principles and processes (18).

In this statement, Corn is primarily referring to the TLVs developed by the ACGIH and their influence on OSHA's permissible exposure limits (i.e. 29 *CFR* 1910.1000). Concern about the utility of TLVs has been expressed both by occupational-health and public-policy academicians (22, 23).

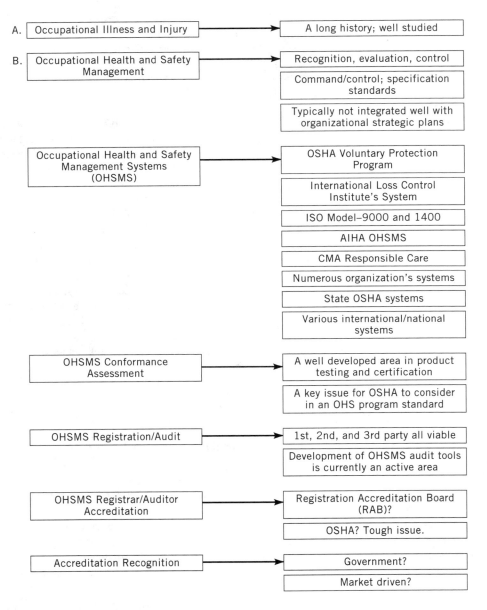

Figure 49.1. An OHS Management Taxonomy.

Although it is crucial to continue with activities in traditional areas of OHS—recognition, evaluation, control—numerous leaders in the field have suggested that actions and research are needed in areas other than the traditional areas to achieve continued gains in improving worker health and safety (24–26). Some of these areas include research in behavioral aspects of OHS, management systems, and intervention effectiveness. Two of these areas are addressed in this chapter, management systems and the measurement of system effectiveness. In terms of management systems as an OHS intervention, this can be referred to as intervention effectiveness.

1.3 Intervention Effectiveness

In the National Institute for Occupational Safety and Health's (NIOSH) 1996 National Occupational Research Agenda (NORA), the need for more research in the area of intervention effectiveness is identified. In NORA, NIOSH states (26):

> The goal of occupational safety and health interventions is to prevent disease and injury through combinations of techniques such as control technologies, exposure guidelines and regulations, worker participation programs, and training. The goal of intervention research is to determine the efficacy and effectiveness of these techniques and programs. New intervention research will assure better use of limited resources in workplace applications of prevention and control strategies. This research uses multidisciplinary approaches and focused field studies. Intervention model development, worker participation, cost effectiveness, hazard identification, and control evaluation are some of the elements of this research.

This research priority was the focus of a NIOSH workshop in 1995 and was expressed in an issue of the *American Journal of Industrial Medicine* (AJIM) (April, 1996) devoted to this subject. Several themes can be identified in NORA publications and the AJIM articles. The first is that OHS intervention research needs to be influenced by program evaluation theory and practice. Based on this, the second is that such intervention studies need to be theory-driven (27, 28). Chen and Rossi define theory-driven evaluations as "those evaluations that systematically integrate aspects of program theory into evaluation design and practice" (29).

These issues are important when considering the development, implementation and evaluation of OHSMSs. To date, research on OHS intervention effectiveness has not provided a good description of the possible causal processes involved between the intervention and expected or desired outcome (30, 31). Rather, such research has adapted more of a try-it-and-see strategy based on the researchers' own intuition and experience. This is a concern because the "...lack of theory-driven research limits our understanding of how an intervention works, which intervening or mediating variables are important, and what interactions may take place between study subjects and the setting characteristics" (30).

2 MANAGEMENT SCIENCE AND SYSTEMS

Considerable attention has been given in management science to organizational arrangements that increase productivity and effectiveness. A central theme in the evolution of organizational thought deals with organizational effectiveness and efficiency. Numerous theories and models have been put forth that present different independent variables which address the dependent variable—efficiency. The dimensions of some of these theories include motivation techniques, working conditions, leadership styles, methods of giving orders, methods of organizing and methods of measuring performance. These dimensions have been articulated within systems theory and systems analysis models and approaches. This section presents some of the basic concepts that provide the foundation upon which OHSMS models and approaches that are presented in this chapter are anchored.

2.1 Classical Organization Theory and Scientific Management

Several phases in the evolution of organizational thought can be identified. The roots of organizational theory, which is commonly known as "classical organizational theory," includes the works of Adam Smith, Charles Babbage, Henri Fayol, Frederick Taylor, and Max Weber, to name a few. Fundamental tenants of the classical school (32) include assertions that:

1. Organizations exist to accomplish production-related and economic goals.
2. There is one best way to organize for production, and that way can be found through systematic, scientific inquiry.
3. Production is maximized through specialization and division of labor.
4. People and organizations act in accordance with rational economic principles.

Although all of the scholars mentioned above have made valuable contributions to the field of organizational theory, the work of Frederick Taylor have had a particularly significant impact on both public and private management in the United States. In his essay "The Principles of Scientific Management," Taylor presented a paradigm that served as the foundation for what would become the "Scientific Management" approach to management (33). Central to Taylor's model was the notion that a manager could determine the "one best way" to fulfill a task. The one best way could, in turn, be determined through: careful observation of the production process; measurement of workers' motions; and, measurement of the time it took to complete a task. Taylor argued that scientific management provided a way to increase profits, eliminate unions, and raise productivity. He presented four points that became central aspects of the scientific management approach: (*1*) reduce management to laws—time motion studies; (*2*) scientifically select workers; (*3*) scientifically motivate workers; and, (*4*) management should have more responsibility for designing work processes and work flow. In this model, output was the most important consideration. The worker's needs were secondary.

Max Weber approached the relationship between workers and the organization from a different perspective than Taylor's. Weber did not operationally define employees in the same way as Taylor, but there was still a distinctly impersonal and dehumanizing aspect to Weber's rule-bound, formalistic, and highly disciplined model. Central to Weber's construct of an organization is that employees must adapt to the organization, not visa-versa, and as such, they tend to fit into a slot or are viewed as simply as a "cog" in the wheel of the organization.

In general, the thoughts and theories of the classical organizational theorists were consistent with the social environment of the early 20th century. The "industrial revolution" began in the late 1800s in both the United States and Europe. These were times of expansion and introduction of new mechanical technologies. Machines and instrumentation ruled. In this environment, it made sense that a model of scientific management would evolve. Industrial engineers were continually seeking ways to make their machines and operations more efficient. In many ways efforts to increase efficiencies made sense, however, problems surfaced. Neglected in the development process were issues related to employees, which, many would say, is the most important factor in any industrial operation.

2.2 The Human Relations Approach in Organization Theory

Mary Parker Follet and Chester Barnard were two early theorists who addressed the human element in organizations. Specifically, in her essay "The Giving of Orders" in 1926, Follet charted new ground through her work on an employee's relationship to the organization and supervision (34). She was the first organizational theorist to apply developments of the psychology schools to organizations. She addressed this relationship between organizations and psychology by looking at the way that supervisors direct employees through the delivery of orders. Follet acknowledged the value in the scientific management notion of "the law of the situation," but she asserted that once the "law" of the situation was discovered, the employee could issue it back to the employer as well as employer to employee.

Along with the findings of Mayo's Hawthorne study, Follet's work provided the foundation of the "human relations movement" and set the foundation for the development of systems approaches in management science. This movement reflects systems thinking and offered strikingly different theories than the classical organizational theorists. The central tenants of the human relations movement are built around the following assumptions (32):

1. Organizations exist to serve human needs.
2. Organizations and people need each other.
3. When the fit between the individual and the organization is poor, one or both will suffer: individuals will be exploited, or will seek to exploit the organization, or both.
4. A good fit between individuals and organizations benefits both: human beings find meaningful and satisfying work, and organizations get the human talent and energy that they need.

The human relations movement initially was viewed as a subset of classical and neoclassical organizational schools of thought. The issues raised and methods developed within the human relations movement were viewed by critics as simply another means with which to control workers and increase efficiency. The humanist theories have continued to evolve from the initial works of Follet, Bernard, and Mayo. Some would say that these theories, as they have evolved, are now accepted in the mainstream of organizational theory. An important modern theorist in the humanist school is Chris Argyris, whose work centers around the general concept of organizational learning. Argyris identifies an important connection between learning about self and others (self-actualization), organizational change, and organizational growth/development. Argyris also emphasizes learning about learning, about the inquiry of inquiry, and asserted that individuals must constantly inquire into their own capacities to learn effectively (35).

People are central to organizations. As the human relations theorists assert, employees cannot be treated like inanimate machines and be expected to perform productively and efficiently. Humans have complex psychological make-ups and are sensitive to their environmental conditions. There is a natural tension between the necessity to organize in society and values that emphasize individualism and individual rights. This tension can be observed in the different organizational approaches of scientific management and the human relations approach. To her credit, Follet acknowledged aspects of scientific management, but suggested modifications. As she identifies, workers do not like to be "watched-over." This notion can be taken a step further and suggest that people in general are adverse to being measured and evaluated, as is central in Taylor's model and many management system approaches. This tension, or paradox can be viewed against the long standing philosophical debate between free-will and determinism. From this perspective, the human relations approach theories emphasis the free-will values of freedom-of-choice, self-expression and self-actualization, while the scientific management and Weberian school of though emphasizes structure, constraints, and authority.

Seeds of thought from Taylor, Follet, and Argyris can be found in the various OHSMSs that evolved in the mid to late 1990s. The formal arrangements and performance measurement aspects are consistent with Taylor. The central role of employee participation is consistent with Follet. And, the role of continual improvement and adapting to external stressors (as performed in OHSMS Management Review) is consistent with Argyris.

2.3 Systems Thinking

Systems thinking and theories have been given significant attention in the management sciences. As early as Chester Barnard's *The Function of the Executive*, he defined the systems view in an organization as (36):

> A cooperative system is a complex of physical, biological, personal, and social components which are in a specific systematic relationship by reason of the cooperation of two or more person for at least one definite end. Such a system is evidently a subordinate unit of larger systems from one point of view; and itself embraces subsidiary systems—physical, biological, etc.—from another point of view. One of the systems comprised within a cooperative system, the one which implicit in the phrase "cooperation of two or more persons," is called an organization.

In the 1960s and 1970 systems theories were pursued and developed by Russel Ackoff, Kenneth Boulding, C. West Churchman, Daniel Katz, Robert Kahn and others. Katz and Kahn were two of the first researchers to take the open system concepts developed by Ludwig von Bertalanffy of biological systems and apply them to organizations (37, 38). In their 1966 book, *The Social Psychology of Organizations*, Katz and Kahn defined organizational open systems as having nine common characteristics (38).

1. *Import of Energy.* Open systems import some from the external environment.
2. *The Through-Put.* Open systems transform the energy available to them.
3. *The Output.* Open systems export some products into the environment.
4. *The Systems as Cycles of Events.* The pattern of activities of the energy exchange has a cyclic character.
5. *Negative Entropy.* To survive, open systems must move to arrest the entropic process; they must acquire negative entropy.
6. *Information Input, Negative Feedback, and the Coding Process.* The inputs into living systems consist not only of energic materials which become transformed or altered in the work that gets done. Inputs are also informative in character and furnish signals to the structure about the environment and about its own functioning in relation to the environment.
7. *The Steady State and Dynamic Homeostasis.* The importation of energy to arrest entropy operates to maintain some constancy in energy exchange, so that open systems which survive are characterized by a steady state.
8. *Differentiation.* Open systems move in the direction of differentiation and elaboration. Diffuse global patterns are replaced by more specialized functions.
9. *Equifinity.* Open systems are further characterized by the principle of equifinity—where the system can reach the same final state from differing initial conditions and by a variety of paths.

In spite of their work on open system models and use general systems theory concepts, Katz and Kahn warned of the danger of applying biological principles to social systems (38).

> There has been no more pervasive, persistent, and futile fallacy handicapping the social sciences than the use of the physical model for the understanding of social structures. The biological metaphor, with its crude comparisons of the physical parts of the body to parts of the social system, has been replaced by more subtitle but equally misleading analogies between biological and social functioning. This figurative type of thinking ignores the essential difference between the socially contrived nature of social systems and physical structure of the machine or the human organism. So long as writers are committed to a theoretical framework based upon the physical model, they will miss the essential social-psychological facts of the highly variable, loosely articulated character of social systems.

2.4 The Systems Approach in OHS Management Systems

The OHSMS approach to OHS management is based on systems theories developed primarily in the natural and social sciences by von Bertalanffy, Katz, Kahn, and others. Although there are numerous definitions provided for a system, the common four elements presented are input; process; output; and, feedback. The relationship between these four elements is depicted in Figure 49.2(**a**).

As suggested by Katz and Kahn, systems can be further characterized as either open or closed systems. In the case of open systems, there are identifiable pathways whereby the

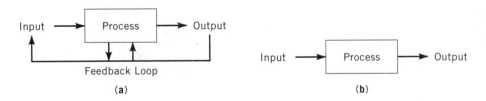

Figure 49.2. (a) System. (b) Program.

system interacts—exchanging information with and gaining energy—from its external environment. This phenomenon is readily observed in biological systems. Conversely, closed systems do not have such pathways, and thus limit their ability to adapt or respond to changing external conditions.

In traditional OHS management approaches, the focus has been on trailing indicators (outcomes or outputs), such as illness, injury, and fatality statistics. In a systems approach, regulatory compliance and trailing indicators are not neglected; however, there is a shift in focus towards performance variables and measurements from the input and process components of the system. These components can be thought of as being "upstream" from the system output.

2.5 Programs vs. Systems

An important distinction to make in an OHSMS approach is that between "programs" and "systems." The distinction is made here between traditional programmatic approaches and newer systems approaches to OHS management. In the paradigm shift suggested by the development and implementation of OHSMSs, a program is operationally defined as singular, vertical, and based on traditional command-control regulations. The focus is on compliance with the program standard/regulation, not the broader impact on OHS performance in the organization. In this conceptualization, programs do not have strong, if any, feedback or evaluation mechanisms whereby the program is adjusted or modified [Fig. 49.2(**b**)].

Conversely, a systems approach. Although not losing sight of programmatic requirements and opportunities for improvement, broadens in perspective to address the manner in which the program affects other programs, and the extent to which the program may or may not improve worker health and safety. Furthermore, a systems approach is driven by OHS improvement, more so than by programmatic regulatory compliance. A key distinction of a systems approach is that there are clear feedback and evaluation mechanisms whereby the system responds to both internal and external events (37, 38).

In this context, an example of program compliance would be with a single standard, such as the Lock-out-Tag-out (**29** *CFR* 1926.417) standard for construction or the Asbestos Standard (**29** *CFR* 1910.1001) for general industry. A systems approach integrates individual programs within the business operations and the external environment, and is thus more comprehensive than any single program. This is demonstrated by the universal OHSMS presented in Figures 49.3, 49.4, and 49.5.

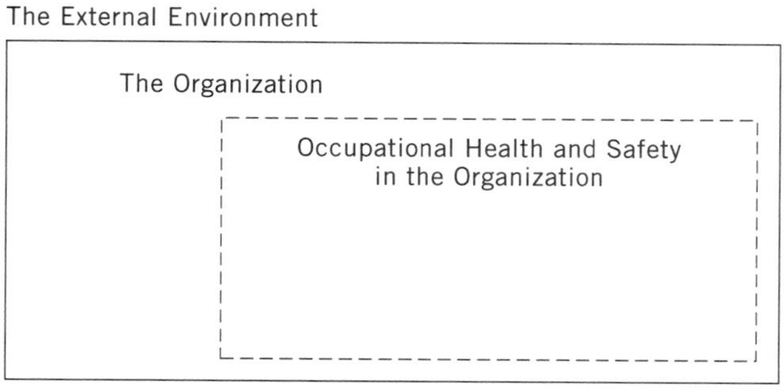

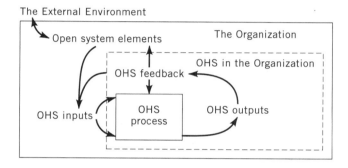

Figure 49.3. OHS in an Organization. Copyright *AIHAJ*.

Figure 49.4. A Systems Approach to OHS Management. Copyright *AIHAJ*.

One could argue that this program/system dichotomy is a potentially weak construct. That is, the programmatic approaches do in fact contain systems qualities and conversely, the systems approaches do in fact contain programmatic qualities. This observation is valid. However, the point of presenting the dichotomy is to elucidate the fact that programmatic OHS management approaches do not reflect or embrace systems concepts. Furthermore, such systems approaches potentially offer previously unrealized opportunities for advancement in OHS performance.

3 OCCUPATIONAL HEALTH AND SAFETY MANAGEMENT SYSTEMS

3.1 Introduction

Since the publication of ISO 9001 and 14001 there has been a proliferation of OHSMS standards, guidance documents, and approaches. Nation-state, professional society, and industry activities in the early to mid-1990s included the development and publication of the following OHSMSs.

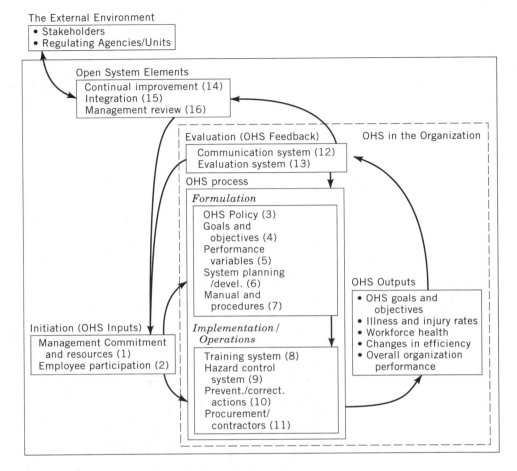

Figure 49.5. A Universal OHSMS Model for OHS Management. Copyright *AIHAJ*.

- Australia, Province of Victoria—SafetyMap
- British Standards Institute—BS 8800
- United States, OSHA—Voluntary Protection Programs
- U.K., Chemical Industries' Association—Responsible Care
- U.S. Chemical Manufacture's—Responsible Care Association
- American Industrial Hygiene Association—OHSMS Guidance Document

Thirty-one different OHSMS standards, guidance documents, approaches, and models have been identified in an analysis prepared by the International Occupational Hygiene Association (IOHA) in 1998. In the IOHA analysis, 24 OHSMSs published or under

OCCUPATIONAL HEALTH AND SAFETY MANAGEMENT SYSTEMS

development by nation-states, standards-development organizations, or professional societies were analyzed and compared.

The IOHA analysis was first discussed between members of the organization's Executive Board and representatives of the Safety and Health Branch of the International Labour Office (ILO) at IOHA's 3rd International Scientific Conference held in Crans Montana, Switzerland in September 1997. The OHSMS issue was a prominent topic at the conference. It was the subject of several keynote presentations and an entire platform session was devoted to various aspects of OHS management. Much discussion centered around ISO's decision to not pursue an international OHSMS standard, and of alternative mechanisms to meet this need.

Provisional agreement was reached between IOHA and the ILO to work together to draft a document on OHSMS. A necessary first step was to survey existing OHSMS documents and those under development to obtain a better overview of OHSMS activities throughout the world. With this knowledge of the OHSMS models and approaches in use, or under development, it would be possible to embark on the development of an international OHSMS that would build on the work already done, minimize conflict with other OHSMSs, and possibly provide a model whereby good OHS management principles are strengthened. In March of 1999 the ILO Board of Governor's elected to proceed with the development of an ILO convention for OHS management systems.

Using the 27 OHSMS variables identified in a OHSMS model presented in Table 49.2 and Figure 49.5, it was found that 19 of the models and approaches analyzed were generally strong in addressing traditional occupational health and safety (OHS) management issues, such as, hazard control, training, evaluation, and risk/hazard assessment. Conversely, there is a general weakness throughout the models and approaches in areas often considered central to management system approaches. These include management commitment, resource allocation, continual improvement, OHSMS integration with other organizational systems, and management review.

A further weakness found throughout what are otherwise strong OHSMSs is the lack of medical surveillance and health programs. In view of the importance of this aspect of preventive health management, this is somewhat surprising. An additional weakness is the manner that employee participation is addressed. While 20 of the 24 models and approaches reviewed contain some level of employee participation language, there is wide variation in the strength of the language.

3.2 OHSMS Standards and Guidelines

Eleven of the OHSMS standards, guidance documents, and approaches identified in the IOHA analysis are summarized and presented here. These standards, guidance documents, and approaches were all published in their final form in 1998. Standards, guidance documents, and approaches that were in draft form in 1998 are not included.

3.2.1 Australia/New Zealand: Joint Standards Australia/Standards New Zealand Committee

This guidance document was published by the Joint Standards Australia/Standards New Zealand Committee SF/1, Occupational Health and Safety Management in 1997 (39). The

Table 49.2. A Universal OHSMS

Initiation (OHS Inputs)

1.0 Management Commitment and Resources
    1.1 Regulatory Compliance and System Conformance
    1.2 Accountability, Responsibility, and Authority
2.0 Employee Participation

Formulation (OHS Process)

3.0 Occupational Health and Safety Policy
4.0 Goals and Objectives
5.0 Performance Variables
6.0 System Planning and Development
    6.1 Baseline Evaluation and Hazard/Risk Assessment
7.0 OHSMS Manual and Procedures

Implementation/Operations (OHS Process)

8.0 Training System
    8.1 Technical Expertise and Personnel Qualifications
9.0 Hazard Control System
    9.1 Process Design
    9.2 Emergency Preparedness and Response System
    9.3 Hazardous Agent Management System
10.0 Preventive and Corrective Action System
11.0 Procurement and Contracting

Evaluation (Feedback)

12.0 Communication System
    12.1 Document and Record Management System
13.0 Evaluation System
    13.1 Auditing and Self-Inspection
    13.2 Incident Investigation and Root Cause Analysis
    13.3 Medical Program and Surveillance

Improvement/Integration (Open System Elements)

14.0 Continual Improvement
15.0 Integration
16.0 Management Review

underlying structure of these guidelines is similar to AS/NZS ISO 14004 but takes into account terms and principles of long-standing use in the field of occupational health and safety.

The objective of the guidelines is to assist in the implementation, development, or improvement of an occupational health and safety management system. The standard is

OCCUPATIONAL HEALTH AND SAFETY MANAGEMENT SYSTEMS

not a specification, but rather aims to encompass the best elements of such systems already widely used in Australia and New Zealand. To do this, it includes examples and practical help guidance.

The standard departs from the format of AS/NZS ISO 9004 and AS/NZS ISO 14004, particularly in those areas covering the application of the risk management processes. This is usually characterized by the three-stage process of: (*1*) hazard identification; (*2*) risk assessment; and, (*3*) risk control.

Similar to AS/NZS ISO 14004, this standard is not intended be used for certification purposes, but it can be used to structure a strong OHSMS. The "practical help" boxes are helpful. They can actually be used to conduct an initial audit or evaluation of an OHSMS.

3.2.2 Australia, Province of Victoria: Health and Safety Organisation

The Safety Management Achievement Program (SafetyMap) has been developed by the Health and Safety Organisation, Victoria (HSO) to present the characteristics of health and safety systems that are effective, comprehensive, and cost efficient (40). Using audit criteria methods common to quality programs, SafetyMap provides organizations with a practical tool to improve their health and safety performance.

SafetyMap is a system that enables organizations of all sizes to:

1. Assess the scope and effectiveness of their health and safety policies and operations.
2. Plan improvements to these operations.
3. Develop benchmarking standards.
4. Gain recognition of the standards achieved.

Continual improvement is identified in SafetyMap's front-end material. SafetyMap is used in an HSO certification program, there are three levels of certification: (*1*) initiation level; (*2*) transition level; and, (*3*) advanced level.

3.2.3 United Kingdom: The Oil Industry International Exploration and Production Form (E&P Forum)

The E&P Forum was founded in 1974 by the International Association of Oil and Gas Companies and Industry Organizations. It is concerned with all aspects of oil and gas exploration and production having international implications, and in particular with safety and health and environmental protection. It represents its members' interests at the United Nations (UN) agencies, European Union (EU) and other international bodies.

The guidelines for an integrated environmental and occupational health and safety management system were prepared in 1994 (41). They were created to assist in the development and application of Health, Safety and Environmental Management Systems (HSEMS) in exploration and production operations. Forum members have participated in the work, to ensure that their collective experience is used and that the Guidelines have wide acceptance.

The HSEMS guidelines have been developed by the E&P Forum to (41):

- Cover relevant Health, Safety and Environment (HSE) issues in a single document.
- Be relevant to the activities of the E and P industry worldwide.
- Be sufficiently generic to be adaptable to different companies and their cultures.
- Recognize, and be applicable to, the role of contractors and subcontractors.
- Facilitate operation within the frame work of statutory requirements.
- Facilitate evaluation of operations to an international standard(s) as appropriate.

This management system is comprehensive and contains all 27 OHSMS variables presented in Table 49.2 and Figure 49.5 except, employee participation and health program/surveillance. Of particular note is the extent to which this management system addressed core structural issues such as direction on procedure development. Another example is the detail provided on requirements for a controlled document system.

3.2.4 International: International Organisation for Standardization Technical Committee 207

ISO 14001 was developed by ISO Technical Committee (TC) 207 and published in 1996 (4). It is an environmental management system that is being used widely as a template for OHS management systems.

Since the promulgation of this standard, many organizations have been using ISO 14001 to integrate environmental, health and safety management systems. In 1995–1996 there were discussions within TC 207 to incorporate health and safety aspects into ISO 14001. This issue was subsequently considered by ISO and tabled for further consideration in 2001.

3.2.5 Japan: Japan Industrial Safety and Health Association

These OHSMS guidelines were published by the Japan Industrial Safety and Health Association (JISHA) in 1997 (42). The guidelines were created to help employers establish and develop an OHSMS that is widely expected to function as a system for ensuring the safety and health of workers with a definite corporate policy and efficient programs created on the employer's responsibility.

Five fundamental concepts are addressed in the JISHA system (42):

1. Self-regulatory OHS Management by Employers.
2. The Guarantee of Implementation and the Maintenance of Records.
3. Twenty Areas of OHS Management.
4. The Clarification of Role Assignment.
5. An Emphasis on Communication Process.

There is strong emphasis on medical surveillance, as partially demonstrated by the requirement for an industrial doctor.

Many of the 27 system variables presented in Table 49.2 and Figure 49.5 are not present in the JISHA OHSMS. There is weak to nonexistent mention of management commitment

OCCUPATIONAL HEALTH AND SAFETY MANAGEMENT SYSTEMS

and resources. There are activities specified that management must do or perform. But it is not explicitly stated that management commitment to OHS is required or recommended. There is no mention of the need for goals, objectives, or performance measures.

3.2.6 The Netherlands: Nederlands Normalisatie-Instituut

This guidance document explicitly states: "This publication contains guidance and recommendations. It should not be quoted as if it were a specification and should not be used for certification purposes (43)." The style of presentation and translation care consistent with these conditions. That is, the document is written with a mixture of shall and should statements that are embedded in general guidance text.

Five OHSMS functions stated in the guide's foreword (43) are

a. it should guarantee effective OH&S management: a planned approach to risks (prevention and management) and control of the implementation of the plans and integration into the business management;
b. it should encourage those involved in the organization to contribute to controlling the risks and step-by-step the level of OH&S management;
c. it should prompt the organization to learn specifically from experience in order to continue improving OH&S (for example: increasingly tackle risks at their source);
d. it should convince important relations outside the organization (clients, Labour Inspectorate) that the organization has arranged its affairs properly in the field of OH&S.
e. it should give an understanding of the way in which the organization meets the statutory obligations in the field of OH&S.

Of particular notice is the attention and strength given to employee participation. It is repeated throughout and in bolder terms than most other OHSMSs published as of 1998.

3.2.7 South Africa: National Occupational Safety Association

The NOSA 5 Star System has been developed and published by the National Occupational Safety Association (NOSA) in South Africa. The stated objective of the OHSMS is to assist "... management in carrying the enormous financial, legal and moral responsibilities which rest on their shoulders in running their business" (44).

The system addresses and measures regulatory compliance issues more strongly than management system issues. It is more compliance based as opposed to management system based. The companion audit workbook includes a section on environmental issues which are based on ISO 14001. In overall emphasis, the Five Star system resembles the DNV International Safety Rating System. This is a valuable tool/system to get started, but will have limited returns as an organization handles the compliance issues and focuses on higher level management system issues.

3.2.8 United Kingdom: British Standards Institute

BS 8800 was published by the British Standards Institute in 1996 and has been widely distributed (45). This OHSMS has received considerable attention outside of the U.K.

because of the role that U.K.-based standards have had in the development of ISO 9001 and ISO 14001. BS 8800 gives guidance on OHS management systems for assisting compliance with stated OHS policies and objectives and on how OHS should be integrated within an organization's overall management system. It gives guidance and recommendations only, and states explicitly that it should not be quoted as if it were a specification, and that it should not be used for certification purposes.

The standard seeks to improve the OHS performance of organizations by providing guidance on how the management of OHS may be integrated with the management of other aspects of business performance, in order to (a) minimize risk to employees and others; (b) improve business performance; and (c) assist organizations to establish and responsible image within the marketplace (45).

The standard shares common management principles with ISO 9001 and ISO 14001, but stresses that these are not prerequisite for using the guide. It gives guidance on two approaches to OHS management, one based on the U.K. Health and Safety Executives booklet HS(G)65, "successful health and safety management" and one based on ISO 14001.

The standard includes several annexes which deal in detail with the practical aspects of implementing the guidance. Those on planning and implementing, risk assessment and measuring performance are especially useful, and contain flowcharts and simple tabulated information to guide users through the process stages.

In 1999 the British Standards Institute published a document titled "Occupational health and safety management systems—Specification." Its reference is OHSAS 18001:1999. This document presents an auditable OHSMS standard that is based on shall statements. This is an intriguing document in that while it is published by BSI and is very close in form and content to BS 8800, the standard's preamble clearly states that it is not an official British Standard. Despite this, at the time of this publication, it appears that OHSAS 18001:1999 will be widely used in the U.K. and beyond (46).

3.2.9 United States: American Industrial Hygiene Association

The American Industrial Hygiene Association's (AIHA) OHSMS was developed in response to the interest and success of ISO 9001 in the mid-1990s. It is a guidance document meant to assist organizations achieve safe working conditions. This OHSMS states that: "It may be used by practicing health and safety professionals in the United States and elsewhere as a basis for designing, implementing, and evaluating OHSMSs . . ." (10).

As an ISO 9001-based OHSMS, the AIHA document contains 20 primary sections that correspond with ISO 9001. This is document can be considered an auditable standard because of its use of "shall" based clauses. In the United States when the American position was being considered on the development of an ISO OHSMS, the AIHA document received considerable attention as a potential template for the development of such an international standard.

The AIHA OHSMS contains all 27 OHSMS variables presented in Table 49.2 and Figure 49.5.

3.2.10 United States: Chemical Manufactures' Association

This Code of Practice has been developed by the Chemical Manufacturers' Association in the United States. Its stated purpose is (47)

The goal of the Employee Health sand Safety Code of Management Practices is to protect and promote the health and safety of people working or visiting member company work sites.

To achieve this goal, the Code provides Management Practices designed to continuously improve work site health and safety. These practices provide a multidisciplinary means to identify and assess hazards, prevent unsafe acts and conditions, maintain and improve employee health, and foster communication on health and safety issues.

Implementation of the Employee Health and Safety Code, together with other Codes of Management Practices, can enable member companies to operate in a manner that further protects and promotes the health and safety of employees, contractors, and the public, and protects the environment.

The CMA document is part of a larger environmental, health and safety management system. While the format is somewhat different than ISO 9001, ISO 14001, or the other popular management systems, this OHSMS approach is comprehensive and contains many of the key OHSMS variables presented in Table 49.2 and Figure 49.5.

3.2.11 United States; Occupational Safety and Health Administration

The United States, Department of Labor, Occupational Safety and Health Administration (OSHA) promulgated the Voluntary Protection Programs (VPP) in the early 1980s as a means to recognize organizations that implemented and maintained strong OHS programs (5). It is a voluntary program and as of August 1998 had over 400 approved sites. This is a popular OHSMS in the United States because of the recognition that an organization receives following certification by OSHA. The VPP model has been used in OSHA as the template for other OHS program activities, including the development of OHS program guidelines in 1989 and consideration of an OHS program standard in 1996. The VPP is strong in the area of employee participation. This is a primary performance variable in the VPP approach.

3.2.12 Development of an International OHSMS Standard

Following a meeting of stakeholders by ISO in September 1996 and the decision of the ISO Technical Management Board early in 1997 not to proceed with an OHSMS standard, there have been a number of developments. The widespread dissatisfaction about the current situation has manifested itself in many ways in the late 1990s. The Dutch Standard NPR 5001 was put forward to the ISO Technical Management Board (TMB) for adoption as an OHSMS guide, but not with certification or accreditation. In 1998, ISO began working towards aligned (as opposed to integrated) standards for quality and environment in the revisions of ISO 9001 and ISO 14001. This integration has rekindled the debate on the role of OHSMS. There is growing feeling that OHSMS should be included in this management systems standards revision and integration process.

AFNOR, the Spanish national standardization body proposed the Spanish standard UNE 81900 for adoption as a European (CEN) standard, though this was rejected by the member countries, mainly on the grounds that it was a "certifiable" standard which they did not want. In Germany, the "social partners" (Government at Federal and State Level, accident funds, employer and employee organization) have developed a common position on OHSMS and are proposing that the European Commission should take action under Articles 117 and 118a of the Treaty of Rome, to make resources available to prepare on OHSMS "Guide." The EC Advisory Committee on OHSMS is also considering whether an OHSMS Guide would be advantageous.

In addition, the importance of, the effectiveness of, and the senior management commitment to, the management system behind occupational health and safety programs is gaining recognition among health and safety professionals. A meeting in 1998 of the WHO PACE program on controlling hazards from dust, the need for an overarching management system, that ensures all the technical and administrative tasks necessary are actually being carried out, was recognized, as was the importance of having measurable performance indices and standards by which to assess its effectiveness.

Similarly, government labor inspectorates have recognized the importance of an effective management system and incorporated management elements in their legislation, codes of practice and guidance. A good example is the U.K., where the Health and Safety Executive's Publication, HS(G)65, "Successful Health and Safety Management" has significantly altered thinking on OHS management in many organizations. Other Governments are in the process of introducing measures, such as Hong Kong, who intend to bring into law a requirement for an OHSMS to be established by the end of 1999 in certain sectors of industry such as construction and ship building.

In addition to the development of comprehensive OHSMSs, others aimed at particular industries or aspects of health and safety have also been produced, such as that by the Dutch Central Committee of Experts (CCE-SCC) who have produced a model for "certification of contractors" OHSMS systems in the engineering and construction industries.

3.3 A New OHSMS Model

In 1996 researchers at the University of Michigan began the development of a universal OHSMS measurement tool (48, 49). A byproduct of this effort has been the development of a new OHSMS model. The development of this OHSMS is potentially very significant because it includes all of the essential aspects of existing models, and consolidates and refines them. And, plausibly facilitates the evolution of a new generation OHSMS. This system can serve as a model for organizations who want to develop/implement an OHSMS but find existing models/guidance documents to be inaccessible. Furthermore, as the international community moves forward in considering an international OHSMS standard, this universal OHSMS can serve as a starting-point for consideration.

3.4 Universal OHSMS Structure

The universal OHSMS structure can be summarized as containing five organizing categories and 27 sections (16 primary and 11 secondary). Considerable attention was given to the superstructure; that is, the five organizing categories and the manner in which the 16 primary sections are distributed among the five categories. The representation of this OHSMS in Figure 49.5 could be presented in a number of ways. However, in an effort to create a more robust model, it was important to arrange the sections in such a way as to

OCCUPATIONAL HEALTH AND SAFETY MANAGEMENT SYSTEMS 2269

facilitate both OHSMS development and diagnostic activities. The five organizing categories are

1. Initiation (OHS Inputs).
2. Formulation (OHS Process).
3. Implementation/Operations (OHS Process).
4. Evaluation (OHS Feedback).
5. Improvement/Integration (Open System Elements).

These categories are partially based on the policy analysis model developed by Brewer and deLeon (50) and a simplified systems model, as depicted in Figures 49.2 and Figure 49.4. The sequence of steps shown in Figures 49.3, 49.4, and 49.5 starts with a very basic construct showing the relationship between OHS management, the organization, and the external environment. Figure 49.4 develops the construct further showing how a systems model and a policy analysis model can fit in the construct. Finally, Figure 49.5 adds the universal OHSMS sections to the construct.

With the universal OHSMS as presented in Figure 49.5, OHSMS developers and evaluators can readily see how the sections relate to one another. This structure assists both the development and implementation of an OHSMS as well as facilitate diagnostic efforts. System evaluation functions are facilitated through this structure since the universal OHSMS components are clearly identified as needed for root-cause-analysis activities (13.2). That is, following preventive and corrective action (10.0) and subsequent root cause analysis activities, evaluators will be able to identify the universal OHSMS sections which may need attention as result of the initial incident or system breakdown. This ability to identify clearly sections and make modifications will facilitate continual improvement (14.0) goals.

3.5 Key System Components

When developing and implementing an OHSMS, particular attention should be given to the following system elements which are key performance variables for OHSMS success.

1. Communication System/feedback channels (12.0).
2. System Evaluation (13.0), specially Auditing/Self-Inspection (13.1), and Root-Cause Analysis (13.2).
3. Continual Improvement (14.0).
4. Integration (15.0).
5. Management Review (16.0).

It is clear that all sections of the universal OHSMS are important to successful OHS management. For instance, many OSHA VPP participants would argue that management commitment (1.0) and employee participation (2.0) can be defined as the essential model components (51). The point of attempting to elucidate the components that are essential to a systems approach is that as future efforts are made to apply OHSMSs to smaller workplaces with limited resources, a thorough understanding of these key systems components will be necessary.

3.5.1 Communication System (12.0)

A well functioning communication system with defined feedback channels is essential for a successful OHSMS. As depicted in the systems model presented in Figures 49.4 and 49.5, this is a basic feature of a system, especially an open system. For the system to survive and potentially grow, there must be mechanisms whereby the system components receive feedback from each other and from the external environment.

This universal OHSMS section provides the means by which all other sections relate and interact. There are any number of ways that the communication system OHSMS principles can be operationally defined. However, in its most basic form, a viable communication system should identify how, and to whom, information for the proper functioning of the OHSMS will be transmitted. The communication system should have mechanisms in place to confirm that information has been received by the intended party and in the prescribed time-frame.

For example, universal OHSMS communication system OHSMS principle COMM01 states "The organization shall ensure that the OHSMS has a clearly defined and functioning communication system whereby information, data, notifications, and other communications required by the OHSMS are transmitted in a timely manner to OHSMS personnel, employees, managers, and supervisors."

3.5.2 Performance Variables (5.0) and Root Cause Analysis (13.2)

In the current OHS lexicon, terms such as "performance measures and metrics" are commonly used but known as variables and measurements here and in the academic literature. In order to add more structure to these discussions, the broader context of scientific inquiry and measurement theory needs to be considered. That is, prior to identifying variables and measurements, it is necessary to identify the concepts and indicators with which they are associated.

Such a measurement hierarchy, from general to specific, can be summarized as follows:

1. It is first necessary to identify the construct to be measured (e.g. OHS performance).
2. This is followed by the identification of the indicator(s) associated with the construct(s) (e.g. employee participation).
3. It is then necessary to identify the variable(s) associated with the construct(s) or indicator(s) (e.g. continual improvement meetings).
4. It is then necessary to quantify operational definitions for these variables (e.g., number of team meetings per year).

Thus, in order to make valid and reliable performance measurements, the indicators, variables, measurement units, and their logical relationships must be established (52, 53).

In terms of the indicators to measure, the distinction has been made in the OHS literature between leading and trailing indicators. As in many disciplines, efforts are underway to identify leading indicators upon which management can rely as predictors of emerging problems. This is seen in the economics field with an emphasis placed on leading economic indicators and in the environmental field with efforts to identify leading environmental health indicators from which environmental management decisions can be made.

Central to the identification of leading OHS indicators are the root-cause-analysis activities required in many OHSMSs. The use of root-cause analysis techniques have been in use for many years in the health and safety field (54). Root-cause analysis has been highlighted in the universal OHSMS because of its central importance in moving up the causal chain to the point of origin in the pursuit of leading OHS indicators. To achieve this, the input models, as well as OHSMSs and EMSs explicitly address root-cause analysis. In VPP and ISO 14001 training courses, root cause analysis is presented in several exercises titled "the four whys" and "follow the string" (51, 55). The point of these exercises is to show OHSMS evaluators several techniques on how to perform root-cause analysis. For example, an evaluator should always ask a set four "why" questions when a nonconformity is found, with each question based on the previous question.

Root-cause analysis may be considered a relatively minor component of an OHSMS, especially in entities in which full compliance is achieved with regulations and in which a numeric rating system is in place. However, compliance with regulations and high scores on a numeric system cannot replace the practice of following a line of inquiry from an unplanned incident, near miss, or regulatory citation to objective evidence that answers the question why? This is central to the philosophy of planning and operating an efficient, effective OHS management system. The lack of procedures for, or documentation of, the use of root-cause analysis may be traceable to, for example, nonconformance in clauses related to policy, management commitment, or training (56).

3.5.3 Continual Improvement (14.0)

Continual improvement is a key concept in OHSMSs and is the central concept reflected in the Deming/Shewhart Plan-Do-Check-Act cycle (57). In an OHSMS context, continual improvement can be defined as the "process of improving the OHS management system to achieve enhancements in overall OHS management performance through continuing reviews of appropriate OHS measures that are in line with the organization's OHS policy (10)."

The basic notion embodied in this concept is that the organization continually seeks ways to improve the OHSMS and thus increase worker health and safety. Continual improvement focuses on problem prevention, corrective action, and performance improvement to affect worker health and safety. Continual improvement does not mean or imply a requirement to attain "better than compliance" conditions as measured against specification regulations or standards. Although "better than compliance" may be a goal of an organization, it is not a requirement of the definition of continual improvement. There are numerous ways that an organization may operationally define continual improvement. However, the ultimate goal of continual improvement should be to reduce the potential for worker injury and illness.

3.5.4 Integration (15.0)

A basic characteristic of OHSMSs is that they are integrated with other business functions and the external environment. As depicted in Figures 49.4 and 49.5, the OHSMS elements are connected through feedback channels. Also depicted in these figures is the manner in which the OHSMS is integrated with both the organization as a whole and the external environment. In order for an OHSMS to succeed, this open-system aspect must be understood and functioning. By definition, the implementation of an OHSMS requires that the OHSMS be connected, or related to other functions in the organization. This means that OHS issues and aspects of the OHSMS will be part of the organizational culture. Furthermore, at a fundamental level, this also means that worker health and safety will be an important value expressed by management and employees alike.

3.5.5 Management Review (16.0)

Management review is the means whereby the overall performance of the OHSMS is evaluated. It provides the link between the OHSMS, the organization, and the environment external to the organization. This involves evaluating the OHSMS's ability to meet the overall needs of the organization, its stakeholders, its employees, and regulating agencies. Management review is different from more specific system-evaluation efforts which address specific aspects of the OHSMS elements.

The distinction between management review and system evaluation can be viewed in terms of how one would plan a long automobile trip. Using this metaphor, the ongoing monitoring of the fuel level, engine temperature, and general performance of the automobile corresponds with the functions performed during OHSMS system evaluation. Management review, on the other hand, corresponds with the ongoing evaluation of whether the car is on the correct highway to reach the intended destination. Continuing with the metaphor in terms of the program/system dichotomy discussed, earlier, it can be said that, checking the tire pressure would correspond with program evaluation and overall trip planning and vehicle performance would correspond with system evaluation.

Management review is the hallmark of a successful system and is a key attribute of strong management commitment to OHS. Without feedback, there can be no strategic planning or continual improvement. Management resistance to participate actively in the OHSMS-review process would be a clear indicator of the lack of management commitment.

3.6 Universal OHSMS Description

The universal OHSMS's 27 sections and five organizing categories are described more fully here. These definitions can aid an organization in taking the initial steps in developing an OHSMS that is unique to their organizational goals and objectives.

3.6.1 Initiation (OHS Inputs)

OHSMS Initiation refers to the act of defining the necessary elements and conditions essential to OHSMS formulation, implementation, and evaluation. These necessary elements include strong management commitment, sufficient resources, and robust employee participation.

1.0 Management Commitment and Resources (COMMIT)

Management commitment to occupational health and safety may be operationally defined in any number of ways. Allocation of sufficient resources for the proper functioning of an OHS program or management system has been identified as a key variable to measure management commitment. Other variables, some of which are found in COMMIT OHSMS principles, are the establishment of organizational structures whereby managers and employees are supported in their occupational health and safety duties and the designation of a management representative who is responsible for overseeing the proper functioning of the OHSMS.

The importance of strong management commitment is reflected in OSHA's VPP and other OHSMSs. In fact, some occupational health and safety professionals assert that management commitment is the *senaquanon* of an OHSMS. The same may be said about employee participation (2.0). Therefore, these two input variables must be present for the development of a robust OHSMS.

1.1 Regulatory Compliance and System Conformance (REG)

Many governmental regulations and nongovernmental standards impose requirements on occupational health and safety management and, therefore, can affect the way an OHSMS is designed, implemented, and operated. Organizations need to understand the governmental regulations and non-governmental standards which impact them. Striving for compliance or conformance with regulations and standards should be a top priority of the organization.

It is not the purpose of REG OHSMS principles to identify for the organization the applicable regulations and standards. Rather, the purpose is to ensure that the organization has a system to identify, document, and implement applicable governmental and nongovernmental requirements.

1.2 Accountability, Responsibility, and Authority (ACCOUNT)

OHSMS accountability, responsibility, and authority addresses the manner in which the organization defines the roles of personnel who are involved in OHSMS management, and the employees, supervisors, and managers who are affected by the system. Crucial to role definition is the manner in which occupational health and safety and OHSMS accountability, responsibility, and authority are defined, supported, and enforced by senior management.

In addition to defining these roles, the organization should ensure that potential discrimination against personnel who have OHSMS management responsibilities is prevented. Repeated and/or willful violations of occupational health and safety procedures should be subject to reprimand.

2.0 Employee Participation (EMPLOY)

Employee participation in occupational health and safety management may be operationally defined in any number of ways. The key issue is that employees have input into occupational health and safety considerations, and that the input is meaningful, valued, and can affect policies and practices. Other important variables, some of which are found in EMPLOY OHSMS principles, include employee participation in OHSMS formulation, implementation, and evaluation activities.

Many OHS professionals have identified employee participation in occupational health and safety management as the variable essential to successful OHS management and illness/injury reduction. Employee participation and management commitment (1.0) are identified as two input variables that must be present in the development of a robust OHSMS.

3.6.2 Formulation (OHS Process)

OHSMS formulation refers to the act of designing the system. This starts with an OHS policy that expresses the organization's commitment to occupational health and safety. Goals, objectives and performance variables are used to implement OHS policy.

Integral to OHSMS formulation is the planning and development of the system in terms of organizational structures. A baseline evaluation that identifies hazards and risks provides essential information. The OHSMS manual is used to present the policies and procedures that define the system.

3.0 Occupational Health and Safety Policy (POLICY)

The OHS policy represents the foundation from which OHS goals and objectives, performance measures, and other system components are developed. The OHS policy should be short, concise, easily understood, and known by all employees in the organization. It can be expressed in terms of organizational mission or vision statements. It is a document that expresses the organization's OHS values.

The OHS policy should demonstrate senior management's commitment to occupational health and safety (1.0), employee participation (2.0), allocation of necessary resources, and continual improvement (14.0). The policy should be evaluated periodically as part of the management review process.

4.0 Goals and Objectives (GOAL)

The development of OHS goals and objectives follows naturally from the OHS policy (3.0) development activities. With the OHS policy established, there is a foundation upon which OHS goals and objectives can be generated.

The establishment of OHS goals and objectives represents the beginning of a progression from the conceptual realm of the OHS policy to an operational realm as expressed in the overall system structure/design (6.0) and OHSMS manual (7.0).

It is important that the organization give careful thought to the way it establishes and develops goals and objectives. They should be measurable and appropriate to the size, nature, and complexity of the organization's activities. The goals and objectives should reflect the organization's commitment to a safe and healthy workplace that is free of known hazards.

5.0 Performance Variables (PERFORM)

The ability to measure OHS performance over time is essential to eliminating occupational injuries and illness. To achieve this, the organization should identify performance variables that are consistent with the OHS variables expressed in the OHS policy (3.0) and goals and objectives (4.0). These variables should be considered, and possibly developed, at the same time that the goals and objectives are being created. Good correspondence between the goals and objectives and the performance variables will enhanced in the organization's ability to perform the diagnostic functions of the evaluation system (13.0) and management review (16.0) components.

Traditional performance variables are trailing indicators of performance. Trailing indicators have been and should continue to be used to evaluate OHS performance. These indicators include, for example, citations for regulatory noncompliance, injury and illness statistics, lost workdays, and insurance costs.

Leading indicators may include percentage of safe behavior, exposure assessment data, employee observations, and percentage of government-mandated training actually delivered. It is not the intent here to require the use of any specific indicators, but rather to provide principles from which the organization can specify, evaluate, and use such indicators. To the extent possible, the performance variables should be leading indicators of workplace health and safety as opposed to trailing indicators.

6.0 System Planning and Development (PLAN)

System planning and development activities address both initial OHSMS development and ongoing revision and modification of the system. This system component addresses the manner in which the overall structure and form of the OHSMS will be developed, implemented, and subsequently modified.

The performance-based nature of OHSMS standards can lead to a wide range of OHSMS structures when adapted and implemented in any given organization. This component is, therefore, one of the most crucial and this stage of the formulation process greatly affects the implementation and ongoing performance of the OHSMS. If this component of the OHSMS is not performed well, the probability of the occurrence of implementation problems will increase.

System planning and development activities represent the implementation of the preceding components in the OHSMS: OHS policy (3.0); goals and objectives (4.0); and, performance measurement (5.0). The planning and development activities must be informed by data from the baseline evaluation (6.1), and the requirements of the hazard control system (9.0).

6.1 Baseline Evaluation and Hazard/Risk Assessment (BASE)

A baseline evaluation of the organization's existing occupational health and safety management practices and OHS hazards is necessary before a robust OHSMS can be completely designed or implemented. The baseline evaluation needs to identify OHS hazards and their associated risks clearly. This information is essential to the development of numerous OHSMS components, including the training system (8.0), hazard control system (9.0), and the emergency preparedness and response system (9.3).

The baseline evaluation can help establish the structure and methods to be used in other evaluation system (Sec. 13.0) functions, such as audits and self-inspections (13.1).

7.0 OHSMS Manual and Procedures (MANUAL)

The OHSMS manual is the document where occupational health and safety and OHSMS policies and procedures are to be found. This document is a key part of OHSMS formulation activities because it compiles and presents the results of the formulation work.

OHSMS management is greatly enhanced by the presence of a written manual that contains OHS procedures to guide the system. It has been argued persuasively that the existence of such a manual is a necessary condition for maintaining a healthy and safe workplace.

The OHSMS manual should be easily accessible to employees, taking into account levels of education and possible language barriers. It should be written in clear language and should use graphic illustrations where possible to communicate the intended information.

3.6.3 Implementation/Operations (OHS Process)

OHSMS implementation represents the act of putting into practical effect the work done in the OHSMS initiation and formulation phases. Implementation involves the activities of the training, hazard control, preventive, and emergency response systems. Also addressed are procurement and contracting activities.

8.0 Training System (TRAIN)

The term training system is used broadly to reflect the importance of knowledge dissemination and skill development in a well functioning OHSMS. Occupational health and safety training has been an integral component of OHS management for many years. It is universally recognized as an essential element in maintaining a healthy and safe workplace.

A successful training system must be based on an understanding of workplace hazards. The system needs to be integrated with the communication system (Sec. 12.0) feedback channels so that, as there are changes in the OHSMS, training efforts can be appropriately modified. A well-functioning training system includes a mechanism to identify training needs and track employee training-status in relation to the identified needs.

8.1 Technical Expertise and Personnel Qualifications (EXPERT)

Successful OHSMS operation requires qualified and competent personnel. This includes personnel in the organization who have direct OHSMS responsibilities as well as external consultants who may provide OHS services to the organization.

The organization can ensure the expertise of OHSMS personnel through preplacement review and postplacement training. In addition, the use of certified professionals can both improve performance and demonstrate an organizational commitment OHS. Taken from examples in the U.S., such professionals could be Certified Industrial Hygienists, Certified Safety Professionals, licensed Professional Engineers, as well as board Certified Occupational Physicians and Health Nurses.

It is not the purpose in EXPERT OHSMS principles to define the OHS qualifications required. Rather, the purpose is to ensure that the organization has a system to set criteria and document their use so that under-qualified OHS persons can be identified and will not have an adverse impact on workplace health and safety.

9.0 Hazard Control System (CONTROL)

The hazard control system is broadly defined to include the various methods used to reduce or eliminate occupational hazards, and the methods through which the control system is modified as workplace conditions may change. Control methods are typically defined in terms of administrative controls, personal protective equipment (PPE), or engineering controls. The intent of these OHSMS principles is not to prescribe the type of OHS controls the organization should use. Rather, the intent is to provide the principles upon which such decisions may be made.

A successful hazard control system is informed by the evaluation (13.0) and communication (12.0) systems. The design of the hazard control system needs to be considered during system planning and development activities (6.0).

A hazard control system goal should be to follow the widely accepted control hierarchy of: (*1*) hazard elimination; (*2*) use of engineering controls; and, (*3*) use of PPE or administrative controls. For hazards which cannot be eliminated, engineering controls should be the first control option considered. Administrative and PPE controls may be considered and used if hazard elimination and engineering controls are not sufficient to protect worker health and safety.

9.1 Process Design (DESIGN)

Process design addresses OHS concerns and issues associated with the installation of new processes or operations. The principles herein can also be applied to modification of existing processes or operations.

In this context, processes and operations represent a wide range of activities. Examples are the installation of new office workstations; modifications made to an existing manufacturing process; or, development of a new medical waste disposal operation.

This system component does not address the overall design of the OHSMS, those activities are addressed in the system planning and development section (6.0).

Successful process and operation design is informed by input from employees who will be assigned to the new process/operation. Such input should be sought, encouraged, valued, and able to affect the design process.

9.2 Emergency Preparedness and Response System (EMERG)

Emergency preparedness and response refers to the manner in which the organization prepares for and responds to OHS emergencies and accidents. As defined here, emergency preparedness and response system actions are initiated and conducted immediately when events occur that can cause serious illness, injuries, or even fatalities. Emergency response covers many possible hazard scenarios including, for example, evacuation of an office building, spill of a flammable liquid, release of a toxic gas, or incapacitation of workers by unknown agents.

Improper planning or execution of the plan can result in consequences that include life-ending, career-ending, or business-unit bankrupting incidents. Details which may seem minor or innocuous, such as a leaky seal on a pump, improper calibration of an LEL meter, or lack of training for procedures to respond to an enunciator, can have catastrophic implications.

9.3 Hazardous Agent Management System (HAZ)

The term hazardous agent refers to chemicals, biological agents, radioactive materials, and hazardous wastes. The hazardous agent management system is an important component of the more broadly defined hazard control system. The key issues addressed in HAZ OHSMS principles are the identification of hazardous agents, understanding of their risks, control of the risks, and establishment of coordinating mechanisms.

Such hazardous agent management system practices typically include: (*1*) an inventory mechanism; (*2*) labeling of containers/areas which contain hazardous substances; (*3*) notification of employees of potential hazards; and, (*4*) employee training.

The hazardous agent management system activities should be coordinated with the following OHSMS activities: system planning/development (6.0); training system (8.0); hazard control system (9.0); and, evaluation system (13.0).

10.0 Preventive and Corrective Action System (ACTION)

Preventive and corrective action refers to actions taken in response to, or in anticipation of, system breakdowns or high hazard/risk events. A key concept of this component is that actions should be as anticipatory as possible. That is, actions should be taken in advance to prevent an incident or other unplanned event that might adversely affect worker health, or that would require emergency or other response actions.

The identification of preventive and corrective actions should be an integral part of the organization's evaluation system (13.0) efforts. Specifically, self-inspection (13.1) efforts which are conducted on an ongoing basis, should identify potential hazards before they can cause losses. As new OHS hazards surface, or breakdowns/deficiencies in the OHSMS occur, preventive and corrective actions should be initiated to eliminate or reduce the effect that these findings may have on workplace health and safety.

11.0 Procurement and Contracting (BUYING)

Products and contractors can impact workplace health and safety. This system component addresses the need to be aware of such impacts and the need for mechanisms to control them.

A successful OHSMS will have at least minimum requirements for the behavior of contractors while on the organization's premises. In some cases, it may be appropriate for contractors to follow all of the organization's safety rules. In other cases, it may be appropriate for the contractors to follow their own safety rules. Regardless of the approach, the contractor's safety record/history should be included as a selection/award criterion.

A mechanism should be in place to evaluate the manner in which all incoming products or items may affect workplace health and safety. In cases where this can not be determined, a product or item should be restricted from use until its impact can be determined.

3.6.4 Evaluation (Feedback)

OHSMS evaluation represents the act of comparing expected and actual performance levels against established criteria. These findings are in turn communicated and used to improve the OHSMS and overall occupational health and safety performance

12.0 Communication System (COMSYS)

The communication system may be defined and implemented in several ways. In its most basic form, a viable communication system should identify how, and to whom, information for the proper functioning of the OHSMS will be transmitted.

A well-functioning communication system with defined feedback channels is essential for a successful OHSMS. This system component provides the means by which all other system components relate and interact.

12.1 Document and Record Management System (DOC)

The document and record management system addresses the way the organization manages and organizes OHSMS documents and records.

A well-functioning document and record management system is essential in organizations that are pursuing OHSMS registration or certification. From a registrar's or auditor's perspective, the document and record management system provides one of the key indicators of whether the OHSMS is currently in conformance, and whether the probability is good that conformance will be maintained over time.

13.0 Evaluation System (EVAL)

The evaluation system is broadly defined and includes baseline evaluations (6.1), auditing (13.1), self-inspection (13.1), incident investigation (13.2), medical surveillance (13.3), and management review (16.0) activities.

In a systems framework these activities are fundamental to the system's ability to function and sustain itself over time. Each of the identified evaluation system activities is addressed in a separate section. General and methods-related evaluation system principles are presented in this section.

13.1 Auditing and Self-Inspection (AUDIT)

Occupational health and safety auditing and self-inspection are specific activities of the evaluation system (13.0) where information is gathered and assessed on individual OHS programs and systems. These activities include an assessment of changes in OHS hazards and the ability of the OHSMS to respond properly to the changes.

Auditing and self-inspection activities provide essential information to other OHSMS components, including the training (8.0), hazard control (9.0), and preventive and corrective action (10.0) systems. AUDIT OHSMS principles are distinct from management review (Sec. 16.0) functions which address the broader issue of the OHSMS structure in relation to changes in the regulatory environment and stakeholder expectations (e.g. employees, regulators, shareholder, community).

13.2 Incident Investigation and Root Cause Analysis (ROOT)

Incident investigation and root-cause analysis refers to the activities conducted to determine the origin and cause(s) of accidents, near miss accidents, injuries, fatalities, or breakdowns in the OHSMS.

An important aspect of incident investigations is the performance of a root-cause analysis to see at what point(s) the OHSMS failed. This includes establishing and moving up a causation-chain from the incident backwards to the point(s) of origin. The goal is to find the source(s) of the breakdown(s) in the OHSMS so that it (they) can be corrected.

13.3 Medical Program and Surveillance (MEDICAL)

Medical program and surveillance refers to the activities associated with providing medical services within the organization, and the development and operation of a medical surveillance program. An occupational medical surveillance program, when workplace hazards dictate, is a key component of an OHS systems approach.

MEDICAL OHSMS principles do not require the presence of a medical surveillance program, only that the need is assessed. In the case where a program is developed and implemented, the principles offer baseline guidance and criteria.

3.6.5 Improvement/Integration (Open System Elements)

OHSMS improvement refers to the act of using evaluation and assessment findings to improve occupational health and safety performance. Integration refers to the act of integrating OHS with management systems and business functions in the organization, as well as the community. Management review provides the mechanism whereby OHSMS performance is broadly assessed in terms of stakeholder expectations (e.g. employees, shareholders, regulators, community).

14.0 Continual Improvement (IMPROVE)

Continual improvement may be operationally defined and implemented in any number of ways. The basic notion is that the organization should seek ways to achieve ongoing improvement of occupational health and safety performance. The primary goal of continual improvement activities should be to eliminate worker injury and illness.

Continual improvement does not mean or imply a requirement to attain better-than-compliance conditions as measured against specification regulations or standards. While better-than-compliance conditions may be a goal of an organization, it is not a requirement of the definition of continual improvement suggested here.

15.0 Integration (INTEGRT)

Integration refers to the actions the organization takes to integrate its occupational health and safety functions with other management system and business processes in the organization and in the community.

A successful OHSMS requires that the OHSMS be connected, or related, to all other key functions in the organization. This means that occupational health and safety issues and aspects of the OHSMS will be part of the organizational culture. In addition, at a fundamental level, this also means that worker health and safety will be an important value expressed by management and employees alike.

16.0 Management Review (REVIEW)

The overall performance of the OHSMS is evaluated through management reviews. It is through this activity that the OHSMS, the organization, and the environment external to the organization are linked. This involves evaluating the OHSMS's ability to meet the overall needs of the organization, its stakeholders, its employees, and regulating agencies. Management review is different from more specific evaluation system (13.0) efforts which address specific aspects of the OHSMS elements.

Management review is a necessary condition for a successful system. Without consideration of feedback, there can be no meaningful planning (6.0) or continual improvement (14.0).

4 DEVELOPMENT AND IMPLEMENTATION

Numerous implementation steps can be identified in the OHSMS process. Initial steps include the consideration of organizational goals and the means by which the goals will be fulfilled. The implementation process and on-going management of an OHSMS has seven basic steps.

1. Goal identification and characterization of the organization's existing systems and structures.
2. System selection/gap analysis.
3. System formulation and development.
4. System implementation.
5. Preregistration/certification evaluation (if registration/certification is sought).
6. System registration/certification (if necessary).
7. System evaluation/continual improvement.

Steps one through four are necessary for the implementation and ongoing maintenance of an OHSMS. Steps five and six are identified for organizations that wish to pursue registration/certification of their OHSMS against an existing conformity-assessment framework.

OCCUPATIONAL HEALTH AND SAFETY MANAGEMENT SYSTEMS 2281

These seven steps follow the general structure of the universal OHSMS presented in Table 49.2 and Figure 49.5. Based on the universal OHSMS model, management commitment and employee participation are necessary steps to consider prior to embarking on the formulation, and implementation of an OHSMS. Without management commitment, resources, and employee participation, the probability is low that the OHSMS can be successfully implemented and maintained.

4.1 Goal Identification and Characterization of the Organization's Existing Systems and Structures

After management commitment, resources, and employee participation have been secured the organization's existing business systems must be characterized. When considering implementation of an OHSMS, many organizations are surprised to find many aspects of the system may already exist in their organization. That is, all organizations have some existing systems in place that affect the organization's outputs and performance variables. All organizations have some level of performance measurement system or set of practices in place. These systems may not be well defined and will need to be characterized before proceeding with the following steps. Unless there is a specific reason for selecting an existing OHSMS approach (e.g. AIHA OHSMS, BS 8800, Victoria, Australia SafetyMap), the organization may find that based on the level of development of existing management systems, it may be most efficient to develop a hybrid of existing models as to not add a new "layer" on top of existing systems that maybe functioning well. As previously indicated, the universal OHSMS presented in Table 49.2 and Figure 49.5 provides a robust template whereby organization specific approaches can be developed.

4.2 System Selection/Gap Analysis

With the organization's existing business systems characterized, it is then necessary to characterize the existing OHS programs and systems. Again, organizations find that many system components are in place based on efforts to maintain regulatory compliance.

In order to perform an efficient gap analysis of the existing systems and programs, the organization must determine which OHSMS model or approach they wish to implement. If ISO 14001 is already being used for environmental management, this approach should be selected to develop the OHSMS. If ISO 9001 or 9002 are being used in the organization, then the AIHA's OHSMS should be selected. If there are country specific, or trade-block specific considerations, than one of the other approaches presented in Section 3 should be considered.

The gap analysis provides the organization with necessary information to meet the stated goals of the system to be implemented. Specifically, the gap analysis can be characterized as follows (58):

> This is the change between how things, organizations, or processes are performing today (baseline) and how they should be according to your vision, a benchmark, and/or customer requirements.

Gap analysis has become an increasingly important part of any study on how an organization or process works. In an organizational assessment, gaps are measured by looking at the differences between actual performance and expected performance. Learning the reasons for such gaps involves a thorough analysis of organizational structure, business financial requirements, customer expectations, process performance, information technology implementation, skills of workers, and similar aspects that affect organizational performance. This analysis can then provide guidelines for closing gaps.

Prior to selecting an OHSMS model, the organization needs to consider the importance of potential third-party certification of the system, and the importance of the ability to audit the selected approach. That is, some OHSMSs are not considered auditable because they are either explicitly developed as guidance documents, or they contain clauses that include "should" statements as opposed to "shall" statements. Strictly speaking, when an OHSMS model is based on should statements, it is considered a non-auditable standard. This is the case with the British OHSMS BS 8800.

This issue of auditability is important since some organizations may implement an OHSMS in order to gain recognition from stakeholders or markets. Such recognition may require external verification by a third-party registrar or certifier. In the case of OSHA's VPP, recognition is provided directly by OSHA. In the case of ISO-based OHSMSs, recognition can be provided by an accredited registrar. More on these points are discussed in the next section of this chapter on conformity assessment.

4.3 System Formulation and Development

An action plan can be developed using the gap analysis information. The first steps include the development of:

- An OHS Policy:
- OHS and OHSMS goals and objectives:
- Performance variables that will be measured and used to manage the OHSMS.
- an OHSMS procedures manual.

There is debate among professionals in the rapidly developing OHSMS field regarding the best way to develop and implement an OHSMS. A common approach is for an organization to secure the services of a professionals to actually write the system, starting with the OHS policy statement continuing through to preparation of procedures and continual improvement forms. The second approach, and the one recommended here, is to have minimal professional support, in the form of coaching and quality control, with the primary development of the system being performed by the organization's employees. For the long term success and functioning of an OHSMS, it is important that its development is done at a "grassroots" level so there is an intimate understanding by personnel of how the system's various components relate.

4.4 System Implementation

OHSMS implementation involves integrating the necessary system components—identified in the formulation stage—into the organization's existing systems and functions. This step typically takes place in increments over a period of time up to several years. That is, following the development of an OHSMS policy statement which is one of the first steps in OHSMS development, the policy statement is typically widely distributed throughout the organization prior to the implementation of "down-stream" components, such as corrective and preventive action procedures.

Key implementation steps include the development of the following system components:

- Training system.
- Hazard control system.
- Preventive and corrective action system.
- Procurement and contracting system.
- Communication system.
- Evaluation system.

An important aspect of successful implementation is to clearly define the necessary actions needed to conform with the selected OHSMS approach and assign responsibility to specific individuals whereby they are held accountable. The process should be managed by the OHSMS Management Representative with oversight provided through annual management reviews.

4.5 Preregistration/Certification Evaluation

Some organizations may elect to have their OHSMS registered or certified by a third-party. This is typically required or requested by stakeholders or markets. In these instances, it is common to have a pre-registration/certification evaluation performed prior to the actual registration/certification audit. The purpose of the pre-registration/certification evaluation is to identify system elements that may need modifications prior to the actual registration/certification audit.

4.6 System Registration/Certification

System registration/certification involves review and auditing of the OHSMS by an independent third-party registrar. Organizations pursue registration for any number of reasons. For instance, registration may be necessary for entry into specific markets, or it may be required by specific customers. In some case is required for second-party registration by clients of suppliers.

Accreditation bodies currently accredit registrars to perform ISO 14001 environmental management system (EMS) registrations. Registration/certification of occupational health and safety management systems (OHSMS) are evolving. At the time of this publication, no accreditions for OHSMS registration/certification exists. However, some EMS registrars are providing OHSMS auditing and certification services.

4.7 System Evaluation/Continual Improvement

Following the implementation of an OHSMS, regular evaluations of the system are necessary. The goal of these on-going evaluations is to ensure the proper functioning of the system and to identify ways to improve it on an on-going basis.

Many organizations are uneasy with the term "continual improvement." There is a sense that a point may be reached where actions are taken in an inefficient manner simply as a means to fulfill continual improvement requirements. This is not the case. The point is that the organization should "keep its eye on the ball" for ways to improve its OHS performance and aspects of the management system. OHSMSs do not dictate how an organization should define its continual improvement activities, but rather, that there is a structured process whereby such definition is generated that is specific to the organization. As such, an organization wants to carefully think through how it will define its continual improvement component, especially if system registration will be sought.

Evaluation of the OHSMS is essential for the proper functioning of the system and to maintain its ongoing relevance to the organization. Evaluations typically include both self-evaluations that are conducted internally by personnel at the operations level (first-party), as well as less frequent evaluations by individuals who are independent of the operation (second- or third-party.)

5 CONFORMITY ASSESSMENT AND THIRD-PARTY CERTIFICATION

Two topics that are closely aligned with the increase of management system standards, models and approaches are third-party certification and conformity assessment. The development of conformity-assessment structures and use of third-party registrars is prominent with ISO 9001 and ISO 14001. In the United States the use of third-party certifiers has been included in proposed OSHA reform legislation. The use of such third parties with existing OSHA structures has sparked heated policy debates in the OHS profession. This section provides an overview of these topics and their relation to OHS management systems.

A closely related topic is the use of second-parties to perform conformity-assessment determinations. Although the focus here is on third parties, second parties play a potentially large role in conformity-assessment structures, and for some organizations, present more challenges than the use of independent third parties. The distinction between second and third parties is made later in this section. The reader should note however, that the in eyes of many organizations second and third parties are viewed the same.

5.1 Introduction

In the broadest terms, conformity assessment refers to the manner in which conformance/compliance to a given standard/regulation is determined. These determinations are important to consider when OHSMS standards-development and effectiveness determinations are made. In the 1990s the role of third party assessment of OHS management systems and programs have been at the forefront of OHS policy debates. Thus, topics related to management systems and discussed here within the context of conformity assessment are first-, second-, and third-party evaluations; system registration/certification.

For clarity and accuracy, several conformity assessment related terms are defined. Figure 49.6 shows the logical relationship between these distinctions in terms of occupational health and safety.

The National Research Council defines *Certification* as (59):

> The process of providing assurance that a product or service conforms to one or more standards or specifications. Some, but not all, certification programs require that an accredited laboratory perform any required tests.

An example are laboratories who certify products or materials.

Registration can be defined as (60):

> The procedure by which a body indicates relevant characteristics of a product, process or service, or particulars of a body or person, and then includes or registers the product, process or service in an appropriate publicly available list (67). An example is the process organizations go through to become ISO 9001 or 14001 registered.

It follows then that *registered* can be defined as (61):

> A procedure by which a body indicates the relevant characteristics of a product, process or service, or the particulars of a body or person, in a published list. [For example] ISO 9000 registration is the evaluation of a company's quality system against the requirements of ISO 9001, 9002, or 9003.

Accreditation can be defined as:

> A procedure by which an authoritative body formally recognizes that a body or person is competent to carry out specific tasks (60). For example, the entities that perform ISO 9001 and 14001 registration need to be accredited by a nationally recognized body, such as the Registration Accreditation Board (RAB) in the United States.

Finally, *conformity assessment* can be defined as:

> The determination of whether a product or process conforms to particular standards or specifications. Activities associated with conformity assessment may include testing, certification, and quality assurance system registration (59).

The terms "certification" and "registration" are often used interchangeably. Prior to the development of the ISO 9001 quality assurance management system, certification was the term used to describe product conformance to a design specification or standard. This is

	First Party	Second Party	Third Party			
	Self-Declaration of Conformity	Customer Review of Supplier Business Unit	Industrial Hygiene/ Safety Testing and Sampling	Product, Equipment and Laboratory Certification	OHSMS and OHS Program Registration or Certification	
Primary Level: Assessment	Internal testing, sampling and quality assurance By: Organization	Direct testing, sampling, and quality assurance of suppliers or business units By: Organization	Traditional IH/Safety sampling and measurement; workplace characterization By: Consultant or independent lab	Certification of OHS products, equipment, and laboratories against a standard or set of standards By: Product, equipment, or laboratory certifier	Audit and registration of occupational health and management systems (OHMS) and OHS Programs By: System Registrar or Program Certifier	
Secondary Level: Accreditation	Acceptance by customer, market, or regulatory authority By: Organization, external body, or government	Acceptance by customer By: Organization, external body, or government	Accreditation of individual's competence By: Private or government accreditation program; i.e., ABIH, BCSP	Accreditation of product, equipment, or laboratory certifier. By: Certifier accreditation program, private or government	Accreditation of quality system registrar By: Registrar/certifier accreditation program, private or government	
Tertiary Level: Recognition	Recognition of first-party accreditation by customer, market, or regulatory authority	Recognition of second-party accreditation by customers within entire industry sector or regulatory authority	Official recognition or certifier accreditation program By: Government, regulatory statute, or market	Official recognition of certifier accreditation program By: Government, regulatory statute, or market	Official recognition of registrar/certifier accreditation program By: Government, regulatory statute, or market	

Figure 49.6. An OHS Conformity Assessment Model. Adapted from Ref. 66.

the case with, for instance, Underwriters Laboratory (UL) certifications. With the advent of management system standards, and the need to make conformity determinations against them, the term registration was introduced to acknowledge system conformance against a specific system standard. For instance, when an organization is found to be in conformance with ISO 9001 or ISO 14001 its name is place on a registry to acknowledge conformance. The organizations is referred to as an ISO 9001 or ISO 14001 registered site, *not* certified site.

Conformity assessment includes a wide range of approaches applicable to sampling, testing, inspection, evaluation, verification, and assurance of conformity, as well as to the certification of quality system assessment and registration, including various combinations of these procedures.

5.2 Conformity Assessment and Traditional OSHA Enforcement

The command-and-control conformance structures of OSHA and the EPA existed prior to the advent of the quality assurance-related conformity-assessment structures presented here. In relation to OHS, the preamble of the Occupational Safety and Health Act (Act) states that (62):

> [This is] an Act to assure safe and healthful working conditions for working men and women; by authorizing enforcement of the standards developed under the Act; by assisting and encouraging the States in their efforts to assure safe and healthful working conditions; by providing for research, information, education, and training in the field of occupational safety and health and for other purposes.

Assessment of conformance with the standards promulgated under the authority granted by the Act is defined in Sections 8, 9, and 10 of the Act, and OSHA's Field Inspection Reference Manual. Conformance assessment and enforcement interpretations are also impacted by decisions of the Occupational Safety and Health Review Commission (OSHRC), and the outcomes of judicial review (62). Nonconformance with OSHA standards and regulations is determined by OSHA inspectors. For instance, when a nonconforming situation is found, a citation is issued.

The agency's means for meeting the Act's enforcement mandate is accomplished through:

1. Traditional enforcement programs.
2. Cooperative assistance and recognition programs.
3. Cooperative enforcement programs.

Traditional enforcement programs include programmed inspections and special-emphasis programs. With programmed inspections, companies are subject to random unannounced inspections scheduled by means of an OSHA list of targeted employers or industries. Candidates for programmed inspections are employers who have declined to participate in a partnership/cooperative program, companies that have poor OHS histories, or industries that have high injury and illness rates as determined by Bureau of Labor

Statistics (BLS) data. Special-emphasis programs may be regional or national, with the focus on specific industries, hazards, or areas. Candidates for special-emphasis programs are specific industries, for example, nursing homes, or hazards, such as silica or lead in construction (63).

Examples of cooperative-assistance and recognition programs are the Voluntary Protection Program (VPP) and the Safety and Health Achievement Recognition Program (SHARP). Examples of cooperative enforcement programs are Local Problem Solving Initiatives (LPSI), such as the GRIP program (Getting Results and Improving Performance) and Cooperative Compliance Programs (CCP), such as the Maine 200-type programs (63).

In the case of OSHA programs which currently use some form of third-party activity, the agency does not delegate its enforcement responsibility. Rather, these programs establish criteria that third parties must meet in order to be permitted to perform the verification function. OSHA staff oversee the performance of the certification to ensure that the third parties meet these criteria.

One of the criticisms of failed OSHA-reform legislation in the 1990s, especially related to the third-party issue, was that the bills would have severely limited OSHA's traditional enforcement capabilities (64).

5.3 Conformity Assessment and the International Organization for Standardization (ISO)

The origins of numerous existing and evolving EHS and OHS conformity-assessment structures can be traced to activities of ISO. The development of conformity-assessment policies and procedures in ISO is the responsibility of ISO's Committee on Conformity Assessment (CASCO) (62), CASCO, in its current form, was created in 1985 as a successor to the former CERTICO—the ISO committee on certification—which was initially established in 1970. The initial focus of CERTICO was on the principles and practices of product certification. The change in name from CERTICO to CASCO explicitly reflected the extension of the field of certification activities from product certification to all conformity-assessment activities, in particular that of quality systems, thereby responding to market demand and to the publication of the ISO 9000 series standards. CASCO's primary task is to develop guidance documents for use by ISO Technical Committees (TC) in the development of specific ISO standards (e.g., ISO 9001 and ISO 14001).

The current trend in CASCO is to prepare generic guidelines whereby TCs can develop conformity-assessment schemes unique to the TCs specific industry/service sector while maintaining some level of uniformity between the conformity-assessment work of different TCs (e.g., accreditation of different kinds of conformity-assessment bodies: laboratories, inspection bodies, certification bodies, etc.) (65).

5.4 An OHS Conformity—Assessment Model

In 1995, the National Research Council's (NRC) Science, Technology, and Economic Policy Board published a report titled "Standards, Conformity Assessment, and Trade (59). The report was issued pursuant to Public Law 102-245, which was enacted to gain a better understanding of the interrelationship between standards, conformity assessment, and international trade.

The NRC report presents a generic conformity-assessment model that can be applied to manufacturing, service delivery, or quality management systems. For use here, the NRC model has been modified specifically for OHS conformity assessment. This new model is presented in Figure 49.6. The model presents a conformity-assessment hierarchy with three levels that includes assessment, accreditation, and recognition.

5.4.1 Primary Level—Assessment

The primary level involves assessment of an organization's conformity with a protocol, standard, or system. An example of a primary-level activity in OHS is the traditional characterization of workplace hazards, which might involve the collection of air samples and noise dosimetry measurements. Additional terms used to describe this level include auditing and evaluation.

5.4.2 Secondary Level—Accreditation

The second level of the hierarchy addresses the accreditation of the person or organization who performs the first-level assessment activities. Examples of this are seen in the certification of health and safety professionals by the American Board of Industrial Hygiene (ABIH) and the Board of Certified Safety Professionals (BCSP). These two bodies provide the certification designations "Certified Industrial Hygienist" (CIH) and "Certified Safety Professional" (CSP), respectively. Another example of a secondary-level activity can be seen in the ISO 9000 series standards where accreditation bodies accredit quality-system registrars. For an ISO 9000 registration to be valid, the primary-level assessment activity must be performed by a registrar accredited by a recognized accreditation body. In the United States, the ISO 9000 accreditation body is the Registration Accreditation Board (RAB), which is an affiliate of the American Society for Quality Control (ASQC).

5.4.3 Tertiary Level—Recognition

Finally, the third level of the conformity-assessment hierarchy involves recognition of the second-level accreditation activities. Recognition can be given either by customers, marketplace demands, or governmental regulating body. An example of such recognition is observed in a number of the asbestos-control regulations that require the use of either CIHs or other certified individuals in clearance activities following asbestos-abatement projects. Still another example of such recognition is observed in OSHA's recognition of the College of American Pathology in the Blood Laboratory Certification Program. With the ISO 9000-series standards, recognition is provided by markets (e.g., European Union) and customers (e.g., U.S. Department of Defense).

5.4.4 First-, Second-, and Third-Parties

The term first-party refers to an organization's internal conformity-assessment activities; sometimes called self-assessment, or self-auditing. This can include all three levels of the hierarchy, assessment, accreditation, and recognition. The organization can have internal assessors who are, in some fashion, internally certified to perform the assessment. The term second-party refers to the activities of a customer assessing supplier organizations. Finally, the term third-party refers to the activities of a person or body who is independent of either the organization or its customers. The third-party columns in Figure 49.6 depicts three categories that include:

- Industrial Hygiene/Safety Testing and Sampling.
- Product, Equipment, and Laboratory Certification.
- Occupational Health and Safety Management System (OHSMS) and OHS Program Registration/Certification.

There is generally good agreement about the description of the conformity-assessment hierarchy presented by the rows in Figure 49.6. However, this agreement begins to break down when the framework's columns are defined; specifically, when describing first-party, second-party, third-party, and regulating-government-body roles and activities in the various levels of the conformity-assessment hierarchy.

Furthermore, there is disagreement among OHS and conformity-assessment professionals regarding the extent to which OSHA's traditional enforcement role can be described in terms of the distinctions in Figure 49.6. Some argue that OSHA's role can be described in terms of the "third party" columns. Others argue that it is not accurate to think of OSHA in terms of a third party since the agency is not truly independent of the process (67). As the following discussion demonstrates, some OSHA programs and standards do fit the model, while others do not.

5.5 Third-Party Certification/Registration in Occupational Health and Safety

The use of third parties in conformity-assessment activities continues to be a prominent topic in OHS policy formulation discussions. With this continuing attention and the implementation of ISO 14001, it is necessary to establish a framework to describe and analyze the third-party issue.

During the mid to late 1990s the use of third parties in OHS has been an issue in legislative debates, OSHA policy discussions, and the OHS community in general (68–71). In these contexts, the term third-party commonly refers to the use of qualified individuals or entities that provide conformity-assessment determinations on aspects of an organization's OHS programs and management systems. The elevation of the third-party issue in these arenas can be attributed to four factors.

The first factor is the proposed but not adopted OSHA-reform legislation of the 104th Congress, which contained requirements for the use of third-parties in activities traditionally performed by OSHA (72, 73). These bills never reached the floors of the House or Senate for a vote. Nevertheless, the hearings conducted in connection with this legislation increased the attention given to the third-party issue (74, 75). The second relates to the conformity-assessment structure developed by ISO, as reflected in ISO 9001 and ISO 14001. Specifically with ISO 9001, third-party registrars play an integral role in this standard's conformity-assessment scheme. The third factor relates to the reference in the National Performance Review (NPR) to the potential use of third parties in the Department of Labor (76). The inclusion of this recommendation in the NPR was referenced often in 104th Congress OSHA-reform hearings. Specifically, the NPR recommended that "the Secretary of Labor issue new regulations for work-site safety and health, relying on private inspection companies or non-management companies (76)." Finally, it should be noted that the origins of the NPR recommendation mentioned above are based on various privatization models which have been discussed in the public management literature since the early 1980s. Such approaches are usually presented as a means of meeting legislated mandates in the face of diminishing resources and in response to calls for increased industry-government cooperation (77, 78).

An analysis of 22 conformity-assessment schemes using the OHS conformity assessment model presented in Figure 49.6 reveals that conformity assessment can be broadly defined and is not restricted to third-party certification; and that, in fact, the development and consideration of OHSMS standards can proceed independently of its conformity-assessment structures. Specifically, the analysis revealed that (79).

1. There are existing conformity-assessment structures upon which a robust occupational health and safety conformity-assessment framework, that incorporates the use of third-party registrars, can be based.
2. There are numerous third-party conformity-assessment models currently in use in the OHS field.
3. An OHSMS conformity-assessment scheme must be based on valid and reliable assessment methods.
4. Despite the recent attention given to OHSMSs, there has been relatively little attention given to the effectiveness of such systems and the means by which they will be evaluated.
5. There has been little, if any, scholarly or systematic examination of the validity or reliability of assessment instruments used in either ISO 9000 or ISO 14000 assessments.
6. No universal assessment instruments for ISO 9001, ISO 14001, or the prominent OHSMSs have been developed (75).

5.6 Conformity Assessment in OSHA and other OHS Programs

Third-party activities can be identified in a number of existing OSHA and other OHS programs managed by non-governmental organizations in the private and not-for-profit sectors. Some of the programs do not have a third-party component, while others can roughly be defined by the terms "OHSMS and OHS Program Registration." No OSHA program includes a third-party certification component that is as comprehensive as the approaches suggested in OSHA-reform legislation proposed in the 104th Congress. Furthermore, OSHA does not have a mechanism to accredit third-party certifiers as proposed in the referenced legislation.

5.6.1 Nationally Recognized Testing Laboratory Program

In this recognition program, OSHA accredits laboratories to test and certify products for use in the workplace. In an effort to overhaul its regulatory procedures related to the testing and certification of select workplace equipment and materials, OSHA published its initial NRTL proposal on March 6, 1984 (80).

The NRTL program can be described by the "product, equipment, and laboratory certification" column depicted in Figure 49.6. In the NRTL program, the primary level (assessment) is performed by the NRTL program laboratories. The secondary level (accreditation) is performed by OSHA's Office of Variance Determination. Recognition of the primary and secondary-level activities is provided by OSHA, state and local governmental bodies, insurance companies, and consumers.

5.6.2 Asbestos Standards (1994 revisions)

On August 10, 1994, OSHA issued final revised standards for occupational exposure to asbestos in general industry (29 CFR 1910.1001) and the construction industry (**29 *CFR* 1926.1101**, formerly **29 *CFR*** 1926.58). In addition, the agency included a separate standard covering occupational exposure to asbestos in the shipyard industry (**29 *CFR*** 1915.1001). Each of the standards contains requirements for the use of industrial hygienists, certified by the American Board of Industrial Hygiene (81).

The revised Asbestos Standards can be described by either the "Industrial Hygiene/Safety Testing/Sampling" or "OHSMS/OHS Program Registration" columns depicted in Figure 49.6 depending upon whether the context is bulk material sampling or work-practice review.

With the Asbestos Standards, the primary level (assessment) is performed by the certified individual (e.g., CIH, PE, or accredited inspector). The secondary level (accreditation) is performed by the certifying/licensing body. In the case of CIHs this would be the American Board of Industrial Hygiene (ABIH). For PEs this would be state licensing boards; and, in the case of accredited inspectors, it would be an accreditation body that is recognized by OSHA or EPA. Recognition of the primary and secondary-level activities is provided by OSHA, state and local governmental bodies, insurance companies, and consumers (typically building owners/managers).

5.6.3 Special Government Employees in Cooperative Programs

The Voluntary Protection Programs (VPP) conformity-assessment process is outlined in OSHA Instruction TED 8.1a, issued May 24, 1996 (82). The process involves: a "desk review" of a site's application package and supporting documentation; an on-site review; analysis of findings and recommendations by the regional OSHA office and the national Office of Cooperative Programs; and, final consideration (approval/disapproval) by the assistant secretary.

Due to the expansion of the VPP at a time of limited OSHA resources, the agency has sought new ways to meet these increased needs. TED 8.1a, appendix J presents requirements whereby volunteers (special government employees) can be used in the VPP conformity-assessment process. The use of such volunteers is primarily for on-site review activities. The use of volunteers is authorized, pursuant to the provisions of Section 7(c)(2) of the OSH Act, which authorizes the Assistant Secretary of Labor for Occupational Safety and Health to "employ experts and consultants or organizations" (62).

As in the case of the use of third parties in corporate-wide settlement agreements, the use of "OSHA volunteers" in the VPP does not fit well in the OHS conformity-assessment framework presented in Figure 49.6. The activities of the volunteers are supervised by OSHA personnel; that is, the role of the volunteer(s) is defined by the VPP team leader. To the extent that this program can be described in terms of the OHS Conformity Assessment framework in Figure 49.6, special governmental employees would be classified in the "OHSMS/OHS Program Registration/Certification" column.

5.6.4 American Industrial Hygiene Association Laboratory Accreditation/ Registration Program/NIOSH PAT program

The American Industrial Hygiene Association (AIHA) provides two accreditation and registration programs for industrial hygiene and environmental laboratories: the Industrial Hygiene Laboratory Accreditation Program (IHLAP); and the Environmental Lead Laboratory Accreditation Program (ELLAP) (83). The AIHA also manages the Asbestos Analysts Registry (AAR). Integral to the AIHA accreditation programs are the use of the NIOSH Proficiency Analytical Testing (PAT) Program results. The PAT program provides quality control reference samples to over 1300 occupational health and environmental laboratories in 18 countries (84).

The AIHA accreditation/registry and NIOSH PAT programs can be described by the "product, equipment, and laboratory certification" column depicted in Figure 49.6. With the AIHA programs, the primary level (assessment) is performed by the accredited laboratory or registered analyst who performs the actual analysis. The secondary level (accreditation) is given by the AIHA. Recognition of the primary and secondary levels of the AIHA programs is provided by federal, state, and local agencies as well as by building owners, employers, unions, and many OHS professionals.

5.7 Conformity Assessment in ISO-Based Approaches

Although it is difficult to identify any single event or trend which can be credited with bringing the third-party issue to the center of OHS policy discussions in the mid 1990s, the activities of ISO can be identified as an international reference point which has provided ideas and models for consideration in the United States.

Beginning with the ISO 9000 series standards, which were first introduced in the early 1980s and present a quality-assurance management-system (QAMS) model, the phenomenon of third-party registration of management systems was rapidly adopted internationally. Third-party registration plays a central role in the ISO 9000 series of quality-assurance-systems standards. Organizations may implement one of the three ISO 9000 standards (ISO 9001, 9002, or 9003) without pursuing registration of the system, but such registration is often needed by the organization for market-entry or is required by individual clients. Registration to one of the ISO 9000 standards is accomplished through the use of accredited third-party registrars.

ISO began consideration of environmental-management system (EMS) standards in the early 1990s following promulgation of the ISO 9000-series standards in the 1980s. As extension of the ISO 9000 model to the environmental management arena was being considered, extension to the occupational health and safety arena was also being explored. In October 1996, ISO published an EMS standard, ISO 14001, which is part of the ISO 14000 series.

ISO management system models, such as ISO 9001 and ISO 14001 are developed by technical committees which follow the conformity-assessment guidelines developed by CASCO. Even though ISO has not developed a specific OHS standard, over the past few years many organizations have been re-structuring their EHS programs/systems using the ISO 9001 and ISO 14001 as guides.

5.7.1 ISO 14001—ISO's Environmental Management System

In October 1996, ISO published five international standards which are part of the overall ISO 14000 series of standards relating to environmental management. ISO 14001:1996 represents the centerpiece of the series standards because it establishes the framework for a comprehensive EMS. As such, it is the only "auditable" standard included in the series (85). The remaining standards published in 1996 and anticipated future ISO 14000 series standards are (will be) for guidance purposes only.

Although ISO 14001:1996 does not contain any requirements to include OHS directly in an ISO 14001 program, an organization may, if it elects to do so, include OHS in its EMS. Such an approach would result in an integrated EHS management system. The ISO 14010:1996, 14011-1:1996, and 14012:1996 documents provide guidance at the conformity-assessment primary level. The secondary and tertiary level structures vary between countries.

In Europe, accreditation programs have been developed by the United Kingdom Accreditation Service (UKAS), formerly the National Accreditation Council for Certifying Bodies (NACCB), and the Raad voor de Accretitatie (RvA) of the Netherlands (accepted throughout Europe and internationally), formerly the Raad voor de Certificate (RvC) (52). In the United States, the American National Standards Institute (ANSI) and the Registration Accreditation Board (RAB) have developed a joint National Accreditation Program (NAP) that is administered by the RAB. The ANSI/RAB program conducted a pilot test in 1996 and formally began its accreditation program in early 1997 (86).

Because it is a relatively new standard, it is not clear at this time how recognition of ISO 14001:1996 will proceed. While the U.S. Environmental Protection Agency (EPA) did actively participate in the ISO 14001 development process, the agency has not taken an official position on the standard (86). Mary McKiel, who was responsible for coordinating the EPA's activities with respect to the standard, has said: "The agency hasn't come out at the political level with any kind of statement about ISO 14000, but more and more in the agency, the 14000 standards are being looked at and sought out as tools that can be used." She continued: "it is likely that the EPA will continue to incorporate the coming international documents in its myriad of voluntary programs, and possibly even into some of its criminal activities (86)."

It is not clear at this time how markets will respond to the ISO 14000 series standards. While it is clear that many organizations have begun to use ISO 14001:1996 as a means to organize their environmental management efforts, and in some cases their OHS efforts, it is not as clear how many organizations will pursue registration of their EMSs. A poll conducted by *Quality Systems Update* in the Summer of 1996 of all ISO 9000 registrars operating in North America found that fewer than 40 percent of the registrars' clients intended to pursue EMS registration (86).

5.7.2 QS 9000—Auto Industry Application of ISO 9000

The discussion of QS 9000 is included here because the standard addresses OHS and environmental management issues in several clauses, and because its conformity-assessment scheme presents several novel aspects not found in ISO 9000 nor ISO 14001 schemes. Unlike its QAMS cousin ISO 9001, QS 9000 has clearly stated OHS requirements.

Quality System Requirements, QS-9000 was developed by the Chrysler, Ford, and General Motors Supplier Quality Requirements Task Force. Previously, each company had developed its own expectations for quality systems and the corresponding assessment documents. This was a second-party auditing system in which the customer audited the supplier. The QS 9000 requirements are mandatory requirements established by two of the big three automotive customers. Chrysler's specific requirement is "all production and service parts suppliers to Chrysler must be third-party registered to QS 9000 by July 31, 1997." General Motor's specific requirement is "all production and service parts suppliers to General Motors must be third-party registered to QS 9000 by December 31, 1997 (87)."

After creating QS 9000, the Supplier Quality Requirements Task Force sought assistance from the Automotive Industry Action Group (AIAG) for document distribution and training coordination. The International Automotive Sector Group (IASG) was given the task of administering various facets of the QS 9000 conformity-assessment system. One of the functions the IASG performs is coordination of accreditation activities. Neither AIAG nor IASG provides registrar accreditation, but both cooperate with recognized accreditation bodies, such as the RvA, RAB, or UKAS in assuring proper interpretation of the standard.

Specifically, the IASG meets regularly to discuss and resolve interpretation issues relative to the QS 9000 criteria and third-party registration of suppliers to QS 9000. The published "agreed upon" interpretations are sanctioned and recognized by Chrysler, Ford, General Motors, the Supplier Quality Requirements Task Force, the participating ISO 9000 accreditation bodies, and QS 9000 qualified registrars (87).

The IASG membership includes representatives from: Chrysler, GM, and Ford; the Supplier Quality Requirements Task Force; QS 9000 qualified registrars; and, automotive suppliers (88). In the QS 9000 model, the final conformity-assessment level, recognition, is provided the Big Three auto makers, as well as other customers, including large and small companies.

One of the most intriguing aspects of the QS 9000 model is the role the IASG has in providing official interpretations of the standard. This is a function unique to the QS 9000's conformity-assessment framework. These unique features address some of the consumer concerns with third-party registration/certification. Specifically, issues of registrar/certifier misconduct, abuse, and possible fraud can potentially be detected by the IASG if a complaint is made from a registrar/certifier client or competitor. The role of the IASG also addresses concerns raised in the literature regarding reliability and validity of measurement of various management systems, whether they are QAMSs, EMSs, or OHSMSs. By virtue of having a body that provides sanctioned interpretations of the standard and its application, it is possible to improve upon the system's overall reliability and validity in ways not possible in other conformity-assessment structures.

The programs/systems of particular interest for consideration in future third-party policy formulation discussions are the AIHA/NIOSH relationship in certifying laboratories, and the QS 9000 conformity-assessment scheme. The laboratory certification program, the QS 9000 conformity-assessment scheme, provide robust models for consideration in future OHS policy formulation discussions.

5.8 The OHSMS Certification Debate

As indicated there has been debate in the 1990s regarding the development of OHSMS standards both at the nation-state and international levels, and potential requirement either by governmental agencies or markets that such standards contain a certification component. To this end, the majority of consensus-standards developed by national-standards bodies of different countries are non-auditable standards or guidance. The main reasons for this are

- The existing, and sometimes comprehensive, legislative framework which many countries have for OHS, and which it has been argued are sufficient in themselves to bring about improvements in health and safety if properly enforced.
- A "backlash" against certifiable specification standards, following introduction of the ISO 9001 and ISO 14001 certifiable standards. This is arguably due mainly to the over-bureaucratic way in which they have been implemented in organizations, and the over zealousness with which paperwork and records have been examined by external assessors, to the detriment of assessment of how 'real-life' health and safety issues are managed.

The reservations of some about the merit of certifiable standards, has not dampened the demand for a certifiable standard to put OHS on an equal footing with environment and quality management systems.

The experience in the UK is a case in point. Over 7000 copies of the OHSMS guidance document, BS8800, were sold in the first 12 months following initial publication. During this period and after, it become apparent that there is a substantial and growing demand for independent verification and recognition of achievement in OHS. The absence of any "officially accredited" certification scheme for those who implement BS8800, has led to a growth in nonaccredited proprietary OHSMS and non-accredited certification schemes, which is seen as undesirable. However, this was an entirely predictable response to a demand from employers who cannot be satisfied by the national standards organizations under present constraints. Similarly, many commercial health and safety consultancies have responded to the demand from their clients by offering OHSMS review and certification services in accordance with BS8800.

In July 1998, the general level of concern about the situation lead to the British Standards Institution (BSI) publishing a draft "Product Assessment Specification" (PAS) against which BS quality assurance auditors will assess the conformity of organizations with the guidance given in BS8800, even though the latter makes it clear than it is not intended for certification purposes. This change in policy between April 1996 when BS8800 was published, and July 1998 is solely a result of the demand from organizations to have their OHS management and performance reviewed by an independent third party, and certificated if it conforms to the relevant standard. It is expected that this change in policy will be a catalyst in the certification marketplace in the UK, and result in many more organizations going for certification to BS8800.

This reflects the fact that certification to a standard is, in general something which organizations strive for, so as to have their achievements recognized and to gain competitive advantage. Attitudes about compliance with legal requirements are often different, and obeying the law is regarded as a duty for which one receives little recognition or credit. These and other arguments were discussed in detail at the session on "International Standard of Occupational Hygiene Management" at the IOHA 3rd International Scientific Conference in September 1997 in Crans Montana, Switzerland.

Ultimately, it is predicted that certification will be available for many OHSMS standards either through "official channels" or through the entrepreneurial activities of quality, environment and health and safety auditors, as this is what many organizations want. There is concern about potential barriers to trade emerging if national standards proliferate. An international standard, which offers internationally recognized certification, is the single most effective way of avoiding this undesirable situation.

6 CONCLUSION

At the time of the writing of this manuscript, the American Industrial Hygiene Association had been given provisional authority to form a Secretariat for ANSI to develop an American national Occupational Health and Safety Management System Standard. The Secretariat would be called Z-10. The active involvement of the appropriate stakeholder groups will result in an American national standard that can be used in the U.S., by U.S.-based multinational organizations, and globally. This standard will also form one of the bases upon which an international standard will ultimately be developed for ISO and ILO.

In conclusion, the practicing Industrial Hygienist, safety engineer, or environmental engineer should understand that Occupational (and Environmental) Health and Safety Management Systems represent a proven tool with which we can effectively protect workers and the environment while increasing organizational productivity and effectiveness.

BIBLIOGRAPHY

1. National Safety Council; *Accident Facts, 1997 Edition*, Itasca, IL, 1997.
2. J. P. Leigh, "Occupational Injury and Illness in the United States," *Arch. Internal. Med.*, **157**(July 28, 1997).

3. International Organization for Standardization, *Quality Systems—Model for Quality Assurance in Design, Development, Production, Installation and Servicing*, International Standard ISO 9001: 1994(E), Geneva, Switzerland, 1995.
4. International Organization for Standardization, *Environmental Management Systems-Specifications with Guidance for Use*, International Standard ISO 14001, Geneva, Switzerland, 1996.
5. U.S. Department of Labor, Occupational Safety and Health Administration, "Voluntary Protection Programs to Supplement Enforcement and to Provide Safe and Healthful Working Conditions," *Fed. Reg.* **53**(133) 26339–26348 (July 12, 1988).
6. United States Department of Labor, Occupational Health and Safety Administration; "Safety and Health Program Management Guidelines," *Fed. Reg.* (Jan. 26, 1989).
7. M. Majewski, Presentation given at the American Industrial Hygiene Conference and Exhibition, Washington, DC, May 22, 1996.
8. Voluntary Protection Program Participants Association, *Benefits of VPP Participation: Data from VPP Sites*, Falls Church, VA, 1996.
9. United States Department of Labor, Occupational Health and Safety Administration; "Proposed Safety and Health System Standard for General Industry, 1910.700" Draft (confidential) agency document, Docket S-027, July 2, 1996.
10. American Industrial Hygiene Association, *Occupational Health and Safety Management System: An AIHA Guidance Document*, AIHA Publications, Fairfax, VA, 1996.
11. American National Standards Institute, "Workshop on International Standardization of Occupational Health and Safety Management Systems: Is there a Need?" Workshop proceedings, Rosemont, IL, May 7–8, 1996.
12. F. Mirer, Presentation given at "Workshop on International Standardization of Occupational Health and Safety Management Systems: Is there a Need?" Rosemont, IL, May 7, 1996.
13. S. P. Levine and D. T. Dyjack, "Development of an ISO 9000-Compatible Occupational Health Standard—II: Defining the Potential Benefits and Open Issues," in C. F. Redinger and S. P. Levine, eds. *New Frontiers in Occupational Health and Safety: A Management Systems Approach and the ISO Model*, AIHA Publications, Fairfax, VA., 1996 p. 122.
14. D. S. Markey and S. P. Levine, "Conformance of ISO OHSMS Standards in Public-Sector Procurement Specifications to GATT/WHO Requirements," in C. F. Redinger and S. P. Levine, eds., *New Frontiers in Occupational Health and Safety: A Management Systems Approach and the ISO Model* AIHA Publications, Fairfax, VA, 1996.
15. G. Pickett and J. Hanlon, *Public Health Administration and Practice*, Times Mirror/Mosby College Publishing, Los Altos, CA, 1990.
16. G. D. Clayton, and F. E. Clayton, eds., *Patty's Industrial Hygiene and Toxicology*, Volume I, *General Principles*, John Wiley & Sons, New York, 1978.
17. World Health Organization *WHO Technical Report Series, No. 66* Geneva, Switzerland, 1963, p. 4.
18. J. K. Corn, *Response to Occupational Health Hazards: A Historical Perspective*, Van Nostrand Reinhold, New York, 1992.
19. J. M. Mendeloff, *The Dilemma of Toxic Substance Regulation: How Overregulation Causes Underregulation at OSHA*, The MIT Press, Cambridge, MA, 1988.
20. D. Dyjack, *Development and Evaluation of an ISO 9000-Harmonized Occupational Health and Safety Management System*, Doctoral Dissertation, University of Michigan, Ann Arbor, MI, 1996.

21. S. Kelman, "Occupational Safety and Health Administration," in J. Q. Wilson, ed., *The Politics of Regulation*, Basic Books, Inc., New York, 1980, pp. 236–266.
22. S. A. Roach and S. M. Rappaport, "But they are not Thresholds: A Critical Analysis of the Documentation of Threshold Limit Values," *Am. J. Industrial Med.* **17** 727–753 (1990).
23. L. Salter, *Mandated Science: Science and Scientists in the Making of Standards*. Kluwer Academic Publishers, Boston, MA, 1988.
24. M. Corn, "Research & the Future of the Profession," Presentation given at the American Industrial Hygiene Conference and Exhibition, Kansas City, MO, May 23, 1995, Forum no. 3 titled, *Setting a National Research Agenda for Industrial Hygiene*.
25. S. P. Levine, Presentation given at the American Industrial Hygiene Conference and Exhibition, Kansas City, MO, May 23, 1995, Forum no. 3 titled, *Setting a National Research Agenda for Industrial Hygiene*.
26. United States Department of Health and Human Services, National Institute of Occupational Safety and Health *National Occupational Research Agenda*, Washington, DC, 1996.
27. M. W. Lipsey, "Key Issues in Intervention Research: A Program Evaluation Perspective," *Am. J. Industrial Med.* **29** 298–302 (1996).
28. C. Needleman and M. Needleman, "Qualitative Methods for Intervention Research," *Am. J. Industrial Med.* **29**, 329–337 (1996).
29. H. T. Chen and P. Rossi, eds., *Using Theory to Improve Program and Policy Evaluations*, Greenwood Press, New York, 1992.
30. L. Goldenhar and P. Schulte, "Methodological Issues for Intervention Research in Occupational Health and Safety," *Am. J. Industrial Med.* **29** 289–294 (1996).
31. L. Goldenhar and P. Schulte, "Intervention Research in Occupational Health and Safety," *J. Occupat. Med.* **36**(7) 763–775 (1994).
32. J. M. Shafritz and S. J. Ott, *Classics of Organizational Theory*, Brooks/Cole Publishing Company, Pacific Grove, CA, 1992.
33. F. Taylor, "Principles of Scientific Management," reprinted in *Classics of Organizational Theory*, Brooks/Cole Publishing Company, Pacific Grove, CA, 1916.
34. M. P. Follet, "The Giving of Orders," reprinted in *Classics of Organizational Theory*, Brooks/Cole Publishing Company, Pacific Grove, CA, 1926.
35. C. Argyris, *On Organizational Learning*, Blackwell Publishers, Inc., Cambridge, MA, 1992.
36. C. Barnard, *The Function of the Executive*, Harvard University Press, Cambridge, MA, 1938.
37. L. von Bertalanffy, "The Theory of Open Systems in Physics and Biology," *Science* **111**, 23–29 (Jan. 13, 1950).
38. D. Katz and R. Kahn, *The Social Psychology of Organizations*, John Wiley & Sons, New York, 1966.
39. *Occupational Health and Safety Management Systems—General Guidelines on Principles, Systems and Supporting Techniques*, AS/NZS 4804:1997 (1997), published jointly by Standards Australia (1 The Crescent, Homebush, NSW 2140, Australia) and Standards New Zealand (Level 10, Radio New Zealand House, 155 The Terrace, Wellington, 6001 New Zealand).
40. *Safety Management Achievement Program (SafetyMap)*, Health and Safety Organisation, Victoria, Australia, 1994.
41. The Oil Industry International Exploration and Production Form (E&P Forum); *Guidelines for the Development and Application of Health, Safety and Environmental Management Systems*, Report No. 6.36/210, 1994.

42. *Occupational Health and Safety Management System (OHS-MS): JISHA Guidelines*. Japan Industrial Safety & Health Association, 5-35-1 Siba Minato-ku, Tokyo, Japan 108-0014, 1998.
43. *Dutch Technical Report: Guide to an Occupational Health and Safety Management System; NPR 5001*, Nederlands Normalisatie-Instituut; Kalfjeslann 2, Postbus 5059, 2600 GB Delft, 1997.
44. *The NOSA 5 Star Safety & Health Management System; Reg. No. 51/00010/08: HB 0.00.50E*, National Occupational Safety Association, P.O. Box 26434, Arcadia, 0007, Republic of South Africa, 1992.
45. *Occupational Health & Safety Management Systems: BS 8800:1996*, British Standards Institute, 389 Chiswick High Road, London, W4 4AL, United Kingdom, 1996.
46. *Occupational Health & Safety Management Systems—Specification; OHSAS 18001:1999* (1999). British Standards Institute, 389 Chiswick High Road, London, W4 4AL, U.K., 1999
47. *Responsible Care: Employee Health and Safety Code of Management Practice*, Chemical Manufacturers' Association, 1300 Wilson Boulevard, Arlington, VA 22209, 1994.
48. C. F. Redinger and S. P. Levine, "Development and Evaluation of the Michigan Occupational Health and Safety Management System Assessment Instrument: A Universal OHSMS Performance Measurement Tool," *Am. Industrial Hyg. J.* **59**, 572–581 (1998).
49. C. F. Redinger and S. P. Levine, *Occupational Health and Safety Management System Performance Measurement: A Universal Assessment Instrument*, AIHA Press, Fairfax, VA, 1999.
50. G. D. Brewer and P. deLeon *Foundations of Policy Analysis*, The Dorsey Press, Chicago, IL, 1993.
51. United States Department of Labor, Occupational Health and Safety Administration *Managing Worker Safety and Health*, Office of Cooperative Programs, Washington, DC Internal OSHA document, Nov. 1994.
52. E. Babbie, *The Practice of Social Science Research*, Wadsworth Publishing Company Belmont, CA, 1992.
53. E. J. Pedhazur and L. P. Schmelkin, *Measurement, Design, and Analysis: An Integrated Approach*, Lawrence Erlbaum Associates, Publishers, Hillsdale, NJ, 1991.
54. F. E. Bird and G. L. Germain, *Practical Loss Control Leadership*, International Loss Control Institute, Loganville, GA, 1990.
55. P. Johnson, *ISO 14000 Environmental Auditor Training Manual*, Perry Johnson, Inc., Southfield, MI, 1996.
56. S. P. Levine and D. T. Dyjack, "Critical Features of an Auditable Management System for an ISO 9000-Compatible Occupational Health and Safety Standard," *Am. Ind. Hyg. J.* **58**, 291–298 (1997).
57. W. E. Deming, *Out of The Crisis*, MIT Press Cambridge, MA, 1986.
58. J. Cortada and J. Woods, *McGraw-Hill Encyclopedia of Quality: Terms and Concepts*, McGraw-Hill, Inc., New York, 1995.
59. National Research Council; *Standards, Conformity Assessment, and Trade: Into the 21st Century*, National Academy Press, Washington, DC, 1995.
60. International Organization for Standardization: *General Terms and Definitions Concerning Standardization and Related Activities, ISO/IEC Guide 2*, Geneva, Switzerland 1991.
61. R. W. Peach, *The ISO 9000 Handbook*, 2nd ed., CEEM Information Services, Fairfax, VA, 1995.
62. "Occupational Safety and Health Act of 1970," Public Law 91-596 (1970).
63. Bureau of National Affairs, Inc. *Occupational Safety & Health Reporter*, (Sept. 11, 1996).

64. F. Mirer and M. Wright, Presentations given at the American National Standards Institute's "Workshop on International Standardization of Occupational Health and Safety Management Systems: Is there a Need?" Workshop Proceedings, Rosemont, IL May 7, 1996.
65. ISO Committee on Conformity Assessment: *Information Provided by the ISO Observer at the WTO/TBT meeting, 1 March 1996, on ISO/IEC Guides in the Conformity Assessment Field Supporting the Reduction of Technical Trade Barriers to Trade*, Document number CASCO/WTO-96-03-01, obtained from the American National Standards Institute, New York, 1996.
66. Ref. 56, p. 67.
67. M. Breitenburg, "Discussion on the Use of Third Parties in Occupational Health and Safety Conformity Assessment," U.S. Department of Commerce, National Institute of Standards and Technology, Office of Standards Services, Gaithersburg, MD, July 23, 1996, personal communication.
68. M. Davidson, "Poll Finds Workers Trust OSHA, Not Employers To Protect Their Safety," *OSHA Week*, Stevens Publishing, Washington, DC (March 6, 1995).
69. S. Graham, "Should OSHA Recruit Outside Help? Certified Third Parties May Help the Agency Do More with Less," *Safety & Health*, 50–55 (Oct. 1995).
70. U.S. Department of Labor, Occupational Safety and Health Administration; *Voluntary Third Party Audits: Pro's and Con's* OSHA VPP Office, Washington, DC, May 16, 1995.
71. U.S. Department of Labor and U.S. Department of Commerce, *Report and Recommendations: Commission on the Future of Worker-Management Relations* The Dunlop Commission. Government Printing Office, Washington, DC, 1994.
72. "Safety and Health Improvement and Regulatory Reform Act of 1995," U.S. Congress, House of Representatives, House Bill, H.R. 1834, introduced June 14, 1995. Congressional Printing Office, Washington, DC, 1995.
73. "Occupational Safety and Health Reform and Reinvention Act," U.S. Congress, Senate Bill, S. 1423, as reported, June 28, 1996, Obtained from LEGI-SLATE, Nov. 8, 1996.
74. U.S. Congress, Senate, *Small Business and OSHA Reform, Senate Hearing 104-316*, Joint Hearing of the Committee on Labor and Human Resources and the Committee on Small Business, 104th Congress, first session, examining certain issues relating to modifications to the Occupational Safety and Health Act of 1970, S.B. 1423, held on Dec. 6, 1995. U.S. Government Printing Office, Washington, DC, 1996.
75. U.S. Congress, Senate, *Occupational Safety and Health Reform and Reinvention Act, Senate Hearing 104-353*, Hearing of the Committee on Labor and Human Resources, 104th Congress, first session, to amend the Occupational Safety and Health Act of 1970 to make modifications to certain provisions, and for other purposes, held on November 29, 1995. U.S. Government Printing Office, Washington, DC, 1996.
76. National Performance Review Commission, *Creating a Government that Works Better and Costs Less*, Government Printing Office, Washington, DC 1993.
77. J. B. Goodman and G. W. Lovement, "Does Privatization Serve the Public Interest," *Harvard Business Review*, 26–38 (Nov.–Dec. 1991).
78. W. T. Gormely, "Privatization Revisited," *Policy Studies Review*, **13** (3/4), 215–234 (Autumn/Winter 1994).
79. C. F. Redinger and S. P. Levine, "Analysis of Third-Party Certification Approaches Using an Occupational Health and Safety Conformity-Assessment Model," *AIHAJ* **59**, 802 (1998).
80. "Definition and Requirements for a Nationally Recognized Testing Laboratory," *Code of Federal Regulations*, Title **29**, Part 1910, Chapter XVII (7-1-95 Edition), 1910.7, pp. 72–77 (1995).

81. U.S. Department of Labor, Occupational Safety and Health Administration; "Occupational Exposure to Asbestos; Final Rule," *Fed. Reg.*, **59** (153) 40964–41158 (Aug 10, 1994).
82. U.S. Department of Labor, Occupational Safety and Health Administration; *VPP Policies and Procedures Manual*, The Office of Cooperative Programs (TED 8.1a) Washington, DC, May 24, 1996.
83. American Industrial Hygiene Association, *AIHA Laboratory Programs*, Fairfax, VA, 1995, (Pamphlet).
84. C. A. Esche and J. H. Groff, "PAT Program, Background and Current Status," *Appl. Occup. Environ. Hyg.* **11**(12), 1376–1377 (Dec. 1996).
85. J. Cascio, *The ISO 14000 Handbook*, CEEM Information Services, Fairfax, VA, 1996.
86. *Quality Systems Update Newsletter*, Irwin Professional Publishing, 1333 Burr Ridge Parkway, Burr Ridge, IL, July 1996.
87. Automotive Industry Action Group, *Quality System Requirements, QS 9000*, 2nd ed., 810-358-3003, Detroit, MI, Feb. 1995.
88. *Quality Systems Update*, Irwin Professional Publishing, 1333 Burr Ridge Parkway, Burr Ridge, Illinois, Sept. 1996.

CHAPTER FIFTY

Business Analysis for Health and Safety Professionals— The State of the Art

Lawrence R. Birkner, CIH and Michael T. Brandt, Ph.D.

1 INTRODUCTION

Increasingly, environmental, health, and safety (EHS) practitioners are reporting that it is becoming more difficult to sell investments in health and safety, and environmental management (1). Global competition, government downsizing—including ongoing efforts to re-engineer Safety and Health regulatory agencies and their underlying legislation (2) changes in technology, and the overall drive by organizations to improve operating efficiency have placed added pressure on EHS practitioners (3, 4). This pressure manifests itself by placing increased financial accountability on all staff functions, including EHS (5, 6).

Business, economics, and environmental management literature have increasingly devoted space for articles concerning such topics as performance measures, quality management, cost control and management, cost and benefit analyses, environmental liabilities, business and environmental management integration, EHS management systems, and EHS performance and the valuation of firms, among many other topics (7–17). It is clear that the expectations of senior executives for EHS performance have changed. EHS practitioners must continue to come to the workplace with exceptional academic education and training in the sciences and engineering, but business and communication tools are also needed to successfully meet the new performance challenges.

Patty's Industrial Hygiene, Fifth Edition, Volume 3. Edited by Robert L. Harris.
ISBN 0-471-29753-4 © 2000 John Wiley & Sons, Inc.

This chapter presents the basic concepts of financial and quality management tools to plan, manage, measure, and control EHS activities, so they may be considered as investments and not pure cost. The goal of this chapter is to demonstrate that EHS practitioners can begin to integrate their technical disciplines into the "business of business". In some cases, the business and financial tools presented are not commonly used by EHS organizations today. In such instances, the discussion focuses on advantages and benefits of using such tools. By using the tools presented in this chapter practitioners will be better able to manage EHS activities to drive improved EHS performance and to facilitate greater line management interest and involvement in this critical business function. These tools are useful to all EHS practitioners, regardless of whether they organizationally reside in private, public or nonprofit sectors of the economy.

2 THE ISSUES

2.1 Driving Forces in Business

According to Rooney of L.L. Bean, when a business spends money, it is either a cost or an investment (18). As such, the EHS practitioner has an opportunity to demonstrate that EHS initiatives reside more on the side of an investment, than on the side, of cost, in the investment equation. In an extensive study of more than 100 corporations, Epstein observed that most companies do not apply the tools used in making capital investment decisions to ES&H departments, programs, or initiatives (19). To further compound this problem, Rahimi, Professor of Ergonomics and Safety at USC, points out that EHS practitioners have difficulty documenting such investment returns (20). For example, Dainoff notes that companies that fail to address cumulative trauma disorders risk factors pay twice—once for the cost of the injury, and again for the lost opportunities for better productivity and product quality (21). Study believes that discussing costs and profit margins has much more impact on executive management than showing incident rates because business managers are much more responsive to impacts on the financial bottom line (22). Finally, as pointed out by Brown and Axelrod, EHS performance can affect the stock value of a company and the overall financial performance of an organization (3, 4).

2.2 The Need for Change in EHS Management

The fundamental issue faced by business and operations managers and the EHS staffs that report to them is determining how to integrate their respective staffs to support organizational goals and objectives effectively. Past misalignments and lack of integration of EHS activities into the business of business has resulted in diminished consideration, if not the complete discounting of this input of EHS organizations in business decision. The recognition of the misalignment by business leaders has motivated executives and EHS practitioners to reassess their working relationship (3, 7–9). In doing so, they are also evaluating new and existing tools to collect, measure, and communicate organizational outputs, performance, and the value of EHS to the organization and its investors (7–15).

Organizations employ specialists like EHS practitioners to buy access to their intellectual capital, and expertise. Successful EHS practitioners in today's knowledge-based economy are developing new skills in business management and financial analysis. They are applying these skills in the emerging leaner and information technology driven organizations. By integrating technical knowledge of hazard anticipation, recognition, evaluation, and control with the information technology and analytic tools used in business decision making, the EHS practitioner can better protect worker health and safety, and the environment.

2.3 Performance Measurement

EHS practitioners have an opportunity to demonstrate the impact that existing or potential EHS risks or hazards have on the financial bottom line of an organization. These impacts can be measured in terms of their incremental impact on operating costs, productivity, product costs stock price, payback periods, and break-even point, among many other metrics (23–30). However, while individual EHS organizations and industries have developed performance targets, few are integrated with traditional cost accounting systems. Without using performance measures common and familiar to business managers and executives, they will not be able to assess the financial impact of EHS investments. Traditional performance measures such as government-required statistics (i.e., OSHA recordable rates), disabling injury rates, workers' compensation costs, emission control costs, regulatory fines, and others represent failures. Further they do not reflect the financial impacts on the business performance (31).

The traditional EHS measures are not sufficiently robust and meaningful enough for management to know how to change business activities and make investments to achieve improved EHS performance (32). Additionally, Wells et al. (33) point out that process improvement, environmental results, and customer satisfaction can and need to be measured to improve and achieve total quality environmental management. Generally accepted models or processes which actually calculate the returns on EHS investments are just now beginning to emerge (34). With these models, EHS activities can be linked, to business functions, and to strategic business and financial objectives, while achieving improvements in EHS processes (29, 30).

Performance measurement and improvement is a well-documented discipline whose details are well beyond the scope of this chapter. However, one cannot discuss the selection of performance measures without a basic introduction to a few critical concepts. W. Edwards Deming identified fourteen points for management (35). He advocated and demonstrated that the implementation of these points is a signal that management intends to stay in business, and protect jobs and shareholder value (35). The Deming approach to quality is the foundation upon which much of the reengineering, business process improvement, customer improvement processes, benchmarking, balanced scorecards, process measurement, cost management, and other business quality improvement initiatives are based.

Deming's fourteen points are readily applicable to service organizations such as internally provided EHS services. Essentially, Deming's method involves a focus on giving customers quality and value for their money. To do so requires that managers view their work as a system or process. One that takes a raw material (an input), for example, and transforms it using a business or industrial process (a system) into something of value (an output), which someone will purchase. Deming's method examines parameters of inputs to and outputs from a process used to produce products or services, including people, material, equipment, methods, and environment (35, 36). These parameters are coupled with Deming's cycle of continuous improvement:

- Recognize the opportunity for improvement.
- Test the theory to achieve the desired result.
- Observe the test results.
- Take action to improve performance based on the results.

Successfully employing Deming's method has contributed to improve worker health and safety, environmental protection, shareholder value, job creation, and increased profitability in many organizations.

Perhaps, one of the reasons for the confusion over performance measures for EHS organizations and programs is the failure to view work activities as a process, or a system for achieving desired results. Each step in Deming's improvement process provides an opportunity to measure performance. The key to measuring performance is selecting meaningful metrics. In order to improve the performance of any process or subprocess, one must understand the sources of variability affecting the process. There are three sources of variability—inputs, process or subprocess, and outputs. The manager must then examine each of Deming's parameters (people, materials, equipment, methods, and environment) to identify the sources of the variability. For example, in a waste management program, the source of input variability may be the inaccurate identification of hazardous waste materials. Such errors will affect the performance of each step in the waste management process downstream which could result, ultimately, in inappropriate treatment and disposal of hazardous waste. These events can result in long-term contingent liabilities to the organization.

Variability in performance can also come from the way in which the process is performed. There are many methods discussed in the literature for analyzing processes for improving performance such as cause and effect diagrams, flow charts, pareto charts, gap analysis, and others (35–40). Through analysis the manager may find measures of process performance that should be monitored to improve the quality of the product or service (outcome) produced by the process. In a hazardous waste management process, it would be important to monitor the number of waste drums packed and labeled to measure and monitor the total waste volumes produced for different waste streams. Alternatively, it may be important to track over time the volume of waste streams targeted for waste minimization. This process measure would be a quality check on the effectiveness of the waste minimization process.

Finally, variability can be assessed by examining the products and services (outcomes) produced by the process. The selection of outcome measures should reflect the needs of the audience, that is, the user of the data. For example, the senior business executives may be interested in the unit and total costs of waste drums disposal but the hazardous waste manager may not only be interested in the number of drums stored, and the unit and total costs associated with the characterization of drum contents. Kaplan and Norton maintain that performance measures should be selected depending on the type of decisions or actions that will be taken, and the measures should reflect the diversity found in business units (31). This view is entirely consistent with Deming's method.

2.4 Communications With Management

The organizational changes that took place in the 1990s and the daily experience of many management consultants suggest that executive decision makers are increasingly frustrated with the dearth of meaningful business metrics to evaluate the EHS function (41). The performance data currently provided business and operations managers is typically not in a form or delivered in timeframe that is congruent with management decision making processes. As a result cost avoidance has been more traditionally the financial component of justifications for EHS investments.

Costs are considered expenses to the organization and are generally viewed as something to be controlled and contained. What is missing from the discussion of EHS investments are the benefits, value, and contribution to the organization from each expenditure (investment) decision. In today's competitive global business environment, cost avoidance no longer is viewed as a particularly meaningful motivator for improved performance or funding of EHS activities. Avoided costs are not likely to out-compete other investment opportunities where the return on investment is more readily quantifiable, tangible, and understood.

Regulatory compliance and an ethical desire not to injure people or the environment are two other factors driving organizations to make EHS investments. Progressive organizations in the twenty-first century will be moving beyond compliance as their driving strategic objective for EHS. EHS will be blended into their ethical desire for operational performance excellence. Consistent with these strategic objectives are EHS investments that minimize failures such as accidents, injuries, illnesses, unusual operating events, regulatory fines, and sanctions. Such investments can also help to maximize production process run-time, and improve productivity, quality, and overall customer satisfaction.

In organizations with such a progressive business philosophy, EHS practitioners will have an opportunity to create partnerships with operations and business managers to demonstrate how EHS investments can support and enhance production activities. Operational efficiencies and improved quality will translate into cost savings, financial benefits, and a competitive advantage in the marketplace. These can be used by managers to help meet or exceed investor expectations for shareholder value and organizational performance.

2.5 Investment–Risk Relationship

Figure 50.1 demonstrates the investment-risk relationship inherent in EHS decisions. Inevitably, these decisions influence the business performance of an enterprise. A significant challenge associated with evaluating EHS investments and risk decisions is to identify and capture the full spectrum of costs and return on investments (ROI). Well managed companies may have a good understanding of their capital investments for mandated environmental pollution controls, but few have the tools, systems, or processes to capture both the direct and indirect costs of EHS issues (19).

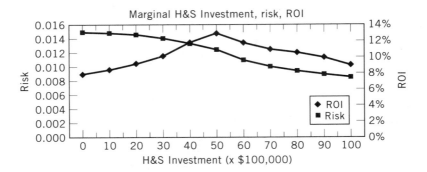

Figure 50.1. Investment-risk relationship.

This figure demonstrates how health and safety investments made to control risks can affect the ability of an operational investment to generate a positive financial return. At some point in the investment in hazard control technologies, incremental risk reductions for each dollar spent become minimal, and the ability of a production operation to return an appropriate level of return is reduced. Similarly, business decision-makers are asking EHS practitioners to determine the Return on Investment (ROI) for each increment of financial investment. The question that must be answered by the EHS executives is: What level of investment should be made to achieve the appropriate level of risk and hazard control that will continue to permit a fair return on investment? Answering this question will begin to move the EHS function closer to the mainstream of business planning, management, and control decisions.

3 FINANCIAL METHODS

Gray reports that EHS investment spending is increasing and he believes that there are significant benefits that can be achieved by evaluating EHS performance and investing in improvements (42). The analysis of EHS performance has not traditionally extended to considering the financial return of these improvements, or the impact of these investments on overall business performance. While this is clearly an important issue, there is limited experience on how to incorporate EHS factors into financial systems. Epstein maintains that a complete measurement of EHS costs will permit a more comprehensive analysis of capital investments and lead to better decision-making (19). Better alignment between the goals of operating managers and those of the EHS function can be achieved by using these methods. The objective of the discussion that follows is to identify methods that can be applied to EHS investment analysis.

3.1 Capital Budgeting

Traditional accounting systems do not identify the financial impact of environmental activities such as waste disposal, pollution prevention, recycling, regulatory compliance, future liabilities, and public relations. Because the accounting systems have not effectively captured these costs, businesses have been unable to integrate environmental and EHS costs fully into the managerial decision-making process. In fact, according to accounting researchers, Kaplan and Thomas, traditional accounting practices and systems provide data and information that is too late and distorted to be of any value in planning and control decisions (43). Kite believes that because of the lack of timely research or leadership in response to the uncertainties within the accounting discipline, U.S. businesses have been making capital investment decisions without critical and relevant data and information on environmental considerations (44). Epstein's study of corporate environmental performance supports this notion by reporting that improvements (capital, process, or product) can only be achieved by measuring all costs and benefits associated with an EHS intervention (19).

EHS managers have an opportunity to present capital intensive hazard control measures before financial executives in their organization's capital budgeting process. This budgeting process is the making of long-term planning decisions for investments in an organization. It can be divided into four distinct phases: awareness, identification, selection, and monitoring, see Figure 50.2. Kite further posits that as applied today, this process cannot address environmental complexities but proposes adjustments to the budget process, shown in Table 50.1, that can help EHS managers have their proposals considered (44). In fact, Epstein reports that EHS issues are being given visibility in the process of capital budgeting decisions (19).

Dixit and Pindyct point out that developing and understanding investment options (alternatives) is critical for making capital investments decisions, particularly for companies operating in the volatile and unpredictable global environment (45). Uncertainty, whatever its cause, requires that manager be much more sophisticated in the ways they assess and account for business risk. This is true for EHS investments, as well, since the regulatory

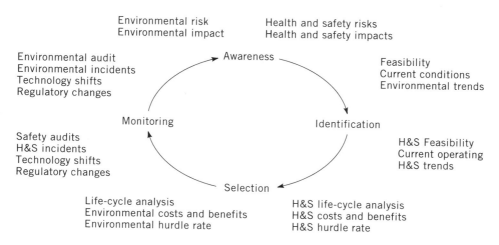

Figure 50.2. Capital budgeting cycle. Adapted from Devaun Kite, Ref. 44.

Table 50.1. EH&S Capital Process Adjustments[a]

Capital Budgeting Process That Considers Environmental Impact

Awareness

The process develops an awareness of the environment as an important variable in strategic planning. This is a two step process. First, environmental risk must be assessed; second, the impact of the risk on organizational goals and objectives must be determined. Environmental risk exists if the project can expose the environment to a source of pollution.

Identification of Feasible Projects

The objective of the project identification stage is to choose projects that support and further organizational goals. The traditional approach must be expanded to assess the environmental climate in which the business operates. This may place additional constraints on business investments. Obviously, environmental regulations must be evaluated as well as community impacts. Future impacts of the project must also be considered by assessing trends.

Project Selection

Discounted Cash Flow techniques are the most popular project selection methods. The Net Present Value assessment must consider environmental issues from an economic life, cash flow, and discount rate which incorporates environmental risk into the hurdle rate.

Economic Life: must consider hazardous wastes or environmental remediation that may have to occur at the end of the projects useful life.

Cash Flow: Direct cash flows associated the environmental impacts frequently get buried inside project accounts making them difficult to discern and ultimately manage. These must be identified and separated out. Indirect cash flows arise from support activities such as waste disposal, material handling and environmental monitoring. Hidden cash flows arise from regulatory compliance, inspecting and training. Contingent cash flows arise from liabilities that may arise from the project including environmental damage, tort actions, and regulatory enforcement.

Discount Rate: is generally the cost of capital or the weighted cost of capital plus some business risk premium for the project. When considering environmental projects an additional risk premium may have to be considered depending on the potential for enevironmental impacts. Thus a 15 percent discount rate that considers the cost of capital and business risk may have to be indexed to a higher rate to fully account for the environmental risk.

Monitoring

Project monitoring is critical to determine if the environmental assumptions and risk assessments were correct. If not, project adjustment may have to be considered to prevent environmental problems on one hand, or to prevent excessive costs if the assumptions turn out to be too conservative. Learning from the monitoring process is key to assessing the risk and issues of future projects.

[a]Adapted from Ref. 44.

and legislative environment, which today drives a great deal of these investments, is also in continuous flux.

3.2 Activity-Based Cost Management (ABCM)

Activity-Based Cost Management (ABCM) is a cost analysis method used to understand the true costs associated with any type of work activity. It traces costs from a budget through the work activities to products, services, and customers. As shown in the example below, the ABCM methodology analyzes costs more precisely than traditional cost accounting to reflect true resource consumption. As a result, many companies are supplementing their existing cost accounting systems with ABCM (46). The basic difference between traditional cost accounting and ABCM lies in the process of tracing overhead costs to the products produced. Traditional systems allocate the resources consumed such as labor and machine hours used to the units produced, but there are resources involved that cannot be allocated to production this way. ABCM traces costs to products based on the product's actual use of activities and on the principle that work activities demand and consume resources. Accordingly, it is not the volume of product produced creating changes in costs, but work activities.

By contrast, overhead costs in an ABCM system are traced into one or more of four cost categories based on cost drivers that reflect a causal relationship between the budget and the work activity. Typical EHS cost drivers might include the number of training classes taught, number of respirators issued, number of illnesses and injuries, and other similar drivers. The cost categories, which represent the final destination for costs in an ABCM analysis, include: unit related, batch related, product sustaining, and facility sustaining. ES&H costs can easily be included in each of these categories. Table 50.2 suggests how costs related to safety could be categorized using ABCM. The key value associated with applying ABCM to EHS is that it directs management's attention to the activities that consume the greatest budget resources and directs management's attention to those areas where resources may not be effectively utilized.

The ABCM methodology plays a central role in identifying critical cost and work activity measures that can be managed to improve EHS performance (47–48). A conceptual

Table 50.2. ABC Applied to Selected Safety Parameters

ABC Applied to Health and Safety Costs	Unit Related	Batch Related	Product Sustaining	Facility Sustaining
Product design/product safety	X		X	
Manufacturing equipment, safety				X
Employee training			X	X
Regulatory tracking			X	
Maintenance				X
Personal protective equipment		X	X	
Legal			X	X
Incident and workers' compensated			X	X
Product transportation		X		

view of ABCM is shown in Figure 50.3. The cost decomposition view is concerned with tracing costs from budget resources through activities to cost objects—a two-stage tracing. In the first stage, activity costs are determined by tracing costs from the budget to activities using resource cost drivers, which traces and measures the use of resources by activities. In the second stage, activity costs are traced to cost objects (products, services, customers) by using activity cost drivers, which measure the frequency and intensity of activity use.

The process view, shown in Figure 50.3, is concerned with the relationship between individual activities linked together as a business process. This view, which represents activity-based management, describes how individual activities linked together as a business process are used in decision making. The budget resources traced to each activity and the relationship among the different activities in a business process are analyzed and evaluated to determine (*1*) why work activities are performed, and (*2*) how and why activities cost what they do.

ABCM can be used to measure and improve the performance of EHS work processes and their outputs. The cost decomposition view yields information about the cost of activities, products, services, and customers; and it yields performance information about the causes of work (resource and activity cost drivers). The cost drivers and activity costs can be used as process performance measures. For example, Table 50.3 shows the activities of a hazardous waste management process with output quantities and cost listed. The

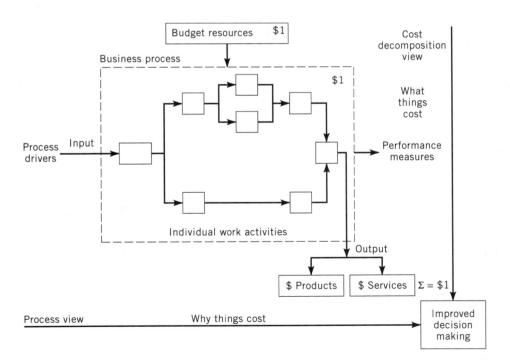

Figure 50.3. Conceptual view of activity-based cost management.

Table 50.3. ABCM Data for Performance Measurement Hazardous Waste Management Example

Activity Transaction	Output Quantity (driver of cost)	Unit Cost ($)
Characterizing waste drums	100 drums	500
Storage drums	900 drums	100
Performing audits	100 reports	500
Performing inspections	200 inspections	250
Manifesting wastes	20 manifests	500
Transporting wastes	150 drums	333
Disposing wastes	150 drums	4,333
Total		$950,000

activity transactions are essential steps in the business process. It is here that a manager can influence the performance of the process. If cost is used as a measure of the performance of EHS process against established cost targets, then the only way to affect cost is by managing and changing the activities that generate them. Intuitively, if the cost is too high, the manager must transform the way the work is performed to reduce the cost. Such transformations will require that the manager challenge the assumptions under which the business process is operating. Another way that a manager can affect performance is by examining the outputs of the business process. The output costs and the volume of output generated, such as the number of waste drums transported is also shown in Table 50.3. The unit cost of $333 per drum transported can be used as a benchmark against which costs of other service providers can be compared. This example shows how process and output data are critical to managing and improving EHS performance.

Traditionally, overhead has been allocated to production using direct labor hours or machine operating hours. Because costs associated with providing a safe work environment have been combined into overhead accounts and allocated based on labor hours, the costs and benefits of health and safety investments have not been visible. However, if safety and health costs and benefits were linked to activities and budget resources were traced to these activities based on risk and risk reduction, a better understanding of EHS impacts on the business is possible.

The process of applying ABCM reveals a level of insight not typically gained in annual budget exercises for several reasons. First, the method forces managers to quantify existing budget resources, identifying critical EHS work activities, and identify products and services produced and customers served systematically. Secondly, The discipline of methodically tracing costs from an existing budget into work activities and finally into products and services forces the manager to examine what things cost, and how and why they cost what they do. Thirdly, the cost data and business process information generated by ABCM is used by decision-makers to better manage and control work activity transactions, which generates costs. Finally, ABCM yields cost data and business process information needed for making decisions and solving complex budgetary and business problems. The data generated includes (*1*) locating the major causes (drivers) of work and their associated costs; (*2*) understanding the drivers of cost; (*3*) optimizing and redefining work activity transactions; and (4) streamlining business processes to increase efficiency and cost effectiveness.

3.3 Break-Even Analysis

Break-even analysis is used by business executives to determine the sales or service volume at which a company is able to recover all its costs without making or losing money. From an accounting perspective, it is the output (products and services) quantity produced when total sales revenues and total costs are equal. It is of particular interest when beginning a new activity such as starting a new business, expanding a business, introducing a new product or service, or selecting a risk reduction or hazard control measure. Some of the questions that EHS practitioners can help business executives answer by using break-even analysis are as follows:

- Would modernization of production facilities by installing state-of-the art pollution abatement technology pay for itself?
- Would an EHS intervention targeting the reduction of employee absenteeism pay for itself?
- Would a safety enhancement to a product generate sufficient sales to justify the cost of the enhancement?
- Would providing services, such as EHS support to customers, generate enough sales through new customers to cost justify providing the support to customers?

Break-even analysis is valuable to business executives because it demonstrates to financial analysts and investors the steps managers are taking to further increase the value of the company through new and innovative initiatives.

Sorine points out that most EHS organizations are not required to offset their costs; instead, they are generally expected to minimize their costs (50). In the future management is going to expect an assessment of the impact of investments in EHS activities on profits. Sorrine specifically raises the question: "How much do degradation expenses (expenses arising out of worker injury, illness, damage to property, and operational ineffectiveness) reduce profits of the company, and to what extent do safety activities prevent loss to profits?" Sorine suggests using a break-even analysis to determine the "nonrealized" profits attributable to degradation costs. Unrealized profits are those that an enterprise was not able to generate due to losses attributable to EHS incidents.

The break-even point is the point at which the total cost and gross income lines cross. In EHS terms it is the point at which the cost of an EHS failure (illness or injury, environmental contamination, regulatory fine, etc.) equals the cost of the investment in hazard or risk controls. It also reflects the direct benefits (such as increased productivity as measured by reduced employee absenteeism, for example) compared to the cost of the intervention (51). Extending this analysis, if the intervention targets bigger cost problems such as absenteeism of higher paid employees or shutdowns of highly profitable production processes, then the benefits achieved by implementing the intervention may outweigh the costs. Break-even analysis has been applied by Greenberg et al. to determine the amount

BUSINESS ANALYSIS FOR HEALTH AND SAFETY PROFESSIONALS—STATE OF THE ART

of improvement in employee productivity that justifies further investments in employee health promotion programs (51). By example they demonstrated that as little as a 2.5% improvement in employee productivity will justify a $500 investment in a smoking cessation program (51).

Some problems identified with the application of break even analysis to EHS are the approximations and assumptions that must be used in the analysis. Some managers may even reject the concept as being "too theoretical" and not addressing practical operating problems. But assumptions are essential for using this analysis. For the results to have any credibility with decision makers, the analyst must construct the scenario for analysis using assumptions that reflect actual conditions (technical and business constraints) as much as possible. An example of how safety and health investments and losses impact the break-even point appears in Table 50.4.

The example indicates that the $50,000 investment actually lowers the breakeven point by $83,612 or $1.67 for each dollar used to train employees and supervisors. This results answers the critical managerial question of: for every incremental dollar spent on an EHS intervention, what is my financial benefit? In this case, $1.67 is saved for every dollar invested up to a total of $83,612.

With an understanding of break even analysis, the EHS practitioner can help to optimize company profitability by identifying and minimizing the nonbeneficial costs associated

Table 50.4. Break Even Analysis.[a]

PKE Inc.	Break Even Point analysis	
Employees	40[b]	
Fixed operating costs	$800,000	
Fixed prevention costs	$50,000	
Operating variable costs	$700,000	
S&H losses	$150,000[c]	
Gross income	$2,000,000	
	BEP w/Prevention	BEP w/No prevention
Total fixed costs	$850,000	$800,000
Total variable costs[d]	$700,000	$850,000
$100\% - \left(\dfrac{\text{Total Variable Costs}}{\text{Gross Income}}\right)$	65.00%	57.50%
BEP[e]	$1,307,692	$1,391,304
Delta (BEP w/o P) − (BEP w/P)	$83,612	

[a] Break Even Point = $\dfrac{\text{Total Fixed Costs}}{100\% - \dfrac{\text{Total Variable Costs}}{\text{Total Gross Income}}}$

[b] Employee and supervisor training.
[c] Average cost for five last time injuries.
[d] 100% − (Tot. Var. Costs/Gr. Income)
[e] Total fixed income/P
P = prevention

with EHS quality failures such as accidents, injuries, illnesses, environmental contamination, and regulatory violations.

3.4 Financial Aspects of Projects

There is no single best method for incorporating EHS considerations into investment decisions. Four methods for analyzing the financial aspects of projects are discussed in this section—discounted cash flow (DCF), net present value (NPV), payback, and economic value added (EVA). It is important to understand, discounting methods force longer term—higher risk projects to be robust in their returns, especially in the later years of the investment, since those cash flows are heavily discounted. Conversely, such analytical methods may make shorter term, lower risk investments look more attractive (42). Nonetheless, EHS managers should understand how these tools can be used to justify a project. These tools are routinely used in business analyses and are elements of the language and decision making processes of executives.

3.4.1 Discounted Cash Flow

Discounted cash flow (DCF) is based on the notion of opportunity cost. When a choice is made by a manager for the use of financial resources, all other alternatives for the use of those resources are forfeited. DCF is the best method to use for making long-term decisions because it specifically weighs the time value of money. This means that a dollar is worth more today than a dollar in the future for three reasons:

- Inflation erodes purchasing power.
- Risk—the uncertainty of receiving a return increases as time goes on.
- Money has a time value since it can be invested to earn interest income.

Epstein advocates the use of DCF analyses to improve the prospects of funding EHS investments (52). DCF is also correlated to the way in which the value of firms are determined by the stock market—the greater the DCF the greater the value of the firm (52). This relationship is based on capital investments made by companies to generate more cash in the future. It is another measure and form of investment.

The cost-effectiveness analysis method used by Birkner and Saltzman to compare health and safety control strategies was first discussed in 1986 (53). The analysis applies DCF/NPV analysis to incremental investments in hazard controls expenses. In this approach, the costs of different hazard control methods (i.e., engineering, personal protective equipment, work practices, or a combination of methods), which assure employee protection and regulatory compliance, are compared for the project lifetime. One of the applications of the method is to help fine tune costs for a given strategy, maximizing the level of employee protection achieved while minimizing expenditures.

Because money has a time value, cash flows from one period cannot simply be added together. Discounting is a process that determines the present value of a future amount of money. DCF is a capital budgeting method that measures cash flow in and out of a project at a single point in time to compare equivalent present dollars. This method takes a future dollar value and reduces it by a discount factor that reflects an available and appropriate interest rate (considering the cost of money plus business risk), see Figure 50.4 below.

The present value, *PV*, of a given amount, *p*, to be received in, *n*, years in the future, discounted at a rate, *i*, can be calculated using the following formula (54):

$$PV = \frac{p}{(1 + i)^n}$$

The first calculation in Figure 50.4 is interpreted as follows. The present value of an investment that yields a 10% annual return promised a $1 return in one year. The value of that dollar is represented by the first calculation. The second calculation shows the present value of a dollar returned after two years for the same investment.

Discounted cash flow (understanding the present-day value of future cash flows) is central to nearly all financial decision making processes. DCF/NPV transforms future cash flows to their present dollar equivalent so that all strategies can be compared on an equal basis (55). Strategies can be ranked on their present value. Cash flows consider the investments, revenues, expenses, and taxes.

To be applied to EHS investment analyses, each control strategy develops cash flows using the elements listed in Table 50.5.

Although DCF/NPV analyses is used extensively for making capital budget decisions, its application to health and safety investments was limited at best, prior to 1986. Other investment analysis methods, such as discounted payback or discounted return on investment can be used to assess the effectiveness of EHS investments. Payback and ROI analyses can also be used without discounting, however for longer investments, their meaning may present an overly optimistic result at best or, at worst, suggest the selection of a bad investment.

3.4.2 Net Present Value

Net present value (NPV) uses the DCF method of calculating the expected net monetary gain or loss from a project. This is done by discounting all expected future cash inflows and outflows to the present time, using a specific discount factor or hurdle rate. The hurdle

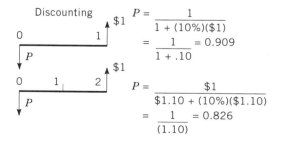

Figure 50.4. Discounted Cash Flow Analysis.

Table 50.5. Drivers for Health and Safety Cash Flows.

Expense	Potential Application
1. Labor	The increment of labor needed to implement each control strategy. The increment can be positive or negative.
2. Training	The costs required to train employees and management in the implementation of the set of strategies being considered.
3. Research and Engineering	Research and engineering costs associated with the development and implementation of the strategies.
4. Production, Selling and Administrative	Lost or increased production or sales associated with each strategy. For example, if the strategy drives different product specifications, additional selling expenses may be incurred.
5. Feedstock	Changes in the use of raw materials.
6. Utilities	Changes in the use of utilities associated with each strategy.
7. Maintenance	Adjustment in maintenance required for each strategy.
8. Other	Other parameters that are peculiar to specific strategies

rate is selected by a company based on the cost of capital and risk factors—in its essence, it is the minimum return that a company would like to receive for its investment. Hurdle rates can be adjusted around a fixed cost of capital to reflect different investment risk. Unless an investment is made for strategic reason, only projects with a positive NPV are acceptable because the return exceeds the cost of capital for a given risk level. This means that the cost of the project will provide a financial return that is greater than if the money used for the project was invested that yielded a 10% return. As a general rule of thumb, projects with a higher NPV are preferred over lower NPVs. The details of calculating NPV are indicated below and further described in the literature (56).

$$NPV = -\text{Initial Investment} + DCF_{\text{year 1}} + DCF_{\text{year 2}} + DCF_{\text{year 3}} + \ldots + DCF_{\text{year } N}$$

Generally, when the Birkner and Saltzman method was applied in the mid 1980s and early 1990s, it yielded negative Net Present Values (NPV) since little effort was made to identify the benefits derived from improvements, which is the revenue resulting from altered safety and health strategies. Thus, strategies were generally ranked based on the negative NPVs with the least negative being considered the most cost-effective.

Ultimately, options create flexibility, which has value that transcends simplistic calculation of NPV. Thus the concept of developing and analyzing alternatives for health and safety investments helps to assure that flexibility and potentially increased investment value by providing an escape route should a selected strategy become less viable for any reason. The bottom line for health and safety managers is that just understanding NPV is insufficient and intelligent choices can only be made when alternatives have been identified, analyzed, and investment escape routes are kept open.

BUSINESS ANALYSIS FOR HEALTH AND SAFETY PROFESSIONALS—STATE OF THE ART

Table 50.6. Categories of Environmental Costs[a]

Regulatory	Upfront	Voluntary
\multicolumn{3}{c}{Potentially Hidden Costs}		

Regulatory	Upfront	Voluntary
Notification	Site studies	Community relations/outreach
Reporting	Site preparation	Monitoring/testing
Monitoring/testing	Permitting	Training
Studies/modeling	R&D	Audits
Remediation	Engineering and procurement	Qualifying suppliers
Recordkeeping	Installation	Reports (e.g., annual environmental reports)
Plans		
Training	*Conventional Company Costs*	Insurance
Inspections		Planning
Manifesting	Capital equipment	Feasibility studies
Labelling	Materials	Remediation
Preparedness	Labor	Recycling
Protective equipment	Supplies	Environmental studies
Medical surveillance	Utilities	R&D
Environmental insurance	Structures	Habitat and wetland protection
Financial assurance	Salvage value	Landscaping
Pollution control		Other environmental projects
Spill response	*Back-End*	Financial support to environmental groups and/or researchers
Stormwater management	Closure/decommissioning	
Waste management	Disposal of inventory	
Taxes/fees	Post-closure care	
	Site survey	

	Contingent Costs	
Future compliance costs	Remediation	Legal expenses
Penalties/fines	Property damage	Natural resource damages
Response to future releases	Personal injury damage	Economic loss damages

	Image and Relationship Costs	
Corporate image	Relationship with professional staff	Relationship with lenders
Relationship with customers	Relationship with workers	Relationship with host communities
Relationship with investors	Relationship with suppliers	Relationship with regulators
Relationship with insurers		

[a] From Ref. 72.

3.4.3 Payback

Payback is another DCF method which determines the time it takes to recover the money invested in a project. The method uses the following formula for calculating the payback period (57):

$$\text{Payback Period (PB)} = \frac{\text{Net Initial Investment}}{\text{Increase in Annual Cash Flow}} \text{ (in years)}$$

Using this method project managers and financial analysts choose a cutoff period for the analysis. Projects with a payback period less than the cutoff period are selected while projects with a payback period greater than the cutoff point may not be selected. Essentially, the payback period calculates a breakeven point at which the financial returns from a project are equal with the cost of the capital invested in the project over a specific timeframe. As suggested above, payback analysis can be discounted or nondiscounted and it is important for a decision-maker to know which type of payback is being used.

Choosing projects based on the shortest payback period does not lead to the best choices. (52). In fact, EHS projects are often expected to recover their financial investments in as little as six months (52). Such unreasonable requirements lead to making the wrong investment decisions (42, 52). The entire context of an investment should be considered. Multiple analytic methods, sometimes called "measures of merit" should be considered in the decision making process.

3.4.4 Economic Value Added (EVA)

EVA is a relatively new financial metric. EVA lies in the intuitive concept that an investment creates value only when its expected financial return exceeds its cost of capital. The cost of capital is the return on a new investment that a company must expect to maintain its stock price (55). Thus a company or company unit creates value for its shareholders only when its operating income exceeds the cost of capital. Epstein states that EVA reflects " . . . a company's net income net the cost of debt and equity capital" as follows (52)

$$\text{EVA} = \text{Net Sales} - \text{Operating Costs} = \text{Operating Profit}$$

$$\text{Operating Profit} - \text{Taxes} = \text{Net Operating Profit}$$

$$\text{Net Operating Profit} - \text{Capital Charges} = \text{EVA}$$

Importantly, Epstein observes that EVA can be used " . . . to evaluate capital investment proposals because the present value of future EVAs for any capital project must equal NPV (52)".

EVA is in an early stage of application in industry, but its methodological rigor has significant implications for EHS practitioners. Although application to health and safety has only been reported anecdotally, an organization that ultimately lives by EVA will create great demands on its EHS practitioners. In the strictest application, no investment will be made unless it can demonstrate returns greater than the cost of capital. EHS professionals will have to demonstrate that EHS interventions and work activities that reduce injuries, illnesses and property or environmental damage create such returns. In an environment where the integration of EHS with business is so critical, EVA provides an approach for bringing capital budgeting, EHS performance drivers and financial incentives into a single metric.

4 COST RECOGNITION, ANALYSIS, AND ACCOUNTING

4.1 Cost Analysis

Both the direct and indirect costs of injury and illness prevention are substantially less than the costs of worker illness/injury treatment and worker rehabilitation. Prevention programs not only help to avoid illnesses, injuries, disability and premature death they reduce the demand for medical practitioner time, which releases practitioner resources to focus on preventive medicine (58). Further, prevention helps management focus on the business of the organization than being diverted into allocating and managing remedial actions—thus one can say that the return on EHS investments is very leveraging with respect to management time.

Historically, programs in preventive medicine, which is rooted in the fundamental principles of public health, must justify their existence in economic terms. Few agreed upon methods for economic analysis have been available. There are references in the literature to methods for assessing the value of health by Sir William Petty in England in 1667, and Adam Smith in the following century. More recently, Weisbrod suggests that the estimates of losses from disease involve questionable, misleading or simply incorrect procedures (59). Real costs and money expenditures are not synonymous terms. There may be expenditures without real costs (e.g. production losses). There may also be social costs without expenditures. Weisbrod suggests that the total cost for disease must be assessed (59).

In 1967, Rice developed an analytic process for assessing the value of a human life expressed in terms of lifetime earnings. She indicated that such a process was a basic tool of economists, program planners, government administrators and others who are interested in measuring the social benefits associated with investments in particular programs. Her analysis included all factors that affected compensation including education, sex, age, race, life expectancy, labor participation and productivity. Additionally she considered the indirect cost of illness to achieve a total cost of health. Her analysis determined that the present value losses to the economy in 1967 associated with occupational illness, disability, health care, and morbidity were $125 billion (60).

Analysis done by the U.S. Environmental Protection Agency (EPA) to evaluate the economic assessment of preventive measures suggests that more than 100,000 deaths a year could be prevented if air pollution were adequately reduced. In 1976, this implied a 7.7% reduction in the total U.S. mortality rate, which translates to $14.5 billion a year. The agency's calculation of the programmatic cost to reduce pollution is $9.5 billion making the net cost (cost effectiveness) $5 billion (61). There are many assumptions in these numbers, nevertheless, governmental agencies do such calculations regularly; in fact, legislative mandates exist to do such analyses.

4.2 Workplace Health and Safety Costs and Benefits

A University of Michigan study in the late 1950s demonstrated that machinery manufacturing firms, which did not have an industrial health program, paid $0.93 for every $100 of payroll on workers' compensation. Firms with such a program paid only $0.56 per $100 of payroll and this cost was more than double that of the plant medical personnel. The study also found the same story among general merchandise stores with $0.54 and $0.37 per $100 of payroll for those without and with industrial health programs, respectively (62).

A study of 1,625 National Association of Manufacturers member companies found that industrial health programs provided a return on the investment. Over 90% of the employers in the survey experienced reductions in accident frequency, in the incidence of occupational disease, in absenteeism, and in insurance premiums. Labor turnover on average dropped 27.3% and workers' compensation costs declined by 28.8% (62).

A more recent analysis by Oxenburgh collected 61 case studies of health and safety interventions at work describe how costs can be assessed and evaluated (63). He presented two models to assist in this analysis: an insurance model and a productivity model. The productivity model appears useful helping to identify parameters for addressing costs and benefits since it addresses calculations associated with productivity, salary (direct and indirect), administrative and personnel costs, absenteeism, turnover, training, and replacement worker costs.

4.3 Workplace Costs and Benefit Recognition

McCallum points out that once the cost of prevention is recognized as infinitesimal when compared to the high price of restoration, disability, or dependent welfare, then economics will provide incentives to assure the needed remedial measures to prevent injury (64). The absence of economic analysis has been the major underlying reason for neglect of emphasis on preventive health and safety measures. EHS practitioners have failed to examine the complete economic picture. Instead, they are inclined to consider safety and health only in terms of initial or acquisition costs, rather than measuring the cost of safeguards against losses that can be incurred by their omission. EHS practitioners are learning that the cost of incorporating safety and health in design is far less expensive than losses experienced when these items are overlooked as fundamental priorities in design (64).

There is a lack of credible and accepted methods for analyzing EHS investments and returns. This is the same problem industry and government first experienced in the early 1900s, at the beginning of the era of workers' compensation. The National Association of Manufacturers recognized as early as 1958 that forward-looking industrial health programs can accomplish the following (62):

- Improve employee health and working conditions.
- Increase efficiency.
- Reduce lost time due to illness.
- Reduce the incidence of accidents.
- Reduce workers' compensation claims.
- Sharply reduce labor turnover.
- Increase productivity.
- Increase customer satisfaction.
- Improve public relations.
- Improve employee–management relationships and employee morale.

The National Safety Council (NSC), under contract to the Occupational Safety and Health Administration (OSHA) to determine if "cost benefit" analysis at the establishment level was feasible (65). After reviewing the responses from 4200 firms, the NSC found that direct costs were generally available but indirect (sometimes called hidden, intangible or hard to quantify) costs were not. Because it is believed that such cost are as large, if not larger than the direct costs, the NSC reported that "cost-benefit analysis was not feasible". In response to this study, OSHA recognized the value of economic analysis as a tool to safeguard against the adoption of standards that might otherwise result in unusually high compliance costs to employers. However, OSHA also recognized the difficulty in doing this analysis due to the difficulty in collecting data and the "difficulties of economic prediction" (65).

The financial analytical tools are still fairly rudimentary as suggested by Simpson in 1988 while looking at the cost-effectiveness of ergonomics (66). He suggested that in order to obtain ergonomic investments, analytic tools must show that ergonomic limitations within a workplace increase the daily operational cost of a given industry and ergonomic change improves the profitability. Although insight on how to assess cost effectiveness is not provided, he does develop a number of empiric arguments using compensation costs. He also develops a metric that sheds some light on the financial impact of economic loss associated with poor workplace and tool design. Referring to work done by Wehrle in 1976 where $217,000 in compensations and outside medical costs were paid over a five-year-period, not discounting, and not adjusting for lost productivity associated with the direct losses, the company had to produce and sell products generating a $43,400 profit per year simply to cover these costs (67). Another study by Druery et al. discusses the financial returns generated by reducing ergonomic stress through a workplace redesign (68). This study indicates a 50% reduction in spinal stress resulting in a $374 saving per employee per year through reduced lost time and a $3000–4000 per person-year improvement in productivity.

Berger contends that prevention-based programs tend to be at least neutrally profitable and, in many cases, extremely profitable (69). However, solutions that are imposed by regulations always cost money and industry tends to lump these costs into a big pot and forget that the costs are related to and driven by the process. If EHS costs can be kept in the forefront in the earliest stages of process R&D, there is an opportunity to keep them low and even see return on the investment. He goes on to describe an EHS cost estimating system, realizing that some costs are visible (direct costs) and some are not (indirect costs) (69). Berger defines four tiers of costs to be quantified as:

Tier 1: Easily observed costs (i.e., training, process design).
Tier 2: Hidden costs (i.e., productivity, product quality, customer satisfaction).
Tier 3: Risk Costs (money to be set aside to hedge operational risks).
Tier 4: The valuation/devaluation associated with the EHS investment.

Berger goes on to describe a cost structure, but the model does not address the investment impact of the costs or "investment."

Haines, learning from the Total Quality Management movement, suggests that industry be challenged to manage EHS costs as an integral part of core business and manufacturing operations (70). He goes on to define the characteristics of performance measures, which management can employ to improve environmental health and safety management as follows

COC_s = (Compliance costs/Total Sales) * 100
(Pure compliance as percent of sales)
COC_e = (EHS expenditures/Total Sales) * 100
(Total EHS expenditures as percent of sales)
COC_p = EHS Expenditures/Total Units of Products
(EHS costs per/unit product manufactured)

Specifically, the measures must be high level to reflect environmental performance trends and help management identify opportunities. With respect to regulatory compliance, which is not optional, compliance indicators must be designed to balance the compliance needs and business economics. He proposes three measures for cost of compliance (COC) that bridge this need. The focus continues to be on cost, not return on EHS investments (70).

4.4 Environmental Accounting

Epstein focusing on environmental accounting, has laid a solid foundation for EHS accounting and investment analysis (19). The Institute of Management Accountants defines Environmental Accounting as the identification, measurement and allocation of environmental costs, the integration of these environmental costs into business decisions, and the subsequent communication of the information to the company's stakeholders (71). After companies identify the impacts on stakeholders as far as possible, they measure those costs and benefits to permit informed management decision making. This analysis can be expressed in physical units or monitized. Some of the applications for environmental accounting are

- Reduce environmental costs.
- Track environmental costs that have been hidden in overhead accounts or overlooked.
- Better environmental costs and performance of processes and products for more accurate costing and pricing.
- Deepen and improve investment analysis and appraisals.
- Support the development and operation of environmental management systems.

Environmental costs and benefits can be categorized as conventional, potentially hidden, contingent and image/relationship. The EPA table (Table 50.6) provides some examples for each of these categories (72).

Some of the tools and techniques used to analyze environmental costs and benefits include: cost allocation, life cycle assessment, hierarchical cost analysis, and activity based costing (19, 24, 29, 32, 48). In cost allocation, past pollution costs, current pollution (prevention costs), and future environmental costs are treated separately. Life cycle analysis looks at the costs that accrue from raw material acquisition, manufacturing to product disposal (15). In hierarchical analysis, costs are tiered as follows: directly related to a project, costs that are hidden, liability costs, and less tangible costs such as goodwill, customer relations, and others (13, 14). All of these techniques can be applied to health and safety, realizing however that traditionally, there have been more intangibilities associated with health and safety costs and benefits. The bottom line here is the ultimate integration of environment, health and safety costs into product and service costs. In that way the true cost of the product or service will be known and the cost structure associated with manufacturing can be managed.

One of the challenges that EHS investments must overcome is their duration. Many of the investments we make today have very long horizons; the cash flows of these investments when discounted to today's dollars are very small, effectively minimizing the level of interest management has in them. To gain interest, the indirect costs and benefits to the company must tell a compelling story, and be founded on good economics.

4.5 Direct and Indirect Costs

In 1931, Henirich did identify the concept of direct and indirect costs associated with safety (73). Direct costs were associated with the costs paid by insurance companies and indirect costs borne by the business enterprise and the employee. He defined total costs as the sum of the direct and indirect costs. Heinrich primarily focused on the cost of the outcome, not the cost of prevention.

$$\text{Total Costs} = \text{Direct} + \text{Indirect Costs}$$

The concept of fixed and variable costs were developed later on. Fixed costs were classified as those costs incurred when no injuries are experienced in a given period due to an EHS illness or injury prevention program. Prevention includes costs associated with professional staff, programs, training, medical services, hardware, recordkeeping, design and engineering, insurance premiums, etc. Variable costs are those costs that occur when an incident occurs, that is response, control and mitigation—incident investigations, medical expenses, employee replacement costs, recordkeeping, lost time, damaged equipment, legal fees, lost productivity, etc. Thus, total EHS cost is the sum of the variable and fixed costs.

$$\text{Total Costs} = \text{Fixed Costs} + \text{Variable Costs}$$

However, this research did set the foundation for work that did recognize the necessity for incorporating prevention into the equation.

4.6 Financial and Operational Ratios

In order to relate the cost of occupational health and safety injuries on the finances of an enterprise, Andrereoi in the summary to his International Labour Office (ILO) monograph (74), develops a number of ratios:

4.6.1 Operational Ratios

$$\text{Average cost of EHS Injuries} = \frac{\text{total cost of EHS injuries}}{\text{number of EHS injuries}}$$

$$\text{Average cost/hour of EHS} = \frac{\text{total cost of EHS injuries}}{\text{man hours}}$$

$$\text{Average cost/unit production} \frac{\text{total cost of EHS injuries}}{\text{quantity of production}}$$

4.6.2 Financial Ratios

$$\text{Cost of EHS injuries as a fraction of wages} = \frac{\text{total cost of EHS injuries}}{\text{total wages}}$$

$$\text{Cost of EHS injuries as a fraction of assets} = \frac{\text{total cost of EHS injuries}}{\text{total assets}}$$

$$\text{Cost of EHS injuries as a fraction of earnings} = \frac{\text{total cost of EHS injuries}}{\text{earnings}}$$

Obviously, many other ratios can be developed based on the specific needs of the enterprise. The point is there may be metrics that may have more business meaning than OSHA recordable and lost time injuries. Clearly, knowing the cost of failure per unit production or earnings can provide management with better tools with which to control operations and set benchmarks and goals.

4.7 Contingent Valuation

Economists are most comfortable when they can measure people's preferences by their behavior in the marketplace. However, the environment, health and safety are rarely sold. It is important to monitize the benefit derived from these intangible attributes of modern society. The questions to be answered include:

- What is the value of an undepleted ozone layer, or what is the value of the ozone layer in its current "damaged" condition?
- How much would the public be willing to pay to prevent further damage?
- What is the value of an unpolluted river?
- Is its value different to the fisherman, the city dweller who drinks the water, or the Sunday hiker who likes to bask in its beauty?
- To what extent is society willing to sacrifice their lifestyles to preserve the river and what are the lost opportunity costs of those sacrifices?
- What is the value of an uninjured worker and how much should an employer invest to keep that worker uninjured and productive?

One approach is to ask people what they are willing to pay. Of course, if you ask the industrialist who needs the water for manufacturing or the hiker, or the city dweller, the answer will be different. Not only will the answers be different, but the expectation of who should bear the costs will be different as well. The question asked when looking at contingent valuation analyses is what are people willing to pay for a benefit, or what compensation would they accept to compensate for its loss? When dealing with the health and safety of a worker, this question is very difficult to answer and when it is asked about the environment, subjectivity lies at the core of the response.

Although the contingent valuation concept has been around since the 1960s, it use was only validated by the courts in 1990. Its scientific and economic validity is still somewhat questionable since when the question is asked about compensation for damages, the answer is always much higher than when the question about preservation is posed. One possible solution is to ask the question, not in terms of prices, but of the quantity of substitutes. If the good has no substitutes, the amount a respondent would be willing to pay at the limit would equal their entire wealth (finite); the amount they would be willing to accept as compensation could well be infinite (75).

4.8 Cost Benefit Analysis

For over five decades, health and safety professions and enlightened business leaders have posited that health and safety regulations and standards are desirable to assure that all businesses are governed by the same set or rules. The government believes they are morally desirable and, where correctly applied to the system of management controls, they positively influence profitability. To date, however, this position has not been actuarially demonstrated to the satisfaction of many and therefore has not been universally accepted by business managers. This skepticism has frequently led management to regard the investments of compliance with safety and health regulations as having a negative net present value (NPV) (76–78). Research has been conducted to examine the application of cost-benefit analysis to EHS (79–82).

Many researchers have investigated the cost of accidents and the influence of EHS management systems on quality (7–9, 23–27). Some have identified links between overall quality performance EHS—the quality model requires that EHS specifications be designed into products and services. This necessarily means making safety and health investments as part of the planning, design and implementation phases of new manufacturing facilities and new products. It also means retrofitting existing operations as scheduling permits. Fisher has classified the costs associated with safety and health functions into prevention, appraisal and failure costs of the quality model. He argues that for even relatively unsophisticated organizations, failure costs can be considerably greater than the combined prevention and appraisal costs (78). The cost model that has been used extensively to help define the relationship between safety and health management system investments and the cost of failure is defined in Figure 50.5.

From the quality perspective, prevention and appraisal costs are captured in the investment curve, when combined with the cost of failure, the total program cost is synthesized. There are three important issues to note—from a management support perspective (programmatic investment) there is an optimal point at which to operate, beyond which,

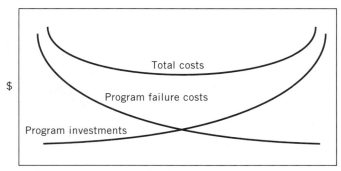

Figure 50.5. Safety and health investment: Failure Model.

the cost-effectiveness of the investment rapidly deteriorates. The model only deals with direct programmatic investments and the direct cost of failure. The model does not address the economic returns, which accrue from the programmatic investments.

At first glance, the direct programmatic investment includes:

- Management time and decision-making.
- Planning and design.
- Hardware.
- Training and communication.
- Inspections and audits.
- Maintenance.
- Supervisory time.
- Safety and health staffing.
- Regulatory compliance.
- Cost of capital.
- Utilities (related to safety and health—i.e. cost to operate ventilation).
- Opportunity costs.

What might be included in the direct cost of failure:

- Medical (ill, injured, and rehabilitated and/or disabled employees).
- Wage compensation.
- Production and product losses.
- Quality deterioration.
- Facility and equipment repair.
- Replacement labor.

- Environmental damage.
- Increased insurance rates or inability to obtain insurance.

In addition, there are also hidden costs associated with failure, and these may include:

- Lost (ill, injured, or disabled) employees.
- Lost business.
- Reduced customer satisfaction.
- Employee morale.
- Reduced productivity.
- Increased regulatory actions.
- Public concern.

Although the economic outcome of any given incident is generally driven by chance, when all incidents are aggregated, patterns emerge that can provide some insight into the power of the programmatic investment. Organizations that fully integrate their safety and health investments into business activities appear to have performance advantages over similar organizations that focus efforts on preventing personal injury and disease (83). A great deal of work in this area has been done by Heinrich (73), Bird (84), and Tye (85).

From the early part of the 20th Century until the mid-1980s, much work had been done to identify, classify, and quantify costs. Very little work had been done to place this analysis on the same financial performance analysis basis as investments that are tied more closely to the core activity of the business. While there is reference to returns on EHS investments, such returns were generally not incorporated in investment decisions.

4.9 Societal Cost Benefit Analysis

With respect to government projects, the principle of economic efficiency was codified in the Flood Control Act of 1936 (86). This act requires that only those projects shall be submitted for Congressional action for which the "benefits to whomsoever they accrue exceed the costs." Effectively, this means that if those who benefit from a project had to bear its entire cost, they would consider it worth paying for. In nearly seventy years, much has changed and governmental agencies with sweeping regulatory powers can impose virtually unlimited costs on the private sector of the economy, frequently with benefits that are intangible, guestimates, or where the benefits accrue to those far removed from those who bear the economic burden. In health and safety terms, the concept has been applied to societal investments that prolong life or reduce mortality. Prolonging life or reducing mortality has some benefit to society as a whole, but it has no discernable benefit to any individual since it is impossible to determine who will be impacted and what the "value of their life" is. The concept of value per life saved is a shorthand way of representing the total benefit accrued by the community. These investments are for the most part driven by politics. Table 50.7 provides some examples of regulatory driven costs in terms of "cost per life saved" by various government programs. What is so striking is the range of investments—from those that have clear benefit and are highly efficient to those that are

Table 50.7. Regulation Driven Costs per Life Saved. Sample Estimates of the Cost per Life Saved in Programs Supported and Operated, or Mandated by the Government[a]

Program	Cost/Life Saved (Dollars)
Medical Expenditure	
Kidney transplant	72,000
Dialysis in hospital	270,000
Dialysis at home	99,000
Traffic Safety	
National Safety Council recommended cost-benefit analysis	37,500
Elimination of all railroad grade crossings (Est.)	100,000
Military policies	
Instructions to pilots when to crash-land planes	270,000
Decision to produce ejection seats	4,500,000
Mandated OSHA Health and Safety Regulations	
Coke oven emissions standard	4,500,000 to 158,000,000
Acrylonitrile exposure standard	1,963,000 to 624,000,000
Consumer Products Safety Commission Standard	240,000 to 1,920,000

highly inefficient with very low social benefit. Most interesting is that the misallocated resources could be channeled into robust investments that yield a savings of many more lives and reduced morbidity. The lesson here is that regulatory mandates or company policies that do not fully assess the economic impact of investments are likely to result in poor decisions that draw limited resources away from where they can be effectively employed and yield significant economic and social value.

Another more operational view of cost-benefit is discussed by Spilling, Eitrheim, and Aaras in their analysis of ergonomics at a Norwegian telephone manufacturing plant, where they concluded that the cost of sick-leave due to musculo-skeletal injury was approximately equal to the cost of lost production (87). Their assessment of previous studies, where investment analysis was used concluded that investments to improve the work environment are relatively easy to assess. Such investments include purchasing and installation of manufacturing equipment, ventilation, lighting, maintenance, and operating costs. What was not considered was the cost of treatment and other health services. They describe an analytical method that looks at alternatives to make business decisions and evaluate the discounted cash flows associated with those alternatives. This clearly demonstrates the application of this approach to health and safety.

5 HIDDEN IMPACTS OF HEALTH AND SAFETY INVESTMENTS

5.1 Quality

Minter clearly sees safety as a business process (88). He notes that the physical and managerial processes that make up a workplace indeed constitute a business process geared to producing a certain product or service. These business processes may be perfectly suitable

BUSINESS ANALYSIS FOR HEALTH AND SAFETY PROFESSIONALS—STATE OF THE ART

at producing the intended product or service, but in many cases prove equally suitable for producing accident or injuries. In this respect, accident or injuries can be considered defects or failures produced by the overall business process.

In order to begin to measure business process performance and ultimately its financial impact, it is important to understand the "safety process" and understand the alignment between the core elements of the business and health and safety. Krause develops a model that solidifies Minter's thinking and drives the overall process toward measurability using concepts developed by Deming (35, 89). The model provides some suggestions as to what factors should be considered in understanding health and safety investments. The model is illustrated in Figure 50.6 and helps us develop many of the relationships that define the hidden impacts. Krause suggests that behavior is the key to effective performance and drives, or at least influences most other business metrics and is closely linked to Deming's quality drivers. Since these factors are observable and measurable, they can be quantified and ultimately incorporated into an analysis of hidden impacts and be considered as a lament of a financial analysis. For this purpose, Krause defines a behavioral inventory that

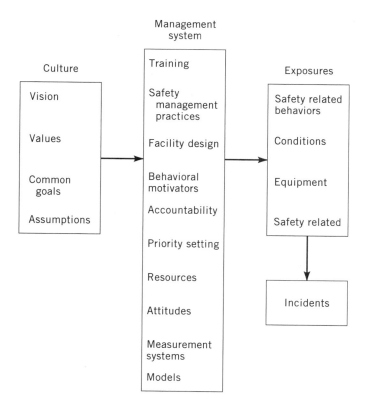

Figure 50.6. The Safety "Business Model" demonstrating the interrelationships between business management systems and safety.

can be associated with most business activities, which identifies correct operational behaviors that drive safety and a sampling method to measure against the inventory.

Hughes and Willis develop the argument that quality control methods can be used to control and reduce environmental expenditures, specifically if costs are characterized into prevention, appraisal, internal failure, and external failure costs (90). Reducing costs must hinge on investing in and effectively managing prevention and appraisal.

5.2 Productivity

One of the major cost drivers for industry today is government regulation. Much of the resistance to regulation stems from regulation that does not recognize the economic environment or does not fully address its impact on business. EHS regulations are no different and may have a greater impact on productivity than regulations in other areas. Robinson in an exhaustive study of the impact of regulation on productivity growth rates in 455 U.S. manufacturing industries between 1974 an 1986, concludes that the costs imposed by regulation are much more significant in terms of diminished productivity than in terms of direct compliance expenditures (91). The study did not support the "technology forcing" interpretations of EPA and OSHA regulations which show that regulations push firms to adopt more efficient products and processes. On the contrary, regulation diverts economic resources and managerial attention away from productivity enhancing innovation. If regulatory initiatives are to continue and society's commitment to environmental quality and occupational health and safety is to be sustained, creative risk and financial management approaches are needed.

Additional drivers of productivity and ultimately health and safety investments are two fundamental changes taking affecting the workplace, described by Parks—global competition and technological change (92). With respect to competition, labor costs appear to be the driver, since they represent 70% of all production costs; thus finding ways to increase productivity is a strategy for competitiveness. Finding ways to make the workplace safe frequently enhances productivity, the problem has been finding ways to evaluate the economic impact of these investments. Another force driving productivity is technology—here the issue is, can technology be used to leverage improved workplace health and safety, and how can that be measured.

Waxler and Higginson point out that both individual and organizational stress and anxiety affect productivity, organizational efficiency, absenteeism, medical costs, and profitability (93). Interestingly, the techniques that organizations can use to reduce stress described by the authors are very similar to the techniques used to enhance safety (see below). Learning how to use the return of investment concepts in these techniques is difficult.

- Improving person–job fit.
- Improving employee training.
- Increasing an employees' sense of control.
- Eliminating punitive management.
- Removing dangerous or hazardous work conditions.
- Improving organizational communications.

6 HEALTH AND SAFETY PERFORMANCE METRICS

EHS performance has a direct effect on employees' productivity and quality of work life, and on company profits and image. Effective programmatic investments in EHS are the key to enhancing performance. The key to enhanced performance is making investments in programs, people and hardware that can in fact drive the process. How does one know when they have an effective program and how can we measure the impact of such effective programs? An answer comes from a tool developed by Kiser and Esler called Performance Indexing (PI) (94). Simply, PI requires managers to set measurable goals, it establishes a process for tracking and measuring performance against the goals, and for communicating the performance outputs.

The fundamental problems with tracking safety performance with traditional tools are that failures—injuries, illnesses, property damage, and business interruption—are the measure of performance. These measures, by their nature, are demotivating and tend to drive people *away* from evaluating the performance of EHS prevention programs. While the goal is to minimize failure, Kiser and Eisler posit that "really effective programs—focus on activities that are positive, proactive and designed to target the underlying causes of failure." The focus is on process measures, not end process metrics. For example, process measures focus on training, inspections, auditing, and behavioral metrics. Selecting the most leveraging measures is the most challenging part of the process that is illustrated in Figure 50.7.

To assist managers in selecting performance objectives, approximately 50 safety performance measures were identified by the authors, including:

- Overall organizational measures (e.g., Lost Time Injury Rates, Responsible Care Score).
- Individual measures (e.g., Incident Investigated, Inspections Performed).
- Departmental and group (e.g., Training Completed, Behavioral Observations Completed).

Although this is an important and current contribution, it is interesting to note that none of the process metrics are financial. Of course most of the process metrics can be transformed into a financial management tool. The bottom line here is that key thinkers in health and safety are not integrating health and safety into the business process.

In 1994, a major symposium on Positive Performance Indicators in Australia addressed the issues of metrics and only briefly addressed the issues of financial performance and its tie to health and performance. Most notable, there was a great realization that measures

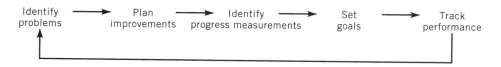

Figure 50.7. The Performance Indexing Process.

of failure cannot be managed and that process measurements must be identified to aid in managing safety in a more integrated manner. Blewett posits that Positive Performance Indicators (PPI) be incorporated in the payment of bonuses for performance pay and that managers be judged on how seriously they examine accidents (95). This approach is not very enlightened and is not a true process indicator.

Greene gets a little closer by tying safety to productivity (96). Specifically, he cites the productivity gains that have been achieved by "intensity of collaboration" between labor and management and suggests that the same can be true for health and safety. Additionally, Greene cites the concept of the Balanced Scorecard discussed by Kaplan and Norton to establish goals to achieve business integration, and that measures of cash flow, sales growth, operating income, market share, etc. could be tied to safety and health and achieving the scorecard goals (97). Through the balance scorecard, health and safety performance could be linked with productivity, internal processes, customer satisfaction, and learning and innovation.

Shaw, in her own review of the literature while addressing the same conference, refers to a work by Carter et al. who argue that a good system of performance measurements needs to include the characteristics found below (98). Clearly a set of financial measures that are linked to health and safety could meet such criteria.

- Controllable or able to be influenced.
- Relevant.
- Assessable or measurable.
- Understandable and clear.
- Accepted as a true indicator of performance.
- Reliable, providing the same measures when assessed by different people; and sufficient to provide accurate information, but not too numerous.

Although financial metrics are important when making EHS investments, they must be balanced with the needs of internal business processes, the customer, the employee and the learning and growth of the organization. The balanced scorecard approach referenced earlier helps to meet this challenge. Managers using the balanced scorecard do not have to rely on short-term financial measures as the sole indicators of performance. The scorecard approach lets them introduce four new management processes that separately and in combination contribute to linking long term strategic objectives with short-term actions (97). This is critical in making EHS investments since they have relatively long time frames to create a return, but they have many short term actions associated with them. Figure 50.8 shows the interrelationship between the scorecards and the details for individual scorecard. The balanced scorecard provides a foundation for the use of systems dynamics to better understand the interrelationship between an investment in health and safety and overall business performance. The parameters of most interest to health and safety professionals are productivity, quality, and customer satisfaction.

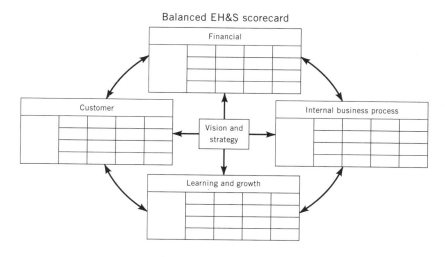

Figure 50.8. The Balanced Scorecard Concept and examples for two of the scorecards.

7 SYSTEMS THINKING AND SYSTEMS DYNAMICS

Senge points out that structures influence behaviors and when different people are placed in the same system, they produce different results (99). This phenomenon is extremely important in understanding how to make health and safety investments that are designed to create a high degree of result uniformity. Senge goes on to point out that the systems perspective tells us that it is necessary to look beyond individual mistakes or bad luck to understand important problems. One must look beyond personalities and events and look to the underlying structures which shape individual actions and create conditions where types of events (in our case safety) become more likely. Finding ways to understand how health and safety investments drive structures, behaviors, and ultimately safety, productivity, quality and customer satisfaction underpins this project. Figure 50.9 is a Systems Thinking model that can be used to help understand the impact of increasing overtime to meet production goal on health and safety.

Seif at Digital Equipment, was able to demonstrate how systems thinking and systems dynamics (SD) was able to position its workforce to meet the demands of a changing marketplace (100). SD was able to create an understanding of hiring, attrition and promotion policies on long term strategies. Kemeny at COPEX, was able to demonstrate the source of recurring financial difficulties using SD and to find the points of leverage to solve these problems (101). Roth explains how Georgia Power used systems thinking tools to face the challenge of a changing regulatory environment and global competition (102). Georgia Power used SD models to better understand the ways in which marketing strate-

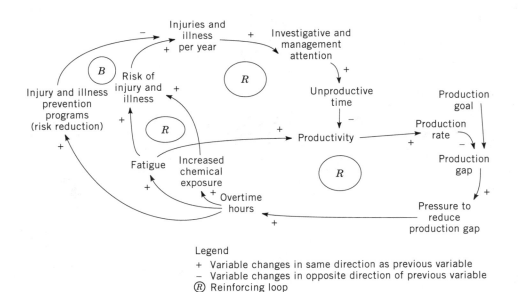

Figure 50.9. Systems Thinking Model of the impact of overtime on health and safety.

gies, customer satisfaction, competition, and price/cost structures would affect their potential role in the industry. These three case studies clearly demonstrate the applicability of systems thinking and systems dynamics to business and the power they have in analyzing health and safety investments that have profound, but frequently subtle impacts on business performance.

One of the key tools that appear to hold promise for analyzing health and safety investments is the causal loop diagram, the central tool in systems dynamics developed by Jay Forrester and his colleagues at MIT in the 1960s and is described by Kim (103).

8 SUMMARY

Clearly, the pressures of competition are driving companies to question the return on investment for every dollar spent on human resources, hardware programs, and people. In EHS, the large number of intangible or hard to measure factors has inhibited our ability to do this analysis. Deming's work in quality improvement is presented to illustrate how EHS practitioners can begin to measure the performance of their work activities and measure the output generated by those activities. Additionally, the growing focus on customer satisfaction, productivity, employee satisfaction and quality are powerful drivers for applying business analysis techniques to EHS.

A central goal of this chapter, is to help those interested in applying business techniques to health and safety a view of the state-of-the-art. Additionally, the chapter provides some examples, tools, concepts and references to help facilitate the application of business techniques to EHS. The goal being the complete integration of EHS into the mainstream of business.

The realization that up until recently, most applications involved understanding the cost of safety and the cost of failure and not the return on investment in EHS projects. Very little work has involved the concept of investing EHS and understanding how those investments impact overall business performance. Whether it was Weisbrod addressing social costs; Rice calculating the value of an illness, injury, or disability; or the work of many others, the focus is generally on cost. Others looking at programmatic or hardware investments, such as the work done by Birkner and Saltzman, and Simpson analyzed the cost-effectiveness of investments, or how much safety can be purchased from each dollar spent. This approach begins to get close to looking at "returns."

The environmental accounting work done by Epstein, however, provides significant guidance in identifying and measuring costs. Of course, understanding how to evaluate cost is the first step in doing investment analysis. This work is a significant contribution to health and safety. The research done by EPA in identifying the categories of environmental costs; hidden, conventional or direct, contingent, and relationship costs is also an important contribution.

Oxenburgh, looking at the impacts of safety investments on productivity, developed cost models that appear to significantly add to health and safety practice. Fisher's cost of failure model begins to address investment, but only empirically in its relationship to the total cost of safety. Heinrich, Bird and Tye addressed the issue of direct, indirect, fixed and variable costs as they relate to safety. And the extensive work done by the International

Labour Office with respect to developing cost ratios has moved the knowledge base forward.

Understanding the hidden impact of health and safety investments is one of the most vexing problems we face. The contributions to this understanding by Minter, Krause, Reichel and Neumann, Robinson, Waxler, and Higginson give us clear direction to the use of systems dynamics and systems thinking for enhancing understanding health and safety investments on overall business performance. Moreover, the Balanced Scorecard concepts promoted by Kaplan and Norton provide further direction for understanding the interaction of financial and nonfinancial metrics. The work done by Seif at Digital Equipment, Kemeny at COPEX, and Roth at Georgia Power provide cases to build on in the area of systems thinking. Finally, the application of ABCM to EHS departments and programs is discussed. ABCM yields cost data and business process information needed for making decisions and solving complex budgetary and business problems.

BIBLIOGRAPHY

1. D. Johnson, "Annual White Paper Report," *Industrial Hygiene and Safety News* Chilton, Dec. 1995.
2. Williams-Steiger Occupational Safety and Health Act of 1970 (84 Stat 1593).
3. H. Brown and T. Larson, "Making Business Integration Work: A Survival Strategy for EHS Managers," *Environ Quality Manage.* 7(3) 1–8 (Spring 1998).
4. R. A. Axelrod, "Ten Years Later: The State Environmental Performance Reports Today," *Environ. Quality Manage.* 7(2) 1–3 (Winter 1998)
5. Matthew J. Kiernan and Jonathan Levinson, "Environment Drives Financial Performance: The Jury Is In," Environ. Quality Manage. 7(1) 1–7 (Winter 1997).
6. P. A. Soyka and S. J. Fledman, "Investor Attitudes toward the Value of Corporate Environmentalism: New Survey Findings," Environ. Quality Manage. 8(1) 1–10 (Autumn 1998)
7. Iannuzzi, A. "The Environmental Quality Business Plan: A Step-by-Step Guide," Environ. Quality Manage. 7(2), 65–69 (Winter 1997)
8. E. D. Weiler, P. G. Lewis, and D. J. Belonger, "Building an Integrated Environmental, Health, and Safety Management System," Environ Quality Manage. 7(3), 59–65 (Spring 1997).
9. G. G. Stock, J. L. Hanna, and M. H. Edwards, "Implementing an Environmental Business Strategy: A Step-by-Step Guide," Environ. Quality Manage. 7(4), 33–41 (Summer 1997).
10. J. P. Roberts, R. H. K. Vietor, and F. Reindardt, "Note on Contingent Environmental Liabilities," *Harvard Business School Case, 9-794-098*, Aug. 1995, pp. 1–23.
11. S. A. Greyser, "Environmental Behavior and Corporate Reputation," *Harvard Business School Case, 9-597-014*, Dec. 1996.
12. R. A. Kopp, Krupnick, and M. Toman, "Cost-benefit Analysis and Regulatory Reform,: Human and Ecological Risk Assessment, 3(5), 787–852, (Nov. 1997).
13. R. H. K. Vietor, and E. Prewitt, "Allied-Signal: Managing the Hazardous Waste Liability Risk," Harvard Business School Case,
14. M. E. Barth, M. J. Epstein, and R. D. Stark, "Polaroid: Managing Environmental Responsibilities and Their Costs," *Harvard Business Review Case*

15. S. M. Datar, M. J. Epstein, and K. White, "Bristol-Myers Squibb: Accounting for Product Life Cycle Costs at Matrix Essentials," *Harvard Business School Case*
16. F. Reinhardt, "Acid Rain: The Southern Co. (B)," *Harvard Business School Case*
17. F. Reinhardt, "Acid Rain: The Southern Co. (A)," *Harvard Business School Case*
18. G. LaBar, *Occupational Hazards* (June 1994).
19. M. J. Epstein, *Measuring Corporate Environmental Performance*, McGraw-Hill, New York, 1996, pp 145, 148, 164, 169, 240
20. G. LaBar, *Occupational Hazards* (June 1994).
21. G. LaBar, *Occupational Hazards* (June 1994).
22. G. LaBar, *Occupational Hazards* (June 1994).
23. C. Fitzgerald, *Selecting Measures for Corporate Environmental Quality: Examples from TQEM Companies*, Executive Enterprises Publications Co., Inc. 1993, pp. 15–24
24. R. S. Greenberg and C. A. Unger, *TQM and the Cost of Environmental Quality*, Executive Enterprises Publications Co., Inc., 1993 pp. 139–144
25. T. J. Larson and H. J. Brown, "Designing Metrics that Fit: Rethinking Corporate Environmental Performance Measurement Systems," *Environ. Quality Manage.* Summer, 1997, pp. 81–88, **6**(3), 81–88 (Summer 1997).
26. W. G. Russell and G. F. Sacchi, "Business-Oriented Environmental Performance Metrics: Building Consensus for Environmental Management Systems," *Environ. Quality Manage.*, **6**(4), 11–19 (Summer 1997).
27. M. G. Schene and J. T. Salmon, "Applying Outcome Evaluation and Measures to Environmental Management Programs," *Environ. Quality Manage.* **6**(4), 71–78 (Summer 1997).
28. M. F. Johnson, M. Magnan, and C. Stinson, "Nonfinancial Measures of Environmental Performance as Proxies for Environmental Risks and Uncertainties," presented at The University of Texas Business School Conference On Activity-Based Costing and Environmental Accounting, Jan. 1997.
29. G. Friend, "EcoMetrics: Integrating Direct and Indirect Environmental Costs and Benefits into Management Information Systems," *Environ. Quality Manage.* 19–29, (Spring 1998).
30. P. A. Marcus, "Using EH&S Management Systems to Improve Corporate Profits," *Environ. Quality Manage.*, Winter, 1996, pp. 11–21, **6**(2), 11–21 (Winter 1996)
31. R. S. Kaplan and D. P. Norton, "Putting the Balanced Scorecard to Work," in Harvard *Business Review on Measuring Corporate Performance*, Harvard Business School Publishing, Boston, MA, 1998, pp. 147–182.
32. L. R. Birkner and L. S. Saltzman, Assessing Control Strategy Cost-effectiveness *Am. Ind. Hyg. Assoc. J.* **47**(1), 50–54 (1986).
33. R. P. Wells, M. N. Hochman, S. D. Hochman, and P. A. O'Connell, *Total Quality Manage.* 315–327 (Summer 1992).
34. G. Labar, *Occupational Hazards Magazine*, 33–36 (June 1994).
35. W. E. Deming, *Out of the Crisis*, Massachusetts Institute of Technology, 1982, pp. 23–24, Chapt. 2.
36. W. W. Scherkenback, *The Deming Route to Quality and Productivity*, Mercury Press, Rockville, MD, 1992.
37. H. J. Harrington, *Business Process Improvement*, McGraw-Hill, Inc., New York, 1991.
38. J. L. Heskett, W. E. Sasseer, Jr., and C. W. L. Hart, *Service Breakthroughs: Changing the Rules of the Game*, The Free Press, New York, 1991.

39. M. Walton, *The Deming Management Method*, The Putnam Publishing Company, New York, 1986.
40. R. C. Whiteley, *The Customer Driven Company: Moving from Talk to Action*, Addison-Wesley Publishing Company, Inc., Reading, MA, 1991.
41. S. DeLorey, *Arthur Andersen Proposal to ORC for the ROHSI Project*, March 1996.
42. R. J. Gray, Bebbington, and D. Walters, *Accounting for the Environment*, Markus Wiener Publishers, Princeton, NJ, 1993, p. 151–152.
43. R. S. Kaplan, and D. P. Norton, "The Balanced Scorecard—Measures that Drive Performance," in *Harvard Business Review on Measuring Corporate Performance*, Harvard Business School Publishing, Boston, MA, 1998.
44. D. Kite, "Capital Budgeting: Integrating Environmental Impact," *Cost Management*, (Summer 1995).
45. A. K. Dixit, and R. S. Pindyct, *Harvard Business Review*, 105–115 (May/June 1995)
46. P. Ainsworth, "ABC Overhead Analysis Beats Traditional Approach," *The Practical Accountant* (July 1995).
47. M. T. Brandt, S. P. Levine, D. G. Smith, and H. J. Ettinger, "Activity-Based Cost Management Part I: Applied to Occupational and Environmental Health Organizations," *Am. Ind. Hyg. Assoc. J.* **59**, 328–334 (1998).
48. M. T. Brandt, S. P. Levine, D. G. Smith, and H. J. Ettinger, "Activity-Based Cost Management Part II: Applied to a Respiratory Protection Program," *Am. Ind. Hyg. Assoc. J.* **59**, 335–345 (1998).
49. M. T. Brandt, "Activity-Based Cost Management (ABCM) Applied to an Environmental, Safety, and Health (ES&H) Department and Program," Dissertation, UMI Dissertation Services, Ann Arbor, MI, 1997.
50. A. Sorine, Accounting for Safety. *Occupat. Hazards* (Sept. 1994).
51. P. E. Greenberg, S. N. Finkelstein, and E. R. Berndt, "Economic Consequences of Illness in the Workplace," *Solan Management Review* 26–38 (Summer 1995).
52. M. J. Epstein and S. D. Young, "Improving Corporate Environmental Performance Through Economic Value Added," *Environ. Quality Manage.* 1–7, (Summer 1998).
53. L. R. Birkner and L. S. Saltzman. Assessing Control Strategy Cost-Effectiveness *Am. Ind. Hyg. Assoc. J.* **47**(1), 50–54 (1986).
54. P. A. Samuelson and W. D. Nordhaus, "Capital, Interest, and Profits," *Economics*, 13th ed, McGraw-Hill Book Company, New York, 1989, Chapt. 30.
55. R. C. Higgins, *Analysis for Financial Management*. Chicago, 1995, pp. 237–277.
56. R. N. Anthony, J. S. Reece, and J. H. Hertenstein, "Short-Run Alternative Choice Decisions," in *Accounting: Text and Cases*, 9th ed, Irwin, Chicago, IL, 1995, Chapt. 26.
57. J. K. Shim and J. G. Siegel, "How to Make Capital-Budgeting Decisions," in *Modern Cost Management & Analysis*, Barron's Educational Services, Inc., 1992, Chapt 16.
58. M. S. Hilbert, Address before the Annual Meeting of the American Public Health Association, 1975.
59. B. A. Weisbrod, "Does Better Health Pay?" *Public Health Reports*, June 1960.
60. D. Rice and P. E. J. MacKenzie. *Cost of Injury in the United States. A Report to Congress*, Center for Disease Control, U.S. Department of Health and Human Services, 1967
61. *The Nation's Health*, (April 1976).
62. HEW Publication, "Small Plant and Medical Programs," 1958.

63. Oxenburgh, M CCH Austrialia Ltd. North Ryde, NWS 1991.
64. D. V. McCallum, "Interdisciplinary Teamwork in the Health and Safety Professions" Paper presented before the American Medical Association, Sept. 21, 1976.
65. Report to the President on Occupational Safety and Health, 1972.
66. G. C. Simpson. Is ergonomics Cost-effective in *Health, Safety and Ergonomics*, Butterworths, London, 1988, p. 154, Chapt. 12.
67. J. H. Wehrle, "Chronic Wrist Injuries Associated with Repetitive Hand Motions in Industry," Ann Arbor Center for Ergonomics MSc thesis, Univ. of Michigan (1976).
68. G. C. Druery, D. P. Roberts, R. Hansgan, and J. R. Baymon, "Evaluation of a Palletizing Aid," *Applied Ergonomics*, **14**, 22 (1985).
69. S. Berger, Estimating Environmental, Safety and Health Costs of Processes During R&D. *Synergist* (Dec. 1995).
70. R. W. Haines, Environmental Performance Indicators: Balancing Compliance with Business Economics. *Total Quality Environ. Manage.* (Summer 1993).
71. *Tools and Techniques of Environmental Accounting for Business Decisions*, Institute of Management Accountants, Montvale, NJ, 1996
72. Environmental Protection Agency, *An Introduction to Environmental Accounting as a Business Management Tool, Key Concepts and Terms*, 1965
73. H. W. Heinrich, *Industrial Accident Prevention*, 4th ed. McGraw-Hill, 1959.
74. D. Andreoni, *The Cost of Occupational Accident and Disease*. Occupational Safety and Health Series 54, International Labour Office, Geneva, 1986.
75. "Economic Focus—A Price on the Priceless," *The Economist*. (Aug. 17, 1991).
76. A. Veltri, "An Accident Cost Impact Model: The Direct Cost Component," *J. Safety Research* **21**, 67–73 (1990).
77. B. Brody, Y. Letourneay, and A. Poirier "An Indirect Cost Theory of Accident Prevention," *J. Occupat. Accidents.* 1990 **13**, 255–270 (1990).
78. T. A. Fisher, Quality' Approach to Occupational Health, Safety and Rehabilitation. *J. Occupat. Health Safety—Australia and New Zealand*, **7**(1), 23–28 (1991).
79. P. Lanoie, and L. Trottier, "Costs and Benefits of Preventing Workplace Accidents: Going from a Mechanical to a Manual Handling System," *J. Safety Research,* **29**(2), 65–75 (1998).
80. R. Kopp, A. Krupnick, and M. Toman, "Cost-benefit Analysis and Regulatory Reform," *Human and Ecological Risk Assessment*, **3**(5), 787–852 (1997).
81. C. R. Harper, and D. Zilberman, "Pesticides and Worker Safety," *Am. J. Agricultural Economics* **74**(1), 68–78 (1992).
82. C. L. Spash, "Reconciling Different Approaches to Environmental Management," *Intern. J. Environ. Pollut.* **7**(4), 497–511 (1997).
83. Health and Safety Executive (HSE). *The Cost of Accident at Work*, London, 1993, pp. 20–23, HS(G)96, ISBN 0 11 886374 6.
84. F. E. Bird, R. G. Loftis, *Loss Control Management*, Institute Publishing, Logenville, GA, 1976.
85. J. Tye, *Accident Ratio Studies 1974–1975*, British Safety Council, 1976.
86. M. J. Bailey, *Reducing Risks to Life: Measurement of Social Benefits*, American Enterprise Institute for Public Policy Research, Washington DC, 1980, pp. 15–27.
87. S. Spilling, Eitrheim, and Aaras, A Cost-Benefit Analysis of Work Environment Investment at STK's Telephone Plant at Kongsvinger, in N. Corlett and J. Wilson, eds, *The Ergonomics of Working Posture*. Tailor and Francis, London 1986, pp. 381–397.

88. S. G. Minter, "A System for Reducing Injury," *Occupat. Hazards*, 6 (1993).
89. T. T. Krause, Safety and Quality: Two Sides of the Same Coin. *Occupat. Hazards*, 47–50 (April 1993).
90. S. B. Hughs and D. M. Willis, "How Quality Control Concepts can Reduce Environmental Expenditures," *Cost Management*, 15–19 (Summer 1995).
91. J. C. Robinson, "The Impact of Environmental and Occupational Health Regulations on Productivity Growth in U.S. Manufacturing," *The Yale Journal on Regulation*, **12**, 387, (1995).
92. S. Parks, "Improving Workplace Performance: Historical and Theoretical Contexts." *Monthly Labor Review*, (May 1995).
93. R. Walxler and T. Higginson, Discovering Methods to Reduce Workplace Stress. *IIE Solutions*, (June 1993).
94. D. M. Kiser, and J. G. Esler, "Kodak's Safety Performance Indexing—A Tool for Environmental Improvement." *Total Quality Environ. Manage.* 15–49 (Autumn 1995)
95. V. Blewett, *Beyond Lost Time Injuries: Positive Performance Indicators of OHS*. Conference Proceedings, Worksafe Australia, Sidney, 1994.
96. R. A. Greene, *Positive Role for OHS in Performance Measurement*, Conference Proceedings, Australia, Sidney, 1994.
97. R. Kaplan and D. Norton, "The Balance Scorecard—Measures that Drive Performance," *Harvard Business Review* (Jan.–Feb. 1992).
98. N. Carter, R. Klein, and P. Day, *How Organizations Measure Success in Government* Routledge, London, 1992.
99. P. M. Senge, *The Fifth Discipline*, Currency Doubleday, New York, 1990, pp. 42–51.
100. N. Seif, *Rethinking Workforce Planning, Managing the Rapids*, Pegasus Communications, Cambridge, MA, 1995 pp. 85–92.
101. J. Kemeny, *Thinking Systemically About Strategy, Planning, Managing the Rapids*, Pegasus Communications, Cambridge, MA 1995 pp. 93–100.
102. W. Roth, *Systems Thinking Applied to the Electric Utility Industry Planning, Managing the Rapids*, Pegasus Communications, Cambridge, MA, 1995 pp. 101–110.
103. D. H. Kim, Learning Laboratories: Practicing Between Performances, *The Systems Thinker*, Pegasus Communications, Cambridge, MA.

CHAPTER FIFTY-ONE

Industrial Hygiene Education, Training, and Information Exchange

Dennis K. George, Ph.D., CIH and Michael R. Flynn, ScD, CIH

1 INTRODUCTION

As defined by the American Industrial Hygiene Association, industrial hygiene is principally an applied discipline dedicated to the anticipation, recognition, evaluation, and control of environmental factors or stresses arising in the occupational environment that may cause sickness, impaired health and well-being, or significant discomfort among workers or among citizens of the community. These factors or stresses may be chemical (e.g., solvents, heavy metals), physical (e.g., noise, heat stress), biological (e.g., bloodborne pathogens, tuberculosis), or ergonomic (e.g., manual materials handling, repetitive motion) in nature and may arise in virtually any sector of the occupational environment. Practitioners in the field of industrial hygiene must be able to:

- Identify situations in the workplace that pose a potential threat of adverse consequences to employee health and well-being,
- Determine the nature and extent of this risk through extensive investigation of all pertinent background information and qualitative and quantitative assessment of the environmental stressors involved,
- Judge the acceptability or unacceptability of the risk upon careful observation of all relevant factors and thorough analysis of all pertinent qualitative and quantitative data.

Patty's Industrial Hygiene, Fifth Edition, Volume 3. Edited by Robert L. Harris.
ISBN 0-471-29753-4   © 2000 John Wiley & Sons, Inc.

- Design and implement adequate measures of remediation for those situations posing unacceptable risk.

Essentially, the industrial hygienist is an individual concerned with the nature and behavior of hazardous agents generated from occupational processes *while these agents are still within the occupational facility* and potentially posing a risk to facility employees.

Whereas "traditional" industrial hygiene focuses on the health and well-being of workers *inside* the plant, today's industrial hygienist is frequently called upon to assess the impact of industrial pollutants upon the health of the surrounding community and the fate of these materials in the neighboring environment. Thus, an industrial hygienist must also be cognizant of the behavior of contaminants *outside* the plant and the assessment and control of hazardous materials escaping into surrounding air, water, or land. The industrial hygienist of today must be prepared to address air quality, water quality (including stormwater and groundwater), and hazardous waste management issues for their facility.

The types of potentially hazardous agents of concern are as many and varied as the industrial processes from which they originate. Extensive knowledge of contaminant generating processes, mechanisms of pollutant behavior and transport, and the means by which the materials cause harm to individuals in both the occupational and outdoor environments is necessary to deal effectively with the associated risks. Thus, managing the occupational health and environmental issues of an industrial concern is, indeed, a highly complex challenge requiring a strong technical background coupled with a very diverse set of applied skills.

In addition to the ethical concerns over the health and well being of employees and surrounding communities, companies are constrained to control employee exposures and environmental releases of contaminants by a plethora of governmental occupational health and environmental regulations. Indeed, dealings with agencies such as the Occupational Safety and Health Administration (OSHA) and the Environmental Protection Agency (EPA) are a routine part of the job for many industrial hygienists. Ensuring compliance with OSHA and EPA regulations, although important, is often a tedious, time-consuming, and, to say the least, frustrating experience for occupational and environmental health professionals.

The necessity of applying a broad base of technical skills and problem solving abilities within such a complicated regulatory framework makes the field of industrial hygiene one of the most complex and challenging of the current scientific disciplines. On the other hand, the opportunity to prevent occupational disease, protect public health and natural resources, and improve the overall quality of life for our society also make this career one of the most rewarding.

The academic preparation of individuals entering industrial hygiene, as well as the continuing education and professional development of those already in practice, are issues critically important to the continuing viability of the industrial hygiene profession. Society is in need of academic programs capable of producing graduates qualified to meet the challenges of this diverse and rapidly expanding field. Continuing education and professional development opportunities are also essential to ensure that industrial hygiene practitioners can expand the scope of their knowledge on an on-going basis and keep abreast of important changes that occur in the practice of their occupation. Additionally, industrial hygienists also need the ability to share ideas and exchange information with peers on a relatively frequent basis. Communication among colleagues is extremely important in terms of sharing lessons learned from field experiences as well as results achieved from basic and applied research projects.

The purpose of this chapter is to introduce the reader to various outlets for industrial hygiene education, training, and information exchange. It is intended to provide guidance for those seeking their first degree in industrial hygiene, those wishing to update their educational credentials, and for those attempting to identify resources enabling them to remain regularly updated and informed on responsible and ethical industrial hygiene practice. For the purposes of this chapter, the word "education" is used to refer to the acquisition of knowledge in a formal, academic setting, typically occurring in the context of a college or university curriculum and ultimately leading to the award of a formal degree. The process of pursing information outside of an academic, "degree-seeking" program of study is termed training, continuing education, or professional development. This avenue encompasses a wide variety of information delivery systems and formats such as seminars, workshops, short courses, self-paced/self-study courses, etc. A brief discussion of various mechanisms and forums of information exchange such as printed materials (e.g., journals), personal interactions (e.g., conferences, meetings), and electronic interfaces (e.g., listserves, web pages, etc.) is also included. Please note that this material is by no means exhaustive. The reader is encouraged to consider this chapter as a beginning point for further exploration.

2 INDUSTRIAL HYGIENE EDUCATION

2.1 The Academic Route into the Industrial Hygiene Profession

An important issue receiving considerable attention in professional industrial hygiene circles of late regards the question: "What formal educational background constitutes the most appropriate academic route into the industrial hygiene profession?" This is perhaps the most important consideration for anyone planning a career in industrial hygiene and contemplating their academic alternatives in preparation for such a vocation. Many different colleges and universities offer degrees in industrial hygiene or similarly named programs such as Occupational Health, Occupational Safety and Health, Environmental and Occupational Health, etc. Currently, some 65 such programs exist in the United States, both at the graduate and undergraduate level. How does one go about evaluating the curriculum offered by a given institution? How does one compare the program offered at one institution to that offered at another?

In a general sense, prospective students must recognize that the term "curriculum" does not simply refer to a collection of individual courses and laboratories in a given major, but also to the manner in which each course in the overall program is woven together into a cohesive unit. The ultimate goal of any academic curriculum should be to deliver a solid, focused, instructional program that offers graduates the best possible opportunity to succeed in their chosen field. Thus, when evaluating any program of study, one should carefully consider the objectives of each specific course, the manner in which it is integrated into the curriculum as a whole, the textbooks and supplemental materials utilized, the nature of the laboratory exercises and the instrumentation used therein, the extent to which field experiences and other relevant extracurricular activities are incorporated into the program, and a host of other similar issues. Although far from complete, the following paragraphs are offered in an attempt to address certain of these questions more specifically. The reader should refer to other, more extensive works for more complete discussions of industrial hygiene educational issues (1–4).

2.2 Graduate versus Undergraduate Education

An emerging industrial hygiene educational issue involves a debate of the advantages of a graduate over an undergraduate industrial hygiene degree. At issue is whether or not a sufficiently rigorous and thorough curriculum can be provided to career-oriented industrial hygiene students at the undergraduate level. Many professional industrial hygienists feel that an undergraduate education alone is not adequate preparation for the competent practice of industrial hygiene and that the master's degree should be the minimum educational "bar" for the profession (3, 5). Some state that graduate training is becoming a prerequisite for a career in industrial hygiene in the United States (1). Others have recommended that only introductory industrial hygiene courses be taught at the undergraduate level as prerequisites for admission to graduate school (6). Sherwood asserts that the required level of education of the professional industrial hygienist in developed countries is becoming two academic years of full-time study after graduation in a branch of science or engineering (3).

Those who strongly advocate this position point to the fact that many sister professional disciplines have already adopted a "professional" school or a graduate degree requirement as the academic route into their respective professions. They feel that a similar requirement is necessary for industrial hygiene to meet the growing challenge and to keep the profession meaningful and relevant to private industry, government agencies, and licensing organizations. Likening the practice of industrial hygiene to that of professional disciplines such as medicine or law, Corn suggested that a pre-professional baccalaureate experience (e.g., chemistry, physics, biology, engineering, etc.) is an essential entree into the more advanced, discipline-specific academic preparation that occurs at the master's level (5).

Another point frequently offered in favor of a graduate degree is the growing expectations placed upon newly graduated industrial hygienists. As corporate industrial hygiene staffs are shrinking in size, fewer opportunities exist for new graduates to develop mentoring relationships with senior industrial hygienists. Many are expected to immediately perform as functioning professionals, making the added maturity and experience gained in graduate school even more of a necessity. Additionally, there is always the concern that inserting industrial hygiene courses into an undergraduate curriculum may force out other important courses that are essential building blocks in the development of fundamental recognition, evaluation, and control skills in the industrial hygienist.

It should be noted, however, that while prelaw or premedicine curricula are widely available at the undergraduate level, a comparable "pre-industrial hygiene" professional avenue has not been defined in academia. Nor is there a "qualifying" examination such as the LSAT or MCAT that one must pass in order to gain acceptance to the professional industrial hygiene school. It is certainly true that at present the majority of the best known and most mature industrial hygiene curricula is found at the graduate level. However, this occurrence may be more a product of happenstance than design. As the practice of industrial hygiene began taking shape as a profession after the turn of the century, persons entering this fledgling field typically did so after having already completed a baccalaureate program or professional degree. Thus, as industrial hygiene academic programs began to develop, the majority were situated in graduate schools, frequently in Schools of Public Health or Engineering. Until recently, few undergraduate industrial hygiene programs existed in the United States. Thus, the discipline-specific academic preparation for the profession has traditionally occurred primarily at the master's level.

However, while the pursuit of a degree in industrial hygiene has historically necessitated a graduate school experience, there are those who contend that the recent development and advancement in the quality of undergraduate industrial hygiene curricula suggest an alternative (7). These individuals propose that properly designed undergraduate industrial hygiene programs can, indeed, provide an excellent educational experience for individuals desiring to directly enter the professional practice of this discipline. They point out that persons entering industrial hygiene graduate schools do so with a variety of academic backgrounds including chemistry, biology, or engineering, noting that although there is much to recommend a bachelor's degree in one of these areas as preparatory to industrial hygiene practice, a significant portion of the advanced courses comprising such degrees relate very little to the understanding of industrial hygiene and might reasonably be replaced with industrial hygiene courses in the curriculum. Supporters of undergraduate industrial hygiene preparation maintain that foundational aspects of the basic sciences and mathematics and the application of these principles to the field of industrial hygiene can be properly addressed at the undergraduate level with sufficient opportunity remaining to cover advanced principles unique to the discipline adequately.

Kortsha states, "At the baccalaureate level, medium and large plants will need industrial hygienists with undergraduate degrees on site to head a variety of health activities. Such individuals are well qualified for the job and are excellent candidates for graduate school. They have a good foundation to build on and many will have had field experience before going on for advanced degrees. It is likely that they will be in much demand" (6).

Answering the question of whether or not an undergraduate degree alone is sufficient preparation for industrial hygiene practice is not an objective of this chapter. In reality, no simple answer exists. At present, there is no clearly defined educational pathway into industrial hygiene; either through graduate or undergraduate academic preparation. At present, approximately 32% of practicing industrial hygienists list a Bachelor's degree as the highest degree held whereas 55% have gone on to obtain a Master's degree (8). Obviously, some feel it worth the effort to pursue an undergraduate degree in a basic science, mathematics, engineering, or perhaps even industrial hygiene, and then move into a more focused industrial hygiene study at the graduate level. Others are comfortable entering industrial hygiene practice after a baccalaureate degree experience. Several excellent academic programs exist at both levels and each prospective student should consider his or her own reasons for selecting a particular route. However, the following paragraphs identify relevant factors that should be considered by anyone attempting to make this decision.

2.2.1 Technical Rigor of Undergraduate Programs

The first issue concerns the selection of courses comprising the industrial hygiene curriculum at the college or university under consideration. Unless carefully conceived, there is a potential for undergraduate programs to contain several industrial hygiene courses without the underlying framework of technical courses necessary to support a thorough understanding of many important industrial hygiene concepts. This danger arises from the fact that undergraduate industrial hygiene curricula must include general education courses as well as foundation/core courses and technical electives that support the upper division courses in the major. To maintain a reasonable number of credit hours for a baccalaureate degree, decisions have to be made regarding the type and number of courses to include in the total program. Each course must be evaluated for its contribution to the overall curriculum, a process that frequently necessitates deletion of one course to accommodate the addition of another that may be deemed more important. Thus, when evaluating a particular baccalaureate curriculum, it becomes critical for the prospective student to ensure that industrial hygiene courses are not substituted into the program at the expense of courses in the basic sciences (chemistry, biology, physics, etc.) and mathematics (1). Development of genuine problem analysis and problem solving skills for an applied field such as industrial hygiene is truly impossible apart from a substantial technical background. Prospective degree-seeking students are cautioned to look suspiciously at programs that do not have substantial preparation in basic mathematics and science as the foundation of their curriculum. A subsequent section of this chapter will address the issue of mathematics and science curricular content more specifically.

2.2.2 Professional Certification Issues

A second issue pertains to the educational qualifications necessary for professional certification in industrial hygiene. The American Board of Industrial Hygiene (ABIH), a not-for-profit corporation organized to improve the practice and educational standards of the profession of industrial hygiene, oversees and administers the professional certification/registration of industrial hygienists. This process culminates in the title of Certified Industrial Hygienist (CIH), a coveted designation widely recognized as the mark of a competent practitioner of the art and science of industrial hygiene. The CIH distinction is frequently utilized as a major determinant in assessing suitability for initial employment as well as subsequent advancement in this position. Many companies who hire outside industrial hygiene consultants specify that the contracted services be performed or at least signed by a CIH. In many states, certification also looms large in title protection issues as well as in other pending legislation. The importance of obtaining this credential should not be underestimated by anyone considering industrial hygiene as a career. Presently, approximately 41% of the industrial hygiene profession are CIHs according to AIHA membership data (8).

Attaining the CIH designation requires a candidate to meet a minimum educational requirement, a minimum field experience requirement, and to achieve a passing score on a two-part examination (a seven-hour core examination and a seven-hour comprehensive examination). The ABIH defines an industrial hygienist as "a person having a baccalaureate or graduate degree from an accredited college or university in industrial hygiene, biology, chemistry, engineering, physics or a closely related physical or biological science who, by virtue of special studies and training, has acquired competence in industrial hygiene (9)". Currently, the minimum educational requirement specified by the ABIH for admission to the CIH examination is stated in the Board's "Regulations for Certification" as follows: "Graduation from a college or university acceptable to the Board with a bachelors degree in industrial hygiene, chemistry, physics, chemical, mechanical, or sanitary engineering, or biology. The Board will consider, and may accept any other bachelors degree from an acceptable college or university which is possessed by an applicant, in consideration of its basic science content. In evaluating the science content of a bachelors degree other than a degree named above, the Board will use the criteria of at least 60 semester credit hours in undergraduate or graduate level courses in science, mathematics, engineering, and technology, with at least 15 of those hours at the upper (junior, senior, or graduate) level. A degree which is heavily comprised of only one of those subject areas, in the absence of others, may be judged to be not acceptable. An applicant who is found to have an unacceptable bachelors degree, may remedy that degree with additional academic science coursework from an acceptable college or university, or by completion of an acceptable cognate graduate degree" (10).

From the above statement, it appears that the ABIH is content, at least for the present, with a bachelor's degree as the minimum educational credential for admission to the CIH examination. However, some who feel strongly that a master's degree is the minimum degree for professional practice are advocating the certification process as a means to encourage this requirement. These individuals are urging the ABIH to require a candidate to have earned at least a Master's degree to be qualified for the CIH exam. This is not a new issue for the Board. This sentiment can be traced at least as far back as 1971 when almost 56% of the respondents to a survey of the American Academy of Industrial Hygiene recommended the Master's Degree as the minimum requirement for eligibility for admission to certification examinations (11). More recently, Corn states, "The desires of senior respected industrial hygienists can only be inferred by younger recruits through deciphering requirements for association member or certification. Curiously, industrial hygiene has taken the very major accreditation step for professional education while not requiring the education" (5).

Prospective students should be aware that the ABIH is currently giving this issue serious consideration. In an article entitled "Educational Requirement" in a 1995 American Academy of Industrial Hygiene Newsletter, the ABIH published the following statement: "In keeping with the purpose of the American Board of Industrial Hygiene "to improve the practice and educational standards of the profession of industrial hygiene," the Board is considering requiring a masters degree in industrial hygiene from an ABET accredited school to meet the minimum educational prerequisite for admission to the certification examinations. The Board believes that many individuals are entering the field with little or no formal industrial hygiene education and limited opportunities for training with experienced professionals. Many other professions require a degree in their discipline prior to certification/licensing/practice. The Board is beginning to debate this issue and would appreciate comments from both Diplomates and applicants/potential applicants of the merits of this proposed change." Although the Board has taken no specific action in this regard to date, it is clear that it is still under consideration. Among the goals of the 1998 ABIH

Strategic Plan is the expressed objective to "Develop a plan to raise the educational eligibility requirements for the CIH" (12). Obviously, the Board continues to debate this issue and even if the minimum educational requirement was raised to the masters degree, there would likely be a phase-in period of several years. However, it would behoove anyone considering a career in industrial hygiene to keep an eye on this situation. Attaining certification is becoming increasingly important for industrial hygiene practice and one would certainly desire to complete the level of education necessary to achieve this credential.

2.2.3 Salary Differential

A final important issue in assessing the comparative advantages of an advanced degree concerns the salary differential between practitioners with a master's as opposed to a bachelor's degree. Figure 51.1 illustrates the salary comparison results of the 1997 AIHA Membership Survey Report (8). Note from this figure that in general, the salary distribution of industrial hygienists favors those with master's degrees in the higher salary ranges as opposed to those with bachelor's degrees, which tend to be more represented in the lower

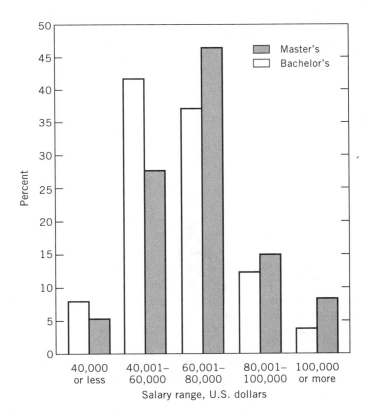

Figure 51.1. Salary Comparison results of the AIHA Membership Survey.

salary ranges. For example, approximately 69% of those with master's degrees make in excess of $60,000 per year while only 52% of those with only a bachelor's have achieved this level. From this data it appears that the advanced degree does make a substantial difference in the salary that can be commanded as well in the advancement potential in a given position.

The salary comparison above does, however, bring to light a significant underlying issue in terms of the relevance of undergraduate programs; namely, competition for jobs. Whether appropriate or not, much of the debate concerning graduate versus undergraduate degrees is likely to be settled in a rather pragmatic fashion outside the walls of academia. It is, in reality, the industrial community, the ultimate consumer of the product of these programs, that will determine the long-term viability of undergraduate programs. If industry managers can secure qualified bachelor's-degreed industrial hygienists that fulfill their technical needs for a salary that is lower than that commanded by master's-degreed industrial hygienists, they will, in all probability, hire them. As long as a market exists for the product of undergraduate programs, these programs will likely continue.

To conclude this discussion, it should be noted that an effort is currently underway within academia to establish a more open and effective line of communication between undergraduate and graduate programs. As of this writing, an Academic Program Special Interest Group is being organized under the auspices of the AIHA. One goal of this endeavor is to more specifically define the role of each type of program in the overall educational preparation of industrial hygienists, and hopefully to facilitate a climate of mutual support and coordination with less competition and duplication of efforts. Optimistically, a clearer academic route into the industrial hygiene profession is on the horizon.

2.3 The Industrial Hygiene Academic Curriculum

The structure and content of the formal academic curriculum is a particularly critical issue when discussing the most "appropriate" industrial hygiene educational background. Questions worthy of consideration in this regard include: "What courses should be mandatory for all students preparing for a career in industrial hygiene?," "What courses should be optional or elective?," and "What courses should be marginally recommended or perhaps altogether deleted?."

Answers to these important questions will obviously vary in accordance with the perspective of the person queried and the issues considered critical in the occupational environment at the time. An individual whose experience has consisted of years of plant-level practice in a heavily industrialized environment might take a different view of what constitutes the most effective educational background than one whose career has been spent in a hospital/clinical setting or one whose most recent experience has been more management related.

Similarly, to illustrate how an industrial hygiene curriculum might change with time, consider the 1971 ABIH survey in which the membership was asked to recommend both required and elective courses for an industrial hygiene curriculum (11). Interestingly, responses to this survey indicated that ergonomics, safety, epidemiology, and biological agents were all preferred as elective rather than required by a significant majority of respondents. In fact, several respondents recommended these courses as worthy of deletion from the curriculum altogether. Due to a variety of factors, these topics have increased in relevance in the last 25 years to the point that it is doubtful that many practitioners would recommend deleting these subjects from today's industrial hygiene educational program. Obviously, as new challenges develop, effective curricula remain flexible enough to incorporate the academic tools necessary to meet those challenges without compromising important existing material.

2.3.1 Science and Mathematics Curriculum Content

When evaluating a baccalaureate curriculum, a prospective student must consider the importance of a solid technical background. As stated previously, industrial hygiene is an applied discipline; that is, the basic concepts of the natural sciences and mathematics are applied to the solution industrial hygiene problems. As such, the industrial hygienist must be equipped with a substantial science/mathematics base that is reflected in the educational preparation. This includes courses in biology, chemistry, and physics, as well as those in basic mathematics. These courses must establish a technical foundation such that the breadth and depth of material necessary for industrial hygiene academic training may be mastered.

Regarding these subjects, Perkins makes the following points (1):

- *Biology*. Inasmuch as the Industrial Hygienist is ultimately concerned with the health of human beings, biology is critical to the practice of the profession. This includes a basic understanding of anatomy and physiology of the human body, molecular and cellular biology, genetics, embryology, biochemistry, microbiology (e.g, biological agents of disease), and ecology.
- *Chemistry*. Industrial Hygiene practice requires fundamental knowledge of both inorganic and organic chemistry including nomenclature, reaction rates, equilibrium, and the ideal gas law, as well as properties and concepts such as vapor pressure, boiling point, solubility, adsorption, absorption, catalytic reaction, etc.
- *Physics*. The study of physics is essential to the understanding of significant physical stressors in the occupational environment such as noise and radiation. Additionally, knowledge of fluid mechanics principles is also important in the study of ventilation as a control of airborne agents.
- *Mathematics*. The bulk of the calculations performed in industrial hygiene utilize mathematics up to the level of algebra and trigonometry. A working knowledge of logarithms is also required as is at least a basic understanding of the application of statistical principles to the design of sampling strategies and the interpretation of industrial hygiene data.

As mentioned previously, the development of problem definition and problem solving skills for any applied science is virtually impossible apart from a solid grounding in the underlying science. Thus, when examining an industrial hygiene curriculum, a prospective student should carefully consider its content in the courses described above.

It is noteworthy that, in addition to the necessity of a strong mathematics/science component in the educational foundation, a number of current practitioners are urging the inclusion of an increased number of engineering courses in the industrial hygiene academic program. Calling for a greater educational emphasis on control of occupational hazards and the prevention of hazardous agent emissions at the source, these individuals suggest that a larger component of engineering courses in the curriculum would well serve the future Industrial Hygienist. Looking forward to the practice of industrial hygiene in the year 2020, Sherwood states "Engineering should provide a broader base for students entering professional education in this field, who will be more concerned with prevention and engineering control of both occupational and environmental hazards, rather than with measurement and epidemiology of the biological and toxicologic sciences" (3).

2.3.2 Industrial Hygiene Curriculum Content

Moving beyond the science and mathematics foundation, the prospective student must also investigate a given curriculum in terms of its slate of courses more specific to professional industrial hygiene practice. Whether bachelors or masters, one should consider the courses in the program of study that apply the underlying science and math foundation to the solution of industrial hygiene problems? An industrial hygienist is widely recognized as somewhat of a "jack-of-all-trades" in terms of applied science; a distinction that implies at least a familiarity, if not a mastery, of a broad spectrum of professional knowledge. That being the case, what types of courses should one look for in an industrial hygiene curriculum? What topics are essential and to what extent should they be covered in a given curriculum? These are critical issues to the potential student.

To gain perspective on the breadth of material with which an Industrial Hygienist should be familiar, it may be useful to consider the topics from which the ABIH draws CIH examination questions. Appendix A, reprinted by permission of the ABIH, lists and amplifies the rubrics of IH knowledge that the Board considers relevant in testing the comprehensiveness of the prospective CIH's knowledge. Although lengthy, this table does give insight to the vast array of subject matter related to industrial hygiene practice.

Reviewing this appendix, one notes subjects traditionally associated with fundamental IH practice. Rubrics with headings such as exposure measurement, noise and vibration, ionizing/nonionizing radiation, biohazards, ergonomics, engineering/nonengineering controls, etc., are part and parcel of the definition of industrial hygiene; i.e., recognition, evaluation, and control of chemical, physical, biological, and ergonomic stressors in the occupational environment. One also finds a rubric on ethics and management covering subjects such as establishment of policy, planning and budgeting, communication, and ethics that relate to the management of occupational stressors in a manner that is consistent with sound ethical practice and current regulatory directives. Additionally, subjects such as community air pollution exposure, biostatistics and epidemiology, toxicology, etc., appear in this list. As Perkins notes, many industrial hygienists are generalists, practicing in the context of an occupational safety and health "team" that includes occupational physicians, occupational health nurses, safety professionals, ergonomists, occupational epidemiologists, occupational toxicologists, and others (1). The industrial hygienist often functions as an interface among these disciplines, and thus, should have at least some training in each of these fields to communicate effectively with, benefit from, and contribute to these professionals.

Extensive coverage of each element in this vast array would be virtually impossible in a single undergraduate or even graduate program. It would be rare to find any given curriculum that addresses each of these topics in great depth. Academic programs, particularly those at the graduate level tend to be focused in one or two specialty areas, depending on the faculty's areas of expertise and research interests. However, regardless of specific emphases, the basic elements related to the recognition, evaluation, and control of the fundamental occupational environmental stressors should be obvious in any industrial hygiene curriculum. Courses and laboratories in the industrial hygiene specialty area must emphasize the fundamental observational techniques and basic survey skills necessary for identifying those potential health hazards associated with specific occupational situations including the types of chemical/physical contaminants generated by these processes. In addition, significant portions of the curriculum should be spent in discussion and demonstration of the principles involved in the accurate measurement of these contaminants and in the proper assessment and interpretation of the results. Similarly, the elements of basic control strategies and their applications to specific situations must be discussed in the classroom along with opportunities for students to design control systems (e.g., local exhaust ventilation) for various unit processes. In this manner, an instructional program delivers the basic industrial hygiene principles of recognition, evaluation, and control. Thus, even though the dynamic nature and broad spectrum of the industrial hygiene profession renders any attempt to pin down the "perfect" curriculum very difficult, certain common denominators do exist in terms of fundamental program content. For example, consider Table 51.1 taken from Perkins, which illustrates a comprehensive master's level industrial hygiene curriculum outline (1).

2.3.3 RAC-ABET Accreditation

Another way to summarize the basic elements of an industrial hygiene program is by discussing the process through which an industrial hygiene curriculum becomes accredited by the Related Accreditation Commission of the Accreditation Board for Engineering and Technology (RAC-ABET) and briefly outlining the ABET curriculum criteria. ABET is a national association that organizes and carries out a comprehensive program of accreditation for those engineering, engineering technology, and engineering-related academic curricula that lead to degrees. A stated objective of the accreditation process is to "identify for prospective students, student counselors, parents, potential employers, public bodies, and officials, engineering-related programs which meet the minimum ABET criteria in engineering-related specialties" (13). Programs considered for accreditation under the RAC of ABET are engineering-related curricula for which a cognizant technical society has submitted program criteria to the ABET Board of Directors. Programs are expected to meet the minimum standards set forth in the general RAC-ABET criteria as well as the specific program criteria submitted by the cognizant technical society. Programs are evaluated on the basis of an extensive self-study questionnaire completed by the institution followed by an on-site visit by a team of RAC-ABET evaluators. Issues addressed include curriculum content, size and qualifications of faculty, adequacy of library holdings, laboratory and other physical facilities, finances, etc.

In the case of industrial hygiene curricula (considered by ABET to be an engineering-related discipline), the cognizant technical society is the American Academy of Industrial

Table 51.1. A Thorough Outline of Study for the Master's Degree in Industrial Hygiene

Fundamental of industrial hygiene
Health physics
Air sampling and analysis
Fundamentals of occupational safety/ergonomics
Biostatistics
Principles of epidemiological research
Noise effects and control
Temperature and pressure in occupational health
Essentials in toxicology
Occupational diseases
Control of occupational hazards
Industrial hygiene case studies
Biological agents
Toxic agents and environmental toxicology
Field interdisciplinary studies
Preceptorship in environmental health
Statistical and computer applications in environmental health
Air pollution
Research thesis
Occupational epidemiology
Hazardous waste management

Hygiene (AAIH). The AAIH developed and submitted accreditation criteria for masters industrial hygiene programs in the mid-1980s with the first round of accreditation taking place in 1987. More recently, the AAIH has developed undergraduate program accreditation guidelines. RAC-ABET began accrediting these programs with the 1997 accreditation cycle. Presently, 26 graduate programs and five undergraduate programs have received RAC-ABET accreditation in industrial hygiene. These programs are listed in Table 51.2.

Briefly, the RAC ABET Criteria for Baccalaureate Programs include (13):

- *Mathematics.* A minimum of 6 semester credit-hours including college algebra or courses more advanced, introductory level calculus through integrals, and statistics.
- *Basic Sciences.* A minimum of 12 semester credit-hours of chemistry courses with laboratories including organic chemistry, 6 semester credit hours of physics with laboratories, and 6 semester credit-hours of biology.
- *Communications, Humanities, and Social Sciences.* A minimum of 21 semester credit-hours in the areas of communications, humanities, and social sciences.
- *Industrial Hygiene Science.* An Industrial Hygiene Science course is one that is designed to expand and apply the basic sciences and mathematics toward professional practice including the solution of closed-form problems and quantitative expression. This segment of the curriculum provides the core Industrial Hygiene Program and

Table 51.2. RAC-ABET Accredited Industrial Hygiene Curricula (November 1998)

Masters Level Programs

University of Alabama at Birmingham
University of California at Berkeley
University of California at Los Angeles
Central Missouri State University
University of Cincinnati
Colorado State University
Harvard University
University of Illinois at Chicago
University of Iowa
Johns Hopkins University
University of Massachusetts at Lowell
Medical College of Ohio
University of Michigan
University of Minnesota
University of North Carolina at Chapel Hill
University of Oklahoma Health Sciences Center
Purdue University
San Diego State University
University of South Carolina
University of South Florida
University of Texas at Houston
Tulane University
University of Utah
University of Washington
Wayne State University
West Virginia University

Bachelors Level Programs

California State University, Northridge
Ohio University
Purdue University
Utah State University
Western Kentucky University

includes at least 15 semester credit-hours in the following: Fundamentals or Principles of Industrial Hygiene, Industrial Hygiene Measurements, including laboratory, Industrial Hygiene Controls, and Toxicology.

- *Industrial Hygiene Practice*. Industrial Hygiene Practice courses apply the Industrial Hygiene Sciences to solve needs of workers and identified clients. This may require written or computed solutions, cost and ethical considerations, or the application of independent judgment. A minimum of 15 semester credit-hours must be in Industrial

Hygiene Practice. The published criteria includes a noncomprehensive list of topics considered appropriate in Industrial Hygiene Practice courses. These topics include epidemiology, industrial ventilation, noise control, ergonomics/human factors, air quality/indoor air quality, environmental health, radiation measurement and control, health physics, laboratory safety, hazardous wastes management, protective equipment, and several others.

- *Additional Technical Subjects.* A minimum of 15 additional semester credit-hours must be taken in additional technical subjects. This may include additional Industrial Hygiene Science or Practice or additional mathematics or science beyond the basic requirement.

A minimum of 120 semester-hour credits are required for the total degree.

The criteria stipulates that candidates for RAC-ABET accredited master's programs must have a bachelor's degree that includes the following minimum course requirements (13):

- 120 semester-hour credits.
- 42 semester-hour credits of technological courses including engineering-related science and engineering-related specialties.
- 21 semester-hour credits in communications, humanities, and social sciences.
- The balance of the program should be designed and sequenced to achieve an integrated and well-rounded engineering-related program.

According to the criteria, master's programs require a minimum of 30 semester hours. These programs must demonstrate an intensive and comprehensive level of *interdisciplinary* instruction and should include special projects, research, and a thesis or internship (13). In terms of curriculum, the master's program must provide a minimum of 18 semester hours in some combination of industrial hygiene science and industrial hygiene practice courses which are defined in the criteria in a manner similar to that of the bachelors programs. Consistent with the interdisciplinary focus of masters degrees, the criteria requires that Industrial Hygiene Science coursework cover principles and practice of industrial hygiene, principles and practice of environmental sciences, and epidemiology and biostatistics. Typical Industrial Hygiene Practice course topics listed in the criteria include control of physical and chemical hazards, environmental health, and occupational safety. The additional 12 semester hours of the masters curriculum are considered to be "unspecified" to allow the institution the flexibility to specialize or focus on some specific aspect of the field. Examples listed include public health, environmental law, and management techniques.

Whatever else might be said about accreditation, it does at least provide some assurance to the perspective student that the institution has taken steps to align its industrial hygiene curriculum with certain fundamental elements deemed appropriate by those practicing in the profession. An industrial hygiene curriculum accredited by RAC-ABET enjoys the credibility associated with this distinction and makes significant advancement toward providing graduates with a recognized and marketable degree.

2.3.4 Coursework in the Allied Professions

A final issue regarding curriculum content concerns the inclusion of courses pertaining to those professions which, in practice, are closely allied with industrial hygiene. At the outset of this chapter it was noted that day-to-day industrial hygiene practice has, in a sense, outgrown traditional boundaries and now encompasses the areas of safety and environmental management as well. In the current climate of corporate downsizing taking place within many organizations, the day of the narrowly-focused specialist with expertise in only one of these technical areas is rapidly giving way to the era of the broad-based generalist whose duties encompass the entire gamut of environmental, health, and safety issues. Increasingly, the plant-level professional is required to possess a varied background and a broad technical knowledge base that includes a combination of each of these areas of professional practice.

This shift in the scope of industrial hygiene practice has been documented in several reliable surveys. For example, a recent survey of industrial hygienists showed that 33% of facilities surveyed operate consolidated Environmental, Health, and Safety departments at the highest level of the organization with only about 25% having separate safety, industrial hygiene, health and environmental departments (14). The 1997 AIHA Membership Survey Report stated that "On average, respondents spend 54% of their time in industrial hygiene, 20% in safety, 14% on other and 12% in environment (8). In 1996, the AIHA Redefinition Task Force, a group created specifically "to examine the changing role of industrial hygienists and to identify the scope and importance of the industrial hygienist's role as it relates to occupational health, safety, and the environment," documented the following among the AIHA constituency (15):

- Only 15% still spend 100% of their time in a traditional industrial hygiene role.
- 54% spend no more than 51% of their time in industrial hygiene.
- 67% spend 10–50% of their time in safety functions.
- 51% spend some time in environmental functions.
- 82% said that in 5 years the profession as a whole would be an integration of environment, safety, and occupational health.
- 62.4% anticipated that their jobs would combine safety, industrial hygiene, and environmental functions.

As of this writing, the AIHA is asking its membership to consider changing the name of the association to the Occupational Health, Safety, and Environment Association (OHSEA) to address the broadening scope of the industrial hygiene profession.

The effect that the expanding scope of industrial hygiene practice will or should have on an academic curriculum is unclear. On one hand, it would not seem prudent to neglect formal educational preparation in areas that are likely to comprise a significant proportion of one's professional practice. Unfortunately however, these additional courses often cannot be easily incorporated into an already burgeoning curriculum, particularly in undergraduate programs, without significantly diluting the industrial hygiene content. The question of adding courses in air quality, water quality, hazardous waste management, etc., to a program at the expense of industrial hygiene courses is one worthy of serious attention. As Sherwood states, "There must be real concern over the continuing rapid and ever-expanding body of knowledge considered necessary for professional practice. Careful balance between superficial scanning of a wide range of topics and in-depth study of critical subjects must be made to maintain intellectual stimulation without leaving blind spots in the field of knowledge" (3). In concluding his 1983 Cummings Memorial Lecture, First comments, "I can foresee no other solution to a mastery of the enormous and growing body of industrial hygiene information, and of the similar expansion of knowledge in the disciplines on which industrial hygiene depends heavily, than to reorient our educational and professional development programs in the direction of enhancing basic understanding of the underlying facts so that we can transform our profession from one of empiricism to one of science" (16).

2.3.5 *Extracurricular Educational Experiences*

In view of this issue, preparation of students for the ambitious roles described above requires creativity in the delivery of the instructional program. As noted previously, providing formal academic course work in each important area is clearly impossible in an undergraduate (or even graduate) program. Thus, many important and interesting areas may be mentioned only briefly or covered at some cursory level during the regular curricular offerings. A significant function of academic programs is to develop innovative ways to integrate this important material into the curriculum so that meaningful topics may be amplified and detailed coverage provided of information that may only be introduced at a superficial level by formal academic courses.

To fulfill this role, industrial hygiene programs can engage students in unique educational experiences outside the normal classroom setting. One method of affording students opportunities to expand their knowledge base in a broad spectrum of industrial hygiene and industrial environmental management subject areas is to sponsor and promote student participation in a variety of professional development and continuing education programs while still in school. These programs (e.g., seminars, short courses, etc.) can be presented by faculty, local professionals, visiting scholars, etc., and can be jointly sponsored by the local industrial community to offset costs. They can be presented during intense sessions over short time periods (ranging from a few hours to a few days) when students are not engaged in regular semester activities. These extracurricular educational activities cannot take the place of formal education, but can certainly supplement the traditional classroom academic experience. Before applying to a given institution, prospective students may want to carefully consider if and in what manner the academic program addresses this issue. More will be said in subsequent sections of this paper regarding professional development courses and continuing education.

2.4 Extending Classroom and Laboratory Experiences into the Real World

Prospective students must also realize that an industrial hygiene academic program is more than what takes place in the classroom. Each of these classroom/laboratory experiences and extracurricular activities described above is an integral part of a student's preparation for a career in industrial hygiene. The importance of these experiences should not be minimized. However, in view of the increasing level of expectations and broad range of capabilities required of entry level industrial hygienists, it is clear that a vital role of an industrial hygiene program is to provide its graduates with "real world" experiences that they can take with them to their initial employment. Today, many companies are hiring inexperienced people, saddling them with heavy responsibilities, and are expecting them to step in and perform immediately. Unfortunately, individuals who have had little or no real-world experience or opportunity to work under the supervision of competent professionals in the field are at an extreme disadvantage in this situation. In view of this trend, providing these opportunities as an integral part of the degree experience becomes even more critical. Adequate academic preparation of these individuals is no accident but is rather the result of a carefully planned curriculum supplemented by real-world activities and experiences.

Many industrial hygiene academic programs are actively seeking to fill this role and create even more unique and effective learning environments for their students. Faculty at these institutions are striving to extend educational opportunities for students beyond routine classroom/laboratory exercises and into problem solving experiences in "real-world" industrial settings. The practice of this philosophy allows students to interact with actual workers, supervisors, and management personnel in genuine industrial workplaces and gives them the excitement and satisfaction of applying problem solving skills in a setting in which their contributions will make a material difference. Combining a solid academic foundation with these real-world experiences is an effective method of producing technically competent and marketable practitioners in the field of industrial hygiene.

Many of the individuals with industrial hygiene responsibility, particularly in smaller industries, are pleased to have students work with them (under faculty supervision) to measure and document the exposure of their workers to a variety of occupational and environmental hazards. This provides a service to these companies that is very difficult for many of them to obtain otherwise. They are very interested at the prospect of having students assist them in carrying out the basic facets of occupational and environmental health for their respective companies. Working with these industries, students can help identify specific areas in which they may be of assistance. As students demonstrate their capabilities to local industrial hygienists, ties between the university and the industrial community are strengthened.

The depth and complexity of these experiences may be highly variable. The more straightforward projects may be accomplished within the context of a regular semester course. Other, more complex issues may require more intensive effort and thus may the subject of a directed research project for a student or a group of students receiving academic credit. An example of a suitable project might be a noise exposure assessment project that would generally proceed as follows: Using techniques learned in the classroom and laboratory, students would work with a company's industrial hygiene program to identify the potentially noise hazardous processes and exposed workers. The students would then assist in designing a survey to evaluate and quantify the noise exposure including a determination of whom to measure, how many measurements to obtain, and what equipment/methods to use in collecting the measurement (i.e. sound level meter, octave band analysis, personal dosimetry, etc.). The students complete the evaluation by actually using the equipment to obtain the specified samples. Working with the industrial hygiene faculty (to insure technical competence) and the company industrial hygiene representatives, the results can be interpreted with a report formulated based on a situation in which the students were personally involved. This process creates a totally new learning environment for students; a unique, project-oriented approach to education that provides students with unparalleled learning opportunities.

The benefits reaped by the students when given the opportunity to practice their industrial hygiene problem solving skills in actual industrial situations cannot be overemphasized. They encounter first hand the reality of interacting with both workers and management personnel in a "real" workplace. They also experience the reality of practicing their profession in an environment where their proposed solutions must not only be effective from a worker protection perspective but they must also be cost effective from a business point of view. Perhaps the most satisfying aspect of these projects is the fact that through conscientious data collection, thoughtful problem analysis, and workable solution development, these students actually make a substantial contribution to the occupational safety and health program of a particular company. Students are understandably excited to participate in studies in which their input is a meaningful part of the decision-making process in a real industrial situation.

Additionally, the industries may also benefit from the academic program's industrial hygiene laboratory monitoring and assessment equipment. Thus, not only do the students benefit from the opportunity to practice their skills in utilizing this equipment in actual industrial settings, but industry benefits from these projects by gaining access to an outstanding array of modern environmental stressor monitoring equipment.

In a similar vein, many academic programs have developed a viable cooperative education/internship program in which students go out into industries for extended periods. During these experiences, the students have an opportunity to spend several weeks working hand in hand with a company's Industrial Hygienist. During this time the student can participate in long-term projects of greater depth and complexity that may require significantly more effort and expertise to manage. The students benefit from the practical experience of the industrial hygiene professional while the company benefits from the technical abilities of the student. In addition to valuable experience in the fundamental practice of industrial hygiene, the student has opportunities to do such things as; practice verbal and written communication skills, sharpen and further develop computer skills, observe how major programs such as Hazard Communication, Hearing Conservation, etc., are implemented in practice, and conduct safety and health training sessions for workers. Also, since internship positions are competitive, students gain valuable experience in the process of preparing a résumé, searching for open positions, and planning for and participating in a job interview. As Sherwood states, "An intern component should also become a standard requirement of professional education, providing it is integrated into the whole educational process and does not lead to exclusion of any required fundamental knowledge" (3).

Due to the significant contribution of these extracurricular opportunities to a student's overall educational experience, perspective students would do well to consider what a given program offers and supports in terms of these activities.

2.5 The Role of Student Organizations

A prospective student may also wish to consider a given program's commitment to foster strong student organizations. Providing an environment that encourages students to interact with other students and professionals is an important function of an industrial hygiene academic program. The formation of social/professional associations is critical to the field of industrial hygiene. As mentioned previously, in many industrial facilities, the industrial hygienist comprises a department of one and may have few opportunities for professional interaction within his/her company. These individuals must depend upon and work through an external network of colleagues. Thus, it is critical that students planning to enter this profession be taught the many benefits associated with participation in professional societies and be made aware of the great networking opportunities afforded by such organizations. Students can be initiated to this practice through participation in an organized student group such as one sponsored by a cognizant national society or local section of a national organization.

For example, through its local sections, the AIHA supports an active outreach program to university-level student associations. This activity provides an excellent forum in which students have the opportunity to work together on various projects related to the field of industrial hygiene and to contribute to the profession and the community in a variety of ways. Student associations can be the focal point for any number of organized student activities and service projects within the program. It brings together students all levels and is an excellent vehicle for communication as well as an effective recruiting tool. Student interaction with the sponsoring AIHA local section lays the groundwork for their future professional organization activities at both the local section and national association levels.

2.6 Nontraditional Academic Programs

For a variety of reasons, many industrial hygiene practitioners are seeking opportunities to complete an advanced degree that would increase their professional knowledge in a way that is directly applicable to their current employment situation. In addition to the direct positive impact this activity would have on the workforce they serve, many industrial hygienists also recognize that a master's degree could significantly enhance their competitiveness for advancement within their company, improve their opportunity for future salary increases, and provide greater job security in their current position. Many others, without bachelor's degrees specifically in industrial hygiene, are drawn into the field with little or no formal education in this discipline. Some have degrees in an engineering-related field or a basic science such as chemistry or biology but have never formally studied the application of these sciences to the solution of industrial hygiene problems. Other industrial hygiene practitioners have educational backgrounds in human resources or other areas of personnel/material management. These individuals find themselves distinctly disadvantaged in terms of their foundation of technical knowledge; never having had an opportunity to study and understand the scientific principles underlying the practice of their profession. Many of these people are currently seeking advanced educational opportunities, such as a graduate degree, in the field of industrial hygiene.

Additionally, professionals in the fields of Industrial Environmental Management and Occupational Safety are also developing an interest in advanced academic preparation in industrial hygiene and are beginning to recognize the advantages of cross-training in industrial hygiene, a discipline in which they are often technically and academically weak. These practitioners are also realizing the importance of recognized credentials, such as advanced degrees from quality institutions, that serve to legitimately document their professional development efforts. The unfortunate reality for many of these individuals is that they simply cannot be absent from their present places of employment for the substantial period of time necessary to complete a graduate program in residence. Thus, they are forced to forego a formal academic experience and focus on short-term professional development activities that have no collectively defined endpoint.

2.6.1 Part-Time Academic Programs

Fortunately, several existing academic programs accept and encourage part-time students. The scheduling format of these programs varies. Some institutions offer courses at night or on weekends when most working professionals would have a better opportunity to attend. Other programs are designed such that students come to campus once or twice each year for intensive training during short sessions of one or two weeks. Opportunities to obtain a degree by working full-time and attending school part-time are beginning to proliferate and more and more reputable institutions are offering this format. Of course, the student still must be in geographical proximity to the program of interest. Also, when working full-time, a student must deal with conflicts that invariably arise between the job and the classroom. However, this prospect is viable and the prospective student may desire to check with the institution to assess these possibilities.

2.6.2 Distributed Learning Academic Programs

The need for scientifically based, technically sound educational opportunities for industrial hygienists and related professionals is encouraging the development of alternative, academically sound instructional programs that can be accessed by students who are not necessarily in close geographic proximity to the institution. These programs (including interactive television and satellite-based programs) aim to provide much-needed educational opportunities in an easily accessible, cost-effective format, to those industrial hygiene professionals desiring to develop additional expertise through formal training in the technical aspects of their career.

A promising avenue for those desiring to further their education is Internet-based academic programs that deliver courses in a format that allows students to complete the bulk of a degree without relocation or prolonged work absences. The Internet provides an ideal platform for the delivery of an instructional program. Advantages of Internet-based instruction have been well documented and include (17):

- The ability to reach students who cannot physically access the classroom.
- The opportunity to enlarge student populations without intensive capital investment in classrooms, buildings, and infrastructure.
- The interaction of students with other students as well as the instructor (e.g., one-to-one, group-to-group, and/or one-to-many interactions) and with the massive amount

of information resources available via the Internet from thousands of Web servers and digital libraries around the world.
- The accessibility of worldwide virtual access with large numbers of executive students and professionals having Internet access at their workplaces or in their homes via a commercial Internet provider.
- The low cost of Internet use with hundreds of Internet Service Providers in the U.S. offering 24-hour access for as little as $15 a month.
- The use of a common and easy user interfaces (e.g., web browsers such as Netscape and Microsoft Explorer) in a controlled and self-operated system.
- The application of full interactive multi-media support through its hyperlink and CD-ROM plug-in capability (text, images, pictures, 3D animations, narration, music, motion video, and virtual reality resources that can be played, replayed, fast forwarded, and rewound).
- The delivery of information in digitally encoded format which virtually eliminates the degradation of information even after many duplications and transmissions.
- The high reliability of Internet workstations.
- The provision of a fully automated gate-keeping authentication function for any information provided on the Internet, thus allowing course contents, tests, and quizzes to be limited only to those who are registered for the course.

By taking advantage of Internet offerings, those individuals discussed above, for whom advanced education has not heretofore been an option, will be able to capitalize on this opportunity.

Another significant advantage of an on-line program is the efficiency with which the most current information on important Industrial Hygiene topics can be made available to a practitioner who desires an updating of his/her knowledge in a particular subject area. Without question, the quantity of consequential information related to the field of Industrial Hygiene, as with any other technical field, is expanding rapidly. Indeed, the responsible practice of the very profession itself is in constant evolution as research efforts and "real-world" experiences continuously yield results and discoveries with important applications to professional practice. Although many excellent graduate programs exist across the country today that prepare individuals to enter the professional practice of industrial hygiene with the most current set of academic skills available, it is essential to recognize that education is a life-long process that does not cease when one leaves the formal academic setting. Unfortunately, upon assumption of a full-time position, the opportunity for an individual to return to a traditional instructional program to update their knowledge base with additional course work is frequently limited. Practitioners with outdated or incomplete knowledge in particular areas are inhibited in their ability to serve the working population they are striving to protect. Even though they may already possess a bachelor's or even a master's degree, these individuals could substantially benefit from an academic forum through which they could receive on-going instruction and professional development opportunities based on the most current information. Internet-based programs can be a significant resource for these individuals.

These Internet-based courses can be continuously revised and updated just as are courses in traditional graduate programs. The difference, of course, is that the information will be more accessible to a much broader segment of the profession. Academically qualified persons desiring to take a course that presents the most current information in a particular area or topic could do so through this program, regardless of where they are currently geographically located (assuming Internet access). The opportunity to participate in a course for academic credit in a graduate program represents an important motivational factor, regardless of whether or not the individual chooses to pursue the entire degree. Taking a rigorous course in an academic context can possibly be more challenging and appealing to many individuals than other forms of self study. (Please note that these courses may also be available as "not-for-credit" offerings as well and thus function in a manner similar to an advanced continuing education course. Indeed, this network will provide a natural outlet for a variety of continuing education/professional development opportunities as will be discussed in a subsequent section.)

Internet-based courses and programs are now being offered by several institutions and are certainly worthy of consideration by those prospective students who may have limited options otherwise. However, students should be aware that for all of their advantages, many issues relating to these programs remain unresolved, at least as of this writing. A chief concern is the regional academic accreditation of Internet-based programs. The ultimate credibility of any academic program is traced to its accreditation by a regional academic accrediting association (e.g., The Southern Association of Colleges and Schools). Although these associations have accredited Internet-based programs in certain disciplines, one must be very careful to check the accreditation status of the specific program of interest. It should also be noted that, as of this writing, no program that is offered principally via the Internet has been accredited by RAC-ABET at either the master's or bachelor's level.

The difficulty of involving students in "hands-on" experiences is another important issue that faces virtually all distance learning programs. Being the applied discipline that it is, it is difficult to conceive how an individual could complete a credible degree in industrial hygiene, at either the master's or bachelor's level, without the benefit of various laboratory and field exercises. The techniques of instrument calibration, equipment utilization, data acquisition, etc., all necessitate an extension beyond the "virtual" environment into at least the laboratory if not into real world situations. This can be an important weakness in Internet-based as well as other distributed learning curricula. This issue may be overcome if some type of periodic laboratory sessions are provided. Prospective students interested in Internet-based or other distance learning programs should consider the types of "hands on" opportunities furnished by the program of interest to ensure that a well-rounded degree is delivered.

3 CONTINUING EDUCATION

As defined early in this chapter, continuing education or professional development is the process of pursing information or knowledge outside of an academic, "degree-seeking" program of study. Continuing education in industrial hygiene can be profitably pursued by those desiring an introduction to the field, by individuals in safety or environmental management desiring cross-training, or by industrial hygiene practitioners who desire the most current information on specific or advanced topics. As Sherwood states "For those who will go on to shape the future profession there is need to strengthen continuing education programs, which must introduce new aspects of technology to meet the continually changing work environment and ensure current awareness of the entire field. Only thus will the profession be able to upgrade the art and science of designing and maintaining safe workplaces for all" (3).

The pursuit of continuing education and professional development offers several advantages. Obviously, the knowledge gained from exposure to the information itself allows the professional to remain fresh and current and to be introduced to new and emerging topics. However, another important advantage afforded course participants is the opportunity to meet and discuss pertinent issues with other professionals in industrial hygiene and related fields. The benefits reaped from these interactions can sometimes even exceed that of the information presented. Also, it is rare for an individual to attend a professional development course without making at least one valuable contact for future networking opportunities.

Continuing education also plays a vital role in the maintenance of the CIH designation. To maintain certification, the CIH must accrue 40 certification maintenance (CM) points per each five years. A portion of these points may be achieved through participation in ABIH approved professional development courses and educational opportunities. In fact the ABIH requires that a minimum of ten CM points per five-year cycle be earned from courses or meetings related to industrial hygiene rubric areas (see Table 51.1). According to the ABIH, "Courses and educational programs must be approved by ABIH before points will be granted (requires submission of sponsor, agenda, dates, and contact ours)" (10).

3.2 Sources of Continuing Education Opportunities

Opportunities to engage in continuing education/professional development courses are provided by various governmental agencies, professional organizations, and colleges and universities. Although the following discussion is by no means exhaustive, it does give the reader some idea of the various outlets for these courses and will hopefully serve as a starting point for further investigation. Please note that the most effective method of obtaining current information regarding short course offerings is to visit the Internet site for the sponsoring organization. Although the web address (i.e., Universal Resource Locator or URL) for a given entity is subject to change, the home page address current as of this writing is given for each source described. The reader is encouraged to visit these sites to obtain the latest information on course offerings, dates, and locations.

3.2.1 Governmental Agencies

Many governmental agencies sponsor continuing education courses either by directly presenting the courses themselves or by funding course development/presentation by other organizations such as colleges or universities. For example, the National Institute for Occupational Safety and Health (NIOSH) supports short-term continuing education courses for occupational safety and health professionals, and others with worker safety and health responsibilities through its 15 university-based Education and Research Centers (ERCs). The scope of these ERC course offerings is truly extensive. A course covering virtually any aspect of the field of industrial hygiene, from introductory to advanced, can likely be found somewhere within the regional ERC network. These courses usually last from one to five days and are offered at a number of different locations throughout the country. A list of ERCs, including contact information, can be found at http://www.niosherc.org. This useful site also provides course listings by center, course listings by topic, and course searching capabilities.

The Occupational Safety and Health Administration (OSHA) Office of Education and Training also offers a number of short-term basic and advanced occupational safety and health training programs to Federal and State OSHA personnel, State consultants, other agency personnel, and private sector employers and employees. These courses cover such areas as OSHA inspection policies and procedures, biological hazards, ergonomics, petrochemical hazards, fire protection, and fall protection. Courses may be offered through the main OSHA Training Institute, located in Des Plaines, Illinois, or through the 18 OSHA Training Institute Education Centers regionally distributed throughout the United States. These Education Centers are located within universities, colleges, labor unions, or other nonprofit organizations. A complete list of courses, dates, locations, registration information, cost, etc. can be found at http://www/osha.slc.gov/Training.

3.2.2 Professional Organizations

Many professional organization also develop and sponsor continuing education courses. The AIHA maintains a standing Continuing Education Committee that lists as its goal to "Provide the membership with a planned system of high quality continuing education instruction, designed and delivered to enhance attainment and retention of technical competence by occupational and environmental health professions" (18). AIHA Professional Development Programs include three to five day courses in "Fundamentals in Industrial Hygiene", "HVAC and Indoor Air Quality", "Comprehensive Industrial Hygiene Review", and "Quantitative Industrial Hygiene", as well as two annual symposia (Toxicology and Risk Assessment) and a Management Institute. These courses are presented in a number of large cities around the United States. The AIHA also sponsors self-study programs in "Classic Safety", "Elemental Industrial Hygiene", and "Effective Risk Communication". In addition to its other duties, the AIHA Continuing Education Committee coordinates the development and presentation of a full slate of professional development courses (half-day to two days in length) at the annual American Industrial Hygiene Conference and Exhibition (AIHCE). Detailed information on all of these activities can be found on the AIHA home page at http://www.aiha.org.

The American Conference of Governmental Industrial Hygienists (ACGIH) also offers or endorses several professional development opportunities for the industrial hygiene community. Courses such as "Bioaerosols: Assessment and Control," "Applied '99: Protecting Today's Worker's—TLVs in Practice," "Special Synopsis in Ventilation," "HVAV/IAQ," "Fundamentals of Industrial Ventilation," and "Occupational Safety and Health Computations" are offered periodically in a number of large cities. Information regarding these opportunities is available on the ACGIH home page at http://www.acgih.org.

In addition to industrial hygiene organizations, other professional groups in areas of related practice offer a number of courses that may be of interest to practicing industrial hygienists. For example, the American Society of Safety Engineers (ASSE, http://www.asse.org) offers an extremely broad spectrum of courses and certifications in many categories such as Safety and Health Management, The Environment, Construction and Facility Management, and Professional Skills Training. These short courses are offered periodically at ASSE Headquarters in Des Plaines, Illinois, as well as several other cities. Detailed course information including instructors, goals, objectives, etc., can be found at http://www.asse.org. The National Safety Council (NSC, http://www.nsc.org) also offers safety and industrial hygiene-related education and training through its Occupational Safety and Health Services.

3.2.3 Other Short-Course Providers

In addition to the programs mentioned above, continuing education courses and seminars may be available from other sources. For example, this training is offered through many colleges or universities other than those associated with the NIOSH ERCs or the OSHA TIs. For example, many of the industrial hygiene academic programs listed in Appendix A offer periodic seminars, workshops, and short courses. Interested readers should contact these programs to determine which courses may be of interest to them. Additionally, many local sections of national organizations sponsor professional development activities for their membership and other interested individuals. National organization Internet home pages can be consulted for local section information.

Finally, the reader is also encouraged to consider Internet-based professional development opportunities. Many of the advantages listed previously as applying to Internet-based graduate programs also apply to professional development and continuing education courses. Although the availability and variety of these offerings is limited at present, it is likely that these opportunities will begin to proliferate in the near future.

4 INFORMATION EXCHANGE

The last section of this chapter deals with the various outlets through which practitioners in industrial hygiene and related disciplines exchange information and ideas. The opportunity to share experiences, insights, opinions, and research results is truly the foundation upon which significant progress in any field of endeavor is built. Important advancements are achieved through communication, collaboration and the collective trial and error process of the entire industrial hygiene profession. In a sense, very few problems are truly unique. Virtually any challenge encountered by an industrial hygienist in daily practice has already been faced and overcome by others in some form or fashion. Access to this information is becoming increasingly important to industrial hygiene practice.

4.1 Professional Journals

Technical journals that focus on the industrial hygiene profession and related fields of practice are rich sources of information and provide an excellent forum through which professionals can share important new discoveries. Journals contain a variety of technical papers, case studies, theoretical/quantitative treatises, and documentation of both basic and applied research results. Many journals also carry reviews of new products, literature, and software and well as classified ads, editorials, and dates/locations of courses or other upcoming events. Although journals are traditionally circulated in printed, hard-copy form, many are now available on-line from the publishers web site. Table 51.3 provides a list of journals that may be of interest to industrial hygienists.

Given the wealth of information recorded over the years, finding data and articles of interest can be quite a challenge. Fortunately, many abstracting services are available to assist in locating the pertinent information. These services are essentially large databases of journal article abstracts marked by keywords. These databases are equipped with search engines that allow the user to identify the abstracts of articles containing the desired information. Once located, the abstracts can be read to determine whether or not the full text of the article is desired. Although several abstracting services are available that may be useful, perhaps the most extensive for occupational health and safety is NIOSHTIC. This resource, developed by NIOSH, is a bibliographic database with comprehensive international coverage of occupational health and safety issues. It contains detailed summaries of over 185,000 significant articles, reports, and publications, spanning over 100 years. The NIOSHTIC data base is available on CD-ROM or by on-line subscription from the Canadian Centre for Occupational Safety and Health (http://www.ccohs.ca).

In addition to the more technical journals, other "trade magazines" exist that also deliver much useful information. *The Synergist*, published by the AIHA, and *ACGIH Today*!, published by the ACGIH, are examples of this type of publication. These periodicals typically contain shorter, less technical articles, as well as current news items, regulatory updates, and information on professional development opportunities such as meetings, conferences, symposia, courses, etc.

4.2 Professional Conferences and Meetings

Professional conferences, symposia, and similar meetings also provide a fruitful environment for meaningful information exchange among colleagues. Perhaps best known of these events, at least with respect to the industrial hygiene profession, is the American Industrial Hygiene Conference and Exhibition (AICHE), held annually in some large city in the United States or Canada. According to the AIHA, "The largest occupational and environmental health meeting in the world, AICHE draws more than 10,000 occupational health and safety experts from all over the globe to discuss and debate the latest issues in technical sessions, roundtables, and forums. The conference, which includes a world-class exhibition, also enables industrial hygienists to study and advance their careers by offering an array of professional development courses" (19). Technical sessions, posters sessions, and case study discussions are included among the many offerings of this conference. Additionally, the personal contacts and networking opportunities are also extensive. Plus, the opportunity to "talk shop" with old friends in the profession can often be an incredibly powerful way to stimulate new ideas and solutions to a variety of problems. AICHE registration information can be found on the AIHA web site.

Table 51.3. Journals of Potential Interest to Industrial Hygienists

American Industrial Hygiene Association Journal
  (American Industrial Hygiene Association)
American Journal of Epidemiology
  (John Hopkins University School of Hygiene and Public Health)
American Journal of Industrial Medicine
  (John Wiley & Sons, Inc., Journals)
American Journal of Public Health
  (American Public Health Association)
Annals of Occupational Hygiene
  (Pergamon Press Journals Division)
Applied Occupational and Environmental Hygiene
  (American Conference of Governmental Industrial Hygienists)
Archives of Environmental Health
  (Heldref Publications)
ASHRAE Journal
  (American Society of Heating, Refrigerating, and Air Conditioning Engineers)
Health Physics
  (Williams & Wilkins)
Job Safety and health
  (The Bureau of National Affairs)
Journal of the American Medical Association
  (American Medical Association)
Journal of Occupational Medicine
  (American Occupational Medicine Association)
Journal of Toxicology, Clinical Toxicology
  (Marcel Dekker)
Journal of Tixocology & Environmental Health
  (Hemisphere)
Noise Control Engineering Journal
  (Institute of Noise Control Engineering)
Occupational and Environmental Medicine
  (British Medical Association)
Occupational Hazards
  (Tenton)
Occupational Health and Safety
  (Occupational Health and Safety)
Occupational Safety & Health Reporter
  (Bureau of National Affairs)
Professional Safety
  (American Society of Safety Engineers)
Safety & Health
  (National Safety Council)
Scandinavian Journal of Work, Environmental & Health
  (Finnish Institute of Occupational Health)

Involvement in local sections of national organizations can also be an effective means of information exchange. Attendance at local section meetings and other sponsored events can offer many of the same benefits of larger, national meetings in terms of networking and exchanging ideas and perspectives with colleagues. The AIHA has an extensive network of local sections distributed regionally throughout the United States. Interested readers should check the AIHA web site to determine the section closest to their location.

4.3 On-line Information Exchange Resources

As a conclusion to this short section on information exchange, the reader is encouraged to explore the world wide web for the vast amount of material relating the industrial hygiene, safety, and environmental issues. Home pages from professional organizations (e.g., AIHA, ACGIH), governmental agencies (e.g., NIOSH, OSHA), consensus standards groups [e.g., American National Standards Institute (ANSI), American Society of Heating, Refrigerating, and Air Conditioning Engineers (ASHRAE)], product manufactures and vendors, universities, labor unions, and even individuals can provide a wealth of practical information for the industrial hygienist. Also, listserves, chat lines, mail rooms, etc., can also be an effective means to communicate with peers. These services essentially provide a platform for running dialogue among all subscribers. An individuals can post a question, concern, or comment and receive replies from all interested participants. This mechanism often opens the door to some very interesting perspectives and can frequently provide some key piece of otherwise elusive information.

APPENDIX A   RUBRICS OF INDUSTRIAL HYGIENE KNOWLEDGE

General Science. General scientific concepts, chemistry, biochemistry, biology, anatomy and physiology, general physics and mathematics through basic differential and integral calculus. Properties of flammable, combustible and reactive materials (compatability) are included. Included are calculations such as those relative to gas laws and unit-of-measure conversions.

Biohazards. Principles of sanitation, personal hygiene, the recognition, evaluation and control of biological agents or materials having the capacity to produce deleterious effects upon other biological organisms, particularly humans (virus, bacteria, fungi, molds, allergens, toxins, recombinant products, bloodborne pathogens, etc.) and infectious diseases that appear in workplaces including industry, agriculture, offices and health care facilities.

Biostatistics and Epidemiology. Principles of epidemiology, techniques used to study the distribution of occupationally induced diseases and physiological conditions in workplaces and factors that influence their frequency. It includes concepts of prospective and retrospective studies, morbidity and mortality and animal experimental studies, data and distribution of data. Also included are basic biostatistics and statistical and nonstatistical interpretation of data in the evaluation of hazards.

Controls/Engineering Controls. Control of chemical and physical exposures through engineering measures. Included are local exhaust ventilation, dilution ventilation, isolation, containment and process change. Also included are mechanics of airflow, ventilation measurements, design principles and related calculations. This rubric also covers inplant recirculation air cleaning technology. Engineering control of ionizing and nonionizing radiation, heat stress, abnormal pressure conditions and noise and vibration sources including principles of isolation, enclosure, absorption and damping are included.

Controls/Nonengineering Controls. Personal protective equipment, including the principles governing selection, use and limitation of respirators, protective clothing and hearing protective devices. Included are respirator fit testing, breathing air specifications, glove permeability, noise reduction characteristics of hearing protective devices, the use of hearing protective devices, eye protection and the use of substitution, source reduction, and administrative controls. Hazard communication and training of employees are also included.

Ergonomics. Application of principles from anthropometry, human factors engineering, biomechanics, work physiology, human anatomy, occupational medicine and facilities engineering to the design and organization of the workplace for the purpose of preventing injuries and illnesses.

Ethics and Management. Acquisition, allocation and control of resources to accomplish industrial hygiene recognition, evaluation and control objectives in an effective and timely manner. Included are such topics as establishment of policy, planning and budgeting, delegation of authority, productivity, accountability, communication, staff versus line authority, organizational structure, performance evaluation and decision making. This rubric also includes ethics.

Exposure Measurement/Analytical Chemistry. Laboratory analytical procedures for workplace environmental samples and related calculations. Included are gas chromatography, infrared, visible and ultraviolet spectrophotometry, high performance liquid chromatography, mass spectroscopy, atomic absorption spectrophotometry, high performance liquid chromatography, mass spectroscopy, atomic absorption spectrophotometry, high performance liquid chromatography, mass spectroscopy, atomic absorption spectrophotometry, wet chemical methods, and microscopy and laboratory quality assurance and chain of custody.

Exposure Measurement/Sampling, Monitoring and Instrumentation. Selection, use, and limitations of field-sampling instruments, full shift and grab samples, including direct reading instruments. Included are the set-up, calibration, and use (including quality assurance practices) of sampling apparatus and direct-reading instruments. Sampling strategy considerations are included. Calculations related to sampling and calibration are included. Included are instruments used to measure exposure to air, noise, ionizing radiation, nonionizing radiation, pressure, and heat.

General Industrial Hygiene Topics.

Community Exposure. Air pollution, air cleaning technology, ambient air quality considerations, emission source sampling, atmospheric dispersion of pollutants, ambient air monitoring, health and environmental effects of air pollutants and related calculations. This also includes other environmental subjects such as chemical emergency planning and response.

Hazardous Wastes/Site Remediation.

Health Risk Analysis, Communication.

Unit Operations/Process Safety/Confined Space. Included are the hazards associated with specific industries and specific industrial or manufacturing processes. Topics include, but are not limited to spray-painting, welding, plating, abrasive-blasting, vapor-degreasing, foundry operations, etc.

Noise and Vibration. Health effects resulting from exposure to noise and vibration. Computations related to combining noise sources and octave band measurements are included as are audiometric testing programs.

Radiation/Ionizing. Physical characteristics and health and biological effects associated with alpha, beta, gamma, neutron, and x-radiation, including source characteristics.

Radiation/Nonionizing. Physical characteristics and health effects associated with electromagnetic fields, static electric and magnetic fields, lasers, radio frequency, microwaves, ultraviolet, visible, near-infrared radiation and illumination.

Regulations, Standards, and Guidelines. Understanding of regulatory principles and requirements including those of the Occupational Safety and Health Administration (OSHA), the Mine Safety and Health Administration (MSHA), the Environmental Protection Agency (EPA), and the Department of Transportation (DOT). Also included are the interpretation and use of American Conference of Governmental Industrial Hygienists (ACGIH) Threshold Limit Values (TLVs), Biological Exposure Indices (BEIs), and industrial ventilation guidelines, American National Standards Institute (ANSI) standards, American Society for Heating, Refrigeration, and Air Conditioning Engineers (ASHRAE) guidelines, American Society for Testing and Materials (ASTM) standards and National Institute for Occupational Safety and Health (NIOSH) Criteria Documents and recommendations.

Thermal and Pressure Stressors. Adverse health effects associated with heat, cold, and nonstandard pressure conditions.

Toxicology. Health effects resulting from exposure to chemical substances. Included are symptomatology, pharmacokinetics, mode of action, additive, synergistic and antagonistic effects, routes of entry, absorption, metabolism, excretion, target organs, toxicity testing protocols and aerosol deposition and clearance in the respiratory tract. Also included are carcinogenic, mutagenic, teratogenic, and reproductive hazards.

ACKNOWLEDGMENTS

The authors wish to acknowledge the contribution of William J. Poppendorf, Ph.D., CIH, who provided a count of the academic institutions offering a degree in Industrial Hygiene or a related field.

BIBLIOGRAPHY

1. J. L. Perkins, *Modern Industrial Hygiene*, Vol. 1, *Recognition and Evaluation of Chemical Agents*, Van Nostrand Reinhold, New York, 1997, pp. 11–45.

2. A. R. Hale, M. Piney, and R. J. Alesbury, "The Development of Occupational Hygiene and the Training of Health and Safety Professionals," *Ann. Occup. Hyg.* **30**, (1) 1–18 (1986).
3. R. J. Sherwood, "Cause and Control: Education and Training of Professional Industrial Hygienists for 2020," *Am. Ind. Hyg. Assoc. J.* **53**(6), 398–403 (1992).
4. "Training and Education in Occupational Hygiene, An International Perspective," *ACGIH Annals* **15**, 3–202 (1988).
5. M. Corn, "Professions, Professionals, and Professionalism," *Am. Ind. Hyg. Assoc. J.* **55**(7), 590–596 (1994).
6. G. X. Kortsha, "Industrial Hygiene at the Crossroads," *Am. Ind. Hyg. Assoc. J.* **51**(9), 453–458 (1990).
7. D. K. George, J. P. Russell, and R. G. Handy, "The Role of Undergraduate Programs in Industrial Hygiene," *Proceedings of the American Society for Engineering Education*, Southeast Section Annual Conference, Clemson, SC 1999.
8. 1997 AIHA Membership Report, *The Synergist*, American Industrial Hygiene Association, Nov. 15, 1997.
9. "Definitions and Function of Industrial Hygiene," *Bulletin of the American Board of Industrial Hygiene*, April 13, 1997, http://www.abih.org/bulletin.htm.
10. "Information about Certification for the Practice of Industrial Hygiene," American Board of Industrial Hygiene, http://www.abih.org/homepage.htm.
11. "Report on Training Requirements—AAIH Committee on Recruitment and Training," *Am. Ind. Hyg. Assoc. J.* 271–273 (April, 1971).
12. Strategic Plan 1998, American Board of Industrial Hygiene, http://www.dcn.davis.ca.us/go/abih/Docs/strategic-plan.htm.
13. *Criteria for Accrediting Engineering-Related Programs, 1998–99 Accreditation Cycle*, Related Education Commission Accreditation Board for Engineering and Technology, Inc., Baltimore, MD, 1998.
14. 13th Annual Environmental Health and Safety White Paper, *Industrial Hygiene and Safety News*, Radnor, PA, 1997.
15. AIHA Redefinition Task Force Report, American Industrial Hygiene Association, *http://www.AIHA.org/name.html#article*.
16. M. W. First, "Engineering Control of Occupational Health Hazards," *Am. Ind. Hyg. Assoc. J.*, **44**(9), 621–626 (1983).
17. A. Jafari, "Issues in Distance Education," *Technical Horizons in Education Online Journal*, http://www.thejournal.com/PAST/OCT/1097exclu3.tml, Oct. 1997.
18. *Who's Who in Industrial Hygiene*, 1997–98 Membership Directory of the American Industrial Hygiene Association.
19. American Industrial Hygiene Conference and Exhibition, American Industrial Hygiene Association, http://www.aiha.org/conf.html.

Index

ABCM (Activity-Based Cost Management), 2311–2314
Access databases, 2127
Accident prevention, 2069–2071
Accommodation, 1928
2-Acetylaminofluorene
 OSHA standards, 2206
Acids
 occupational exposure limits for organic, 1923–1924
 OSHA standards for inorganic, 2206
Acrylonitrile
 lifetime cancer risk for persons exposed, 1936
 OSHA standards, 1622, 2206
Activity-Based Cost Management (ABCM), 2311–2314
Administrative Procedures Act of 1946, 1596
Aerosols
 unusual work schedule exposure, 1865
Aesthetically displeasing agents
 occupational exposure limits, 1927–1928
Agricultural pesticides
 worker protection standard, 1771
Air contaminants
 unusual work schedule exposure limits, 1810–1811
Air monitoring
 biological monitoring compared, 2048–2050
Air pollution regulations
 odors, 1730–1731
Air quality
 proposed standards for indoor, 1628–1629
Alcohol consumption
 effect on biological monitoring, 2046–2047
American Industrial Hygiene Conference and Exhibition, 2369
Americans with Disabilities Act of 1990 (ADA), 2212
4-Aminodiphenyl
 OSHA standards, 2206
Application programs, 2118
Argentina
 occupational exposure limits, 1959
Arsenic, inorganic
 lifetime cancer risk for persons exposed, 1936
 OSHA standards, 1627
Asbestos
 lifetime cancer risk for persons exposed, 1936
 OSHA standards, 1617–1619, 2206
Atomic Energy Act of 1954, 1604–1605
Attorney–client privilege, 1713–1714
Audits
 hazard communication program, 1776–1785
 industrial hygiene data, 2106–2109
 odors, 1731–1733
Australia
 occupational exposure limits, 1959
 occupational health and safety management systems standards, 2261–2263
Austria
 occupational exposure limits, 1959–1960
Average daily doses, 2168–2169

2375

Bases
  occupational exposure limits, 1923–1924
BATs (Biologische Arbeitsstoff-Toleranz-Werte), 2217–2222
BEIs (Biological Exposure Indices), 1919, 2002
Belgium
  occupational exposure limits, 1960
Benzene
  hazard communication requirements, 1770
  lifetime cancer risk for persons exposed, 1936
  OSHA standards, 1622–1623, 2206
  reproductive toxicity, 2182–2186
  risk analysis, 2173–2181
Benzidine
  OSHA standards, 2206
Biological Exposure Indices (BEIs), 1919, 2002
Biological monitoring, 2001–2003, 2199
  air monitoring compared, 2048–2050
  analytical method availability, 2015–2016
  biomarker selection, 2014–2016
  biomarker stability, 2016
  biomarker uptake, distribution, and elimination, 2003–2008
  blood, 2012–2013
  ethical issues, 2028–2029
  exhaled air, 2008–2010
  exposure monitoring, 2008–2014
  hair exposure to heavy metals, 2014
  of inorganic gases, 2042
  of inorganic pollutants, 2042
  of metals, 2039–2042
  mixture exposure, 2044–2048
  organic compounds, 2029–2039
  purposes, 2015
  recordkeeping requirements, 2096, 2101
  reference values, 2017–2029
  sensitivity, 2015
  urine, 2010–2012
Biological threshold, 2161
Biologic half-life, 1791–1793
  determining for chemicals, 1848–1849
  estimated for various chemicals, 1849
Biologische Arbeitsstoff-Toleranz-Werte (BATs), 2017–2022
Biomarkers, 2002–2008
  biochemical changes as, 2043–2044
  DNA adducts as, 2034–2035
  glutathione conjugates as, 2032–2034
  half-life of, 2004–2006
  protein adducts as, 2035–2039
  reference values, 2017–2029

  selection of, 2014–2016
  specificity, 2014–2015
  stability, 2016
Blood
  biological monitoring, 2012–2013
Bloodborne pathogens
  OSHA standards, 1624, 2206
Body burden, 2001
Brazil
  occupational exposure limits, 1960
Break-even analysis, 2314–2316
Brief and Scala model, for adjusting OELs for unusual work schedules, 1812–1815, 1949–1952
Business analysis, 2303–2308
  cost recognition, analysis, and accounting, 2321–2330
  financial methods, 2308–2320
  health and safety performance metrics, 2333–2335
  hidden impacts of health and safety investments, 2330–2332
  systems thinking, 2236–2237
Butadiene
  OSHA standards, 1627, 2206

Cadmium
  OSHA standards, 1626, 2206
Canada
  occupational exposure limits, 1960–1961
Cancer
  benzene risk analysis case study, 2173–2181
  biochemical markers of, 2211–2212
  dose-response curves, 2161–2164
  early detection of, 2210–2211
  hazard identification, 2155–2156
  risk characterization, 2170–2171
  WellWorks program, 2234
Capital budgeting, 2308–2311
Carbon monoxide
  occupational exposure limits, 1904
Carcinogens
  adjusting limits for unusual work schedules, 1872–1877
  German biomarker reference values, 2023–2024
  occupational exposure limits, 1931–1937
  OSHA health standards, 1619–1620
CD-ROMs
  commercially available products, 2145–2147
  delivery systems, 2138–2139

INDEX   2377

hardware and software requirements, 2138
suitability of information for, 2139–2140
Ceiling values, 1919
Central processing unit (CPU), 2116
Certified Industrial Hygienist, 2348–2349
Chain of custody, of evidence, 1720–1721
Chemical carcinogens
  occupational exposure limits, 1931–1937
Chile
  occupational exposure limits, 1961–1962
bis-Chloromethyl ether
  OSHA standards, 2206
p-Chlorotoluene (PCT)
  reproductive toxicity risk assessment, 2188–2192
Citations, under OSHA law, 1657
  contesting, 1662–1664
Civil lawsuits, 1699–1703
Clean Air Act, 1603
Clean Air Act Amendments
  and hazard communication, 1771–1772
Coke oven emissions
  OSHA standards, 1620, 2206
Communication problems
  between the legal community and industrial hygienists, 1697–1699
Compliance and projection
  challenges to OSHA enforcement scheme, 1664–1666
  contesting citations and penalties, 1662–1664
  criminal sanctions, 1661
  imminent danger situations, 1664
  investigations and inspections, 1651–1655
  recordkeeping and reporting, 1655–1657
  sanctions for violations, 1657–1666
  warrants, 1652–1654
Compressed workweeks, 1789
Computer programs, 2117–2119
Computers, 2114, 2116
Confidentiality, of industrial hygiene data, 2101–2102
Confined spaces
  OSHA health standards, 1627
Contingent valuation, 2326–2327
Continuing education, industrial hygiene, 2365–2368
Contracts, 1692
Cooperative Compliance Program, 1665–1666
Corporate occupational exposure limits, 1946–1949
Cost analysis

  of injury and illness prevention, 2321
Cost benefit analysis, 2327–2330
Cotton dust
  OSHA standards, 1621–1622, 2206
Criminal lawsuits, 1703
Criminal sanctions
  under environmental laws, 1693–1695
  under OSHA law, 1661
Cumulative trauma disorder, 2204–2205

Data, metadata, 2119–2122
Data, private, 2123–2125
Data, public domain, 2122–2123, 2125
Data automation, 2113–2115
  business considerations, 2127–2138
  CD-ROMs, 2138–2140, 2145–2147
  conversion problems, 2127
  costs, 2127–2128
  legal requirements, 2115–2116
  long term data storage, 2125–2126
  metadata, 2119–2222
  private data, 2123–2125
  public domain data, 2122–2123, 2125
  related technologies, 2138–2140
Data storage, 2118–2119
  long term, 2125–2126
Daubert decision
  effect of expert witness testimony, 1705–1710
DBCP. *See* 1,2-Dibromo-3-chloropropane.
DBIII databases, 2127
DCF. *See* Discounted cash flow.
De minimus violations, under OSHA law, 1658–1659
Denmark
  occupational exposure limits, 1962
Depositions, 1715–1716
Dermal occupational exposure limits, 1938–1939
Developmental toxicants
  occupational exposure limits, 1924–1926
1,2-Dibromo-3-chloropropane (DBCP)
  lifetime cancer risk for persons exposed, 1936
  OSHA standards, 1622, 2206
3,3'-Dichlorobenzidine
  OSHA standards, 2206
4-Dimethylamino azobenzene
  OSHA standards, 2206
Dioxins
  occupational exposure limits, 1929–1930
Disability management, 2074–2075
Discounted cash flow (DCF), 2316–2317

Discovery, 1714–1715
Distributed learning programs, 2363–2365
DNA adducts
  as biomarkers, 2034–2035
DNA repair products
  as biomarkers, 2035
Dose-response assessment, 2159–2166
  benzene, 2178–2179, 2184
  p-chlorotoluene, 2190
  toluene, 2187
DRAW (direct read after write) optical drives, 2138

Economic Value Added (EVA), 2320
Ecuador
  occupational exposure limits, 1962
Edman degradation method
  biological monitoring application, 2038
Egregious case policy, under OSHA law, 1665
Electronic data processing systems, 2113–2114
  application selection development, 2128–2138
  concepts of, 2116–2119
Electrophilic metabolite biomarkers, 2031–2034
Emergency Planning and Community Right-to-Know Act of 1986 (EPCRA), 1737
Employee assistance programs, 2074
Environmental, health, and safety (EHS) practitioners, 2303
Environmental accounting, 2324–2325
Environmental impact of standards, 1615
Environmental Pesticide Control Act of 1972, 1600
Environmental protection, 1693–1695
Environmental Protection Agency (EPA)
  hazard communication, 1770–1772
Epidemiology studies, 2154–2155
Equal Employment Opportunity Commission (EEOC)
  health surveillance programs, 2212
Ergonomic standards
  proposed OSHA health standard, 1628
Ethical issues
  biological monitoring, 2028–2029
  industrial hygienists, 1686–1688
Ethylene dibromide
  lifetime cancer risk for persons exposed, 1936
Ethyleneimine
  OSHA standards, 2206
Ethylene oxide
  labeling requirements, 1769

lifetime cancer risk for persons exposed, 1936
  OSHA standards, 1623, 2206
  recordkeeping requirements, 2100
EVA. See Economic Value Added.
Evidence
  and expert witness testimony, 1710
Excel databases, 2127
Exhaled air
  biological monitoring, 2008–2010
Ex parte warrants, under OSHA law, 1653
Expert witnesses, 1703–1704
  depositions, 1715–1716
  discovery in pretrial, 1714–1715
  evidence as testimony, 1710
  hypothetical questions as testimony, 1711–1712
  opinion questions as testimony, 1710–1711
  preparation by attorney, 1716–1717
  qualifications for, 1717–1718
  recordkeeping, 1718
  reports, 1719–1720
  types of testimony by, 1704–1712
Exposure assessment, 2166–2170
  benzene, 2178–2179, 2185
  p-chlorotoluene, 2190–2191
  records, 2086–2088, 2097–2099
  toluene, 2187–2188
Exposure limits. See Occupational exposure limits.
Exposure monitoring, 2008–2014
Extractable metabolite biomarkers, 2030–2031
Extracurricular educational experiences, 2359

Federal Aviation Act of 1958, 1601
Federal Coal Mine Health and Safety Act of 1969, 1600
  health surveillance requirements, 2205–2207
Federal Consumer Product Safety Act, 1603–1604
Federal Mine Safety and Health Act of 1977, 1600
  compliance and projection, 1651
Federal Noise Control Act of 1972, 1601–1602
Federal Railroad Safety Act of 1970, 1600
Federal Rule of Evidence 702, 1704
Federal Toxic Substances Control Act of 1976, 1602–1603
Fertility
  reproductive toxicants occupational exposure limits, 1926–1927
  reproductive toxicity case study, 2181–2192

tests for effects on, 2209–2210
Finland
  occupational exposure limits, 1962
Formaldehyde
  OSHA standards, 1624, 2206
Fraudulent misrepresentation, 1691–1692
Frye Rule, 1705–1710

Gases
  biological monitoring of inorganic, 2042
  occupational exposure limits, *See*
    Occupational exposure limits
  pharmacokinetic models, 1866
General duty clause, OSHA health standards,
  1608–1611
Genetic monitoring, 2201
Germany
  biomarker reference values, 2017, 2028
  occupational exposure limits, 1910, 1962
Glutathione conjugates
  as biomarkers, 2032–2034
Graduate industrial hygiene education,
  2346–2351
Group health care, 2063

Haber's Law Model, 1893, 1953–1955
Hair
  biological monitoring for heavy metals
    exposure, 2014
Half-life, of biomarkers, 2004–2006
Hazard communication, 1735–1737. *See also*
  Material Safety Data Sheets; Threshold
  limit values.
  EPA Regulations, 1770–1772
  hazard determination, 1740–1741
  interpretation, 1772–1776
  labeling, 1743, 1745–1746, 1750–1751
  material information systems, 1751–1753
  OSHA Regulations, 1768–1770
  requirements, 1737–1747
  training for, 1746, 1747–1750
  written programs for, 1743–1745
Hazard communication programs
  audit, 1776–1785
  costs, 1759–1762
  management, 1753–1768
Hazard Communication Standard (HCS),
  1736–1737
  applications, 1737–1739
  functions for typical departments, 1757–1758
  hazardous material information systems,
    1751–1753
  interpretation, 1772–1776
  labeling, 1750–1751
  program audit, 1776–1785
  program elements, 1740–1747, 1756,
    1760–1761
  program management, 1753–1768
  recordkeeping, 2088, 2100
  related regulations, 1768–1772
  training, 1746, 1747–1750
Hazard communication standards, 1624–1625
Hazard identification, 2153–2159
  benzene, 2173–2178, 2182–2184
  p-chlorotoluene, 2188–2190
  toluene, 2186–2187
Hazardous Materials Transportation Act, 1601
Hazardous material tracking, 1752–1753
Hazardous Substances Act, 1604
Hazardous waste operations
  hazard communication requirements, 1770
  OSHA standards, 2206
HCS. *See* Hazard Communication Standard.
Health and safety performance metrics,
  2333–2335
Health Belief Model, 2223
Health care
  cost-effective loss prevention, 2067–2092
  costs, 2061–2063
  group costs, 2063
  increased costs of, 2063–2067
  statistical data, unreliability of, 2066
Health Check program, 2234–2235
Health promotion, 2221–2224
  ecology, 2225–2226
Health Promotion Model, 2223–2224
Health promotion programs
  design, 2226–2228
  effect of diversity in workplace, 2236–2237
  effect of ethnicity on, 2237–2238
  effect of gender on, 2237
  future trends, 2235–2243
  measurement of success, 2229–2235
  work-site programs, 2225–2228
Health records
  linking to industrial hygiene data, 2096–2097
Health Risk Analyses, 2230
Health surveillance programs, 2199–2200
  and EEOC, 2212
  and exposure data, 2212
  for general health maintenance, 2204
  hazard-oriented medical examinations,
    2204–2212

Health surveillance programs (*Continued*)
  legal issues, 2215–2216
  objectives, 2200–2203
  OSHA requirements, 2205–2206
  problem areas, 2214–2215
  recordkeeping, 2213–2214
Health surveillance systems, 2097, 2102
Hearing conservation programs, 2075–2077
Heavy metals
  biological monitoring of hair for, 2014
Hickey and Reist Model, for adjusting OELs for unusual work schedules, 1835–1846, 1850–1854, 1880–1884
Hypothetical questions, asked of expert witnesses, 1711–1712

Imminent danger situations, 1664
Individual liability, 1692–1693
Indoor air quality
  proposed standards, 1628–1629
Industrial hygiene
  defined, 2343–2345
  knowledge areas, 2371–2373
  liability issues, 1685–1695
  litigation practice, 1697–1721
Industrial hygiene data
  auditing, 2106–2109
  computerization of records, 2102–2104
  confidentiality of, 2101–2102
  linking to health records, 2096–2097
  related governmental recordkeeping, 2104–2106
Industrial hygiene education, 2343–2346
  academic curriculum, 2351–2359
  continuing education, 2365–2368
  distributed learning, 2363–2365
  extracurricular experiences, 2359
  graduate *vs.* undergraduate, 2346–2351
  information exchange, 2368–2371
  nontraditional academic programs, 2362–2365
  part-time academic programs, 2363
  professional organizations, 2367–2368
  RAC-ABET accreditation, 2354–2357
  real world application, 2359–2361
  student organizations, 2362
Industrial hygiene programs
  administration of, 2085–2086
  budgeting for, 2109–2110
  OSHA recordkeeping requirements, 2099–2101
  records, 2086–2088, 2097–2099

  reports of surveys and studies, 2088–2096
Industrial hygienists
  ethical responsibilities, 1686–1688
  role in health care, 2068–2069
  salaries, 2350–2351
Inorganic acids
  OSHA standards, 2206
Inorganic arsenic
  lifetime cancer risk for persons exposed, 1936
  OSHA health standards, 1627
Inorganic gases
  biological monitoring, 2042
Inorganic pollutants
  biological monitoring, 2042
Input devices, 2116–2117
Inspections, under OSHA law, 1651–1655
Intentional misrepresentation, 1691–1692
Internet
  distributed learning programs over, 2363–2365
  information exchange resources, 2371
Intervention effectiveness, 2253
Investigations, under OSHA law, 1651–1655
Investment-risk relationship, 2307–2308
Ionizing radiation
  recordkeeping requirements, 2100
Ireland
  occupational exposure limits, 1962–1963
ISO 9001, 2248
ISO 14001, 2248, 2264
Iuliucci model, for adjusting OELs for unusual work schedules, 1825–1826

Japan
  biomarker reference values, 2025–2026, 2028
  occupational exposure limits, 1963
  occupational health and safety management systems standards, 2264–2265
Job safety and health law, 1595–1598. *See also* OSHA health standards.
  employee rights and duties under OSHA, 1629–1631
  employer duties under OSHA, 1608–1629
  Federal regulations other than OSHA, 1598–1605
  future of, 1666–1667
  investigations and inspections, 1651–1655
  recordkeeping and reporting, 1655–1657
  regulation by states, 1605–1608
  sanctions for violating, 1657–1666
Joint application design (JAD) process, 2143–2144

INDEX

Labels
  hazardous chemicals in workplace, 1743, 1750–1751
  quality control, 1765
Laboratory chemicals
  hazard communication, 1769
  OSHA standards, 2206
Lawsuits, 1699–1703
Lead
  OSHA standards, 1620–1621, 2206
Liability, 1685–1688
  criminal sanctions, 1693–1695
  individual, 1692–1693
  potential types of, 1688–1693
Lifetime average daily dose, 2168–2169
Litigation. *See also* Expert witnesses.
  attorney–client privilege, 1713–1714
  chain of custody, 1720–1721
  depositions, 1715–1716
  discovery, 1714–1715
  and health care costs, 2066–2067
  industrial hygiene practice, 1697–1721
  lawsuits, 1699–1703
  recordkeeping, 1718–1720
  role of industrial hygienist, 1698–1699
  and standardized procedures, 1721–1722
  witness preparation, 1716–1718
  work product, 1712–1713
LIVE FOR LIFE Program, 2231–2232
Local area networks (LANs), 2119
Loss prevention and control programs, 2061–2067
  case studies, 2077–2082
  cost-effectiveness, 2067–2082
  design, 2067
  hearing conservation programs, 2075–2077
  implementation, 2067–2075
  management and evaluation, 2077
  role of industrial hygienist, 2068–2069
  role of occupational health clinician, 2071–2075
  role of safety professional, 2069–2071
Lotus Notes databases, 2127

Management science and systems, 2253–2259
Manufacturer's labels, 1750
Mason and Dershin Model, for adjusting OELs for unusual work schedules, 1826–1835
Material information systems, 1751–1753
Material Safety Data Sheets, 1741–1743, 1745
  management, 1751–1752
  and material tracking, 1752–1753
  preparation, 1751
  quality control, 1765
  record keeping, 1766–1767
Maximum Allowable Concentrations (MACs), 1905–1906
Medical benefit system, 2064–2065
Medical examinations
  compulsory and voluntary, 2214
  hazard-oriented, 2204–2212
Medications
  effect on biological monitoring, 2047–2048
Mental health programs
  preventive, 2241–2242
Mercapturic acid analysis, 2032–2034
Metadata, 2119–2122
Metal and Non-Metallic Mine Safety Act of 1966, 1600
Metals
  biological monitoring, 2039–2042
  heavy metals in hair, 2014
Methyl chloromethyl ether
  OSHA standards, 2206
Methylene chloride
  OSHA standards, 1626–1627, 2206
4,4′-Methylenedianiline
  OSHA standards, 2206
Methyl ethyl ketone
  Reference Concentration, 1915–1916
Michaelis-Menton kinetics, 1807–1808
Microbial aerosols
  unusual work schedule exposure, 1865
Mine safety, 1600, 2205–2207
Mine Safety and Health Administration (MSHA), 1600
Mine safety and health legislation, 1600
Mixtures
  biological monitoring of exposure to, 2044–2048
  occupational exposure limits, 1937–1938
  reproductive toxicity risk assessment, 2191–2192
  risk characterization, 2172–2173
Modifying factors, with OELs, 1913–1914
Moonlighting, and environmental exposures, 1877–1882

α-Naphthylamine
  OSHA standards, 2206
β-Naphthylamine
  OSHA standards, 2206

National Advisory Committee on Occupational Safety and Health Administration, 1597–1598
National Institute for Occupational Safety and Health (NIOSH), 1597
  continuing education programs, 2366–2367
  health surveillance requirements, 2207
Natural Gas Pipeline Safety Act of 1968, 1601
Negligence, 1689–1691
Netherlands
  occupational exposure limits, 1963
  occupational health and safety management systems standards, 2265
Net present value (NPV), 2317–2319
Networking, 2119
Neurotoxic agents
  occupational exposure limits, 1927
New Zealand
  occupational health and safety management systems standards, 2261–2263
NIOSH. *See* National Institute for Occupational Safety and Health.
4-Nitrobiphenyl
  OSHA standards, 2206
N-Nitrosodimethylamine
  OSHA standards, 2206
Noise exposure
  OSHA health standards, 1625–1626
  recordkeeping requirements, 2100
  unusual work schedule OELs, 1809–1810
Noncancer risk characterization, 2171–2172
Nonlinear pharmacokinetic models, 1807–1808
Nonserious violations, under OSHA law, 1658
No Observed Adverse Effect Level (NOAEL), 2165
No Observed Effect Level (NOEL), 1913
North American Free Trade Agreement, 1605
NPV (net present value), 2317–2319
Nuisance law
  odors, 1728–1730

Occupational exposure limit models, 1811–1812
  for adjusting OELs, 1949–1952
  biologic half-life determination, 1848–1849
  Brief and Scala model, 1812–1815, 1949–1952
  Iuliucci model, 1825–1826
  model comparison, 1849–1855
  OSHA model, 1815–1825, 1954–1955
  pharmacokinetic models, 1826–1848, 1855–1872

Occupational exposure limits. *See also* Risk analysis.
  adjusting for carcinogens, 1872–1877
  adjusting for seasonal occupations, 1882–1886
  adjusting for unusual work schedules, 1808–1811
  approaches to setting, 1910–1939
  for chemical carcinogens, 1931–1937
  corporate, 1946–1949
  data used for developing, 1948
  dermal, 1938–1939
  for developmental toxicants, 1924–1926
  history and biological basis, 1903–1965
  intended uses, 1906–1908
  mixtures, 1937–1938
  models for, 1921–1924
  for neurotoxic agents, 1927
  for odors, 1927–1928
  outside U.S., 1957–1965
  for persistent chemicals, 1928–1930
  philosophy of, 1908–1909
  physiologically based pharmacokinetic approach to adjusting, 1868–1872, 1956–1957
  for reproductive toxicants, 1926–1927
  for respiratory sensitizers, 1930–1931
  for sensory irritants, 1919–1924
  for systems toxicants, 1917–1919
  uncertainty factors, 1913–1917
  in U.S., 1909–1910
Occupational health and safety management systems, 2247–2253
  conformity assessment, 2284–2297
  implementation, 2280–2284
  management science and systems, 2253–2259
  standards and guidelines, 2261–2280
  universal structure, 2268–2280
Occupational health clinicians
  role in health care costs, 2071–2075
Occupational noise
  OSHA health standards, 1625–1626
Occupational Safety and Health Act (OSHA). *See also* OSHA health standards.
  agencies responsible for, 1597–1598
  beyond-compliance strategies, 2247–2249
  compliance and projection, 1651–1667
  employee duties under, 1629
  employee rights under, 1629–1631
  employer duties under, 1608–1629
  health surveillance requirements, 2205, 2206

INDEX

legislative history, 1595–1597
recordkeeping requirements, 1655–1657, 2099–2101
Occupational Safety and Health Administration (OSHA), 1597
  chemical classification by primary adverse effect, 1819–1823
  continuing education programs, 2367
  model for adjusting OELs for unusual work schedules, 1815–1825, 1954–1955
Occupational Safety and Health Review Commission, 1597
Odors
  air pollution regulations, 1730–1731
  audit, 1731–1733
  checklist for air pollution situations, 1731–1733
  human response to, 1727
  legal issues, 1725–1733
  nuisance law, 1728–1730
  occupational exposure limits, 1927–1928
OELS. See Occupational Exposure Limits.
One-compartment pharmacokinetic models, 1800–1801
Opinion questions, asked of expert witnesses, 1710–1711
Organic acids
  occupational exposure limits, 1923–1924
Organic bases
  occupational exposure limits, 1923–1924
Organic xenobiotics
  biological monitoring, 2029–2039
Organization of Russian States
  occupational exposure limits, 1910, 1964–1965
Organization theory, 2254–2256
OSHA. See Occupational Safety and Health Act; Occupational Safety and Health Administration.
OSHA health standards
  2-acetylaminofluorene, 2206
  acrylonitrile, 1622, 2206
  4-aminodiphenyl, 2206
  asbestos, 1617–1619, 2206
  benzene, 1622–1623, 2206
  benzidine, 2206
  bloodborne pathogens, 1624, 2206
  butadiene, 1627, 2206
  cadmium, 1626, 2206
  carcinogenic chemicals, 1619–1620
  challenging validity of, 1612
  bis-chloromethyl ether, 2206
  coke oven emissions, 1620, 2206
  confined spaces, 1627
  cotton dust, 1621–1622, 2206
  1,2-dibromo-3-chloropropane, 1622, 2206
  3,3′-dichlorobenzidine, 2206
  4-dimethylamino azobenzene, 2206
  economic and technological feasibility, 1612–1615
  environmental impact, 1615
  ethyleneimine, 2206
  ethylene oxide, 1623, 2206
  formaldehyde, 1624, 2206
  general duty clause, 1608–1611
  hazard communication, 1768–1770
  for hazard communication, 1624–1625
  inorganic acids, 2206
  inorganic arsenic, 1627
  laboratory chemicals, 2206
  lead, 1620–1621, 2206
  methyl chloromethyl ether, 2206
  methylene chloride, 1626–1627, 2206
  4,4′-methylenedianiline, 2206
  naphthylamines, 2206
  4-nitrobiphenyl, 2206
  N-nitrosodimethylamine, 2206
  occupational noise, 1625–1626
  overview, 1616–1617
  for personal protective equipment, 1628
  promulgation process, 1611–1612
  β-propiolactone, 2206
  proposed ergonomic standard, 1628
  proposed standards, 1628–1629
  for respirators, 1627–1628
  specific duty clause, 1611
  variances, 1615–1616
  vinyl chloride, 1619, 2206
Outer Continental Shelf Lands Act, 1605
Output devices, 2117
Overtime, and environmental exposures, 1877–1882

Particulates
  pharmacokinetic models, 1864–1865
Payback, 2319–2320
Penalties, under OSHA law, 1657–1662
  contesting, 1662–1664
Performance Indexing, 2333
Performance measurement, 2305–2307
Periodic medical examinations, 2202–2203

Permissible Exposure Limits (PELs), 1909
Persistent chemicals
    occupational exposure limits, 1928–1930
Personal protective equipment
    OSHA health standards, 1628
Pharmacokinetic models, 1797–1800
    chemical accumulation in body, 1804–1807
    generalized approach to use of, 1855–1859
    Hickey and Reist Model, 1835–1846,
        1850–1854, 1880–1884
    Mason and Dershin Model, 1826–1835
    for mixtures, 1867–1868
    nonlinear, 1807–1808
    for OELs, 1955–1956
    one-compartment model, 1800–1801
    for particulates, 1864–1865
    for radioactive material, 1866
    for reactive gases and vapors, 1866
    Roach Model, 1846–1848
    two-compartment model, 1801–1804
Pharmacokinetics, 1790–1797
Philippines
    occupational exposure limits, 1963–1964
Positive Performance Indicators, 2333–2334
Preassignment medical examinations, 2202
Preplacement medical examinations,
    2072–2073, 2201–2202
Prevention programs. *See* Loss prevention and
    control programs.
Preventive mental health programs, 2241–2242
Private data, 2123–2125
Process Safety Management (PSM) standard,
    1770
Productivity
    as hidden impact of health and safety
        investments, 2332
Professional certification, 2348–2350
Professional conferences, 2369–2371
Professional journals, 2368–2369, 2370
Professional liability, 1685
Professional organizations, 2367–2368
β-Propiolactone
    OSHA standards, 2206
Protein adducts
    as biomarkers, 2035–2039
Public domain data, 2122–2123, 2125

Quality
    as hidden impact of health and safety
        investments, 2330–2332

RAC-ABET accreditation, 2354–2357
Radiation Control for Health and Safety Act of
    1968, 1605
Radiation control standards, 1604–1605
Radioactive material
    pharmacokinetic models, 1866
Reasoned Action theory, 2223–2224
Recommended Exposure Limits (RELs), 1909
Recordkeeping
    expert witnesses, 1718–1720
    exposure assessment, 2086–2088, 2097–2099
    hazard communication program, 1766–1768
    health surveillance programs, 2213–2214
    industrial hygiene programs, 2085–2088,
        2097–2099
    noise exposure, 2100
    OSHA law requirements, 1655–1657,
        2099–2101
    respirators, 2100
    Worker's compensation, 2098
Reference Concentrations (RfC), 1913–1917
Reference Doses (RfD), 1913
Reference values, for biological monitoring,
    2017–2029
Renal clearance, 2011
Repeated violations, under OSHA law,
    1659–1661
Reports, 1655–1657
    expert witnesses, 1719–1720
    of surveys and studies, 2088–2096
Reproductive toxicants
    occupational exposure limits, 1926–1927
    risk analysis case study, 2181–2192
Resource Conservation and Recovery Act of
    1976 (RCRA), 1771
Respirators
    health surveillance requirements, 2205, 2216
    OSHA health standards, 1627–1628
    recordkeeping requirements, 2100
Respiratory sensitizers
    occupational exposure limits, 1930–1931
Responsible care programs, 2108
Risk analysis, 2151–2152
    benzene case study, 2173–2181
    elements of, 2152–2173
    reproductive toxicity case study, 2181–2192
Risk characterization, 2170–2173
    benzene, 2179–2181, 2186
    p-chlorotoluene, 2191–2192
    toluene, 2188
Roach Model, for adjusting OELs for unusual
    work schedules, 1846–1848

INDEX

Russian States, Organization of
　occupational exposure limits, 1910, 1964–1965

Safety professionals
　role in accident prevention, 2069–2071
Salaries, 2350–2351
Sanctions, for violating OSHA law, 1657–1666
Scientific management, 2254–2255
Seasonal occupations, 1882–1886
Sensory irritants
　occupational exposure limits, 1919–1924
Serious violations, under OSHA law, 1657–1658
Shift work, 1787–1789. *See also* Unusual work schedules.
Short term exposure limits (STELs), 1919
Skin
　occupational exposure limits, 1938–1939
South Africa
　occupational health and safety management systems standards, 2265
Specific duty clause, OSHA health standards, 1611
Steady state, pharmacokinetics, 1793–1797
Structure-activity relationship (SAR) evaluation, 1917
Student organizations, 2362
Subclinical effects, 2208
Superfund Amendments and Reauthorization Act of 1986 (SARA), 1737, 1770–1771
Surveillance. *See* Health surveillance programs.
Surveys
　industrial hygiene programs, 2088–2096
Systems thinking, 2256–2259
　and business analysis, 2336–2337
Systems toxicants
　occupational exposure limits, 1917–1919

Take Heart program, 2232–2234
Termination medical examinations, 2203
Testimony
　by expert witnesses, 1704–1712
　witness preparation, 1716–1717
Threshold limit values (TLVs), 1906
　approaches to setting, 1910–1939
　future of, 1943–1945
　intended uses of, 1906–1908
　philosophy of, 1908–1909
　protection of workers with, 1939–1945
TLVs. *See* Threshold Limit Values.

Toluene
　reproductive toxicity risk assessment, 2186–2187
Torts, 1688–1692
Toxicity
　interspecies differences, 2165–2166
Trade secrets
　and hazard communication, 1746–1747, 1767
Training
　hazard communication, 1746, 1747–1750, 1767
Two-compartment pharmacokinetic models, 1801–1804

Uncertainty factors, with OELs, 1913–1917
Undergraduate industrial hygiene education, 2346–2351
United Kingdom
　occupational health and safety management systems standards, 2263–2264, 2265–2266
United States
　biomarker reference values, 2017, 2023–2024, 2026–2028
　occupational exposure limits, 1909–1910
　occupational health and safety management systems standards, 2266–2267
Unusual work schedules, 1787–1790. *See also* Pharmacokinetic models.
　adjustment of exposure limits for, 1808–1811 1949–1952
　biologic studies, 1886–1893
　moonlighting, overtime, and environmental exposures, 1877–1882
　noise exposure limits, 1809–1810
　seasonal occupations, 1882–1886
　uncertainties in predicting toxicological response, 1893–1895
Urine
　biological monitoring, 2010–2012
　DNA repair products in, 2035
Vapors
　occupational exposure limits, *See* Occupational exposure limits
　pharmacokinetic models, 1866

Variances, for OSHA health standards, 1615–1616
Vinyl chloride
　OSHA standards, 1619, 2206

Warrants, under OSHA law, 1652–1654
WellWorks program, 2234

Wide area networks (WANs), 2119
Willful violations, under OSHA law, 1659–1661
Worker Right-to-Know Programs, 1737
Workers' compensation, 1692–1693
  direct costs, 2062
  indirect costs, 2063
  recordkeeping, 2098
Workplace Environment Exposure Limits (WEELs), 1909

Work product
  general principles governing discoverability of, 1712–1713
Work-related injuries
  management and treatment of, 2073–2074
World Wide Web. *See* Internet.
WORM (write one-read many) optical drive, 2138

Yearly average daily dose, 2168–2169